HOLT
ALGEBRA
WITH TRIGONOMETRY

Eugene D. Nichols
Mervine L. Edwards
E. Henry Garland
Sylvia A. Hoffman
Albert Mamary
William F. Palmer

Holt, Rinehart and Winston, Inc.

Harcourt Brace Jovanovich, Inc.

Austin • Orlando • San Diego • Chicago • Dallas • Toronto

About the Authors

Eugene D. Nichols
Distinguished Professor of Mathematics Education
Florida State University
Tallahassee, Florida

Mervine L. Edwards
Chairman, Department of Mathematics
Shore Regional High School
West Long Branch, New Jersey

E. Henry Garland
Head of Mathematics Department
Developmental Research School
DRS Professor
Florida State University
Tallahassee, Florida

Sylvia A. Hoffman
Resource Consultant in Mathematics
Illinois State Board of Education
State of Illinois

Albert Mamary
Superintendent of Schools for Instruction
Johnson City Central School District
Johnson City, New York

William F. Palmer
Professor of Education and Director
Center for Mathematics and Science Education
Catawba College
Salisbury, North Carolina

Printed in the United States of America

56 032 765   ISBN 0-03-005433-8

Acknowledgments

Reviewers

Jeanette Gann
Mathematics Supervisor
High Point Schools
High Point, North Carolina

Patrice Gossard, Ph.D.
Mathematics Teacher
Cobb County School District
Marietta, Georgia

Linda Harvey
Mathematics Teacher
Reagan High School
Austin, Texas

Gerald Lee
Chairman, Mathematics Department
McArthur High School
Lawton, Oklahoma

Janet Page
Mathematics Teacher
Hoffman Estates High School
Hoffman Estates, Illinois

Photo Credits

Illustration

Chapter Contents

Symbol List

$\lvert x \rvert$	the absolute value of x	$\overrightarrow{AB}$	ray AB
$\overset{?}{=}$	is possibly equal to	$\vec{V}$	vector V
$\approx$	is approximately equal to	$\lVert \vec{v} \rVert$	the norm of the vector $\vec{v}$
$\neq$	is not equal to	$\angle A$	angle A
$\{\ \}$	set enclosures	$m \angle A$	the measure of $\angle A$
$\mid$	such that	$\circ$	degrees
$\varnothing$	empty set	$'$	minutes of arc
$<$	is less than	$\triangle ABC$	triangle ABC
$\leq$	is less than or equal to	i	the imaginary number $i = \sqrt{-1}$
$>$	is greater than		
$\geq$	is greater than or equal to	$\overline{a + bi}$	the conjugate of the complex number $a + bi$
$-x$	the additive inverse of x	Σ	summation symbol
$\frac{1}{x}$	the multiplicative inverse of x	$!$	factorial
$\pm$	plus-or-minus symbol	$\binom{n}{r}$	the number of combinations of n things taken r at a time; $\binom{n}{r} = \frac{n!}{r!(n-r)!}$
(x,y)	ordered pair		
(x,y,z)	ordered triple	$P(E)$	the probability of event E
$f(x)$	f of x; the value of the function f at x	σ	standard deviation
$g(f(x))$	composition; g of f of x	$\log_b x$	the base-b logarithm of x
f^{-1}	the inverse of the function f	A^{-1}	the inverse of matrix A
$\sqrt{x}$	the square root of x	$\det$	determinant
$\sqrt[n]{x}$	the nth root of x	$\mathrm{Sin}^{-1} x$	the principal values of the inverse of sin x
x^n	the nth power of x		
$x^{\frac{p}{q}}$	the pth power of the qth root of x	π	the Greek letter pi
		θ	the Greek letter theta
$\overline{AB}$	segment AB	φ	the Greek letter phi
$\overleftrightarrow{AB}$	line AB		

High-speed computers make it possible to display
information graphically in mathematics, science, art, and
design. The image above is an example of a purely
mathematical object called a *fractal*. This particular fractal is
known as the Mandelbrot set.

1.1 Operations with Real Numbers

Objective   To perform basic operations with pairs of real numbers

Listed below are examples of several kinds of real numbers.

Natural numbers	1, 2, 3, 4, . . .
Whole numbers: natural numbers and zero	0, 1, 2, 3, . . .
Integers: whole numbers and their opposites	. . ., -2, -1, 0, 1, 2, . . .
Rational numbers: quotients of integers	$\frac{3}{4}$, or 0.75; $\frac{-8}{1}$, or -8
(terminating decimals)	$\frac{-17}{5}$, or $-3\frac{2}{5}$, or -3.4
(repeating decimals)	$\frac{1}{3}$, or 0.33 . . ., or $0.\overline{3}$
Irrational numbers: nonterminating, nonrepeating decimals	$\sqrt{2}$, or 1.4142135 . . . 5.1511511151 . . . ; π

A **rational number** is a number that can be written as the quotient $\frac{a}{b}$ of two integers, a and b, where $b \neq 0$. The decimal form of a rational number either terminates (such as 0.75) or repeats (such as 0.33 . . .).

An **irrational number** is a nonterminating, nonrepeating decimal.

The set of **real numbers** is the set of all rational and irrational numbers. The sign rules for adding real numbers can be stated in terms of the *absolute values* of the numbers. The **absolute value** of a *positive* real number or zero is the number itself. The absolute value of a *negative* real number is its opposite. The absolute value of any real number x, symbolized by $|x|$, is never negative, as shown by the following examples.

$$|9| = 9 \qquad |-3| = 3 \qquad |0| = 0$$

Sign Rules for Addition
To add two nonzero real numbers with *like signs*, add their absolute values, and give the sum the sign of the original numbers.

To add two nonzero real numbers with *unlike signs*, subtract the lesser absolute value from the greater and give the difference the sign of the original number with the greater absolute value.

EXAMPLE 1   Simplify.   **a.** $-4 + (-2.6)$ **b.** $2.4 + (-7.5)$

Solutions   **a.** $-4 + (-2.6) = -(|-4| + |-2.6|) = -(4 + 2.6) = -6.6$
b. $2.4 + (-7.5) = -(|-7.5| - |2.4|) = -(7.5 - 2.4) = -5.1$

The operation of subtraction is defined as follows.

Definition

$a - b = c$ means that $c + b = a$.

For example, $3 - 7 = -4$ because $-4 + 7 = 3$.

Finding $a - b$ can be done by finding the opposite of b, that is $-b$, and then adding the real numbers a and $-b$.

Rule of Subtraction
$a - b = a + (-b)$ for all real numbers a and b.

EXAMPLE 2   Simplify.   **a.** $7.2 - 9.6$ **b.** $-12 - 3.5$

Solutions   **a.** $7.2 - 9.6 = 7.2 + (-9.6)$ **b.** $-12 - 3.5 = -12 + (-3.5)$
$\qquad\qquad\qquad\quad = -2.4 \qquad\qquad\qquad\qquad\qquad\qquad = -15.5$

The sign rules for multiplying or dividing a pair of nonzero real numbers fall into two cases, like signs and unlike signs, as illustrated below.

like $\begin{cases} 10(2.5) = 25 \\ -10 \div (-2.5) = 4 \end{cases}$ unlike $\begin{cases} 6.8(-2) = -13.6 \\ -6.8 \div 2 = -3.4 \end{cases}$
signs   signs

Recall that *division by zero is undefined*. This can be explained by the definition of division that follows.

Definition

$a \div b = c$ means that $c \cdot b = a$.

For example, $15 \div 3 = 5$ because $5 \cdot 3 = 15$. However, $8 \div 0 = x$ has no solution because $x \cdot 0 = 8$ has no solution.

Sign Rules for Multiplication and Division
For all nonzero real numbers a and b with like signs, the product $a \cdot b$ and the quotient $a \div b$ are positive numbers.

For all nonzero real numbers a and b with unlike signs, the product $a \cdot b$ and the quotient $a \div b$ are negative numbers.

$a \cdot 0 = 0$ and $0 \cdot a = 0$ for each real number a.
$\frac{0}{a} = 0$ $(a \neq 0)$ and $\frac{a}{0}$ is undefined for each real number a.

A **variable** is a symbol, usually a letter, that represents a number. An expression such as $c - d$ can be evaluated for given values of the variables c and d, such as 15 for c and -8 for d. To do this, substitute the values for the variables, and simplify the result as shown.

$$c - d = 15 - (-8) = 15 + 8 = 23$$

EXAMPLE 3 Evaluate each expression for $a = -16$, $b = 8$, $c = -4$, and $d = 0$, if possible.

Solutions

Expression	Result	Answer
$a + c$	$-16 + (-4)$	-20
$b - c$	$8 - (-4)$	12
$c - b$	$-4 - 8$	-12
$d - b$	$0 - 8$	-8
bc, or $b \cdot c$	$8(-4)$	-32
$-5c$, or $-5 \cdot c$	$-5(-4)$	20
$\dfrac{a}{b}$	$\dfrac{-16}{8}$	-2
$\dfrac{a}{c}$	$\dfrac{-16}{-4}$	4
$\dfrac{a}{d}$	$\dfrac{-16}{0}$	undefined
$\dfrac{d}{b}$	$\dfrac{0}{8}$	0
ad	$-16 \cdot 0$	0

Operations occur more than once in expressions such as $23 - 52 - (-14) - 7$ and $-5 \cdot 4(-10)(-3)$. Such expressions can be simplified by combining pairs of numbers in several steps.

EXAMPLE 4 Simplify. **a.** $23 - 52 - (-14) - 7$ **b.** $-5 \cdot 4(-10)(-3)$

Solutions

a. Change subtracting to *adding the opposite*. Group addends with the same sign and add.

$$23 - 52 - (-14) - 7$$
$$23 + (-52) + 14 + (-7)$$
$$(23 + 14) + [-52 + (-7)]$$
$$37 + (-59)$$
$$-22$$

b. Multiplication is the only operation. Work with pairs of factors.

$$-5 \cdot 4(-10)(-3)$$
$$-20 \cdot 30$$
$$-600$$

Classroom Exercises

Simplify, if possible. If not possible, write *undefined*.

1. $-15 + 45$
2. $-17 + (-11)$
3. $-18 + 18$
4. $-11 - 30$
5. $23 - (-5)$
6. $-35 - (-15)$
7. $-5(-7)$
8. $-15 \div 3$
9. $-24 \div (-6)$
10. $0 \div \pi$
11. $6 \div 0$
12. $-\sqrt{7} + 0$
13. $-5 - 3 - 3$
14. $-7(-2)(-10)$
15. $-5 \cdot 3(-4)$

Written Exercises

Simplify, if possible. If not possible, write *undefined*.

1. $-45 + 175$
2. $-73 + (-52)$
3. $-62 - 32$
4. $-18 - (-12)$
5. $-8(-11)$
6. $-5 \cdot 12$
7. $-6 \cdot 0$
8. $\frac{-84}{-21}$
9. $\frac{0}{9}$
10. $\frac{9}{0}$
11. $-9 + 9$
12. $18 - 38 - 12$
13. $-6 - (-31) - 15$
14. $-9.3 + 5.7$
15. $-18.2 - (-15.9)$

Evaluate for $a = -24$, $b = 6$, $c = -3$, and $d = 0$, if possible.

16. $a + c$
17. $b - c$
18. bc
19. $-4c$
20. ad
21. $\frac{a}{b}$
22. $\frac{a}{c}$
23. $\frac{b}{d}$
24. $a - b - c$
25. $b - c - a$
26. $c - a - b$
27. $c - b - 9$

Simplify, if possible.

28. $-8(-15)(-2)$
29. $-6 \cdot 7 \cdot 5$
30. $15(-6.5)$
31. $-8.25(-400)$
32. $-27 \div (-3.6)$
33. $-5(-5)(-5)(-5)$
34. $\frac{1}{2} - \frac{5}{6} - \frac{3}{4}$
35. $-3 - \frac{2}{5} - \left(-\frac{3}{10}\right)$
36. $-\frac{3}{2}\left(-\frac{8}{5}\right)\left(\frac{10}{9}\right)$

37. $a \div b = c$ if and only if $c \cdot b = a$. Thus, $3 \div 0 = c$ has no solution, and $\frac{3}{0}$ is undefined. Explain why $\frac{0}{0}$ must also be undefined.

In Exercises 38 and 39, x and y are any two *positive* real numbers with x less than y.

38. Is $(x - y)^{12}$ positive or negative?

39. If n is a positive integer, is $(x - y)^{2n}$ positive or negative?

1.2 Order of Operations

Objective

To apply the rules for the order of operations

More than one operation may appear in a numerical expression, as in $18 - 12 \div 6 \cdot 2 + 4$. The value of such an expression will depend on the order in which the operations are performed. Thus, *only one* of the following can be a correct interpretation of $18 - 12 \div 6 \cdot 2 + 4$.

A. $18 - 12 \div 6 \cdot 2 + 4$
$\quad\ 18 - 2 \cdot 2 + 4$
$\quad\quad\ 18 - 4 + 4$
$\quad\quad\quad\ 14 + 4$
$\quad\quad\quad\ 18$ (correct)

B. $18 - 12 \div 6 \cdot 2 + 4$
$\quad\quad\ 6 \div 6 \cdot 2 + 4$
$\quad\quad\quad\ 1 \cdot 2 + 4$
$\quad\quad\quad\quad 2 + 4$
$\quad\quad\quad\ 6$ (incorrect)

In the second interpretation (B), all operations are performed from left to right, but this is incorrect. The first interpretation (A) is the one mathematicians accept as correct. The rule is as follows.

> Multiply and divide in order from left to right. Then add and subtract, also in order from left to right.

On a calculator, enter 18 ⊖ 12 ⊘ 6 ⊗ 2 ⊕ 4 ⊜ and read the display. A *scientific* calculator will show 18, which is correct. Such a calculator automatically obeys the rule for the order for operations when there are no grouping symbols. If the calculator displays 6, which is incorrect, you will need to enter the data according to the rule for the order of operations:

$$⊖ \ 12 \ ⊘ \ 6 \ ⊗ \ 2 \ ⊕ \ 18 \ ⊕ \ 4 \ ⊜$$

EXAMPLE 1

Simplify: $-8 \div 4 - (-3)16 \div 4 + 5(-8)$

Solution

Multiply and divide, left to right.
$\quad -8 \div 4 - (-3)16 \div 4 + 5(-8)$
$\quad\quad -2 - (-48) \div 4 + (-40)$
$\quad\quad -2 - (-12) + (-40)$

Use $a - b = a + (-b)$.
The answer is -30.
$\quad\quad -2 + 12 + (-40)$, or -30

The order-of-operations rule above does not include the operation of raising a number to a power. *Raising a number to a power* takes priority over all other operations, as illustrated below.

$$3 + 4^2 = 3 + (4^2) \qquad\qquad -5 \cdot 2^3 = -5(2^3)$$
$$= 3 + 16, \text{ or } 19 \qquad\qquad = -5 \cdot 8, \text{ or } -40$$

Parentheses and brackets are *grouping symbols*. They are often used to show an intended order of operations when the usual rules are not to be observed. Thus,

$$2 \cdot 5 - 7 = 10 - 7 = 3, \text{ but } 2(5 - 7) = 2(-2) = -4$$

When one expression is enclosed within another, simplify the innermost expression first. Thus,

$$5[7 - 3 - (2 + 6)] = 5[7 - 3 - 8] = 5[-4] = -20$$

EXAMPLE 2   Simplify each expression.

Solutions
a. $(3 + 4)^2$
  7^2
  49

b. $-5 + 3(2 - 7)(4 - 6)^3$
  $-5 + 3(-5)(-2)^3$
  $-5 + (-15)(-8)$
  $-5 + 120, \text{ or } 115$

A horizontal fraction bar serves as a grouping symbol. The intended order of operations is (1) simplify the numerator, (2) simplify the denominator, and (3) simplify the quotient, as shown below.

$$\frac{-18 - 3(-2)}{4 \cdot 6 - (-8)} = \frac{-18 + 6}{24 + 8} = \frac{-12}{32} = \frac{-3}{8} = -\frac{3}{8}$$

Rules for Order of Operations
1. First, simplify expressions within pairs of grouping symbols, beginning with the innermost pair.
2. Next, simplify all powers.
3. Then, multiply and divide in order from left to right.
4. Finally, add and subtract in order from left to right.

Classroom Exercises

Simplify, if possible. If not possible, write *undefined*.

1. $3 + 4 \cdot 2$
2. $(3 + 4)2$
3. $4 \cdot 2 - 3 \cdot 5$
4. $4(2 - 3) \cdot 5$
5. -5^2
6. $(-5)^2$
7. $6 \cdot 8 \div 4 \cdot 2$
8. $6[8 \div (4 \cdot 2)]$
9. $7 - 3^2$
10. $(7 - 3)^2$
11. $-5 \cdot 3^2$
12. $(-5 \cdot 3)^2$
13. $\dfrac{4 - 6 \cdot 2}{-2 \cdot 5 - 6}$
14. $\dfrac{-2 \cdot 3}{14 - 7 \cdot 2}$

Written Exercises

Simplify, if possible. If not possible, write *undefined*.

1. $-7 + 5 \cdot 3$

2. $6(-8) - 10$

3. $10(-4) - 5(-8)$

4. $-12(-6) + 8(-4)$

5. $\dfrac{30 - (-6)}{-5 - (-14)}$

6. $\dfrac{8 + 2 \cdot 5}{3 \cdot 4 - 6 \cdot 2}$

7. $4^2 + 16 \cdot 9 - 2^4$

8. $5(-3 - 4) + 35$

9. $40 - 10^2$

10. -10^2

11. $-6 \cdot 2^3$

12. $(-3 \cdot 2)^3$

13. $7(-6 - 5^2)$

14. $3[-18 - (6 - 14)]$

15. $40 - 36 \div 18 \cdot 2$

16. $\dfrac{-2 - 2[12 - 2(2.6 - 4.1)]}{(2.7 - 3.2)(4.4 - 8.4)^2}$

17. $\dfrac{-7.16 - 5.42 \cdot 3}{3^2 - 6(1.5)}$

18. $(12 - 3^2)(12 - 3)^2$

19. $-15 - 5[12 - 3(8.4 - 12.4)^2]$

For each statement, determine whether it is true or false.

20. $(2^4)^2 = (2^2)^4$

21. $2^{(4^2)} = (2^4)^2$

22. $(5^2)^3 = 5^{(2^3)}$

Mixed Review

Evaluate each expression for $a = 6$, $b = -2$, and $c = -3$. *1.1*

1. bc

2. $a - b$

3. $5ac$

4. $b - c - a$

Using the Calculator

The example below shows how to use a *scientific* calculator with a minimum number of keystrokes.

$$\dfrac{85}{3.9 - 5.6}: \qquad 85 \; \boxed{\div} \; \boxed{(} \; 3.9 \; \boxed{-} \; 5.6 \; \boxed{)} \; \boxed{=} \; -50$$

Second method: $3.9 \; \boxed{-} \; 5.6 \; \boxed{=} \; \boxed{1/x} \; \boxed{\times} \; 85 \; \boxed{=} \; -50$

Simplify using a calculator and a minimum number of keystrokes.

1. $-23 + 54 \cdot 83 - 31 \cdot 47$

2. $75 - 15 \cdot 12 \div 30 + 36$

3. $\dfrac{-75 - 5(65)}{50(3.2)}$

4. $\dfrac{28.8 - 43.2}{-5.1 + 3.9}$

1.3 Algebraic Expressions

Objectives

To identify properties of operations with real numbers
To simplify algebraic expressions
To evaluate algebraic expressions

In Lessons 1.1 and 1.2, you reviewed the four major operations with real numbers. Real numbers are of great practical importance to scientists, engineers, and others who make and interpret measurements.

The set of real numbers is also of great interest to professional mathematicians. A main reason for this interest is that real numbers (together with the operations of addition and multiplication) provide one of the best examples of the mathematical structure known as a *field*.

Any set F is a **field** if the eleven properties listed below are true for the set. Unless otherwise stated, each property applies to all numbers a, b, and c in F.

Name of Property	Statement of Property
Closure for Addition	$a + b$ is in F.
Closure for Multiplication	$a \cdot b$ is in F.
Commutative of Addition	$a + b = b + a$
Commutative of Multiplication	$a \cdot b = b \cdot a$
Associative of Addition	$(a + b) + c = a + (b + c)$
Associative of Multiplication	$(a \cdot b) \cdot c = a \cdot (b \cdot c)$
Distributive of Multiplication over Addition	$a(b + c) = a \cdot b + a \cdot c$ and $(b + c)a = b \cdot a + c \cdot a$
Identity for Addition	There is a unique number 0 in F such that for all a in F, $a + 0 = a.$
Identity for Multiplication	There is a unique number 1 in F such that for all a in F, $a \cdot 1 = a.$
Additive Inverse	For each a in F, there is a unique number $-a$ in F such that $a + (-a) = 0$. a and $-a$ are **additive inverses**, or **opposites**.
Multiplicative Inverse	For each nonzero a in F, there is a unique number $\frac{1}{a}$ in F such that $a \cdot \frac{1}{a} = 1$. a and $\frac{1}{a}$ are **multiplicative inverses**, or **reciprocals**.

There are fields other than the set of real numbers. For example, the set of rational numbers (see Lesson 1.1) is a field. Also, the set of complex numbers (see Chapter 9) is a field.

EXAMPLE 1 Identify the property of operations with real numbers illustrated by the sentence.

Solutions

Sentence	Property
a. $v(4t) = (4t)v$	**a.** Commutative of Mult
b. $-a^2b + a^2b = 0$	**b.** Add Inverse
c. $7 \cdot \pi$ is a real number.	**c.** Closure for Mult
d. $-6r + 0 = -6r$	**d.** Identity for Add
e. $(8 + n) + (-n) = 8 + [n + (-n)]$	**e.** Associative of Add
f. $m - n = 1(m - n)$	**f.** Identity for Mult
g. $m(n^2 + n) = m \cdot n^2 + m \cdot n$	**g.** Distr of Mult over Add
h. $\dfrac{w}{4} \cdot \dfrac{4}{w} = 1$	**h.** Mult Inverse
i. $2c + 3c^2 = 3c^2 + 2c$	**i.** Commutative of Add
j. $-5(3t) = (-5 \cdot 3)t$	**j.** Associative of Mult
k. $8 + \sqrt{2}$ is a real number.	**k.** Closure for Add

Three examples of *algebraic expressions* are given below.

$$-4(5x) \qquad 7a^2 + 5a^2 \qquad -n + (3m + n)$$

An **algebraic expression** contains one or more variables (see Lesson 1.1) and one or more operations. The expression $-4(5x)$ contains three **factors**, -4, 5, and x. Factors are multiplied. The expression $7a^2 + 5a^2$ has two **terms**, $7a^2$ and $5a^2$. Terms are added. For the term $7a^2$, 7 is called the numerical **coefficient** of a^2.

All of the expressions above can be simplified. Steps in their simplification are shown in the table below.

Steps	Property or reason
$-4(5x) = (-4 \cdot 5)x$ $= -20x$	Associative Property of Multiplication (The factors -4, 5, and x were regrouped.)
$7a^2 + 5a^2 = (7 + 5)a^2$ $= 12a^2$	Distributive Property (The like terms, $7a^2$ and $5a^2$, were combined by adding their coefficients.)
$-n + (3m + n)$ $= (3m + n) + (-n)$ $= 3m + [n + (-n)]$ $= 3m + 0$ $= 3m$	Commutative Property of Addition Associative Property of Addition Additive Inverse Additive Identity (The three terms were reordered and regrouped.)

The expression $5x^2y + 3ab^2 + 2x^2y + 4a^2b$ contains four terms. $5x^2y$ and $2x^2y$ are called **like terms** because they differ only in their coefficients. Like terms can be combined using the Distributive Property.

$$5 \cdot x^2y + 2 \cdot x^2y = (5 + 2)x^2y = 7x^2y$$

The terms $3ab^2$ and $4a^2b$ are **unlike terms**. They differ in their variable factors, ab^2 and a^2b. Unlike terms cannot be combined.

The Distributive Property of Multiplication over Addition can be extended to subtraction.

For all real numbers a, b, and c,
$$a(b - c) = a \cdot b - a \cdot c \text{ and } (b - c)a = b \cdot a - c \cdot a$$

This statement is proved as a theorem in Lesson 1.8.

EXAMPLE 2   Simplify each expression.

Solutions
a. $-8x^2 + 5(4x^2 - 6)$
$= -8x^2 + 5 \cdot 4x^2 - 5 \cdot 6$ Distribute 5 to $4x^2$ and 6.
$= -8x^2 + 20x^2 - 30$
$= 12x^2 - 30$ Combine like terms.

b. $5n - (3n + 6)$ Use the multiplicative identity:
$= 5n - 1(3n + 6)$ $(3n + 6) = 1(3n + 6)$.
$= 5n + (-1 \cdot 3n) + (-1 \cdot 6)$ Distribute -1 to $3n$ and 6.
$= 2n - 6$ Combine like terms.

Because in Example 2a, $-8x^2 + 5(4x^2 - 6)$ is simplified to $12x^2 - 30$, the two expressions are said to be *equivalent*. Similarly, in Example 2b, $5n - (3n + 6)$ and $2n - 6$ are shown to be equivalent expressions. Two algebraic expressions are **equivalent expressions** if they are equal for *all* values of the variables for which the expressions have meaning.

EXAMPLE 3   Evaluate $2x^2 - 6xy + 4x^2 - xy$ for $x = -3$ and $y = 2$.

Plan   Find a simpler equivalent expression for $2x^2 - 6xy + 4x^2 - xy$. Then evaluate the simpler expression.

Solution
$$\begin{aligned} 2x^2 - 6xy + 4x^2 - xy &= 2x^2 + 4x^2 - 6xy - xy \\ &= 6x^2 - 7xy \\ &= 6(-3)^2 - 7(-3)2 \\ &= 6 \cdot 9 - (-42) = 96 \end{aligned}$$

The value is 96.

Summary (Simplifying Expressions)	Example
Factors are *multiplied* to give a product. Factors can be rearranged to simplify a product.	four factors $(3 \cdot x)(5 \cdot y) = (3 \cdot 5)(x \cdot y)$ $\qquad\qquad\qquad = 15xy$
Terms are *added* to give a sum. Terms can be rearranged, and like terms combined, to simplify a sum.	four terms $7a^2 + 5b^2 + 2a^2 + 8b$ $= (7a^2 + 2a^2) + (5b^2 + 8b)$ $= 9a^2 + 5b^2 + 8b$

Classroom Exercises

Identify the property of operations that is illustrated. All variables represent real numbers.

1. $-6.3 + 6.3 = 0$

2. $6c + (5 - 2c) = (5 - 2c) + 6c$

3. $8.2 + \sqrt{10}$ is a real number.

4. $3c + c = 3c + 1c$

5. $(7a)r = 7(ar)$

6. $(5x + y)z = 5x \cdot z + y \cdot z$

Simplify each expression. Then evaluate for $x = 3$ and $y = 4$.

7. $4x^2 - y + x^2$

8. $2y - 3(y + 1)$

9. $4 - 2(x - 1)$

Written Exercises

Simplify.

1. $8x + 1 - 2x$

2. $xy^2 + x^2y - xy^2$

3. $3m - n + 15m - 14n + m - 12$

4. $16r - 11 - 18t - 15r + 17t + 10$

5. $9x^2 - 2y^2 + xy + 3x^2 + 2xy - y^2$

6. $4x^3 + 5x^2 - 6x^3 + 7x^2$

7. $14 - 4(5x + 3) + 10x$

8. $23 - (14 - 2n) - 12n$

Evaluate each expression for $x = -2$ and $y = 5$, if possible.

9. $x - y + 2$

10. $3(x + 2) + y$

11. $8x + 3y + 4x - 5y - 6$

12. $5(4x - y) - (15x + 2y) + 4$

13. $x^2 - y^2 + xy$

14. $3x^2y - 3xy - 3xy^2$

15. $8.4x - 7.5y + 0.8 - 3.2x + 4.3y$

16. $5.7xy - 3.2x + 2.4y - 3.5xy + x - y$

17. $y(-5x) - 2(7 - 3x) - (4x + y)$

18. $3.4x + 4.5(2x - y) - (10x - 3.3y)$

Identify the rule or property of operations with real numbers that is illustrated. All variables represent real numbers.

19. $8(x - y) = 8[x + (-y)]$

20. $12t - (3t + 4) = 12t - 1(3t + 4)$

21. $6 + 3(c + d) = 6 + (3c + 3d)$

22. $d^2(-7c) = (-7c)d^2$

23. $(-7c)d^2 = -7(cd^2)$

24. $5m + [3n + (-3n)] = 5m + 0$

25. $15 + \sqrt{2}$ is a real number.

26. $(7a + 0) - 5b = 7a - 5b$

In Exercises 27–29, show that the given generalization is *false*. (HINT: A generalization is false if it is possible to find one *counterexample*— that is, one example for which the generalization is not true.)

27. Subtraction is commutative $(a - b = b - a)$ for all real numbers a and b.

28. Subtraction is associative $[(a - b) - c = a - (b - c)]$ for all real numbers a, b, and c.

29. Division is commutative for all real numbers.

30. Use the definition of subtraction in Lesson 1.1 to prove the Rule of Subtraction, $a - b = a + (-b)$.

Mixed Review

Simplify, if possible.   *1.2*

1. $-7(-13) - 3(-4)$

2. $8 + (-2)(-3)(-4)$

3. $36 - (-3)^2$

Application: *Insect Songs*

Scientists have noticed a relationship between the temperature and the number of times a cricket chirps in a minute.

For one kind of cricket, a formula for the temperature T in degrees Fahrenheit in terms of the number of chirps C that a cricket makes in a minute is

$$T = 50 + \frac{C - 92}{4.7}.$$

1. What is the temperature if a cricket chirps 186 times per minute?

2. What is the value of the fractional term in the formula for $C = 92$? What temperature would this be?

1.4 Linear Equations

To solve linear equations in one variable

The equation $6x - 15 - 3x + 5 = 9x + 6 - 4x$ is a **first-degree** or **linear equation** in one variable. (*Second-degree equations*, such as $6x^2 - 15 = 9x$, are discussed in Chapters 6 and 9.) Any value of x that makes the equation true is called a **solution** or **root** of the equation.

You can solve this equation by:
simplifying each side,
subtracting $5x$ from each side,
adding 10 to each side, and then
dividing each side by -2, the
coefficient of x. The root is -8.

$$6x - 15 - 3x + 5 = 9x + 6 - 4x$$
$$3x - 10 = 5x + 6$$
$$-2x - 10 = 6$$
$$-2x = 16$$
$$\frac{-2x}{-2} = \frac{16}{-2}$$
$$x = -8$$

In the above example, each of the equations ($6x - 15 - 3x + 5 = 9x + 6 - 4x$, $3x - 10 = 5x + 6$, and so on) has -8 as its only solution. Equations that have the same solution (or solutions) are called **equivalent equations**. A linear equation in one variable is solved by writing a sequence of equivalent equations, each one simpler than the one preceding it. To do this, one or more of the following properties are used.

Properties of Equality
For all real numbers a, b, and c,
if $a = b$, then $a + c = b + c$
and $a - c = b - c$; ← (Addition/Subtraction Properties)
if $a = b$, then $a \cdot c = b \cdot c$
and $\frac{a}{c} = \frac{b}{c}$ ($c \neq 0$). ← (Multiplication/Division Properties)

EXAMPLE 1   Solve $2(3c + 19) - 9c = 7 - 3(11 - 3c)$.

Solution

$$6c + 38 - 9c = 7 - 33 + 9c \quad \leftarrow \text{First, the Distributive}$$
$$-3c + 38 = -26 + 9c \qquad \text{Property is used.}$$
$$-12c = -64$$
$$c = \frac{-64}{-12} = \frac{16}{3}$$

The solution is $\frac{16}{3}$, or $5\frac{1}{3}$.

EXAMPLE 2   Solve $2.1x + 3 = 51.3 - 1.35x$.

Plan   To eliminate the decimals, multiply each side by an appropriate power of 10. Since the greatest number of decimal places is 2 (in 1.35), multiply each side by 10^2, or 100.

Solution
$$100(2.1x + 3) = 100(51.3 - 1.35x)$$
$$210x + 300 = 5{,}130 - 135x$$
$$345x = 4{,}830$$
$$x = 14$$

Check
$$2.1x + 3 = 51.3 - 1.35x$$
$$2.1(14) + 3 \overset{?}{=} 51.3 - 1.35(14)$$
$$29.4 + 3 \overset{?}{=} 51.3 - 18.9$$
$$32.4 = 32.4 \text{ True}$$

The solution is 14.

The technique demonstrated in Example 2 can also be used to solve an equation such as $\frac{3}{4}x + \frac{1}{2} = \frac{7}{6}x - 3$.

Use the Multiplication Property of Equality to eliminate the fractions. Recall that the *least common denominator* (LCD) of a set of fractions is the *least common multiple* (LCM) of the denominators. The least common denominator of $\frac{3}{4}$, $\frac{1}{2}$, and $\frac{7}{6}$ is 12. So, we multiply each side of the equation by 12.

EXAMPLE 3   Solve $\frac{3}{4}x + \frac{1}{2} = \frac{7}{6}x - 3$.

Solution   Multiply each side by 12, the LCD of the fraction.
$$12\left(\tfrac{3}{4}x + \tfrac{1}{2}\right) = 12\left(\tfrac{7}{6}x - 3\right)$$
$$9x + 6 = 14x - 36$$
$$-5x = -42$$
$$x = \frac{42}{5}, \text{ or } 8\tfrac{2}{5}$$

The root is $8\tfrac{2}{5}$.

A **proportion** is an equation of the form $\frac{a}{b} = \frac{c}{d}$. Both $\frac{2}{3} = \frac{10}{15}$ and $2 \cdot 15 = 3 \cdot 10$ are true equations. This suggests a property of proportions that can be used to solve an equation such as $\frac{2x + 8}{3x - 2} = \frac{5}{4}$.

Proportion Property

If $\frac{a}{b} = \frac{c}{d}$, then $a \cdot d = b \cdot c$.

EXAMPLE 4   Solve $\dfrac{2x+8}{3x-2} = \dfrac{5}{4}$.

Solution   Use the Proportion Property. $\dfrac{2x+8}{3x-2} = \dfrac{5}{4}$

$$(2x+8)4 = (3x-2)5$$
$$8x + 32 = 15x - 10$$
$$-7x = -42$$
$$x = 6$$

The solution is 6.

An equation with more than one variable, such as $px - 5 = n$, is called a **literal equation**. It is "solved for x" when x is alone on one side.

EXAMPLE 5   Solve each equation for x.

Solutions

a. $px - 5 = n$

$px = n + 5$

$x = \dfrac{n+5}{p}$

b. $y = \dfrac{7}{3}x + z$

$y - z = \dfrac{7}{3}x$

$\dfrac{3}{7}(y - z) = \dfrac{3}{7} \cdot \dfrac{7}{3}x$

$\dfrac{3}{7}(y - z) = x$

c. $a(x + b) = d$

$x + b = \dfrac{d}{a}$

$x = \dfrac{d}{a} - b$

Classroom Exercises

Solve each equation for x.

1. $\dfrac{1}{3}x = 6$

2. $7 - x = 9$

3. $2 = 5 + x$

4. $5 - 10x = 0$

5. $\dfrac{1}{5} = \dfrac{2}{x}$

6. $\dfrac{x}{4} = \dfrac{x}{6}$

7. $ax = b$

8. $x + n = m$

9. $5x + 14 = 6x$

10. $x - c - d = 0$

11. $\dfrac{x}{a} = b$

12. $5 - x = -2$

Written Exercises

Solve each equation. Check.

1. $4a - 2 = 10$

2. $10x = 2x - 32$

3. $7y + 11 = 8 - 5y$

4. $-43 = 7 - 8t$

5. $2(x - 4) = -5$

6. $15 = 4(6 - 3c)$

Solve each equation.

7. $18x - 14 - 21x = 17 - x + 7$

8. $14 + 5(6n - 2) = 25n - 16$

9. $8y - 5.4 = 3y - 1.2$

10. $2.9n - 5 = 3 - 0.3n$

11. $\frac{3}{5}x + 11 = 31 - \frac{1}{5}x$

12. $\frac{2}{7}y - 15 = \frac{6}{7}y - 9$

13. $12 - (7c - 11) = 3(5 - 3c)$

14. $2.4(20 - x) + 1.8x = 43.8$

15. $\frac{x - 3}{2} = \frac{x}{5}$

16. $\frac{x + 2}{x - 4} = \frac{7}{4}$

Solve each equation for x.

17. $ax - b = c$

18. $5x - y = 3x + 3y$

19. $2tx + 7 = 8 - 3tx$

20. $a(2x - b) = 7$

21. $4r(x + t) = 3rx + n$

22. $\frac{x - a}{2} = \frac{x + b}{3}$

Solve each equation.

23. $p + p(0.08)3 = 3,100$

24. $p(0.09)4 + p(0.11)4 = 1,200$

25. $3(5.4 + r) = 5.4r$

26. $4.2(r + 8.7) = 8.7r$

27. $540 = v \cdot 6 - 5 \cdot 6^2$

28. $242 = v \cdot 4 - 5 \cdot 4^2$

29. $\frac{3}{4}n - \frac{7}{8} = \frac{3}{2}n - \frac{3}{4}$

30. $\frac{2}{3}y + \frac{5}{2} = \frac{4}{5}y + \frac{7}{6}$

31. $\frac{8x - 3}{6x + 9} = \frac{-2}{3}$

32. $\frac{5x + 6}{3} = \frac{3x - 14}{5}$

33. $3(2y - 5) - (7y - 6) = -10$

34. $-6(3 - 3x) = -9(2x - 30)$

35. $\frac{3}{5}(10x - 15) + \frac{5}{4}(16x + 24) = 86$

36. $\frac{5}{6}(18 - 12t) - \frac{2}{3}(12t + 15) = 14$

Equations such as $x + 3 = 3 + x$ or $2(y - 5) = y - 10 + y$ that
have all real numbers as solutions are called *identities*. The equation
$2x = 2x + 6$, however, has no root because $2x$ is less than $2x + 6$ for
each real number x. Determine whether each equation is an identity or
has no roots.

37. $5x + 2 = 2 + 5x$

38. $3a + 1 = 3a$

39. $5(4n + 6) - 10 = 2(15 + 10n)$

40. $\frac{1}{3}y + \frac{1}{4} = \frac{1}{2}y + 0.25 - \frac{1}{6}y$

41. $2[8 - (15 - 2z)] = 4z - 15$

42. $5(4y - 2) - 7 = 15y - 17 + 5y$

Mixed Review

Identify the property of operations with real numbers that is
illustrated.   *1.3*

1. $(m - n)3 = 3(m - n)$

2. $-2(3x) = (-2 \cdot 3)x$

3. $-12c + 12c = 0$

4. $\frac{2}{7} \cdot \frac{7}{2}t = 1t$

1.5 Problem Solving: Using Formulas

Objective

To solve word problems using given formulas

Sometimes an equation is found to be useful in solving problems from business, industry, and everyday life. In such a case, the equation is often called a **formula**. Shown below is one formula from banking.

If you borrow $800 (principal) at 9% (rate) for 3 years (time), then at the end of three years you will owe money for the simple interest, which will be included in the total amount due (principal plus interest).

> You will owe $800(0.09)3, or $216 in simple interest, and
> $800 + $216, or $1,016 as the total amount.

In general, if you borrow or loan p dollars at an annual simple interest rate of r (expressed as a decimal) for t years, then

$$i = prt \qquad \text{and} \qquad A = p + prt$$

where p is the principal, r is the rate, t is the time, i is the amount of interest, and A is the total amount due.

The general rule, or *formula*, $A = p + prt$ contains four variables. The value of one of the variables can be found when values for the other variables are given, as shown in the example below.

EXAMPLE 1

At the end of 2 years 6 months, Sylvia owed a total of $3,637.50 on her junior-college loan. If she was being charged a simple interest rate of 8.5% per year, how much money did she originally borrow? Use the formula $A = p + prt$. (THINK: Did she borrow more or less than $3,637.50?)

Solution

What are you to find?
What is given?

$\qquad$ The principal, p
$\qquad$ t = 2 yr 6 mo, or 2.5 yr
$\qquad$ A = $3,637.50

$\qquad$ r = 8.5%, or 0.085

Write the formula.
Substitute in the formula for A, r, and t.

$\qquad$ $A = p + prt$
$\qquad$ $3,637.5 = p + p(0.085)2.5$
$\qquad \qquad \quad = 1p + 0.2125p$
$\qquad \qquad \quad = 1.2125p$

Solve the equation.

$\qquad$ $\dfrac{3,637.5}{1.2125} = p$
$\qquad \qquad 3,000 = p$

State your answer.

$\qquad$ Sylvia borrowed $3,000.

Suppose an object is projected upward from the earth with an initial vertical velocity of 40 meters per second (40 m/s). While aloft, the object's height h in meters at the end of t seconds is given approximately by the formula $h = 40t - 5t^2$.

You can use this formula to verify the height for each time shown in the figure. For example, for $t = 3$ s,
$h = 40(3) - 5(3)^2 = 75$ m,
and for $t = 5$ s,
$h = 40(5) - 5(5)^2 = 75$ m.

In general, if an object is shot upward with an initial vertical velocity of v meters per second, then its height h in meters at the end of t seconds will be given by the formula $h = vt - 5t^2$.

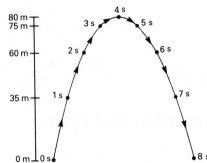

Rise and Fall of an Object When Initial Vertical Velocity is 40 m/s

EXAMPLE 2

Find the initial vertical velocity v (in meters per second) necessary for an object launched upward to reach a height of 800 m at the end of 4 s. Use the formula $h = vt - 5t^2$.

Solution

What are you to find?	The initial vertical velocity, v
What is given?	$h = 800$, $t = 4$
Write the formula.	$h = vt - 5t^2$
Substitute in the formula for h and t.	$800 = v \cdot 4 - 5 \cdot 4^2$
Solve the equation.	$880 = 4v$
State your answer.	$220 = v$

The necessary velocity is 220 m/s.

A formula can be solved for any of its variables in terms of the other variables. (Review Example 5 in Lesson 1.4.)

EXAMPLE 3

A trapezoid has an area of 144 square feet, or 144 ft^2. Find the length of one base given that the length of the other base is 7 ft and the height is 16 ft.

Plan

From the *Formulas from Geometry* at the back of the book, find the formula for the area of a trapezoid. Solve the formula for b_1 (read: "b-sub-one"). Then in the new formula, substitute 144 for A, 7 for b_2, and 16 for h. Compute the value of b_1, and label with the correct unit.

a.

$$A = \tfrac{1}{2}h(b_1 + b_2)$$
$$2 \cdot A = 2 \cdot \tfrac{1}{2} \cdot h(b_1 + b_2)$$
$$2A = h(b_1 + b_2)$$
$$2A = hb_1 + hb_2$$
$$2A - hb_2 = hb_1$$
$$\frac{2A - hb_2}{h} = b_1$$

b.

$$b_1 = \frac{2A - hb_2}{h}$$
$$= \frac{2 \cdot 144 - 16 \cdot 7}{16}$$
$$= \frac{288 - 112}{16}$$
$$= \frac{176}{16}, \text{ or } 11$$

The first base is 11 ft long.

Classroom Exercises

Change each percent to a decimal.

1. 7%   **2.** 12%   **3.** $9\tfrac{1}{2}\%$ **4.** $11\tfrac{3}{4}\%$

Use a decimal to express each time period in years only.

5. 5 years 3 months   **6.** 4 years 9 months   **7.** May 15, 1987 to Nov 15, 1993

Find the simple interest earned in one year for the given principal (p) and annual interest rate (r).

8. $p = \$500, r = 8\%$ **9.** $p = \$900, r = 10\%$ **10.** $p = \$1,000, r = 12\%$

11. Find p in dollars given that $A = \$7,625$, $r = 10\tfrac{1}{2}\%$, and $t = 5$ yrs. Use $A = p + prt$.

12. Find v in meters per second given that $h = 2,080$ m and $t = 8$ s. Use $h = vt - 5t^2$.

Written Exercises

Use the formula $i = prt$ to find each of the following.

1. i given that $p = \$6,000$, $r = 7\%$, and $t = 4$ years

2. p given that $i = \$500$, $r = 12\tfrac{1}{2}\%$, and $t = 5$ years

3. r given that $i = \$990$, $p = \$1,200$, and $t = 10$ years

Solve each problem. Use the formula $i = prt$.

4. Edgar borrowed $3,500 for 4 years 9 months, and was charged $1,330 in simple interest. What was the annual interest rate?

5. Angela owes $1,806 in simple interest on a loan of $4,300 at 12% per year. Find the time for which Angela has held the loan in years and months.

Solve each problem. Use the formula $A = p + prt$.

6. Find p in dollars given that $A = \$9,990$, $r = 9\frac{1}{2}\%$, and $t = 7$ years.

7. Mr. Worthy took out a loan on June 1, 1986, at $7\frac{1}{2}\%$ per year. How much did he borrow if he owed a total of $\$3,030$ on December 1, 1989?

Solve each problem. Use the formula $h = vt - 5t^2$.

8. Find v in meters per second given that $h = 300$ m and $t = 6$ s.

9. Given that $h = 85$ m and $t = 5$ s, find v in meters per second.

10. At the end of 4 seconds, what will be the height of an object shot upward with an initial vertical velocity of 90 meters per second?

11. What initial vertical velocity is required to boost an object to a height of 168 m at the end of 4 s?

For each exercise, select the appropriate formula from the *Formulas from Geometry* on page 788. Then solve the formula for the appropriate variable. Use the new formula to solve the problem.

12. Find the length of a rectangle that is 6.5 cm wide and has a perimeter of 32 cm.

13. The area of a triangle is 34.5 ft^2. Find the length of its base given that its height is 6 ft.

14. A trapezoid has an area of 111 cm^2. Find its height given that its bases are 17 cm and 20 cm in length.

15. A cube has a volume of 50.653 m^3. Find the length of an edge given that the area of a face is 13.69 m^2.

Solve each problem. Use an appropriate formula.

16. If $h = vt - 5t^2$, then the maximum value of h is $\frac{v^2}{20}$. What initial vertical velocity is needed to launch an object to a maximum height of 4,500 m?

17. Find, to the nearest hundred meters, the maximum height reached by a projectile launched upward with an initial vertical velocity of 240 m/s (see Exercise 16).

18. Find t in years and months given that $A = \$11,400$, $p = \$7,500$, and $r = 8\%$ (simple interest).

19. Find the annual simple interest rate r given that $A = \$1,044$, $p = \$600$, and $t = 8$ years.

20. At the end of 4 years 3 months, Mr. Benjamin owed a total of $\$4,594$ on a loan made at $10\frac{1}{4}\%$ (simple interest). What was the amount of the loan?

Determine the time needed for each investment to be doubled in total value (investment plus simple interest earned).

21. $3,000 at 8% per year

22. p dollars at 10% per year

If an object is sent upward from d meters above the surface of the earth with an initial vertical velocity of v meters per second, then its height h in meters at the end of t seconds is determined by the formula

$$h = d + vt - 5t^2.$$

Use the formula $h = d + vt - 5t^2$ for Exercises 23–25.

23. Find h for $d = 120$ m, $v = 80$ m/s, and $t = 6$ s.

24. Find d for $h = 275$ m, $v = 70$ m/s, and $t = 5$ s.

25. At the end of 4 s, what is the height above the earth of an object shot upward with an initial vertical velocity of 85 m/s from the bottom of a shaft that extends 40 m below the earth's surface?

Midchapter Review

Simplify, if possible. *1.1, 1.2*

1. $-7 - 3 \cdot 2(-5) + 6 - 21 \div 3$

2. $-5 \cdot 2^3$

3. $(-5 \cdot 2)^3$

4. $\dfrac{6 - 9 \cdot 2}{-24 + 8 \cdot 3}$

Identify the property of operations with real numbers that is illustrated. *1.3*

5. $9 + \sqrt{2}$ is a real number.

6. $(3m + 2)n = 3m \cdot n + 2 \cdot n$

7. $(3m + 2)n = n(3m + 2)$

8. $-4c + 4c = 0$

9. Simplify $3(6x - 2y) - (8x - y) + xy$, and evaluate for $x = -5$ and $y = -2$.

Solve. *1.4*

10. $3(2x + 11) - x = 7 - 2(12 - 5x)$

11. $3.5x + 5 = 32.5 - 1.5x$

12. Solve $t(3x - 4) = v$ for x.

13. Use the formula $A = p + prt$ to find p in dollars given that $A = \$1,120$, $r = 8\%$, and $t = 5$ years. *1.5*

1.6 Problem Solving: One or More Numbers

Objective   To solve word problems about one or more numbers

When solving word problems, it is often necessary to translate word phrases and sentences into algebraic expressions or equations.

Word phrase or sentence	Algebraic expression or equation
Twelve decreased by a number	$12 - n$
Twelve less than a number	$n - 12$
Twice a number, increased by 12	$2a + 12$
Five decreased by the sum of a number and 12	$5 - (y + 12)$
Nine less than a number is 8 more than twice the number.	$n - 9 = 2n + 8$
Three divided by a number is the same as 6 divided by 2 more than the number.	$\dfrac{3}{y} = \dfrac{6}{y + 2}$

In some word problems, you are asked to find more than one number. To do this, begin by representing each number in terms of the same variable.

EXAMPLE 1   The second of three numbers is 4 times the first number. The third number is 5 less than the second number. If twice the first number is decreased by the third number, the result is the same as 23 more than the second number. What are the three numbers?

Solution

What are you to find?   Three numbers
Choose a variable.   Let f = the first number.
What does it represent?   Then $4f$ = the second number, and
$4f - 5$ = the third number.

What is given?   Twice the 1st, minus the 3rd,
equals the 2nd plus 23.

Write an equation. $2f - (4f - 5) = 4f + 23$
Solve the equation. $2f - 4f + 5 = 4f + 23$
$-6f = 18$
$f = -3 \rightarrow$ the first number $= -3$
$f = -3$ $4f = -12$ $4f - 5 = -17$

Check in the original problem.   "Twice the 1st, minus the 3rd" is
$2(-3) - (-17)$, or 11.
"23 more than the 2nd" is $-12 + 23$, or 11.

State the answer.   The three numbers are -3, -12, and -17.

Many problems require information that is not given directly in the problem. For example, a problem may involve consecutive integers or consecutive multiples of an integer. Four such word phrases, along with numerical examples and algebraic expressions, are given in the table below.

Word Phrase	Example	Algebraic Expression
Consecutive integers	$-2, -1, 0, 1$	$n, n + 1, n + 2, \ldots$ (n is an integer.)
Consecutive odd integers	$7, 9, 11, 13, 15$	$x, x + 2, x + 4, \ldots$ (x is odd.)
Consecutive even integers	$-4, -2, 0, 2$	$y, y + 2, y + 4, \ldots$ (y is even.)
Consecutive multiples of 3	$12, 15, 18, 21$	$t, t + 3, t + 6, \ldots$ (t is a multiple of 3.)

EXAMPLE 2

Find three consecutive odd integers such that twice the sum of the first two integers, decreased by the third integer, is the same as 20 less than the second integer.

Solution

Let x, $x + 2$, and $x + 4$ represent the three odd integers.
$$2[x + (x + 2)] - (x + 4) = x + 2 - 20$$
$$4x + 4 - x - 4 = x - 18$$
$$2x = -18$$
$$x = -9 \leftarrow \text{the first odd integer} = -9$$

$$x = -9 \qquad x + 2 = -7 \qquad x + 4 = -5$$

Check

Twice (the 1st plus the 2nd) minus the 3rd is $2[-9 + (-7)] - (-5)$, or -27.

20 less than the 2nd is $-7 - 20$, or -27.

The three consecutive odd integers are -9, -7, and -5.

Classroom Exercises

Represent each number in terms of one variable.

1. The second of three numbers is 5 times the first number. The third number is 4 less than the second number.

2. The greatest of three numbers is 40 more than the smallest. The remaining (middle) number is 5 less than 6 times the smallest number.

3. Write an equation for the three numbers of Classroom Exercise 1, given that their sum is 18.

4. Write an equation for the three numbers of Classroom Exercise 2, given that their sum is 3 times the middle number.

5–6. Solve the equations of Classroom Exercises 3 and 4.

7. Eight less than 4 times a number is the same as 8 times the sum of the number and 2. What is the number?

8. Fifteen more than 7 times a number is twice the sum of the number and -6. What is the number?

Written Exercises

1. If 18 is decreased by the sum of a number and 6, the result is 4 less than 3 times the number. Find the number.

2. If 3.5 times a number is increased by 2, the result is the same as 7.5 times 4 less than the number. Find the number.

3. The greater of two numbers is 10 less than 3 times the smaller. If the greater is increased by twice the smaller, the result is 8 less than 3 times the greater. What are the two numbers?

4. A Brand X bottle contains one dozen fewer capsules than a Brand Y bottle. Five Brand X and 8 Brand Y bottles contain 486 capsules. How many capsules are in a Brand X bottle?

5. The second of three numbers is 4 times the first number. The third number is 12 less than the first. Find the three numbers given that their sum is 42.

6. Find three consecutive integers whose sum is 37 less than 5 times the third integer.

7. There are three consecutive even integers such that the sum of the second and third integers is 30 more than the first integer. Find these three even integers.

8. There are three consecutive odd integers such that twice their sum is 30 less than 8 times the third integer. Find these odd integers.

9. The greatest of three numbers is 7 times the smallest, and the middle number is 8 more than the smallest. Six more than the greatest number, decreased by 3 times the middle number, is 9 less than the smallest number. Find the three numbers.

10. The second of three numbers is twice the first, and the third is 5 less than the second. If 12 less than the first is decreased by the third, the result is -15. What are the three numbers?

11. Find three consecutive integers such that 6 times the first, decreased by the third, is 398.

12. Find three consecutive odd integers such that 5 times the second, decreased by twice the third, is 155.

24 Chapter 1 Linear Equations

13. Find three consecutive multiples of 15 such that twice their sum is 360.

14. One half of the sum of three consecutive multiples of 10 is 90. What are these multiples of 10?

15. A company's sales for June increased $600 over its May sales figure. July sales were twice June sales, and the sales for August doubled the figure for July. The sales total was $20,200 for the four months. Find the sales figure for each month.

16. A size *A* box holds twice as many oranges as a size *C* box, and a size *B* box holds one dozen fewer oranges than a size *A* box. Six size *A* boxes, 5 size *B* boxes, and 8 size *C* boxes will hold a total of 45 dozen oranges. Find the number of oranges that a size *B* box will hold.

17. Find all sets of three consecutive even integers such that 6 less than twice the third integer is the sum of the first and second integers.

18. A wire 50 in. long is bent to form a rectangle. Represent the rectangle's length given that $x - 3$ inches represents the width.

19. A first angle of measure *x* and a second angle of measure *y* are *complementary*—that is, the sum of their measures is 90. The second angle measures 4 times the measure of the first angle less $5a$. Two angles are *supplementary* if the sum of their measures is 180. Represent the measure of the supplement of the first angle in terms of *a*.

Mixed Review

Solve each equation for *x*.   *1.4*

1. $\dfrac{3x - 6}{x + 5} = \dfrac{3}{8}$

2. $\dfrac{2}{3}x + \dfrac{1}{2} = \dfrac{5}{6}x$

3. $a(5x + b) = 7$

▰/ *Brainteaser*

The sum of two numbers is 1. Which is larger: the sum of the larger number and the square of the smaller number, or the sum of the smaller number and the square of the larger number?

Application: **Equilibrium**

A sculptor is commissioned to create a large mobile for a public hall. For the mobile (shown below) to be balanced, the support cables must be attached at the precise points of equilibrium (*A* and *B*).

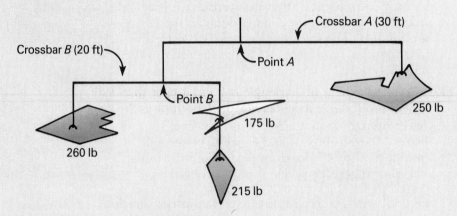

To calculate its point of equilibrium, think of each crossbar as a simple lever. A lever system is balanced when $F_1 \times d_1 = F_2 \times d_2$, where F_1 and F_2 are the forces (weights) and d_1 and d_2 are the distances.

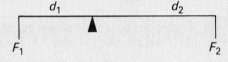

EXAMPLE   Find the point of equilibrium for Crossbar *B* (point *B*). Assume that the weight of the crossbars and the cables can be disregarded.
Let x = distance from the 260-lb shape to the fulcrum (d_1).
Therefore, $d_2 = 20 - x$.

Solution
$$F_1 \times d_1 = F_2 \times d_2$$
$$260 \cdot x = (175 + 215)(20 - x)$$
$$x = 12 \leftarrow \quad \text{distance from 260-lb shape to fulcrum} = 12 \text{ ft}$$

The point of equilibrium for Crossbar *B* is 12 ft from the 260-lb shape.

1. Find the point of equilibrium for Crossbar *A*.

2. If a 150-lb shape were added to the cable holding the 260-lb shape, how far would point *B* move? Which way?

3. The construction team mistakenly attaches the main support cable to a point on Crossbar *A* 12 ft from the 250-lb shape. For the mobile to be balanced, how much weight must be added? To which end?

4. The sculptor resolves the imbalance in Exercise 3 by adding another shape to Crossbar *A* at a point 4 ft from the cable holding the 250-lb shape. How much must this extra shape weigh in order to restore equilibrium?

1.7 Problem Solving: Perimeter and Area

Objective   To solve word problems involving perimeter and area

If a word problem involves a geometric figure, you can represent the data by drawing and labeling the figure. Area and perimeter formulas can be found in the *Formulas from Geometry* at the back of the book.

EXAMPLE 1   Side *a* of a triangle is 9 cm longer than side *b*. Side *c* is 1.5 times as long as side *a*. Find the length of each side given that the perimeter of the triangle is 75 cm.

What are you to find?   The length of each side

What is given?
Represent the data as
    shown at the right.

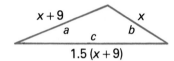

The **perimeter** is the sum of the lengths of the sides.

Write an equation. $(x + 9) + x + 1.5(x + 9) = 75$
Solve the equation. $x + 9 + x + 1.5x + 13.5 = 75$
$$3.5x = 52.5$$
$$x = 15$$

The three lengths are: $x = 15$
$x + 9 = 24$ $1.5(x + 9) = 36$

Check in the   The perimeter is $15 + 24 + 36 = 75$.
    original problem.
State the answer.   The lengths are 15 cm, 24 cm, and 36 cm.

The lengths of the sides of one geometric figure may be related to the lengths of the sides of another. In such cases, draw and label both figures to represent the data as shown below.

EXAMPLE 2   Rectangle I is 4 m longer than it is wide. Rectangle II is 3 m wider than, and twice as long as rectangle I. The difference between their perimeters is 24 m.

a. Find the length of rectangle II.   **b.** Find the area of rectangle I.

Plan   Draw and label the two rectangles. Use w to represent the width of rectangle I.

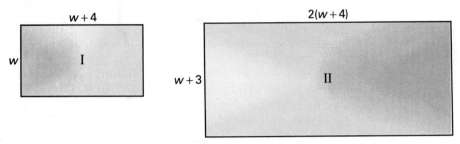

Solutions   **a.** Perimeter I $= 2w + 2(w + 4)$, or $4w + 8$
Perimeter II $= 2(w + 3) + 2[2(w + 4)]$, or $6w + 22$
Perimeter II $-$ Perimeter I $= 24$
$$6w + 22 - (4w + 8) = 24$$
$$6w + 22 - 4w - 8 = 24$$
$$2w = 10$$
$$w = 5 \leftarrow \text{width of rectangle I} = 5 \text{ m}$$
The length of rectangle II, $2(w + 4)$, is 18 m.

b. The area of rectangle I, $w(w + 4)$, is $5 \cdot 9$, or 45 m^2.

The formulas for the circumference and area of a circle include the irrational number π.
An approximate value of π is 3.14. This can be written as $\pi \approx 3.14$, where the symbol $\approx$ means "is approximately equal to."

Some scientific calculators have a key for π. Use such a calculator to find π to six decimal places.

$C = 2\pi r$
$A = \pi r^2$

EXAMPLE 3   The circumference of a circle is 47.1 in. Find the area of the circle to the nearest in^2.

Solution   Let $r =$ the length of the radius.
Use $C = 2\pi r$ and $\pi \approx 3.14$ to find r.
$$C = 2\pi r$$
$$47.1 \approx 2(3.14)r, \text{ or } 6.28r$$
$$\frac{47.1}{6.28} \approx r \text{ Thus, } 7.50 \approx r, \text{ or } r \approx 7.50$$
Use $A = \pi r^2$ to find the area with $r \approx 7.5$ and $\pi \approx 3.14$.
$$A \approx 3.14(7.5^2) = 3.14(56.25) = 176.625$$

The area is 177 in^2, to the nearest in^2.

Classroom Exercises

Find the perimeter or circumference of each figure. Include the appropriate unit of measure. Use $\pi \approx 3.14$.

1.

3 ft | rectangle
9 ft

2.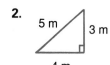

5 m   3 m
4 m

3.

5 cm

4–6. Find the area of each figure of Exercises 1–3. Include the appropriate unit of measure. Use $\pi \approx 3.14$.

7. Side a of a triangle is 5 ft longer than side b. Side c is twice as long as side b. Find the length of each side given that the perimeter of the triangle is 45 ft.

Written Exercises

Write an expression for the perimeter of the figure shown below.

1.

rectangle | $x - 1$
x

2.

square | $3x$

3.

x
4 | trapezoid | 4
$x + 4$

4. Side c of a triangle is 4 m longer than side a, and side b is 5 m shorter than side a. The perimeter is 59 m. What are the lengths of the sides?

5. The length of a rectangle is 6 yd more than 3 times its width. Find the length and the area of the rectangle given that its perimeter is 44 yd.

6. The hypotenuse of a right triangle is 15 in. long and one leg is 3 in. longer than the other leg. Given that the perimeter is 36 in., find the area of the right triangle in square inches.

7. A square and a rectangle have the same width. The rectangle is 8 m longer than the square, and one perimeter is twice the other perimeter. Find the dimensions of the rectangle.

8. One side of an equilateral triangle is 6 ft shorter than one side of the square. The difference between the perimeters of the two figures is 28 ft. Find the perimeter of each figure.

9. The circumference of a circle is 31.4 cm. Find its area to the nearest 0.1 cm^2. Use $\pi \approx 3.14$.

10. One base of a trapezoid is 1 in. shorter than the height, and the other base is 1 in. longer than twice the height. Find the area given that each leg is 5 in. long and the perimeter is 22 in.

11. The two longest sides of a pentagon (5-sided figure) are each 3.5 times as long as the shortest side. The remaining sides are each 9 m longer than the shortest side. Find the length of each side given that the perimeter is 78 m.

12. Each of the three longer sides of a hexagon (6-sided figure) is 2 in. less than 4 times the length of each of the three shorter sides. Find the length of a longer side given that the perimeter is 114 in.

13. One leg of a right triangle is 4 cm longer than twice the length of the other leg. The hypotenuse is 2 cm longer than the longer leg. Find the area given that the perimeter is 60 cm.

14. A longer side of a parallelogram is 3 ft shorter than 4 times the length of a shorter side. One side of a rhombus (equilateral quadrilateral) and a longer side of the parallelogram are equal in length. The difference between the two perimeters is 24 ft. What is the perimeter of the rhombus?

15. Find, to the nearest square meter, the area of a circular roof whose circumference is 78.5 m. Use $\pi \approx 3.14$.

16. Describe in your own words how to find the area of a rectangle, if its length is described in terms of its width, and its perimeter is given in feet.

17. The lengths of the sides of a quadrilateral have the ratio of 2 to 4 to 5 to 7. What is the length of each side if the perimeter is 117 in.?

18. One side of an equilateral triangle is 4 cm longer than one side of a regular hexagon. The difference between their perimeters is 10.5 cm. Find the perimeter of the triangle.

19. A right triangle has a perimeter of 30 ft, and one leg is 5 ft long. Find the area of the triangle. (HINT: $(m - n)^2 = m^2 - 2mn + n^2$)

20. The perimeters x, y, and z of three figures are integers with the ratios $x:y = 2:3$ and $y:z = 5:4$. Find the three perimeters given that their sum is 74 ft.

Mixed Review

Name the property of operations with real numbers that is illustrated. *1.3*

1. $-4.2 + \sqrt{6}$ is a real number.
2. $(a + 3) + 5 = a + (3 + 5)$
3. $6t + [-(6t)] = 0$
4. $8x + x = 8x + 1x$

1.8 Proof in Algebra

Objectives
To prove two algebraic expressions equivalent
To determine whether relations are equivalence relations

Statements that are assumed to be true are called **axioms** or **postulates**. The eleven properties of a field listed in Lesson 1.3 are postulates because they are accepted as true without proof. A **theorem** is a statement that has been logically deduced from a set of axioms or postulates.

The following additional postulates are needed to prove some theorems for all real numbers a, b, and c.

Zero Property for Multiplication
$a \cdot 0 = 0$

Properties of Equality

If $a = b$, then $a + c = b + c$.	(Add Prop of Eq)
If $a + c = b + c$, then $a = b$.	(Subt Prop of Eq)
If $a = b$, then $a \cdot c = b \cdot c$.	(Mult Prop of Eq)
If $a \cdot c = b \cdot c$ and $c \neq 0$, then $a = b$.	(Div Prop of Eq)
$a = a$	(Reflexive Prop of Eq)
If $a = b$, then $b = a$.	(Symmetric Prop of Eq)
If $a = b$ and $b = c$, then $a = c$.	(Transitive Prop of Eq)

Substitution
If $a = b$, then a can be replaced by b, and vice-versa, in any equation without changing the *truth-value* (truth or falsity) of the equation.

It was stated in Lesson 1.3 that the terms of an expression can be reordered and regrouped. One such rearrangement is stated in the following theorem. Notice that each statement in the proof of the theorem is justified by a postulate for operations with real numbers.

Theorem 1.1

$(a + b) + (c + d) = (a + c) + (b + d)$ for all real numbers a, b, c, and d.

Proof

	Statement	Reason
1.	$(a + b) + (c + d)$ $= [(a + b) + c] + d$	Associative Prop of Add
2.	$= [a + (b + c)] + d$	Associative Prop of Add
3.	$= [a + (c + b)] + d$	Commutative Prop of Add
4.	$= [(a + c) + b] + d$	Associative Prop of Add
5.	$= (a + c) + (b + d)$	Associative Prop of Add
6.	Thus, $(a + b) + (c + d) = (a + c) + (b + d)$	

Statement 6 follows from the Transitive Property of Equality used several times in succession. (If $x = y$ and $y = z$, then $x = z$.)

Theorem 1.2

$-x = -1x$ for each real number x.

Proof

Statement	Reason
1. x is a real number.	Given
2. $-1x$ is a real number.	Closure Prop for Mult
3. $-1x + x$ is a real number.	Closure Prop for Add
4. $-1x + x = -1x + 1x$	Identity for Mult
5. $ = (-1 + 1)x$	Distr Prop
6. $ = 0 \cdot x$	Add Inverse Prop
7. $ = 0$	Zero Prop for Mult
8. $-1x + x = 0$	Trans Prop of Eq (Steps 4–7)
9. $-x + x = 0$	Add Inverse Prop
10. $-1x + x = -x + x$	Sub (Steps 8–9)
11. $ -1x = -x$	Subt Prop of Eq
12. $ -x = -1x$	Sym Prop of Eq

It was stated in Lesson 1.3 that multiplication is distributive over subtraction: $(a - b)c = ac - bc$. To prove this, it is helpful to prove first that $(-a)b = -(ab)$.

Theorem 1.3

$(-a)b = -(ab)$ for all real numbers a and b.

Proof

Statement	Reason
1. $(-a)b = (-1a)b$	Theorem 1.2 $[-a = -1a]$
2. $ = -1(ab)$	Associative Prop of Mult
3. $ = -(ab)$	Theorem 1.2 $[-(ab) = -1(ab)]$
4. $(-a)b = -(ab)$	Trans Prop of Eq

Theorem 1.4

$(a - b)c = ac - bc$ for all real numbers a, b, and c.

Proof

Statement	Reason
1. $(a - b)c = [a + (-b)]\, c$	Rule of Subt
2. $ = ac + (-b)c$	Distr Prop
3. $ = ac + [-(bc)]$	Theorem 1.3 $[(-b)c = -(bc)]$
4. $ = ac - bc$	Rule of Subt
5. $(a - b)c = ac - bc$	Trans Prop of Eq

A real-number fact, such as $-1 - 2 = -3$, can be used to replace one numeral with an equivalent numeral in a proof, as shown below.

Theorem 1.5

$-z - 2z = -3z$ for each real number z.

Proof

Statement	Reason
1. $-z - 2z = -1z - 2z$	Theorem 1.2 $[-z = -1z]$
2. $\quad\quad = (-1 - 2)z$	Theorem 1.4 $[(-1 - 2)z = -1z - 2z]$
3. $\quad\quad = -3z$	Number fact $[-1 - 2 = -3]$
4. $-z - 2z = -3z$	Trans Prop of Eq

You are asked to prove Theorems 1.6 and 1.7 below in Exercises 7 and 10.

Theorem 1.6

$-(-a) = a$ for each real number a.

Theorem 1.7

$(ab)(cd) = (ac)(bd)$ for all real numbers a, b, c, and d.

There are other relations besides $=$ that may be reflexive, symmetric, or transitive. In general, a relation $\Re$ is
  (1) reflexive when: $a \,\Re\, a$ is true for all a.
  (2) symmetric when: If $a \,\Re\, b$, then $b \,\Re\, a$ is true for all a and b.
  (3) transitive when: If $a \,\Re\, b$ and $b \,\Re\, c$, then $a \,\Re\, c$ is true for all a, b, and c.

Definition

Any relation that is reflexive, symmetric, and transitive is called an **equivalence relation.**

EXAMPLE

Is the relation $<$ *(is less than)* for real numbers (1) reflexive, (2) symmetric, (3) transitive, or (4) an equivalence relation?

Solution

(1) $<$ is *not* reflexive: "$8 < 8$" is a false statement.
(2) $<$ is *not* symmetric: "If $5 < 6$, then $6 < 5$" is not true.
(3) $<$ *is* transitive: "If $7 < 9$ and $9 < 11$, then $7 < 11$" is true and "If $a < b$ and $b < c$, then $a < c$" is true for all real numbers a, b, and c.
(4) $<$ is *not* an equivalence relation because it does not satisfy all three conditions.

Classroom Exercises

Justify each step in the proof.

1. $-(a + b) = -a - b$
1. $-(a + b) = -1(a + b)$
2. $\qquad\qquad = -1a + (-1)b$
3. $\qquad\qquad = -a + (-b)$
4. $\qquad\qquad = -a - b$
5. $-(a + b) = -a - b$

2. $(x + y) - y = x$
1. $(x + y) - y = (x + y) + (-y)$
2. $\qquad\qquad = x + [y + (-y)]$
3. $\qquad\qquad = x + 0$
4. $\qquad\qquad = x$
5. $(x + y) - y = x$

Determine whether the relation is (1) reflexive, (2) symmetric, (3) transitive, or (4) an equivalence relation.

3. *is a multiple of,* for all numbers in $\{1, 2, 3, 4, 5, 6, 7, 8, 9, 10\}$

4. *is the same age as,* for all people

Written Exercises

Determine whether each relation is (1) reflexive, (2) symmetric, (3) transitive, or (4) an equivalence relation. If not (1), (2), or (3), supply a counterexample.

1. $\geq$, for real numbers

2. $\neq$, for real numbers

3. $\not<$, for real numbers

4. $\not\geq$, for real numbers

5. *is a divisor of,* for positive integers

6. *is a multiple of,* for positive integers

Supply a reason to justify each step in the following proofs. (Exercises 7–10)

7. $-(-a) = a$ (Theorem 1.6)
1. $-(-a) = -1(-1 \cdot a)$
2. $\qquad\quad = [-1(-1)]a$
3. $\qquad\quad = 1a$
4. $\qquad\quad = a$
5. $-(-a) = a$

8. $(a - b) + b = a$
1. $(a - b) + b = [a + (-b)] + b$
2. $\qquad\qquad\quad = a + [-b + b]$
3. $\qquad\qquad\quad = a + 0$
4. $\qquad\qquad\quad = a$
5. $(a - b) + b = a$

9. $-(a - b) = b - a$

1. $-(a - b) = -1(a - b)$
2. $ = -1a - (-1)b$
3. $ = -a - (-b)$
4. $ = -a + [-(-b)]$
5. $ = -a + b$
6. $ = b + (-a)$
7. $ = b - a$
8. $-(a - b) = b - a$

10. $(ab)(cd) = (ac)(bd)$ (Theorem 1.7)

1. $(ab)(cd) = [(ab)c]d$
2. $ = [a(bc)]d$
3. $ = [a(cd)]d$
4. $ = [(ac)b]d$
5. $ = (ac)(bd)$
6. $(ab)(cd) = (ac)(bd)$

11. Use Theorems 1.2 and 1.7, a real-number fact, the Identity Property for Multiplication, and the Transitive Property of Equality to prove that:

$(-x)(-y) = xy$, for all real numbers x and y.

Prove that each statement is true for all real numbers a, b, c, and d. Supply a reason for each step.

12. $(a + b) + c = (c + b) + a$

13. $a(-b) = -(ab)$

14. $a(b + c) + d = (d + ac) + ab$

15. $(ab + c) + ad = a(b + d) + c$

Determine whether each relation is (1) reflexive, (2) symmetric, (3) transitive, or (4) an equivalence relation.

16. *is due west of*, for all places in the Northern Hemisphere

17. *is a bisector of*, for all segments in a plane

18. *is the brother of*, for all people

19. *is perpendicular to*, for all the lines in a plane

20. Is the set of *irrational numbers* a field? Prove your answer.

Mixed Review

Evaluate each expression for $x = -3$ and $y = 5$, if possible. If not possible, write *undefined*. *1.3*

1. $4(5x - 3y + 2) - 5(3x + 2y) + 4$

2. $2x^3 - y^2 + xy - x$

3. $\dfrac{y^2 - x^2}{(y - x)^2}$

4. $\dfrac{8}{x + y - 2}$

Chapter 1 Review

Key Terms

absolute value (p. 1)
Additive Inverse (p. 8)
algebraic expression (p. 9)
Associative Property (p. 8)
axiom (p. 31)
Closure Property (p. 8)
coefficient (p. 9)
Commutative Property (p. 8)
Distributive Property (p. 8)
equivalence relation (p. 33)
equivalent equations (p. 13)
equivalent expressions (p. 10)
factor (p. 9)
field (p. 8)
first-degree equation (p. 13)
formula (p. 17)
Identity for Addition (p. 8)
Identity for Multiplication (p. 8)
irrational number (p. 1)
like terms (p. 10)

linear equation (p. 13)
literal equation (p. 15)
Multiplicative Inverse (p. 8)
postulate (p. 31)
Properties of Equality (pp. 13, 31)
proportion (p. 14)
Proportion Property (p. 14)
rational number (p. 1)
real number (p. 1)
reciprocal (p. 8)
Reflexive Property (p. 31)
root (p. 13)
Substitution (p. 31)
Symmetric Property (p. 31)
term (p. 9)
theorem (p. 31)
Transitive Property (p. 31)
unlike terms (p. 10)
variable (p. 3)
Zero Property (p. 31)

Key Ideas and Review Exercises

1.1, 1.2 To simplify a numerical expression, follow the order of operations (summarized in Lesson 1.2), and recall that division by 0 is undefined.

Simplify, if possible. If not possible, write _undefined_.

1. $8 - 12 \cdot 3 + 4(-7) - (2 - 12)$

2. $\dfrac{18 - 5(-4)}{2 \cdot 6 - 4 \cdot 3}$

1.3 The eleven properties of operations with real numbers are listed in Lesson 1.3. The rule of subtraction states that $a - b = a + (-b)$.

Identify the rule or property of operations with real numbers that is illustrated.

3. $4y - 7y = 4y + (-7y)$

4. $3(7t) = (3 \cdot 7)t$

5. $5(4c + 3) = 5(4c) + 5 \cdot 3$

6. $3(7t) = (7t)3$

7. $(6a - 2b) + 0 = 6a - 2b$

8. $4 \cdot \frac{1}{4}n = 1n$

9. Simplify $xy + x^2y - 3xy + 9x^2y$, and evaluate for $x = -4$ and $y = 1.5$.

1.4 To solve a linear equation in one variable, first simplify all algebraic expressions. Then use the Properties of Equality listed in Lesson 1.4.

Solve each equation.

10. $3(4x - 6) - (2x + 7) =$
$10 - 5(4 - 3x)$

11. $2.6y - 3.4 = 3.6 - 4.4y$

12. $\frac{3}{4}w - \frac{1}{2} = \frac{5}{6}w + 2$

13. $\frac{9x - 5}{6} = \frac{7x - 4}{5}$

14. Solve $8a(x - 2) = 5ax + 3$ for x.

1.5   Substitution in a given formula can be used to solve some problems.

15. Bernie borrowed some money at $8\frac{1}{2}\%$ (simple interest) per year. At the end of 4 years 3 months, he owed a total of $7,623 (in principal and interest). How much money did Bernie borrow? Use the formula $A = p + prt$.

1.6   To solve a word problem about more than one number, begin by representing the numbers in terms of one variable.

16. The greatest of three numbers is -8 times the smallest, and the remaining number is 10 more than the smallest. Find the three numbers, given that the greatest is 16 times the sum of the other two numbers.

17. Write in your own words how you would represent the numbers for a word problem in which you are to find three consecutive integers.

1.7   To solve a word problem involving geometric figures, begin by sketching and labeling the figures.

18. A rectangle is 4 m wider than a square and 3 times as long. The difference between the two perimeters is 36 m. Find the area of the square and the length of the rectangle.

1.8   To prove algebraic theorems, use the field postulates in Lesson 1.3 and the postulates and theorems in Lesson 1.8.

19. Justify each step in the proof.
Theorem: $(ab)c + d = d + b(ca)$

Proof:
1. $(ab)c + d = d + (ab)c$
2. $\qquad\qquad = d + c(ab)$
3. $\qquad\qquad = d + (ca)b$
4. $\qquad\qquad = d + b(ca)$
5. $(ab)c + d = d + b(ca)$

20. The properties of an equivalence relation are listed in Lesson 1.8. Determine whether, for all triangles, the relation $\cong$ (*has the same size and shape as*) is **a.** reflexive, **b.** symmetric, **c.** transitive, or **d.** an equivalence relation.

Chapter 1 Test

Simplify, if possible. If not possible write *undefined*.

1. $12 - 6 \cdot 3 + 5(-2) - (-4)$

2. $\dfrac{5 \cdot 8 - 4(-3)}{-4 \cdot 6 - (-24)}$

Identify the property of operations with real numbers that is illustrated.

3. $7 + 5x = 5x + 7$ **4.** $5(8y) = (5 \cdot 8)y$ **5.** $-7t + 7t = 0$

6. Simplify $2xy + 3x^2y - xy$, and evaluate for $x = 5$ and $y = -4$.

Solve each equation.

7. $3(2.4x - 1.2) - (5x + 4) =$
$7.2 - 2(x - 1)$

8. $\frac{2}{3}y - \frac{5}{4} = \frac{1}{2}y + 4$

9. Solve $5(mx + 3) = 2mx + c$ for x.

10. Andrew borrowed some money at $9\frac{1}{2}\%$ (simple interest) per year. At the end of 4 years 6 months, he owed a total of $11,420. How much money did Andrew borrow? Use the formula $A = p + prt$.

11. The greatest of three numbers is -4 times the smallest number, and the remaining number is 14 more than the smallest number. Find the three numbers given that the greatest is 5 times the sum of the other two numbers.

12. A rectangle is 5 ft wider than a square. The rectangle is twice as long as it is wide. Find the area of the square given that the difference between the two perimeters is 70 ft.

13. Justify each step in the proof.

Theorem: $d + c(ba) = a(bc) + d$

Proof:
1. $d + c(ba) = c(ba) + d$
2. $\quad\quad\quad\quad = (ba)c + d$
3. $\quad\quad\quad\quad = (ab)c + d$
4. $\quad\quad\quad\quad = a(bc) + d$
5. $d + c(ba) = a(bc) + d$

14. Determine whether the relation $>$ is
  a. reflexive
  b. symmetric
  c. transitive
  d. an equivalence relation.

15. Use a numerical counterexample to prove that multiplication is *not* distributive over multiplication.

16. Solve $3 - 2\,|4x - (3 - x)| = x - 2 - 11(x - 1)$.

17. Prove $(x - y) - x = -y$, for all real numbers x and y.

Strategy for Achievement in Testing

Try to obtain the following information before taking any important test.

(1) Is the test of the type that is intentionally made longer than most people can complete in the time provided? If so, knowing that fact may remove some of the tension and help your concentration.

(2) Is there a greater penalty for giving a wrong answer than for leaving a question unanswered? If not, it may pay to guess.

Directions: Choose the *one* best answer to each question or problem.

1. If $3a + 6b = 90$, then $2a + 4b = \underline{\ ?\ }$.
   (A) 120   (B) 75
   (C) 60   (D) 45
   (E) None of these

2. If x^* is defined to be $x + 2$, find the value of $(3^* + 5^*)^*$.
   (A) 8   (B) 10
   (C) 12   (D) 14
   (E) None of these

3. A rectangle measures 9 in. by 12 in. Find the length of a diagonal.
   (A) 21 in.   (B) 15 in.
   (C) $10\frac{1}{2}$ in.   (D) 6 in.
   (E) None of these

4. Each angle in the figure below is a right angle. Find the perimeter of the figure.
   (A) 77 m (B) 36 m (C) 29 m (D) 18 m
   (E) Not enough data to answer

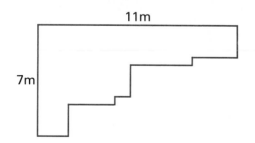

11m

7m

5. Which of the following rectangles has the greatest perimeter?
   (A) 2 yd by 1 yd   (B) 7 ft by 2 ft
   (C) 50 in. by 58 in.   (D) 5 ft 6 in. by 3 ft 6 in.   (E) They all have the same perimeter.

6. If $\dfrac{(x - 3)(y + 3)z}{3w} = 60$, which of the numbers x, y, z, and w *cannot* be 3?
   (A) x (B) y (C) z
   (D) w (E) All of these

7. If $\dfrac{15k}{3kx + 36} = 1$ and $x = 4$, what is the value of k?
   (A) 2   (B) 3   (C) 4
   (D) 8   (E) 12

8. Which number is a factor (divisor) of the product $15 \times 26 \times 77$?
   (A) 4   (B) 9   (C) 36
   (D) 55   (E) None of these

9. Which number is the sum of three consecutive integers?
   (A) 158   (B) 258   (C) 358
   (D) 458   (E) None of these

10. Find the number of fractions with two-digit numerators and two-digit denominators that are equivalent to $\frac{2}{9}$.
    (A) 10   (B) 9   (C) 8
    (D) 7   (E) 6

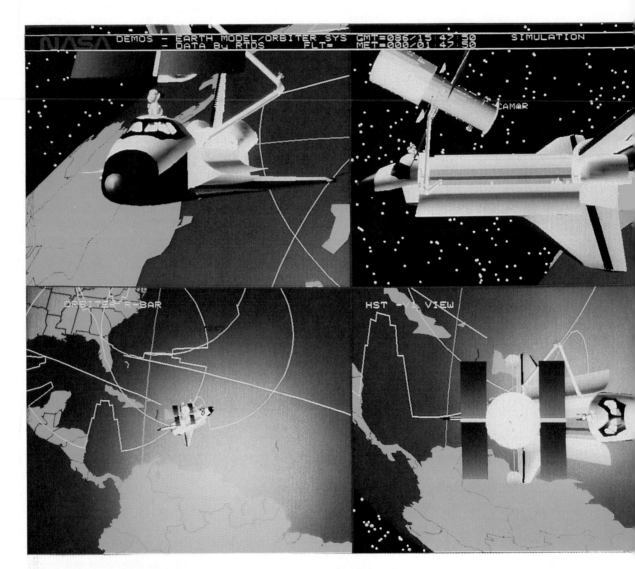

Computer-aided design, known as CAD, uses computer graphics to create a three-dimensional image of an object. CAD enables NASA engineers to gather information about the space shuttle to predict how it will perform under a given set of conditions.

2.1 Linear Inequalities

Objective

To find and graph the solution sets of linear inequalities

In Chapter 1, you learned how to solve problems with exactly one answer. You did this by writing a linear equation and finding its only solution, or root. Sometimes problems have a *range* of answers. To solve these problems, the relations $<$ *(is less than)* and $>$ *(is greater than)* are often used. These relations are called **order relations.**

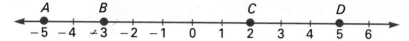

Use the number line above to verify the statements below.

$$-3 < 2 \text{ because } B \text{ is to the } left \text{ of } C$$
$$5 > -5 \text{ because } D \text{ is to the } right \text{ of } A$$

Several properties are developed below using numerical examples.

1.
$$-9 < 6$$
$$-9 + 4 < 6 + 4$$
$$-5 < 10$$

2.
$$-9 < 6$$
$$-9 - 2 < 6 - 2$$
$$-11 < 4$$

3.
$$-9 < 6$$
$$\frac{-9}{3} < \frac{6}{3}$$
$$-3 < 2$$

4.
$$-9 < 6$$
$$-9 \cdot 2 < 6 \cdot 2$$
$$-18 < 12$$

In the two cases below, the order of the relation changes.

5.
$$-9 < 6$$
$$-9(-2) > 6(-2)$$
$$18 > -12$$

6.
$$-9 < 6$$
$$\frac{-9}{-3} > \frac{6}{-3}$$
$$3 > -2$$

Multiplying or dividing each side of an inequality by the same *negative* number *reverses* the order of the inequality.

Properties of Inequality
For all real numbers a, b, and c,

if $a < b$, then $a + c < b + c$ and $a - c < b - c$
(Addition/Subtraction)

if $c > 0$ and $a < b$, then $a \cdot c < b \cdot c$ and $\dfrac{a}{c} < \dfrac{b}{c}$

if $c < 0$ and $a < b$, then $a \cdot c > b \cdot c$ and $\dfrac{a}{c} > \dfrac{b}{c}$
(Multiplication/Division)

The properties stated for $a < b$ can easily be restated for $a > b$. For example, if $c < 0$ and $a > b$, then $a \cdot c < b \cdot c$.

You can use inequality properties to find the set of all solutions, or the **solution set,** of a *linear inequality,* as shown below.

EXAMPLE 1   Find the solution set of $5x - 4 < 8x + 2$.

Solution

$$5x - 4 < 8x + 2$$

Add 4 to each side. $5x < 8x + 6$

Subtract $8x$ from each side. $-3x < 6$

Divide each side by -3 $x > -2$
and reverse the order.

The solution set is the set of all real numbers greater than -2.

As Example 1 illustrates, a solution set may have an infinite number of solutions. For this reason, such sets are called **infinite sets**. The solution set of Example 1 is an infinite set written as follows.

$$\{x \mid x > -2\} \leftarrow \text{Read: ``The set of all } x \text{ such that } x > -2.\text{''}$$

The number-line graph of this solution set is shown below. The open dot indicates that -2 is not a solution.

The order relations can be combined with the *equals* relation ($=$) to form the relations $\leq$ *(is less than or equal to)* and $\geq$ *(is greater than or equal to).* Notice that the following inequalities are both true.

$$2 \leq 5 \text{ (because } 2 < 5) \qquad -3 \geq -3 \text{ (because } -3 = -3)$$

EXAMPLE 2   Find the solution set of $2(15 - 3x) \geq 4x - 5$. Graph it on a number line.

Solution

$$2(15 - 3x) \geq 4x - 5$$

Use the Distributive Property. $30 - 6x \geq 4x - 5$

Subtract 30 from each side. $-6x \geq 4x - 35$

Subtract $4x$ from each side. $-10x \geq -35$

Divide each side by -10. $x \leq 3.5 \leftarrow$ NOTE: "$\geq$" changes
                                                            to "$\leq$."

The solution set is $\{x \mid x \leq 3.5\}$. The graph is the ray to the left of 3.5, including its endpoint. The endpoint is represented by a closed dot.

If $5 > 3$, then $3 < 5$, and in general, if $a > b$, then $b < a$. This property allows an inequality such as $4 > n$, with the variable on the right, to be written as $n < 4$, with the variable on the left. When the variable is written on the left, the symbols $<$ and $>$ will "match" the arrows on the graphs of the relations as shown below.

EXAMPLE 3 Find and graph the solution set of $\frac{1}{6}x - \frac{3}{2} < \frac{2}{3}x$.

Solution Multiply each side of the inequality by 6, the LCD.
Subtract x from each side.
Divide each side by 3, and write the variable on the left side.

$$6\left(\frac{1}{6}x - \frac{3}{2}\right) < 6 \cdot \frac{2}{3}x$$
$$x - 9 < 4x$$
$$-9 < 3x$$
$$-3 < x$$
$$x > -3$$

The solution set is $\{x \mid x > -3\}$. The graph is the ray (excluding its endpoint) that extends to the right of -3.

Focus on Reading

Match each sentence with exactly one graph.

1. $6 < x$
2. $6 \le x$
3. $x < 6$
4. $x \le 6$

a.
b.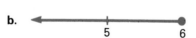
c.
d.

Classroom Exercises

For each statement, determine whether it is true or false.

1. $-18 < -10$ **2.** $-12 > 10$ **3.** $5 \le 7$ **4.** $8 \ge 8$

State the solution set of each inequality. Graph it on a number line.

5. $-4y \le 12$ **6.** $x + 7 > 5$ **7.** $28 > 7n$ **8.** $-t < -3$
9. $\frac{1}{2}c < 6$ **10.** $-6a \ge -24 + 2a$ **11.** $x - 4 < 21$ **12.** $2x + 5 \ge 1$

Written Exercises

Find the solution set. Graph it on a number line.

1. $2 \geq x - 3$

2. $4y - 12 \leq 8y$

3. $7 - 2n < 3n - 3$

4. $11c + 4 > 9c - 3$

5. $18y + 36 > 8y - 4$

6. $9x + 10 - x \leq 2x + 40$

7. $3(4x - 5) \leq 8x + 3$

8. $8 - 2(3x + 7) \geq 12$

9. $5x > 12x - (x - 24)$

10. $30 - (12 - x) < 7x$

11. $7x - 15 - 3x < 14 - 6x + 11$

12. $5(2 - 3x) \geq 4 - 3(4x + 7)$

13. $5p - (7p + 2) < 29 + 3(2p - 5)$

14. $-x > 0$

15. $-\frac{2}{3}n > 6$

16. $\frac{5}{3} < \frac{1}{6}y$

17. $\frac{3}{4}a - \frac{1}{2} > a + \frac{2}{3}$

18. $\frac{3x - 8}{5} \leq \frac{x + 4}{3}$

19. $\frac{2x + 7}{3} \leq \frac{x + 6}{2}$

20. $\frac{5}{6}a - \frac{3}{8}a \geq \frac{1}{2}a - 2$

21. $\frac{3x + 1}{-4} < 5$

22. $\frac{2y + 3}{-5} > 3 - y$

23. $-3(a + 5) > \frac{a - 5}{-2}$

Find a numerical counterexample that proves each statement false.

24. If $a < b$ and $b \neq 0$, then $a^2 < b^2$ for *all* real numbers a and b.

25. If $a < b$, then $\frac{1}{a} > \frac{1}{b}$ for *all* real numbers $a \neq 0$ and $b \neq 0$.

For each statement, determine whether it is true or false. If it is false, provide a numerical counterexample to show it is false.

26. If $a < b$ and $b < c$, then $a < c$ for all real numbers a, b, and c.

27. If $a < b$ and $b < c$, then $a + b < c$ for all real numbers a, b, and c.

28. If $\frac{a}{b} < \frac{c}{d}$, then $ad < bc$ for all positive real numbers a, b, c, and d.

29. If $\frac{a}{b} < \frac{c}{d}$, then $\frac{a}{b} < \frac{a + c}{b + d}$, and $\frac{a + c}{b + d} < \frac{c}{d}$ for all positive real numbers a, b, c, and d.

Mixed Review

Evaluate each expression for $x = -4$ and $y = 3$.   *1.3*

1. $8(3x - 4y) - (20x - 30y)$

2. $5xy - x^2y + 2xy^2$

Solve each equation for x.   *1.4*

3. $\frac{3x - 2}{4} = \frac{2x + 5}{3}$

4. $2a(x - m) = am - c$

Statistics: Stem-and-Leaf Plots

The data below are values of the Pollution Index for Austin, Texas for 23 weekdays of May, 1990. Each value is the highest average hourly concentration of the air pollutant ozone in parts per million (ppm) for that day. A score of 0–50 is considered good, 50–100 moderate, and 100–199 unhealthful. Readings approaching or exceeding 200 warrant advisories or warnings.

37 27 36 38 63 47 45 39 46 33 30 38
54 24 74 49 62 36 33 29 47 37 33

In their present form, the numbers appear to be haphazard. To organize them, construct a **stem-and-leaf plot** as shown below.

1. Divide each number into two parts: a stem and a leaf. For the given data, it is convenient to take the tens digit as the stem and the units digit as the leaf.

2. Write the stem of the smallest number. Then in a column below it, list all of the possible stems up to the stem of the largest number, as shown at the right above. Draw a vertical line to separate the stems from the leaves.

2	4, 7, 9
3	0, 3, 3, 3, 6, 6, 7, 7, 8, 8, 9
4	5, 6, 7, 7, 9
5	4
6	2, 3
7	4

3. The first ozone reading, 37, has a leaf of 7. Write this leaf opposite its corresponding stem, 3. Do this for every reading in the table, listing each leaf in the row opposite its stem.

It is now easy to make observations about the data. For example, the plot shows at a glance that the greatest number of readings lies in the thirties.

Exercises

For Exercises 1–3, refer to the stem-and-leaf plot above.

1. Give the three readings represented by the first row of the plot.
2. For how many days was the air quality moderate? unhealthful?
3. What was the most common quality assessment of the Pollution Index?
4. Construct a stem-and-leaf plot of the following test scores.

    78   55   94   89   92   85   69   98   86   89
    86   67   74   46   100   48   71   79   67   87

    Which of the stems contain most of the test scores?

2.2 Compound Sentences

Objective

To find and graph the solution sets of compound sentences

Two **simple sentences** such as "John is shopping" and "John is at the supermarket" can be combined into a **compound sentence** such as "John is shopping *and* John is at the supermarket."

For a mathematical example of a compound sentence, consider first the simple sentences "$4 + 1 = 5$" and "$6 < 3$." These can be connected by *and* or by *or* to form the two compound sentences below.

$$4 + 1 = 5 \text{ and } 6 < 3 \qquad 4 + 1 = 5 \text{ or } 6 < 3$$

The first compound sentence, which contains *and,* is called a **conjunction** of two simple sentences. The second sentence, which contains *or,* is called a **disjunction** of the simple sentences.

The conjunction (*and* sentence) "$4 + 1 = 5$ and $6 < 3$" is *false* because at least one of the simple sentences ($6 < 3$) is false.

The disjunction (*or* sentence) "$4 + 1 = 5$ or $6 < 3$" is *true* because at least one of the simple sentences ($4 + 1 = 5$) is true.

Truth-Value of a Compound Sentence

If p and q are a pair of statements, then
1. the conjunction p *and* q is true if and only if both p is true and q is true.
2. the disjunction p *or* q is false if and only if both p is false and q is false.

Next, consider the conjunction $x > 3$ *and* $x < 6$. Note that $x > 3$ *and* $x < 6$ is true only for numbers between 3 and 6, as shown by the table below.

x	$x > 3$ *and* $x < 6$	
1	$1 > 3$ *and* $1 < 6$	False
3	$3 > 3$ *and* $3 < 6$	False
4	$4 > 3$ *and* $4 < 6$	True (4 is between 3 and 6.)
8	$8 > 3$ *and* $8 < 6$	False

The graph of the solution set of a conjunction is the intersection (set of all points in common), or *overlap* of the graphs of its parts.

Graph of $x > 3$:

Graph of $x < 6$:

Graph of $x > 3$ *and* $x < 6$:

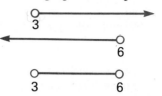

The solution set is $\{x \mid 3 < x < 6\}$, where $3 < x < 6$ means that $3 < x$ *and* $x < 6$. Read $3 < x < 6$ as: "x is between 3 and 6."

Finally, consider the disjunction $x > 3$ *or* $x < 6$. This compound sentence is true for *all* real numbers, as suggested by the table.

x	$x > 3$ or $x < 6$	
1	1 > 3 or 1 < 6	True
3	3 > 3 or 3 < 6	True
4	4 > 3 or 4 < 6	True
8	8 > 3 or 8 < 6	True

The graph of the solution set of a disjunction is the *union* of the graphs of its parts.

Graph of $x > 3$:

Graph of $x < 6$:

Graph of $x > 3$ *or* $x < 6$:

The solution set is the set of all real numbers.

EXAMPLE 1   Graph the solution set of $5x - 10 \leq 3x$ *and* $7 < 4x + 15$ on a number line. State the solution set.

Plan   Simplify and graph each part. Then graph their intersection.

Solution
$$
\begin{array}{ccc}
5x - 10 \leq 3x & and & 7 < 4x + 15 \\
2x \leq 10 & and & -8 < 4x \\
x \leq 5 & and & -2 < x
\end{array}
$$

Graph $x \leq 5$.

Graph $x > -2$.

Graph the intersection.

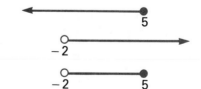

Graph of the solution set.

The solution set is $\{x \mid -2 < x \leq 5\}$.

EXAMPLE 2 Graph the solution set of $3x + 11 > 2$ *or* $8 - x > 4$ on a number line. State the solution set.

Plan Simplify and graph each part. Then graph their union.

Solution

$$
\begin{array}{ccc}
3x + 11 > 2 & or & 8 - x > 4 \\
3x > -9 & or & -x > -4 \\
x > -3 & or & x < 4
\end{array}
$$

Graph $x > -3$.

Graph $x < 4$.

Graph the union.

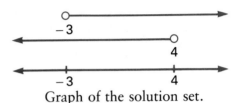

Graph of the solution set.

The solution set is the set of all real numbers.

Sometimes the graphs of the parts of a compound sentence have no points in common, or no overlap. For example, consider the disjunction $4x < 12$ *or* $2x > 12$ and its solution set.

Simplify each part. $x < 3$ *or* $x > 6$

Graph $x < 3$.

Graph $x > 6$.

Graph their union.

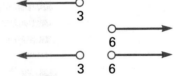

The solution set is $\{x \mid x < 3 \; or \; x > 6\}$.

The conjunction $x < 3$ *and* $x > 6$, however, has *no* solutions. Because there is no overlap, the graph of the conjunction has no points, and the solution set has no numbers. This solution set is called the **empty set**. It is represented by the symbol $\varnothing$.

EXAMPLE 3 Graph the solution set of $5x + 21 \le 6$ *or* $17 - 3x < 5$ on a number line. State the solution set.

Solution

$$
\begin{array}{ccc}
5x + 21 \le 6 & or & 17 - 3x < 5 \\
5x \le -15 & or & -3x < -12 \\
x \le -3 & or & x > 4
\end{array}
$$

Graph the union because the disjunction is used.

The solution set is $\{x \mid x \le -3 \; or \; x > 4\}$.

A compound inequality such as $-5 < 2y - 9 \le 5$ can be solved using the Properties of Inequality on each of the three parts: -5, $2y - 9$, and 5.

EXAMPLE 4   Find the solution set of $-5 < 2y - 9 \le 5$. Graph it on a number line.

Solution
$$-5 < 2y - 9 \le 5$$
$$4 < 2y \le 14$$
$$2 < y \le 7$$

The solution set is $\{y \mid 2 < y \le 7\}$. The graph is shown below.

Classroom Exercises

Match each compound sentence with the graph of its solution set.

1. $x < -4$ or $x < 2$

2. $x < -4$ and $x < 2$

3. $x < -4$ or $x > 2$

4. $x < -4$ and $x > 2$

5. $x > -4$ or $x > 2$

6. $x > -4$ and $x > 2$

7. $x > -4$ or $x < 2$

8. $x > -4$ and $x < 2$

a. ←————○
          -4

b. ←————————○
                  2

c. ○————————→
     -4

d. ○————————→
     2

e. ○————————○
     -4 2

f. ←————————→

g. ←——○   ○——→
        -4 2

h. ∅

For each compound sentence, determine whether it is true or false. Justify your answer.

9. $-5 < 3$ and $(-5)^2 < 3^2$

10. $-2 \cdot 6 > 2 \cdot 5$ or $-2(-1) > 0$

Graph the solution set of each compound sentence on a number line. State the solution set.

11. $x < 1$ or $x > 2$

12. $x > -1$ and $x < 2$

13. $x < 2$ or $x < 5$

14. $x < 2$ and $x < 5$

Written Exercises

Graph the solution set of each compound sentence on a number line.
State the solution set.

1. $x + 9 < 5 \text{ or } 4x > 12$

2. $x - 3 \le 2 \text{ or } 5x \le 10$

3. $5y \ge 15 \text{ and } y + 8 > 2$

4. $2y > 6 \text{ and } y - 2 < 4$

5. $c - 8 < 2 \text{ or } 6c > -18$

6. $7x < -14 \text{ and } x + 5 > 8$

7. $3 \le a + 7 < 9$

8. $-12 < 4t \le 20$

9. $-15 < 2x - 9 < 15$

10. $3x + 11 \le 20 \text{ or } 4x \ge 20$

11. $5x + 12 < 2 \text{ or } 5x - 12 < 3$

12. $14 - 3x < 2 \text{ or } 5 - 4x > 17$

13. $7 - 6x \le 19 \text{ and } 14 - 5x \le 4$

14. $-2 \le 3x + 10 \text{ and } 5 > 2x - 3$

15. $2x < 7x - 10 \text{ or } 8x \le 3x - 15$

16. $2x - 7 < 5x + 8 \text{ or } 8 - 2x > 0$

17. $-7.5 < 4x + 2.5 < 11.3$

18. $-0.7 \le 3x - 4.3 \le 8$

19. Write in your own words how to draw the graph of the solution set of the disjunction (*or* sentence) of two inequalities.

Find the solution set.

20. $8 < 4x - 7 < 22 \text{ and } x$ is an odd integer.

21. $18 < 5y + 6 < 50 \text{ and } y$ is an even integer.

22. $2w + 2(3w - 6) \le 40 \text{ and } w > 0 \text{ and } 3w - 6 > 0$

23. $4a + 2(6a - 8) < 100 \text{ and } a > 0 \text{ and } 4a - 8 > 0$

24. $-8 < -4x < 12$

25. $1 \le 6 - 5x \le 16$

26. $3x - 10 < 5x + 2 < 3x + 4$

27. $-5 < 3x - 2 < 4 \text{ or } 3x - 2 > 10$

Mixed Review

1. A rectangle is 6 ft longer than it is wide. A second rectangle is twice as wide and 3 times as long as the first rectangle. The sum of the two perimeters is 104 ft. Find the area of the first rectangle and the length of the second rectangle. **1.7**

2. Supply a reason for each step in the proof shown below. **1.3, 1.8**

1. $a(x + b) + c = c + a(x + b)$
2. $\qquad\qquad = c + (ax + ab)$
3. $\qquad\qquad = c + (ax + ba)$
4. $\qquad\qquad = (c + ax) + ba$
5. $a(x + b) + c = (c + ax) + ba$

2.3 Sentences with Absolute Value

To solve equations involving absolute value

To find and graph the solution sets of inequalities involving absolute value

Definition

The **absolute value** of any real number x, written as $|x|$, is the distance on a number line between its location and the origin. Thus, the absolute value of 4 is 4 ($|4| = 4$), and the absolute value of -4 is also 4 ($|-4| = 4$), as shown in the figure below.

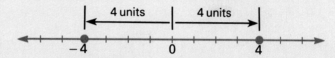

$|x| = -x$, if and only if x is a negative number ($x < 0$)

$|x| = x$, if and only if x is a nonnegative number ($x \geq 0$)

It follows from this definition that if $|y| = 10$, then $y = -10$ or $y = 10$.

Equation Property for Absolute Value
If $|x| = k$ and $k \geq 0$, then $x = -k$ or $x = k$.

If $|x| = k$ and $k < 0$, then there is no solution.

EXAMPLE 1

Solve the equation $|3n + 4| = 19$.

Plan

Use the Equation Property above. Replace x with $3n + 4$ and k with 19.

Solution

$$|3n + 4| = 19$$
$$3n + 4 = -19 \text{ or } 3n + 4 = 19$$
$$3n = -23 \text{ or } \qquad 3n = 15$$
$$n = \frac{-23}{3} \text{ or } \qquad n = 5$$

Checks: $|3 \cdot \frac{-23}{3} + 4| \overset{?}{=} 19$
$$19 = 19 \text{ True}$$
$$|3 \cdot 5 + 4| \overset{?}{=} 19$$
$$19 = 19 \text{ True}$$

The roots are $-7\frac{2}{3}$ and 5.

The solutions of $|x| < 4$ are the numbers whose graphs are *less than* 4 units from the origin.

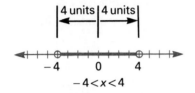

$-4 < x < 4$

The solutions of $|x| \geq 4$ are those numbers whose graphs are *at least* 4 units from the origin.

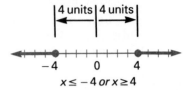

$x \leq -4 \text{ or } x \geq 4$

Inequality Properties for Absolute Value
If $|x| \leq k$ and $k > 0$, then $-k \leq x \leq k$.

If $|x| \geq k$ and $k \geq 0$, then $x \leq -k \text{ or } x \geq k$.

EXAMPLE 2   Find the solution set of $|4y - 2| \leq 10$. Graph it.

Plan   Use the property: If $|x| \leq k$, then $-k \leq x \leq k$.

Solution
$$|4y - 2| \leq 10$$
$$-10 \leq 4y - 2 \leq 10$$
$$-8 \leq 4y \leq 12$$
$$-2 \leq y \leq 3$$

The solution set is $\{y \mid -2 \leq y \leq 3\}$. The graph is shown below.

-2 3

EXAMPLE 3   Find the solution set of $|2n + 1| > 7$. Graph it.

Solution   Restate the property above as: If $|x| > k$, then $x < -k \text{ or } x > k$.

$$|2n + 1| > 7$$
$$2n + 1 < -7 \text{ or } 2n + 1 > 7$$
$$2n < -8 \text{ or } \qquad 2n > 6$$
$$n < -4 \text{ or } \qquad n > 3$$

The solution set is $\{n \mid n < -4 \text{ or } n > 3\}$.

-4 3

Classroom Exercises

Match each sentence on the left with a sentence at the right.

1. $|x + 5| = 3$
2. $|x + 5| < 3$
3. $|x + 5| < -3$
4. $|x + 5| > 3$

a. $x + 5 = -3$ *and* $x + 5 = 3$
b. $x + 5 = 3$ *or* $x + 5 = -3$
c. $x + 5 < -3$ *or* $x + 5 > 3$
d. $x + 5 > -3$ *and* $x + 5 < 3$
e. There is no solution.

Solve each equation.

5. $|n + 4| = 7$
6. $|y - 3| = 4$
7. $|12 - x| = 12$
8. $|4 + c| = 22$
9. $|11 - r| = 3$
10. $|s - 6| = 17$

Written Exercises

Solve each equation.

1. $|x - 4| = 9$
2. $|a + 5| = 3$
3. $|6 - y| = 8$
4. $|2t - 17| = 11$
5. $|3n + 12| = 18$
6. $|10 - 4a| = 32$

Find the solution set of each inequality. Graph it.

7. $|x - 4| > 1$
8. $|2c| > 6$
9. $|y + 3| \le 7$
10. $|n - 2| \le 5$
11. $|3c| \le 12$
12. $|t + 8| > 4$

Solve each equation.

13. $|4y + 5| = 17$
14. $|5x - 7| = 3$
15. $|10 - 3n| - 5 = 0$
16. $|2x + 7.5| = 19.5$
17. $|2.5x - 8| = 10$
18. $|7.5 - 5n| = 2.5$

Find the solution set of each inequality. Graph it.

19. $|2y - 3| < 11$
20. $|4t + 6| \le 14$
21. $|3y + 6| > 15$
22. $|4n - 5| \ge 15$
23. $|2x + 3.5| > 11.5$
24. $14 > |4z - 2|$

Solve each equation. (HINT: $|a - b| = |-a + b| = |b - a|$)

25. $|n - 3| + |3 - n| = 14$
26. $|2y + 7| + |3 - 2y| = 20$
27. $|x + 5| + |-x - 9| = 30$
28. $|3 - 2c| + |-c - 9| = 27$

Mixed Review

Solve for x. *1.4*

1. $3x - 6 = 0$
2. $400(0.09)x = 180$
3. $2[x + (x + 2)] = 23$

2.4 Problem Solving: Using Inequalities

Objective

To solve word problems involving a compound sentence

In Chapter 1 you used algebra to solve word problems with exactly one answer. As you have learned since, however, some problems have several answers, and others none. The problem below, for example, has many answers, yet algebra can still be used to solve it.

Find the time period during which $400 invested at 9% per year will earn from $135 to $180, inclusive, in simple interest.

$$i = prt \leftarrow \text{formula for simple interest}$$
$$135 \leq prt \leq 180$$
$$135 \leq 400(0.09)t \leq 180$$
$$135 \leq 36t \leq 180$$
$$3.75 \leq t \leq 5$$

Thus, for any time between 3 years 9 months and 5 years, inclusive, $400 invested at 9% per year will earn between $135 and $180, inclusive, in simple interest.

Sometimes, the realistic scope, or *domain*, of a variable must also be considered. In the problem below, the length $(3x - 6)$ and width $(2x)$ of the rectangle must always be positive.

EXAMPLE 1

A rectangle is twice as wide as a square. The rectangle's length is 6 ft less than 3 times the square's length. The sum of the two perimeters is less than 86 ft. Find the set of all possible lengths for the square.

Solution

Perimeter of square: $4x$
Perimeter of rectangle: $10x - 12$
Sum of perimeters: $4x + (10x - 12)$

The length of each side must be positive.

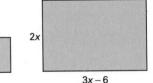

$$4x + 10x - 12 < 86 \text{ and } 3x - 6 > 0 \text{ and } 2x > 0 \text{ and } x > 0$$
$$14x < 98 \text{ and } \qquad 3x > 6 \text{ and } \quad x > 0$$
$$x < 7 \text{ and } \qquad x > 2$$
So, $2 < x < 7$.

Check

Check a number between 2 and 7:

If $x = 5$, then $4x = 20$, $10x - 12 = 38$, $20 + 38 = 58$, and $58 < 86$. The length of the square is between 2 ft and 7 ft.

In checking problems, such as Example 1, it is also a good idea to check numbers that are not between 2 and 7 ($x \leq 2$ or $x \geq 7$):
If $x = 8$, then $4x = 32$, $10x - 12 = 68$, $32 + 68 = 100$, and $100 \nless 86$ (100 is not less than 86). If $x = 1$, then the length of the rectangle $(3x - 6)$ would be -3 ft, which is not possible.

EXAMPLE 2   Find three consecutive odd integers such that twice the sum of the first two is greater than 23, and 3 times the sum of the last two is less than 61.

Solution   Let x, $x + 2$, $x + 4$ be the three consecutive odd integers.

x is odd *and* $2[x + (x + 2)] > 23$ *and* $3[(x + 2) + (x + 4)] < 61$
$$4x + 4 > 23 \text{ and} \qquad\qquad 6x + 18 < 61$$
$$x > 4\tfrac{3}{4} \text{ and} \qquad\qquad x < 7\tfrac{1}{6}$$

Thus, x is odd *and* $x > 4\tfrac{3}{4}$ *and* $x < 7\tfrac{1}{6}$. In other words, x is an odd integer between $4\tfrac{3}{4}$ and $7\tfrac{1}{6}$. So, x is either 5 or 7.

There are two answers: 5, 7, 9 and 7, 9, 11.

Recall (pg. 18) that $vt - 5t^2$ is the height h an object will reach at the end of t seconds if it is launched upward with an initial vertical velocity of v meters per second.

EXAMPLE 3   What is the range of the initial vertical velocity that will be needed to launch an object to a height of 810 m $\pm$ 30 m at the end of 6 s? Use the formula $h = vt - 5t^2$.

Solution

$$h = vt - 5t^2$$
$$810 - 30 \leq vt - 5t^2 \leq 810 + 30 \leftarrow 810 \pm 30 \text{ refers to}$$
$$780 \leq 6v - 5 \cdot 6^2 \leq 840 \qquad \text{numbers } between \text{ 810} - 30$$
$$960 \leq 6v \leq 1{,}020 \qquad\qquad \text{and } 810 + 30 \text{, inclusive.}$$
$$160 \leq v \leq 170$$

An initial vertical velocity between 160 m/s and 170 m/s, inclusive, will be needed.

Classroom Exercises

1. Find three consecutive even integers given that x is the first even integer and $6\tfrac{2}{3} < x < 9\tfrac{1}{2}$.

2. Find three consecutive multiples of 5 given that y is the first multiple and $12 < y < 24$.

In Exercises 3 and 4, the interval for n is $6 < n < 10$.

3. What is the interval for $n - 4$?   4. What is the interval for $5(n - 4)$?

In Exercises 5–7, a certain rectangle has a width x and a length $4(x - 5)$. Can x be:

5. equal to -2?   6. equal to 2?   7. equal to 5?

Written Exercises

1. Find the time period during which $600 invested at 8% per year will earn between $120 and $300, inclusive, in simple interest.

2. How much principal must be invested at 10% per year for 3 years 6 months to earn between $525 and $700, inclusive, in simple interest?

3. Find the set of real numbers for which 18 more than 5 times the sum of a number and 4 is greater than 98.

4. If 6 less than a positive number is multiplied by 2, the result is less than 17. Find the set of all such positive numbers.

5. A rectangle's length is 12 in. less than 3 times its width. Find the set of all the possible widths given that the perimeter must be less than 48 in.

6. Each leg of an isosceles triangle is 5 cm shorter than twice the length of the base. Find the set of possible lengths of a leg given that the perimeter is less than 40 cm.

7. Find three consecutive integers such that the sum of the first two is greater than 12, and the sum of the last two is less than 18.

8. Find three consecutive even integers for which the sum of the first two integers is greater than 9 more than the third integer, and the sum of the first and third integers is less than 17 more than the second integer.

9. What is the range of the initial vertical velocity needed for an object to reach a height of 760 m $\pm$ 120 m at the end of 4 s? (Use the formula: $h = vt - 5t^2$.)

10. A rectangle is 3 times as wide as a square. The rectangle's length is 15 ft shorter than 5 times the square's length. Given that the sum of the two perimeters must be less than 150 ft, find the set of all possible lengths for the rectangle.

11. Mrs. McTier wants to invest some money at 11% per year in order to have a total amount (investment plus interest) between $27,900 and $37,200, inclusive, at the end of 5 years. Find the amount of money she must invest. (HINT: Let $p + prt$ represent the total amount.)

12. Mr. McTier invests $600 at 9% per year, and $800 at 12% per year. During what time period will the $1,400 earn between $525 and $975, inclusive, in simple interest?

13. Find three consecutive multiples of 5 such that 3 times the sum of the first two multiples is less than 200, and twice the sum of the last two is greater than 125.

14. The second of three negative numbers is 6 less than the first negative number. The third negative number is 3 times the second number. Find the set of all possible third numbers given that twice the sum of the first two numbers is greater than the third number.

15. Given that the base of a triangle is 5 ft, find all the whole-number values for its height that will result in an area between 8 ft^2 and 18 ft^2.

A rectangular solid (box) has a height h of 4 in. and a width w that is 8 in. less than the length l. Use this information for Exercises 16 and 17.

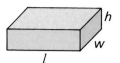

16. Find all the combinations of whole-number values of l, w, and h for which the resulting volume V will be less than 140 in^3. ($V = lwh$)

17. Find all the combinations of whole-number values of l, w, and h for which the resulting surface area T will be between 636 in^2 and 936 in^2. ($T = 2lw + 2lh + 2wh$)

Midchapter Review

Find the solution set. **2.1, 2.3**

1. $3(4 - 2x) < 18 - 3x$
2. $\frac{3}{4}n - \frac{1}{6}n \geq \frac{1}{3}n + 3$
3. $|4n - 2| = 14$
4. $|4n - 2| < 14$
5. $|4n - 2| > 14$

Graph the solutions. State the solution set. **2.2**

6. $x + 5 > 3x - 5$ and $4x + 4 > x - 2$
7. $3x - 4 > x + 6$ or $4x > x - 6$

Application: *Circumference of the Earth*

About 240 B.C., Eratosthenes, then the head of the Library at Alexandria, made the first scientific measurement of the earth's circumference. He knew that in the city of Syene at noon on Midsummer's Day, the sun shone directly into a well, while in Alexandria, a camel's ride of 5,000 stadia (1 stadium = 606 ft) to the north, the sun cast a shadow of 7.2 degrees from the vertical.

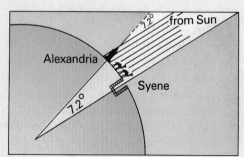

Using alternate interior angles to show that 7.2 was also the central angle between Alexandria and Syene, Eratosthenes then concluded that the measure 7.2 was to a circle as 5,000 stadia is to the earth's circumference.

1. Use this data to find the earth's circumference to the nearest mile. (Use 1 stadium = 606 ft.)

2. Explain why Eratosthenes had to assume that all the sun's rays are parallel to one another.

Problem Solving Strategies

Making a Table

Objective

To construct tables to represent the data in word problems

EXAMPLE

José is 9 years younger than Maria. In 4 years, she will be twice as old as he will be. Construct a table that represents their ages, both now and in 4 years, in terms of one variable.

Plan

The table must distinguish between:
(1) José and Maria, and
(2) ages now and ages in 4 years.

Table: Let m = Maria's age now.

Ages	José	Maria
Now	$m - 9$	m
In 4 yr	$m - 9 + 4$	$m + 4$

Notice that the information "she will be twice as old as he" has not yet been used in representing the data. This fact would be used in the next problem-solving step, that is, writing an equation.

Exercises

1. Bus A left a terminal and averaged 35 mi/h. Two hours later, Bus B left the terminal and traveled the same route at 45 mi/h. Construct a table that represents the rate, time, and distance for each bus. Let t = the time in hours for Bus A. (Use the formula: $r \times t = d$.)

	rate	time	distance
Bus A	35	t	
Bus B	45		

2. Some corn worth 75¢/lb is mixed with oats worth $1.10/lb to make 25 lb of an animal feed that is worth 96¢/lb. Construct a table that represents the weight, value per pound, and total value for the corn, oats, and mixture. Let x = number of pounds of corn.

	No. of lb	¢/lb	Value in ¢
corn	x	75	
oats		110	
mixture	25	96	

3. Amos is 3 times as old as Beth and Carlo is 4 years older than Beth. Five years ago, the sum of their ages was 29 years. Construct a table that represents their ages now and 5 years ago.

2.5 Problem Solving: Ages

Objective	To solve word problems involving ages

You will often find that a word problem becomes easier to solve once its data has been represented in a table. Note that in the example below, each entry in the bottom row of the table is clearly 6 greater than the corresponding entry in the row above it.

EXAMPLE

Charlene is 19 times as old as Enrique. Sonya is 30 years younger than Charlene. In 6 years, Charlene's age will be twice the sum of Enrique's and Sonya's ages then. How old is each person now?

Solution

Let E = Enrique's age now. Then, Charlene's age now is represented by $19E$ and Sonya's age by $19E - 30$.

Ages	Enrique	Charlene	Sonya
Now	E	$19E$	$19E - 30$
In 6 years	$E + 6$	$19E + 6$	$(19E - 30) + 6$, or $19E - 24$

In 6 years, Charlene's age	will be	twice the sum of the others' ages (in 6 years)
$19E + 6$	=	$2[(E + 6) + (19E - 24)]$

$$19E + 6 = 40E - 36$$
$$42 = 21E$$
$$E = 2 \qquad 19E = 38 \qquad 19E - 30 = 8$$

Check

So, Enrique is 2 years old now, Charlene is 38 years old now, and Sonya is 8 years old now. In 6 years, Charlene will be 38 + 6, or 44, and twice the sum of the others' ages then will be $2[(2 + 6) + (8 + 6)]$, or 44.

An age problem can also involve an inequality and have many answers in its solution set. In such a problem, be sure to consider the realistic domain of the variable. For example, if you are x years old and 6 years ago you were $x - 6$ years old, then $x - 6 > 0$ and $x > 6$. So, you must now be more than 6 years old.

Classroom Exercises

Represent each age now in terms of one variable.

1. A horse is 15 years older than twice the age of a colt.
2. A redwood tree is 100 years older than 3 times the age of a pine tree.
3. A second city was chartered 15 years before a first city. A third city is 3 times as old as the first city.
4. A bald eagle is twice as old as a falcon. A sparrow was hatched 10 years after the eagle was hatched.

In the table shown, find the entry for:

5. Mary's age 10 yr ago.
6. Sue's age now.

Ages	Mary	Sue
now		
10 years ago	x	$4x$

7. If Sue is now 2 times as old as Mary, how old is Sue?
8. In Classroom Exercise 2, suppose that in 115 years the redwood tree will be twice as old as the pine tree. Find the age of each now.
9. Mrs. Adams is 26 years older than her son. Ten years ago, she was 3 times as old as he was. How old is each now?

Written Exercises

1. Today, Selma is 5 times as old as Frank, but in 10 years she will be only 3 times as old. How old are they now?

2. Andre is 4 years younger than Charlene. Six years ago, the sum of their ages was 16. Find their ages now.

3. Barbara is 6 years older than Matt, and Christi is twice as old as Barbara. Three years ago, the sum of all their ages was 41. Find their present ages.

4. An oak tree is 30 years older than a pine tree. In 8 years, the oak will be 3.5 times as old as the pine. Find their present ages.

5. Truck 1 is 3 times as old as Truck 2, and Truck 3 is 16 years younger than Truck 1. One year ago, the age of Truck 1 was twice the sum of the ages of Trucks 2 and 3. Find their present ages.

6. Building A was built 30 years before Building B, and 20 years after Building C. In 10 years, Building C will be 40 years younger than the combined ages of Buildings A and B. What is the present age of each building?

7. City B is 3 times as old as City A. Six years ago, the sum of their ages was less than 32 years. Find the possibilities for the present age of City B.

8. Mary is 10 years younger than her brother. In 3 years, the sum of their ages will be less than 24. Find their possible ages now.

9. The ratio of the present ages of three paintings is 1 to 3 to 4. In three years, the sum of their ages will be less than 81 years. Find the possible present ages of each painting. (HINT: The ratio 1:3:4 can be written more generally as $1x:3x:4x$.)

10. The ratio of the ages of three companies is 2 to 4 to 5. Twenty years ago, the age of the oldest company was greater than the combined ages of the other two companies. Find the possibilities for the present age of the oldest company.

11. A person who is n years old now was p years old 8 years ago. Represent the person's age f years from now in terms of p and f.

12. In x years, a painting will be y years old. Represent the painting's age z years ago in terms of x, y and z.

13. Company A was formed n years after Company B was started, and p years before Company C was organized. Two years ago, the combined age of the companies was 50 years. Express the present age of Company B in terms of n and p.

14. City A is half as old as City B, and City B is one-third as old as City C. City B is now b years old. In 6 years, the combined age of Cities A and B will be $3t$ years less than the age of City C. Express the present age of City C in terms of t.

Mixed Review

Solve each sentence for x.

1. $5.3x - 1.2(5x - 6) = 2.3x - 1.8$ **1.4**

2. $\frac{2}{3}x - \frac{3}{4} = \frac{5}{8}x - \frac{7}{4}$ **1.4**

3. $-5 < 3x + 1 < 16$ **2.2**

4. $-6x + 4 > -8 \ or \ 2 < x - 3$ **2.2**

▬▬/Application: *Etching*

In one method of etching, artists use acid baths to cut images into metal plates. Once etched, the plates are inked and used to print the artwork. Control of the bath's acid content is crucial. For example, a printer has 64 fluid ounces (fl oz) of 10% nitric acid solution. To cover some zinc plates, she adds 32 fl oz of water. The calculations below show, to the nearest tenth of an ounce, the amount of acid (x) she now has to add to bring the acid content up to 8%.

$$10\% \ of \ 64 \ fl \ oz + x \ fl \ oz = 8\% \ of \ (64 + 32 + x) \ fl \ oz$$
$$0.1(64) + x = 0.08 \ (64 + 32 + x)$$
$$x = 1.4 \ fl \ oz \ of \ acid$$

1. To the nearest tenth, how much acid must be added to 128 fl oz of a 12% solution to make it a 15% solution?

2. A printer mixed a 40% acid solution with water to make 64 fl oz of an 8% solution. How much of each did he use?

2.6 Problem Solving: Two Objects in Motion

Objectives

To solve word problems involving two objects moving in opposite directions

To solve word problems involving two objects moving in the same direction

A bus driven at an average rate of 45 mi/h for 4 h travels $45 \cdot 4$, or 180 mi. A sprinter who runs at an average rate of 11 m/s for 5 s travels $11 \cdot 5$, or 55 m. The general relationship among distance, rate, and time for an object in motion is stated below.

The distance d traveled by a moving object is equal to its average rate of speed r multiplied by its time t in motion: $d = rt$.

Sometimes two objects travel in *opposite* directions, as shown in Example 1 below.

EXAMPLE 1

Two trains are moving toward each other on parallel tracks from stations 332 mi apart. One train is averaging 48 mi/h and the other 56 mi/h. If the faster train leaves 1 h 30 min after the slower train, how long will it take the faster train to meet the slower one?

Solution

Change 1 h 30 min to $1\frac{1}{2}$ h, or 1.5 h.

Let t = time for the faster train.
Then $t + 1.5$ = time for the slower train.

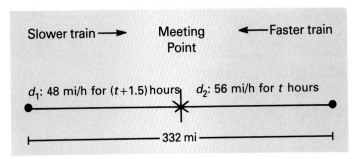

Use the formula $d = rt$.

Substitute for r_1, t_1, r_2, and t_2.

$$d_1 + d_2 = 332$$
$$r_1 t_1 + r_2 t_2 = 332$$
$$48(t + 1.5) + 56t = 332$$
$$48t + 72 + 56t = 332$$
$$104t = 260$$
$$t = 2.5$$

It will take the faster train $2\frac{1}{2}$ h (2 h 30 min) to meet the slower one.

In the next example, the two objects start from the same point and move in the *same* direction.

EXAMPLE 2 An ocean liner, traveling at 30 km/h, leaves port at 7:00 A.M. At 10:00 A.M., a helicopter leaves the same port and travels the same route at 66 km/h. At what time will the helicopter overtake the liner? (THINK: Do the liner and the helicopter travel at the same speed? for the same time? for the same distance?)

Solution Let t = time for the helicopter. Then $t + 3$ = time for the liner.

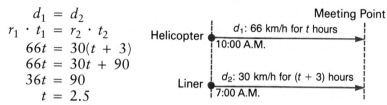

$$d_1 = d_2$$
$$r_1 \cdot t_1 = r_2 \cdot t_2$$
$$66t = 30(t + 3)$$
$$66t = 30t + 90$$
$$36t = 90$$
$$t = 2.5$$

The helicopter will fly for $2\frac{1}{2}$ h, or 2 h 30 min, and overtake the liner at 10:00 A.M. plus 2 h 30 min, or 12:30 P.M.

EXAMPLE 3 Marcia averaged 8 mi/h and André averaged 10 mi/h during a long-distance race. If they began the race at 7:30 A.M., at what time were they 2.5 mi apart?

Solution Let t = both Marcia's time and André's time in hours after 7:30 A.M.

$$d_1 + 2.5 = d_2$$
$$r_1 \cdot t_1 + 2.5 = r_2 \cdot t_2$$
$$8t + 2.5 = 10t$$
$$2.5 = 2t$$
$$1.25 = t$$

Marcia was 2.5 mi behind André at $1\frac{1}{4}$ h, or 1 h 15 min, after 7:30 A.M. So it was at 8:45 A.M. that they were 2.5 mi apart.

Classroom Exercises

Match each sentence on the left with a diagram on the right.

1. $d_1 = d_2$
2. $d_1 + d_2 = 100$
3. $d_1 + 100 = d_2$
4. $r_1 t_1 + 100 = r_2 t_2$
5. $r_1 t_1 + r_2 t_2 = 100$
6. $r_1 t_1 = r_2 t_2$

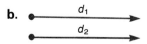

Mr. Adams left his house at 7:00 A.M., and traveled for t hours. Mrs. Adams left the house at 9:00 A.M., and traveled for $t - 2$ hours along the same road until she overtook Mr. Adams. (Exercises 7–10)

7. If $t = 5$ h, for how many hours did Mr. Adams travel?

8. If $t = 5$ h, for how many hours did Mrs. Adams travel?

9. If $t = 8$ h, for how many hours did Mrs. Adams travel?

10. If $t = 6$ h, at what time did Mrs. Adams overtake Mr. Adams?

Written Exercises

1. Two buses start toward each other at the same time from towns 190 mi apart. If one bus travels at 40 mi/h and the other travels at 55 mi/h, in how many hours will they meet?

2. From a buoy in a lake, Sharon swims toward the near shore at 4.2 m/s, while George swims in the opposite direction at 2.8 m/s. If they begin at the same time, in how many seconds will they be 105 m apart?

3. A ship leaves port traveling at 15 km/h. Two hours later, a motor boat leaves the same port traveling at 75 km/h in the same direction as the ship. How long will it take the motor boat to overtake the ship?

4. Two trains leave a terminal at the same time traveling in the same direction on parallel tracks. One train averages 57 mi/h, and the other averages 42 mi/h. In how many hours will they be 30 mi apart?

5. A bus left the city at 8:00 A.M., traveling at 70 km/h. Another bus, traveling in the opposite direction at 55 km/h, left at 11:00 A.M. At what time were the buses 460 km apart?

6. A van left City A at 7:30 A.M., and traveled toward City B at 65 km/h. At 11:30 A.M., a bus left City B and headed toward the van at 75 km/h. At what time did the bus pass the van given that the cities are 470 km apart?

7. An aircraft carrier and a destroyer leave the same port at 6:00 A.M., sailing in the same direction. The carrier averages 16 mi/h and the destroyer averages 40 mi/h. At what time will they be 54 mi apart?

8. Amy drove her car the same distance in 6 h that Juan drove his car in $5\frac{1}{2}$ h. Find the speed (rate) of Amy's car, given that she drove 5 mi/h slower than Juan.

9. It took a truck 5 h 30 min to cover the same distance that a car covered in 4 h 15 min. Find the speed of the car, given that the car was driven 10 mi/h faster than the truck.

10. Cities A and B are 360 mi apart on an east-west highway. A bus left City A for City B at 10:00 A.M.. Two hours later, a car left City B for City A. The bus's speed was 8 mi/h slower than the car's and the car passed the bus at 3:00 P.M. Find the speed of the bus.

11. A plane left an airfield at 9:00 P.M. cruising at 240 mi/h. At 11:30 P.M., a pursuit plane left the same airfield flying at 440 mi/h in the same direction. At what time did the pursuit plane overtake the slower plane?

12. A cruiser and a submarine leave the same port at the same time and travel eastward. The cruiser averages 18 mi/h, and the submarine 42 mi/h. For what period of time will the cruiser be 12 mi or less behind the submarine?

13. A plane left an airfield and flew eastward at 250 mi/h. Two hours later, a second plane left the same airfield and flew 450 mi/h in pursuit of the first plane. For which period after take-off was the second plane 100 mi or less behind the first plane?

14. An ocean freighter left port at 8:00 A.M. traveling 20 mi/h. At 10:00 A.M. a speedboat left the same port traveling the same route at 70 mi/h. As the speedboat caught and then passed the freighter, how long were they 5 miles or less from each other?

15. An automobile traveled x miles in a hours, then y miles in b hours, and finally z miles in c hours. Represent the average speed for the total trip in terms of the variables.

Mixed Review

Find the solution set of each sentence. *2.2, 2.3*

1. $3x - 2 < 16$ or $11 > 21 - x$

2. $5x - 4 < 7x$ and $6 - 2x > -4$

3. $|2x - 3| = 9$

4. $|2x - 3| > 9$

/ Using the Calculator

Use a calculator to solve each equation or inequality. Round each answer to the nearest hundredth.

Example   Solve. $|150x + 1.1| = 5.6$

$$150x + 1.1 = -5.6 \quad or \quad 150x + 1.1 = 5.6$$

$$x = \frac{-5.6 - 1.1}{150} \quad or \quad x = \frac{5.6 - 1.1}{150}$$

The roots are -0.04 and 0.03.

Exercises

1. $5.43x - 72.8 = 6.92$

2. $\frac{2}{3.75}y + 4.03 < 10.54$

3. $\frac{3.6}{4.5}t = \frac{48}{7.5}$

4. $\left| \frac{2.4}{3.2}n - \frac{6.2}{4.5} \right| = \frac{58}{4.5}$

5. $\frac{x}{350} = \frac{9}{140}$

6. $\frac{5.4}{1.5} = \frac{63}{y}$

Key Terms

absolute value (p. 51)
compound sentence (p. 46)
conjunction (p. 46)
disjunction (p. 46)
empty set (p. 48)
infinite set (p. 42)

linear inequality (p. 42)
order relations (p. 41)
Properties of Inequality (p. 41)
simple sentence (p. 46)
solution set (p. 42)

Key Ideas and Review Exercises

2.1   When solving a linear inequality, recall that multiplying or dividing each side by a *negative* number reverses the order.

Find the solution set. Graph it on a number line.

1. $3x - 2(4x - 7) < 8 - 2x$

2. $\frac{5}{6}a - \frac{2}{3} \leq \frac{1}{2}a - 4$

3. $-3x > 6$

4. $21 - 6y \leq 8 - 4y$

5. $6(2n - 4) \geq 5(3n - 4) - 10$

6. $\frac{3}{4}a < 24$

2.2   To graph the solution set of the disjunction (*or* sentence) or conjunction (*and* sentence) of two simple sentences, first graph the two simple sentences on separate number lines. Then for the disjunction, graph the *union* of the parts. For the conjunction, graph the *intersection* of the parts.

Graph the solution set on a number line.

7. $6x + 3 < 15$ or $2x < 12 - x$

8. $3 - 4x \leq 11$ *and* $19 \geq 7 - 2x$

9. Write in your own words how to draw the graph of the solution set of the conjunction of two linear inequalities.

2.3   To solve a sentence involving absolute value, use one of the following properties.

If $|x| = k$ and $k \geq 0$, then $x = -k$ or $x = k$.

If $|x| \geq k$ and $k \geq 0$, then $x \leq -k$ or $x \geq k$.

If $|x| \leq k$ and $k > 0$, then $-k \leq x \leq k$.

10. Solve $|3y - 5| = 22$.

11. Solve $|x - 8| = 12$.

Find the solution set. Graph it on a number line.

12. $|2y + 7| < 15$ **13.** $|4t - 6| \geq 10$

2.4   Solving a word problem can sometimes involve writing and solving a compound sentence.

14. Find three consecutive odd integers such that the sum of the first two is greater than 28, and the sum of the last two is less than 42.

2.5   To solve a word problem involving the ages of several objects, represent the ages now and at another time using one variable.

15. An oak tree is 5 times as old as a pine tree, and an elm tree is 6 years younger than the oak. In 3 years, the age of the elm will be twice the age of the pine. How old is the oak tree now?

16. The ratio of the present ages of two houses is 2 to 5. Ten years ago, the sum of their ages was less than 43 years. Find the possibilities for the present age of the younger house.

17. Building B was built 10 years before Building A and 20 years after Building C. Twenty years ago, the age of Building C was the same as the combined ages of Buildings A and B. What is the present age of each building?

2.6   To solve a problem involving objects A and B traveling distances d_1 and d_2 at rates r_1 and r_2 for times t_1 and t_2, respectively:

   (1) If $d_1 + d_2 = k$, use $r_1 t_1 + r_2 t_2 = k$.
   (2) If $d_1 = d_2$, use $r_1 t_1 = r_2 t_2$.
   (3) If $d_1 = d_2 + k$, use $r_1 t_1 = r_2 t_2 + k$.

18. A car left City A at 9:00 A.M. driving 52 mi/h toward City B. A truck left City B at 11:00 A.M. traveling 43 mi/h toward City A. At what time did the car and truck meet given that Cities A and B are 294 mi apart?

19. A cargo plane left an airfield at 8:30 A.M. flying due west at 150 mi/h. At 11:30 A.M. a pursuit plane left the same field flying 350 mi/h to overtake the cargo plane. At what time did the pursuit plane overtake the cargo plane?

20. In 2 h, Alice drove her car the same distance that Gene drove his car in 2 h 30 min. Find the speed of Gene's car given that he drove 10 mi/h slower than Alice drove.

21. A bus left City A at 9:00 A.M. heading toward City B at 50 mi/h. Then, a car left City B at 10:00 A.M. traveling 40 mi/h toward City A. At what time did they meet if the cities are 230 miles apart?

22. Sam began a marathon run at 6:30 A.M. and averaged 12 km/h, while Pam began the run at 7:00 A.M. and averaged 15 km/h. At what time did Pam pass Sam?

Chapter 2 Test

For Exercises 1–7, find the solution set and graph it on a number line.

1. $5x - 3(5x - 8) \geq 16 - 6x$

2. $3a > 18$ *and* $a - 2 > 1$

3. $4y + 6 < -10$ *or* $2y + 5 > 9 - 2y$

4. $5x - 3 < 7$ *or* $8 - x > 3$

5. $-4 \leq 3x - 7 \leq 8$

6. $6x - 1 < 17$ *and* $2 - x < 4x + 12$

7. $|a - 5| < 3$

8. Solve $|3n - 15| = 6$.

9. Find three consecutive even integers such that the sum of the first two is greater than 27 and the sum of the last two is less than 39.

10. An oak tree is 6 times as old as a pine tree and an elm tree is 6 years younger than the oak tree. In 3 years, the age of the elm will be 3 times the age of the pine. How old is the oak tree now?

11. The ratio of the present ages of two museums is 3 to 5. Twelve years ago, the sum of their ages was less than 56 years. Find the possibilities for the present age of the younger museum.

12. Cities A and B are located on an east-west highway at a distance of 244 mi from each other. A truck left City A at 8:00 A.M. driving 47 mi/h toward City B. A car left City B at 10:00 A.M. traveling 53 mi/h toward City A. At what time did the car pass the truck?

13. What is the range of the initial vertical velocity that will be needed to launch an object to a height of 800 m $\pm$ 40 m at the end of 8 s? Use the formula $h = vt - 5t^2$.

14. A ship leaves port at 9:30 A.M. sailing 50 km/h due east. At 11:00 A.M., a helicopter leaves the same port flying 80 km/h to overtake the ship. At what time does the helicopter reach the ship?

15. In 3 h, Andrea drove her car the same distance that Covey drove his car in 3 h 30 min. Find the speed of Covey's car given that he drove 8 mi/h slower than Andrea drove.

16. Find the solution set of $|a - 4| < 7$ *or* $|a - 4| > 10$. Graph it on a number line.

17. Solve $|3y + 2| + |8 - 3y| = 30$.

18. An antique automobile is now n years old. In 5 years, it will be f years old. Represent its age p years ago in terms of f and p.

College Prep Test

In each item, you are to compare a quantity in Column 1 with a quantity in Column 2. Write the letter of the correct answer from these choices.

A—The quantity in Column 1 is greater than the quantity in Column 2.
B—The quantity in Column 2 is greater than the quantity in Column 1.
C—The quantity in Column 1 is equal to the quantity in Column 2.
D—The relationship cannot be determined from the given information.

NOTE: Information centered over both columns refers to one or both of the quantities to be compared.

Column 1	Column 2	
Sample Question 1		**Answer**
$x = 3$ *and* $y = 4$		The answer is B, since $3 \cdot 4 > 3 + 4$.
$x + y$	xy	
Sample Question 2		**Answer**
$5a$	$7a$	The answer is D, since $5a > 7a$ if $a = -1$, $7a > 5a$ if $a = 1$, and $5a = 7a$ if $a = 0$.

Column 1	Column 2
1. $5 \cdot 629 \cdot 6$	$10 \cdot 629 \cdot 3$
2. $a \cdot a = 16$ *and* $b \cdot b = 25$	
a	b

3.

Column 1	Column 2
c	d
4. $58 \cdot 2^3 \cdot 47$	$47 \cdot 3^2 \cdot 58$
5. $a < c$ *and* $b < c$	
$a + b$	$2c$

Column 1	Column 2
6. $\|x\| = 3$ *and* $\|y\| = 4$	
$4x$	$3y$
7. $p < q$ *and* $q < r$	
$p - r$	2

Items 8–9 use the following definition:

For all positive integers x,
$\bigcirc\!\!\!\!x = x$, if x is even, and
$\bigcirc\!\!\!\!x = x + 1$, if x is odd.

Column 1	Column 2
8. $\textcircled{61} + \textcircled{65}$	$\textcircled{62} + \textcircled{64}$
9. $\textcircled{y} + \textcircled{y} + \textcircled{1}$	$\textcircled{2y} + \textcircled{1}$

Cumulative Review *(Chapters 1 and 2)*

Simplify, if possible. Choose the correct response among the choices A, B, C, D, and E.

1. $-5 - (-4) + (-8)$　　　　*1.1*

(A) -17　(B) -7　(C) 7
(D) -9　(E) 9

2. $\dfrac{5 - 4 \cdot 3 + 9}{-9 \cdot 4 + (-6)^2}$　　　　*1.2*

(A) $-\frac{1}{36}$　(B) $-\frac{1}{6}$　(C) $\frac{2}{9}$
(D) 0　(E) Undefined

3. $8 - 3(4x - 6) - (x - 2) + 5$　*1.3*

(A) $19x - 23$　(B) $19x - 3$
(C) $-13x + 5$
(D) $-13x + 33$
(E) None of these

Solve the equation. Choose the correct solution(s) among the choices A, B, C, and D.

4. $\frac{2}{3}x - \frac{5}{2} = \frac{3}{4}x - 4$　　　　*1.4*

(A) $\frac{48}{31}$　(B) $9\frac{1}{2}$
(C) 12　(D) 18

5. $3(5x - 4) - 4(4x + 3) =$　*1.4*
$8 - (5 - 2x)$

(A) -1　(B) -4
(C) -9　(D) 3

6. $\dfrac{6}{3x - 2} = \dfrac{4}{4x + 3}$　　　　*1.4*

(A) $-2\frac{1}{6}$　(B) $-\frac{5}{12}$
(C) $-\frac{5}{6}$　(D) 15

7. $|2x + 5| = 11$　　　　*2.3*

(A) $3, -3$　(B) $-8, 8$
(C) $8, -3$　(D) $3, -8$

For Exercises 8–14, match the sentence with a property or rule (A–G) listed below the sentences. All the variables represent real numbers.　*1.3*

8. $3t + (4t + 6) = (3t + 4t) + 6$

9. $-5 - 8 = -5 + (-8)$

10. $-5.4 + \sqrt{3}$ is a real number.

11. $2.4(5n + 15) =$
$2.4(5n) + 2.4(15)$

12. $(x + 3)4 = 4(x + 3)$

13. $-5y + 5y = 0$

14. $8c + c = 8c + 1c$

(A) Multiplicative Identity
(B) Additive Inverses
(C) Associative of Addition
(D) Closure for Addition
(E) Rule of Subtraction
(F) Distributive
(G) Commutative of Multiplication

Match the sentence with its solution set. Choose the correct response among the choices A–D.

15. $4x - 3(2x - 6) < 24$　　　　*2.1*

(A) $\{x \mid x < -3\}$
(B) $\{x \mid x > -3\}$
(C) $\{x \mid x < -21\}$
(D) $\{x \mid x > -21\}$

16. $3x < 18$ *or* $5 > x - 4$　　　　*2.2*

(A) $\{x \mid x < 6\}$
(B) $\{x \mid x < 9\}$
(C) $\{x \mid 6 < x < 9\}$
(D) $\{x \mid x < 6$ *or* $x > 9\}$

17. $6(x - 4) < x + 6$ *and*   **2.2**
$5 - 3x < x - 7$

   (A) $\{x \mid x < 3\}$
   (B) $\{x \mid x < 6\}$
   (C) $\{x \mid 3 < x < 6\}$
   (D) $\{x \mid 6 < x < 3\}$

18. $|x - 5| < 6$ **2.3**

   (A) $\{x \mid x > 11\}$
   (B) $\{x \mid x > -1\}$
   (C) $\{x \mid -1 < x < 11\}$
   (D) $\{x \mid x < -1 \ or \ x > 11\}$

Simplify, if possible.   **1.2**

19. $\dfrac{17 + 3 \cdot 5}{6 + 2 \cdot 4 + 2}$

20. $\dfrac{3 \cdot 8 - 7 \cdot 8}{(-6)^2 + 3(-12)}$

21. $-8 - 7(-2)$

22. $6(9 - 11)^3$

23. $-8 + 6 \div 2 - 4(-3) - (-20)$

24. $9 - 2[3(6 - 10) + 8]$

Simplify.   **1.3**

25. $8c - 4 - d + 12 + 9d - c$

26. $12x - 3(8x - 5) + 6$

27. $17 - (6 - 11n) - 8n$

Solve. (Exercises 28–31)   **1.4**

28. $5(3x + 4) - (x + 12) =$
$15 - 3(8 - 4x)$

29. $\frac{5}{6}m - \frac{7}{2} = \frac{5}{4}m + 4$

30. $3.2y - 5.6 = 7.9 - 5.8y$

31. $\dfrac{8x - 4}{5} = \dfrac{7x - 3}{4}$

32. Solve $6ax + 3c = 5c - 3ax$ **1.4**
for x.

33. Justify each step in the proof.   **1.8**

Theorem: $a(bc + d) =$
$\qquad\qquad ad + b(ca)$
Proof:

1. $a(bc + d) = a(bc) + ad$
2. $\qquad\quad = (bc)a + ad$
3. $\qquad\quad = b(ca) + ad$
4. $\qquad\quad = ad + b(ca)$
5. $a(bc + d) = ad + b(ca)$

34. Ann borrowed some money   **1.5**
at $9\frac{1}{2}\%$ per year in simple
interest. At the end of 5
years 6 months, she owed a
total amount of $6,090,
including principal and
interest. Use the formula
$A = p + prt$ to find how
much money Ann borrowed.

35. The greatest of three num-   **1.6**
bers is -3 times the smallest
number, and the remaining
(middle) number is 15 more
than the smallest. Find the
three numbers given that the
greatest is 3 times the sum of
the other two numbers.

36. A rectangle is 5 in. wider   **1.7**
than a square and 4 times as
long. The difference between
the two perimeters is 52 in.
Find the area of the square.

37. The age of an antique auto-   **2.5**
mobile is 10 years less than
the age of an antique table.
In 15 years the table's age
will be 45 years less than
twice the auto's age. How
old is the auto now?

38. Mario drove his racing car   **2.6**
the same distance in 6 h that
Bobby drove his car in 5 h 45
min. Find the speed at which
Mario drove, given that he
drove 10 mi/h slower than
Bobby.

Research and development engineers use computer graphics to create new designs without building expensive models. The image above shows an aircraft being tested for wind resistance. The red lines simulate the movement of air around the hull and wings.

3.1 Relations and Functions

Objectives

To graph relations and determine their domains and ranges
To determine relations from their graphs
To determine whether a relation is a function

Sentences with one variable, such as $3x - 1 < x$, are graphed on a number line. Sentences with two variables, such as $x + y = 1$, are graphed on a *coordinate plane*.

The figure at the right shows a vertical number line (**y-axis**) perpendicular to a horizontal number line (**x-axis**). The two axes intersect in a point called the **origin,** where they form a **coordinate plane** separated into four **quadrants**. The quadrants are numbered counterclockwise beginning with Quadrant I on the upper right.

Each point in a coordinate plane corresponds to exactly one **ordered pair** of real numbers, (x, y). The point is the graph of the ordered pair. The x-coordinate is called the **abscissa.** The y-coordinate is called the **ordinate.**

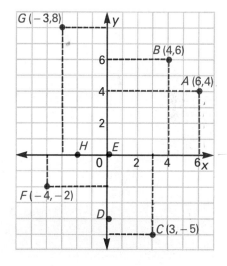

The coordinates of point A in the graph at the right are written as the ordered pair $(6, 4)$. Order is important because $A(6, 4)$ and $B(4, 6)$ are not the same point.

To graph, or *plot*, the point $C(3, -5)$, proceed as follows: Start at the origin. Move 3 units to the *right* and 5 units *down*. Place a dot in the coordinate plane. Label the point. C is the graph of the ordered pair $(3, -5)$.

The sign of the coordinate indicates the direction in which to move. The absolute value of the coordinate indicates the distance. The x- coordinate (abscissa) indicates the direction and distance from the origin along the x-axis. The y-coordinate (ordinate) indicates the direction and distance from the origin along the y-axis.

The graph at the right consists of four points. It can be described by a set of ordered pairs called a *relation*:

$R = \{(-3, 1), (-3, 5), (0, -5), (4, -2)\}$

This relation can also be shown by a **mapping**.

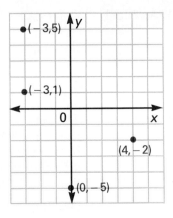

x-values   y-values

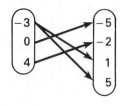

The set of x-values is called the *domain* of the relation.
Domain: $\{-3, 0, 4\}$

The set of y-values is called the *range* of the relation.
Range: $\{-5, -2, 1, 5\}$

Notice that -3 is listed only once in the domain.

Definitions

A **relation** is a set of ordered pairs.

The **domain** of a relation is the set of all first coordinates of the ordered pairs.

The **range** of a relation is the set of all second coordinates of the ordered pairs.

EXAMPLE 1 Write the relation P whose graph is shown at the right. List its domain and range.

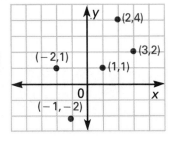

Solution $P = \{(-2, 1), (-1, -2), (1, 1), (2, 4), (3, 2)\}$

Domain of P: $\{-2, -1, 1, 2, 3\}$

Range of P: $\{-2, 1, 2, 4\}$

(In the range, list the 1 only once.)

In Example 1, the domain and the range are finite sets because the relation consists of only a finite number of ordered pairs. For many relations, however, this is not the case. As illustrated in Example 2, some relations consist of an infinite number of ordered pairs.

EXAMPLE 2   Determine the domain and range of the relation Q, whose graph is shown at the right.

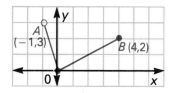

Solution   The open circle at $A(-1, 3)$ indicates that this point is not a point of the graph. Therefore, -1 is not included in the domain of the relation, and 3 is not in the range.

Domain: $\{x \mid -1 < x \le 4\}$

Range: $\{y \mid 0 \le y < 3\}$

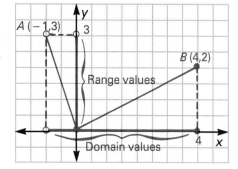

Look back at the relation P in Example 1. Notice that no two first coordinates are the same. This is not true for all relations, but when it is true, the relation is called a *function*.

Definition   A **function** is a relation such that no two ordered pairs have the same first coordinate.

EXAMPLE 3   Determine whether each of the following relations is a function.

a. $S = \{(-5, 3), (4, 3), (11, -1)\}$ b.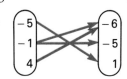

Solutions   a. No two ordered pairs have the same first coordinate. The relation is a function.

b. The ordered pairs $(-1, -6)$ and $(-1, -5)$ have the same first coordinate. The relation is not a function.

You can also determine whether a relation is a function by applying the following **vertical-line test** to its graph: *If a vertical line can be drawn that crosses the graph more than once, then the relation is not a function. Otherwise, the relation is a function.*

Classroom Exercises

State the domain and range of each relation.

1. $\{(2, -1), (4, 2)\}$ **2.** $\{(-3, -3), (8, -6), (1, -6)\}$ **3.** $\{(1, 1), (-1, 3), (1, -3)\}$

4–6. Graph each relation in Exercises 1–3.

7–9. Tell which relations in Exercises 1–3 are functions.

Written Exercises

Graph each relation and determine its domain and range. Is it a function?

1. $\{(2, 1), (1, 2)\}$ **2.** $\{(0, 0), (-1, 4)\}$

3. $\{(-1, 6), (-1, 2)\}$ **4.** $\{(3, 1), (-1, 1)\}$

5. $\{(4, 2), (3, -2), (-2, 4), (-3, -3), (0, 3)\}$

6. $\{(9, 8), (-6, 9), (-9, 8), (6, -1)\}$

7. $\{(-1, -1), (7, -1), (8, -1), (-5, -1)\}$

8. $\{(6, 2), (6, -1), (6, -5), (6, 0)\}$

Write the relations whose graphs are given. List the domain and range of each. Determine whether each relation is a function.

9.
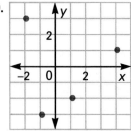

10.
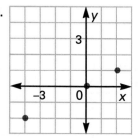

11.
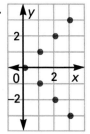

Determine whether each mapping represents a function.

12.
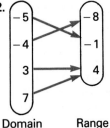
Domain   Range

13.
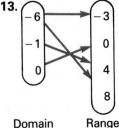
Domain   Range

14.
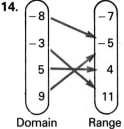
Domain   Range

Determine the domain and range of each relation. Use the vertical-line test to determine whether the relation is a function.

15.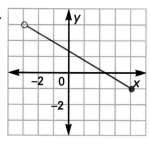

16.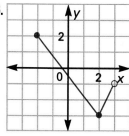

17.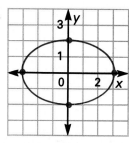

18.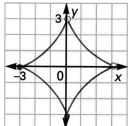

19.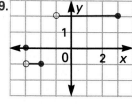

20.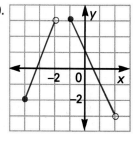

Give the domain and range of each relation. Is the relation a function?

21. $\{\ldots(-7,-5),(-4,-3),(-1,-1),(2,1),(5,3),\ldots\}$

22. $\{(1,5),(4,10),(9,17),(16,26),(25,37),\ldots\}$

Mixed Review

Find the solution set. Graph it on a number line.   *2.1, 2.3*

1. $2x - 15 < 8x + 3$ **2.** $\dfrac{3x - 8}{5} < \dfrac{x + 4}{3}$ **3.** $|4x - 2| = 8$ **4.** $|2x - 3| < 11$

Brainteaser

Three partners in an appliance store appear before a bankruptcy judge. Each partner makes four statements. Says Mr. Houlihan: "Washington owes me $5,000. Steinberg owes me $10,000. Everything Steinberg says is true. Nothing Washington says is true." Mr. Washington replies: "Everything Houlihan says is wrong. I don't owe him anything. Steinberg owes me $5,000. I should know because I was the bookkeeper." Finally, Mr. Steinberg says: "Washington was not the bookkeeper. And only two of his statements are true. Furthermore, I don't owe anyone anything. And I always tell the truth." Only one partner is telling the complete truth. Which one is it?

3.2 Graphs of Functions

Objectives

To find functions described by equations
To find values of functions
To find the ranges of functions for given domains

A function composed of an infinite number of points cannot be described by listing all the ordered pairs corresponding to those points. This type of function is usually described by writing an *equation in two variables*.

A function that has a line as its graph is called a **linear function**. Its equation is called a *linear equation in two variables*. The graph at the right describes a linear function. Its equation, $2x - y = 2$, is a linear equation in two variables. Each solution is an ordered pair of real numbers.

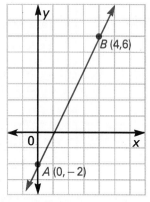

Definition

The *standard form* of a **linear equation in two variables** is $ax + by = c$, where a and b are *not both* equal to 0.

EXAMPLE 1 Graph $x - 2y = -8$.

Solution

1. Solve for y in terms of x:
   $y = \frac{1}{2}x + 4$.
2. Find two ordered pairs by choosing two values for x; find the corresponding values for y.
3. Plot the points; draw the graph.

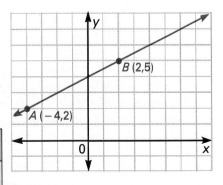

x	$\frac{1}{2}x + 4$	y	ordered pair
2	$\frac{1}{2} \cdot 2 + 4$	5	$(2, 5)$
-4	$\frac{1}{2} \cdot (-4) + 4$	2	$(-4, 2)$

The y-values in Example 1 can also be found using a scientific calculator. For example, if $x = -4$, the steps are as shown below.

$$0.5 \; \boxed{\times} \; 4 \; \boxed{+/-} \; \boxed{+} \; 4 \; \boxed{=} \; 2$$

In Example 2, the equation that is graphed is *not* linear.

EXAMPLE 2   Graph the equation $x^2 - y = 4$. Use the following values for x to make a table of ordered pairs: $\{-3, -2, -1, 0, 1, 2, 3\}$.

Solution   Since the equation is not linear, more than 2 points need to be plotted.

1. Solve the equation for y: $y = x^2 - 4$.

2. Make a table of ordered pairs using the given values for x.

3. Plot the points and draw a smooth curve through them.

x	$x^2 - 4$	y	ordered pair
-3	$(-3)^2 - 4$	5	$(-3, 5)$
-2	$(-2)^2 - 4$	0	$(-2, 0)$
-1	$(-1)^2 - 4$	-3	$(-1, -3)$
0	$(0)^2 - 4$	-4	$(0, -4)$
1	$(1)^2 - 4$	-3	$(1, -3)$
2	$(2)^2 - 4$	0	$(2, 0)$
3	$(3)^2 - 4$	5	$(3, 5)$

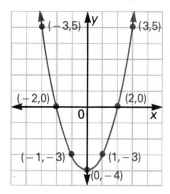

Recall that a function may be described by a mapping, as shown at the right.

Each member of the range is a *value* of the function. The statement, "The value of f at -4 is -1," can be written as $f: -4 \rightarrow -1$ or, more commonly, as $f(-4) = -1$. Similarly, $f(-2) = 0$ is read as "The value of f at -2 is 0."

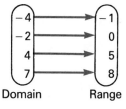

Domain   Range

In general, the notation $f(x)$ means "the value of f at x," and it is read either as "f of x" or as "f at x." Thus, $f(-2)$ can be read either as "f at -2," or as "f of -2."

Definition

For any ordered pair (x, y) of a function f, the **value of the function** at x is equal to y—that is, $f(x) = y$.

EXAMPLE 3   Given that $f(x) = -4x + 8$, find each of the following values.

   a. $f(-2)$ **b.** $f\left(\frac{3}{8}\right)$ **c.** $f(3a + 1)$

Solutions   **a.** $f(x) = -4x + 8$ **b.** $f(x) = -4x + 8$
                $f(-2) = -4(-2) + 8$ $f\left(\frac{3}{8}\right) = -4\left(\frac{3}{8}\right) + 8$
                $= 16$ $= -1\frac{1}{2} + 8$

                                                                          $= 6\frac{1}{2}$

               c. $f(x) = -4x + 8$
                        $f(3a + 1) = -4(3a + 1) + 8$
                                      $= -12a - 4 + 8$
                                      $= -12a + 4$

EXAMPLE 4   Given that $f(x) = 6x - 2$, find $f(a + h) - f(a)$.

Plan   Substitute $a + h$ for x in $f(x)$. Then substitute a for x in $f(x)$. Subtract.

Solution
$$f(x) = 6x - 2 \qquad\qquad f(x) = 6x - 2$$
$$f(a + h) = 6(a + h) - 2 \qquad f(a) = 6a - 2$$

$$f(a + h) - f(a) = [6(a + h) - 2] - [6a - 2]$$
$$= 6a + 6h - 2 - 6a + 2 = 6h$$

EXAMPLE 5   A function g is defined for all numbers in its domain by $g(x) = -x^2 - 4$. The domain of g is $\{-3, 0, 1.2\}$. Find the range of g.

Solution
$$g(x) = -x^2 - 4 \qquad g(x) = -x^2 - 4 \qquad g(x) = -x^2 - 4$$
$$g(-3) = -(-3)^2 - 4 \quad g(0) = -(0)^2 - 4 \quad g(1.2) = -(1.2)^2 - 4$$
$$= -9 - 4 \qquad\qquad = 0 - 4 \qquad\qquad = -1.44 - 4$$
$$= -13 \qquad\qquad\quad = -4 \qquad\qquad\quad = -5.44$$

The range is $\{-13, -5.44, -4\}$.

Focus on Reading

Match each item at the left with its corresponding item at the right.

1. set of first coordinates of a relation   **a.** $f(x)$

2. the value of f at x **b.** range

3. set of second coordinates of a relation   **c.** $3x - 2y = 8$

4. set of ordered pairs of real numbers   **d.** relation

5. linear equation   **e.** domain

Classroom Exercises

Find the values of y for $x = -3, -2, -1, 0, 1, 2, 3$.

1. $y = 3x - 2$ **2.** $y = -x + 1$ **3.** $y = -2x + 4$ **4.** $y = 5x + 3$

5–8. Graph each equation in Exercises 1–4.

Written Exercises

Graph each equation.

1. $3x - y = 6$ **2.** $6x - y = 3$ **3.** $f(x) = -x - 1$

Make a table of ordered pairs using the following values for x: $-3, -2, -1, 0, 1, 2, 3$. Graph each equation.

4. $x^2 - y = 8$ **5.** $x^2 - y = 3$ **6.** $f(x) = 3x^2 - 10$

Let $f(x) = -6x + 8$ and $g(x) = -x^2 + 6$. Find each of the following values.

7. $f(-4)$ **8.** $f(0.2)$ **9.** $g(2a)$

Find the range of each function for the given domain D.

10. $f(x) = -3x + 5; D = \{-4, 6, 8\}$ **11.** $g(x) = \frac{2}{3}x - 4; D = \{-6, 2, 3\}$

Let $f(x) = 4x - 3$ and $g(x) = x^2 - 2$. Find each of the following values.

12. $f(4) - f(8)$ **13.** $g(2) - g(-3)$ **14.** $f(4) - g(5)$

15. $f(3a - 4)$ **16.** $g(a) - g(3)$ **17.** $f(3a) - g(2a)$

18. If $h(a + b) = h(a) + h(b)$ for all real numbers a and b, prove that $h(2a) = 2h(a)$.

19. Let W be a function such that for all positive real numbers x, y, and z, $W(xyz) = W(x) + W(y) + W(z)$. Prove that $W(t^3) = 3W(t)$ for all positive real numbers t.

Mixed Review

Identify the property of operations with real numbers. *1.3*

1. $(x - 7)4 = 4(x - 7)$ **2.** $-8y + 8y = 0$ **3.** $a(b + c) = ab + ac$

4. Simplify $xy^2 + xy - 7xy^2 + 3xy$. *1.3* **5.** Solve $\dfrac{5x - 2}{3} = \dfrac{3x - 5}{2}$. *1.4*

Application: *The Doppler Effect*

The next time a train passes by, stop for a moment and listen. Doesn't the pitch of its whistle seem higher as the train moves toward you and lower as it moves away? The frequency of the whistle never changes, of course, but the pitch of the sound reaching you does. This change in observed pitch, caused by the motion of the source, is known as the Doppler Effect.

The Doppler Effect for a moving source and a stationary observer can be described using the formula shown below, where f' is the frequency (in vibrations per second) of the sound heard, f is the frequency of the sound produced, and v is the velocity of the source of sound (in mi/h). Use positive values for v when the source is moving toward you, and negative values when it is moving away from you.

$$f' = \frac{f}{1 - \dfrac{v}{760}}$$

EXAMPLE

If a train were moving away from you at 19 mi/h, and the frequency of its whistle was 820 vib/s, what frequency would you hear?

Use -19 for v because the train is moving away.

$$f' = \frac{f}{1 - \dfrac{v}{760}} = \frac{820}{1 - \dfrac{(-19)}{760}} = \frac{820}{1.025} = 800$$

You would hear a frequency of 800 vib/s.

1. A train is moving toward you at 87.4 mi/h. The frequency of its whistle is 3,540 vib/s. What frequency do you hear?

2. A race car is leaving the pit at 171 mi/h. To the pit crew, the pitch of its engine seems to be 440 vib/s. What is the engine's actual pitch?

3. The frequency produced by a siren is 1,540 vib/s. You hear a frequency of 1,400 vib/s. What is the speed of the source? Is it coming or going?

4. In the formula above, what does the 760 represent? (HINT: In any formula, all the measures must cancel out. So, 760 must be a measure of miles per hour.)

The table at the right lists the 1989 batting averages for each of the New York Mets with at least 30 "at bats". A **box-and-whisker** plot gives a quick graphical display of the data.

Batting Averages

.183	.228	.233	.270	.291
.183	.229	.247	.272	.308
.205	.231	.256	.286	
.225	.231	.258	.287	

A. List the values in order, as in the table above. Take the first and last numbers. These 2 numbers determine the **range** of the batting averages.

B. Find the **median** of the values. For an *odd* number of values, the median is the middle value. For an *even* number of values, as in the table above, the median is the average of the 2 middle values. The median here is the average of the 9th and 10th values: $(.233 + .247) \div 2 = .240$.

C. Find the first and third *quartiles*. The **first quartile** is the median of the lower half of the values, and the **third quartile** is the median of the upper half. (For an odd number of values, the middle value is counted in both halves.) Taking the medians of the upper and lower 9 values gives a first quartile of .228 and a third quartile of .272.

You are now ready to draw a box-and-whisker plot for the data.

D. Select a scale appropriate for the range of the values, and draw a box from the first quartile to the third quartile. Draw a bar at the median.

E. Add the "whiskers" by drawing segments to the boxes from the highest and lowest values as shown.

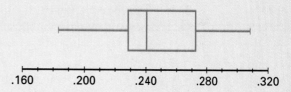

Exercises

1. Referring to the plots on the right, which team had the higher median?

2. Estimate each of the team's best and worst batting averages.

3. What was the range of the middle half of the averages for each team?

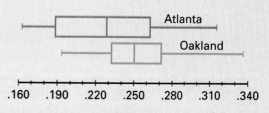

4. In 1989, the Oakland Athletics won 61% of their games. The Atlanta Braves won only 39% of theirs. Do the plots show greater batting strength for the Athletics? Discuss.

3.3 Slope

Objectives

To find the slope of a line given two of its points
To determine whether three given points lie on the same line

A mountain road that rises rapidly is said to have a steep *slope*. Linear functions can also be described using this same concept of slope.

The *slope*, or degree of steepness, of a line is described by the ratio slope $= \dfrac{\text{rise}}{\text{run}}$, as illustrated in the figures below.

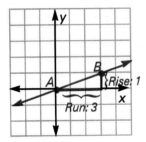 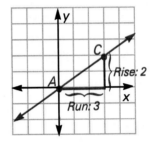 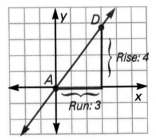

$$\text{slope of } \overset{\leftrightarrow}{AB} = \tfrac{1}{3} \qquad \text{slope of } \overset{\leftrightarrow}{AC} = \tfrac{2}{3} \qquad \text{slope of } \overset{\leftrightarrow}{AD} = \tfrac{4}{3}$$

The steeper the line, the greater is the absolute value of $\dfrac{\text{rise}}{\text{run}}$.

In the figure at the right, a right triangle is drawn to illustrate the slope of $\overset{\leftrightarrow}{PQ}$.

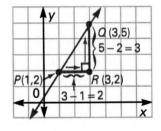

$\overline{RQ}$: rise $= 5 - 2 = 3$ (change in y)

$\overline{PR}$: run $= 3 - 1 = 2$ (change in x)

So, the slope $= \dfrac{\text{rise}}{\text{run}} = \dfrac{5 - 2}{3 - 1} = \dfrac{3}{2}$.

Notice that the slope of $\overset{\leftrightarrow}{PQ}$ is

$$\frac{\text{change in } y}{\text{change in } x} = \frac{5 - 2}{3 - 1} = \frac{3}{2}, \text{ or } \frac{2 - 5}{1 - 3} = \frac{-3}{-2} = \frac{3}{2}.$$

Definition

The **slope**, *m*, of a nonvertical line containing the points $A(x_1, y_1)$ and $B(x_2, y_2)$ is given by the following formula.

$$\text{slope of } \overset{\leftrightarrow}{AB} = m(\overset{\leftrightarrow}{AB}) = \frac{\text{change in } y}{\text{change in } x} = \frac{y_2 - y_1}{x_2 - x_1}, \text{ or } \frac{y_1 - y_2}{x_1 - x_2}$$

EXAMPLE 1   Find the slope of $\overset{\leftrightarrow}{AB}$.

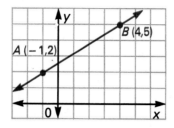

Solution

$$m(\overset{\leftrightarrow}{AB}) = \frac{y_2 - y_1}{x_2 - x_1} = \frac{5 - 2}{4 - (-1)} = \frac{3}{5}$$

The slope is $\frac{3}{5}$.

In Example 1, the slope of $\overset{\leftrightarrow}{AB}$ is $\frac{3}{5}$, a positive number, and the line slants up to the right.

The sign of the slope of a line tells whether the line slants up to the right or down to the right. In the figure at the right, the slope of $\overset{\leftrightarrow}{AB}$ is

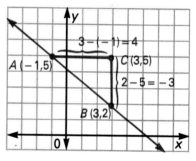

$$\frac{2 - 5}{3 - (-1)} = \frac{-3}{4}, \text{ or } -\frac{3}{4}.$$

Notice that the slope of the line is negative and that the line slants down to the right.

In general, a line with a *positive* slope slants *up* to the right. A line with a *negative* slope slants *down* to the right.

When different pairs of points on a line are used to compute the slope of the line, the slope is the same. This suggests the following theorem which can be proved using the properties of similar triangles from geometry.

Theorem 3.1   The slope of a nonvertical line is the same for any two points on the line.

Slope can also be used to determine whether three points are *collinear*. Three points are **collinear** if they all lie on the same line.

EXAMPLE 2   Determine, without graphing, whether the following three points lie on the same line: $A(0, 1)$, $B(-4, 4)$, and $C(8, 3)$.

Plan   If the three points lie on the same line, then the slopes for any two of them must be the same. Determine whether $m(\overleftrightarrow{AB}) = m(\overleftrightarrow{BC})$.

Solution   For $A(0, 1)$, $B(-4, 4)$, $m(\overleftrightarrow{AB}) = \dfrac{4 - 1}{-4 - 0} = -\dfrac{3}{4}$ ⟵ different slopes

For $B(-4, 4)$, $C(8, 3)$, $m(\overleftrightarrow{BC}) = \dfrac{3 - 4}{8 - (-4)} = -\dfrac{1}{12}$

Because the slopes are different, the three points do not lie on the same line. They are not collinear.

In Example 3, the slopes of horizontal or vertical lines are considered.

EXAMPLE 3   Find the slope of each line.

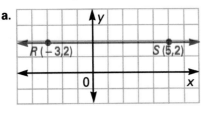

a.   **b.**

Solutions   **a.** $m(\overleftrightarrow{RS}) = \dfrac{2 - 2}{5 - (-3)} = \dfrac{0}{8} = 0.$ **b.** $m(\overleftrightarrow{AB}) = \dfrac{3 - (-1)}{-1 - (-1)} = \dfrac{4}{0}$

The slope is 0.   Since division by zero is undefined, the slope is undefined.

Theorem 3.2   The slope of any horizontal line is 0.

Theorem 3.3   The slope of any vertical line is undefined.

Some important ideas about slope are summarized below.
(1) Lines with positive slopes slant up to the right.
(2) Lines with negative slopes slant down to the right.
(3) The steeper the line, the greater is the absolute value of its slope.

Classroom Exercises

Find the slope of each line whose graph is shown.

1.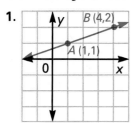

2.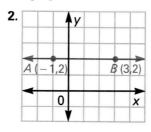

3.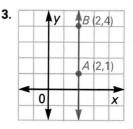

Describe the slant of the line with the given slope.

4. $\frac{2}{3}$ **5.** -4 **6.** $\frac{0}{4}$ **7.** Undefined

Written Exercises

Find the slope of $\overleftrightarrow{PQ}$. Describe the slant of the line.

1. $P(0, 0)$, $Q(3, 6)$ **2.** $P(0, 0)$, $Q(-2, 11)$ **3.** $P(4, 2)$, $Q(6, 8)$

4. $P(7, -5)$, $Q(-4, -5)$ **5.** $P(-3, 6)$, $Q(-3, 10)$ **6.** $P(7, 5)$, $Q(-5, 14)$

7. $P(9, 2)$, $Q(5, 8)$ **8.** $P(-8, -2)$, $Q(-4, 8)$ **9.** $P(-3.5, -7)$, $Q(-1.5, -7)$

10. $P(-6, -2)$, $Q(-6, -8)$ **11.** $P(-7, -5)$, $Q(-3, -6)$ **12.** $P(31, 1)$, $Q(-2, -8)$

Determine, without graphing, whether points A, B, and C are collinear.

13. $A(0, -7)$, $B(2, -3)$, $C(5, 3)$ **14.** $A(-4, 4)$, $B(8, -5)$, $C(0, -4)$

Find the slope of $\overleftrightarrow{PQ}$ for the given coordinates.

15. $P(-2, -3)$, $Q\left(-\frac{1}{2}, -\frac{1}{3}\right)$ **16.** $P\left(-\frac{2}{3}, \frac{1}{2}\right)$, $Q\left(\frac{5}{6}, 5\right)$ **17.** $P\left(\frac{1}{7}, \frac{1}{4}\right)$, $Q\left(\frac{5}{7}, \frac{3}{4}\right)$

18. $P(3a, -b)$, $Q(-a, 2b)$ **19.** $P(7a, -3b)$, $Q(-a, -b)$ **20.** $P(-10a, 12b)$, $Q(-2a, 7b)$

21. A line contains $A(6, y)$ and $B(9, -1)$. Find y given that $m(\overleftrightarrow{AB}) = \frac{2}{3}$.

22. The points $P(-2, 4)$ and $Q(0, 2)$ determine a line. Suppose that the point $R(3, a - 1)$ is a point on the same line. Find the value of a.

Mixed Review

Solve each equation for x. **1.4**

1. $6x - 1 = 9 - 2x$ **2.** $\frac{1}{2}x + \frac{1}{3} = x$ **3.** $ax - 2 = c$

3.4 Equation of a Line

Objective

To write an equation of a line, given its slope and one of its points, or given two of its points

Recall that the *standard form* of a linear equation in two variables is $ax + by = c$. Also, recall that the slope of a nonvertical line is the same for any two points on that line. (Theorem 3.1)

The line in the figure at the right contains the point $A(x_1, y_1)$. Let $G(x, y)$ represent any other point on the line. Let the slope of the line be m. Then, the slope of $\overleftrightarrow{AG} = m$.

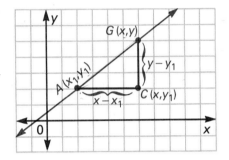

$$\frac{y - y_1}{x - x_1} = m$$

$$y - y_1 = m(x - x_1)$$

Equation of a Line (*Point-Slope Form*)

If a nonvertical line has slope m and contains the point $A(x_1, y_1)$, then the **point-slope form** of the equation of that line is given by:

$$y - y_1 = m(x - x_1).$$

EXAMPLE 1

Write an equation, in standard form, of the line containing the point $P(2, -3)$ and having the slope $-\frac{4}{3}$.

Solution

Write the point-slope form.

$$y - y_1 = m(x - x_1)$$

$$y - (-3) = -\frac{4}{3}(x - 2)$$

$$y + 3 = -\frac{4}{3}(x - 2)$$

Multiply each side by 3.

$$3y + 9 = -4(x - 2)$$

$$3y + 9 = -4x + 8$$

Write in standard form.

$$4x + 3y = -1$$

EXAMPLE 2   Write an equation, in standard form, of the line containing the points $P(3, -1)$ and $Q(-6, -13)$.

Plan   First find the slope of the line $\overleftrightarrow{PQ}$.

Solution
$$m = \frac{-13 - (-1)}{-6 - 3} = \frac{-12}{-9} = \frac{4}{3}$$

Then write the point-slope form of the equation. Use either point $P(3, -1)$ or $Q(-6, -13)$.

$$
\begin{aligned}
y - y_1 &= m(x - x_1) \\
y - (-1) &= \tfrac{4}{3}(x - 3) \\
y + 1 &= \tfrac{4}{3}(x - 3) \\
3y + 3 &= 4x - 12 \\
-4x + 3y &= -15
\end{aligned}
\qquad\text{or}\qquad
\begin{aligned}
y - y_1 &= m(x - x_1) \\
y - (-13) &= \tfrac{4}{3}[x - (-6)] \\
y + 13 &= \tfrac{4}{3}(x + 6) \\
3y + 39 &= 4x + 24 \\
-4x + 3y &= -15
\end{aligned}
$$

An equation in standard form is $-4x + 3y = -15$, or $4x - 3y = 15$.

Recall that the slope of a vertical line is undefined (Theorem 3.3). Therefore, the point-slope form cannot be used to write an equation of a vertical line, for example, one containing the points (3, 3) and (3, 1). The value of the x-coordinate of every point on this line remains *constant* (always 3). Thus, an equation of the line is $x = 3$.

Equation of a Vertical Line
The standard form of the equation of a vertical line is $x = c$, where c is a constant.

Recall that the slope of any horizontal line is 0 (Theorem 3.2).

Although the point-slope form could be used to write an equation of the line containing the points $(-1, 3)$ and $(5, 3)$, it is not needed because the y-coordinate of every point on the line remains *constant* (always 3). So, an equation of the line is $y = 3$.

Equation of a Horizontal Line
The standard form of the equation of a horizontal line is $y = c$, where c is a constant.

A function whose graph is a horizontal line is called a *constant linear function*, or a **constant function**.

Classroom Exercises

Tell whether the line containing the points A and B is vertical or horizontal. Also, give an equation of the line.

1. $A(-3, 6)$, $B(-3, 8)$ **2.** $A(-4, -6)$, $B(11, -6)$ **3.** $A(3, 7)$, $B(3, 0)$

Write an equation, in standard form, of the line containing the given point and having the given slope.

4. $P(4, 1)$, $m = \frac{2}{3}$ **5.** $P(1, 3)$, $m = 3$ **6.** $P(-1, -3)$, $m = -\frac{2}{3}$

Written Exercises

In Exercises 1–18, you are given either the coordinates of two points on a line or the slope of a line and one point on it. Write an equation for the line in standard form.

1. $R(1, 3)$, $m = \frac{2}{3}$ **2.** $A(4, -3)$, $m = 4$ **3.** $L(2, 7)$, $m = -\frac{3}{4}$

4. $P(6, 0)$, $Q(-3, -6)$ **5.** $S(8, -3)$, $T(-4, 3)$ **6.** $G(-4, 1)$, $H(8, 8)$

7. $A(-4, 7)$, $B(-4, -1)$ **8.** $M(9, 3\frac{1}{4})$, $N(-5, 3\frac{1}{4})$ **9.** $W(2.3, -5)$, $U(2.3, -4)$

10. $T(-1, -6)$, $U(6, -8)$ **11.** $R(\frac{1}{6}, -\frac{1}{2})$, $C(0, 1)$ **12.** $Y(2, -\frac{1}{3})$, $Z(6, 1)$

13. $G(0.2, -0.4)$, $H(1.2, 2.6)$ **14.** $S(0.1, -0.19)$, $T(0, -0.4)$

15. $R(2.1, 14.42)$, $S(-1.2, 4.52)$ **16.** $J(1.03, 6.3)$, $K(-0.07, 1.9)$

17. $I(5, -4)$, undefined slope **18.** $A(-4, -3)$, horizontal line

19. Write how to find an equation of a line containing any pair of given points. Discuss these cases: (1) lines that are neither vertical nor horizontal, (2) vertical lines, (3) horizontal lines.

20. Write an equation of the line containing the points $A(4, 5)$ and $B(12, -1)$. Then use the equation to find y when $x = 2$.

21. Write an equation of the line containing the points $A(4, 5)$ and $B(0, 8)$. The point $(t, -7)$ is a point on the line. Find t.

Mixed Review

Find the solution set. 2.1, 2.2, 2.3

1. $4x - 3(2x - 6) < 24$ **2.** $3x - 2 > 7$ *and* $-x - 2 > -8$

3. $|2x - 6| = 8$ **4.** $|x + 6| \leq 4$

3.5 Graphing Linear Relations

Objectives

To write equations of lines given their slopes and y-intercepts
To find the slopes and y-intercepts of lines given their equations
To graph lines using the slope-intercept method
To determine whether given points lie on given lines

The line $\overleftrightarrow{PQ}$ at the right crosses, or *inter-cepts*, the y-axis at $P(0,4)$. The y-*intercept* is 4. The slope of $\overleftrightarrow{PQ}$ is

$$m = \frac{8 - 4}{6 - 0} = \frac{4}{6} = \frac{2}{3}.$$

An equation of $\overleftrightarrow{PQ}$ can be written using the point-slope method.

$$y - y_1 = m(x - x_1)$$
$$y - 4 = \tfrac{2}{3}(x - 0)$$
$$y - 4 = \tfrac{2}{3}x$$

Now solve for y.
$$y = \tfrac{2}{3}x + 4$$

Notice the pattern.   slope y-intercept

When you solve an equation of a nonvertical line for y, the result is the *slope-intercept form* of the equation. The y-*intercept* is the y-coordinate of the point at which the line intersects the y-axis.

In general, if a curve crosses the y-axis at the point $(0, b)$, then b is called the **y-intercept** of the curve.

Equation of a Line *(Slope-Intercept Form)*
If a nonvertical line has a slope m and y-intercept b, then the **slope-intercept form** of the equation of that line is given by $y = mx + b$.

EXAMPLE 1

Write an equation, in slope-intercept form, of the line whose slope is $-\frac{4}{5}$ and whose y-intercept is 2.

Plan

Use the slope-intercept form.

Solution

$$y = mx + b \qquad y = -\tfrac{4}{5}x + 2$$

EXAMPLE 2   Find the slope and *y*-intercept of the line whose equation is given.
     a. $5x + 3y = 4$ **b.** $-x - y = 8$

Plan   Solve each equation for *y* to obtain the form $y = mx + b$.

Solutions

a. $5x + 3y = 4$
$$3y = -5x + 4$$
$$y = -\tfrac{5}{3}x + \tfrac{4}{3}$$
slope $= -\tfrac{5}{3}$ *y*-intercept $= \tfrac{4}{3}$

b. $-x - y = 8$
$$-1y = 1x + 8$$
$$y = -1x - 8$$
slope $= -1$ *y*-intercept $= -8$

The slope-intercept form of an equation of a line can also be used to graph a line such as $2x - 5y = 20$.

Slope-Intercept Method of Graphing a Linear Equation

1. Solve the equation for *y*. Identify the slope and *y*-intercept.

$$2x - 5y = 20$$
$$-5y = -2x + 20$$
$$y = \tfrac{2}{5}x - 4$$

slope $= \dfrac{2}{5}$ *y*-intercept $= -4$

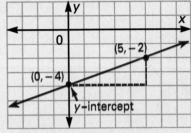

2. Place a dot at -4 on the *y*-axis [the point $(0, -4)$]. Use the slope, $\tfrac{2}{5}$, to find a second point as follows: From the point $(0, -4)$, move 5 units to the right and 2 units up, to the point $(5, -2)$.
3. Place a dot at the point $(5, -2)$. Draw a line through the two points.

EXAMPLE 3   Graph the function $f(x) = -\tfrac{3}{4}x + 6$.

Solution

1. Find the slope and *y*-intercept.
$$f(x) = -\tfrac{3}{4}x + 6$$
$$m = -\tfrac{3}{4}\qquad b = 6$$

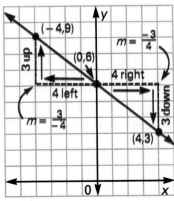

2. Plot the point $(0, 6)$. From this point, move 4 units to the right and 3 units down $\left(m = \dfrac{-3}{4}\right)$, or 4 units left and 3 units up $\left(m = \dfrac{3}{-4}\right)$.

3. Plot the second point. Draw a line through the two points. The graph is shown above.

Recall that $y = c$ is the standard form of an equation of a horizontal line, and that $x = c$ is the standard form of an equation of a vertical line. Vertical and horizontal lines can be graphed without making tables or using special methods.

EXAMPLE 4   Graph $-7x - 14 = 0$.

Plan   Solve for x. Compare to the standard form $x = c$.

Solution
$$-7x - 14 = 0$$
$$-7x = 14$$
$$x = -2$$

The graph is shown at the right.

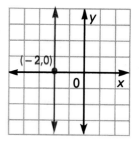

(−2,0)

You can determine whether a point such as $P(15,2)$ lies on the line $2x - 5y = 20$ by substituting the coordinates of the point into the equation and determining whether the result is a true equation. The procedure is illustrated in Example 5.

EXAMPLE 5   Determine whether the given point lies on the line described by the given equation.
   a. $P(15, 2)$, $2x - 5y = 20$ **b.** $R(7, -3)$, $5y - x - 12 = 0$

Solutions   **a.** Substitute 15 for x and 2 for y.

$$2 \cdot 15 - 5 \cdot 2 \overset{?}{=} 20$$
$$30 - 10 \overset{?}{=} 20$$
$$20 = 20 \text{ (True)}$$

The pair $(15, 2)$ *satisfies* the equation. So, P is a point on the line.

b. Substitute 7 for x and -3 for y.

$$5(-3) - 17 - 12 \overset{?}{=} 0$$
$$-15 - 7 - 12 \overset{?}{=} 0$$
$$-34 = 0 \text{ (False)}$$

The pair $(7, -3)$ does *not satisfy* the equation. So, R is not a point on the line.

Classroom Exercises

Find the slope and *y*-intercept, if any, of the line whose equation is given.

1. $y = \frac{3}{4}x - 7$ **2.** $y = -\frac{4}{7}x + 9$ **3.** $y = 3x + 7$

4. $x = 8$ **5.** $y = -1$ **6.** $x = 4$

7. $y = \frac{3}{2}x - 1$ **8.** $y = -5$ **9.** $-3x = 12$

10. $y = -x - 7$ **11.** $y = 4 + x$ **12.** $3x = -15$

13–24. Graph each equation in Classroom Exercises 1–12.

Written Exercises

Write an equation of the line with the given slope and *y*-intercept.

1. slope: $-\frac{2}{3}$, *y*-intercept: 5

2. slope: -2, *y*-intercept: -5

Find the slope and *y*-intercept of the line whose equation is given.

3. $y = -\frac{4}{9}x + 1$

4. $y = \frac{5}{4}x - 9$

5. $y = -\frac{2}{3}x - 1$

6. $y = -x$

7. $4x + 5y = -3$

8. $y = 4$

Graph each equation.

9. $y = \frac{2}{3}x - 1$

10. $y = -\frac{4}{5}x + 1$

11. $y = -x + 5$

12. $y = x - 6$

13. $2x - 5y = 15$

14. $4x + 3y = 18$

15. $2x - 7y = 21$

16. $-x - y = 6$

17. $f(x) = 2x - 6$

18. $f(x) = -x - 3$

19. $f(x) = \frac{3}{4}x - 3$

20. $f(x) = -\frac{5}{4}x + 8$

21. $-5y = -20$

22. $8 = -4y + 20$

23. $-14 = -7x$

24. $12 = 9 + 3x$

25. $-6x - 12y = 36$

26. $f(x) = -2x - 1$

Determine whether the point lies on the line described by the equation.

27. $P(1, 4)$; $3x + 4y = 19$

28. $P(3, 4)$; $x - 4y = -13$

29. $P(6, 8)$; $\frac{2}{3}x - \frac{1}{4}y = 2$

30. $P(10, -6)$; $\frac{2}{5}x + \frac{5}{3}y = 6$

Graph the line described by each equation.

31. $3(x - 4) - (4 - 4y) = 8$

32. $-(x - 5y) - 2(3x + y) = 12$

33. For what value of *a* will the graph of $3ax - 4y = 8$ have the same slope as the line whose equation is $3x - 2y = 14$?

34. Point *P* is a point on the line whose equation is $4x - 3y = 6$. The abscissa of *P* is 1 less than twice its ordinate. Find the coordinates of *P*.

Midchapter Review

Graph each relation or equation.   *3.1, 3.2, 3.5*

1. $\{(-2, 0), (-1, 2), (0, 1), (1, 0)\}$

2. $\{(2, 3), (-1, 4), (0, 4), (2, 4)\}$

3. $x + y = 2$

4. $f(x) = -x + 4$

5. $y = -\frac{1}{4}x + 2$

6. $2x - y = 3$

Find the slope of $\overleftrightarrow{PQ}$. Describe the slant of the line.   *3.3*

7. $P(5, 1)$, $Q(0, 2)$

8. $P(-3, 4)$, $Q(-2, 8)$

9. Write an equation of the line through $P(4, -1)$ that has a slope of $\frac{3}{5}$.   *3.4*

Application: *Familiar Linear Relationships*

The linear relations you have been studying are readily observed in many familiar situations in your own life. The relation between degrees Fahrenheit and degrees Celsius, for example, is linear. Constant rates, such as constant distance per time (speed), are also linear relations.

Use the following exercises to explore linear relationships.

1. Use the slope-intercept form $y = mx + b$ and the two points given in the figure at the right to show that $C = \frac{5}{9}(F - 32)$, the relationship between Fahrenheit and Celsius, is truly a linear relation.

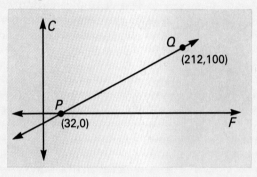

2. Using the points in the second figure, show that $F = \frac{9}{5}C + 32$ is a linear relationship.

3. At what temperature (point) do the Celsius and Fahrenheit scales register the same (intersect)? Show your derivation.

4. Thermometers sell for \$3.50 each. Let n represent the number of thermometers and C the cost of n thermometers. Write an equation for representing C in terms of n. Draw the graph showing this relationship.

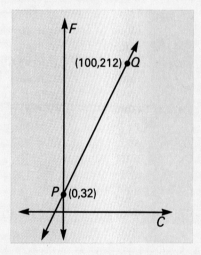

5. A car traveled at an average speed of 70 km/h. How far did it travel in 3 h? in 5 h? Draw the graph showing this relation.

3.6 Linear Models

Objectives

To construct linear models for sets of data
To use the equations of linear models to predict new data values

Scientists, economists, sociologists, and business managers all record data in order to find patterns by which future results may be predicted. For example, a naturalist might record the number of ants in an anthill at the end of a number of months in order to predict insect population changes over time. Below is a sample of such data and its graph.

Months: x	2	3	4	5	6	7
Ant population: y	8,000	9,500	10,000	11,000	13,000	13,500

Notice that a line can be drawn through two of the given points, say $A(2, 8,000)$ and $D(5, 11,000)$, that passes very close to the other plotted points. This line is called a **linear model**, and it approximates the relationship in the study between time and population. Its equation can be used to predict the ant population at any time for which recorded data is not available.

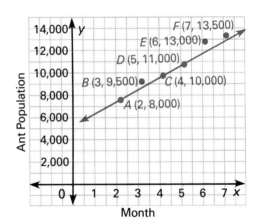

EXAMPLE 1

Use points $A(2, 8,000)$ and $D(5, 11,000)$ to write the equation of the linear model above.

Plan

Find the slope. Use the point-slope form to write the equation.

$$m = \frac{11,000 - 8,000}{5 - 2} = \frac{3,000}{3} = 1,000$$

Solution

$$y - y_1 = m(x - x_1)$$
$$y - 8,000 = 1,000(x - 2)$$
$$y - 8,000 = 1,000x - 2,000$$
$$y = 1,000x + 6,000$$

The equation of the linear model $y = 1{,}000x + 6{,}000$ describes a linear function where y, the number of ants, is a function of x, the number of months. $y = 1{,}000x + 6{,}000$ can therefore be written as

$$f(x) = 1{,}000x + 6{,}000.$$

This function can be used to predict the approximate number of ants at any time, as shown in Example 2.

EXAMPLE 2   Use the formula above to predict the number of ants after:

    a. 4 months.   **b.** 8 months.

Solutions   **a.** 4 months

    Find $f(4)$.
    $f(x) = 1{,}000x + 6{,}000$
    $f(4) = 1{,}000 \cdot 4 + 6{,}000$
    $f(4) = 10{,}000$ (This confirms the value in the table.)

    b. 8 months

    Find $f(8)$.
    $f(x) = 1{,}000x + 6{,}000$
    $f(8) = 1{,}000 \cdot 8 + 6{,}000$
    $f(8) = 14{,}000$

So, there would be 10,000 ants after 4 months and 14,000 ants after 8 months.

The linear model used in Examples 1 and 2 also illustrates the following physical interpretation of slope.

$A(2, 8{,}000)$ corresponds to 8,000 ants after 2 months. $D(5, 11{,}000)$ corresponds to 11,000 ants after 5 months.

change of 3,000
in ant population
↓

$$\text{Slope of } \overleftrightarrow{AD} = \frac{11{,}000 - 8{,}000}{5 - 2} = \frac{3{,}000}{3} = \frac{1{,}000}{1}$$

↑
change of 3 months

EXAMPLE 3 A researcher constructs a linear model for a set of data that shows how dosages of x grams of an appetite stimulant caused rats to gain y grams of weight. Over a restricted dosage range, the equation of the linear model is found to be $y = 2x + 50$, or $f(x) = 2x + 50$.

Predict each of the following:
a. the weight gain for 30 g of stimulant,
b. the dosage necessary for a weight gain of 120 g, and
c. the average gain in weight with respect to the dosage of stimulant.

Solutions

a. $y = 2x + 50$
$y = 2 \cdot 30 + 50$
$y = 60 + 50$, or 110

b. $y = 2x + 50$
$120 = 2x + 50$
$70 = 2x$
$35 = x$

The weight gain is 110 g. The necessary dosage is 35 g.

c. $y = 2x + 50$, so the slope is 2.

The average gain in weight is 2 g for every 1 g of stimulant.

Classroom Exercises

For Exercises 1–8, suppose that the equation $y = -5x + 100$ represents the number of bacteria y (in thousands) left x hours after an antibacterial spray is initiated. Predict the number of bacteria left after the given number of hours.

1. 2

2. 5

3. 10

4. 7

5. 8

6. 12

7. If the linear model is correct, in how many hours will there be no bacteria left?

8. Give the average rate of change in the number of bacteria with respect to time (in hours).

Written Exercises

A linear model for the cost y (in dollars) of producing x radios contains the points $A(5, 160)$ and $B(10, 220)$.

1. Write an equation for the linear model.

2. Find the cost of producing 15 radios.

A linear model for the total number y of a particular brand of television set sold over x years contains the points $A(2, 2,600)$, $B(5, 3,500)$, and $C(10, 5,000)$.

3. How many television sets were sold after 2 years?

4. How many years did it take to sell 5,000 sets?

5. Write the equation of the linear model.

6. Find the average rate of change in the total number of sets sold with respect to time (in years).

The table at the right shows the number of chirps per minute that a cricket makes at various temperatures. Use the table for Ex. 7–13.

Temperature (°C): x	6	8	10	15	20
Number of chirps per minute: y	12	28	40	74	110

7. Graph a linear model using the data for 10°C and 20°C.

8. Write the equation of the linear model.

9. Use the equation to predict the number of chirps at 40°C.

10. Write the average rate of change in the number of chirps with respect to the temperature.

11. Predict at what temperature the number of chirps will be 180.

12. Predict the number of chirps at a temperature of 4°C. Explain the significance of your answer.

13. Predict the number of chirps at temperature a°C.

Mixed Review

1. Simplify $5xy - 7x - 6xy - 4x$.   *1.3*   **2.** Solve $0.06x = 1.02$.   *1.4*

3. Solve $|x - 2| = 5$.   *2.3*   **4.** Given that $f(x) = 4 - x - x^2$, find $f(-3)$.   *3.2*

Application: *Möbius Strip*

Look at the bands shown below. The band on the left is the familiar ring-shaped sort with two sides and two edges. The band on the right, however, has a twist. This type of band is called a **Möbius strip.**

1. How many sides does a Möbius strip have? How many edges?

2. If a "Möbius twist" were added to an endless answering-machine cassette, what would happen to the time available for a message?

3.7 Parallel and Perpendicular Lines

Objectives

To determine whether two given lines are perpendicular, parallel, or neither

To write an equation of a line that contains a given point and that is parallel (or perpendicular) to a given line

In the figure at the right, $\overleftrightarrow{PQ}$ is *parallel* to $\overleftrightarrow{RS}$. (Parallel lines are lines in the same plane that never meet.) Notice that their slopes are the same.

$$m(\overleftrightarrow{PQ}) = \frac{5 - 3}{3 - 0} = \frac{2}{3}$$

$$m(\overleftrightarrow{RS}) = \frac{1 - (-1)}{5 - 2} = \frac{2}{3}$$

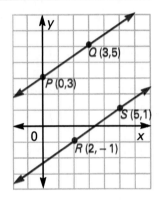

Theorem 3.4

Slopes of Parallel Lines: In a plane, if two nonvertical lines are **parallel**, then they have the same slope. Conversely, two lines in a plane that have the same slope are parallel.

EXAMPLE 1

Write an equation, in standard form, of the line containing the point $P(8, 3)$ that is parallel to the line whose equation is $4y - 3x = 8$.

Plan

Because the lines are parallel, their slopes are the same.

Solution

(1) Solve $4y - 3x = 8$ for y.
$$4y = 3x + 8$$
$$y = \tfrac{3}{4}x + 2$$
The slope of the given line is $\tfrac{3}{4}$.

(2) Write an equation of the parallel line.
$$y - y_1 = m(x - x_1)$$
$$y - 3 = \tfrac{3}{4}(x - 8)$$
$$12 = 3x - 4y$$

An equation of the line in standard form is $3x - 4y = 12$.

In the figure, $\overleftrightarrow{PQ}$ is *perpendicular* to $\overleftrightarrow{RS}$. (Perpendicular lines meet at right angles.)

$$m(\overleftrightarrow{PQ}) = \frac{-1 - 3}{4 - (-2)} = \frac{-4}{6}, \text{ or } -\frac{2}{3}$$

$$m(\overleftrightarrow{RS}) = \frac{4 - 1}{3 - 1} = \frac{3}{2}$$

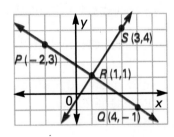

Notice that the product of the two slopes is $-\frac{2}{3} \cdot \frac{3}{2}$, or -1.

Therefore, the slope of one line is the opposite of the reciprocal of the slope of the other. This suggests the following theorem.

Theorem 3.5

Slopes of Perpendicular Lines
Given two lines that are neither vertical nor horizontal, if the two lines are **perpendicular**, then the product of their slopes is -1.

Conversely, if the product of the slopes of two lines is -1, then the lines are perpendicular. Therefore, Theorem 3.5 can be rewritten as follows: Two nonvertical, nonhorizontal lines are perpendicular *if and only if* the product of their slopes is -1.

EXAMPLE 2   Write an equation, in standard form, of the line containing the point $(-5, 4)$ that is perpendicular to the line with equation $3x - 2y = 8$.

Solution

1. Find the slope of the given line.

Solve for y. $3x - 2y = 8$
$$-2y = -3x + 8$$
$$y = \frac{3}{2}x - 4$$

The slope is $\frac{3}{2}$. So, the slope of a perpendicular line is $-\frac{2}{3}$.

2. Use the point-slope form to write an equation.

$$y - y_1 = m(x - x_1)$$
$$y - 4 = -\frac{2}{3}(x + 5)$$
$$3y - 12 = -2x - 10$$
$$2x + 3y = 2$$

An equation in standard form is $2x + 3y = 2$.

You can tell whether two lines are parallel or perpendicular by comparing their slopes. For example, consider the slopes of $y = \frac{5}{2}x + 2$ and $y = -\frac{2}{5}x + \frac{7}{10}$. The lines are perpendicular because the product of their slopes, $\frac{5}{2}$ and $-\frac{2}{5}$, is -1.

Classroom Exercises

Tell whether the given lines are parallel, perpendicular, or neither.

1. $y = \frac{2}{3}x - 1$; $y = -\frac{3}{2}x + 5$ **2.** $y = 3x - 5$; $y = 3x + 4$ **3.** $x = 6$; $y = 8$

Give the slope of a line parallel to the given line.

4. $y = 3x - 5$ **5.** $y = -3x$ **6.** $y = 4$

Written Exercises

Write an equation, in standard form, of the line that contains the given point and is parallel to the given line.

1. $P(1, 1); y = 3x + 2$ **2.** $Q(0, -2); y = -\frac{1}{2}x - 1$ **3.** $R(-1, 2); y = -x$

4. $P(4, 3); x + 2y = 7$ **5.** $A(5, 1); 4x + 3y = 15$ **6.** $C(-2, -3);$
$$3x - 5y = 10$$

7. $D(3, -1); x = 4$ **8.** $F(-3, -6); 4y - 6 = 2$ **9.** $P(-5, 3); 3x - 7 = 14$

Write an equation, in standard form, of the line that contains the given point and is perpendicular to the given line.

10. $P(4, 2);$ **11.** $A(-5, -1);$ **12.** $J(3, 5);$
$\quad y = \frac{2}{3}x - 4$ $4y + 3x = 8$ $y - x = 11$

13. $R(4, -3);$ **14.** $A(-5, -4);$ **15.** $T(3, 2);$
$\quad 8 = -4y - 4$ $3x - 4 = 5$ $-x - 3y = 15$

Determine whether the given lines are parallel, perpendicular, or neither.

16. $y = 2x;$ **17.** $y = 3x + 4;$ **18.** $4x - 5y = 10;$
$\quad y = -2x$ $y = 3x - 5$ $15x + 12y = 36$

Write an equation of the line satisfying the given conditions.

19. perpendicular to the line $-x - y = 9$ at $P(-6, -3)$

20. containing $T(3, 0)$ and parallel to the line through $R(5, 4)$ and $S(3, 1)$.

21. perpendicular to the line containing $A(2, -1)$ and $B(3, 1)$ at B

22. For what value of k will the line $3x + ky = 8$ be perpendicular to the line $4x - 3y = 6$?

23. For what value of k will the line $2x - ky = 4$ be parallel to the line $x - y = -1$?

Mixed Review

Let $f(x) = 3x - 4$ and $g(x) = x^2 - 2$. Find each of the following.   **3.2**

1. $f(-4)$ **2.** $g(-3)$ **3.** $f(-5)$ **4.** $f(2) - g(-3)$

◢/ Brainteaser

Find a natural number n such that $n(n - 1)(n - 2)(n - 3)(n - 4) = 95{,}040$.

Extension

Logical Reasoning

When the equation $2x + 6 = 14$ is solved, the result is $x = 4$. Recall from Chapter 1 that if you give a reason for each step and write them in statement-reason format, you are presenting the proof of your conclusion. Thus, if you are given $2x + 6 = 14$, you can *conclude* $x = 4$ from the following proof. It is *assumed* that the given is true.

Statement	Reason
1. $2x + 6 = 14$	1. Given
2. $2x = 8$	2. Subt Prop of Eq
3. $\therefore \ x = 4$	3. Div Prop of Eq

A statement written in the following form is a **conditional** statement.

"If p, then q."

hypothesis or given   conclusion

Theorems can also be stated in the form of a conditional. For example, Theorem 3.2 of this chapter states that the slope of a horizontal line is 0.

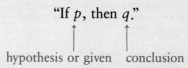

If *a line is horizontal*, then *the slope of the line is 0.*

hypothesis or given   conclusion

EXAMPLE 1 Write as a conditional: The slope of a vertical line is undefined.

Solution Hypothesis: a line is vertical.   Conclusion: slope is undefined.

Conditional: If *a line is vertical*, then *its slope is undefined.*

Interchanging the hypothesis and the conclusion of a conditional produces another conditional, called the *converse*. The converse of a true conditional may or may not be true.

EXAMPLE 2 Conditional: If you live in Ohio, then you live in the United States. (True)

Converse: If you live in the United States, then you live in Ohio. (False)

Definition The **converse of a conditional** is the statement formed by interchanging the hypothesis and conclusion.

Recall that when $2x + 6 = 14$ was solved, the result was $x = 4$. However, you should not conclude that 4 is the solution of the equation without checking.

Substitute 4 for x.

$$2x + 6 = 14$$
$$2 \cdot 4 + 6 \stackrel{?}{=} 14$$
$$14 = 14 \text{ True}$$

Thus, 4 is a solution of $2x + 6 = 14$, since the conditional and its converse are *both* true.

In this case the conditional and its converse can be combined into a single statement called a *biconditional*.

Conditional: If $2x + 6 = 14$, then x = 4.   True
Converse: If $x = 4$, then $2x + 6 = 14$.   True
Biconditional: $2x + 6 = 14$ *if and only if* $x = 4$.

Definition

When a conditional statement and its converse are combined by "if and only if," the resulting statement is called a **biconditional**. The biconditional, "*p if and only if q*," is true only when the conditional, "*if p, then q*," and its converse, "*if q, then p*," are *both* true.

A conditional or its converse can be shown false by finding one example where it is false. Such an example is called a **counterexample**.

Exercises

Write the following statements as conditionals.

1. The slope of a horizontal line is 0.
2. The slopes of parallel lines are the same.
3. Supplementary angles are angles the sum of whose measures is 180.

Identify the hypothesis and conclusion of each conditional.

4. If $3x - 7 = 8$, then $x = 5$.
5. If $a > b$ and $b > c$, then $a > c$.
6. If $x = 2$, then $x^2 = 4$.

Determine whether each conditional is true. Write the converse of the conditional. Is it true? Write a biconditional for the conditional and its converse.

7. If $6x - 6 = 4x + 14$, then $x = 10$.
8. If $x = -4$, then $|x| = 4$.
9. Write the conditional and converse that produces the following biconditional: The square of a number is 9 if and only if the number is 3. Is the biconditional true? Why?

3.8 Direct Variation

Objectives

To determine whether relations are direct variations
To find missing values in direct variations
To solve word problems involving direct variations

Recall that the function $f(x) = mx + b$ (or $y = mx + b$) is a linear function. If the y-intercept of the graph of a linear function (b) is 0, then the function is a special type of function called a *direct variation*.

The figure at the right is the graph of the direct variation $f(x) = \frac{2}{3}x$, or $y = \frac{2}{3}x$. Notice that for any pair (x, y) other than $(0, 0)$, the ratio $\frac{y}{x}$ is *constant*.

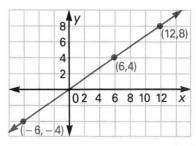

The table below illustrates this.

x	y
−6	−4
6	4
12	8

$$\frac{y}{x} = \frac{-4}{-6} = \frac{4}{6} = \frac{8}{12} = \frac{2}{3}$$

The ratio of y to x is always the same, or constant: $\frac{2}{3}$.

Definition

A linear function defined by an equation of the form $y = kx \ (k \neq 0)$ is called a **direct variation.** The constant k is called the **constant of variation.**

The equations $\frac{y}{x} = k$ and $y = kx$ can be read as "y varies directly as x."

EXAMPLE 1

Determine whether the table below expresses a direct variation. If so, find the constant of variation and an equation of the function.

Solution

x	y
2	−8
5	−20
6	−24

$$\frac{y}{x} = \frac{-8}{2} = \frac{-20}{5} = \frac{-24}{6} = -4$$

$$\frac{y}{x} = -4, \text{ or } y = -4x$$

So, y varies directly as x. The constant of variation is -4, and an equation of the function is $y = -4x$.

EXAMPLE 2   Determine whether the equation represents a direct variation. If so, find the constant of variation.

   a. $-9y = 27x$

   b. $y = 7x - 4$

Solutions   **a.** Solve for y. $y = -3x$

The equation is of the form $y = kx$, where $k = -3$. So, $y = -3x$ represents a direct variation, and -3 is the constant of variation.

   b. The equation is not of the form $y = kx$.

So, $y = 7x - 4$ does not represent a direct variation.

EXAMPLE 3   The bending of a beam varies directly as the mass of the load it supports. A beam is bent 20 mm by a mass of 40 kg. How much will the beam bend when supporting a mass of 100 kg?

Solution   1.   Write a formula to find k.

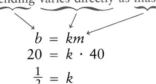
bending varies directly as mass

$$b = km$$
$$20 = k \cdot 40$$
$$\frac{1}{2} = k$$

   2.   Substitute $\frac{1}{2}$ for k in the formula.
   $b = km$ becomes $b = \frac{1}{2}m$.

   3.   Find b for $m = 100$.
   $b = \frac{1}{2}m = \frac{1}{2} \cdot 100 = 50$

So, the beam will bend 50 mm when supporting a mass of 100 kg.

A direct variation problem can also be solved using a proportion (see Lesson 1.4). If (x_1, y_1) and (x_2, y_2) are any two ordered pairs of a direct variation, and neither is $(0, 0)$, then $\frac{y_1}{x_1} = k$, $\frac{y_2}{x_2} = k$, and by the Transitive Property of Equality, $\frac{y_1}{x_1} = \frac{y_2}{x_2}$.

In Example 3, the bending of the beam *is directly proportional* to the mass of the load. The solution below shows how a proportion can be used to solve the problem.

The beam is bent 20 mm by a mass of 40 kg.

(b_1, m_1)
$(20, 40)$

Find the bending for a mass of 100 kg.

(b_2, m_2)
$(b_2, 100)$

$$\frac{b_1}{m_1} = \frac{b_2}{m_2}$$
$$\frac{20}{40} = \frac{b_2}{100}$$

Solve the proportion for b_2. $40b_2 = 2{,}000$
$$b_2 = 50 \text{ mm}$$

Furthermore, direct variation need not be restricted to linear functions, as shown in the next example.

EXAMPLE 4 A nursery charges $48 for enough plants to cover a square patch 8 ft on a side. At that rate, how much would it cost to cover a square patch 6 ft on a side?

Solution The cost of the ground cover (C) is directly proportional to the area of the patch, which is the square of one side (s).

$$\frac{C_1}{(s_1)^2} = \frac{C_2}{(s_2)^2}$$

$$\frac{48}{8^2} = \frac{C_2}{6^2}$$

$$64C_2 = 1,728$$

$$C_2 = 27$$

Calculator Solution

Solve the proportion for C_2.

$$C_2 = \frac{(48)(6^2)}{8^2}$$

Then use a calculator as follows.

48 $\boxed{\times}$ 6 $\boxed{x^y}$ 2 $\boxed{\div}$ 8 $\boxed{x^y}$ 2 $\boxed{=}$ 27

So, it would cost $27 to cover a square patch 6 ft on a side.

Classroom Exercises

Give an equation that describes each direct variation. Use k for the constant of variation.

1. y varies directly as z.

2. m is directly proportional to t.

3. p varies directly as w.

4. a varies directly as the square of g.

5. e is directly proportional to f.

6. g varies directly as h.

Find the constant of variation for each direct variation.

7. $\frac{y}{x} = \frac{4}{7}$

8. $y = \frac{2}{3}x$

9. $-4y = -20x$

10. $-5y = -15x$

11. $\frac{y}{2x} = 7$

12. $\frac{3y}{x} = -6$

Written Exercises

Determine whether the table or equation expresses a direct variation. If so, find the constant of variation, and for the tables, write an equation of the function.

1.

x	y
5	1
15	3
25	5
45	9

2.

x	y
4	7
12	21
-8	-14
-16	-28

3.

x	y
-18	6
12	-4
-45	15
54	-18

4.

x	y
5	4
9	8
-3	-4
-9	-10

5.

x	y
7	7
-8	-8
13	13
-3	-3

6.

x	y
4	5
6	7
-3	-2
0	1

7.

x	y
2	3
8	12
14	21
-4	-6

8.

x	y
2	-1
4	1
9	6
11	8

9.

x	y
-4	5
8	-10
24	-30
-12	15

10. $y = -4x$

11. $y = -x$

12. $y = 3x - 5$

13. $\dfrac{y}{x} - 4 = 0$

14. $y = -3x$

15. $y = 2x$

16. $7y = 4x$

17. $-4y = -8x$

18. $y = 4 - 3x$

19. $y = \dfrac{3}{x}$

20. $y = x + 3$

21. $xy = 4$

In Exercises 22–25, t varies directly as u. Find the value as indicated.

22. Given that $t = 21$ when $u = 7$, find t when $u = 4$.

23. Given that $t = -27$ when $u = 3$, find t when $u = 18$.

24. Given that $t = 16$ when $u = -8$, find t when $u = 4$.

25. Given that $t = 2.6$ when $u = 20.8$, find u when $t = 3.9$.

26. Given that y varies directly as x, and $y = 0.18$ when $x = -6$, find y when $x = -9$.

27. Given that v varies directly as w, and $v = 24$ when $w = -6$, find v when $w = 9$.

28. Given that p is directly proportional to q, and $p = -81$ when $q = 9$, find p when $q = -3$.

29. Given that y is directly proportional to x, and $y = -6$ when $x = -30$, find x when $y = 22.2$.

30. A number y varies directly as the square of the number x. If $y = 32$ when $x = 4$, find y when $x = 6$.

31. A number y varies directly as the cube of a second number x. If $y = 54$ when $x = 3$, find y when $x = 5$.

32. The current I in amperes in an electric circuit varies directly as the voltage V. When 24 volts are applied, the current is 8 amperes. Find the current when 36 volts are applied.

33. The cost of gold varies directly as its mass. Given that 6 g of gold cost $102, find the cost of 14 g of gold.

34. Gas consumption of a car is directly proportional to the distance traveled. A car uses 20 gallons of gas to travel 300 mi. How much gas will the car consume on a trip of 650 mi?

35. The distance required to stop a car varies directly as the square of its speed. It requires 144 ft to stop a car at 60 mi/h. What distance is required to stop a car at 45 mi/h?

36. The height reached by a ball thrown vertically upward is directly proportional to the square of its initial velocity. If a ball reaches a height of 46 m when it is thrown upward with an initial velocity of 30 m/s, what height will the ball reach if the initial vertical velocity is 40 m/s?

37. Hooke's Law says that the distance a spring is stretched by a hanging object is directly proportional to the weight w of the object. If the distance is 80 cm when the weight is 6 kg, what will be the distance when the weight is 10 kg?

38. Given that A is directly proportional to B, and $(A_1, B_1) = (36, 5)$, find B_2 when $A_2 = -9$.

39. If y varies directly as x, and the value of y is positive and held constant, what happens to the value of k as x increases?

40. If y varies directly as x, what effect will tripling x have on y?

41. If y varies directly as the square of x, what effect will doubling x have on y?

42. Write an equation of a function f given that f is a direct variation whose constant of variation is k. Then show that $f(a) + f(b) = f(a + b)$ for all real numbers a and b.

Mixed Review

Find the solution set. *2.1, 2.2*

1. $3x - 2 < 5x + 6$
2. $3x - 8 > x + 6$ *and* $2x - 5 < 23$
3. $x + 5 \leq 8$ *or* $2x - 3 > 1$
4. $-5 < 2x - 9 \leq 5$

For each compound sentence, determine whether it is true or false. *2.2*

5. $2^4 = 4^2$ *or* $4 - 7 = -(7 - 4)$
6. $8 > 4 \cdot 2$ *and* $-6 \leq -3 \cdot 2$

3.9 Graphing Inequalities

Objective

To graph linear inequalities in two variables

In Chapter 2, you graphed linear inequalities in *one* variable as *rays*.

The graphs of linear inequalities in two variables are *half-planes*.

The graph of $y = \frac{1}{2}x - 2$ separates the coordinate plane into two regions called **half-planes**. One half-plane is above the line, and the other is below the line. The line is called the **boundary** of the two half-planes. One half-plane is described by $y > \frac{1}{2}x - 2$ and the other by $y < \frac{1}{2}x - 2$.

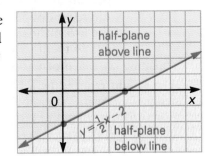

EXAMPLE 1

Graph the inequality $-x - 5y \geq 15$. Check.

Solution

1. Solve the inequality for y.
$$-x - 5y \geq 15$$
$$-5y \geq x + 15$$
$$y \leq -\frac{1}{5}x - 3$$

2. Graph the equation $y = -\frac{1}{5}x - 3$. Because the inequality sign is $\leq$, the points on the line are part of the graph. Use a *solid* line to show this.

3. Shade the region *below* the line.

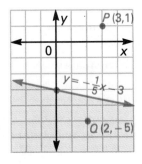

Check

Pick one point, such as $Q(2, -5)$, in the region below the line, and another point, such as $P(3, 1)$, above the line. Substitute.

$$-x - 5y \geq 15$$
$$-2 - 5(-5) \overset{?}{\geq} 15$$
$$23 \geq 15 \text{ True}$$

$$-x - 5y \geq 15$$
$$-3 - 5 \cdot 1 \overset{?}{\geq} 15$$
$$-8 \geq 15 \text{ False}$$

Because $23 \geq 15$ is true, the point $Q(2, -5)$ is a part of the graph. Thus, the graph of $-x - 5y \geq 15$ is the line and the shaded region below the line.

EXAMPLE 2   Graph the inequality $3x - 4 < 11$.

Solution   Solve for x. $3x - 4 < 11$
$$3x < 15$$
$$x < 5$$

Graph the equation $x = 5$. Use a *dashed* line because the inequality symbol is $<$, *not* $\leq$. The points on the line are not part of the graph. Shade the region to the left of the line because the x-coordinate of every point to the left of the line is less than 5. The graph is the shaded region, not including the vertical dashed line.

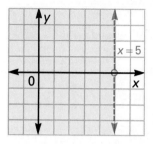

Classroom Exercises

Solve each inequality for y. Then, graph each inequality.

1. $2y > 4x - 6$
2. $-3y < 6x - 3$
3. $-y > x - 2$
4. $-5y \leq -10x + 20$

Written Exercises

Graph each inequality. Check.

1. $y < 3x - 4$ **2.** $x > 4$ **3.** $y > -4x - 8$
4. $y < 5$ **5.** $-4x \leq 8$ **6.** $y \leq -4x - 3$
7. $3y > 18$ **8.** $y > -x + 6$ **9.** $y < 2x - 5$
10. $y > \frac{2}{3}x - 1$ **11.** $x \leq -4$ **12.** $y < -4$
13. $3x - 2y \leq 8$ **14.** $5x + 2y < 10$ **15.** $3x - 7y \geq 21$
16. $4 < y < 6$ **17.** $8 > 2x - 6 > 4$ **18.** $4 < -2x - 8 \leq 10$

Mixed Review

Write an equation, in standard form, of the line satisfying the given conditions. *3.4, 3.7*

1. containing the points $A(0, -5)$ and $B(9, 1)$
2. containing the point $T(-6, 8)$ and having slope $-\frac{2}{3}$
3. containing $P(1, 2)$ and parallel to the line $3x - 2y = 6$

Problem Solving Strategies

Insufficient Data and Contradictions

Sometimes, a problem may not provide enough data to obtain an answer.

Example 1

How much change would you receive from a 10-dollar bill if you purchased 6 quarts of motor oil?

Example 1, as stated, cannot be solved. There is *insufficient data*. That is, to solve this problem, you would need to know the price of 1 quart and the amount of sales tax, if any.

On the other hand, there can be enough data to solve a problem but the solution leads to a contradiction. This is illustrated in Example 2.

Example 2

A rectangle and a square have the same width and the rectangle's length is 5 ft more than twice its width. Find the perimeter of the rectangle if the difference between the two perimeters is 8 ft.

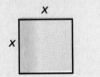 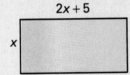

$$2x + 2(2x + 5) - 4x = 8$$
$$2x + 10 = 8$$
$$2x = -2$$
$$x = -1$$

The width x of the square cannot be -1 ft because the width of a square must be a positive number. Thus, there is *no solution*.

Exercises

Solve each problem if possible. If there is insufficient data, state this and tell what additional data is needed. If the solution leads to a contradiction, state "no solution" and describe the contradiction.

1. How much change would you receive from $20 if you purchased 3 neckties at $2.20 each, 2 belts at $5.80 each, and paid a 5% sales tax.
2. The sum of three consecutive integers is 112. Find the three integers.
3. A rectangle is twice as long as it is wide. Find the length of the rectangle.
4. Side a of a triangle is 1 in. shorter than side b, and side c is 7 in. longer than side b. Find the length of each side if the triangle's perimeter is 30 in.
5. A rectangular floor, 18 ft long, is to be covered by a carpet that costs $45 per sq yd, tax included. Find the cost of the carpet.

Mixed Problem Solving

Solve each problem.

1. Find the two numbers whose sum is 85, given that the greater number is 16 times the smaller number.

2. Karen is 7 years older than Bob. Three years ago, Karen's age was twice Bob's age then. How old is Karen now?

3. Rita is 24 years older than Manuel, and 2 years from now she will be 5 times as old as he will be. How old is Rita now?

4. Find three consecutive multiples of 5 such that the sum of the two smaller numbers is 50 more than the greatest number.

5. Find three consecutive odd integers such that twice the sum of the first two integers is 73 more than 3 times the third integer.

6. What initial vertical velocity is required to boost an object to a height of 375 m at the end of 3 s? Use the formula
   $h = vt - 5t^2$.

7. Find three consecutive multiples of 9 such that 5 times the sum of the first two multiples is greater than 600, and 4 times the sum of the last two multiples is less than 660.

8. The perimeter of a triangle is 71 ft. The length of the first side is 11 ft less than twice the length of the second side. The third side is 14 ft longer than the second side. Find the length of the third side.

9. An aircraft carrier traveling 25 km/h left a port at 6:30 A.M. At 10:00 A.M., a helicopter left the same port, traveling the same route at 60 km/h. At what time did the helicopter overtake the carrier?

10. The stock of women's shoes in a shoe store is 40 pairs more than twice the stock of men's shoes. The stock of children's shoes is 80 pairs less than twice the stock of women's shoes. Given that the total stock is 600 pairs of shoes, find the number of pairs of children's shoes.

11. Machine A is 3 times as old as Machine B. Machine C was constructed 20 years after Machine A was built. In 12 years, the age of the oldest machine will be the same as the combined ages of the other two machines now. How old is each machine now?

12. The length of a rectangle is 5 cm less than 3 times its width, and one side of an equilateral triangle is as long as the length of the rectangle. Given that the sum of the two perimeters is 60 cm, find the length of the rectangle.

13. At 8:00 A.M., a van left City A and traveled 60 km/h toward City B. At 10:00 A.M., an automobile left City B and traveled at 80 km/h toward City A. At what time did the automobile and van meet, given that the cities are 330 km apart?

Key Terms

abscissa (p. 73)
collinear (p. 85)
constant function (p. 89)
constant of variation (p. 105)
coordinate plane (p. 73)
direct variation (p. 105)
domain (p. 74)
function (p. 75)
half-plane (p. 110)

linear function (p. 78)
linear model (p. 96)
mapping (p. 74)
ordered pair (p. 73)
ordinate (p. 73)
origin (p. 73)
point-slope form (p. 88)
quadrant (p. 73)
range (p. 74)

relation (p. 74)
slope (p. 84)
slope-intercept form (p. 91)
value of a function (p. 79)
vertical-line test (p. 75)
x-axis (p. 73)
y-axis (p. 73)
y-intercept (p. 91)

Key Ideas and Review Exercises

3.1 To verify that a relation is a function, show either that (1) no two ordered pairs have the same first coordinate, or that (2) no vertical line crosses the graph of the relation more than once.

Graph each relation and determine its domain and range. Is it a function?

1. $\{(5,3), (4,-1), (-1,5), (-2,-2), (0,3)\}$ 2. $\{(5,3), (-6,0), (-1,-1), (5,-4)\}$

3. Determine the domain and range of the relation whose graph is given at the right. Use the vertical-line test to determine whether the relation is a function.

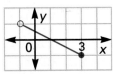

3.2 To graph a function given its equation and domain, solve for y, and construct a table of values for x and y. Plot the points; draw the graph.

4. Graph $x^2 - y = 4$ for the following values of x: $-3, -2, -1, 0, 1, 2, 3$.

Let $f(x) = x^2 - 4x - 2$ and $g(x) = -5x + 2$. Find each of the following.

5. $f(-4)$ 6. $g(-3)$
7. The range of f for the domain $\{-6, -2, 5\}$. 8. $g(3 + h)$

3.3 To determine the slant of a line, find the slope. If the slope > 0, then the line slants up to the right. If the slope < 0, then the line slants down to the right. If the slope $= 0$, then the line is horizontal. If the slope is undefined, then the line is vertical. (See p. 84 for the slope formula.)
To verify that three points are collinear, show that the slope is the same for any two pairs of the points.

Find the slope of $\overleftrightarrow{PQ}$. Describe the slant of the line.

9. $P(-4, 2), Q(-3, -2)$ 10. $P(-6, -5), Q(-6, -1)$ 11. $P(-3, 4), Q(7, 4)$
12. Determine whether the points $A(2, 5), B(0, -1),$ and $C(1, 2)$ are collinear.

3.4   To find an equation of a line,

1. Use $y - y_1 = m(x - x_1)$ if the line is nonvertical and m and (x_1, y_1) are given. If two points are given, first use the slope formula to find m.
2. Use $y = c$ if the line is horizontal and contains $P(x, c)$.
3. Use $x = c$ if the line is vertical and contains $P(c, y)$.

Find an equation, in standard form, of the line containing the given points.

13. $P(0, -1)$, $Q(-3, 1)$ **14.** $P(4, 5)$, $Q(4, -2)$ **15.** $P(7, 2)$, $Q(1, -7)$

3.5   To graph a line using the slope-intercept method, write an equation in $y = mx + b$ form. Plot the point $(0, b)$, where b is the y-intercept. Use this point and the slope, m, to plot a second point. Draw the graph.

Graph each equation.

16. $f(x) = -2x + 1$ **17.** $3x - 5y = 15$ **18.** $3y - 4 = 11$

3.6   To write an equation of a linear model for a given set of data, first plot the points corresponding to the data. Then draw the straight line that best fits the points. Finally, use two points on the line to write the equation.

19. Use the table to construct a linear model. Write an equation of the model.

Number of months: x	2	3	4	5	6
Ants (thousands): y	9	10	13	18.5	19

20. Describe in your own words the procedure for writing an equation of a linear model corresponding to a given set of data.

3.7   To find an equation of a nonvertical line parallel to a given line, use the fact that parallel lines have the same slope.

To find an equation of a nonvertical, nonhorizontal line perpendicular to a given line, use the fact that the product of the slopes of two perpendicular lines is -1.

Find an equation of the line satisfying the given conditions.

21. containing $P(-5, 1)$ and parallel to the line whose equation is $4x - y = 8$

22. containing $Q(-2, 5)$ and perpendicular to the line whose equation is $-3x - 5y = 20$

3.8   To verify that a function is a direct variation, show that for all $x \neq 0$ and for all y, $\frac{y}{x} = k$ or $y = kx$, where k is a nonzero constant.

Determine whether the equation expresses a direction variation.

23. $y = 4x + 2$ **24.** $c = 5s$ **25.** $6y = 11x$

Graph each relation and determine its domain and range. Is each a function?

1. $\{(-4, 1), (3, -2), (-4, 6), (-1, 1)\}$ **2.** $\{(-5, 0), (2, -4), (6, -3), (8, 0)\}$

Let $f(x) = -x^2 - x + 4$ and $g(x) = -4x + 5$. Find each of the following.

3. $g(-3)$ **4.** $f(4)$

5. the range of f for the domain $\{-2, 0, 5\}$ **6.** $g(2 + h)$

Find the slope of $\overleftrightarrow{AB}$. Describe the slant of the line.

7. $A(-2, 5), B(3, -1)$ **8.** $A(-4, -1), B(-4, 6)$ **9.** $A(-6, -3), B(2, -3)$

10. Determine whether the points $A(0, 1)$, $B(2, -3)$, $C(4, 5)$ are points on the same line.

Find an equation, in standard form, of the line containing the given points.

11. $P(8, 1), Q(-4, 10)$ **12.** $A(-3, -1), B(4, -1)$ **13.** $A(5, -3), B(3, -11)$

Graph each equation.

14. $f(x) = -\frac{2}{3}x + 5$ **15.** $-4x - 3y + 12 = 0$ **16.** $-4 = 6 - 5y$

17. Use the table to construct a linear model. Write an equation of the model.

Number of hours: x	4	6	8	10	12	14
Thousands of bacteria: y	80	70	58	52	36	33

18. Predict the number of bacteria after 18 h.

Write an equation, in standard form, of the line satisfying the given conditions.

19. containing $P(-1, 2)$ and parallel to the line whose equation is $-2x - 4y = 8$

20. containing $T(3, -1)$ and perpendicular to the line whose equation is $x - y = 5$

Determine whether each equation represents a direct variation. If so, find the constant of variation.

21. $-6y = 12$ **22.** $x - y = 4$ **23.** $\frac{x}{y} = -4$

Graph each inequality.

24. $y < -3x + 1$ **25.** $4x + 2y \le 8$ **26.** $3x - 1 > 5$

27. Graph $-4 < 2x - 8 \le 6$ in a coordinate plane.

College Prep Test

Choose the *one* best answer to each question or problem.

1. Which ordered pair (x,y) is not a solution of the equation
$$-13 + 28 = xy - 13?$$
(A) (4,7)   (B) (14,2)   (C) (14,14)
(D) $\left(\frac{1}{2}, 56\right)$ (E) $(-7, -4)$

2. If y varies directly as the square of x, what will be the effect on y of doubling x?
(A) y will double.
(B) y will be half as large.
(C) y will be four times as large.
(D) y will decrease in size.
(E) None of these.

3. If $5x + 4y - xy + 8 = 0$ and $x + 3 = 9$, then $3 - y = $ ___?___
(A) -19 (B) -16 (C) 8
(D) 19   (E) 22

4. Segments of the lines $x = 4$, $x = 9$, $y = -5$, and $y = 4$ form a rectangle. What is the area of this rectangle?
(A) 6   (B) 10   (C) 20
(D) 18   (E) 45

5. In the figure below, what is the area of the shaded region of the rectangle?

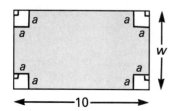

(A) $10w$ (B) $4a^2$ (C) $10w - 4a$
(D) $10w - a^2$ (E) $10w - 4a^2$

6. For any integer k, which of the following represents three consecutive even integers?
(A) $2k, 4k, 6k$
(B) $k, k + 1, k + 2$
(C) $k, k + 2, k + 4$
(D) $4k, 4k + 1, 4k + 2$
(E) $2k, 2k + 2, 2k + 4$

7. Six oranges cost as much as 3 pears. If 2 oranges cost 50¢, what will 2 pears cost?
(A) $1.00   (B) 50¢   (C) $3.00
(D) 75¢   (E) $1.50

8. Dividing a number by $\frac{1}{5}$ gives the same result as multiplying by which one of the following?
(A) $\frac{1}{5}$ (B) 0.5   (C) 20%
(D) 3   (E) 5

9. What is the arithmetic mean (average) of $\frac{1}{3}$ and $\frac{1}{4}$?
(A) $\frac{1}{6}$ (B) $\frac{1}{7}$ (C) $\frac{1}{12}$
(D) $\frac{7}{12}$ (E) $\frac{7}{24}$

10. Given that $x = 5a$ and $y = \dfrac{1}{15a + 2}$, what is y in terms of x?
(A) $\dfrac{1}{3x + 2}$ (B) $\dfrac{1}{3x}$
(C) $\dfrac{1}{x + 2}$ (D) $\dfrac{5}{x + 2}$
(E) None of these

HURRICANE GILBERT
12 SEPTEMBER 1988 1200 UTC

A three-dimensional model of Hurricane Gilbert is created from two sets of data transmitted via satellite. The computer uses the data to calculate the height of cloud tops and to track the movement of the storm. Images can be generated to view the hurricane from any height or angle.

4.1 Graphing Linear Systems

Objectives

To solve systems of two linear equations in two variables by graphing

To determine whether a system of two linear equations is consistent, inconsistent, dependent, or independent

Recall that equations such as $2x + y = 3$ and $2y - x + 4 = 0$ describe *linear* functions. The graphs of the two functions are shown below. Notice that their *point of intersection*, $P(2, -1)$, is the only point on both lines. Therefore, $(2, -1)$ is the only ordered pair (x, y) that satisfies *both* equations.

Recall (Lesson 2.2) that the conjunction *p and q* is true if and only if both *p* is true and *q* is true. Therefore, the conjunction

$$2x + y = 3 \quad and \quad 2y - x + 4 = 0$$

is true for $(2, -1)$ because $(2, -1)$ is a solution of both equations.

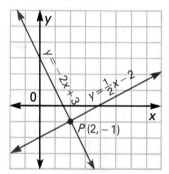

$$
\begin{array}{ll}
2x + y = 3 & 2y - x + 4 = 0 \\
2 \cdot 2 - 1 \stackrel{?}{=} 3 & 2(-1) - 2 + 4 \stackrel{?}{=} 0 \\
4 - 1 \stackrel{?}{=} 3 & -2 - 2 + 4 \stackrel{?}{=} 0 \\
3 = 3 & -4 + 4 \stackrel{?}{=} 0 \\
\text{True} & 0 = 0 \\
& \text{True}
\end{array}
$$

So, the ordered pair $(2, -1)$ is the only solution of both parts of the conjunction

$$2x + y = 3 \quad and \quad 2y - x + 4 = 0,$$

and of the following **system of equations:** $\begin{aligned} 2x + y &= 3 \\ 2y - x + 4 &= 0 \end{aligned}$

To solve a system of two equations in two variables by graphing:
1. Graph both equations on the same coordinate plane.
2. Find the coordinates of the point of intersection of the two lines.

Definitions

If a system of linear equations has *at least* one solution, the system is called a **consistent system.**

If a system of linear equations has *at most* one solution, the equations are called **independent**, and the system is called an **independent system.**

For example, the system $\begin{matrix} 2x + y = 3 \\ 2y - x + 4 = 0 \end{matrix}$ is both *consistent* and *independent* because it has exactly one solution, $(2, -1)$.

EXAMPLE 1   Solve the system $\begin{matrix} 5y - 2(x + y) = 15 \\ 2x - 3y = 12 \end{matrix}$ by graphing.

Solution   First, write each equation in slope-intercept form, then graph it.

$$5y - 2(x + y) = 15$$
$$5y - 2x - 2y = 15$$
$$3y = 2x + 15$$
$$y = \tfrac{2}{3}x + 5$$

$$2x - 3y = 12$$
$$-3y = -2x + 12$$
$$y = \tfrac{2}{3}x - 4$$

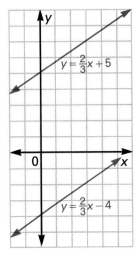

Notice that the slopes of the two lines are the same, but their *y*-intercepts are different. Therefore, the lines are parallel, and because parallel lines never meet, there is no point of intersection. So, the system has no solution.

Definition   A system of equations that has *no* solution is an **inconsistent system**.

EXAMPLE 2   Solve by graphing. $0.2x - 0.3y = 0.9$ *and* $\dfrac{y}{2} = \dfrac{x}{3} - \dfrac{3}{2}$

Solution   First, write each equation in slope-intercept form and graph it.

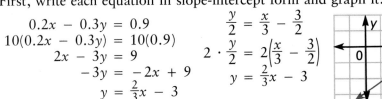

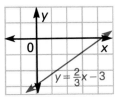

$$0.2x - 0.3y = 0.9$$
$$10(0.2x - 0.3y) = 10(0.9)$$
$$2x - 3y = 9$$
$$-3y = -2x + 9$$
$$y = \tfrac{2}{3}x - 3$$

$$\dfrac{y}{2} = \dfrac{x}{3} - \dfrac{3}{2}$$
$$2 \cdot \dfrac{y}{2} = 2\left(\dfrac{x}{3} - \dfrac{3}{2}\right)$$
$$y = \tfrac{2}{3}x - 3$$

The slopes and *y*-intercepts of the two equations are the same. Therefore, the graphs coincide, and *all* ordered pairs (x,y) such that $y = \tfrac{2}{3}x - 3$ are solutions of the system. The solution set, which is infinite, can be written as $\{(x,y) \mid y = \tfrac{2}{3}x - 3\}$.

Definition   If a system of equations has an infinite number of solutions, the equations are **dependent**, and the system is a **dependent system**.

As the examples of this lesson have shown, a system of linear equations may have one solution, no solution, or infinitely many solutions.

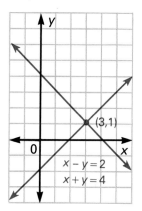
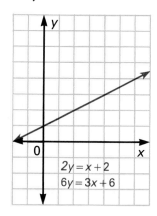

One solution	No solution	Infinitely many solutions
Consistent	*Inconsistent*	*Consistent*
Independent	*Independent*	*Dependent*

Classroom Exercises

Determine whether the given ordered pair is a solution of the given system of equations.

1. $(3, -1)$;
$2x + y = 5$
$x - y = 1$

2. $(3, 4)$;
$x + y = 7$
$2x - y = 1$

3. $(5, 1)$;
$x + 2y = 7$
$2x - y = 9$

4. $(-3, 1)$;
$2x + y = -5$
$y = -\frac{1}{3}x$

Solve each equation for y.

5. $2y = 4x - 8$

6. $-3y = 6x - 12$

7. Solve Exercise 1 by graphing.

Written Exercises

Solve each system by graphing. Indicate whether the system is *consistent* or *inconsistent* and *dependent* or *independent*.

1. $y = x + 4$
$y = -x + 2$

2. $y = 2x - 4$
$y = 2x + 5$

3. $y = 3x - 2$
$2y = 6x - 4$

4. $y = 2x$
$y = -2x + 4$

5. $-x + y = 6$
$-2x + 2y = 10$

6. $x + y = 6$
$-3x - 3y = -18$

Solve each system by graphing.

7. $y = 2 - 3x$
$\dfrac{x}{2} = \dfrac{1}{3} - \dfrac{y}{6}$

8. $y = 8 - x$
$\dfrac{y}{2} = x + \dfrac{5}{2}$

9. $2x = 4 - y$
$-\dfrac{3}{2}x + \dfrac{1}{4}y = -1$

10. $5x - 3(y + x) = -6$
$2x - 3y = 3$

11. $6x - 2(5y - 2x) + 12y = 10$
$-5x - (y - 8x) + 2y - 3 = 0$

12. $0.4x - 0.2y = 0.6$
$0.6x + 0.3y = -0.9$

13. $0.2x + 0.2y = 0.4$
$0.4x - 0.4y = 0.8$

14. Write, in your own words, how to solve a system of linear equations by graphing.

Find the value(s) of k that satisfy the given condition for each system.

15. $kx - 5y = 8$
$7x + 5y = 10$ is consistent.

16. $5x - 2y = 7$
$-20x + ky = -28$ has infinitely many solutions.

Mixed Review

Find the solution set.   *2.1, 2.3, 3.5*

1. $3x - 2(4x + 1) \geq x - 8$

2. $|2x - 4| = 10$

3. $|x - 6| < 1$

4. Graph $3x - 5y = 15$.

Application: *Supply and Demand*

Supply and demand equations are used by economists to describe, for a particular time, the relationship between price y and quantity x. Supply equations have positive slopes because, as price increases, so does production. Demand equations, however, have negative slopes because, as price increases, demand decreases. The solution of a system of supply and demand equations is the equilibrium price for that item.

1. Supply of and demand for rag rugs can be described by $2y - x = 18$ and $2y + 2x = 36$, where $x =$ thousands of rugs and $y =$ price in dollars. Find the equilibrium price.

2. What are some of the factors that could cause supply to fall below the equilibrium point? What are some of the factors that could cause the equilibrium point to shift?

The table below shows gas mileage ratings for different cars reported in a recent year. The engine size *x* is in liters of displacement. The mileage *y* is in highway miles per gallon.

x	1.0	1.3	1.5	1.6	1.8	1.9	2.0	2.2	2.3	2.5
y	58	47	49	38	41	32	40	32	33	31

Statisticians sometimes use a method called the **median line of fit** to determine a line through a set of data points like the one above.

A. Graph the points on a coordinate plane. This is called a **scatter diagram.**

B. Partition the data into three nearly equal sets such that there is an odd number of points in the first and third sets. In this example, the obvious choice is sets of 3, 4, and 3 points.

C. Locate the *median x* and *y* values in the first and third sets. *The median is the value with as many values above it as below it.*

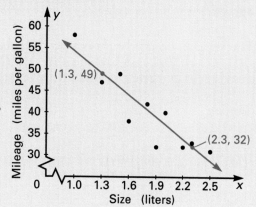

The median of the *x* values in the first group of 3 points is 1.3. The median of the *y* values is 49. Combine these 2 values as the point (1.3, 49) and add it to the diagram. Do the same with the median *x* and *y* values from the 3rd group to locate a second point, (2.3, 32). Connect the 2 points with a straight line.

Exercises

1. Find the slope of the line using the points (1.3, 49) and (2.3, 32).
2. Using the slope from Exercise 1 and one of the two points, determine the equation of the line.
3. Predict the mileage for an engine with a 1.7 l engine.
4. Use the data in the table below to construct a scatter diagram, using "pages" for the *x*-variable. Draw the median line of fit and determine its equation. Predict the number of hours to type 70 pages.

Pages	5	8	14	20	23	29	36	40	44	49
Hours to Type	.5	1.1	1.4	2.8	1.8	3.8	4.7	3.9	5.0	4.8

4.2 The Substitution Method

Objectives

To solve systems of two linear equations using the substitution method
To solve word problems using systems of equations

Because solving a system of equations by graphing involves estimating the coordinates of the point of intersection, the graphing method works poorly when the coordinates are not integers. There are, however, algebraic methods for solving systems of equations that produce exact results. One of these is the *substitution method*.

EXAMPLE 1

Solve the system $\begin{matrix} 2x - y = 8 \\ -4x = -23 - 3y \end{matrix}$ by substitution.

Plan

First, solve one of the equations for one of its variables. Notice that it is easier to solve for a variable whose coefficient is already 1 or -1, such as y in $2x - y = 8$.

Solution

Solve $2x - y = 8$ for y.

$$2x - y = 8$$
$$-y = 8 - 2x$$
$$y = -8 + 2x$$

Substitute $-8 + 2x$ for y in the other equation.

$$-4x = -23 - 3y$$
$$-4x = -23 - 3(-8 + 2x)$$
$$-4x = -23 + 24 - 6x$$
$$2x = 1$$
$$x = \frac{1}{2}$$

To find the value of y, substitute $\frac{1}{2}$ for x in $y = -8 + 2x$. (This equation is used because it is already solved for y.)

$$y = -8 + 2x$$
$$= -8 + 2 \cdot \frac{1}{2}$$
$$= -8 + 1$$
$$= -7$$

Check

$$2x - y = 8$$
$$2(\tfrac{1}{2}) - (-7) \stackrel{?}{=} 8$$
$$8 = 8 \quad \text{True}$$

$$-4x = -23 - 3y$$
$$-4(\tfrac{1}{2}) \stackrel{?}{=} -23 - 3(-7)$$
$$-2 = -2 \quad \text{True}$$

The solution is $\left(\frac{1}{2}, -7\right)$.

In Example 1, the system $\begin{matrix} 2x - y = 8 \\ -4x = -23 - 3y \end{matrix}$ was simplified to the system: $\begin{matrix} x = \frac{1}{2} \\ y = -7 \end{matrix}$

Because these two systems have the same solution set, $\{(x,y) \mid x = \frac{1}{2}$ *and* $y = -7\}$, the two systems are referred to as **equivalent systems.**

Sometimes the equations of a linear system contain fractions or decimals. When this occurs, it helps to eliminate the fractions or decimals first.

To accomplish this, multiply each side of the equation by the LCM of the denominators or, for decimals, the appropriate power of 10, as illustrated in Example 2 below.

EXAMPLE 2   Solve the conjunction $\frac{x}{3} - \frac{y}{6} = \frac{1}{2}$ and $0.3x + 0.7y = 4.7$ using the substitution method.

Solution   First, eliminate the fractions. Multiply each side of the first equation by the LCD, 6.

$$6\left(\frac{x}{3} - \frac{y}{6}\right) = 6 \cdot \frac{1}{2}$$
$$6 \cdot \frac{x}{3} - 6 \cdot \frac{y}{6} = 6 \cdot \frac{1}{2}$$
$$2x - y = 3$$

Next, eliminate the decimals. Multiply each side of the second equation by 10.

$$10(0.3x + 0.7y) = 10(4.7)$$
$$10(0.3x) + 10(0.7y) = 10(4.7)$$
$$3x + 7y = 47$$

Now solve the equivalent system.

(1) $2x - y = 3$
(2) $3x + 7y = 47$

Solve Equation (1) for y, because it has the convenient coefficient -1.

(1) $2x - y = 3$
$-y = 3 - 2x$
$y = 2x - 3$

Substitute $2x - 3$ for y in Equation (2).

(2) $3x + 7y = 47$
$3x + 7(2x - 3) = 47$
$3x + 14x - 21 = 47$
$17x = 68$
$x = 4$

To find the value of y, substitute 4 for x in $y = 2x - 3$, because this equation is already solved for y.

$y = 2x - 3$
$= 2 \cdot 4 - 3$
$= 8 - 3$, or 5

Check

$$\frac{x}{3} - \frac{y}{6} = \frac{1}{2}$$
$$\frac{4}{3} - \frac{5}{6} \stackrel{?}{=} \frac{1}{2}$$
$$\frac{1}{2} = \frac{1}{2} \text{ True}$$

$$0.3x + 0.7y = 4.7$$
$$0.3(4) + 0.7(5) \stackrel{?}{=} 4.7$$
$$1.2 + 3.5 \stackrel{?}{=} 4.7$$
$$4.7 = 4.7 \text{ True}$$

The solution is $(4, 5)$.

Sometimes a word problem can be solved using a system of equations.

EXAMPLE 3 At a senior-citizens center, 35 tables were set up for games of checkers (2 players per table) and bridge (4 players per table). There were 92 players occupying 35 tables. Find the number of checker players and the number of bridge players.

Solution What are you to find? the number of each type of player

Choose two variables. Let c = the number of checker tables.
What do they represent? Let b = the number of bridge tables. Then
$2c$ = the number of checker players, and
$4b$ = the number of bridge players.

What is given? The number of The number of
tables is 35. players is 92.
Write a system. (1) $c + b = 35$ (2) $2c + 4b = 92$

Solve the system. Solve (1) for one variable. $c = 35 - b$

Substitute $35 - b$ for c in (2).
(2) $2(35 - b) + 4b = 92$

Solve for b. $70 - 2b + 4b = 92$
$b = 11$ ←number of
bridge tables

Substitute 11 for b in (1).
(1) $c + 11 = 35$
Solve for c. $c = 24$ ←number of
checker tables

Number of checker players = $2c = 2 \cdot 24 = 48$

Number of bridge players = $4b = 4 \cdot 11 = 44$

Check in the original 35 tables were set up: $11 + 24 = 35$
problem.

There were 92 players: $48 + 44 = 92$

State the answer. There were 48 checker players and 44 bridge players.

Classroom Exercises

Tell which equation is easier to solve. Which variable would you solve for?

1. $2x + 3y = 7$
$3x - y = 5$

2. $2y = 6 - x$
$4x + 5y = 18$

3. $7x = 6 - 2y$
$4y = 9 - x$

4. $3x - y = 9$
$5x + 2y = 4$

Solve using the substitution method.

5. $y = 3x$
$4x + 3y = 26$

6. $x = 4y$
$2x + 4y = 24$

7. $y = x - 4$
$y + x = 6$

8. $x = y + 1$
$3x + y = 11$

Written Exercises

Solve using the substitution method.

1. $y = 2x$
$x + y = 9$

2. $x = y + 1$
$x + 4y = 11$

3. $y = 3x - 1$
$2x + y = 14$

4. $x = 4y$
$2x + 3y = 22$

5. $2x - y = 5$
$-x + y = -3$

6. $15x + 4y = 23$
$10x - y = -3$

7. $3x - y + 2 = 0$
$5x - 3y + 4 = 0$

8. $9m + 8n = 21$
$2m = 7 - n$

9. $5a + 4b = 7$
$2a - b = 8$

Solve each conjunction.

10. $\frac{1}{2}x + \frac{1}{4}y = \frac{3}{4}$ and $0.5x - 0.2y = 0$ **11.** $0.12x + 0.15y = 132$ and
$x = 1,000 - y$

Solve each problem by using a system of two linear equations and the substitution method.

12. The sum of two numbers is 35. Twice the first number is equal to 5 times the second number. Find the two numbers.

13. Martha's age is twice that of her brother Ned. The sum of their ages is 24. Find the age of each.

14. An apartment building contains 200 units. Some of these are one-bedroom units that rent for $435 each month. The rest are two-bedroom units that rent for $575 per month. When all the units are rented, the total monthly income is $97,500. How many apartments are there of each type?

15. The units digit of a two-digit number is $\frac{1}{2}$ the tens digit. If the digits are reversed, the sum of the new number and the original number is 132. Find the original number. (HINT: Write the number as $10t + u$, where t and u are the tens and units digits, respectively.)

16. The graph of $y = ax^2 + bx$ contains the points $P(2,5)$ and $Q(1,-1)$. Find a and b.

17. Find the coordinates of the vertices of the triangle formed by the graphs of the lines $2x - y = -1$, $x - y = -1$, and $2x - 5y = -29$.

Mixed Review

1. Find $f(-3)$ for $f(x) = -x^3$. **3.2**

2. Graph $f(x) = 2x - 4$. **3.5**

3. Find the slope of $\overline{AB}$ for $A(5, -1)$ and $B(-2,3)$. **3.3**

4. If y varies directly as x, and if $y = 12$ when $x = 3$, what is y when $x = 15$? **3.8**

4.3 The Linear Combination Method

Objective

To solve systems of two linear equations using the linear combination method

Another algebraic method for solving a system of linear equations, the *linear combination method*, applies the following theorem.

Theorem 4.1

If $a = b$ and $c = d$, then $a + c = b + d$.

Proof

Statement	Reason
1. $a = b$, $c = d$	1. Given
2. $a + c = b + c$	2. Add Prop of Eq (Lesson 1.4)
3. $a + c = b + d$	3. Subst (Lesson 1.8)

In the system at the right, the y-terms, $2y$ and $-2y$, are *additive inverses*. So, when the equations are added, the resulting equation has only one variable term, $7x$. Next, solving for x, and then solving for y by substituting for x in either equation, gives the solution $(2, 1)$.

$$(1) \quad 3x + 2y = 8$$
$$(2) \quad 4x - 2y = 6$$
$$\overline{ 7x = 14}$$
$$x = 2$$

$$(1) \quad 3x + 2y = 8$$
$$3 \cdot 2 + 2y = 8$$
$$6 + 2y = 8$$
$$2y = 2$$
$$y = 1$$

Verify that the ordered pair $(2, 1)$ is a solution of the system: $\begin{array}{l} 3x + 2y = 8 \\ 4x - 2y = 6 \end{array}$

Also, $(2, 1)$ is a solution of the equation formed from the sum of *any* two constant multiples of these equations in the manner illustrated below (see Mult Prop of Eq in Lesson 1.4).

Use 5 as a multiplier for (1).

(1) $3x + 2y = 8$
$5(3x + 2y) = 5 \cdot 8$
(3) $15x + 10y = 40$

Use 3 as a multiplier for (2).

(2) $4x - 2y = 6$
$3(4x - 2y) = 3 \cdot 6$
(4) $12x - 6y = 18$

Add the corresponding sides of the resulting equations.

(3) $15x + 10y = 40$
(4) $\underline{12x - 6y = 18}$
(5) $27x + 4y = 58$

Verify that $(2, 1)$ is a solution of Equation (5).

$27 \cdot 2 + 4 \cdot 1 \overset{?}{=} 58$
$54 + 4 = 58$ True

Thus, if $(2, 1)$ is a solution of (1) $3x + 2y = 8$ and (2) $4x - 2y = 6$, then $(2, 1)$ is also a solution of (3) and (4). The resulting equation,

(5) $27x + 4y = 58$, is called a *linear combination* of Equations (1) and (2).

A **linear combination** of two equations may be obtained in the following manner.
1. Multiply the first equation by a nonzero constant.
2. Multiply the second equation by a nonzero constant.
3. Add the two resulting equations to obtain a new equation.

If the constants in Steps (1) and (2) above are chosen properly, the new equation obtained in Step (3) can be solved for one of the variables. The other variable will have been eliminated in the addition step. This is illustrated in Example 1.

EXAMPLE 1   Solve the system $\begin{aligned} 8x + 5y &= 22 \\ 6x + 2y &= 13 \end{aligned}$ using the linear combination method.

Plan   Solve the system by eliminating either the x- or y-terms. For example, to make the y-terms additive inverses of each other, multiply each side of the first equation by 2, and each side of the second equation by -5.

The result is the equivalent system: $\begin{aligned} 16x + 10y &= 44 \\ -30x - 10y &= -65 \end{aligned}$

Solution
$$\begin{aligned} 8x + 5y &= 22 \\ 6x + 2y &= 13 \end{aligned} \rightarrow \begin{aligned} 2(8x + 5y) &= 2(22) \\ -5(6x + 2y) &= -5(13) \end{aligned} \rightarrow \begin{aligned} 16x + 10y &= 44 \\ \underline{-30x - 10y = -65} \end{aligned}$$

Add the resulting equations. $-14x = -21$

Solve for x. $x = \dfrac{3}{2}$

To find the value of y, substitute $\frac{3}{2}$ for x in either equation. Equation (1) is used here.

$$\begin{aligned} (1) \quad 8x + 5y &= 22 \\ 8\left(\tfrac{3}{2}\right) + 5y &= 22 \\ 12 + 5y &= 22 \\ 5y &= 10 \\ y &= 2 \end{aligned}$$

So, the solution is $\left(\frac{3}{2}, 2\right)$. The check is left for you.

In Example 1, the system $\begin{aligned} 8x + 5y &= 22 \\ 6x + 2y &= 13 \end{aligned}$ was reduced to the equivalent system: $\begin{aligned} x &= \tfrac{3}{2} \\ y &= 2 \end{aligned}$

At the right, the two systems are represented together graphically, as is their common solution $\left(\frac{3}{2}, 2\right)$.

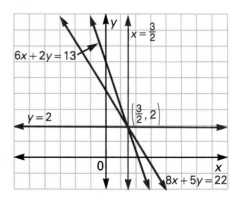

EXAMPLE 2 Solve each system using the linear combination method.

  a. $3x - 6y = 7$
  $-x + 2y = 6$

  b. $10x - 6y = 4$
  $-5x + 3y = -2$

Solutions

a. (1) $3x - 6y = 7$
(2) $-x + 2y = 6$
  $3(-x + 2y) = 3 \cdot 6$
(3) $-3x + 6y = 18$

Add Equations (1) and (3).
(1) $3x - 6y = 7$
(3) $\underline{-3x + 6y = 18}$
  $0 + 0 = 25$ False

Thus, there is *no* solution.
The system is *inconsistent*.
The solution set is $\varnothing$.

b. (1) $10x - 6y = 4$
(2) $-5x + 3y = -2$
  $2(-5x + 3y) = 2(-2)$
(3) $-10x + 6y = -4$

Add Equations (1) and (3).
(1) $10x - 6y = 4$
(3) $\underline{-10x + 6y = -4}$
  $0 + 0 = 0$ True

Thus, there are an *infinite number* of solutions. The system is
dependent. The solution set is
$\{(x,y) \mid -5x + 3y = -2\}$.

The graph of the system in Example 2a is a pair of parallel lines (see
Example 1 of Lesson 4.1). The graph of the system in Example 2b is a
pair of coincident lines (see Example 2 of Lesson 4.1).

Classroom Exercises

For each given system, select a multiplier for one of the equations so
that the sum of the equations of the new system is an equation in one
variable.

1. $3x - 5y = 1$
  $x + y = 7$

2. $4x - 5y = 13$
  $3x + y = 6$

3. $3x + 7y = 4$
  $-x + 5y = 2$

4. $3x + y = 4$
  $x + 2y = 3$

Solve using the linear combination method.

5. $x + y = 4$
  $x - y = 2$

6. $x + y = 2$
  $x - y = 8$

7. $-x + y = 5$
  $x + y = 7$

8. $3x - y = 8$
  $2x + y = 2$

Indicate whether each system has exactly one solution, no solution, or
infinitely many solutions.

9. $3x - 4y = -1$
  $x + 4y = 5$

10. $3x - 5y = 9$
  $-3x + 5y = -9$

11. $7x - 8y = 5$
  $-7x + 8y = 5$

12. $3x = 2y + 3$
  $6x + 4y = 3$

Solve using the linear combination method. If the system is inconsistent
or dependent, indicate this.

13. $3x + 2y = 4$
  $5x - 2y = 28$

14. $5x - 3y = 8$
  $10x - 6y = 18$

15. $6x + y = 12$
  $x - 2y = 2$

16. $4x - 2y = 1$
  $-8x + 4y = -2$

Written Exercises

Solve using the linear combination method. If the system is inconsistent or dependent, indicate this.

1. $5x - 2y = 30$
   $x + 2y = 6$

2. $-7x + y = 10$
   $7x + 2y = -1$

3. $-8x + 7y = 17$
   $8x + 11y = -35$

4. $5x + 3y = 4$
   $4x - 3y = 14$

5. $6x - y = 5$
   $8x + 2y = 10$

6. $4x - 3y = 1$
   $2x + y = 3$

7. $7y - 2x = -4$
   $4y = x - 1$

8. $2x = 1 + 3y$
   $5x + 6y = 16$

9. $12x + 10y = 4$
   $9x = 21 - 12y$

10. $3x + 2y = 5$
    $-6x - 4y = -10$

11. $2x - 4y = 5$
    $-x + 2y = 8$

12. $3x + 2y = 5$
    $4x = 22 + 5y$

13. $5x - 2y = 3$
    $2x + 7y = 9$

14. $3x + 2y = 12$
    $2x + 5y = 8$

15. $3x = -2y + 13$
    $2x + 3y = 2$

16. $\dfrac{x - 2y}{8} = \dfrac{1}{2}$
    $3x + 2y = 4$

17. $\dfrac{6}{7}x - y = \dfrac{3}{7}$
    $\dfrac{9}{7}x + 2y = \dfrac{1}{7}$

18. $\dfrac{x}{2} = \dfrac{y + 4}{3}$
    $\dfrac{x - y}{6} = \dfrac{1}{2}$

19. $10t + u = 7(t + u) + 3$ *and* $3u = t + 2$
20. $2(b + c) = 16$ *and* $3(b - c) = 16$
21. $12x - 11y = 7$ *and* $13x - 10y = 21$
22. $27x - 165y = 105$ *and* $11x + 55y = 165$

Solve each problem using a system of linear equations and the linear combination method.

23. The larger of two numbers is 3 more than twice the smaller. If twice the larger is decreased by 5 times the smaller, the result is 0. Find the two numbers.

24. Twice one number is 15 less than a second number. When 13 is added to the second number, the result is 7 less than 9 times the first number. Find the numbers.

25. The perimeter of a rectangle is 22 m. The length of the rectangle is 1 m less than 3 times the width. Find the length and the width.

26. Twice one number is 20 more than a second number. One-fourth of the first number is 2 less than $\frac{1}{4}$ of the second. Find each number.

27. If a theater added 5 seats to each row, it would need 20 fewer rows to seat the same number of people. If each row had 3 fewer seats, it would take 20 more rows to seat the same number. How many people does the theater seat now? [HINT: $(a + b)(c + d) = ac + bc + ad + bd$]

Solve for x and y.

28. $ax + by = n$
    $2ax + 3by = n$

29. $\dfrac{x}{y} = \dfrac{a}{b}$

    $\dfrac{x + 1}{2a} = \dfrac{y + 1}{b}$

30. $\dfrac{8}{x} + \dfrac{15}{y} = 33$

    $\dfrac{4}{x} + \dfrac{5}{y} = -1$

Midchapter Review

Solve each system by graphing. Indicate whether the system is consistent or inconsistent and dependent or independent. *4.1*

1. $y = x - 5$
   $y = 3 - x$

2. $y = -x$
   $y = x + 2$

3. $x - y = 4$
   $y - x = 4$

4. $3x - y = 6$
   $2y = 6x - 12$

Solve using either the substitution method or the linear combination method. *4.2, 4.3*

5. $y = -3x$
   $x - y = 12$

6. $\dfrac{1}{3}t + \dfrac{1}{5}u = \dfrac{8}{15}$
   $3t = 4 - u$

7. $2x - y = -4$
   $4x + 2y = 12$

8. $7x - 2y = 4$
   $3x - 3y = 21$

9. The total annual cost T of operating a new car is $f + cm$, where f is the fixed cost (such as depreciation and insurance), c is the operating cost per mile, and m is the number of miles driven. The total cost for 10,000 mi is $1,800, and the total cost for 15,000 mi is $2,300. Find the fixed cost and the operating cost per mile. *4.3*

Application: Profit and Loss

For any business, profit (or loss) is simply the difference between costs and revenues. The equations below show costs and revenues for a blanket production run of x blankets. Fixed costs are $30,000; variable costs are $4 per blanket; revenues are $7 per blanket.

Costs = fixed + variable
    $y = 30,000 + 4x$

Revenues = price × sales
    $y = 7x$

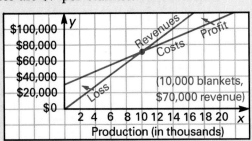

The solution to any cost-revenue system is the business's break-even point (where revenues = costs). In this example, the revenue from sales of 10,000 blankets equals the cost of their production.

Calculate the break-even point for The Bagel Bakery. Fixed costs are $24,000; variable costs are $1/dozen bagels; revenues are $5/dozen.

4.4 Problem Solving: Money Problems/ Mixture Problems

Objectives

To solve word problems involving coins
To solve word problems involving unit price
To solve word problems about investments
To solve word problems involving mixtures

A system of linear equations can be used to solve such problems as finding the numbers of coins having a given total value, finding the number of tickets sold at a given price, or finding the amount of money invested at a given rate of return.

These problems involve *numbers* of items and *monetary values* of those items. For example, in problems involving coins, it is often necessary to distinguish the *number* of coins from the *total value* of each in cents. Thus, the total *number* of coins in 3 dimes and 8 nickels is $3 + 8$, or 11, while the total *value* of these 11 coins is $3 \cdot 10 + 8 \cdot 5 = 30 + 40$, or 70 cents.

Similarly, the total value in cents of q quarters and d dimes is $q \cdot 25 + d \cdot 10$, or $25q + 10d$.

EXAMPLE 1

Renee has $3.75 in dimes and quarters. The number of quarters is 2 less than 3 times the number of dimes. Find the number of coins of each type.

What are you to find?	the number of dimes and quarters
Choose two variables.	Let d = the number of dimes.
What do they represent?	Let q = the number of quarters.
	Then $10d$ = the value of dimes in cents,
	and $25q$ = the value of quarters in cents.

What is given? No. of quarters is 2 less than 3 times number of dimes.

$$q = 3d - 2$$

Value of quarters plus value of dimes is $3.75.

$$25q + 10d = 375$$

Write a system. (1) $q = 3d - 2$ (2) $25q + 10d = 375$

Solve the system. Substitute $3d - 2$ for q in (2). Solve for d.
(2) $25(3d - 2) + 10d = 375$
$75d - 50 + 10d = 375$
$d = 5 \leftarrow$ number of dimes

Substitute 5 for d in (1). Solve for q.
$$\begin{aligned}(1)\ q &= 3 \cdot 5 - 2 \\ &= 13 \leftarrow \text{number of quarters}\end{aligned}$$

Check in the original problem.

13 quarters is 2 less than 3 times 5 dimes.
$$13 = 3 \cdot 5 - 2$$
Renee has $3.75:
$$\begin{aligned}10 \cdot\ \ 5 &= 50 \text{ cents in dimes} \\ 25 \cdot 13 &= \underline{325 \text{ cents in quarters}} \\ &\ \ \ \ 375 \text{ cents, or } \$3.75\end{aligned}$$

Calculator check using decimals:
$$0.1\ \boxed{\times}\ 5\ \boxed{+}\ 0.25\ \boxed{\times}\ 13\ \boxed{=}\ 3.75$$

State the answer.

Renee has 5 dimes and 13 quarters.

EXAMPLE 2   The receipts from a ball game were $154,500, paid by 30,000 specta-
tors. Bleacher seats sold for $4.75 each, and reserved seats sold for
$5.95 each. How many seats of each type were sold?

Solution   What are you to find?

the number of bleacher seats and reserved seats sold

Choose two variables. What do they represent?

Let x = the number of bleacher tickets.
Let y = the number of reserved tickets.

Then $4.75x$ = the value of the bleacher tickets and $5.95y$ = the value of the reserved tickets.

What is given?

Value of bleacher tickets + value of reserved tickets = $154,500

(1) $4.75x + 5.95y = 154,500$

No. of bleacher tickets + no. of reserved tickets = 30,000

(2) $x + y = 30,000$

Write a system.   (1) $4.75x + 5.95y = 154,500$ (2) $x + y = 30,000$

Solve the system.   Multiply each side of (1) by 100 and (2) by -595.

(3) $475x + 595y = 15,450,000 \leftarrow (1) \cdot 100$
(4) $-595x - 595y = -17,850,000 \leftarrow (2) \cdot (-595)$

Add (3) and (4). $-120x = -2,400,000$
$$x = 20,000 \leftarrow \text{number of bleacher seats}$$

Substitute 20,000 for x in (2).
(2) $20,000 + y = 30,000$
$$y = 10,000 \leftarrow \text{number of reserved seats}$$

State the answer.   20,000 bleacher seats and 10,000 reserved seats were sold. The check is left for you.

EXAMPLE 3   Julio invested some money at $8\frac{1}{2}\%$ interest, and a second amount at $6\frac{3}{4}\%$. The amount at $6\frac{3}{4}\%$ was \$1,200 more than the amount at $8\frac{1}{2}\%$. Find the amount invested at each rate, given that the total income after 2 years was \$711. Use the simple-interest formula $i = prt$.

Solution   Let x = the amount invested at $6\frac{3}{4}\%$, or 6.75%.

Let y = the amount invested at $8\frac{1}{2}\%$, or 8.5%.

Write an equation relating the amounts invested.

Amount at $6\frac{3}{4}\%$ was \$1,200 more than $8\frac{1}{2}\%$ amount.

$$(1)\ x = y + 1,200$$

Equation for total income using $i = prt$:

income on x dollars at $6\frac{3}{4}\%$ for 2 years		income on y dollars at $8\frac{1}{2}\%$ for 2 years		total income
$x \cdot 0.0675 \cdot 2$	$+$	$y \cdot 0.085 \cdot 2$		$= 711$
$0.135x$	$+$	$0.17y$		$= 711$

Multiply each side by 1,000. $(2)\ 135x + 170y = 711,000$

Solve by substitution. $135\,(y + 1,200) + 170y = 711,000$
Substitute $y + 1,200$ for x $135y + 162,000 + 170y = 711,000$
in Equation (2).

$$305y = 549,000$$
$$y = 1,800$$
$$\uparrow$$
amount at 8.5%

Substitute 1,800 for y $x = 1,800 + 1,200$
in Equation (1). $= 3,000 \leftarrow$ amount at 6.75%

The check is left for you.

The amounts invested were \$3,000 at $6\frac{3}{4}\%$ and \$1,800 at $8\frac{1}{2}\%$.

EXAMPLE 4   A chemist mixes a solution containing 18% alcohol with a second solution containing 45% alcohol. The result is 12 oz of solution that is 36% alcohol. How many ounces of each solution are mixed? (THINK: Which is greater, the number of ounces of 18% solution or of 45% solution? How do you know?)

Solution   Let x = the number of ounces containing 18% alcohol.

Let y = the number of ounces containing 45% alcohol.

	Number of ounces of solution	Number of ounces of alcohol
18% solution	x	$0.18x$
45% solution	y	$0.45y$
Total	12	$0.36(12) = 4.32$

Write two equations.

Total ounces of solution = 12

$$x + y = 12$$

Total ounces of alcohol = 4.32

$$0.18x + 0.45y = 4.32$$
$$18x + 45y = 432$$

Solve the system $\begin{matrix}(1)\ x + y = 12 \\ (2)\ 18x + 45y = 432\end{matrix}$ using the substitution method.

The above system can be simplified to the equivalent system: $\begin{matrix} x = 4 \\ y = 8 \end{matrix}$

Thus, there are 4 ounces of the 18% solution and 8 ounces of the 45% solution. The solution and check are left for you.

Classroom Exercises

Give an algebraic expression for the income on each investment. Use $i = prt$.

1. x dollars at 5% for 3 years

2. y dollars at 4% for 5 years

3. x dollars at $9\frac{1}{2}$% for 6 months

4. The Hayakawas invested $20,000, part at $11\frac{1}{4}$% and the rest at $12\frac{1}{2}$%. At the end of one year, the return was $2,350. How much money was invested at each rate?

Written Exercises

Use a linear system of two equations to solve each problem.

1. Jon has $1.15 in dimes and nickels. The number of nickels is 5 less than twice the number of dimes. Find the number of each type of coin.

2. Megan has twice as many quarters as half-dollars. The total value is $5.00. How many of each type of coin are there?

3. The cost of an adult ticket to a football game is $2.00, and a student ticket is $1.50. The total receipts from 300 tickets were $550. How many tickets of each kind were sold?

4. Hank invested one sum of money at 8.5%, and another at 10.75%. The amount invested at 8.5% was $4,000 less than twice the amount at 10.75%. If his total income from simple interest for one year was $1,325, how much did he invest at each rate?

5. A chemist has one solution containing 20% acid, and a second containing 30% acid. How many liters of each solution must be combined to obtain 80 liters of a mixture that is 28% acid?

6. A scientist wants to combine two metal alloys into 20 kg of a third alloy which is 60% aluminum. He plans to use one alloy with 45% aluminum content, and a second alloy with 70% aluminum content. How many kilograms of each alloy must be combined?

7. Julio borrowed $6,000, part at $15\frac{1}{2}\%$ and the rest at $14\frac{1}{4}\%$. At the end of 3 years and 6 months, he had paid $3,150 in simple interest. How much did he borrow at each rate?

8. An $8.50 assortment of pads contains $0.65 pads and $0.50 pads. If the number of $0.50 pads is 1 less than $\frac{1}{2}$ the number of $0.65 pads, how many of each type of pad are in the assortment?

9. Jake invested one-third of his money at $9\frac{3}{4}\%$, and the rest at $11\frac{1}{2}\%$. At the end of 18 months he had earned $4,912.50 in simple interest. Find the total amount of money invested.

10. Bill invested a total of $8,000, part at 10% and the remainder at 15%. The total simple interest income for one year is $12\frac{1}{2}\%$ of the total investment. Find the amount of each investment.

11. The Easy Kar Rental Agency charges a fixed amount for each day a car is rented and a fixed amount for each mile the car is driven. Mr. Wong spent $159 to rent a car for 5 days and 600 mi. Ms. Weintraub spent $130 to rent a car for 4 days and 500 mi. Find the daily rental charge and the charge per mile.

12. A metallurgist has 30 grams of a metal alloy that is 60% silver. How much pure silver must she add to the alloy to obtain a mixture that is 70% silver?

Mixed Review

1. Solve $\frac{4}{5}x - 2 = \frac{1}{3}x + 5$. *1.4*

2. Solve $\frac{3x - 2}{2x + 1} = \frac{2}{3}$. *1.4*

3. Write an equation of the line containing the point $P(4,10)$ and parallel to the line described by $3x + 2y = 4$. *3.7*

4. Suppose y varies directly as the square of x. If $y = 4$ when $x = 2$, what is y when $x = 5$? *3.8*

4.5 Problem Solving: Wind and Current Problems/Angle Problems

Objectives

To solve motion problems involving wind or current

To solve word problems involving complementary or supplementary angles

Recall (see Lesson 2.6) that the distance d traveled by an object is given by the formula $d = rt$, where r is the object's rate (speed) and t is the elapsed time. The rate of a plane in still air is called the *airspeed* of the plane. The rate of the plane with respect to the ground is its *ground speed*. Ground speed is affected by the wind as shown below.

Rate of plane *with* wind is airspeed + tail wind = ground speed.

Rate of plane *against* wind is airspeed − head wind = ground speed.

tail wind   head wind

EXAMPLE 1

A plane flies 900 mi in 3 h with a tail wind, and returns the same distance in $4\frac{1}{2}$ h against the wind. Find the wind speed. Assume a constant airspeed and a constant wind speed.

Solution

Let p = the airspeed of the plane.

Let w = the wind speed.

Then $p + w$ = the ground speed with the wind, and $p - w$ = the ground speed against the wind.

rate (r) × time (t) = distance (d)		
rate	time	distance
With wind	$p + w$	3	$3(p + w)$
Against wind	$p - w$	$4\frac{1}{2}$	$4\frac{1}{2}(p - w)$

Write two equations. Use the fact that both distances are 900 mi.

$$3(p + w) = 900 \qquad \qquad \frac{9}{2}(p - w) = 900$$

$$\frac{1}{3} \cdot 3(p + w) = \frac{1}{3} \cdot 900 \qquad \frac{2}{9} \cdot \frac{9}{2}(p - w) = \frac{2}{9} \cdot \frac{900}{1}$$

$$p + w = 300 \qquad \qquad p - w = 200$$

Now solve the system of equations. $(1)\ p + w = 300$
$(2)\ p - w = 200$

Add equations (1) and (2). $2p = 500$
$p = 250 \leftarrow$ airspeed = 250 mi/h

To find the value of w,
substitute 250 for p in (1).

(1) $250 + w = 300$
$w = 50 \leftarrow$ wind speed $= 50$ mi/h

Check

Distance *with* the wind:
$3(250 + 50) = 3(300) = 900$

Distance *against* the wind:
$4\frac{1}{2}(250 - 50) = 4\frac{1}{2}(200) = 900$

So, the wind speed is 50 mi/h.

Systems of equations can also be applied to problems involving relationships between angles. The diagrams below illustrate the properties of complementary and supplementary angles.

Two angles are *complementary* if the sum of their degree measures is 90.

Two angles are *supplementary* if the sum of their degree measures is 180.

$x + y = 90$

$x + y = 90$

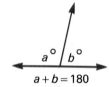
$a + b = 180$

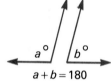
$a + b = 180$

EXAMPLE 2

The degree measure of the larger of two complementary angles is 30 less than twice the measure of the smaller. Find the measure of each angle.

What are you to find?

The measures of two angles.

Choose two variables.
What do they represent?

Let x = the measure of the smaller angle.
Let y = the measure of the larger angle.

What is given?
Write two equations.

Larger is 30 less than twice smaller.
(1) $y = $ 30 less than $2x$, or $2x - 30$

Angles are complementary.
(2) $x + y = 90$

Write a system.

(1) $y = 2x - 30$
(2) $x + y = 90$

Solve the system.

Substitute $2x - 30$ for y in (2).

(2) $x + (2x - 30) = 90$
$3x - 30 = 90$
$3x = 120$
$x = 40 \leftarrow$ measure of smaller angle

Substitute 40 for x in (1).

(1) $y = 2x - 30$
$= 2 \cdot 40 - 30$
$= 50 \leftarrow$ measure of
larger angle

Check in the original problem.

Larger is 30 less than twice smaller.
$50 = 2 \cdot 40 - 30$ True
Angles are complementary.
$50 + 40 = 90$ True

State the answer.

The measures are 40 and 50.

Classroom Exercises

Using x and y to represent the measures of the two angles, write a system of equations to represent each pair of sentences.

1. Two angles are complementary. The measure of one is twice the measure of the other.

2. Two angles have equal measure. The angles are supplementary.

Find the ground speed of each plane.

3. airspeed = 500 mi/h, head wind = 50 mi/h

4. airspeed = 660 mi/h, tail wind = 40 mi/h

5. airspeed = A mi/h, head wind = H mi/h

6. airspeed = A mi/h, tail wind = T mi/h

7. A plane flew 2,400 mi in 5 h with a tail wind. The return trip took 6 h against the wind. Find the wind speed. Assume a constant airspeed and a constant wind speed.

Written Exercises

1. A plane flew 1,800 mi in 4 h with a tail wind, and flying at the same airspeed, returned the same distance in $4\frac{1}{2}$ h against the wind. Find the wind speed.

2. A plane flew 720 mi in 3 h against the wind, and flying at the same airspeed, returned the same distance in 2 h with the wind. Find the airspeed of the plane.

3. Flying with the wind, a plane can travel 1,500 mi in $2\frac{1}{2}$ h. Flying at the same airspeed, the return trip against the wind requires $\frac{1}{2}$ h more to make the same trip. Find the airspeed.

4. The measure of the larger of two complementary angles is twice the measure of the smaller. Find the measure of each angle.

5. Two angles are supplementary. The measure of the larger is 60 less than twice the measure of the smaller. Find the measure of each angle.

6. Two angles are supplementary. The measure of the larger is 60 more than 3 times the measure of the smaller. Find the measure of each angle.

7. Two angles are complementary. The measure of one is $\frac{2}{3}$ the measure of the other. Find the measure of the supplement of the larger angle.

8. The measure of the supplement of angle A is 3 times that of its complement. Find the measure of angle A.

9. An Everglades air boat moves 78 mi downstream (with the current) in the same amount of time it requires to move upstream (against the current) a distance of 48 mi. The boat's engines drive it in still water at an average rate of 16 mi/h greater than the speed of the current. Find the speed of the current. (HINT: Current is to water as wind is to air.)

10. Traveling downstream (with the current), a boat covers 54 mi in 3 hours. Traveling upstream (against the current), it takes the boat twice as long to cover $\frac{2}{3}$ of the distance. Find the rate of the boat in still water and the speed of the current. (See the HINT offered in Exercise 9).

11. Four times the measure of the complement of angle B is 12 more than twice the difference between the measures of its supplement and complement. Find the measure of angle B.

12. A woman can row 11 mi downstream in the same time it takes her to row 7 mi upstream. She rows downstream for 3 h, then turns and rows back for 4 h, but finds that she is still 5 mi from where she started the trip. How fast does the stream flow (see the HINT offered in Exercise 9)?

Mixed Review

Identify the property of operations with real numbers.   *1.3*

1. $3x + (5x + 2) = (3x + 5x) + 2$

2. $-7x(-4y) = -4y(-7x)$

3. Find $f(-2)$ for $f(x) = -x^5$.   *3.2*

4. Write an equation of the line that contains $A(4,3)$ and $B(4,-2)$.   *3.4*

Brainteaser

A ticket to a chamber-music concert costs twice as much as a ticket to a band concert. Suppose that you have a choice of inviting 40 friends to one band concert or inviting two friends to 10 chamber music concerts. Which is the more expensive choice?

4.6 Systems of Three Linear Equations

Objectives

To solve systems of three linear equations in three variables
To solve word problems involving systems of three linear equations

At the right, direct substitution shows that the given equation is true when $x = 2$, $y = 1$, and $z = 3$.

$$2x + 3y - z = 4$$
$$2 \cdot 2 + 3 \cdot 1 - 3 = 4$$
$$4 + 3 - 3 = 4 \text{ True}$$

The **ordered triple** $(2, 1, 3)$ is a solution of the equation $2x + 3y - z = 4$. Any equation of the form $ax + by + cz = d$ (with a, b, and c not all zero) is a **first-degree equation in three variables.** Such equations are also called *linear* even though their graphs are not lines but *planes* in three-dimensional space.

As with systems of two linear equations in two variables, systems of three linear equations in three variables may have exactly one solution, no solution, or infinitely many solutions. You can also use the same methods you learned earlier to solve systems of three equations in three variables.

EXAMPLE 1

Solve for x, y, and z.

$$(1) \quad 2x - 3y - z = 12$$
$$(2) \qquad\quad y + 3z = 10$$
$$(3) \qquad\qquad\quad z = 4$$

Plan

Since Equation (3) is already solved for z, substitute 4 for z in Equation (2) to find y. Then the values of y and z can be substituted in Equation (1) to find x.

Solution

Substitute 4 for z in Equation (2).

$$(2) \qquad y + 3z = 10$$
$$y + 3 \cdot 4 = 10$$
$$y = -2$$

Substitute -2 for y and 4 for z in Equation (1).

$$(1) \qquad 2x - 3y - z = 10$$
$$2x - 3(-2) - 4 = 12$$
$$2x + 6 - 4 = 12$$
$$2x = 10$$
$$x = 5$$

Check

$$2x - 3y - z = 12 \qquad\qquad y + 3z = 10 \qquad\qquad z = 4$$
$$2 \cdot 5 - 3(-2) - 4 \overset{?}{=} 12 \qquad -2 + 3 \cdot 4 \overset{?}{=} 10 \qquad 4 = 4 \text{ True}$$
$$10 + 6 - 4 \overset{?}{=} 12 \qquad\qquad -2 + 12 \overset{?}{=} 10$$
$$12 = 12 \text{ True} \qquad\qquad 10 = 10 \text{ True}$$

Thus, $(5, -2, 4)$ is the solution of the system.

EXAMPLE 2   Solve.   (1) $-x + 3y + z = -10$
(2) $3x + 2y - 2z = 3$
(3) $2x - y - 4z = -7$

Plan   Choose any two equations, say the first two, and eliminate one of the variables, say x, by linear combination. Then repeat the procedure, eliminating the same variable x from a different pair of equations. This leaves two equations in two variables.

Solution   Multiply (1) by 3 and add to (2) to eliminate x.

(1) $-x + 3y + z = -10 \rightarrow \quad -3x + 9y + 3z = -30$
(2) $3x + 2y - 2z = 3 \rightarrow \quad \underline{3x + 2y - 2z = 3}$
$(4) \qquad\qquad\qquad 11y + z = -27$

Multiply (1) by 2 and add to (3).

(1) $-x + 3y + z = -10 \rightarrow \quad -2x + 6y + 2z = -20$
(3) $2x - y - 4z = -7 \rightarrow \quad \underline{2x - 1y - 4z = -7}$
$(5) \qquad\qquad\qquad 5y - 2z = -27$

Solve the resulting system, (4) and (5). Multiply (4) by 2 and add.

(4) $11y + z = -27 \rightarrow \quad 22y + 2z = -54$
(5) $5y - 2z = -27 \rightarrow \quad \underline{5y - 2z = -27}$
$\qquad\qquad\qquad\qquad 27y = -81$
$\qquad\qquad\qquad\qquady = -3$

Substitute -3 for y in either (4) or (5) to find the value of z.

(4) $\qquad 11y + z = -27$
$\qquad 11(-3) + z = -27$
$\qquad\qquad\qquad z = 6$

Substitute 6 for z and -3 for y in one of the original equations to find the value of x.

(1) $\qquad -x + 3y + z = -10$
$\qquad -x + 3(-3) + 6 = -10$
$\qquad\qquad -x - 9 + 6 = -10$
$\qquad\qquad\qquad -x = -7$, or $x = 7$

Check   A check of Equation (1) is shown below using a scientific calculator. The rest of the check is left for you.

Calculator check of (1): 7 $\boxed{+/-}$ $\boxed{+}$ 3 $\boxed{\times}$ 3 $\boxed{+/-}$ $\boxed{+}$ 6 $\boxed{=}$ -10

Thus, the solution of the system is $(7, -3, 6)$.

EXAMPLE 3   Joe has $4.70 in half-dollars, quarters, and dimes. The number of quarters is 10 less than twice the number of dimes. The number of half-dollars decreased by the number of quarters is 2. Find the number of coins of each type.

Solution   Let h = number of half-dollars, q = number of quarters and d = number of dimes. Total value is 470 cents: $50h + 25q + 10d = 470$.

The number of quarters is 10 less than twice the number of dimes: $q = 2d - 10$. The number of half-dollars decreased by the number of quarters is 2: $h - q = 2$.

So, the system is: (1) $50h + 25q + 10d = 470$
(2) $q = 2d - 10$
(3) $h - q = 2$

Equation (2) is already solved for q in terms of d. Substitute $2d - 10$ for q in each of the other two equations.

When we substitute $2d - 10$ for q in (1) and (3) we obtain:

(4) $50h + 60d = 720$ and (5) $h - 2d = -8$.

When we solve the system (4) and (5) by multiplying Equation (5) by 30 and adding, we obtain $h = 6$. Substituting 6 for h in Equation (5) we get $d = 7$. Substituting 7 for d in Equation (2), we get $q = 4$.

Thus, there are 6 half-dollars, 4 quarters, and 7 dimes.

Classroom Exercises

State whether $(3, 4, -1)$ is a solution of the given equation.

1. $x + y + z = 6$ **2.** $2x + 3y - z = 13$ **3.** $x - y + 3z = 4$

4. $3x + y - 2z = 15$ **5.** $-x - y - 2z = -5$ **6.** $3x = 2y - z$

Solve each system of equations for x, y, and z.

7. $x + 2y + z = 14$
$\quad\ y + z = 8$
$\quad\quad\quad z = 3$

8. $x = 3$
$\quad 2x + y = 10$
$\quad 2x + y + z = 9$

9. $y = 6$
$\quad x - y + 3z = 20$
$\quad y - x = 8$

Written Exercises

Solve each system of equations.

1. $5x - 4y - z = 25$
$\quad\ 2y + 7z = 17$
$\quad\quad\quad z = 3$

2. $3x + 2y - z = 16$
$\quad x = 4$
$\quad 2x - y = 5$

3. $-3x - 2y = 14$
$\quad x - 2y + z = -1$
$\quad y = -4$

4. $x + 2y - z = 1$
$\quad 2x + y + 3z = 5$
$\quad 3x + y + 2z = 8$

5. $a + b + c = 6$
$\quad 3a - b + c = 8$
$\quad 2a + 3b - 2c = 10$

6. $2x + 3y - 2z = 4$
$\quad 3x - 2y + 2z = 16$
$\quad -x - 12y + 8z = 5$

7. $x + 2y = 2$
$2y - 3z = 4$
$2x + 3z = -1$

8. $x + 3y - z = 8$
$2x - y + 2z = -9$
$2x + z = -5$

9. $3x + 5y = -3$
$10y - 2z = 2$
$x + 4z = 6$

10. Moira has $2.95 in half-dollars, dimes, and nickels. The number of half-dollars is 1 less than twice the number of dimes. The sum of the number of half-dollars and the number of nickels is 8. Find the number of coins of each type.

11. Find three numbers in decreasing order such that their sum is 20, the difference of the first two numbers is 1, and the sum of the smallest number and the greatest number is 13.

12. A three-digit number is 198 less than the number when its digits are reversed. Twice the sum of the digits is 5 more than 7 times the tens digit. The tens digit is 5 less than twice the hundreds digit. Find the original number. (HINT: Write the number as $100h + 10t + u$, where h, t and u are the hundreds, tens, and units digits, respectively.)

13. The tens digit of a three-digit number is 3 less than 5 times the units digit. Three times the sum of the digits is 2 more than 4 times the hundreds digit. If the digits are reversed, the new number is 594 less than the original number. Find the original number (see the HINT offered in the previous exercise).

Solve each system.

14. $\dfrac{2x + 4}{3} - (y + 4) + \dfrac{z + 1}{3} = 0$

$\dfrac{x - 4}{6} + \dfrac{y + 1}{8} - \dfrac{z - 2}{4} = -\dfrac{1}{2}$

$\dfrac{x + y + 1}{2} = \dfrac{3}{4} - \dfrac{z - 1}{4}$

15. $0.2x - 0.3y + 0.1z = 0.3$
$0.5x - 0.5y + 0.1z = 0.8$
$0.6x - 0.7y + 0.2z = 0.9$

16. $\dfrac{1}{x} + \dfrac{2}{y} + \dfrac{2}{z} = 16$

$\dfrac{1}{y} + \dfrac{2}{z} + \dfrac{2}{x} = 15$

$\dfrac{1}{z} + \dfrac{2}{x} + \dfrac{2}{y} = 14$

17. $x + y + z + w = 10$
$2x - y + z - 3w = -9$
$3x + y - z - w = -2$
$2x - 3y + z - w = -5$

18. Find a, b, and c such that the solution of the equation $ax + by + cz = 6$ contains the ordered triples $\left(\frac{1}{2}, \frac{3}{8}, \frac{3}{2}\right)$, $\left(\frac{2}{3}, \frac{1}{2}, 1\right)$, and $\left(1, \frac{1}{2}, \frac{1}{2}\right)$.

Mixed Review

Solve. *1.4, 2.1, 2.3*

1. $5 - \frac{2}{3}(6 - 9x) = \frac{1}{2}x + 4$

2. $0.6t - 0.2 = 0.12t - 0.08$

3. $4(y + 1) < 7y - 2(3 - y)$

4. $8 - 2|2 - a| = -2$

Problem Solving Strategies

Guess and Check

A problem can sometimes be solved by a guess-and-check method. This method can also be used to suggest an equation whose solution(s) will help in solving the problem. Both of these objectives are achieved in the example below and practiced in the exercise that follows.

Example

The sum of the areas of a square and a rectangle is 240 ft^2. The rectangle is 3 ft wider than and twice as long as the square. Find the length of a side of the square.

Plan

Choose (guess) a value of s for a side of the square. Check in the problem as shown below. Repeat the process until the sum of the areas is 240 ft^2.

Solution

Square		Rectangle			Sum of Areas	?	Sum of Areas
side	area	width	length	area			
s	s^2	w	l	lw	$s^2 + lw$		240
5	25	5 + 3	2 · 5	80	25 + 80 = 105	<	240
10	100	10 + 3	2 · 10	260	100 + 260 = 360	>	240
8	64	8 + 3	2 · 8	176	64 + 176 →	=	240
x	x^2	$x + 3$	$2x$	$2x(x + 3)$	$x^2 + 2x(x + 3)$	=	240

Notice above that 5 for s is too small, since $105 < 240$, and 10 is too large, since $360 > 240$. Try 7 for s; then try 9 for s. This process continues until the value of s narrows down to 8, the correct value of s. Also, the patterns developed in each column are generalized, using the variable x, to provide the equation $x^2 + 2x(x + 3) = 240$, whose positive root, 8, is the length of the side in feet. Now, try the following exercise in a similar way. Include the equation and its solution(s).

Find three consecutive multiples of 4 such that twice the product of the first two, increased by 16, is the square of the third.

1st	2nd	3rd	2 · 1st · 2nd + 16	?	$(3rd)^2$
4	8	12	2 · 4 · 8 + 16 = 80	<	$12^2 = 144$
16	20	24			
.					
.					
.					
x					

4.7 Solving Systems of Linear Inequalities

Objectives
To solve systems of two linear inequalities by graphing
To graph polygonal regions described by conjunctions of several linear inequalities

To solve a system such as $2x + 3y > 12$ *and* $-3y \geq 6 - 4x$, first graph both inequalities on the same coordinate plane. The solution will be the region where the two graphs overlap, as shown in Example 1.

EXAMPLE 1   Solve by graphing. $2x + 3y > 12$
$-3y \geq 6 - 4x$

Solution   First, solve each inequality for y.

$$2x + 3y > 12 \qquad\qquad -3y \geq 6 - 4x$$
$$3y > -2x + 12 \qquad\qquad -3y \geq -4x + 6$$
$$y > -\tfrac{2}{3}x + 4 \qquad\qquad y \leq \tfrac{4}{3}x - 2 \leftarrow \text{Change} \geq \text{to} \leq.$$

Now, graph each inequality on the same coordinate plane.

First graph $y = -\tfrac{2}{3}x + 4$.   Graph $y = \tfrac{4}{3}x - 2$ on the same
↗   ↑   ↑   ↑ coordinate plane.
slope   y-intercept   slope   y-intercept

Use a dotted line because $>$ means *is greater than*. Shade the region above the line.

Use a solid line because $\leq$ means *is less than or equal to*. Shade the region below the line.

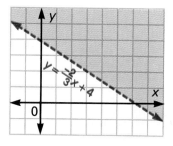   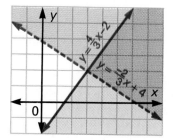

The solution is the set of points in the double-shaded region, including the upper boundary of that region.

Check   Pick a point in the double-shaded region, say $P(6, 2)$. Show that its coordinates satisfy both inequalities.

$$2x + 3y > 12$$
$$2 \cdot 6 + 3 \cdot 2 \overset{?}{>} 12$$
$$12 + 6 \overset{?}{>} 12$$
$$18 > 12 \text{ True}$$

$$-3y \geq 6 - 4x$$
$$-3 \cdot 2 \overset{?}{\geq} 6 - 4 \cdot 6$$
$$-6 \overset{?}{\geq} 6 - 24$$
$$-6 \geq -18 \text{ True}$$

The pair $(6, 2)$ satisfies both inequalities.

Thus, the graph of the solution set is the double-shaded region *below* and *including* the graph of $-3y = 6 - 4x$ and *above* the graph of $2x + 3y = 12$.

Example 1 illustrates the *intersection* of the graphs of the two given inequalities. The **intersection** of two graphs is the set of all points in *both* the first graph *and* the second graph. This set of points is the graph of the solution set of Example 1.

The **union** of two graphs, such as those of the inequalities in Example 1, is the set of all points in either the first graph *or* the second graph, including all points in both graphs.

EXAMPLE 2   Graph the solution set of the following conjunction. Then find the coordinates of the vertices of the polygonal region that is formed.

$$3 \leq x \leq 6 \text{ and } y \geq 0 \text{ and } x + y \geq 5 \text{ and } 21 - 3y \geq 2x$$

Solution   1. Rewrite $3 \leq x \leq 6$ as the conjunction $x \geq 3$ *and* $x \leq 6$. Then graph both inequalities on the same coordinate plane.

Graph of $x \geq 3$:

Graph the vertical line $x = 3$. Use a solid line. Shade the half-plane to the right.

Graph of $x \leq 6$:

Graph the vertical line $x = 6$. Use a solid line. Shade the half-plane to the left.

Graph of $3 \leq x \leq 6$:

The graph is the *intersection* of the graphs of $x \geq 3$ and $x \leq 6$.

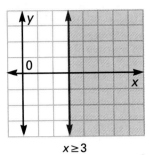

$x \geq 3$

$x \geq 3 \text{ or } x \leq 6$
Union of the graphs

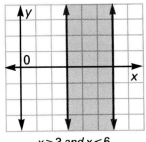

$x \geq 3 \text{ and } x \leq 6$
Intersection of the graphs

2. Graph $y \geq 0$ on the same coordinate plane as $3 \leq x \leq 6$. Graph the horizontal line $y = 0$. Use a solid line. Shade the half-plane above the line.

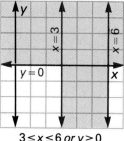

$3 \leq x \leq 6$ or $y \geq 0$

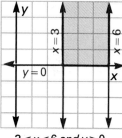

$3 \leq x \leq 6$ and $y \geq 0$

3. Continue in a similar fashion with the two remaining inequalities.

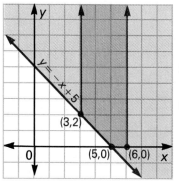

$3 \leq x \leq 6$
$y \geq 0$
$x + y \geq 5$

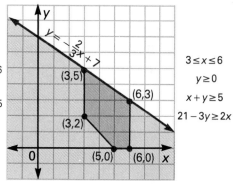

$3 \leq x \leq 6$
$y \geq 0$
$x + y \geq 5$
$21 - 3y \geq 2x$

Thus, the graph of the solution set of the conjunction is the polygonal region with vertices $A(3, 2)$, $B(5, 0)$, $C(6, 0)$, $D(6, 3)$, and $E(3, 5)$. The graph of the region is shown at the right.

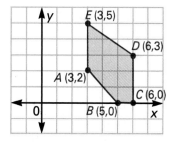

There is, however, a shorter and more efficient way to graph the system of linear inequalities in Example 3.

First graph the boundary lines defined by each inequality.

$$3 \leq x \leq 6 \rightarrow x = 3,\ x = 6$$
$$y \geq 0 \rightarrow \qquad\qquad y = 0$$
$$x + y \geq 5 \rightarrow \qquad\quad y = -x + 5$$
$$21 - 3y \geq 2x \rightarrow \quad y = -\tfrac{2}{3}x + 7$$

Then shade the interior of the polygonal region bounded by the lines.

Classroom Exercises

State each inequality as a conjunction.

1. $3 \leq x \leq 7$

2. $3x - 2 \leq y \leq 3x + 8$

Describe each step in solving the following systems of linear inequalities.

3. $y \geq x$
$\quad y \leq -x + 4$

4. $-x + 2y \geq 4$
$\quad y \geq 3$

5. $x + y \leq 6$
$\quad x - y \geq 4$

6. $4 \leq x \leq 8$
$\quad y + 2x \geq 6$

7. Graph the solution set of Classroom Exercise 3.

Written Exercises

Graph the solution set.

1. $y > x + 5$
$\quad y < -x + 1$

2. $y \leq 3x - 4$
$\quad y \geq 6 - x$

3. $y < x + 4$
$\quad x \leq 4$

4. $y \geq \frac{1}{2}x - 1$
$\quad y < 2$

5. $x + y > -1$
$\quad 3x - 2y > 4$

6. $4x - y \leq -3$
$\quad y \leq -1$

7. $2x - 1 \leq y < 5$

8. $4 \leq x \leq 8$

9. $6 \leq y < -2x + 1$

Graph the solution set of each conjunction of inequalities. Then find the coordinates of the vertices of the polygonal region formed.

10. $y \leq 2x - 4$ and $x \leq 6$ and $y \geq 2$ **11.** $2x + y \leq 5$ and $y \geq 1$ and $x \geq 1$

12. $x \geq 0$ and $y \geq 0$ and $x + 2y \leq 7$ and $2x + y \leq 8$

Write a system of inequalities describing each polygon and its interior.

13.

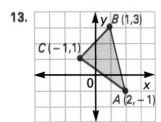

14.

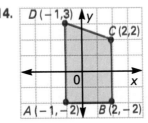

15.

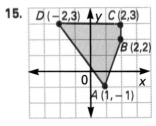

Mixed Review

1. Solve $\dfrac{2x - 4}{x - 3} = \dfrac{2}{3}$. *1.4*

2. Solve $|2x - 4| < 8$. *2.3*

3. Given that $f(x) = -2x^3$, find $f(-4)$. *3.2*

4. Solve $\begin{array}{l} 2x - y = 4 \\ 3x + y = 1 \end{array}$. *4.3*

Extension

Introduction to Linear Programming

A branch of mathematics called **linear programming** deals with the graphs of systems of linear inequalities.

Suppose that the sum S of two numbers, x and y, is to be maximized. Suppose, moreover, that x and y must also be the coordinates of a point in the convex polygonal region defined in Example 2 of Lesson 4.7.

Because the polygonal region contains an infinite number of points, it would be impossible to find the maximum value of S by checking all of these points. The following theorem, however, reduces to a manageable number the list of points that must be checked.

$$3 \le x \le 6$$
$$y \ge 0$$
$$x + y \ge 5$$
$$21 - 3y \ge 2x$$

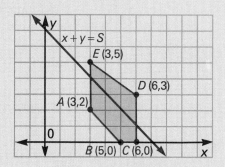

Theorem 4.2

Suppose that x and y are the coordinates of points in a convex polygonal region defined by a system of linear inequalities. Let P represent the linear expression $ax + by + c$ (with a and b not both zero). Then P attains its maximum value at one or more of the vertices of the region, and its minimum value at one or more of the remaining vertices.

The sum-of-two-numbers formula, $S = x + y$, is an example of an expression that can be maximized or minimized using Theorem 4.2. To find the maximum and minimum values of S, calculate the sum of the coordinates for each vertex of the polygonal region.

$$x + y = S$$

Test $(3, 2)$ $3 + 2 = 5$ ⟍
Test $(5, 0)$ $5 + 0 = 5$ ⟋ minimum of 5
Test $(6, 0)$ $6 + 0 = 6$
Test $(6, 3)$ $6 + 3 = 9$ ← maximum of 9
Test $(3, 5)$ $3 + 5 = 8$

So, S has a maximum value of 9 and a minimum value of 5 under the given constraints on x and y.

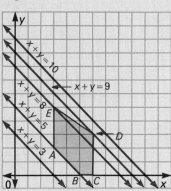

Example 1   Find the maximum and minimum values of P for $P = 2x + 7y$ given the following constraints: $x + y \le 4$, $x - y \le 2$, $x \ge 0$, and $y \ge 0$.

Solution
1. Graph the polygonal region and label the vertices.
2. Test the coordinates of each vertex in the formula $2x + 7y = P$.
   $(2, 0)$: $2 \cdot 2 + 7 \cdot 0 = 4 + 0 = 4$
   $(3, 1)$: $2 \cdot 3 + 7 \cdot 1 = 6 + 7 = 13$
   $(0, 4)$: $2 \cdot 0 + 7 \cdot 4 = 0 + 28 = 28$
   $(0, 0)$: $2 \cdot 0 + 7 \cdot 0 = 0 + 0 = 0$

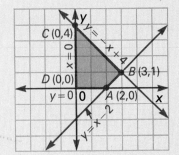

So, the maximum value of P is 28, and the minimum value is 0.

Example 2   A company manufactures television sets and video cassette recorders (VCRs). It must produce at least 20 TV sets per month, but cannot make more than 60 of them. The company also cannot produce more than 100 VCRs per month. Total production cannot exceed 140 TV sets and VCRs combined. The profit for a TV set is \$45, and \$175 for a VCR. Find the number of each that should be manufactured in order to maximize profit.

Solution   Let x = the number of TV sets produced each month.

Let y = the number of VCRs produced each month.

At least 20 TV sets means *20 or more TV sets.*	$x \ge 20$
No more than 60 TV sets means *60 or fewer TV sets.*	$x \le 60$
No more than 100 VCRs means *100 or fewer VCRs.*	$y \le 100$
The number of VCRs cannot be negative.	$y \ge 0$
Total production *is less than or equal to 140.*	$x + y \le 140$

Total profit $P = 45x + 175y$.

Now find the maximum value of P under the above constraints.

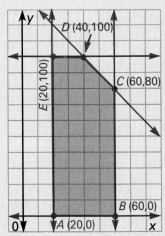

1. Graph the polygonal region and label the vertices.
2. Test the coordinates of each vertex in $45x + 175y = P$.

The maximum value occurs at $D(40, 100)$:
$45 \cdot 40 + 175 \cdot 100 = 19{,}300$.

So, the maximum profit is \$19,300, which corresponds to $D(40, 100)$, or 40 TV sets and 100 VCRs.

Exercises

The formula $P = 2x + 3y$ has constraints on x and y. The graph of these constraints is shown at the right. Find the value of P at the given vertex.

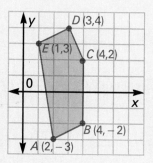

1. A **2.** B **3.** C

4. D **5.** E

6. Give the maximum value of P.

7. Give the minimum value of P.

8. In the graph of the constraints, find a point (x, y) for which $P = 0$.

Find the maximum and minimum values of P under the given constraints. Begin by graphing a polygonal region and determining the coordinates of the vertices of the polygon.

9. $P = 4x + 7y$
   $x + y \leq 8$
   $x - y \leq 2$
   $x \geq 0$
   $y \geq 0$

10. $P = 3x + 5y$
    $x - 2y \geq 0$
    $x + 2y \leq 8$
    $1 \leq x \leq 6$
    $y \geq 0$

11. $P = 2x + 7y$
    $x \geq 1$
    $0 \leq y \leq 4$
    $4x - 2y \leq 8$

12. $P = 3x + 2y$
    $x + y \leq 5$
    $y - x \geq 5$
    $4x + y \geq -10$

Solve each problem using linear programming.

13. A farmer uses two types of fertilizer on his bean field. Each 100-lb bag of 5-10-10 fertilizer ($10 per bag) contains 5 lb of nitrogen. Each 100-lb bag of 10-10-10 fertilizer ($12 per bag) contains 10 lb of nitrogen. The bean field needs at least 160 lb of nitrogen, and the farmer's fertilizer budget for the field is $240. If he already has 6 bags of 5-10-10 fertilizer on order, how many more bags of each type should he order to minimize the cost?

14. At radio station QXYZ, 6 min of each hour are devoted to news and the remainder to music, commercials, and other programming. Station policy requires at least 30 min of music per hour, as well as at least 3 min of music for each minute of commercial time. Find the maximum number of minutes of commercial time available each hour.

15. The manager of the Heinz bicycle factory, which has to make 50 frames per day to show a profit, has 240 man-hours available to him each day. It takes 2 h for one man, or 2 man-hours, to make a frame. It takes $\frac{1}{3}$ man-hour to make each wheel, and at least two wheels must be made for each frame. Find the maximum number of completed bicycles that the Heinz factory can make in one day.

Chapter 4 Review

Key Terms

consistent system (p. 119)
dependent system (p. 120)
equivalent systems (p. 124)
first-degree equation in three variables (p. 142)
inconsistent system (p. 120)
independent system (p. 119)

intersection (p. 148)
linear combination (p. 129)
linear programming (p. 151)
ordered triple (p. 142)
system of equations (p. 119)
union (p. 148)

Key Ideas and Review Exercises

4.1   To solve a system of two linear equations in two variables by graphing, graph both equations on the same coordinate plane.

Three situations can occur. (1) There is one point of intersection. The coordinates of the point are the solution. The system is *consistent* and *independent*. (2) The lines are parallel. There is no solution. The system is *independent* and *inconsistent*. (3) The lines coincide. There are infinitely many solutions. The system is *consistent* and *dependent*.

Solve each system by graphing. Indicate whether the system is *consistent* or *inconsistent* and *dependent* or *independent*.

1. $x + y = 4$
$x - y = 6$

2. $y = \frac{2}{3}x - 4$
$2x - 3y = 9$

3. $4x + 2y = 12$
$6x + 3y = 18$

4.2   To solve a system of two linear equations by substitution, look for an equation with a variable whose coefficient is 1 or -1. Solve the equation for that variable. Then substitute for that variable in the remaining equation.

Solve using the substitution method.

4. $y = 2x - 6$
$3x + 2y = 16$

5. $2x - y = 5$
$3x + 5y = 14$

6. $\dfrac{y}{2} + \dfrac{x - 3}{3} = 3$
$2x - y = 8$

4.3   To solve a system of two linear equations using the linear combination method, choose multipliers for one or both equations so that one pair of terms will be additive inverses of each other. Add both equations, and solve using substitution.

Solve using the linear combination method. If the system is inconsistent or dependent, indicate this.

7. $2x + 5y = 16$
$4x + 3y = 18$

8. $6x + 3y = 9$
$4x + 2y = 5$

9. $3x - 2(y - 4) = 2$
$\dfrac{x}{2} = \dfrac{y - 1}{4}$

4.4, 4.5 To solve special types of word problems, use systems of equations as follows. *Coin, Unit price, Investment*: One equation relates *numbers* of coins (items) or *amounts* invested. The second relates cent or dollar *values*.

Mixture: One equation relates total amounts of a substance. The second relates amounts of a particular ingredient.

Wind: Equations have the form *rate* × *time* = *distance*, where
(1) ground speed = airspeed + tail wind, and
(2) ground speed = airspeed − head wind.

Complementary/Supplementary Angles: For one of the equations, use the sum of the degree measures. The sum of the measures of two complementary angles is 90. The sum of the measures of two supplementary angles is 180.

Use a system of equations to solve each problem.

10. Mr. Ruiz has $1.10 in dimes and nickels. The number of nickels is 2 more than $\frac{1}{2}$ the number of dimes. Find the number of coins of each type.

11. Tina invested $4,000, part at 8% and part at 11%. The total amount of simple interest earned on her investments in two years was $790. How much did she invest at each rate?

12. A small plane flew 800 mi in 5 h against the wind, and returned the same distance with the wind in 4 h. Find the plane's average airspeed.

4.6 To solve a system of three linear equations in three variables, you can use the substitution or the linear combination method, or both methods together.

Solve each system of equations for x, y, and z.

13. $x = 4$
$\quad 2x - 3y = 11$
$\quad 3x - 4y - 2z = 8$

14. $x + 2y = -6$
$\quad y + 2z = 11$
$\quad 2x + z = 16$

15. $2x + 3y - 2z = 4$
$\quad 3x - 3y + 2z = 16$
$\quad 6x - 2y + 8z = 10$

4.7 To solve a system of linear inequalities, graph both inequalities on the same coordinate plane and shade the region common to both.

Graph the solution set.

16. $y > x - 4$ *and* $y < -x + 2$

17. $y < x - 6$ *and* $x \geq -3$

18. Graph the solution set of $2x - y \leq 4$ *and* $y \leq 2$ *and* $x \geq 1$. Then find the coordinates of the vertices of the polygonal region formed.

19. Describe in a paragraph the difference in what is being represented by each of the two equations written to solve a problem dealing with coins. How is the strategy for solving coin problems similar to that for solving investment problems?

Solve each system by graphing. Indicate whether the system is consistent or inconsistent and dependent or independent.

1. $x + y = 6$
$-2x + y = -6$

2. $6x - 2y = 6$
$y = 3x - 4$

3. Solve by substitution.
$x = 8 - y$
$2x + 3y = 22$

4. Solve by linear combination.
$4x + 2y = 10$
$3x + 5y = 11$

Solve the conjunction by any method.

5. $3x + y = -5$ and $\frac{x}{3} + \frac{y}{2} = -\frac{1}{6}$

6. $2x - (3 - 2y) = 5$ and $\frac{3x + 4y}{2} = 9$

Use a system of equations to solve each problem.

7. The perimeter of a rectangle is 38 cm. The length is 9 cm less than 3 times the width. Find the length and the width.

8. Bill has $3.00 in nickels and quarters. The number of nickels is 4 more than twice the number of quarters. Find the number of coins of each type.

9. Alicia invested a total of $6,400, part at 9% per year and the rest at 8% per year. At the end of 2 years 6 months, the money had earned a total of $1,320 in simple interest. How much did she invest at each rate?

10. The ratio of the measures of two complementary angles is 4:5. Find the measure of each angle.

11. Flying with a tail wind, a plane can travel 3,600 mi in 6 h. Flying against the wind, the plane can return in 7.2 h. Find the average airspeed of the plane.

Solve each system of equations for x, y, and z.

12. $2x + 3y + z = 11$
$3x - y - 2z = -5$
$4x + 3y + 3z = 19$

13. $3x + 4y = 3$
$5y - 2z = -3$
$x + 3z = -13$

14. Graph the solution set of $2x - y \leq -3$ and $x \leq -2$.

15. Graph the solution set of $x + 2y \leq 8$ and $3x + 2y \leq 12$ and $x \geq 0$ and $y \geq 0$. Then find the coordinates of the vertices of the polygonal region formed.

16. Find the value(s) of k for which the following conjunction has infinitely many solutions: $3x - 4y = 8$ and $-6(x - 3) + ky = -10$.

College Prep Test

In each item, you are to compare a quantity in Column 1 with a quantity in Column 2. Write the letter of the correct answer from the following choices.

> A—The quantity in Column 1 is greater than the quantity in Column 2.
> B—The quantity in Column 2 is greater than the quantity in Column 1.
> C—The quantity in Column 1 is equal to the quantity in Column 2.
> D—The relationship cannot be determined from the given information.

> NOTE: Information centered over both columns refers to one or both of the quantities to be compared.

Sample Items	Answers
Column 1 — **Column 2**	
S1. 4,570 rounded to the nearest hundred — 4,523 rounded to the nearest hundred	The answer is A: $4,600 > 4,500$.
S2. $x > -6$ and $y < 6$ x — y	The answer is D: For example, if $x = 7$ and $y = 5$, then $x > y$. If $x = 0$ and $y = 5$, then $x < y$.

	Column 1	Column 2
1.	$\dfrac{8}{12}$	$\dfrac{23}{24}$
2.	$-1 < y < 1$, and $y \neq 0$ $3y$	$\dfrac{3}{y}$
3.	$\frac{1}{3}(30 + 90)$	$\frac{1}{6}(60 + 180)$
4.	$6 \leq x \leq 12$ and $4 \leq y \leq 11$ Maximum value of $y - x$	5
5.	$x < y < 0$ $\dfrac{y}{x}$	x
6.	k is a positive integer. $(-1)^{2k}$	$(-1)^{2k+1}$

	Column 1	Column 2
	The figure below shows a circle of diameter x in a circle of radius x.	
7.	Area of the shaded region	$\dfrac{3\pi x^2}{4}$
8.	$x > 0$ x plus an increase of 75% of x	$0.75x$
9.	$4(x + y) = 24$ and $3x - y = 6$ x	y

Simplify, if possible. *1.2*

1. $7(6 - 5^2)$

2. $\dfrac{-8.34 - 1.53 \cdot 2}{3 \cdot 0.27 - (0.9)^2}$

Identify the property of operations with real numbers. (Exercises 3–6) *1.3*

3. $5a + (3a + 6) = (5a + 3a) + 6$

4. $\frac{2}{3}(6x + 12) = \frac{2}{3} \cdot 6x + \frac{2}{3} \cdot 12$

5. $-4a + 4a = 0$

6. $9c + 5d = 5d + 9c$

7. Simplify $5ab - a^2b - ab + 2a^2b$, and then evaluate for $a = 3$, $b = 2$.

For Exercises 8–11, choose the root of the equation from among the five choices. *1.4*

8. $2(3x + 8) - 3(x - 2) = 9x - (4x - 6)$
   (A) 8   (B) -8 (C) 14
   (D) -2 (E) 2

9. $x - 0.01 = 12.2 - 2.3x$
   (A) 37   (B) 0.37   (C) 3,700
   (D) 370   (E) 3.7

10. $\frac{3}{2}x - 1 = \frac{4}{5}x + 6$
    (A) -10 (B) 10   (C) 1
    (D) -1 (E) $-\frac{50}{7}$

11. $\dfrac{2x + 8}{3x - 2} = \dfrac{5}{4}$
    (A) -6 (B) $\frac{10}{7}$ (C) 6
    (D) $\frac{6}{7}$ (E) $3\frac{1}{7}$

Find the solution set.

12. $5x - 2(3x - 1) \geq x - 10$ *2.1*

13. $3x - 2 > 7$ *and* $2x - 4 \leq 18$ *2.2*

14. $|2x - 8| = 12$ *2.3*

15. $|x - 7| < 4$

Let $f(x) = -3x - 5$ and $g(x) = -2x^2 - x$. Find each value.   *3.2*

16. $f(-4)$

17. $g(-4)$

Find the slope of $\overleftrightarrow{PQ}$. Describe its slant as *up to the right, down to the right, horizontal,* or *vertical.* *3.3*

18. $P(-4,5)$, $Q(-8,7)$

19. $P(3,-1)$, $Q(-5,-1)$

20. $P(-6,-7)$, $Q(-6,8)$

Write an equation of the line satisfying the given conditions.

21. The line containing the points $A(0,1)$ and $B(-6,-3)$.   *3.4*

22. The line containing $P(3,-4)$ and parallel to the line $x - y = 4$.   *3.7*

Graph each equation or inequality on a coordinate plane. (Exercises 23–26).

23. $3x - 4y = 12$ *3.5*

24. $4x - 3 = 13$

25. $6y - 4 \leq 14$ *3.9*

26. $4x + 3y \leq -9$

27. If y varies directly as the square of x, and $y = 6$ when $x = 4$, what is y when $x = 2$?   *3.8*

Solve each system by graphing. Determine whether the system is *consistent* or *inconsistent* and *dependent* or *independent*. *4.1*

28. $3x - 2y = 4$
$x + y = 3$

29. $4x - 2y = 6$
$-6x + 3y = 9$

Solve algebraically.

30. $2x - y = 2$ *4.2*
$3x + 2y = 10$

31. $4x + 6y = 6$ *4.3*
$6x + 9y = 5$

32. $3x - 2(y - 1) = 7$
$\dfrac{2x}{3} = \dfrac{y + 1}{2}$

33. $x = 4$ *4.6*
$2x + y = 11$
$3x - y + z = 10$

34. $x + y + z = 2$
$2x + y + 2z = 3$
$3x - y + z = 4$

35. Graph the solution set. *4.7*
$y > 2x - 4$
$y < -x + 2$

36. Jack borrowed $4,000, and *1.5* paid $585 in simple interest after 1 year and 6 months. Find the annual rate of interest. Use the formula $i = prt$.

37. The sum of three consecutive *1.6* odd integers is 69 more than twice the third odd integer. Find the three odd integers.

38. A rectangle and a square *1.7* have the same width. The rectangle is 6 cm longer than a side of the square. One perimeter is twice the other. Find the dimensions of the rectangle.

39. Two trains started toward *2.6* each other from stations 732 km apart. One train traveled at 96 km/h, and the other at 148 km/h. After how many hours did they meet?

40. Al is 4 times as old as Bill. *2.5* Ten years from now, he will be twice as old as Bill will be. Find the age of each now.

41. The sum of two numbers is *4.2* 14. Twice the larger number, increased by 3 times the smaller number, is 34. Find the two numbers.

42. Mr. Brown invested a total of *4.4* $6,400, part at 9% per year and the rest at 8% per year. At the end of 2 years 6 months, he had earned a total of $1,380 in simple interest. How much was invested at each rate?

43. Angelo has $1.90 in dimes and nickels. The number of nickels is 4 fewer than 5 times the number of dimes. How many of each type does he have?

44. The cost of an adult ticket to a football game is $1.75. The cost of a student ticket is $1.25. Total receipts last week from ticket sales were $1,700. If the number of student tickets sold was twice the number of adult tickets, how many of each type were sold?

45. An airplane flies 840 mi in 3 h *4.5* with a tail wind. The return trip at the same airspeed takes $3\frac{1}{2}$ h. Find the wind speed and the airspeed.

This computer-generated drawing of a fern is an example of a fractal. Fractals are *self-similar structures,* containing patterns within patterns. Fractal-like structures are found in nature in clouds, mountain ranges, coastlines—and in the "fern" shown here.

5.1 Positive Integral Exponents

To simplify expressions containing positive integral exponents
To solve exponential equations

Over three hundred years ago, the mathematician René Descartes (1596–1650) economized by writing $b \cdot b$ as b^2, $b \cdot b \cdot b$ as b^3, $b \cdot b \cdot b \cdot b$ as b^4, and so on. This notation is now universally used.

In the expression b^3, 3 is the **exponent**, and b is the **base**. Exponents that are positive integers signify the number of times the base is used as a factor. Thus, b^3 means $b \cdot b \cdot b$. Read b^3 as "the third power of b," or "b-cubed." When no exponent is written, the exponent is understood to be 1.

The meaning of the exponent suggests laws of exponents that can be used to simplify products and quotients of powers with the same base.

Product or Quotient	Expanded Form	Simplified Form	Suggested Relationship
$a^4 \cdot a \cdot a^2$	$aaaa \cdot a \cdot aa$	a^7	$a^4 \cdot a^1 \cdot a^2$ $= a^{4+1+2} = a^7$
$\dfrac{a^6}{a^2}$	$\dfrac{\cancel{aa} \cdot aaaa}{\cancel{aa}}$	a^4	$\dfrac{a^6}{a^2} = a^{6-2} = a^4$
$\dfrac{b^2}{b^5}$	$\dfrac{\cancel{bb}}{\cancel{bb} \cdot bbb}$	$\dfrac{1}{b^3}$	$\dfrac{b^2}{b^5} = \dfrac{1}{b^{5-2}} = \dfrac{1}{b^3}$
$\dfrac{c^3}{c^3}$	$\dfrac{\cancel{c} \cdot \cancel{c} \cdot \cancel{c}}{\cancel{c} \cdot \cancel{c} \cdot \cancel{c}}$	1	$\dfrac{c^3}{c^3} = 1$

Laws of Exponents
For all positive integers m and n and all real numbers b,
1. $b^m \cdot b^n = b^{m+n}$ (Product of powers)

2. $\dfrac{b^m}{b^n} = \begin{cases} b^{m-n}, & \text{where } m > n \text{ and } b \neq 0 \\ \dfrac{1}{b^{n-m}}, & \text{where } n > m \text{ and } b \neq 0 \\ 1, & \text{where } m = n \text{ and } b \neq 0 \end{cases}$ (Quotient of powers)

EXAMPLE 1   Simplify.

a. $-8a^5b \cdot 4a^2b^3$

b. $\dfrac{-12a^5b^4c}{18a^2b^4c^6}$

Solutions   a. $-8 \cdot a^5 \cdot b^1 \cdot 4 \cdot a^2 \cdot b^3$

$= -8 \cdot 4 \cdot a^5a^2 \cdot b^1b^3$

$= -32a^7b^4$

b. $\dfrac{-12}{18} \cdot \dfrac{a^5}{a^2} \cdot \dfrac{b^4}{b^4} \cdot \dfrac{c^1}{c^6}$

$= \dfrac{-2}{3} \cdot \dfrac{a^3}{1} \cdot \dfrac{1}{1} \cdot \dfrac{1}{c^5} = \dfrac{-2a^3}{3c^5}$

Notice in the answers for Example 1 that $a^7 \cdot b^4$ and $\dfrac{a^3}{c^5}$ cannot be simplified because the bases are different.

Exponents can also be used to show the power of a power, such as $(b^5)^3$, the power of a product, such as $(5b)^3$, and the power of a quotient, such as $\left(\dfrac{5}{b}\right)^3$.

$(b^5)^3 = b^5 \cdot b^5 \cdot b^5$

$= b^{15}$, or $b^{5 \cdot 3}$

$(5b)^3 = 5b \cdot 5b \cdot 5b$

$= 125b^3$, or $5^3 \cdot b^3$

$\left(\dfrac{5}{b}\right)^3 = \dfrac{5}{b} \cdot \dfrac{5}{b} \cdot \dfrac{5}{b}$

$= \dfrac{125}{b^3}$, or $\dfrac{5^3}{b^3}$

Laws of Exponents

For all positive integers m and n and all real numbers a and b,

(1) $(a^m)^n = a^{m \cdot n}$ (Power of a power)

(2) $(ab)^n = a^n \cdot b^n$ (Power of a product)

(3) $\left(\dfrac{a}{b}\right)^n = \dfrac{a^n}{b^n}$, where $b \ne 0$ (Power of a quotient)

EXAMPLE 2   Simplify.

a. $8(a^3)^5$

$= 8 \cdot a^{3 \cdot 5}$

$= 8a^{15}$

b. $2a^2(5a^3b)^4$

$= 2a^2 \cdot 5^4(a^3)^4b^4$

$= 2a^2 \cdot 625a^{12}b^4$

$= 1,250a^{14}b^4$

c. $\left(\dfrac{-2a^2}{5b^4c}\right)^3$

$= \dfrac{(-2a^2)^3}{(5b^4c)^3}$

$= \dfrac{(-2)^3(a^2)^3}{5^3(b^4)^3c^3}$

$= \dfrac{-8a^6}{125b^{12}c^3}$

Equations such as $3^x = 81$, $2^{3x} = 32$, and $5^{4x-2} = 125$ are called **exponential equations.** To solve these equations, you need to know the various powers of 3, 2, and 5.

EXAMPLE 3 Solve $5^{4x-2} = 125$.

Plan The left side, 5^{4x-2}, is in exponential form and the base is 5. So, express the right side, 125, as a power of 5.

Solution $5^{4x-2} = 5^3$

Because the bases are equal, the exponents must also be equal.

$$4x - 2 = 3$$
$$4x = 5$$
$$x = \frac{5}{4} \qquad \text{Check: } 5^{4x-2} = 5^{4\cdot\frac{5}{4}-2} = 5^3 = 125$$

Thus, the solution is $\frac{5}{4}$.

Classroom Exercises

Simplify.

1. $a^3 \cdot a^3$

2. $\dfrac{b^{16}}{b^4}$

3. $(c^2)^3$

4. $(de)^3$

5. $(f^2g)^4$

6. $\dfrac{10x^3y^5}{-2xy^2}$

7. Solve $2^x = 128$ for x.

Written Exercises

Simplify each expression.

1. $5x^2 \cdot x^4$

2. $y(6y^3)$

3. $2z(-6yz)$

4. $7xy^3 \cdot 4x^2y^4$

5. $-11mn^6 \cdot 2m^3n$

6. $-6c^2d \cdot 4c^4d^6 \cdot 10c^3d^2$

7. $\dfrac{18a^3b^7}{-9a^9b^2}$

8. $\dfrac{8x^6y^4z^3}{6x^2y^4z^7}$

9. $\dfrac{-12a^5bc^2d^5}{-8ac^{10}d^5}$

10. $n(n^4)^3$

11. $(2x)^4$

12. $(-4c)^3$

13. $5(-3a)^2$

14. $(5c^2)^3$

15. $5x(-3x^3)^4$

16. $\left(\dfrac{-2a}{5}\right)^4$

17. $\left(\dfrac{-7c^3}{10d^5}\right)^2$

18. $\left(\dfrac{3b^3}{2c^2}\right)^3$

Solve each equation.

19. $3^{x-1} = 81$

20. $10^{n+2} = 1{,}000$

21. $2^{3y} = 64$

22. $5^{-x} = 25$

23. $3^{2n+1} = 243$

24. $5^{3x-1} = 625$

25. $4^{6-y} = 256$

26. $2^{6x+6} = 1{,}024$

27. $4^{5-3x} = 64$

Simplify each expression.

28. $(-5a)^3 b^7 c \cdot (2a)^4 bc^5$

29. $(10x)^2 y^2 \cdot (5x)^2 z^4 (-4yz^2)$

30. $8m(-5m^3 n^4)^3$

31. $4c^2 d(2a^3 cd^4)^4$

32. $\left| \dfrac{5a^2 b^3}{-2c^4 d^2} \right|^2$

33. $\left| \dfrac{-2mnp^3}{3x^3 y^2} \right|^3$

34. $x^a y^{2b} \cdot x^{3c} y^{4d}$

35. $x^{3a+2} y^{3b} \cdot x^{4a} y^{2b-1}$

36. $(x^a y^{b+2})^2$

37. $(x^{2a} y^{b-1})^c$

38. $x^a (x^b y^c)^d$

39. $x^{2a} y^b (x^{a+3} y^b)^3$

40. $\dfrac{x^{8a} y^{4b}}{x^{2a} y^b}$

41. $\dfrac{x^{5a+2b} y^{3a-5b}}{x^{a+b} y^{2a-3b}}$

Mixed Review

Find the solution set.   *2.1, 2.2, 2.3*

1. $8 - 5(4x - 3) < 6x - (7 - 4x)$

2. $|2x - 7| < 11$

3. $|4x + 6| > 14$

4. $3x - 2 < 10 \text{ or } 3 - 5x > 18$

5. $3x + 6 > 2x - 4 \text{ and } -2x + 2 < 2x - 10$

6. $x + 3 > 2x - 6 \text{ or } -x - 3 > 2x - 9$

▰/Brainteaser

A group of blocks is stacked in a corner as shown at the right. There are no empty spaces behind any of the blocks.

a. How many blocks cannot be seen?

b. How many blocks will not be seen if another layer of blocks is added to the bottom of the stack?

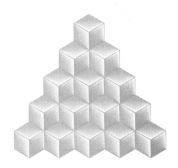

5.2 Zero and Negative Integral Exponents

To evaluate expressions containing negative integral exponents
To simplify expressions containing negative integral exponents
To use scientific notation

The domain of the exponent x in the function $y = 2^x$ can be expanded to include zero and negative integers. That is, expressions such as 2^0 and 2^{-3} can be defined so that the Laws of Exponents (see Lesson 5.1) cover all integral exponents—positive, negative, and zero.

First, consider 2^0.

$$\frac{2^5}{2^5} = \frac{32}{32} = 1 \text{ and } \frac{2^5}{2^5} = 2^{5-5} = 2^0.$$

Therefore, for the Laws of Exponents to hold for all integral exponents, 2^0 must equal 1.

This suggests the following definition of *zero exponent*.

Definition

Zero Exponent: $b^0 = 1$, for each real number b where $b \neq 0$.

Similarly, 2^{-3} can be defined to mean $\frac{1}{2^3}$, and $\frac{1}{5^{-4}}$ defined to mean 5^4.
$\frac{2^0}{2^3} = 2^{0-3} = 2^{-3}$ and $\frac{2^0}{2^3} = \frac{1}{2^3}$. Thus, $2^{-3} = \frac{1}{2^3}$.

(By calculator, 2 $\boxed{x^y}$ 3 $\boxed{+/-}$ $\boxed{=}$ 0.125 and also 2 $\boxed{x^y}$ 3 $\boxed{=}$ $\boxed{1/x}$ = 0.125.)

$\frac{1}{5^{-4}} = \frac{5^0}{5^{-4}} = 5^{0-(-4)} = 5^4$. Thus, $\frac{1}{5^{-4}} = 5^4$.

Definition

Negative Exponent: For each real number b, $b \neq 0$, and each positive integer n, $b^{-n} = \frac{1}{b^n}$ and $\frac{1}{b^{-n}} = b^n$.

When simplifying expressions such as $5a^{-2}$ and $(5a)^{-2}$, note that the base in $5a^{-2}$ is a while the base in $(5a)^{-2}$ is $5a$.

Thus, $5a^{-2} = 5 \cdot a^{-2} = 5 \cdot \frac{1}{a^2} = \frac{5}{a^2}$ and $(5a)^{-2} = \frac{1}{(5a)^2} = \frac{1}{25a^2}$.

EXAMPLE 1 Find the value of each expression.

a. $4 \cdot 5^{-2} = \dfrac{4}{5^2} = \dfrac{4}{25}$

b. $\dfrac{3}{(4 \cdot 5)^{-2}} = 3(4 \cdot 5)^2$

$= 3 \cdot 20^2 = 1{,}200$

EXAMPLE 2 Simplify and write each expression using positive exponents.

a. $-6x^3 \cdot y^{-4} \cdot 5x^{-7} \cdot y^6$

$= -6 \cdot 5 \cdot x^3 x^{-7} \cdot y^{-4} y^6$

$= -30 \cdot x^{3+(-7)} \cdot y^{-4+6}$

$= -30x^{-4} y^2$

$= \dfrac{-30y^2}{x^4}$

b. $\dfrac{-10a^7 b^{-2} c^{-3}}{15a^{-2} b^{-5} c^5}$

$= \dfrac{-10}{15} \cdot \dfrac{a^7}{a^{-2}} \cdot \dfrac{b^{-2}}{b^{-5}} \cdot \dfrac{c^{-3}}{c^5}$

$= \dfrac{-2}{3} \cdot \dfrac{a^7 a^2}{1} \cdot \dfrac{b^5}{b^2} \cdot \dfrac{1}{c^3 c^5}$

$= \dfrac{-2a^9 b^3}{3c^8}$

The numbers 7.62×10^5 and 7.62×10^{-3} are written in *scientific notation*. They are rewritten in ordinary notation below.

$7.62 \times 10^5 = 762{,}000.0$ $\qquad$ $7.62 \times 10^{-3} = 0.00762$

Notice that the exponents 5 and -3 tell the number of places and the direction in which to move the decimal point in order to obtain ordinary notation. A number is written in **scientific notation** when it is in the form $a \times 10^c$, where $1 \le a < 10$ and c is an integer.

EXAMPLE 3 Find the value of the expression and state it in ordinary notation.

$(8.2 \times 10^4) \times (1.5 \times 10^{-7})$

$= (8.2 \times 1.5) \times (10^4 \times 10^{-7})$

$= 12.3 \times 10^{-3}$

$= 0.0123$

Classroom Exercises

Simplify each expression.

1. 3^{-2}

2. $5 \cdot 2^{-3}$

3. $(5 \cdot 2)^{-3}$

4. 2.34×10^2

5. $6c^{-3}$

6. $(6c)^{-3}$

7. $(7 \cdot 4)^0$

8. $a^{-4} \cdot a^7$

Written Exercises

Find the value of each expression.

1. 5^{-3} **2.** 4^{-3} **3.** $6 \cdot 3^{-2}$ **4.** $(4 \cdot 3)^{-2}$

5. $8^4 \cdot 8^{-6}$ **6.** $8^{-4} \cdot 8^6$ **7.** 3.8×10^{-3} **8.** 842×10^{-2}

9. $\dfrac{4}{5^{-3}}$ **10.** $\dfrac{2^{-3}}{3^{-2}}$ **11.** $\dfrac{4.6}{10^{-3}}$ **12.** $\dfrac{5.79}{10^2}$

Simplify each expression. Use positive exponents.

13. $8x^{-3}$ **14.** $(5y)^{-4}$ **15.** $-2a^{-3}$ **16.** $(-2a)^{-4}$

17. $x^{-3} \cdot x^5$ **18.** $5y^2 \cdot 2y^{-5}$ **19.** $-2n^{-3} \cdot 7n^{-5}$ **20.** $5c^{-4} \cdot 2c^0$

21. $(x^{-3})^4$ **22.** $(5a^3)^{-2}$ **23.** $(2c^{-4})^{-3}$ **24.** $2(-3x^{-2})^4$

25. $\dfrac{12c^6}{15c^{-2}}$ **26.** $\dfrac{10n^{-6}}{15n^{-8}}$ **27.** $\left(\dfrac{x^2}{y^{-3}}\right)^5$ **28.** $\left(\dfrac{2a^{-3}}{d^2}\right)^4$

Find the value of each expression and state it in ordinary notation.

29. $\dfrac{8.8 \times 10^2}{4.4 \times 10^{-2}}$ **30.** $\dfrac{3.3 \times 10^{-3}}{2.2 \times 10^{-5}}$ **31.** $\dfrac{1.4 \times 10^{-1}}{2.8 \times 10^3}$ **32.** $\dfrac{2.1 \times 10^{-8}}{8.4 \times 10^{-6}}$

33. $(4.6 \times 10^{-2}) \times (1.5 \times 10^{-1})$ **34.** $(5.6 \times 10^{-3}) \times (3.8 \times 10^7)$

Simplify each expression. Use positive exponents.

35. $\dfrac{2a^6b^{-5}}{6c^{-3}d^4}$ **36.** $\dfrac{15x^7y^{-12}}{10x^{-3}y^{10}}$ **37.** $\left(\dfrac{12x^{-4}}{y^3z^{-3}}\right)^2$ **38.** $\left(\dfrac{3a^{-2}b^3}{2c^{-1}}\right)^4$

39. $3x^{-8}y^2 \cdot 5x^3y^{-4}$ **40.** $(-3a^3d^{-2})^4$ **41.** $\left(\dfrac{-4x^{-3}y}{3z^2w^{-4}}\right)^3$ **42.** $\left(\dfrac{2a^3}{3b^2}\right)^{-2}$

43. x^{m-n}, where $m < 0 < n$ **44.** y^{a-b}, where $0 < a < b$

Mixed Review

Find each value given that $f(x) = 6x - 2$. *3.2*

1. $f(7)$ **2.** $f(-2c^2)$ **3.** $f(a) + f(5)$ **4.** $f(c + 3) - f(c)$

5. Find the set of real numbers for which 3 less than 3 times the sum of a number and 6 is greater than 12. *2.4*

Extension

Estimating Products and Quotients

Scientific notation can be used to express very large or very small numbers in a compact form. For example, 982,000,000 ft/s, the speed of light, can be written as 9.82×10^8 ft/s. When a measurement is written in scientific notation, $a \times 10^c$, the *significant* digits are placed in the first factor, a, where $1 \le a < 10$.

Each of the nine nonzero digits, 1 through 9, in a measurement is a **significant digit**. Zero (0) is a significant digit when its *only* purpose is *not* to place the decimal point as shown below.

Number	Number of Significant Digits	
0.0507	3	0.0<u>507</u>
5.007	4	<u>5.007</u>
5.70	3	<u>5.70</u>
57,000	2	<u>57</u>,000
57,000 to the nearest hundred	3	<u>570</u>,00

Example 1 Round 0.0006080 to two significant digits, and write the result in scientific notation.

$$0.0006080 = 0.00061 \text{ to two significant digits, and}$$
$$0.00061 = 6.1 \times 10^{-4} \text{ in scientific notation.}$$

One strategy to be used when solving problems is to determine if your answer is reasonable. Scientific notation can be used to *estimate* a quotient or product to be compared with a calculated answer.

Example 2 Estimate, to one significant digit, the time needed for light to travel 10 mi at a rate of 982,000,000 ft/s.

Plan Convert 10 mi to feet. Round the distance and rate to two significant digits, and write each in scientific notation.

Use $\dfrac{distance}{rate} = time \left(\dfrac{d}{r} = t\right)$ to find the time in seconds.

Solution $d = 10 \text{ mi} = 10 \times 5{,}280 \text{ ft} = 52{,}800 \text{ ft} \approx 53{,}000 \text{ ft} = 5.3 \times 10^4 \text{ ft}$
$r = 982{,}000{,}000 \text{ ft/s} \approx 980{,}000{,}000 \text{ ft/s} = 9.8 \times 10^8 \text{ ft/s}$

$$t = \frac{d}{r} \approx \frac{5.3 \times 10^4}{9.8 \times 10^8} = \frac{5.3}{9.8} \times 10^4 \times 10^{-8}$$

$$\approx 0.5 \times 10^{-4} = 5 \times 10^{-5}$$

Thus, it will take approximately 5×10^{-5}, or 0.00005 s. This compares favorably with the calculated answer, 0.00005376782 s.

Example 3 If $r = 0.0000718$ km/h and $t = 232$ h, estimate $r \cdot t$ to one significant digit.

$$(7.2 \times 10^{-5}) \times (2.3 \times 10^2) = (7.2 \times 2.3) \times (10^{-5} \times 10^2)$$

$$\approx 20 \times 10^{-3} = 2 \times 10^{-2}$$

Thus, $r \cdot t$ is approximately 2×10^{-2}, or 0.02 km.

This compares favorably with the calculated answer, 0.0166576 km.

Exercises

Find the number of significant digits. Then round the number to two significant digits, and write the result in scientific notation.

1. 78,350   **2.** 457,000   **3.** 37,800,000   **4.** 0.00462

5. 0.7506   **6.** 20.64×10^5 **7.** 32.4×10^{-3} **8.** 0.0718×10^{-2}

9. the velocity of light: 29,979,300,000 cm/s

Estimate the time t in seconds to one significant digit. Use $\frac{d}{r} = t$.

10. $d = 564,000,000$ mi, $r = 28,100$ mi/s

11. $d = 1,000$ mi, $r = 982,000,000$ ft/s

12. $d = 38,600,000$ km, $r = 640$ m/s

Estimate the distance d to one significant digit. Use $r \cdot t = d$.

13. $r = 0.00818$ mi/s, $t = 72,000$ s

14. $r = 186,000$ mi/s, $t = 0.0023$ s

Choose the correct estimate of the value for each expression to one significant digit. Do not use a calculator.

15. $382,124 \times 5,012$
     a. 20,000,000
     b. 200,000,000
     c. 2,000,000,000

16. $39.46 \div 81,460$
     a. 0.05
     b. 0.0005
     c. 0.00005

5.3 Polynomials

Objectives

To classify polynomials by the number of terms and their degree
To simplify sums and differences of polynomials
To multiply polynomials

A **monomial** is a numeral, or a variable, or the product of a numeral and one or more variables. Five examples of monomials are shown below.

$$\frac{2}{3} \qquad z \qquad -4t \qquad x^3y^2 \qquad \frac{1}{2}c^2d$$

A **polynomial** is a monomial or the sum of two or more monomials. The monomials in a polynomial are called the **terms** of the polynomial. The expressions below are examples of polynomials.

$$-5c^4 \text{ (monomial)}$$
$$3x^2 + \tfrac{1}{2}x \text{ (binomial)}$$
$$2x^3y - 5xy - 3y^2 \text{ (trinomial)}$$

Polynomials can be classified by the number of terms they contain. A **binomial** has exactly two terms; a **trinomial** has exactly three terms. A polynomial such as $5x - 3y + 2z - 1.5$ is classified as a polynomial with four terms. Expressions such as $\frac{3y}{x}$ (division by a variable) and $\sqrt{x}$ (the square root of a variable) are not polynomials.

The **degree of a monomial** is the sum of the exponents of its variables. For example, x^2yz is of degree 4 because $x^2yz = x^2y^1z^1$ and $2 + 1 + 1 = 4$. The degree of a nonzero number such as 6 is 0, because $6 = 6 \cdot 1 = 6x^0$.

The **degree of a polynomial** with more than one term is the same as that of its term with the greatest degree. The polynomial $5xy + 6x^2yz^2 - 2x + 6$ is of degree 5. This can be determined as shown below.

$$\underset{\substack{\text{degree} \\ 2}}{5x^1y^1} + \underset{\substack{\text{degree} \\ ⑤}}{6x^2y^1z^2} - \underset{\substack{\text{degree} \\ 1}}{2x^1} + \underset{\substack{\text{degree} \\ 0}}{6x^0}$$

First-degree polynomials such as $3x - 7$ and $4x + 5y - 6$ are also called **linear polynomials** (see Lesson 3.2). Second-degree polynomials such as $5y^2 - 3y + 2$ and $x^2 - 6xy + 9y^2$ are called **quadratic polynomials**.

EXAMPLE 1   Classify each polynomial by number of terms, degree, and whether the polynomial is linear, quadratic, or neither.

Polynomial	Number of terms	Degree	Linear/Quadratic
8	Monomial	0	Neither
$-3x^2$	Monomial	2	Quadratic
$5x + 7$	Binomial	1	Linear
$x^2 - 4xy + 4y^2$	Trinomial	2	Quadratic
$a^2 + 3ab^3 + a^3bc - a$	Four terms	5	Neither

Polynomials can be added, subtracted, multiplied, and divided. To simplify the sum of two or more polynomials, combine like terms.

EXAMPLE 2   Find the simplified form of the sum of the polynomials $5a^2b - 2ab + b^2$ and $7ab - 8a^2b + 2b^2 - 6$.

Plan   Use the Rule of Subtraction, $x - y = x + (-y)$, to convert all subtractions to additions. Then use the Commutative and Associative Properties of Addition to regroup the terms.

Solution
$$(5a^2b - 2ab + b^2) + (7ab - 8a^2b + 2b^2 - 6)$$
$$= 5a^2b + (-2ab) + b^2 + 7ab + (-8a^2b) + 2b^2 + (-6)$$
$$= 5a^2b - 8a^2b - 2ab + 7ab + b^2 + 2b^2 - 6$$
$$= -3a^2b + 5ab + 3b^2 - 6$$

The product of a monomial and a polynomial can be simplified using the Distributive Property.

$$-2x^2(3x - 4x^3 + 5) = -2x^2 \cdot 3x - 2x^2(-4x^3) - 2x^2 \cdot 5$$
$$= -6x^3 + 8x^5 - 10x^2$$
$$= 8x^5 - 6x^3 - 10x^2$$

To simplify $(a + b)(c + d)$, the product of two binomials, you can use the Distributive Property three times as shown below.

$$(a + b)(c + d) = a(c + d) + b(c + d) = ac + ad + bc + bd$$

The product $ac + ad + bc + bd$ can be found using the "First-Outer-Inner-Last Terms" method (FOIL method).

$$(a \quad + \quad b) \quad (c \quad + \quad d) = ac + ad + bc + bd$$

F O I L

EXAMPLE 3 Simplify $(5x + 4)(2x - 3)$. Use the FOIL method.

Solution

$$\begin{aligned}
(5x + 4)(2x - 3) &= 5x \cdot 2x + 5x(-3) + 4 \cdot 2x + 4(-3) \\
&= 10x^2 - 15x + 8x - 12 \\
&= 10x^2 - 7x - 12
\end{aligned}$$

To simplify the product of a monomial and two binomials, it is usually easier to begin with the binomials. For example,

$$\begin{aligned}
7y(2y^2 - 5)(3y^2 - 2) &= 7y[2y^2 \cdot 3y^2 + 2y^2(-2) - 5 \cdot 3y^2 - 5(-2)] \\
&= 7y(6y^4 - 4y^2 - 15y^2 + 10) \\
&= 7y(6y^4 - 19y^2 + 10) \\
&= 42y^5 - 133y^3 + 70y
\end{aligned}$$

The product $(3a - b)(2a + 4b - 5)$ can be simplified by distributing each term of the binomial to each term of the trinomial, as shown in the next example.

EXAMPLE 4 Simplify $(3a - b)(2a + 4b - 5)$.

Solution

$$\begin{aligned}
(3a - b)(2a + 4b - 5) &= 3a(2a + 4b - 5) - b(2a + 4b - 5) \\
&= 6a^2 + 12ab - 15a - 2ab - 4b^2 + 5b \\
&= 6a^2 + 10ab - 15a - 4b^2 + 5b
\end{aligned}$$

Classroom Exercises

Classify each polynomial by the number of its terms.

1. $x^2 - 4$ **2.** $x^2 - 2xy + y^2$ **3.** $9a^2y^2$ **4.** $x - y + z - w$

Simplify.

5. $(2x^2 + x - 5) - (x - y^2 + 2)$ **6.** $3ab(b^2 - ab + 2)$

Written Exercises

Give the degree of each polynomial.

1. $x^2 + x - 6$ **2.** $2x^3y^2$ **3.** $-4abc$ **4.** $a + b + c$

5. 8   **6.** $7x$ **7.** $x^3 - 2x$ **8.** $x^2y^2 + xy^2$

Simplify.

9. $(x^2 - 4x) + (2x^2 + 1)$ **10.** $(3m^2 - 8) - (m^2 - 3)$

11. $(6x^2 + 4x - 3) - (3x^2 - 3x - 4)$ **12.** $(3n - 2n^2 + 10) - (6 + 2n^2 - 7n)$

13. $-3x(7x + 2y - 5)$ **14.** $5x^2(3x^2 - 2x - 4)$

15. $(x + 1)(x - 2)$ **16.** $(y - 3)(y - 4)$

17. $(3t + 4)(2t - 3)$ **18.** $(5n - 3)(2n - 1)$

19. $(5x - y)(2x + 3y)$ **20.** $(3m + 4n)(5m + 2n)$

21. $(x^2 + 8)(x^2 + 10)$ **22.** $(n^3 - 5)(n^3 - 7)$·

23. $(2y^3 - 7)(5y^3 - 2)$ **24.** $(8x^2 + 3)(3x^2 + 5)$

25. $4(x + 3)(2x - 5)$ **26.** $-4(3x + 10)(3x - 10)$

27. $(3x + 4)(x^2 - 3x + 2)$ **28.** $(4y - 3)(2y^2 + 5y - 1)$

29. $(3x^2y^2 - 2xy - xy^2) + (5xy - 3x^2y^2 - xy^2)$

30. $(6x^3y - 8x^2y^2) - (4xy^3 - x^2y^2 + 6x^3y)$

Simplify.

31. $(5x^2 + y^2)(3x^2 - 4y^2)$ **32.** $(c^3 - 2d^2)(2c^3 + 6d^2)$

33. $(2a - 3b)(4a - 5b + 3)$ **34.** $(5x + 2y)(x - y + 4)$

35. $-3y(4y - 3)(5y + 2)$ **36.** $(3.2x + 0.4)(0.7x + 0.3)$

37. $(x^n + 4)(x^n - 2)$ **38.** $(2x^{2a} + 3)(x^{2a} + 1)$

39. $(x^a + y^{2a})(x^a + 3y^{2a})$ **40.** $(5x^{2a} + 2y^b)(4x^{2a} - 3y^b)$

Mixed Review

Write an equation in standard form, $ax + by = c$, for the line that contains the given point(s) and/or satisfies other conditions that are stated. *3.4, 3.7*

1. $P(0,3)$; slope $= \frac{4}{5}$

2. $Q(-2,4)$; slope $= -\frac{2}{3}$

3. $R(4,-2)$; $S(-2,6)$

4. $T(5,-4)$; parallel to the line described by $2x + 3y = 6$

5.4 Factoring

To factor a trinomial into two binomials
To factor a polynomial whose terms contain a common factor
To solve literal equations by factoring

Sometimes you need to find the *factors* of a whole number in order to solve a problem in arithmetic. In algebra, too, it is often necessary to factor a polynomial in order to solve a problem.

You can reverse the FOIL method to help you factor $3x^2 + 14x - 5$ by trial-and-error. Try $3x$ and $1x$ for the first terms because $3x \cdot 1x$ is $3x^2$.

$$3x^2 + 14x - 5 = (\underline{\quad?\quad})(\underline{\quad?\quad})$$

$$3x^2 + 14x - 5 = (3x \underline{\quad?\quad})(1x \underline{\quad?\quad})$$

To find the last terms, notice first that because their product is negative, their signs must be opposite. So, the last terms are either $+5$ and -1, or -5 and $+1$. After trying several possible pairs of factors, you may find a pair that works. In this case, the pair $3x - 1$ and $x + 5$ give the correct middle term, $+14x$. So, $3x^2 + 14x - 5 = (3x - 1)(x + 5)$.

EXAMPLE 1 Factor $5n^2 - 42n + 16$ into two binomials, if possible.

Solution

$$5n^2 - 42n + 16 = (5n - \underline{\quad})(n - \underline{\quad}) \leftarrow (\text{THINK: Why must the missing terms be negative?})$$

Possible Factors	Middle Term	
$(5n - 1)(n - 16)$	$-80n - 1n = -81n$	
$(5n - 16)(n - 1)$	$-5n - 16n = -21n$	
$(5n - 4)(n - 4)$	$-20n - 4n = -24n$	
$(5n - 8)(n - 2)$	$-10n - 8n = -18n$	
$(5n - 2)(n - 8)$	$-40n - 2n = -42n$	$\leftarrow$ (Correct middle term)

So, $5n^2 - 42n + 16 = (5n - 2)(n - 8)$.

Some polynomials cannot be factored when only integers are allowed as coefficients of the factors. In $12y^2 + 9y + 1$, no pair of factors gives the correct middle term.

Possible Factors	Middle Term
$(12y + 1)(y + 1)$	$13y$
$(6y + 1)(2y + 1)$	$8y$
$(4y + 1)(3y + 1)$	$7y$

Thus, $12y^2 + 9y + 1$ cannot be factored using integers.

The product of a monomial and a polynomial is a polynomial whose terms contain a *common monomial factor*.

$$4x^2(6x^2 - 11x - 10) = 24x^4 - 44x^3 - 40x^2$$

Each monomial term of $24x^4 - 44x^3 - 40x^2$ has $4x^2$ as a common factor. So, $24x^4 - 44x^3 - 40x^2$ can be factored into a monomial and a trinomial.

$$24x^4 - 44x^3 - 40x^2 = 4x^2(6x^2 - 11x - 10)$$

Because $4x^2$ is a common factor, 2, 4, x, x^2, $2x$, $4x$, and $2x^2$ must also be common factors. However, $4x^2$ is the *greatest common factor* (GCF). The **greatest common factor** of a polynomial is the common factor that has the greatest degree and the greatest constant factor.

The factoring of $24x^4 - 44x^3 - 40x^2$ into $4x^2(6x^2 - 11x - 10)$, however, is not "complete" because the factor $6x^2 - 11x - 10$ can be factored into two binomials. The *complete* factoring is shown below.

$$\begin{aligned} 24x^4 - 44x^3 - 40x^2 &= 4x^2(6x^2 - 11x - 10) \\ &= 4x^2(3x + 2)(2x - 5) \end{aligned}$$

To factor a polynomial completely: **1.** factor out the GCF of its terms, if any, and **2.** factor any resulting factor, if possible.

EXAMPLE 2   Factor $18a^3b^4 + 24a^2b^2 - 12a^2b$ completely.

Solution   Find the GCF of the terms.

The GCF of 18, 24, and 12 is 6.

The GCF of a^3, a^2, and a^2 is a^2.

The GCF of b^4, b^2, and b is b.

So, the GCF of the terms is $6a^2b$.

Thus, $18a^3b^4 + 24a^2b^2 - 12a^2b = 6a^2b(3ab^3 + 4b - 2)$.

EXAMPLE 3   Factor $45x^2y - 21xy^2 - 6y^3$ completely.

Solution   The GCF of the terms is $3y$. Factor out $3y$. $45x^2y - 21xy^2 - 6y^3$
$3y(15x^2 - 7xy - 2y^2)$

Factor $15x^2 - 7xy - 2y^2$, if possible. $3y(5x + y)(3x - 2y)$

$5x + y$ and $3x - 2y$ cannot be factored using integers.

Thus, $45x^2y - 21xy^2 - 6y^3 = 3y(5x + y)(3x - 2y)$.

Recall that an equation such as $a(x - b) = 2 - cx$ is called a *literal equation* (see Lesson 1.4). Sometimes factoring can be used to solve such equations.

EXAMPLE 4   Solve $a(x - b) = 2 - cx$ for x.

Solution

$$a(x - b) = 2 - cx$$

Use the Distributive Property. $$ax - ab = 2 - cx$$

Rewrite the equation with the x-terms alone on one side. $$ax + cx = ab + 2$$

Factor out the common factor. $$(a + c)x = ab + 2$$

Solve for x. $$x = \frac{ab + 2}{a + c}$$

Classroom Exercises

Factor into two binomials, if possible.

1. $x^2 - 12x + 20$ **2.** $y^2 - 5y - 24$ **3.** $45c^3 + 5c$
4. $x^2 + 2x - 3$ **5.** $x^2 + 3x + 1$ **6.** $x^2 - 5x + 4$
7. $4d^3 + 2d^2$ **8.** $x^2 - x - 6$ **9.** $3n^2 + 8n + 4$
10. $5t^2 - 17t + 6$ **11.** $4y^2 - 5y - 6$ **12.** $4y^2 + 2y - 12$

Find the GCF of each group of expressions.

13. $12x^2, 36x, 18$ **14.** $15x^2, 20x^3$
15. a^3b^2, a^2b^3, a^2b **16.** $x(a - b), 2y(a - b)$

Factor completely.

17. $a^3 + 2a^2 - 8a$ **18.** $m^3 - 4m^2 + 3m$ **19.** $y^4 + 7y^3 + 10y^2$

Written Exercises

Factor into two binomials, if possible.

1. $x^2 + x - 20$
2. $x^2 - 8x - 20$
3. $y^2 - 5y + 6$
4. $y^2 - 5y - 6$
5. $a^2 + 10a + 36$
6. $b^2 - 7b - 24$
7. $3n^2 - 5n - 1$
8. $2a^2 - 8a + 1$
9. $7y^2 - 36y + 5$
10. $3c^2 + 14c - 5$
11. $10t^2 - 11t - 6$
12. $4n^2 + 7n - 15$

Factor completely, if possible.

13. $9m^2 - 18m + 8$
14. $8c^2 - 34c + 21$
15. $3n^2 + 6n - 45$
16. $4x^2 - 28x + 48$
17. $6t^3 - 11t^2 - 10t$
18. $8a^3 - 18a^2 + 9a$
19. $6x^3 - 28x^2 - 10x$
20. $24y^3 - 44y^2 - 40y$
21. $12x^2 + 4xy - y^2$

Solve each equation for x.

22. $tx + vx = 7$
23. $mx - 5x = f$
24. $rx - a = tx + 5$
25. $c - 4x = d - ax$
26. $c(x + 4) = 8x + 7c$
27. $a(6x + 5) = 2nx - 4a$

28. Find two integers for k such that $x^2 + kx + 3$ can be factored into two binomials.

29. There are four integers for k such that $y^2 + ky - 10$ can be factored. Find these four integers.

30. Find one positive integer and at least three negative integers for k such that $x^2 + 2x + k$ can be factored.

Factor completely.

31. $12y^2 - 8y - 15$
32. $24x^2 + 5x - 36$
33. $15x^3y + 5x^2y^2 - 10xy^2$
34. $15c^3 - 9c^2d + 3c^3d$
35. $36y^5 - 46y^3 - 12y$
36. $30n^5 + 93n^3 + 72n$
37. $x^{6m} - 8x^{3m} + 12$
38. $6y^{4m} + 19y^{2m} + 15$
39. $5x^{2a} - 23x^a y^b + 12y^{2b}$

Mixed Review

Find the solution set. Graph it on a number line. (Exercises 1 and 2)
2.2, 2.3

1. $3x - 2 < 10$ *or* $2 - 3x > 14$
2. $|2x - 7| < 15$
3. Solve $\dfrac{6}{2x + 3} = \dfrac{10}{3x - 5}$. *1.4*
4. Solve $m(tx - 6) = m + 5$ for x. *1.4*

5.5 Special Products

To simplify products of the form $(x + y)(x - y)$, $(x + y)^2$, and $(x - y)^2$
To evaluate quadratic functions

Products of the form $(a + b)(a - b)$ are readily simplified.

$$(a + b)(a - b) = a^2 - ab + ab - b^2, \text{ or } a^2 - b^2$$

Product of Sum and Difference of Two Terms
For all real numbers a and b, $(a + b)(a - b) = a^2 - b^2$. That is, the product of the sum and difference of two terms is the difference of the squares of the two terms.

EXAMPLE 1 Simplify.

Solutions

a. $(4x + 3y)(4x - 3y)$
$= (4x)^2 - (3y)^2$
$= 16x^2 - 9y^2$

b. $(5n^2 - 8)(5n^2 + 8)$
$= (5n^2)^2 - 8^2$
$= 25n^4 - 64$

The second special product, $(a + b)^2$, is the square of a binomial. Notice that $(a + b)^2 = (a + b)(a + b) = a^2 + ab + ab + b^2 = a^2 + 2ab + b^2$. So, the square of a binomial is the sum of (1) the square of the first term, (2) twice the product of the terms, and (3) the square of the last term. In a similar way, you can find that $(a - b)^2 = a^2 - 2ab + b^2$.

Square of a Binomial
For all real numbers a and b, $(a + b)^2 = a^2 + 2ab + b^2$ and $(a - b)^2 = a^2 - 2ab + b^2$.

The trinomials $a^2 + 2ab + b^2$ and $a^2 - 2ab + b^2$ are called **perfect-square trinomials**.

EXAMPLE 2 Simplify.

Solutions

a. $(3n + 5)^2$
$= (3n)^2 + 2 \cdot 3n \cdot 5 + 5^2$
$= 9n^2 + 30n + 25$

b. $(4x^2 - 5)^2$
$= (4x^2)^2 - 2 \cdot 4x^2 \cdot 5 + 5^2$
$= 16x^4 - 40x^2 + 25$

Recall that if $g(x) = 4x - 7$, then $g(-2) = 4(-2) - 7 = -15$ and $g(a + h) = 4(a + h) - 7 = 4a + 4h - 7$. The function g is called a *linear function* because $4x - 7$ is a linear (first-degree) polynomial. Now, let $f(x) = 3x^2 - 2x + 5$. Then the function f is a *quadratic function* because $3x^2 - 2x + 5$ is a quadratic (second-degree) polynomial. To find $f(a - 4)$, substitute the binomial as shown below.

$$\begin{aligned} f(x) &= 3x^2 - 2x + 5 \\ f(a - 4) &= 3(a - 4)^2 - 2(a - 4) + 5 \\ &= 3(a^2 - 8a + 16) - 2a + 8 + 5 \\ &= 3a^2 - 24a + 48 - 2a + 13 \\ &= 3a^2 - 26a + 61 \end{aligned}$$

EXAMPLE 3 Let $f(x) = 2x^2 - 5x + 12$. Find the following values.
a. $f(-3)$ **b.** $f(4c)$ **c.** $f(x + h) - f(x)$

Solutions Given that $f(x) = 2x^2 - 5x + 12$:
a. $f(-3) = 2(-3)^2 - 5(-3) + 12 = 2 \cdot 9 + 15 + 12 = 45$
b. $f(4c) = 2(4c)^2 - 5 \cdot 4c + 12 = 2 \cdot 16c^2 - 20c + 12$
    $= 32c^2 - 20c + 12$
c. $f(x + h) - f(x)$
    $= [2(x + h)^2 - 5(x + h) + 12] - [2x^2 - 5x + 12]$
    $= 2(x^2 + 2hx + h^2) - 5x - 5h + 12 - 2x^2 + 5x - 12$
    $= 2x^2 + 4hx + 2h^2 - 5h - 2x^2$
    $= 4hx + 2h^2 - 5h$

Classroom Exercises

Choose one of the expressions a–d that is equivalent to the given expression.

1. $(3x - 5)(3x + 5)$
 a. $3x^2 - 10$
 b. $9x^2 - 30x + 25$
 c. $3x^2 - 25$
 d. $9x^2 - 25$

2. $(2y - 7)^2$
 a. $4y^2 - 49$
 b. $2y^2 - 14y - 49$
 c. $4y^2 - 28y + 49$
 d. $4y^2 - 14y + 49$

3. $(2c^2 + d)(2c^2 - d)$
 a. $4c^2 - d^2$
 b. $2c^4 - d^2$
 c. $4c^4 + 4c^2d + d^2$
 d. $4c^4 - d^2$

Simplify.

4. $(a + 2)(a - 2)$

7. $(b - 2)(b - 2)$

5. $(x^2 - 1)(x^2 + 1)$

8. $(6x + 3)(6x - 3)$

6. $(y + 2)^2$

9. $(4x - 3)^2$

Written Exercises

Simplify.

1. $(n + 11)(n - 11)$
2. $(7x - 6)(7x + 6)$
3. $(2x + 5)^2$
4. $(4y - 5)^2$
5. $(3 - 7a)^2$
6. $(6c + 2)^2$
7. $(x + c)^2$
8. $(2a^2 - 6)(2a^2 + 6)$
9. $(3x + 2a)(3x - 2a)$
10. $(x - 2y)^2$

Let $f(x) = 4x^2 - 10$ and $g(x) = 3x^2 - 4x$. Evaluate.

11. $f\left(-\frac{1}{2}\right)$
12. $g\left(\frac{1}{3}\right)$
13. $f(3c)$
14. $g(-2a)$
15. $f(2x - 5)$
16. $g(x + h) - g(x)$

Simplify.

17. $(3m^2 + 4n)(3m^2 - 4n)$
18. $(6c - 5d^2)(6c + 5d^2)$
19. $(x + 0.1)^2$
20. $(5x - 4y)^2$
21. $(3x + 2b^2)^2$
22. $(4a^2 - k^2)^2$

Let $f(x) = 5x^2 - 2x - 15$ and $g(x) = -2x^2 + 6x + 10$. Evaluate.

23. $f(4)$
24. $f(-5c) + f(c)$
25. $f(c + 6) - f(2c)$
26. $f(x + h) - f(x)$
27. $g(2a - 1) + g(-a)$
28. $g(x + h) - g(x)$
29. $(x^m + y^n)(x^m - y^n)$
30. $(x^{3a} - y^{2b})(x^{2a} + y^b)$

Midchapter Review

Simplify each expression. Use positive exponents.   *5.1, 5.2, 5.3, 5.5*

1. $\dfrac{-10a^3b^2c^7}{15a^6b^2c^2}$
2. $\left(\dfrac{-3a}{2c^2}\right)^4$
3. $\dfrac{10a^6c^{-3}d^{-1}}{5a^{-2}c^2d^{-3}}$
4. $(-2x^{-2})^3$

5. $(8x^3 - 5x^2 + 4x) - (5x^3 - 2x^2 - 3x - 2)$
6. $3c(5c + 4)(2c - 3)$
7. $(4x - 3y)(4x + 3y)$
8. Solve $3^{2x-1} = 81$ for x. *5.1*

Factor completely.   *5.4*

9. $6n^2 - 14n - 12$
10. $60m^2n - 9mn^2 - 6n^3$

5.6 Special Factors

To factor the difference of two squares
To factor perfect-square trinomials
To factor the sum or difference of two cubes

When the special products $(a + b)(a - b)$ and $(a + b)^2$ are simplified, the results are $a^2 - b^2$, a difference of two squares, and $a^2 + 2ab + b^2$, a perfect-square trinomial. These procedures can be reversed to find the special factors of $a^2 - b^2$ and $a^2 + 2ab + b^2$.

Special Factors
For all real numbers a and b,
$$a^2 - b^2 = (a + b)(a - b) \qquad \text{(Difference of two squares)}$$
$$\left.\begin{array}{l} a^2 + 2ab + b^2 = (a + b)^2 \\ a^2 - 2ab + b^2 = (a - b)^2 \end{array}\right\} \quad \text{(Perfect-square trinomials)}$$

The trinomial $16n^2 + 40n + 25$ is a perfect-square trinomial because it can be written as the square of a binomial.

$$16n^2 \quad + \quad \overbrace{40n}^{2 \cdot 4n \cdot 5} \quad + \quad 25$$
$$(4n \quad + \quad 5)^2$$

Check: The square of $4n$ is $16n^2$, the first term. The square of 5 is 25, the third term. Twice $4n \cdot 5$ is $40n$, the middle term.

EXAMPLE 1   a. Factor $16c^2 - 49$.   b. Factor $16c^4 - 8c^2d + d^2$.

Solutions   a. $16c^2 - 49 = (4c)^2 - 7^2$
$$= (4c + 7)(4c - 7)$$
        b. $16c^4 - 8c^2d + d^2 = (4c^2)^2 - 2 \cdot 4c^2 \cdot d + d^2$
$$= (4c^2 - d)^2$$

The special products $a^2 - b^2$ and $(x + y)^2$ may sometimes occur in the same expression, as in $(c + d)^2 - 5^2$. Factor as follows.

Use $a^2 - b^2$ to factor $(c + d)^2 - 5^2$. $\leftarrow a^2 - b^2 = (a + b)(a - b)$
$(c + d)^2 - 5^2 = [(c + d) + 5][(c + d) - 5]$
$\qquad\qquad = (c + d + 5)(c + d - 5)$

EXAMPLE 2   Factor.   **a.** $25x^2 - 40x + 16 - 9y^2$ **b.** $16x^2 - 4y^2 - 12y - 9$

Solutions   **a.** Use $25x^2 - 40x + 16 = (5x - 4)^2$.

$$(25x^2 - 40x + 16) - 9y^2 = (5x - 4)^2 - (3y)^2$$
$$= (5x - 4 + 3y)(5x - 4 - 3y)$$

b. Use $4y^2 + 12y + 9 = (2y + 3)^2$.

$$16x^2 - 4y^2 - 12y - 9 = 16x^2 - (4y^2 + 12y + 9)$$
$$= (4x)^2 - (2y + 3)^2$$
$$= [4x + (2y + 3)][4x - (2y + 3)]$$
$$= (4x + 2y + 3)(4x - 2y - 3)$$

The sum of two cubes, $a^3 + b^3$, and the difference of two cubes, $a^3 - b^3$, can be factored into the product of a binomial and a trinomial. To understand this factoring, start with the two factors.

$$(a + b)(a^2 - ab + b^2) = a^3 - a^2b + ab^2 + a^2b - ab^2 + b^3$$
$$= a^3 + b^3$$
$$(a - b)(a^2 + ab + b^2) = a^3 + a^2b + ab^2 - a^2b - ab^2 - b^3$$
$$= a^3 - b^3$$

Special Factors
For all real numbers a and b,
$a^3 + b^3 = (a + b)(a^2 - ab + b^2)$ (Sum of two cubes)
$a^3 - b^3 = (a - b)(a^2 + ab + b^2)$ (Difference of two cubes)

EXAMPLE 3   Factor $64x^3 - 27y^3$.

Solution $64x^3 - 27y^3$ can be written as $(4x)^3 - (3y)^3$, a difference of cubes.

$$a^3 - b^3 = (a - b)(a^2 + ab + b^2)$$

$$64x^3 - 27y^3 = (4x)^3 - (3y)^3 = (4x - 3y)(16x^2 + 12xy + 9y^2)$$

Classroom Exercises

Determine whether each polynomial is a *difference of two squares*, a *perfect-square trinomial*, a *sum of two cubes*, a *difference of two cubes*, or *none of these*.

1. $4c^2 - 9$ **2.** $y^2 + 16$ **3.** $x^2 + 8x + 16$

4. $x^2 - 12x - 36$ **5.** $8x^3 + 1$ **6.** $a^2 + 2a - 1$

7–12. Factor each polynomial in Classroom Exercises 1–6, if possible.

Written Exercises

Factor each polynomial, if possible. Write each perfect-square trinomial as the square of a binomial.

1. $4c^2 - 25$
2. $49 - 36d^2$
3. $x^2 - 8x + 16$
4. $100a^2 + 20a + 1$
5. $c^2 + 1$
6. $x^2 + 10x + 16$
7. $y^4 - 49$
8. $25 - d^4$
9. $16n^2 + 24n + 9$
10. $4a^2 - 20a + 25$
11. $(x + y)^2 - 16$
12. $(c - d)^2 - 25$
13. $x^2 - (y - 2)^2$
14. $m^2 - (n + 3)^2$
15. $x^2 + 6x + 9 - y^2$
16. $m^2 - 2mn + n^2 - 9$
17. $a^4 - 16a^2 + 64$
18. $y^3 - 8$
19. $x^3 + 27$
20. $t^2 - v^2 + 8v - 16$
21. $64d^3 + 1$

22. Find the two values of k for which $x^2 + kx + 16$ is a perfect-square trinomial.
23. For what value of k is $y^2 - 12y + 9k$ a perfect-square trinomial?

Factor each polynomial, if possible.

24. $49a^2b^2 - 100$
25. $36x^6 - y^4$
26. $4c^2 - 20cd + 25d^2$
27. $25a^4 + 60a^2b^2 + 36b^4$
28. $25c^2 + 9$
29. $4x^2 - 10xy + 25y^2$
30. $9x^2 + 30xy + 16y^2$
31. $m^3 - 64n^3$
32. $125c^3 + d^3$
33. $8t^3 + 125v^3$
34. $27x^6 - 1{,}000y^3$
35. $4x^2 + 12xy + 9y^2 - 25$
36. $9x^2 - 6xy + y^2 - 16$
37. $x^{2n} - y^{4n+6}$
38. $4x^{6m+4} + 12x^{3m+2}y^n + 9y^{2n}$
39. Is $x^2 + y^2 \geq 2xy$ true for all real numbers x and y? Explain.
40. Prove that $[(a + b) + c]^2 = a^2 + b^2 + c^2 + 2ab + 2ac + 2bc$ for all real numbers a, b, and c.

Mixed Review

1. Betty is 3 years older than Alex, and Carlo is twice as old as Betty. In 5 years, Carlo's age will be 7 times the difference in Betty's and Alex's ages. Find Carlo's present age.   *2.5*

2. Two twins leave their house at the same time, traveling in the same direction. One drives a moped at 30 mi/h and the other rides a bike at 20 mi/h. In how long will they be 5 mi apart?   *2.6*

Find the slope and describe the slant of the line determined by the two points.   *3.3*

3. $P(-3,5)$, $Q(2,-4)$
4. $M(2,7)$, $N(-3,7)$
5. $A(6,1)$, $B(6,-4)$

5.7 Combined Types of Factoring

To factor polynomials completely
To factor polynomials by grouping pairs of terms

Some integers can be factored into the product of more than two integers. For example, $18 = 2 \cdot 9 = 2 \cdot 3 \cdot 3$. In a similar way, some polynomials can be factored into the product of more than two polynomials. This can be illustrated by factoring $75x^4 - 27x^2$ completely as shown below.

Recall that the first step in factoring a polynomial completely is to find the GCF of the terms of the polynomial. The GCF of the terms in $75x^4 - 27x^2$ is $3x^2$. So, $75x^4 - 27x^2 = 3x^2(25x^2 - 9)$.

Second, factor $25x^2 - 9$, if possible. Recall that $25x^2 - 9$ is a difference of squares. So, $3x^2(25x^2 - 9) = 3x^2(5x + 3)(5x - 3)$.

Some fourth-degree trinomials can be factored into two second-degree binomials. Then one or both of the binomials can be factored into the product of two first-degree binomials. This case is shown in Example 1.

EXAMPLE 1   Factor $4y^4 - 17y^2 + 4$ completely.

Plan   The terms of $4y^4 - 17y^2 + 4$ have no common factor, and $4y^4 - 17y^2 + 4$ is not a special product. Therefore, try to factor it into two binomials.

Solution
$$4y^4 - 17y^2 + 4 = (4y^2 - 1)(y^2 - 4)$$

Each factor, $4y^2 - 1$ and $y^2 - 4$, is a difference of squares.

$$(4y^2 - 1)(y^2 - 4) = (2y + 1)(2y - 1)(y + 2)(y - 2)$$

Thus, $4y^4 - 17y^2 + 4 = (2y + 1)(2y - 1)(y + 2)(y - 2)$.

Because the coefficient of the first term in $-18x^2 + 24x - 8$ is negative, it is helpful to factor out -2 rather than 2.

$$\begin{aligned} -18x^2 + 24x - 8 &= -2(9x^2 - 12x + 4) \\ &= -2(3x - 2)(3x - 2) \\ &= -2(3x - 2)^2 \end{aligned}$$

EXAMPLE 2 Factor $-24y^3 + 81$ completely.

Plan Factor out the GCF, -3. Then factor the resulting polynomial.

Solution
$$-24y^3 + 81 = -3(8y^3 - 27)$$

Factor $8y^3 - 27$ as a difference of cubes: $(2y)^3 - 3^3$.

$$8y^3 - 27 = (2y)^3 - 3^3 = (2y - 3)(4y^2 + 6y + 9)$$

Thus, $-24y^3 + 81 = -3(2y - 3)(4y^2 + 6y + 9)$.

Some polynomials with four terms can be factored by grouping the terms into pairs of terms. The *common monomial factor z* can be factored out in $5xz - 3z$. In a similar way, the *common binomial factor* $y - 2$ can be factored out in $5x(y - 2) - 3(y - 2)$.

This technique can be used to factor $5xy - 10x - 3y + 6$ by grouping pairs of terms as follows.

$$(5xy - 10x) + (-3y + 6) = 5x(y - 2) - 3(y - 2)$$
$$= (5x - 3)(y - 2)$$

EXAMPLE 3 Factor $6x^2 + 8x - 15xy - 20y$ by grouping pairs of terms.

Solution
$$6x^2 + 8x - 15xy - 20y = (6x^2 + 8x) + (-15xy - 20y)$$
$$= 2x(3x + 4) - 5y(3x + 4)$$
$$= (2x - 5y)(3x + 4)$$

Here are the steps for factoring a polynomial completely.
1. Factor out the greatest common factor (GCF).
2. Factor the resulting polynomial, if possible.
3. Factor each polynomial factor, if possible.

Classroom Exercises

Determine whether each expression is factored completely. If not, complete the factoring.

1. $11 \cdot 17 \cdot 21$
2. $19 \cdot 29 \cdot 37$
3. $7(3x - 6y)$
4. $-7(a^3 + 2ab)$
5. $4(x^2 - 2x + 3)$
6. $5c(9c^2 + 1)$

Factor out a negative integer; then factor completely.

7. $-12a + 10b - 6c$
8. $-6x^2 + 45xy - 21y^2$
9. $-4x^2 + 9$

Written Exercises

Factor each polynomial completely.

1. $4x^2 - 28x + 48$
2. $3n^2 + 6n - 45$
3. $12y^2 - 3$
4. $50c^2 - 18$
5. $4n^2 - 40n + 100$
6. $5x^2 + 40x + 80$
7. $x^4 - 13x^2 + 36$
8. $y^4 - 5y^2 + 4$
9. $x^4 - 5x^2 - 36$
10. $y^4 - 3y^2 - 4$
11. $2y^3 + 54$
12. $3n^3 - 375$
13. $-a^2 + 10a - 25$
14. $-x^2 - 8x - 16$
15. $-4n^2 - 4n + 3$
16. $-6m^2 + 11m - 3$
17. $-y^3 + 1$
18. $-x^3 - 64$

Factor each polynomial by grouping pairs of terms.

19. $8xy - 20x + 6y - 15$
20. $6cd + 14c - 15d - 35$
21. $8x^2 - 2xy + 12x - 3y$
22. $20c^2 - 15c - 4cd + 3d$
23. $3x^3 - 4x^2 + 15x - 20$
24. $10y^3 - 25y^2 - 16y + 40$
25. $6c^3 - 4c^2 - 3cd + 2d$
26. $8a^4 + 28a^2 - 6a^2b - 21b$

Factor each polynomial completely.

27. $10a^3 - 5a^2 - 50a$
28. $2a^3 - 50ab^2$
29. $3x^3y - 6x^2y + 3xy$
30. $18a^3b + 60a^2b^2 + 50ab^3$
31. $9y^4 - 40y^2 + 16$
32. $25y^4 - 101y^2 + 4$
33. $9y^4 + 32y^2 - 16$
34. $25y^4 - 21y^2 - 4$
35. $24x^3 + 375$
36. $-54y^3 + 128$
37. $-y^4 + 64y$
38. $2x^4 + 2{,}000x$
39. $-8ac^2 - 8ac - 2a$
40. $-3xy^2 + 27x$

Factor completely.

41. $a^3 - a^2b - a^2b^2 + ab^3$
42. $6c^3 + 12c^2d - 6cd^2 - 12d^3$
43. $4c^8 - 13c^4 + 9$
44. $36d^8 - 13d^4 + 1$
45. $x^{a+5} + x^3$
46. $x^{6c} - x^{5c}$
47. $y^{n+6} + y^{n+4}$
48. $x^{6c} - 9x^{4c}$
49. $x^{4n+2} - 8x^{2n+2} + 16x^2$
50. $y^{4n} - y^n$

Mixed Review

Find the slope of the line described by each equation.   *3.5*

1. $3x - 5y = 4$
2. $8y - 16 = 0$
3. $-3x = 6$
4. $\frac{1}{2}x + \frac{1}{3}y = 1$

5.8 Dividing Polynomials

Objectives

To divide polynomials in one variable by binomials
To determine whether a binomial is a factor of a polynomial

A divide-multiply-subtract cycle can be used to divide 678 by 32.

Ordinary form	Divide-multiply-subtract (D-M-S) cycle		Expanded form
$\begin{array}{r} 21 \\ \hline 32\overline{)678} \\ 640 \\ \hline 38 \\ 32 \\ \hline 6 \end{array}$	$600 \div 30 = 20$	(D)	$\begin{array}{r} 20 + 1 \\ \hline 30 + 2\overline{)600 + 70 + 8} \\ 600 + 40 \\ \hline 30 + 8 \\ 30 + 2 \\ \hline 6 \end{array}$
	$20(30 + 2) = 600 + 40$	(M)	
	$(600 + 70) - (600 + 40) = 30$	(S)	
	$30 \div 30 = 1$	(D)	
	$1(30 + 2) = 30 + 2$	(M)	
	$(30 + 8) - (30 + 2) = 6$	(S)	

So, $678 \div 32 = 21$, remainder 6. Check: $32 \times 21 + 6 = 678$.

32 is not a factor of 678 because there is a nonzero remainder.

The divide-multiply-subtract cycle shown in expanded form above can be used to divide a polynomial in one variable by a binomial.

EXAMPLE 1 Divide $5x^2 - 11x - 21$ by $x - 3$ to find the polynomial quotient and remainder. Is $x - 3$ a factor of $5x^2 - 11x - 21$?

Solution

$$\begin{array}{r} 5x + 4 \\ \hline x - 3\overline{)5x^2 - 11x - 21} \\ 5x^2 - 15x \\ \hline 4x - 21 \\ 4x - 12 \\ \hline -9 \end{array}$$

$$
\begin{aligned}
5x^2 \div x &= 5x & \text{(D)} \\
5x(x - 3) &= 5x^2 - 15x & \text{(M)} \\
(5x^2 - 11x) - (5x^2 - 15x) &= 4x & \text{(S)} \\
4x \div x &= 4 & \text{(D)} \\
4(x - 3) &= 4x - 12 & \text{(M)} \\
(4x - 21) - (4x - 12) &= -9 & \text{(S)}
\end{aligned}
$$

The polynomial quotient is $5x + 4$, and the remainder is -9.

Check

$$(x - 3)(5x + 4) + (-9) = 5x^2 - 11x - 21$$

So, $x - 3$ is not a factor of $5x^2 - 11x - 21$ because there is a nonzero remainder.

EXAMPLE 2   Divide $6x^3 - 8 - 7x^2 - 26x$ by $3x + 4$. Is $3x + 4$ a factor of the dividend?

Plan   Arrange the terms of the dividend in descending order of the exponents.

Solution

$$
\begin{array}{r}
2x^2 - 5x - 2 \\
3x + 4\overline{)6x^3 - 7x^2 - 26x - 8} \\
\underline{6x^3 + 8x^2} \\
-15x^2 - 26x \\
\underline{-15x^2 - 20x} \\
-6x - 8 \\
\underline{-6x - 8} \\
0
\end{array}
$$

Check $(3x + 4)(2x^2 - 5x - 2) + 0 = 6x^3 - 8 - 7x^2 - 26x$

Thus, the polynomial quotient is $2x^2 - 5x - 2$, the remainder is 0, and $3x + 4$ is a factor of the dividend.

If there are "missing terms" in the dividend, use the coefficient 0 and replace the missing terms. To divide $4x^3 - 5x - 35$ by $2x - 5$, first replace the missing x^2-term with $0x^2$.

$$
\begin{array}{r}
2x^2 + 5x + 10 \\
2x - 5\overline{)4x^3 + 0x^2 - 5x - 35} \\
\underline{4x^3 - 10x^2} \\
10x^2 - 5x \\
\underline{10x^2 - 25x} \\
20x - 35 \\
\underline{20x - 50} \\
15
\end{array}
$$

Classroom Exercises

Simplify each quotient.

1. $(16x^5 - 24x^3) \div (8x)$

2. $(7cd^2 + 9c^2d) \div (cd)$

3. $(15a^2b^2 - 12a^2b) \div (3a^2b)$

4. $\dfrac{25m^4 - 20m^3 - 15m^2}{5m^2}$

5. $\dfrac{-8cd^4 + 10c^5d^2 - 14c^3d^3}{2cd}$

6. $\dfrac{28x^3y^3 + 16x^2y^2 - 24xy^3}{4xy^2}$

7. Divide $(2c^3 - 3c^2 + 2c - 1)$ by $(c - 1)$.

8. Is the divisor in Classroom Exercise 7 a factor of the dividend?

Written Exercises

Divide to find the polynomial quotient and the remainder. Is the divisor a factor of the dividend?

1. $(3c^2 + c - 3) \div (c + 1)$
2. $(y^2 + 5y + 5) \div (y + 2)$
3. $(8x^2 - 6x - 20) \div (x - 2)$
4. $(2n^2 + 23n + 45) \div (2n + 5)$
5. $(4y^3 + 9y^2 + y + 36) \div (y + 3)$
6. $(9x - 17x^2 - 9 + 5x^3) \div (x - 3)$
7. $(16n^3 - 33n + 20) \div (4n - 5)$
8. $(9y^4 + 5y^2 - 12) \div (3y + 2)$
9. $(3a^4 + a + 20 - 49a^2) \div (a + 4)$
10. $(8y^4 - 22y^2 + 9) \div (2y - 3)$
11. $(y^3 - 8) \div (y - 2)$
12. $(125x^3 + 64) \div (5x + 4)$
13. $(3c^4 - 2c^3 + 8c - 48) \div (c^2 - 4)$
14. $(4x^4 + 6x^3 - 15x - 5) \div (2x^2 - 5)$
15. $(y^6 - 3y^4 + y^2 - 8) \div (y^4 + 1)$
16. $(9x^6 + 3x^5 + x^2 - 4) \div (3x^3 + 1)$
17. $\left(3x^3 + \frac{1}{2}x^2 - 5x + 2\right) \div \left(x - \frac{1}{2}\right)$
18. $\left(x^3 - \frac{7}{3}x^2 - x + \frac{1}{2}\right) \div \left(x + \frac{2}{3}\right)$
19. $(x^5 + y^5) \div (x + y)$
20. $(x^7 - y^7) \div (x - y)$
21. $(x^{3m} - 4x^{2m} + 3x^m - 12) \div (x^m - 4)$
22. $(24x^3y + 7x^2y^2 - 6x^2y - 6xy^3 - 4xy^2) \div (3x + 2y)$
23. Find the value of m that will make $x - 2$ a factor of
$3x^4 + 10x^3 - 40x^2 + mx - 12$.

Mixed Review

1. A stack of 20 coins contains dimes and quarters. Given that the stack has a face value of $2.75, find the number of coins of each type. **4.4**
2. One of two supplementary angles measures 45 less than twice the measure of the other angle. Find the measure of each angle. **4.5**
3. If $f(x) = 3x + 7$, what is $f\left(a + \frac{2}{3}\right)$? **3.2**
4. If $g(x) = 2x^2 + 1$, what is $g(c - 3)$? **5.5**

▰▰▰ Brainteaser

The sum of the volumes of two cubes is equal to the sum of the lengths of all their edges. For each cube, the length of an edge is a whole number. Find the dimensions of each cube.

5.9 Composition of Functions

To find composite functions
To evaluate composite functions

Two functions can often be combined into one. The resulting function is called a *composite function.*

Consider a function f: *adding* 6, followed by a second function g: *multiplying* by 3. Apply the function f to the set $\{1, 2, 3\}$ and then apply the function g to the resulting set.

x	f: adding 6	$f(x)$	g: multiplying by 3	$g(f(x))$
1	$1 + 6 = 7$	7 $\longrightarrow$	$7 \times 3 = 21$	21
2	$2 + 6 = 8$	8 $\longrightarrow$	$8 \times 3 = 24$	24
3	$3 + 6 = 9$	9 $\longrightarrow$	$9 \times 3 = 27$	27

Notice that the result of applying g to $f(x)$ is written as $g(f(x))$. Another way of writing the combined function is $(f \circ g)(x)$, or $f \circ g$.

Definition

> If f and g are functions such that the range of g is in the domain of f, then the function whose value at x is $g(f(x))$ is called the **composite** of the functions f and g. The operation of combining the two is called **composition.**

It is important that the results of applying the first function be values that will work for the second function. For instance, if the second function is the square root function, then negative values will not work for it. In other words, the *range* of the first function must be in the *domain* of the second function.

EXAMPLE 1 Let $f(x) = x + 5$ and $g(x) = x^2 - 10$.
 a. Define the composite function $g(f(x))$ as an expression in terms of the variable x.
 b. Evaluate the expression for $x = 5$.

Solutions
 a. $g(f(x)) = g(x + 5)$ $\longleftarrow$ Substitute $x + 5$ for $f(x)$.
 $\quad g(f(x)) = (x + 5)^2 - 10$ $\longleftarrow$ THINK: The function g
 $\quad\quad\quad = x^2 + 10x + 25 - 10$ squares a number and
 $\quad\quad\quad = x^2 + 10x + 15$ subtracts 10.

b. $g(f(x)) = x^2 + 10x + 15$ ⟵   Substitute $x = 5$.

$\quad g(f(5)) = 5^2 + 10(5) + 15$

$\quad\quad\quad\quad = 25 + 50 + 15 = 90$

EXAMPLE 2   Let $f(x) = x - 10$ and $g(x) = 2x + 3$. Define the composite functions $g(f(x))$ and $f(g(x))$. Evaluate each for $x = 4$. Are the results the same?

Solutions

$g(f(x)) = g(x - 10)$ $f(g(x)) = f(2x + 3)$

$\quad\quad\quad = 2(x - 10) + 3$ $\quad\quad\quad = (2x + 3) - 10$

$\quad\quad\quad = 2x - 20 + 3$ $\quad\quad\quad = 2x - 7$

$\quad\quad\quad = 2x - 17$

$g(f(4)) = 2(4) - 17$ $f(g(4)) = 2(4) - 7$

$\quad\quad\quad = 8 - 17$ $\quad\quad\quad = 8 - 7$

$\quad\quad\quad = -9$ $\quad\quad\quad = 1$

The results in this case are not the same.

Classroom Exercises

Given the functions $t = \{(7,5), (5,7), (4,-6), (-3,-2)\}$, $s = \{(5,-6), (4,-3), (-3,8), (-6,5)\}$, and $f(x) = 2x$, find each value if it exists.

1. $s(t(4))$ **2.** $t(s(4))$ **3.** $s(t(7))$

4. $t(s(-6))$ **5.**   **6.** $t(s(-3))$

7. $s(s(4))$ **8.** $t(t(7))$ **9.** $f(t(5))$

Written Exercises

Given $f(x) = 3x - 5$ and $g(x) = x^2 + 2$, find each value.

1. $g(f(2))$ **2.** $f(g(2))$ **3.** $g(f(-3))$ **4.** $f(g(-3))$

5. $f(g(\frac{1}{3}))$ **6.** $f(f(-\frac{1}{3}))$ **7.** $g(g(2))$ **8.** $g(f(a))$

9. $f(g(a))$ **10.** $f(f(a))$ **11.** $g(g(-a))$ **12.** $f(g(5c))$

Given $s(x) = x^2 + 4x$ and $t(x) = -2x + 3$, find each value.

13. $s(t(2))$ **14.** $t(s(-5))$ **15.** $t(s(2.5))$ **16.** $s(t(4.5))$

17. $t(s(a))$ **18.** $s(t(a))$ **19.** $t(s(3c))$ **20.** $s(t(-5c))$

21. $t(s(c - 5))$.   **22.** $s(s(-4))$ **23.** $s(s(2))$ **24.** $t(t(a + b))$

For each pair of functions, f and g, determine g(f(x)) and f(g(x)).

25. $f(x) = x + 8$, $g(x) = 3x$
26. $f(x) = x + 7$, $g(x) = x - 4$
27. $f(x) = 5x - 4$, $g(x) = 7x + 9$
28. $f(x) = 12 - x$, $g(x) = -x - 12$
29. $f(x) = x + 3$, $g(x) = x^2 - 1$
30. $f(x) = 2x - 1$, $g(x) = x^2 + x$
31. $f(x) = 2x^2 + 6$, $g(x) = 3x - 2$
32. $f(x) = 3x^2 - 4x$, $g(x) = 2x + 5$
33. $f(x) = 6x$, $g(x) = \frac{1}{3}x$
34. $f(x) = 4x + 12$, $g(x) = \frac{1}{4}x - 3$
35. $f(x) = \frac{1}{5}x - 7$, $g(x) = 5x + 35$
36. $f(x) = \frac{2}{3}x + 6$, $g(x) = \frac{3}{2}x - 9$

Determine s(t(x)) and t(s(x)). Then use s(t(x)) and t(s(x)) to find s(t(10)) and t(s(10)).

37. $s(x) = x^2 - x - 12$, $t(x) = x + 3$
38. $s(x) = x + 2$, $t(x) = (2x - 3)^2$

For Exercises 39–43, $f(x) = x^2 - 3$, $g(x) = 5x + 2$, and $h(x) = 4 - x$. Write an equation for each composition.

39. $f \circ (g \circ h)$ (HINT: Find $f(g(h(x)))$.)
40. $(f \circ g) \circ h$ (HINT: Find $f \circ g$ first.)
41. $f \circ (h \circ g)$
42. $h \circ (g \circ f)$
43. $(h \circ g) \circ f$

Mixed Review

Find the solution set for each sentence. *2.2, 2.3*

1. $-7 \le 5x - 27 < 18$
2. $|x + 5| > 2$
3. $2x - 1 < 5$ or $1 - x < 5$
4. $2x - 1 < 5$ and $1 - x < 5$

Application: *Measures of Motion*

Scientists often express measures of motion using negative exponents. For instance, velocity in units of meters per second (m/s) can be expressed as ms^{-1}, and acceleration can be written as units of ms^{-2}. When solving motion problems, exponential notation makes it easy to find the units of the answer. For example, if velocity is divided by acceleration, the result is time (in seconds).

$$\frac{m}{s} \div \frac{m}{s^2} = \frac{m}{s} \cdot \frac{s^2}{m} = ms^{-1} \cdot m^{-1}s^2 = s$$

Find the units of the answer. Use meters and seconds.

1. acceleration divided by velocity multiplied by the square of time
2. velocity squared multiplied by time divided by distance

192 Chapter 5 Polynomials

1. How many liters of water must be added to 30 liters of a 28% salt solution to dilute it to a 12% solution?

2. How much water must be evaporated from 15 liters of an 18% salt solution to obtain a 27% solution?

3. Find four consecutive even integers such that 8 times the smallest integer is the same as 5 times the largest integer.

4. Timothy was 7 years old when Ada was born. Hal, who is 5 years older than Timothy, is now 3 times as old as Ada. How old is each person now?

5. How many milliliters of water must be evaporated from 200 ml of an 18% iodine solution to obtain a 30% solution?

6. Skim milk contains no butterfat. How many liters of skim milk must be added to 800 liters of milk that is 3.2% butterfat to obtain milk that is 2.0% butterfat?

7. If y varies directly as the square of x, and $y = 12$ when $x = 0.6$, what is y when $x = 0.9$?

8. A triangle has a perimeter of 58 cm, and one side is 22 cm long. Find the lengths of the other two sides given that their ratio is 4:5.

9. The measures of two supplementary angles are in the ratio of 3:5. Find the measure of the complement of the smaller angle.

10. If an object is sent upward, what initial vertical velocity is required to reach a height of 275 m at the end of 5 s? Use the formula $h = vt - 5t^2$.

11. An airplane flew 480 mi in 2 h aided by a tailwind. It would have flown the same distance in $2\frac{1}{2}$ h against the same wind. Find the wind speed.

12. A car left a rest area and drove south on a highway at 45 mi/h. Ten minutes later, a motorcycle left the same rest area and headed south at 60 mi/h. How long did it take the motorcycle to overtake the car?

13. A collection of 35 coins contains dimes and quarters. If there were 7 more dimes, the face value of the dimes would equal the face value of the quarters. How many quarters are in the collection?

14. A dealer planned to buy 30 radios of a popular model. When the price of each radio was reduced by $3, the dealer was able to buy 3 more radios for the same total cost. What was the total cost of the radios?

15. If $8,000 were invested, part at 9% and the remainder at 11%, the amount of simple interest earned at the end of 5 years would be $4,150. Find the amount invested at 9%.

16. Some almonds worth $5.40/kg are mixed with some peanuts worth $3.60/kg to make 24 kg of a mixture worth $4.80/kg. How many kilograms of peanuts are in the mixture?

Chapter 5 Review

Key Terms

base (p. 161)
binomial (p. 170)
composite function (p. 190)
degree of a monomial (p. 170)
degree of a polynomial (p. 170)
exponent (p. 161)
exponential equation (p. 163)
greatest common factor (p. 175)

linear polynomial (p. 170)
monomial (p. 170)
perfect-square trinomial (p. 178)
polynomial (p. 170)
quadratic polynomial (p. 170)
scientific notation (p. 166)
trinomial (p. 170)

Key Ideas and Review Exercises

5.1,   *Laws of Exponents*   (For restrictions, see Lessons 5.1 and 5.2.)
5.2

$$b^m \cdot b^n = b^{m+n} \qquad (a^m)^n = a^{m \cdot n} \qquad \left(\frac{a}{b}\right)^n = \frac{a^n}{b^n}$$

$$\frac{b^m}{b^n} = b^{m-n} \qquad (ab)^n = a^n \cdot b^n$$

For negative exponents and the zero exponent:

$$b^{-n} = \frac{1}{b^n} \qquad\qquad \frac{1}{b^{-n}} = b^n \qquad\qquad b^0 = 1$$

Simplify each expression. Use positive exponents.

1. $10c(2c^2)^3$ **2.** $5x^{-5} \cdot 3x^3 \cdot 2x^0$ **3.** $\dfrac{18x^{-5}y^4}{12x^{-2}y^{-1}}$

Find the value of each expression and state it in ordinary notation.

4. $\dfrac{7.5 \times 10^2}{2.5 \times 10^{-3}}$ **5.** $(8.4 \times 10^3) \times (2.5 \times 10^{-7})$

5.3   Polynomials can be added, subtracted, and multiplied.

Simplify. Then give the degree of the polynomial in the answer.

6. $(7xy - xy^2 - x^2y) - (2xy + 5xy^2 - 6x^2y)$

7. $(7x^2 + 3y)(2x^2 - y)$

5.4   Some trinomials can be factored into binomials by reversing the FOIL method for multiplying.

Factor each trinomial into two binomials, if possible.

8. $x^2 - 5xy - 36y^2$ **9.** $12x^4 + x^2 - 20$ **10.** $c^2 - 3c - 16$

11. Solve $ax - t = 5x + 6$ for x.

5.5,   *Special Products and Special Factors*
5.6,
5.7 $a^2 - b^2 = (a + b)(a - b)$ $a^3 + b^3 = (a + b)(a^2 - ab + b^2)$
   $a^2 + 2ab + b^2 = (a + b)^2$ $a^3 - b^3 = (a - b)(a^2 + ab + b^2)$

Common binomial factor: $a(x + y) + b(x + y) = (a + b)(x + y)$

Simplify.

12. $(5m + 3n)(5m - 3n)$ **13.** $(6c - 5d)^2$ **14.** $(a + 1)(a^2 - a + 1)$

Let $f(x) = 3x^2 - 2x$. Evaluate.

15. $f(-4)$ **16.** $f(3c - 2)$ **17.** $f(x + h) - f(x)$

Factor, if possible. Write each perfect-square trinomial as the square of a binomial.

18. $25x^2y^2 - 36$ **19.** $16c^2 - 40c + 25$ **20.** $9a^2 + 4$
21. $x^3 + 8$ **22.** $125y^3 - 27$ **23.** $x^4 - 9$

The steps for *factoring completely* are listed on p. 185.

Factor each polynomial *completely*.

24. $27y^2 - 12$ **25.** $2n^3 - 24n^2 + 72n$ **26.** $x^4 - 26x^2 + 25$

Factor each polynomial by grouping pairs of terms.

27. $4xy + 2x - 6y - 3$ **28.** $6x^3 - 15x^2 - 4x + 10$

5.8   If $P(x) \div B(x) = Q(x) + R$ for polynomials $P(x)$, $B(x)$, $Q(x)$, and some number R, then $Q(x)$ is the *polynomial quotient* and R is the *remainder*. If $R = 0$, then $B(x)$, the *divisor*, is a factor of $P(x)$, the *dividend*.

Divide to find the polynomial quotient and the remainder. Is the divisor a factor of the dividend?

29. $(15c^3 - 28c^2 + 15c - 8) \div (3c - 2)$
30. $(2y^4 + 2y + 15 - 19y^2) \div (y + 3)$

5.9   To find $f(g(x))$ and $g(f(x))$, given equations for $f(x)$ and $g(x)$: substitute $g(x)$ for x in $f(x)$ and substitute $f(x)$ for x in $g(x)$.

Let $f(x) = 2x + 1$ and $g(x) = x^2 - 4$.

31. Find $g(f(-3))$.   **32.** Find $f(g(3c))$.
33. Write an equation that determines the function $g(f(x))$, the composition of g with f.

Simplify each expression. Use positive exponents.

1. $-5ac \cdot 7a^3bc^2 \cdot 2b^2c^3$

2. $\dfrac{15x^8y^3z^5}{10x^4y^3z^7}$

3. $(6a^{-3})^2$

4. $(2b^0c^2)^{-3}$

Simplify.

5. $(5a^2 - 2b)(3a^2 + b)$

6. $(2t - 5)(7t^2 + 4t - 2)$

7. $(7ab^2 - a^2b - 5ab) - (6a^2b - ab^2 - 2ab)$

8. $3y(4y - 5)(4y + 5)$

9. $(8m + 3n)^2$

10. $(x - 1)(x^2 + x + 1)$

Find the value of each expression and state it in ordinary notation.

11. $\dfrac{8.8 \times 10^2}{2.2 \times 10^{-3}}$

12. $(4 \times 10^6) \times (2.1 \times 10^{-8})$

Give the degree of each polynomial.

13. $-7x^3y$

14. $5x^3 - 4x + 6$

Factor completely, if possible.

15. $5x^2 + 10x - 40$

16. $9c^3 - 30c^2 + 25c$

17. $25x^4 - y^2$

18. $4x^4 + 1$

19. $n^3 - 64$

20. $y^4 - 29y^2 + 100$

21. Solve $nx + 4 = px + r$ for x.

22. Factor $15xy + 6y - 20x - 8$ by grouping pairs of terms.

23. Solve $2^{4x-3} = 32$ for x.

Let $f(x) = x^2 + 6$ for Exercises 24–26.

24. Find $f(-4)$. **25.** Find $f(c + 3)$. **26.** Find $f(c + d) - f(d)$.

27. Divide $(2x^3 - 15x + 15)$ by $(x + 3)$ to find the polynomial quotient and the remainder. Is the divisor a factor of the dividend?

For Exercises 28–30, let $f(x) = 3x - 2$ and $g(x) = x^2 + 4$.

28. Find $g(f(-1))$. **29.** Find $f(g(-1))$. **30.** Find $f(g(c))$.

31. Simplify $(3x^{2a} - y^b)(x^{2a} + y^b)$.

32. Factor $16x^{2n} + 24x^n + 9$.

Strategy for Achievement in Testing

If guessing is not penalized on the test, you may be able to improve your score by eliminating the obviously wrong choices and guessing one of the remaining choices.

Sample: If $3^{x+1} = 27^{x-1}$, then $x =$ __?__
(A) 0 (B) -2 (C) 1 (D) 3 (E) None of these

Even if you do not know how to solve the problem directly, you might still notice that choices (A) and (C) are incorrect. By eliminating these two choices, you have improved your chances of guessing the correct answer, (E).

Choose the *one* best answer to each question or problem.

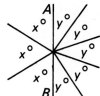

1. $\overline{AB}$ is a line segment in the figure above. Find $\dfrac{x+y}{x-y}$.

(A) $\dfrac{5}{3}$ (B) $\dfrac{8}{5}$ (C) $\dfrac{8}{3}$
(D) 4   (E) 24

2. What is the greatest monomial factor of $2(6x^2y - 9xy^3)(15a^3x + 10ay^2)$?

(A) 2   (B) $6xy$ (C) $10a$
(D) $30axy$ (E) None of these

3. If $4^{x-1} = 8^{2x+2}$, then $x =$ __?__

(A) 2   (B) 1   (C) 0
(D) -1 (E) -2

4. If $f(x) = (3x - 1)^2 - 5x + 2x^2$, then $f(0.1) =$ __?__

(A) 0.01   (B) 0.1   (C) 1
(D) 1.21   (E) None of these

5. $x\star$ means $4(x - 2)^2$. Find the value of $(3\star)\star$.

(A) 8   (B) 12   (C) 16
(D) 36   (E) None of these

6. A function f is described by $f(x) = 3x - 6$ and a function g is described by $g(x) = 12 - 6x$. Which statement is true?

(A) $g(f(x)) = f(g(x))$
(B) $g(f(x)) = 2 \cdot f(g(x))$
(C) $g(f(x)) = -2 \cdot f(g(x))$
(D) $g(f(x)) = f(g(x)) + 18$
(E) None of these is true.

7. If $f(x) = 2x - 6$ and $g(f(x)) = x$, then __?__

(A) $g(x) = \dfrac{x+6}{2}$

(B) $g(x) = \dfrac{1}{2}(x - 6)$

(C) $g(x) = -2x + 6$

(D) $g(x) = -2x - 6$

(E) $g(x) = \dfrac{1}{2x - 6}$

8. If $f(x) = x^{-2}$, then $f(f(2)) =$ __?__

(A) $\dfrac{1}{16}$ (B) $\dfrac{1}{4}$ (C) 4
(D) 16   (E) None of these

9. $9[4^{-2}(-2)^4 - 3^{-2}]^{-1} =$ __?__

(A) 8   (B) $\dfrac{1}{8}$ (C) $-\dfrac{1}{8}$
(D) -8 (E) None of these

This is a magnification of a region of the Mandelbrot Set, the
fractal that appears on page XVI. Notice that the shape of the
dark region is very much like the overall shape of the set.
The basic shape occurs infinitely many times in the set, an
example of the self-similarity of fractals.

6.1 The Zero-Product Property

Objective

To solve equations of degree 2, 3, or 4 for their rational roots by factoring

You already know that the solution, or *root*, of a linear (first-degree) equation in one variable can be found by solving for that one variable (see Lesson 1.4). Thus, $2x - 1 = 0$ can be solved in two steps to give $x = \frac{1}{2}$. Equations of a higher degree, however, such as $x^2 + x = 0$ and $x^4 = 3x^3 - 8$, can rarely be solved in this same direct manner. In fact, higher-degree equations can often be solved only approximately. In this chapter, though, you will work with higher-degree equations that *can* be solved exactly. The simplest of these is the second-degree, or *quadratic* equation.

Equations such as $3x^2 - 10x - 8 = 0$, $0 = 9x^2 - 16$, and $5x^2 + 15x = 0$ are called **quadratic equations** in one variable because each equation contains a quadratic (second-degree) term.

Standard Form of a Quadratic Equation in One Variable
The standard form of a quadratic equation in one variable is either $ax^2 + bx + c = 0$ or $0 = ax^2 + bx + c$, where a, b, and c are real numbers and $a \neq 0$.

Some quadratic equations can be solved by writing the equation in standard form, factoring the polynomial, and then setting each factor equal to zero. This method is based on the *Zero-Product Property*.

Zero-Product Property
If $m \cdot n = 0$, then $m = 0$ or $n = 0$, for all real numbers m and n.

EXAMPLE 1   Solve $3x^2 = 6 - 7x$.

Plan

Write the equation in standard form and factor the polynomial. Then use the Zero-Product Property.

Solution

$$3x^2 = 6 - 7x$$
$$3x^2 + 7x - 6 = 0$$
$$(3x - 2)(x + 3) = 0$$
$$3x - 2 = 0 \quad or \quad x + 3 = 0$$
$$x = \frac{2}{3} \quad or \quad x = -3$$

Checks

$x = \frac{2}{3}$:
$$3x^2 = 6 - 7x$$
$$3\left(\frac{2}{3}\right)^2 \stackrel{?}{=} 6 - 7 \cdot \frac{2}{3}$$
$$3 \cdot \frac{4}{9} \stackrel{?}{=} \frac{18}{3} - \frac{14}{3}$$
$$\frac{4}{3} = \frac{4}{3} \text{ True}$$

$x = -3$:
$$3x^2 = 6 - 7x$$
$$3(-3)^2 \stackrel{?}{=} 6 - 7(-3)$$
$$3 \cdot 9 \stackrel{?}{=} 6 + 21$$
$$27 = 27 \text{ True}$$

The solutions, or *roots*, are $\frac{2}{3}$ and -3.

Sometimes it is more convenient to use the form $0 = ax^2 + bx + c$.

EXAMPLE 2 **a.** Solve $-5x^2 = 20x$.   **b.** Solve $16 = 9n^2$.

Solutions

a.
$$-5x^2 = 20x$$
$$0 = 5x^2 + 20x$$
$$0 = 5x(x + 4)$$
$$5x = 0 \quad or \quad x + 4 = 0$$
$$x = 0 \quad or \quad x = -4$$
The roots are 0 and -4.

b.
$$16 = 9n^2$$
$$0 = 9n^2 - 16$$
$$0 = (3n + 4)(3n - 4)$$
$$3n + 4 = 0 \quad or \quad 3n - 4 = 0$$
$$n = -\frac{4}{3} \quad or \quad n = \frac{4}{3}$$
The roots are $-\frac{4}{3}$ and $\frac{4}{3}$.

A fourth-degree equation can be solved for its *rational roots*.

EXAMPLE 3 Find the rational roots of each equation.
a. $x^4 - 13x^2 + 36 = 0$ **b.** $4x^4 + 35x^2 - 9 = 0$

Plan Factor each trinomial completely.

Solutions **a.**
$$x^4 - 13x^2 + 36 = 0$$
$$(x^2 - 4)(x^2 - 9) = 0$$
$$(x + 2)(x - 2)(x + 3)(x - 3) = 0$$

Use the Zero-Product Property extended to four factors.

$$x + 2 = 0 \quad or \quad x - 2 = 0 \quad or \quad x + 3 = 0 \quad or \quad x - 3 = 0$$
$$x = -2 \qquad\qquad x = 2 \qquad\qquad x = -3 \qquad\qquad x = 3$$

There are four rational roots: $-2, 2, -3$, and 3.

b.
$$4x^4 + 35x^2 - 9 = 0$$
$$(4x^2 - 1)(x^2 + 9) = 0$$
$$(2x + 1)(2x - 1)(x^2 + 9) = 0$$
$$2x + 1 = 0 \quad or \quad 2x - 1 = 0 \quad or \quad x^2 + 9 = 0$$
$$x = -\frac{1}{2} \qquad\qquad x = \frac{1}{2} \qquad x^2 + 9 \text{ cannot be factored.}$$
$$x^2 = -9 \text{ has no real roots.}$$

There are two rational roots: $-\frac{1}{2}$ and $\frac{1}{2}$.

Recall the Division Property of Equality (see Lesson 1.4).

If $a = b$, then $\dfrac{a}{c} = \dfrac{b}{c}$ $(c \neq 0)$.

To solve $4x^2 - 36 = 0$, divide each side of the equation by 4, and factor. This is done at the right.

$$4x^2 - 36 = 0$$
$$x^2 - 9 = 0$$
$$(x + 3)(x - 3) = 0$$
$$x + 3 = 0 \text{ or } x - 3 = 0$$
$$x = -3 \qquad x = 3$$

The roots are -3 and 3.

However, to solve $x^3 - 9x = 0$, do *not* divide each side by x because x may equal 0.

<table>
<tr><th>Correct method</th><th>Incorrect method</th></tr>
<tr><td>$x^3 - 9x = 0$</td><td>$\dfrac{x^3}{x} - \dfrac{9x}{x} = \dfrac{0}{x}$</td></tr>
<tr><td>$x(x^2 - 9) = 0$</td><td>$x^2 - 9 = 0$</td></tr>
<tr><td>$x(x + 3)(x - 3) = 0$</td><td>$(x + 3)(x - 3) = 0$</td></tr>
<tr><td>$x = 0 \text{ or } x = 3 \text{ or } x = -3$</td><td>$x = -3 \text{ or } x = 3$</td></tr>
<tr><td>The roots are 0, -3, and 3.</td><td>The roots are -3 and 3.</td></tr>
</table>

Dividing each side by x caused the loss of the root 0.

Focus on Reading

For each statement, determine whether it is true or false.

1. $0 = 3x^2 + 6x$ is a quadratic equation.

2. $4x - 8 = 0$ is a quadratic equation.

3. A quadratic equation may have two rational roots.

4. $x^2 - 6x + 9 = 0$ is a quadratic equation with exactly one root.

5. A fourth-degree equation may have four rational roots.

6. A fourth-degree equation may have exactly two rational roots.

Classroom Exercises

For each statement, determine whether it is true or false.

1. If $x(x - 5) = 0$, then $x = 0$ or $x - 5 = 0$.

2. If $(3n - 2)(n - 4) = 0$, then $n = 2$ or $n = 4$.

3. If $(y - 6)(y - 6) = 0$, then $y = 6$.

4. $x^2 + 4 = 0$ has no real-number roots.

Find the rational roots of each equation.

5. $-3x = 0$

6. $x^2 = 0$

7. $(x - 1)x = 0$

8. $(x - 2)(x - 3) = 0$

9. $(5n - 3)(n^2 + 4) = 0$

10. $y^2 + 15y + 54 = 0$

Written Exercises

Find the rational roots, if any, of each equation.

1. $(3x - 1)(x + 2) = 0$ **2.** $(2a + 5)(a - 4) = 0$ **3.** $y(y + 5) = 0$

4. $(3c - 2)(2c + 3) = 0$ **5.** $x^2 - 13x + 40 = 0$ **6.** $0 = x^2 + 14x + 40$

7. $y^2 = y + 12$ **8.** $6a = 16 - a^2$ **9.** $0 = 3c^2 - 15c$

10. $7x^2 = -28x$ **11.** $z^2 - 49 = 0$ **12.** $16 = y^2$

13. $x^2 + 25 = 0$ **14.** $0 = n^2 + 36$ **15.** $0 = a^2 - 4a + 4$

16. $x^2 + 8x + 16 = 0$ **17.** $y^4 - 26y^2 + 25 = 0$ **18.** $a^4 = 29a^2 - 100$

19. $y^4 - 24y^2 - 25 = 0$ **20.** $a^4 = 21a^2 + 100$ **21.** $6x^2 - 42x + 60 = 0$

22. $0 = 5n^2 + 20n - 60$ **23.** $0 = 4n^2 + 40n + 100$ **24.** $60c - 180 = 5c^2$

25. $a^3 - 9a = 0$ **26.** $0 = c^3 - 16c$ **27.** $x^3 + 3x^2 - 10x = 0$

28. $2x^2 + 8x = x^3$ **29.** $2c^2 = 7c - 6$ **30.** $15 - 4y = 3y^2$

31. $16n^2 - 9n = 0$ **32.** $30c = 12c^2$ **33.** $30t^2 = 125t - 120$

34. $2m + 70 = 24m^2$ **35.** $9y^4 - 37y^2 + 4 = 0$ **36.** $25b^4 - 34b^2 + 9 = 0$

37. $9y^4 + 35y^2 - 4 = 0$ **38.** $25b^4 + 16b^2 - 9 = 0$ **39.** $(n - 7)^2 = 29 - n^2$

40. $2(24 - n^2) = (8 - n)^2$ **41.** $5x^3 = 30x - 25x^2$ **42.** $3y^3 + 36y^2 = 3y^4$

Solve each equation for x in terms of a.

43. $x^2 - ax - 6a^2 = 0$ **44.** $3x^2 + 24a^2 = 18ax$ **45.** $x^4 - 10a^2x^2 + 9a^4 = 0$

46. Solve $|x^2 - 5| = 4$. (HINT: There are four roots.)

47. Find the two roots of $|a^2 + 2a - 8| = |a^2 - 4|$.

48. Find the three roots of $|2y^2 - 11y + 5| = |y^2 - 3y - 10|$.

49. The Zero-Product Property is stated in "If . . ., then . . ." form (see page 199). Write the converse of this property. Is the converse true? (See Extension, pages 103–104.)

Mixed Review

1. Find three consecutive even integers such that 2 more than 5 times the third integer is the same as 3 times the sum of the first two integers. **1.6**

2. Supply a reason for each step in the proof that $a(b + c) + d$ is equivalent to $d + (ba + ca)$. **1.8**

    1. $a(b + c) + d = d + a(b + c)$
    2. $\qquad\qquad\quad = d + (ab + ac)$
    3. $\qquad\qquad\quad = d + (ba + ca)$
    4. $a(b + c) + d = d + (ba + ca)$

Problem Solving Strategies

Solving a Simpler Problem

A service engineer is asked to install telephone lines connecting 10 buildings. Every pair of buildings is to have a separate line. How many lines are needed?

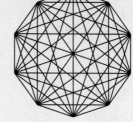

Use a decagon (10-sided polygon) to **model** the situation. Let each vertex represent a building. The problem is to find the total number of *sides and diagonals* of the figure.

Consider a number of **simpler problems.** Draw several polygons and find the sum of the diagonals and sides for each.

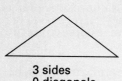

3 sides
0 diagonals

4 sides
2 diagonals

5 sides
5 diagonals

6 sides
9 diagonals

Make a **table** to organize the results. Look for a **pattern**. Use the pattern to extend the table.

Number of Sides	3	4	5	6	7	8	9	10
Number of Diagonals	0	2	5	9	14	?	?	?

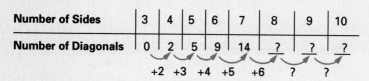

+2 +3 +4 +5 +6   ?   ?

A decagon will have 35 diagonals and 10 sides. So, 45 lines will be needed to connect the 10 buildings.

Exercises

1. There are 7 students at a party. Each student shakes hands once with every other student in the room. How many handshakes are there?

2. How many different squares are there in the 7 by 7 grid at the right?

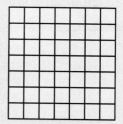

6.2 Problem Solving: Using Quadratic Equations

Objective	To solve word problems using quadratic equations

In the following example, you see that a given word problem may have two sets of answers.

EXAMPLE 1 The product of two consecutive integers is 12. Find all such consecutive integers.

What are you to find?	Two consecutive integers
Choose a variable.	Let m = the first consecutive integer.
What does it represent?	Then $m+1$ is the second consecutive integer.
What is given?	The product of m and $m+1$ is 12.
Write an equation.	$m(m+1)=12$
Solve the equation.	$m^2+m=12$
	$m^2+m-12=0$
	$m+4=0$ or $m-3=0$
	$m=-4$ $m=3$
	If $m=-4$, then $m+1=-3$.
	If $m=3$, then $m+1=4$.
Check in the original problem.	The check is left for you.
State the answer.	There are two pairs of consecutive integers, $-4, -3$ and $3, 4$.

When a quadratic equation is used to solve a problem, it is necessary to determine whether both solutions provide reasonable answers to the problem. For example, if x apples are placed in each of $x-6$ boxes and there is a total of 40 apples, then $x(x-6)=40$ and the roots are 10 and -4.

If $x=10$, then $x-6=4$ and there are 10 apples in each of 4 boxes. If $x=-4$, then $x-6=-10$, and there are -4 apples in each of -10 boxes, which is not reasonable. The root -4 is therefore not a reasonable solution of the problem.

EXAMPLE 2 One hundred cubes are placed in rows so that the number of cubes in each row is 1 more than 6 times the number of rows. Find the number of cubes in each row.

Plan

| Number of rows | × | Number of cubes in one row | = | Total number of cubes |

Let r = the number of rows. Then $6r + 1$ = the number of cubes per row.

Solution

$r(6r + 1) = 100$ ← (THINK: The answer must be a factor of what number?)

$$6r^2 + r = 100$$
$$6r^2 + r - 100 = 0$$
$$(6r + 25)(r - 4) = 0$$

$6r + 25 = 0$ or $r - 4 = 0$
$r = -\frac{25}{6}$ or $r = 4$

The number of rows cannot be $-\frac{25}{6}$, so $r = 4$ and $6r + 1 = 6 \cdot 4 + 1$, or 25. Check: 4 rows × 25 cubes per row = 100 cubes. There are 25 cubes in each row.

Classroom Exercises

Tell which of the following values substituted for x in Exercises 1–8 give results that are *not* reasonable: -3, $2\frac{1}{2}$, 8.

1. a first number x and a second number $4x - 2$
2. three consecutive integers beginning with x
3. four consecutive odd integers beginning with x
4. x nickels, $3x$ quarters, $3x - 2$ dimes
5. x males and $x - 10$ females
6. a rectangle, x units wide and $2x + 3$ units long
7. a triangle with a base x units long and a height of $2x - 5$ units
8. x pounds of rice in each of $4x$ bags
9. Find three consecutive multiples of 5 so that the square of the third number, decreased by 5 times the second number, is the same as 25 more than twice the product of the first two numbers.
10. Some light bulbs are placed in boxes, and the boxes are packed in cartons. The number of bulbs in each box is 4 less than the number of boxes in each carton. Find the number of bulbs in each box given that a full carton contains 60 light bulbs.

Written Exercises

1. Find three consecutive even integers such that the product of the first two integers is 48.

2. Find three consecutive integers so that the product of the second and third integers is 56.

3. Seventy chairs are placed in rows so that the number of chairs in each row is 3 less than the number of rows. What is the number of chairs in each row?

4. Thirty bulbs are placed in some boxes where the number of bulbs in each box is 7 more than the number of boxes. Find the number of bulbs in each box.

5. One number is 5 less than another number, and their product is 24. Find all such pairs of numbers.

6. The product of two integers is 44. One is 3 more than twice the other. Find both numbers.

7. Find three consecutive multiples of 5 so that the product of the first two numbers is 150.

8. Find three consecutive multiples of 10 so that the product of the first two numbers is 15 times the third number.

9. Eighty peaches were packed in boxes so that the number of boxes was 6 less than twice the number of peaches in each box. Find the number of boxes used.

10. A complete album holds 480 stamps. How many pages are in the album given that the number of pages is 8 more than 4 times the number of stamps on each page?

11. Find three consecutive integers so that the square of the first, increased by the square of the third, is 74.

12. Find three consecutive odd integers such that the sum of the squares of the second and third integers is 130.

13. The product of two numbers is 4, and one number is 9 times the other number. Find all such pairs of numbers.

14. Find two numbers whose product is -9 given that one number is -16 times the other number.

15. Some tiles are arranged in rows so that the number of tiles in each row is 8 more than the number of rows. The same number of tiles can be arranged in 3 more rows than in the first pattern with 16 tiles in each row. Find the total number of tiles.

16. Some chairs are placed in rows so that the number of rows is 2 less than the number of chairs in each row. The same number of chairs can be placed in 6 more rows than in the first case with 15 chairs per row. Find the total number of chairs.

17. Find four consecutive multiples of 0.5 for which the product of the first and third numbers is 0.25 less than 3 times the second number.

18. Find four consecutive multiples of π in which the product of the second and third multiples is $2\pi^2$ more than 3π times the fourth multiple.

Mixed Review

1. The length of a rectangle is 6 units more than its width. Represent the perimeter and the area of the rectangle in algebraic terms. *1.7*

2. The base of a triangle is 6 units shorter than 4 times its height. Represent the area of the triangle in algebraic terms. *1.7*

3. What is the perimeter of the right triangle whose sides are $5x$, $12x$, and $13x$ and whose area is 30? *1.7*

4. Find the area of the rectangle whose dimensions are $6a$ by $8a$, and whose perimeter is 70. *1.7*

Application: *Projectile Motion*

From observation, Aristotle (384-322 B.C.) and Galileo (1564-1642) knew that an object thrown horizontally off a cliff took a curved path to the ground. Galileo was able to show that the projectile traveled forward and downward simultaneously, and that the horizontal and vertical components of its motion were independent of each other.

The distance x (in feet) traveled horizontally can be expressed using $d = rt$. If the initial horizontal velocity is 8 ft/s, then $x = 8t$. The height y (in feet) can also be expressed in terms of time (in seconds): $y = h_o - 16t^2$, where h_o is the original height. If the cliff is 100 ft high, then $y = 100 - 16t^2$. At $t = 1$ s for example, $x = 8$, $y = 84$, and the object is at a point 8 ft away from the cliff and 84 ft high.

To find the spot where this projectile hits the ground ($y = 0$), set the polynomial $100 - 16t^2$ equal to zero and solve for t by factoring. Then substitute this value for t into $x = 8t$ to find how far the projectile will travel horizontally in that time.

$$100 - 16t^2 = 0 \qquad\qquad x = 8t$$
$$(10 + 4t)(10 - 4t) = 0 \qquad\quad = 8(\tfrac{5}{2})$$
$$t = \frac{10}{4} = \frac{5}{2}s \qquad\qquad = 20 \text{ ft}$$

1. A cannon is fired from a fort 144 ft high. The shell is shot horizontally with an initial velocity of 900 ft/s. How far away from the fort does the shell hit the ground?

2. A motorcycle daredevil prepares to jump a canyon from the North Rim to the South Rim. The canyon is 352 ft wide, and the South Rim is 64 ft lower than the North Rim. What is the minimum speed (in ft/s) necessary, given a horizontal takeoff, for the jump to work?

6.3 Problem Solving: Area and Length

Objectives

To find the lengths of sides of right triangles using the Pythagorean relation

To solve area problems using quadratic equations

For each right triangle ABC with right angle C, the lengths of the two legs are a and b and the length of the hypotenuse is c. For any such right triangle, the Pythagorean relation states that

$$a^2 + b^2 = c^2.$$

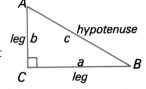

EXAMPLE 1 One leg of a right triangle is 14 m longer than the other leg and 2 m shorter than the hypotenuse. Find the length of each side of the triangle.

Solution

What are you to find?   the length of each side

What is given?

Represent the data as shown at the right.

Write an equation.

Substitute in the formula for a, b, and c.

Solve the equation.

$$a^2 + b^2 = c^2$$
$$x^2 + (x + 14)^2 = (x + 16)^2$$
$$x^2 + x^2 + 28x + 196 = x^2 + 32x + 256$$
$$x^2 - 4x - 60 = 0$$
$$(x - 10)(x + 6) = 0$$
$$x = 10 \quad or \quad x = -6$$

The length of a side cannot be -6 m.
If $x = 10$, then $x + 14 = 24$ and $x + 16 = 26$.

Check in the original problem.

24 m is 14 m longer than 10 m.

24 m is 2 m shorter than 26 m.

$$10^2 + 24^2 \stackrel{?}{=} 26^2$$
$$676 = 676 \text{ True}$$

State the answer.

The legs measure 10 m and 24 m. The hypotenuse is 26 m long.

Recall that the area of a triangle is found by multiplying half the length of the base by the height of the triangle.

EXAMPLE 2   The height of a triangle is 9 cm less than 3 times the length of its base. Find the length of the base and the height of the triangle given that the area of the triangle is 60 cm^2.

Solution   Let b = the length of the base.
Then $3b - 9$ = the height.

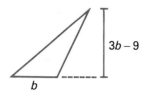

$$\tfrac{1}{2} \times \text{base} \times \text{height} = \text{area}$$
$$\tfrac{1}{2}b(3b - 9) = 60$$
$$b(3b - 9) = 120$$
$$3b^2 - 9b - 120 = 0$$
$$b^2 - 3b - 40 = 0$$
$$(b - 8)(b + 5) = 0$$
$$b = 8 \quad \text{or} \quad b = -5 \qquad \text{(The base cannot be } -5 \text{ cm.)}$$

If $b = 8$, then $3b - 9 = 15$.

Check $\tfrac{1}{2} \times 8$ cm $\times 15$ cm $= 60$ cm^2

The base is 8 cm long, and the height is 15 cm.

EXAMPLE 3   The length of a rectangular floor is 4 m shorter than 3 times its width. The width of a rectangular carpet on the floor is 2 m shorter than the floor's width, and the carpet's length is 2 m more than twice its own width. Find the area of the floor given that 31 m^2 of the floor is not covered by the carpet.

Solution   Draw and label both rectangles.

Let x = the width of the floor.

Then $3x - 4$ = the length of the floor; $x - 2$ = the width of the carpet; and $2(x - 2) + 2$, or $2x - 2$ = the length of the carpet.

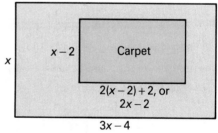

Use the formula for the area of a rectangle: $A = lw$.

$$(\text{Area of floor}) - (\text{Area of carpet}) = \text{Area not covered}$$
$$x(3x - 4) - (x - 2)(2x - 2) = 31$$
$$3x^2 - 4x - (2x^2 - 6x + 4) = 31$$
$$x^2 + 2x - 35 = 0$$
$$(x - 5)(x + 7) = 0$$
$$x = 5 \quad \text{or} \quad x = -7 \qquad \text{(Width cannot be } -7 \text{ m.)}$$

If $x = 5$, then $3x - 4 = 11$, and $x(3x - 4) = 55$.

The area of the floor is 5 m $\times$ 11 m, or 55 m^2.

Classroom Exercises

Which figure has the greater area?

1. a square 8 m wide or a rectangle 10 m by 6 m?

2. a triangle with base 7 m and height 9 m or a rectangle 4 m by 8 m?

3. a rectangle x units by $4(x - 2)$ units or a rectangle $4x$ units by x units?

4. a square $4x$ units wide or a rectangle $8x$ units by $2x - 2$ units?

Write an expression for the area of each figure.

5.

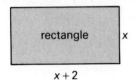

6.

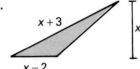

7. In a right triangle, the hypotenuse is 8 in. longer than one leg and 4 in. longer than the other leg. Find the length of each side.

Written Exercises

1. One leg of a right triangle is 2 ft shorter than the other leg. The hypotenuse is 10 ft long. Find the length of each leg.

2. The length of a rectangle is 4 cm less than twice its width. Find the length and the width given that the area is 70 cm^2.

3. The length of a rectangle is 5 m more than 4 times its width. Find the length and the width given that the area is 51 cm^2.

4. The height of a triangle is 8 times the length of its base. Find the length of the base and the height given that the area is 36 ft^2.

5. The base of a triangle is 3 yd shorter than its height, and the area of the triangle is 27 yd^2. Find its height and the length of its base.

6. One square is 3 times as wide as another square. Find the area of each square given that the difference between the two areas is 200 cm^2.

7. The width of one square is twice the width of another square. The sum of their areas is 180 m^2. Find the area of the larger square.

8. The length of a rectangle is 3 cm more than 3 times the width and 1 cm less than the length of a diagonal. Find the length of the rectangle.

9. A square and a rectangle have the same width. The rectangle's length is 4 times its width. Find the area of each figure given that the sum of the two areas is 80 ft^2.

10. The area of a triangle is 57 cm². Find the height of the triangle given that the height is 5 cm less than 4 times the length of the base.

11. A rectangle is 7 m longer than it is wide, and each diagonal is 8 m longer than the rectangle's width. What is the length of each diagonal?

12. The length of a rectangle is 5 in. more than twice its width. The width of a square is the same as the length of the rectangle. Find the area of each figure given that one area, decreased by the other area, is 88 in².

13. A rectangular wall is 3 m longer than twice its height. The width of a picture on the wall is 1 m less than the wall's height. The picture is 2 m longer than it is wide. Find the area of the wall given that 19 m² of the wall are not covered by the picture.

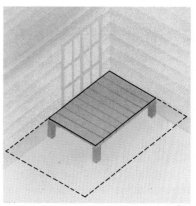

14. A rectangular patio floor has a length that is 1 m less than twice its width. The floor is extended by adding an additional 3 m to the original patio's length and an additional 2 m to its width. Find the area of the original floor given that the total area of the patio floor is 40 m² after the extension is built.

15. A rectangle's length is 4 times its width. A triangle's height is 4 times the length of its base. The triangle's height is the same as the rectangle's length. Find the area of each figure given that the sum of the areas is 150 m².

16. A triangle's base is 2 ft shorter than a rectangle's width. The rectangle's length is 6 times its width, and the triangle's height is the same as the rectangle's length. Find the area of each figure given that the difference between the areas is 105 ft².

17. A rectangular lawn measures 80 ft by 104 ft and is surrounded by a uniform sidewalk. The outer edge of the sidewalk is a rectangle with an area of 10,260 ft². Find the width of the uniform sidewalk.

18. A rectangular picture measures 30 cm by 52 cm. It is surrounded by a uniform frame whose outer edge is a rectangle with an area of 2,280 cm². Find the width of one side of the frame.

19. A rectangular piece of cardboard is 4 in. longer than it is wide. A 3-inch square is cut from each corner, and the four flaps are turned up to form an open box with a volume of 351 in³. Find the length and the width of the original piece of cardboard.

20. Find the value of x given that the *area* of the block-letter H at the right is 5,400 square units.

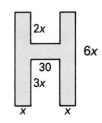

Midchapter Review

Solve. *6.1*

1. $2x^2 = 5x + 12$
2. $12x = 4x^2$
3. $16 = 25x^2$
4. $9y^4 - 40y^2 + 16 = 0$
5. $5n^2 + 40n + 80 = 0$
6. $a^3 - 25a = 0$
7. Find three consecutive even integers such that the square of the third integer decreased by 5 times the first integer is 7 times the third integer. *6.2*
8. A rectangle is 3 m longer than it is wide. The rectangle's length is 3 m less than a diagonal's length. Find the length of a diagonal. *6.3*
9. The base of a triangle is 3 ft shorter than twice the triangle's height. Find the length of the base given that the area of the triangle is 27 ft². *6.3*

Statistics: *Population Sampling*

Biologists often use **tag sampling** to estimate the total population of fish in a lake. The fish from a *random sample* are counted, tagged, and released. At a later time, another random sample is taken and the following formula is used to predict the total fish population, T.

$$\underset{\text{2nd sample}}{\text{Tagged}} \rightarrow \frac{s_1}{S} = \frac{F}{T} \begin{matrix} \leftarrow \text{1st Sample} \\ \leftarrow \text{Population} \end{matrix}$$

A random sample of 500 fish is taken from a lake, tagged, and released. Later, another random sample is taken. This sample contains 800 fish, 38 of which have tags. Substituting in the formula,

$$s_1 = 38, S = 800, \text{ and } F = 500.$$

$$\frac{38}{800} = \frac{500}{T} \qquad 38T = 400{,}000$$

So $T = 10{,}526.31$, or about 10,500 fish in the lake.

Exercises

Compute an estimate for the total population T to the nearest hundred.

1. $F = 900, S = 400, s_1 = 73$
2. $F = 850, S = 250, s_1 = 25$

6.4 Inequalities of Degree 2 and Degree 3

Objective

To find and graph the solution sets of second- or third-degree inequalities in one variable

In Chapter 2, you solved and graphed linear inequalities by using your knowledge of linear *equalities* as well as that of compound sentences involving *and* and *or*. Now you can solve *quadratic inequalities* in a similar fashion, using your knowledge of quadratic equalities.

An inequality such as $x^2 - 2x - 8 > 0$ is called a **quadratic inequality** in one variable. The standard form of a quadratic inequality is either

$$ax^2 + bx + c > 0 \quad \text{or} \quad ax^2 + bx + c < 0, \text{ where } a \neq 0.$$

When $ax^2 + bx + c$ can be factored as in $x^2 - 2x - 8 > 0$, or $(x + 2)(x - 4) > 0$, the following property involving the positive product of two factors can be used to solve the inequality.

Positive Product of Two Factors
If $m \cdot n > 0$, then $[m < 0 \text{ and } n < 0]$ or $[m > 0 \text{ and } n > 0]$.
(If the product of two numbers is positive, then the numbers must be either both negative or both positive.)

EXAMPLE 1

Find the solution set of $x^2 - 2x - 8 > 0$. Graph the solutions.

Plan

Factor the polynomial. Use the property of the Positive Product of Two Factors.

Solution

$$x^2 - 2x - 8 > 0$$
$$(x + 2)(x - 4) > 0$$

| $(x + 2 < 0 \text{ and } x - 4 < 0)$ | or | $(x + 2 > 0 \text{ and } x - 4 > 0)$ |
| $(x < -2 \text{ and } x < 4)$ | or | $(x > -2 \text{ and } x > 4)$ |

$$x < -2 \qquad \text{or} \qquad x > 4$$

The solution set is $\{x \mid x < -2 \text{ or } x > 4\}$.

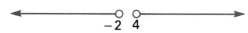

The quadratic inequality $x^2 - 9 < 0$, or $(x + 3)(x - 3) < 0$, can be solved using the property of the *Negative Product of Two Factors*.

Negative Product of Two Factors
If $m \cdot n < 0$, then [$m < 0$ *and* $n > 0$] *or* [$m > 0$ *and* $n < 0$].
(If the product of two numbers is negative, then one of the numbers must be positive and the other must be negative.)

If the coefficient of the square term in a quadratic inequality is negative, it may be helpful to divide each side by -1 as shown below.

EXAMPLE 2   Find the solution set of $-x^2 + 9 > 0$. Graph the solutions.
Divide each side by -1 and reverse the order of the inequality.

Solution

$$-x^2 + 9 > 0$$
$$x^2 - 9 < 0$$
$$(x + 3)(x - 3) < 0$$

Next, use the property of the Negative Product of Two Factors.

$$(x + 3 < 0 \ and \ x - 3 > 0) \quad or \quad (x + 3 > 0 \ and \ x - 3 < 0)$$
$$(x < -3 \ and \ x > 3) \quad\quad or \quad\quad (x > -3 \ and \ x < 3)$$

$$\underbrace{\phantom{(x < -3 \ and \ x > 3)}}$$ $$\underbrace{\phantom{(x > -3 \ and \ x < 3)}}$$

$$\text{no solution} \quad\quad\quad\quad\quad -3 < x < 3$$

The solution set is $\{x \mid -3 < x < 3\}$.

For a *polynomial inequality* of degree 3 or higher, it is usually easier to graph the solutions *first*, and then write the solution set. Begin by factoring the polynomial completely. Then solve the equation that corresponds to the inequality, as shown below.

EXAMPLE 3   Graph the solutions of $x^3 - x^2 - 20x > 0$. Then write the solution set.

Solution   Factor the polynomial completely.
$$x^3 - x^2 - 20x > 0$$
$$x(x^2 - x - 20) > 0$$
$$x(x + 4)(x - 5) > 0$$

Let $x(x + 4)(x - 5) = 0$. The roots of this *equation* are 0, -4, and 5. Plot these numbers with *open* dots on a number line. (The open dots indicate that these numbers are *not* solutions of the inequality $x(x + 4)(x - 5) > 0$.)

I   II   III   IV

$$x < -4 \quad {-4} \quad -4 < x < 0 \quad 0 \quad 0 < x < 5 \quad 5 \quad x > 5$$

The points at -4, 0, and 5 separate the preceeding number line into four parts: I, II, III, and IV. Choose a number for x in each part and evaluate $x(x + 4)(x - 5)$ to find where its value is positive.

Part	x	$x(x + 4)(x - 5)$	Value	Positive(?)
$x < -4$	-5	$-5(-1)(-10)$	-50	No
$-4 < x < 0$	-2	$-2(+2)(-7)$	$+28$	Yes
$0 < x < 5$	$+1$	$+1(+5)(-4)$	-20	No
$x > 5$	$+6$	$+6(+10)(+1)$	$+60$	Yes

Notice in the above table that the value of the polynomial *changes sign* whenever one of the boundary points, -4, 0, or 5, is crossed.

The graph of the solutions is parts II and IV, a segment and a ray.

The solution set is $\{x \mid -4 < x < 0 \quad or \quad x > 5\}$.

If the inequality $x^3 - x^2 - 20x > 0$ were combined with its related equation to form the compound sentence $x^3 - x^2 - 20x \geq 0$, then the numbers 0, -4, and 5, would be solutions of the compound sentence.

Classroom Exercises

Match each inequality to exactly one graph.

1. $6 < x < 9$

2. $x < 6 \; or \; x > 9$

3. $x < 6 \; or \; 9 < x < 12$

4. $6 < x < 9 \; or \; x > 12$

5. $x^2 + 9 > 0$

6. $x^2 + 6 < 0$

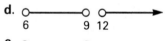

a.

b.

c. no points

d.

e.

f.

Match each sentence at the left with one compound sentence at the right.

7. $(x + 3)(x - 5) < 0$

8. $(x + 3)(x - 5) > 0$

9. $(x - 1)(x - 4) \leq 0$

a. $[x - 1 \leq 0 \; and \; x - 4 \geq 0] \; or$
$[x - 1 \geq 0 \; and \; x - 4 \leq 0]$

b. $[x + 3 < 0 \; and \; x - 5 > 0] \; or$
$[x + 3 > 0 \; and \; x - 5 < 0]$

c. $[x + 3 < 0 \; and \; x - 5 < 0] \; or$
$[x + 3 > 0 \; and \; x - 5 > 0]$

Written Exercises

Find the solution set of each inequality. Graph the solutions.

1. $n^2 - 8n + 12 > 0$ 2. $x^2 - 3x - 10 < 0$ 3. $c^2 + 7c + 10 \leq 0$
4. $n^2 - 7n + 6 \geq 0$ 5. $x^2 - 16 \geq 0$ 6. $y^2 - 36 \leq 0$
7. $-y^2 + 25 < 0$ 8. $-a^2 + 4 > 0$ 9. $n^2 - 4n > 0$
10. $c^2 + 6c < 0$ 11. $-3c^2 - 12c \geq 0$ 12. $-4x^2 + 12x \leq 0$

13. Find the set of all numbers whose squares, decreased by 49, result in a negative number.

14. Find the set of all numbers such that 5 times the square of a number, decreased by 20 times the number, is a positive number.

Graph the solutions of each inequality. Then write the solution set.

15. $y^3 - y^2 - 12y < 0$ 16. $x^3 + 6x^2 - 7x > 0$ 17. $x^3 - 16x \geq 0$
18. $y^3 - 25y \leq 0$ 19. $-y^3 - y^2 + 20y > 0$ 20. $-x^3 + 49x < 0$

Find the solution set of each inequality. Graph the solutions.

21. $2x^2 + x < 15$ 22. $4a^2 > 9a + 9$ 23. $2y^2 \geq 9y$
24. $4c^2 \leq 10c$ 25. $-c^2 - c + 12 < 0$ 26. $-x^2 + 6x \geq 8$
27. $0 < -a^2 + 4a - 3$ 28. $0 \geq 35 - 2y - y^2$ 29. $2x^3 - x^2 - 21x > 0$
30. $3a^3 + 5a^2 - 28a \leq 0$ 31. $a^2 < 4$ or $a^2 > 36$ 32. $c^2 \geq 4$ and $c^2 \leq 36$

33. Write in your own words the property of the Negative Product of Two Factors.

Graph the solutions and write the solution set.

34. $4x^2 + 1 > 3x^2$ 35. $y^2 - 3 > 2y^2 + 3$ 36. $|x^2 - 5| < 4$
37. $|y^2 - 10| > 6$ 38. $a^4 - 10a^2 + 9 < 0$ 39. $x^4 - 26x^2 + 25 > 0$
40. $a^4 - 16 \geq 0$ 41. $n^4 - 81 \leq 0$ 42. $y^4 - 20y^2 + 64 > 0$

Mixed Review

Divide to find the polynomial quotient and the remainder. Is the divisor a factor of the dividend?   *5.8*

1. $(2x^3 - 4x - 10) \div (x - 2)$ 2. $(y^3 + 4y^4 - y - 2 - 2y^2) \div (y + 1)$
3. $(3x^3 + 8x^2 - 4x - 3) \div (x + 3)$ 4. $(y^3 - 10) \div (y - 2)$
5. An 18% alcohol solution is mixed with a 45% alcohol solution to produce 12 ounces of a 36% alcohol solution. How many ounces of each solution must be used? *4.4*

6.5 Synthetic Substitution and the Remainder Theorem

Objectives

To divide a polynomial $P(x)$ by $x - a$ using synthetic division

To evaluate a polynomial $P(x)$ using synthetic substitution

To determine whether a binomial $x - a$ is a factor of a polynomial $P(x)$

For brevity, a polynomial such as $3x^3 - 10x^2 + 14x - 7$ is sometimes represented as $P(x)$ (read "P of x" or "P at x"). Similarly, a *polynomial function* can be represented as $y = P(x)$. Thus, the two polynomial functions $y = 3u^2 + 1$ and $y = 2w^5 + w^2 - 1$ might be represented as $y = P_1(u)$ and $y = P_2(w)$, respectively.

Recall from Lesson 5.8 the long-division technique for dividing a polynomial by any binomial. In this lesson, a polynomial $P(x)$ is divided by a binomial in the special form $x - a$. It is also determined whether the divisor $x - a$ is a factor of $P(x)$.

Here, the polynomial

$P(x) = 3x^3 - 10x^2 + 14x - 7$

is divided by $x - 2$ using long division.

$$
\require{enclose}
\begin{array}{r}
3x^2 - 4x + 6 \\
x - 2 \enclose{longdiv}{3x^3 - 10x^2 + 14x - 7} \\
\underline{3x^3 - 6x^2 } \\
-4x^2 + 14x \\
\underline{-4x^2 + 8x } \\
6x - 7 \\
\underline{6x - 12} \\
5
\end{array}
$$

Shown below is a more compact form of this division called *synthetic division*. Note that the divisor $x - a = x - 2$, and that $a = 2$. To begin, write 2 and the coefficients of $P(x)$.

Bring down the 3. Then, use the "multiply by 2 and add" cycle as shown.

Quotient: $3x^2 - 4x + 6$ Remainder: 5

Notice that both methods yield the same results: (1) the polynomial quotient $Q(x)$ is $3x^2 - 4x + 6$, (2) the remainder R is 5, and (3) the divisor $x - 2$ is *not* a factor of the dividend $P(x)$ because the remainder is not 0.

It can also be shown that the value of $P(x) = 3x^3 - 10x^2 + 14x - 7$ is 5 when $x = 2$. That is, $P(2) = 5$.

$$\text{Dividend} = \text{Divisor} \times \text{Quotient} + \text{Remainder}$$

$$
\begin{aligned}
P(x) &= (x - 2)(3x^2 - 4x + 6) + 5 \\
P(2) &= (2 - 2)(3 \cdot 2^2 - 4 \cdot 2 + 6) + 5 \\
&= 0 + 5 \\
&= 5
\end{aligned}
$$

Thus, the value of the polynomial at $x = 2$ is 5, which is the same as the remainder when the polynomial is divided by $x - 2$. This suggests the following theorem.

Theorem 6.1

Remainder Theorem: When a polynomial in x is divided by $x - a$, the remainder equals the value of the polynomial when $x = a$.

One consequence of the Remainder Theorem is that $x - a$ and $Q(x)$ are factors of $P(x)$ when $R = 0$. This suggests the **Binomial Factor Theorem** (also referred to as the *Factor Theorem*).

Theorem 6.2

Binomial Factor Theorem: The binomial $x - a$ is a factor of the polynomial $P(x)$ if and only if $P(a) = 0$.

If $x - a = x + 5$, then $x - a = x - (-5)$ and $a = -5$. Thus, if a polynomial $P(x)$ is divided by $x + 5$, the remainder will be the value of $P(-5)$.

EXAMPLE 1 Given $P(x) = 2x^4 + 11x^3 - 27x - 10$, use synthetic division to divide $P(x)$ by $x + 5$. (1) Find the polynomial quotient $Q(x)$ and the remainder R. (2) Find the value of $P(x)$ for $x = -5$. (3) Determine whether $x + 5$ is a factor of $P(x)$.

Plan There is no x^2 term. Write 0 for the coefficient of the missing x^2 term.

Solution

$x + 5 = x - (-5)$

$$
\begin{array}{r|rrrrr}
 & 2 & 11 & 0 & -27 & -10 \\
 & & (-10) & (-5) & (25) & (10) \\
\hline
-5 & 2 & 1 & -5 & -2 & 0 \\
\end{array}
$$

(1) $Q(x) = 2x^3 + x^2 - 5x - 2$ and $R = 0$
(2) $P(-5) = 0$, because $P(-5) = R$ and $R = 0$
(3) $x + 5$ is a factor of $P(x)$ because $R = 0$.

Finding a value of a polynomial by synthetic division is called **synthetic substitution.** For example, $P(x) = 6x^3 + 7x^2 - 8x - 9$ is evaluated for $x = 2$ by synthetic substitution as follows.

$$
\begin{array}{c|rrrr}
 & 6 & 7 & -8 & -9 \\
 & & (12) & (38) & (60) \\
\hline
2 & 6 & 19 & 30 & \boxed{51}
\end{array}
$$

The last number, 51, is the value of $P(x)$ when $x = 2$. $P(2) = 51$.

A more compact form of the synthetic substitution above is shown below. In this form, some of the operations are performed mentally. Thus, the multiples of 2 (12, 38, and 60) are not shown.

$$
\begin{array}{c|rrrr}
 & 6 & 7 & -8 & -9 \\
\hline
2 & 6 & 19 & 30 & \boxed{51}
\end{array}
$$

The "multiply by 2 and add" cycle above may be better understood by seeing $P(x)$ expressed in a different way. First, rewrite $P(x)$ as follows.

$$
\begin{aligned}
P(x) = 6x^3 + 7x^2 - 8x - 9 &= (6x^2 + 7x - 8)x - 9 \\
&= [(6x + 7)x - 8]x - 9
\end{aligned}
$$

Next, use direct substitution to see the "multiply by 2 and add" cycle.

$$
P(2) = [(6 \cdot 2 + 7)2 - 8]2 - 9 = \boxed{51}
$$

Multiply by 2. Then add.

EXAMPLE 2   Evaluate $P(x) = 2x^4 - x^3 + 4x - 12$ for $x = -2$ and for $x = 3$.

Solution

$$
\begin{array}{c|rrrrr}
 & 2 & -1 & 0 & 4 & -12 \\
 & & (-4) & (10) & (-20) & (32) \\
\hline
-2 & 2 & -5 & 10 & -16 & \boxed{20} \\
 & & (6) & (15) & (45) & (147) \\
\hline
3 & 2 & 5 & 15 & 49 & \boxed{135}
\end{array}
$$

Compact form

$$
\begin{array}{c|rrrrr}
 & 2 & -1 & 0 & 4 & -12 \\
\hline
-2 & 2 & -5 & 10 & -16 & \boxed{20} \\
\hline
3 & 2 & 5 & 15 & 49 & \boxed{135}
\end{array}
$$

Thus, $P(-2) = 20$ and $P(3) = 135$.

EXAMPLE 3   Evaluate $P(x) = 3x^4 + 3x^3 - 20x^2 - 2x + 12$ for $-3, -2, 2$, and 3. Name the binomial factors found, if any.

Plan   To evaluate $P(x)$ for a or $-a$, write $a\rfloor$ or $-a\rfloor$, respectively. If $P(a) = 0$, then $x - a$ is a factor of $P(x)$. If $P(-a) = 0$, then $x + a$ is a factor of $P(x)$.

Solution

$$
\begin{array}{r|rrrrr}
 & 3 & 3 & -20 & -2 & 12 \\
 & & (-9) & (18) & (6) & (-12) \\
\hline
-3 & 3 & -6 & -2 & 4 & \boxed{0} \\
 & & (-6) & (6) & (28) & (-52) \\
\hline
-2 & 3 & -3 & -14 & 26 & \boxed{-40} \\
 & & (6) & (18) & (-4) & (-12) \\
\hline
2 & 3 & 9 & -2 & -6 & \boxed{0} \\
 & & (9) & (36) & (48) & (138) \\
\hline
3 & 3 & 12 & 16 & 46 & \boxed{150} \\
\end{array}
$$

Compact form

$$
\begin{array}{r|rrrrr}
 & 3 & 3 & -20 & -2 & 12 \\
\hline
-3 & 3 & -6 & -2 & 4 & \boxed{0} \\
\hline
-2 & 3 & -3 & -14 & 26 & \boxed{-40} \\
\hline
2 & 3 & 9 & -2 & -6 & \boxed{0} \\
\hline
3 & 3 & 12 & 16 & 46 & \boxed{150} \\
\end{array}
$$

Thus, $P(-3) = 0$, $P(-2) = -40$, $P(2) = 0$, and $P(3) = 150$. Two factors of $P(x)$ are $x + 3$ and $x - 2$.

Focus on Reading

Use the example of synthetic substitution shown at the right and the list of expressions below to complete the paragraph.

$$
\begin{array}{r|rrrr}
 & 2 & 9 & 0 & -6 \\
-4 & 2 & 1 & -4 & \boxed{10} \\
\end{array}
$$

The dividend is ___, the divisor is ___, 10 is the ___, and the polynomial quotient is ___. The value of $P(x)$ is ___ when the value of x is ___. The binomial $x + 4$ is not a ___ of $2x^3 + 9x^2 - 6$.

a. remainder
b. -10
c. 10
d. $x + 4$
e. $x - 4$
f. $2x^3 + x^2 - 4x + 10$

g. factor
h. -4
i. $2x^2 + x - 4$
j. $2x^2 + 9x - 6$
k. $2x^3 + 9x^2 - 6$

Classroom Exercises

Determine whether each statement is true or false.

1. If $P(x) = (x + 3)(x^2 - 2x - 5) + 4$, then $x + 3$ is a factor of $P(x)$.

2. If $P(x) = (x - 5)(2x^2 + 7)$, then $x - 5$ is a factor of $P(x)$.

3. If $P(x) = (x + 4)(x^2 - 5x + 6)$, then $P(x) = 0$ when $x = -4$.

4. If $P(x) = (x - 6)(x^2 + 4x + 4) + 5$, then $P(6) = 5$.

Evaluate $P(x) = x^2 + x - 2$ for each value of x.

5. $x = 0$ **6.** $x = -1$ **7.** $x = 1$ **8.** $x = -2$ **9.** $x = 2$

10. Use the results of Exercises 5–9 to name the binomial factors of $x^2 + x - 2$.

Written Exercises

Use synthetic division to find the polynomial quotient and the remainder. Is the divisor a factor of the dividend?

1. $(2x^3 - 11x^2 + 18x - 15) \div (x - 3)$

2. $(3x^3 + 17x^2 - 8x - 12) \div (x + 6)$

3. $(y^4 + 2y^3 - 3y^2 + 2y + 16) \div (y + 2)$

4. $(3y^4 - 12y^3 - 20y^2 + 30y + 10) \div (y - 5)$

5. $(4n^5 - 2n^3 + 6n^2 - 9n + 1) \div (n - 1)$

Evaluate each polynomial for the given values using synthetic substitution. Name the binomial factors found, if any.

6. $x^3 + 2x^2 - 8x - 21$ for $x = 3$ and again for $x = -2$

7. $x^3 - 4x^2 - 4x - 5$ for $x = -2$ and again for $x = 5$

8. $3y^4 + y^3 - 14y^2 + 10y - 60$ for $y = 2$ and again for $y = -3$

9. $2x^5 - 6x^4 - 3x^2 + 44$ for $x = 3$ and again for $x = 2$

Evaluate each polynomial for $-3, -2, -1, 0, 1, 2,$ and 3. Name the binomial factors found, if any.

10. $2x^4 - 4x^3 - 5x^2 - 2x - 3$ **11.** $x^4 + x^3 - 4x - 16$

Use synthetic division to find the polynomial quotient and the remainder. Is the divisor a factor of the dividend?

12. $(c^3 - 8{,}000) \div (c - 10)$ **13.** $(c^3 + 8{,}000) \div (c + 20)$

14. $(x^6 + x^4 + x^2 - 84) \div (x + 2)$ **15.** $(x^5 - x^3 - 210) \div (x - 3)$

Use synthetic substitution to find the value of k in each polynomial $P(x)$, given a factor of $P(x)$.

16. $x - 2$ is a factor of $P(x) = x^3 + 4x^2 - 15x + k$.

17. $x + 3$ is a factor of $P(x) = 2x^4 + 5x^3 + kx^2 + 10x - 6$.

Use synthetic division to find the polynomial quotient and the remainder. Is the divisor a factor of the dividend?

18. $(2x^3 - 7x^2 + 7x - 2) \div (x - \frac{1}{2})$ **19.** $(2y^3 - \frac{5}{2}y^2 + \frac{15}{4}y - \frac{3}{2}) \div (y - \frac{1}{2})$

Mixed Review

Write an equation in standard form, $ax + by = c$, for each line described below.

1. crossing the y-axis at $P(0, -2)$ with a slope of $\frac{2}{3}$ *3.5*

2. passing through $P(4, -2)$ and parallel to the line described by $2x - y = 5$ *3.7*

3. passing through $P(4, -2)$ and perpendicular to the line described by $2x - y = 5$ *3.7*

Using the Calculator

You can use synthetic substitution and a calculator with the $\boxed{+/-}$ function to evaluate any polynomial $P(x)$ for a given value of x. For example, to evaluate $P(x) = 5x^4 + 4x^3 - 6x^2 - 7x + 12$ for $x = 8$, a positive number, use the sequence of keystrokes below.

$5 \boxed{\times} 8 \boxed{+} 4 \boxed{=} \boxed{\times} 8 \boxed{-} 6 \boxed{=} \boxed{\times} 8 \boxed{-} 7 \boxed{=} \boxed{\times} 8 \boxed{+}$
$12 \boxed{=} 22{,}100$

Thus, $P(8) = 22{,}100$.

Notice that the keystrokes use the multiply-and-add cycle of synthetic substitution.

The keystrokes for $x = -8$, a negative number, are as follows.

$5 \boxed{\times} 8 \boxed{+/-} \boxed{+} 4 \boxed{=} \boxed{\times} 8 \boxed{+/-} \boxed{-} 6 \boxed{=} \boxed{\times} 8 \boxed{+/-} \boxed{-}$
$7 \boxed{=} \boxed{\times} 8 \boxed{+/-} \boxed{+} 12 \boxed{=} 18{,}116$

Thus, $P(-8) = 18{,}116$.

Evaluate $P(x) = 6x^4 - 5x^3 + 4x^2 - 7x - 9$ for the given value of x.

1. $x = 2$ **2.** $x = 3$ **3.** $x = -2$ **4.** $x = -3$

6.6 The Integral Zero Theorem

To find the rational zeros of integral polynomials

To factor integral polynomials into first-degree factors

To find the rational roots of integral polynomial equations of degree greater than 2

The polynomial $P(x) = 2x^4 - 7x^3 - 5x^2 + 28x - 12$ is an **integral polynomial** because all of its coefficients are integers. By synthetic substitution, or otherwise, you can show that $P(3) = 0$. The number 3 is called a *zero* of $P(x)$.

Definition

A number r is called a **zero of a polynomial function** $P(x)$ if and only if $P(r) = 0$.

A *zero of a polynomial function* may be referred to more briefly as a *zero of a polynomial*.

EXAMPLE 1 Find the four rational zeros of $P(x) = 2x^4 - 7x^3 - 5x^2 + 28x - 12$, given that $x - 3$ and $x + 2$ are factors.

Plan Use synthetic division and the Factor Theorem.

1. Divide by one of the factors, say $x - 3$, to obtain a quotient, $Q_1(x)$, with a zero remainder ($R_1 = 0$).

2. Divide $Q_1(x)$ by $x + 2$ to find $Q_2(x)$ with $R_2 = 0$. Because $x + 2$ is a factor of $Q_1(x)$, it must also be a factor of $P(x)$.

3. Factor $Q_2(x)$.

Solution

1.
$$\begin{array}{r|rrrrr} & 2 & -7 & -5 & 28 & -12 \\ 3 & 2 & -1 & -8 & 4 & \enclose{circle}{0} \end{array}$$

$Q_1(x) = 2x^3 - x^2 - 8x + 4$; $R_1 = 0$

2.
$$\begin{array}{r|rrrr} & 2 & -1 & -8 & 4 \\ -2 & 2 & -5 & 2 & \enclose{circle}{0} \end{array}$$

$Q_2(x) = 2x^2 - 5x + 2$; $R_2 = 0$

The *second-degree* factor $2x^2 - 5x + 2$ can be factored into two first-degree factors: $2x^2 - 5x + 2 = (2x - 1)(x - 2)$.

Thus, $P(x) = 2x^4 - 7x^3 - 5x^2 + 28x - 12$
$= (x - 3)(2x^3 - x^2 - 8x + 4)$
$= (x - 3)(x + 2)(2x^2 - 5x + 2)$
$= (x - 3)(x + 2)(2x - 1)(x - 2)$

Thus, the four rational zeros are 3, -2, $\frac{1}{2}$, and 2 because $P(3) = 0$, $P(-2) = 0$, $P(\frac{1}{2}) = 0$, and $P(2) = 0$.

In Example 1, three of the four *rational zeros* $(3, -2, 2)$ are **integral zeros**. The four zeros are also the roots of the polynomial equation
$$2x^4 - 7x^3 - 5x^2 + 28x - 12 = 0.$$

For any given integral polynomial, the number of *potential* integral zeros is finite. The positive integral zeros are integral factors of the constant term. From this, the **Integral Zero Theorem** follows:

Theorem 6.3

Integral Zero Theorem: If an integer r is a zero of the integral polynomial $P(x)$, then r is a factor of the constant term in $P(x)$. That is, if r is an integral zero of
$$a_0 x^n + a_1 x^{n-1} + a_2 x^{n-2} + \cdots + a_{n-1} x^1 + a_n,$$
then r is a factor of a_n.

An integral polynomial may have *no* integral zeros. For example, the potential integral zeros of $x^3 - x + 1$ are 1 and -1, but neither of these numbers is a zero of $x^3 - x + 1$.

EXAMPLE 2   Solve the equation $2x^4 - 5x^3 - 12x^2 - x + 4 = 0$.

Plan   Let $P(x) = 2x^4 - 5x^3 - 12x^2 - x + 4$. The roots of the equation are the zeros of $P(x)$. The potential integral zeros are ± 1, ± 2, and ± 4. Test each potential zero until a second-degree factor is found.

Solution

$$
\begin{array}{r|rrrrr}
 & 2 & -5 & -12 & -1 & 4 \\
\hline
-1 & 2 & -7 & -5 & 4 & \enclose{circle}{0} \\
\hline
-1 & 2 & -9 & 4 & \enclose{circle}{0} &
\end{array}
$$

$2x^2 - 9x + 4 = (x - 4)(2x - 1)$

The zeros of $P(x)$ are -1, 4, and $\frac{1}{2}$.

Successive Factorizations of $P(x)$

$(x + 1)(2x^3 - 7x^2 - 5x + 4)$
$(x + 1)(x + 1)(2x^2 - 9x + 4)$
$(x + 1)(x + 1)(x - 4)(2x - 1)$

In the example above, the number -1 is called a zero with a **multiplicity** of 2 because the factor $x + 1$ appears *twice* in the factorization of $P(x)$. Each of the remaining zeros, 4 and $\frac{1}{2}$, has a multiplicity of 1.

Classroom Exercises

State the potential integral zeros of each polynomial.

1. $2x^2 + 14x + 3$ **2.** $2x^5 + 9$

Given that -3 is a zero of the polynomial $P(x)$, find each of the following.

3. a first-degree factor of $P(x)$ **4.** a root of $P(x) = 0$

5. Find each root of $(x + 7)(x - 5)(3x - 1)(2x + 3) = 0$.

Find each zero and its multiplicity for the given polynomial $P(x)$.

6. $P(x) = (x + 3)(x - 5)(x + 3)$ **7.** $P(x) = (x - 1)^3(x + 2)$

Written Exercises

Factor $P(x)$ into first-degree factors. Find the zeros of $P(x)$.

1. $P(x) = x^3 - 2x^2 - x + 2$ **2.** $P(x) = x^3 + 3x^2 - x - 3$

3. $P(x) = 2x^3 + 3x^2 - 17x + 12$ **4.** $P(x) = 6x^3 + 25x^2 + 21x - 10$

5. $P(x) = 10x^3 - 41x^2 + 2x + 8$ **6.** $P(x) = 10x^3 + 51x^2 + 3x - 10$

7. $P(x) = x^4 - 15x^2 + 10x + 24$ **8.** $P(x) = 6x^4 + 23x^3 - 36x^2 - 3x + 10$

Solve each equation. Find the multiplicity m for each root whose $m > 1$.

9. $x^3 - 6x^2 + 12x - 8 = 0$ **10.** $3x^3 + 29x^2 + 65x - 25 = 0$

11. $4a^4 - 4a^3 - 43a^2 - 26a + 24 = 0$ **12.** $5c^4 + 6c^3 - 79c^2 + 84c + 20 = 0$

13. Write a fourth-degree polynomial $P(x)$ whose zeros are 2, -2, -1, and $\frac{1}{2}$.

14. Write a fourth-degree polynomial $P(x)$ whose zeros are 3 and $-\frac{1}{2}$, each with a multiplicity of 2.

15. The length of a rectangular solid (box) is 3 ft longer than twice its width. The height is 2 ft less than the width. Find the 3 dimensions given that the volume is 195 ft^3. Use $V = lwh$.

16. Solve the equation $x^6 - 3x^4 + 3x^2 - 1 = 0$.

Mixed Review

Find the slope of $\overleftrightarrow{PQ}$ and describe its slant.   *3.3*

1. $P(-3, -4)$, $Q(1,2)$ **2.** $P(7,3)$, $Q(7, -2)$ **3.** $P(1,5)$, $Q(-2,5)$

Key Terms

Binomial Factor Theorem (p. 218)
integral polynomial (p. 223)
integral zero (p. 224)
Integral Zero Theorem (p. 224)
multiplicity (p. 224)
polynomial function (p. 217)
polynomial inequality (p. 214)

quadratic equation (p. 199)
quadratic inequality (p. 213)
Remainder Theorem (p. 218)
synthetic division (p. 217)
synthetic substitution (p. 219)
zero of a polynomial (p. 223)
Zero-Product Property (p. 199)

Key Ideas and Review Exercises

6.1 To find the rational roots of an equation such as $16x^4 = 17x^2 - 1$, write the equation in standard form. Then factor the polynomial completely, set each factor equal to 0, and solve.

Find the rational roots of each equation.

1. $6x^2 = 5x + 4$ **2.** $16x^4 = 17x^2 - 1$ **3.** $20y^3 + 48y^2 = 36y$

6.2 To find a pair of numbers or consecutive multiples of a number, represent the numbers in terms of one variable and follow the six-step procedure for problem solving.

4. Find three consecutive multiples of 3 so that twice the square of the second number is 90 more than the product of the first and third numbers.

6.3 To solve problems involving the areas of rectangles, triangles, and right triangles, first sketch and label each figure. Then use an appropriate formula: for a rectangle, $A = lw$; for a triangle, $A = \frac{1}{2}bh$; for a right triangle, $a^2 + b^2 = c^2$ and $A = \frac{1}{2}ab$.

5. The length of the base of a triangle is 10 cm less than 4 times its height. Find the length of the base and the height given that the area of the triangle is 25 cm^2.

6.4 To solve a second-degree inequality such as $-y^2 + 4 \geq 0$, divide each side by -1 and reverse the order. Then use the appropriate property of the product of two factors:

If $m \cdot n > 0$, then $[m < 0 \text{ and } n < 0]$ or $[m > 0 \text{ and } n > 0]$.
If $m \cdot n < 0$, then $[m < 0 \text{ and } n > 0]$ or $[m > 0 \text{ and } n < 0]$.

Find the solution set of each inequality. Graph the solutions.

6. $x^2 - 6x - 16 > 0$ **7.** $-y^2 + 4 \geq 0$

6.4   To find the solution set of a third-degree inequality such as $a^3 - 25a > 0$:
1. factor the polynomial completely,
2. plot the roots of the related equation on a number line to separate the line into parts,
3. from each part, test a number in the factorization, and
4. graph the parts for which the test is true.

8. Graph the solutions of $a^3 - 25a > 0$. Then write the solution set.

6.5   To divide a given polynomial $P(x)$ by a binomial of the form $x - a$ using synthetic division:
1. Use synthetic substitution to evaluate $P(x)$ for $x = a$.
2. Use the resulting row of numbers to obtain the coefficients of the quotient and the remainder.

Use synthetic division to find the polynomial quotient and the remainder. Is the divisor a factor of the dividend?

9. $(x^6 + 2x^4 + x^2 - 900) \div (x + 3)$

6.5   To evaluate a given polynomial $P(x)$ for $x = -m$ and again for $x = n$, and to name any binomial factors found:
1. include 0 for the coefficient of any missing terms,
2. use synthetic substitution with $\underline{-m}$ and $\underline{n}$,
3. then use the Binomial Factor Theorem: if $P(-m) = 0$, then $x + m$ is a factor of $P(x)$; if $P(n) = 0$, then $x - n$ is a factor of $P(x)$.

10. Use synthetic substitution to find the value of $P(x) = 2x^4 - 17x^2 + x - 12$ for $x = -2$ and again for $x = 3$. Name the binomial factors found, if any.

6.6   To factor a given polynomial $P(x)$ into first-degree factors and find the zeros of $P(x)$:
1. list the integral factors of the constant term to find the potential integral zeros,
2. then use synthetic substitution to factor $P(x)$ completely.

11. Factor $P(x) = 3x^3 + 2x^2 - 37x + 12$ into first-degree factors. Find the zeros of $P(x)$.

12. Solve $4x^4 + 15x^3 + 9x^2 - 16x - 12 = 0$. Find the multiplicity m for each root whose $m > 1$.

13. Write in your own words what is meant by a *zero* of a polynomial.

Find the rational roots of each equation.

1. $5x^2 + 18x = 8$

2. $4y^4 - 37y^2 + 9 = 0$

3. $6a^3 - 33a^2 + 36a = 0$

4. Find three consecutive multiples of 5 so that twice the square of the second number is 250 more than the product of the first and third numbers.

5. The length of the base of a triangle is 2 m less than twice its height. Find the length of the base and the height given that the area of the triangle is 56 m^2.

6. A rectangle is twice as wide as a square. The rectangle is 5 ft longer than it is wide. The sum of the areas of the two figures is 240 ft^2. What is the area of the square?

Find and graph the solution set of each inequality.

7. $x^2 + x - 12 < 0$

8. $-y^2 + 25 \leq 0$

9. Graph the solutions of $a^3 - 36a > 0$. Then write the solution set.

Use synthetic division to find the polynomial quotient and the remainder. Is the divisor a factor of the dividend? (Exercise 10)

10. $(3x^4 + 10x^3 + 2x^2 - 5x - 3) \div (x + 3)$

11. Use synthetic substitution to find the value of $P(x) = 2x^4 - 33x^2 + 3x + 4$ for $x = 4$ and again for $x = -1$. Name the binomial factors found, if any.

12. Factor $P(x) = 2x^3 + 7x^2 + 2x - 3$ into first-degree factors. Find the zeros of $P(x)$.

13. Solve $2x^4 - 5x^3 - 21x^2 + 45x + 27 = 0$. Find the multiplicity m for each root whose $m > 1$.

14. Solve $|x^2 - 20| = 16$.

15. Solve $x^4 - 10a^2x^2 + 9a^4 = 0$ for x in terms of a.

16. Graph the solutions of $x^4 - 29x^2 + 100 \leq 0$. Write the solution set.

In each item, you are to compare a quantity in Column 1 with a quantity in Column 2. Write the letter of the correct answer from these choices:

A—The quantity in Column 1 is greater than the quantity in Column 2.
B—The quantity in Column 2 is greater than the quantity in Column 1.
C—The quantity in Column 1 is equal to the quantity in Column 2.
D—The relationship cannot be determined from the given information.

NOTE: Information centered over both columns refers to one or both of the quantities to be compared.

Sample Question	Answer
$x \neq 2$ Column 1 — Column 2 $\dfrac{x^2 - 4}{x - 2}$ — x	$\dfrac{x^2 - 4}{x - 2} = \dfrac{(x - 2)(x + 2)}{x - 2} = x + 2$, if $x \neq 2$, $x + 2 > x$ for each value of x. The answer is A.

Column 1	Column 2

1. $y > 0$

$\dfrac{y^2 - 25}{y + 5}$ — $y - 1$

2. $P(x) = 18x^6 + 12x^4 - 8x^2$
$P(7)$ — $P(-7)$

3. $(x - 6)^2$ — $(x + 6)^2$

4. $(x + 1)^2$ — $x(x + 2)$

5.

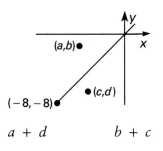

$a + d$ — $b + c$

Column 1	Column 2

6. Number of minutes in two weeks — Number of seconds in 7 hours

7. x is an integer.

$\dfrac{x}{-2}$ — $2x$

8.

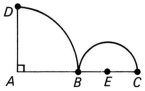

B is the midpoint of $\overline{AC}$.
$\overparen{BC}$ and $\overparen{BD}$ are arcs of circles with centers at E and A, respectively.

Length of $\overparen{BC}$ — Length of $\overparen{BD}$

9. x^* means x^{-2}.

$(3^*)^*$ — 3^*

Cumulative Review (Chapters 1–6)

Identify the property of operations with real numbers. (Exercises 1–3) *1.3*

1. $(a + 5)c = ac + 5c$

2. $-7t + 7t = 0$

3. $n = 1n$

4. Simplify *1.3*
$5(3x + 2) - (7x - 4y^3)$
and evaluate the result for
$x = 1.5$ and $y = -2$.

For Exercises 5–8, solve the *1.4*
equation. Then, select the
root from the choices
A, B, C, D, and E.

5. $5x + 7 - 4(2 - x) =$
$7x - (2x + 9)$
(A) $-\frac{5}{7}$ (B) -2 (C) 2
(D) -6 (E) 6

6. $x - 0.04 = 8.6 - 2.2x$
(A) 0.27 (B) 2.7 (C) 27
(D) 270 (E) 2,700

7. $\dfrac{3x + 8}{5x - 2} = \dfrac{2}{5}$
(A) $-8\frac{4}{5}$ (B) $-\frac{5}{44}$ (C) $\frac{25}{36}$
(D) $1\frac{11}{25}$ (E) $-1\frac{11}{25}$

8. $\frac{1}{2}x - 2 = \frac{3}{5}x + 4$
(A) -6 (B) -60 (C) $-\frac{6}{11}$
(D) 20 (E) -20

Solve each equation for x.
(Exercises 9–16)

9. $10x - (8 - 2x) =$ *1.4*
$12 + 4(x - 7)$

10. $\dfrac{5x - 2}{5} = \dfrac{2x + 6}{3}$

11. $\frac{2}{3}x - \frac{1}{2} = \frac{3}{4}x + 2$

12. $a(2x - b) = 6ab - 5$

13. $|2x - 7| = 3$ *2.3*

14. $5^{2x-1} = 125$ *5.1*

15. $m(x - a) = 3x + 4am$ *5.4*

16. $2x^2 - 7x = 15$ *6.1*

17. Supply the reason for each *1.8*
step in the proof of the theorem.

Theorem:
$(ab + c) + ad = a(b + d) + c$

Proof:
1. $(ab + c) + ad$
$= ab + (c + ad)$
2. $= ab + (ad + c)$
3. $= (ab + ad) + c$
4. $= a(b + d) + c$
5. $(ab + c) + ad$
$= a(b + d) + c$

Find the solution set.

18. $7 - 5x < 2(x - 7)$ *2.1*

19. $-4 \le 3x - 7 \le 8$ *2.2*

20. $|2x + 5| < 11$ *2.3*

21. $|x - 3| > 5$

22. $x^2 - 2x - 8 < 0$ *6.4*

For Exercises 23–25, $f(x) = 3x - 2$
and $g(x) = x^2 + 4$. Find the following.

23. $g(-4) - f(4)$ *3.2*

24. $g(c + 5) - g(c)$ *5.5*

25. $f(g(2))$ *5.9*

Find the slope and describe the *3.3*
slant of the line determined by
the points P and Q.

26. $P(5,2)$, $Q(-3,4)$

27. $P(-2,4)$, $Q(-5,4)$

Draw the graph on a coordinate plane.
(Exercises 28–31)

28. $2x - 3y = 6$　　　　*3.5*

29. $4y + 2 = 14$

30. $2x + 3 \leq 9$　　　　*3.9*

31. $3x + 4y > 12$

32. If y varies directly as x, and　　*3.8*
$y = 84$ when $x = 3.5$, what
is x when $y = 60$?

33. Given that y varies directly as
the square of x and $y = 12$
when $x = 3$, find y when
$x = 6$.

Solve each system for (x, y).
(Exercises 34–35)

34. $3x - 4y = 7$　　　　*4.2*
$y = 2x - 8$

35. $5x - 3y = 1$　　　　*4.3*
$3x + 4y = 18$

Simplify. Use positive exponents.

36. $(4n^{-2})^3$　　　　*5.2*

37. $\dfrac{7.2 \times 10^{-6}}{3.6 \times 10^{-8}}$

38. $5a(3a + 4)(a - 6)$　　　　*5.3*

39. $(5x - 3)^2$　　　　*5.5*

Factor completely.　　　　*5.7*

40. $9x^4 - 37x^2 + 4$

41. $2x^3 + 16$

42. $16x^3 - 100x$

For Exercises 43 and 44, $P(x) = $　　*6.6*
$6x^4 - 11x^3 - 22x^2 + x + 6.$

43. Factor $P(x)$ into first-degree
factors.

44. List the zeros of $P(x)$.

45. The second of three numbers　　*1.6*
is 5 times the first number,
and the third number is 4
less than the second number.
Twice the first number, decreased by the third number,
is the same as 20 less than
the second number. Find the
three numbers.

46. Maria is 5 years older than　　*2.5*
Andre. In 4 years, the sum of
their ages will be less than
45 years. Find the possible
ages for Maria.

47. A rectangle is 6 in. longer　　*6.3*
than twice its width. A sec-
ond rectangle is 3 times as
wide and one-half as long as
the first rectangle. The sum
of the areas of the two rectangles is 140 in^2. Find the
length of the first rectangle.

Write a system of two equations with
two variables to solve each problem.

48. A collection of 35 coins con-　　*4.4*
tains nickels and dimes. Find
the number of dimes given
that the total face value of
the coins is $2.55.

49. Teresa invested $600, part at　　*4.4*
7% and the rest at 9%. At
the end of 4 years, the total
simple interest earned was
$184. How much money
did she invest at 9%?

50. A plane flies 800 m in 4 h　　*4.5*
with a tail wind, and returns
the same distance at the same
airspeed in 5 h against the
wind. Find the wind speed.

51. The ratio of the measures of　　*4.5*
two supplementary angles is
4:5. Find the measure of
each angle.

Architects use CAD technology to create three-dimensional drawings of a building project. The image above shows the central space of a building in exact detail. The computer enables architects to view the interior from any perspective or change the style of any design element.

7.1 Rational Expressions and Functions

Objectives

To find the domains of rational functions
To find the zeros of rational functions
To simplify rational expressions
To find the values of rational functions

Each expression below is a quotient of two polynomials.

$$\frac{3x + 2}{12x + 7} \qquad \frac{8}{x^2 - x - 6} \qquad \frac{x^2 - 2}{9}$$

Such expressions are called *rational expressions*. Thus, the expression $\frac{x^2 + 4x + 3}{\sqrt{3x - 4}}$ is not a rational expression because $\sqrt{3x - 4}$ is not a polynomial.

Definition

A **rational expression** is a quotient of two polynomials.

Recall that the quotient $\frac{a}{b}$ of two numbers is undefined if the denominator b is 0. Similarly, a rational expression is undefined if its denominator is 0.

EXAMPLE 1 For what values of x is $\dfrac{2x - 3}{x^2 - 8x + 12}$ undefined?

Plan Find the values of x for which the denominator, $x^2 - 8x + 12$, is 0.

Solution Solve the quadratic equation $x^2 - 8x + 12 = 0$.
$$(x - 6)(x - 2) = 0$$
$$x - 6 = 0 \qquad or \qquad x - 2 = 0$$
$$x = 6 \qquad\qquad\qquad x = 2$$

Thus, $\dfrac{2x - 3}{x^2 - 8x + 12}$ is undefined for $x = 6$ and for $x = 2$.

A **rational function** is a function described by a rational expression. For example, the function described by the rule $f(x) = \dfrac{2x - 3}{x^2 - 8x + 12}$ is a rational function.

Recall that the domain of a function f is the set of all values of x for which the function f is defined. For example, the domain of f in Example 1, where $f(x) = \dfrac{2x - 3}{x^2 - 8x + 12}$, is the set of all real numbers except 2 and 6; that is, $\{x \mid x \neq 2 \ and \ x \neq 6\}$.

Unless otherwise stated, the domain of any rational function is assumed to be the set of all real numbers for which the value of any denominator is not zero.

Recall that $\dfrac{a}{b} = c$ means that $c \cdot b = a$ (see Lesson 1.1). This definition is the basis of the following theorem.

Theorem 7.1

For all real numbers a and b, where $b \neq 0$, if $\dfrac{a}{b} = 0$, then $a = 0$.

The **zero of a function** f is a number x such that $f(x) = 0$ (see Lesson 6.6). In the next example, Theorem 7.1 is used to find the zeros of a rational function.

EXAMPLE 2   Find the zeros of the function $f(x) = \dfrac{2x - 3}{x^2 - 8x + 12}$.

Plan   Find the values of x for which $2x - 3 = 0$.

Solution $2x - 3 = 0, \ 2x = 3, \ x = \dfrac{3}{2}$

Thus, $\dfrac{3}{2}$ is the only zero of f.

A useful property of fractions is the *Cancellation Property*.

Cancellation Property of Fractions
For all real numbers a, b, and c, where $b \neq 0$ and $c \neq 0$, $\dfrac{ac}{bc} = \dfrac{a}{b}$.

For example, $\dfrac{6}{10} = \dfrac{3 \cdot 2}{5 \cdot 2} = \dfrac{3}{5}$

This property can be used to simplify rational expressions. A rational expression is said to be in *simplest form* when it is written as a polynomial or the quotient of two polynomials whose greatest common factor is 1. A rational expression can be simplified by:
1. factoring the numerator and denominator completely, and
2. applying the Cancellation Property of Fractions.

EXAMPLE 3   Simplify $\dfrac{2x}{6x^4 - 18x^3}$.

Solution   Factor the numerator and denominator completely.

$$\frac{2x}{6x^4 - 18x^3} = \frac{2x}{6x^3(x - 3)}$$

$$= \frac{2x \cdot 1}{2x \cdot 3x^2(x - 3)}$$

Use the Cancellation Property of Fractions.

$$= \frac{1}{3x^2(x - 3)} \leftarrow \frac{ac}{bc} = \frac{a}{b}$$

A shorter method of simplification, *dividing out common factors*, is based on the Cancellation Property of Fractions. It is illustrated in Example 4 below. In order to divide out *common* factors, it is sometimes useful to change the numerator or denominator to a more convenient form.

EXAMPLE 4   Simplify $\dfrac{y^2 - y - 42}{14 + 5y - y^2}$.

Plan   First rewrite $14 + 5y - y^2$ in a more convenient form as follows.
(1) Write in descending order of exponents. $-1y^2 + 5y + 14$
(2) Factor out -1. $-1(y^2 - 5y - 14)$

Solution

$$\frac{y^2 - y - 42}{14 + 5y - y^2} = \frac{y^2 - y - 42}{-1(y^2 - 5y - 14)}$$

$$= \frac{\overset{1}{(y + 6)(\cancel{y - 7})}}{-1(y + 2)(\cancel{y - 7})} = \frac{y + 6}{-1(y + 2)}, \text{ or } -\frac{y + 6}{y + 2}$$

EXAMPLE 5   Simplify $f(t) = \dfrac{t^3 - t^2 - 4t + 4}{t^2 + t - 2}$. Then find $f(9.3)$.

Solution   Factor the numerator by grouping pairs of terms.

$$\frac{t^2(t - 1) - 4(t - 1)}{t^2 + t - 2} = \frac{(t - 1)(t^2 - 4)}{t^2 + t - 2} = \frac{\overset{1}{(\cancel{t - 1})}\overset{1}{(\cancel{t + 2})}(t - 2)}{\underset{1}{(\cancel{t - 1})}\underset{1}{(\cancel{t + 2})}} = t - 2$$

So, $f(t) = t - 2$ and $f(9.3) = 9.3 - 2$, or 7.3.

Rational expressions of the form $\dfrac{f(a) - f(b)}{a - b}$ or $\dfrac{f(x + h) - f(x)}{h}$ are encountered often in the study of calculus. When the function f is a polynomial function, such rational expressions can be simplified.

EXAMPLE 6   **a.** Given that $f(x) = 5x + 4$, simplify $\dfrac{f(a) - f(7)}{a - 7}$.

b. Given that $g(x) = x^2 - 2$, simplify $\dfrac{g(x + 3) - g(x)}{3}$.

Solutions   **a.** $\dfrac{f(a) - f(7)}{a - 7}$

$= \dfrac{(5a + 4) - (5 \cdot 7 + 4)}{a - 7}$

$= \dfrac{5a - 35}{a - 7}$

$= \dfrac{5(a - 7)}{a - 7} = 5$

b. $\dfrac{g(x + 3) - g(x)}{3}$

$= \dfrac{[(x + 3)^2 - 2] - (x^2 - 2)}{3}$

$= \dfrac{x^2 + 6x + 9 - 2 - x^2 + 2}{3}$

$= \dfrac{6x + 9}{3} = 2x + 3$

Classroom Exercises

Determine whether each expression is a rational expression. If not, explain.

1. $\dfrac{4x^2 - 7x + 12}{x - 8}$

2. $\dfrac{0}{7x - 5}$

3. $\dfrac{\sqrt{3x + 5}}{2x - 9}$

For what value(s) of x is the rational expression undefined?

4. $\dfrac{5x^2 - 25x}{2x}$

5. $\dfrac{3x - 15}{x^2 - 25}$

6. $\dfrac{3x - 6}{10 - 5x}$

7-9. Simplify each rational expression in Classroom Exercises 4–6.

Written Exercises

Find the domain of the function with the given rule.

1. $f(x) = \dfrac{3x - 15}{2x - 6}$

2. $f(x) = \dfrac{x^2 - 9}{3x + 15}$

3. $f(x) = \dfrac{x^2 - 9x}{x^2 - 7x}$

4. $f(x) = \dfrac{x^2 - 5x}{x^2 - 7x + 12}$

5-8. Find the zeros of each function in Written Exercises 1–4.

Simplify each rational expression.

9. $\dfrac{x^5(x + 4)}{x^7(x + 4)}$

10. $\dfrac{(2x + 5)(x - 4)}{(x - 4)(5 + 2x)}$

11. $\dfrac{x}{x^2 + 5x}$

12. $\dfrac{3x - 15}{6x - 30}$

13. $\dfrac{x^2 - x}{x^2 - 1}$

14. $\dfrac{4x}{8x^2 - 8x}$

15. $\dfrac{5x}{10x^2 - 20x}$

16. $\dfrac{2x^3 - 4x^2}{6x^3}$

17. $\dfrac{6x^2 - 12x}{8x^3}$

18. $\dfrac{a^2 + 4a - 12}{3a^2 - 12a + 12}$

19. $\dfrac{6m^2 - 4m}{9m^2 - 12m + 4}$

20. $\dfrac{27a^4 - 75a^2}{27a^3 - 18a^2 - 45a}$

21. $\dfrac{x^2 - 8x - 20}{12x - x^2 - 20}$

Simplify the expression that describes each function. Then find the indicated value of the function.

22. $f(x) = \dfrac{x^2 - 25}{2x^2 - 7x - 15}$

Find $f(-4)$.

23. $f(x) = \dfrac{-x^2 + 8x - 12}{3x^2 - 2x - 8}$

Find $f(-3)$.

In Exercises 24–26, $f(x) = 3x - 8$ and $g(x) = x^2 + 5$. Simplify the following.

24. $\dfrac{f(c) - f(10)}{c - 10}$

25. $\dfrac{g(2a) - g(3)}{2a - 3}$

26. $\dfrac{f(x + 12) - f(x)}{12}$

Simplify each rational expression.

27. $\dfrac{5a^4 - a^4m}{a^4m^2 - 7a^4m + 10a^4}$

28. $\dfrac{x^4 - 10x^2 + 9}{27 - 3x^2}$

29. $\dfrac{x^2a^2 - 9a^2 - 4x^2 + 36}{xa - 3a + 2x - 6}$

30. $\dfrac{-3x - x^2 + 4}{x^3 + 4x^2 - x - 4}$

31. $(3 + 2a - a^2)(a^4 - 10a^2 + 9)^{-1}$

32. $(ax + ay - xb - yb)(a^2 - b^2)^{-1}$

33. Prove Theorem 7.1.

Mixed Review

Solve each equation or system of equations.   *1.4, 2.3, 4.2*

1. $\dfrac{2x - 3}{2} = \dfrac{4x + 1}{3}$

2. $|2x + 4| = 8$

3. $\begin{array}{l} y = 2x - 6 \\ 3x - 2y = 2 \end{array}$

7.2 Products and Quotients of Rational Expressions

Objective

To find products and quotients of rational expressions

The product of two rational expressions is a rational expression of the form $\dfrac{\textit{product of the numerators}}{\textit{product of the denominators}}$.

Rule for Multiplying Rational Expressions
If $\dfrac{a}{b}$ and $\dfrac{c}{d}$ are rational expressions, where $b \neq 0$ and $d \neq 0$,

then $\dfrac{a}{b} \cdot \dfrac{c}{d} = \dfrac{a \cdot c}{b \cdot d}$.

The above rule is illustrated below.

1. Apply the Rule for Multiplying Rational Expressions.

$$\frac{6}{5x + 20} \cdot \frac{x^2 + 4x}{9} = \frac{6(x^2 + 4x)}{(5x + 20) \cdot 9}$$

2. Factor the numerator and denominator.

3. Divide out common factors.

$$= \frac{\overset{1}{\cancel{3}} \cdot 2 \cdot x \overset{1}{\cancel{(x + 4)}}}{5 \cancel{(x + 4)} \cdot \underset{1}{\cancel{3}} \cdot 3}$$

4. Simplify.

$$= \frac{2x}{15}$$

It is usually convenient to divide out common factors *before* using the Rule for Multiplying Rational Expressions. This is shown in Example 1.

EXAMPLE 1

Simplify $\dfrac{6a^3b}{15xy^4} \cdot \dfrac{10x^2y^3}{9ab^4}$.

Solution

$$\frac{6a^3b}{15xy^4} \cdot \frac{10x^2y^3}{9ab^4} = \frac{\overset{1}{\cancel{3}} \cdot 2 \cdot \overset{a^2}{\cancel{a^3}} \cdot \overset{1}{\cancel{b}}}{\underset{1}{\cancel{5}} \cdot \underset{1}{\cancel{3}} \cdot \underset{1}{\cancel{x}} \cdot \underset{y}{\cancel{y^4}}} \cdot \frac{\overset{x}{\cancel{5}} \cdot 2 \cdot \overset{x}{\cancel{x^2}} \cdot \overset{1}{\cancel{y^3}}}{3 \cdot 3 \cdot \underset{1}{\cancel{a}} \cdot \underset{b^3}{\cancel{b^4}}} = \frac{4a^2x}{9yb^3}$$

EXAMPLE 2   Simplify $\dfrac{2x^2 - 12x - 14}{x^3 - 16x} \cdot \dfrac{-16 - 4x}{6x - 42}$.

Plan   Write $-16 - 4x$ in a more convenient form: $-16 - 4x = -4x - 16$ $= -4(x + 4)$. Then factor the other numerators and denominators completely, divide out the common factors, and multiply.

Solution $\dfrac{2x^2 - 12x - 14}{x^3 - 16x} \cdot \dfrac{-16 - 4x}{6x - 42} = \dfrac{2(x^2 - 6x - 7)}{x(x^2 - 16)} \cdot \dfrac{-4(x + 4)}{6(x - 7)}$

$$= \dfrac{\overset{1}{\cancel{2}}(\cancel{x - 7})\overset{1}{(x + 1)}}{x(\cancel{x + 4})(x - 4)} \cdot \dfrac{-2 \cdot \cancel{2}(\cancel{x + 4})}{\cancel{2} \cdot 3(\cancel{x - 7})}$$

$$= \dfrac{-4(x + 1)}{3x(x - 4)}$$

To divide a rational number by a nonzero rational number, multiply the first number by the *reciprocal*, or *multiplicative inverse*, of the second number. Rational expressions are divided in a similar manner.

Rule for Dividing Rational Expressions
If $\frac{a}{b}$ and $\frac{c}{d}$ are rational expressions where $b \neq 0$, $c \neq 0$, and $d \neq 0$, then $\frac{a}{b} \div \frac{c}{d} = \frac{a}{b} \cdot \frac{d}{c}$.

EXAMPLE 3   Simplify $\dfrac{3x^2 - 2xy}{8x^3 y} \div \dfrac{9x^2 - 4y^2}{12x^2 y^5}$.

Solution $\dfrac{3x^2 - 2xy}{8x^3 y} \div \dfrac{9x^2 - 4y^2}{12x^2 y^5} = \dfrac{3x^2 - 2xy}{8x^3 y} \cdot \dfrac{12x^2 y^5}{9x^2 - 4y^2}$

$$= \dfrac{\overset{1}{x}(\overset{1}{\cancel{3x - 2y}})}{\underset{2}{\cancel{8}} \cdot \underset{1}{\cancel{x^3}} \cdot \underset{1}{\cancel{y}}} \cdot \dfrac{\overset{3}{\cancel{12}} \cdot \overset{1}{\cancel{x^2}} \cdot \overset{y^4}{\cancel{y^5}}}{(3x + 2y)(\cancel{3x - 2y})}$$

$$= \dfrac{3y^4}{2(3x + 2y)}$$

EXAMPLE 4   Simplify $\dfrac{-2x-5}{x^2-1}\cdot\dfrac{x^2-1}{2x^2-x-15}\div\dfrac{x+1}{x-3}$.

Solution $\dfrac{-2x-5}{x^2-1}\cdot\dfrac{x^2-1}{2x^2-x-15}\cdot\dfrac{x-3}{x+1}$

$=\dfrac{-1(2x+5)}{(x-1)(x+1)}\cdot\dfrac{(x-1)(x+1)}{(x-3)(2x+5)}\cdot\dfrac{(x-3)}{(x+1)}=\dfrac{-1}{x+1}$, or $-\dfrac{1}{x+1}$

Classroom Exercises

Express each quotient as the product of rational expressions.

1. $\dfrac{14}{2b}\div\dfrac{7x}{2y}$ **2.** $\dfrac{5}{x-y}\div\dfrac{x+y}{4}$ **3.** $\dfrac{x-2}{x+3}\div\dfrac{2x-4}{x-3}$ **4.** $\dfrac{5}{6x-3}\div\dfrac{15x}{1-2x}$

5–8. Simplify the quotients in Classroom Exercises 1–4, if possible.

Written Exercises

Simplify.

1. $\dfrac{x^3}{5}\cdot\dfrac{10}{x^4}$

2. $\dfrac{4}{x^5}\cdot\dfrac{x^2}{12}$

3. $\dfrac{x-3}{14x^2}\cdot\dfrac{7x^3}{x-3}$

4. $\dfrac{x^3}{x^5(2x+5)}\cdot(2x+5)$

5. $\dfrac{2a}{3b}\cdot\dfrac{b^2}{4a^3}$

6. $\dfrac{10x^3}{9y^5}\cdot\dfrac{6y^4}{15x^4}$

7. $\dfrac{3}{4x+8}\cdot\dfrac{x^2+2x}{9}$

8. $\dfrac{x^2+6x}{10}\cdot\dfrac{4}{x^2-36}$

9. $\dfrac{7}{a^2-9a+20}\cdot\dfrac{a^2-4a}{14}$

10. $\dfrac{x^2y^4}{6a^2}\div\dfrac{x^4y}{10a^2}$

11. $\dfrac{8x+40}{6x^3}\div\dfrac{4x+20}{8x^5}$

12. $\dfrac{x^2-9}{6}\div\dfrac{3-x}{8}$

13. $\dfrac{x^2-y^2}{5x^3y^2}\cdot\dfrac{15x^2y^5}{4x+4y}$

14. $\dfrac{4a+8}{5a-20}\div\dfrac{10+3a-a^2}{a^2-4a}$

15. $\dfrac{3x^2+10x-8}{3x^2-17x+10}\cdot\dfrac{5+9x-2x^2}{x^2+3x-4}$

16. $\dfrac{2y^2-18}{24-6y}\div\dfrac{3y^2+24y+45}{2y^2-9y+4}$

17. $\dfrac{4ab^3}{3a^2-a-10}\div\dfrac{6a^5b^7}{3a^2+17a+20}$

18. $\dfrac{3x^2+14x-5}{x^2+2x-15}\div\dfrac{3x^2-25x+8}{8+15x-2x^2}$

19. $\dfrac{6y^2-13y-5}{-3x-x^2}\cdot\dfrac{2xy+6y+5x+15}{4y^2-25}$

20. $\dfrac{x^2+2x-3}{x^2-2x+1}\div\dfrac{x^2+5x+6}{2-x-x^2}$

21. $\dfrac{x^3 - 27}{8x^3} \cdot \dfrac{20x - 4x^2}{x^2 - 8x + 15}$

22. $\dfrac{a^2 + 3a - 18}{3 + 2a - a^2} \cdot \dfrac{a^3 + 1}{2a^2 + 7a - 30}$

23. $\dfrac{x^2 + 4x - 32}{x^2 - 12x + 35} \cdot \dfrac{x^2 - 4x - 21}{16x - 4x^2} \cdot \dfrac{x^2 - 10x}{x^2 + 11x + 24}$

24. $\dfrac{2x^2y - x^3}{a^3 - 27b^3} \cdot \dfrac{x^3 + 8y^3}{x^2 - 4y^2} \div \dfrac{x^5 - 2x^4y + 4x^3y^2}{ax + 6by - 2ay - 3bx}$

25. $\dfrac{x^3y^3 - 8y^3 - x^3 + 8}{x^2 - 10x + 25 - 49y^2} \cdot \left(\dfrac{xy + 2 - 2y - x}{x^2 - 5x - 7yx} \right)^{-1}$

Mixed Review

Find the solution set. *2.1, 2.2, 2.3, 6.4*

1. $6 - 2x \geq 4(6 - x)$

2. $-6 \leq 2x - 4 \leq 8$

3. $|x - 5| > 4$

4. $x^2 + 4x - 12 < 0$

Application: *Photography*

When light passes through a camera lens, the curvature of the lens causes the light rays to bend so that they all pass through a point F, called the **focal point** of the lens. The distance from the center of the lens to the vertical plane through F is f, the *focal length* of the lens.

A photographic image is in focus when

$$\frac{1}{p} + \frac{1}{q} = \frac{1}{f},$$

where p is the distance of the object from the lens, q is the image distance, and f is the focal length.

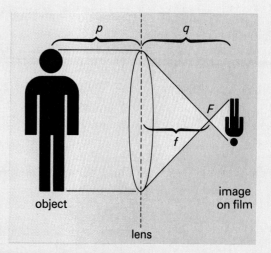

For a lens of focal length 50 mm, an image distance of 52 mm, and an image size of 12 mm, find the following.

1. the distance of the object from the lens in meters

2. the size of the object given that
$$\frac{\text{image size}}{\text{object size}} = \frac{\text{image distance}}{\text{object distance}}$$

7.3 Dimensional Analysis

Objective

To convert from one unit of measurement to another using dimensional analysis

The equation 12 in = 1 ft relates feet to inches. This relation can be written as a fraction that is equal to 1 in two ways.

Divide each side by 1 ft.

$$\frac{12 \text{ in}}{1 \text{ ft}} = 1$$

Divide each side by 12 in.

$$1 = \frac{1 \text{ ft}}{12 \text{ in}}$$

To convert 18 in to feet, multiply 18 in by $\frac{1 \text{ ft}}{12 \text{ in}}$

so that the common factor *in.* will divide out and leave the new unit *ft.*

$$18 \text{ in} = \frac{18 \ \cancel{\text{in}}}{1} \cdot \frac{1 \text{ ft}}{12 \ \cancel{\text{in}}} = \frac{18}{12} \text{ ft} = 1\frac{1}{2} \text{ ft}$$

To convert 18 ft to inches, multiply 18 ft by $\frac{12 \text{ in}}{1 \text{ ft}}$.

This conversion procedure is called **dimensional analysis.**

Fractions such as $\frac{12 \text{ in}}{1 \text{ ft}}$ that are used to make the conversions are called *conversion factors* or **conversion fractions.**

EXAMPLE 1

Convert 60 miles per hour (mi/h) to feet per second (ft/s).

Plan

Write 60 mi/h as a fraction: $\frac{60 \text{ mi}}{1 \text{ h}}$. Then note that 5,280 ft = 1 mi, 60 min = 1 h, and 60 s = 1 min. Use these relationships to write conversion fractions.

Solution

$$\frac{60 \ \cancel{\text{mi}}}{1 \ \cancel{\text{h}}} \cdot \frac{5,280 \text{ ft}}{1 \ \cancel{\text{mi}}} \cdot \frac{1 \ \cancel{\text{h}}}{60 \ \cancel{\text{min}}} \cdot \frac{1 \ \cancel{\text{min}}}{60 \text{ s}} = \frac{60 \cdot 5,280 \text{ ft}}{60 \cdot 60 \text{ s}}$$

$$= \frac{88 \text{ ft}}{\text{s}}, \text{ or } 88 \text{ ft/s}$$

Thus, 60 mi/h is equivalent to 88 ft/s.

Conversion fractions can be used with area data to convert measurements with square units. Recall that 3 ft = 1 yd and 12 in = 1 ft.

Square each side. $9 \text{ ft}^2 = 1 \text{ yd}^2$ and $144 \text{ in}^2 = 1 \text{ ft}^2$

This leads to the following conversion fractions.

$$\frac{9 \text{ ft}^2}{1 \text{ yd}^2} = 1 \qquad \frac{1 \text{ yd}^2}{9 \text{ ft}^2} = 1 \qquad \frac{144 \text{ in}^2}{1 \text{ ft}^2} = 1 \qquad \frac{1 \text{ ft}^2}{144 \text{ in}^2} = 1$$

EXAMPLE 2   A rug measures 12 ft by 10 ft and costs $27.90 per square yard. Find the cost of the rug.

Solution   The area of the rug is 12 ft · 10 ft, or 120 ft^2. Use a conversion fraction.

$$\frac{120 \text{ ft}^2}{1} \cdot \frac{1 \text{ yd}^2}{9 \text{ ft}^2} \cdot \frac{\$27.90}{1 \text{ yd}^2} = \frac{120 \cdot \$27.90}{9} = \$372$$

Thus, the cost is $372.

Classroom Exercises

Determine the conversion fraction that should be used for each conversion indicated.

1. 18 yd to feet

2. 26 pt to quarts

3. 90 in. to yards

4. 4.5 lb to ounces

5. 300 g to kilograms

6. 18 qt to gallons

Written Exercises

1–6. Convert each measure in Classroom Exercises 1–6 as indicated.

Use dimensional analysis to convert each measure.

7. 12 oz/pt to pounds per gallon

8. 972 lb/ft^2 to ounces per square inch

9. A test car is driven for 5 h at an average speed of 40 mi/h. The car averages 24 mi/gallon. Find the cost of the gas used given that gas costs $1.35/gallon.

10. A construction crew uses 125 ft^3 of sand each work week. Given that sand weighs 1.215 tons per cubic yard (T/yd^3), find the number of pounds of sand the crew uses in 3 work weeks.

11. A city has 320 youths involved in its softball program. Twenty players are assigned to each team. Each team is provided with 30 bats for the season. The cost is $42.60 per carton, with a dozen bats in each carton. Find the total cost of the bats given that the city receives a 15% discount.

Mixed Review

Find the slope and describe the slant of the line.   *3.5*

1. $3x - 2y = 10$ **2.** $5x - 10 = 0$ **3.** $2y - 5 = 3$ **4.** $y = 2 - x$

Problem Solving Strategies

Using a Table to Find a Pattern

An important problem solving strategy is to use a table to organize data and find a pattern. The data in the table may help you to discover a pattern so that you can solve the problem.

The Italian astronomer Galileo (1564–1642) discovered the relationship between the length of a pendulum and the time required for it to make a complete swing. Thus, given the swing time, he could predict the length of the pendulum. A student, trying an experiment similar to Galileo's, constructs the following table based on her results.

Time of Swing	Length of Pendulum
1 s	0.25 m
2 s	1 m
3 s	2.25 m
5 s	6.25 m

Using this data, the student discovers a pattern and then writes an equation relating the time of the swing to the length of the pendulum.

Exercises

1. By noting a pattern in the data, predict the length of a pendulum with a swing time of 4 seconds. (HINT: Express the length of the pendulum in fourths of a meter, as $\frac{1}{4}$, $\frac{4}{4}$, and so on.)
2. Write a formula for the length given the time for one swing.

Tables can also be used to indicate possible solutions to problems involving geometric patterns. For example, the three circles below show the number of regions formed by drawing all possible segments between a given number of points on a circle.

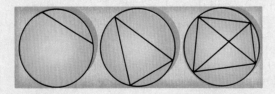

3. Use the pattern to guess the missing numbers in the table.

Number of points connected	2	3	4	5	6
Number of regions	2	4	8	__	__

4. Draw all possible segments between 5 points on a circle and then 6 points on a circle and count the number of regions formed. Are the predictions made from the table accurate in these cases?

7.4 Sums and Differences of Rational Expressions

Objective

To find sums and differences of rational expressions

$$x(y + z) = xy + xz \qquad x(y - z) = xy - xz$$

The Distributive Properties above can be used to simplify the sum $\frac{3}{7} + \frac{2}{7}$ and the difference $\frac{3}{7} - \frac{2}{7}$.

$$\frac{3}{7} + \frac{2}{7} = \frac{1}{7} \cdot 3 + \frac{1}{7} \cdot 2 \qquad\qquad \frac{3}{7} - \frac{2}{7} = \frac{1}{7} \cdot 3 - \frac{1}{7} \cdot 2$$

$$= \frac{1}{7}(3 + 2) = \frac{1}{7} \cdot 5, \text{ or } \frac{5}{7} \qquad = \frac{1}{7}(3 - 2) = \frac{1}{7} \cdot 1, \text{ or } \frac{1}{7}$$

The above results can be generalized into the following theorem.

Theorem 7.2

For all real numbers x, y, and z, where $z \neq 0$,

$$\frac{x}{z} + \frac{y}{z} = \frac{x + y}{z} \text{ and } \frac{x}{z} - \frac{y}{z} = \frac{x - y}{z}.$$

EXAMPLE 1

Simplify $\dfrac{3a}{a^2 - 16} - \dfrac{a + 8}{a^2 - 16}$.

Plan

Use $\dfrac{x}{z} - \dfrac{y}{z} = \dfrac{x - y}{z}$ and simplify the resulting rational expression.

Solution

$$\frac{3a}{a^2 - 16} - \frac{a + 8}{a^2 - 16} = \frac{3a - (a + 8)}{a^2 - 16} \quad \leftarrow \text{Be sure to include the}$$
$$\qquad\qquad\qquad\qquad\qquad\qquad \text{parentheses in the numerator.}$$

$$= \frac{3a - a - 8}{a^2 - 16}$$

$$= \frac{2a - 8}{a^2 - 16} = \frac{2(a - 4)}{(a + 4)(a - 4)}, \text{ or } \frac{2}{a + 4}$$

In the sum $\dfrac{3y}{10} + \dfrac{7y - 2}{5}$, the denominators are not the same. The denominators would be the same, however, if you multiplied the numerator and denominator of $\dfrac{7y - 2}{5}$ by 2.

$$\frac{3y}{5 \cdot 2} + \frac{(7y - 2)2}{5 \cdot 2} = \frac{3y}{10} + \frac{14y - 4}{10} = \frac{17y - 4}{10}$$

For the sum $\dfrac{3y}{10} + \dfrac{7y - 2}{5}$, 10 is the *Least Common Denominator* (LCD) of the fractions, or the *Least Common Multiple* (LCM) of the denominators.

To find the LCM of several algebraic expressions such as

$$x^2 - 6x + 9, \; 2x^2 - 6x, \; \text{and} \; 4x + 12:$$

1. factor the expressions completely, and

2. form the LCM by writing the product of the factors, with each factor counted only the greatest number of times it occurs in any one expression.

EXAMPLE 2   Find the LCM of $x^2 - 6x + 9$, $2x^2 - 6x$, and $4x + 12$.

Solution   Factor each expression completely.

$$
\begin{array}{lll}
x^2 - 6x + 9 & 2x^2 - 6x & 4x + 12 \\
(x - 3)(x - 3) & 2 \cdot x(x - 3) & 2 \cdot 2(x + 3)
\end{array}
$$

There are at most *two* $(x - 3)$s, *one* x, *two* 2s, and *one* $(x + 3)$ in any expression.

The LCM is $(x - 3)(x - 3) \cdot x \cdot 2 \cdot 2(x + 3)$, or $4x(x - 3)^2(x + 3)$.

EXAMPLE 3   Simplify $\dfrac{5}{4x - 16} + \dfrac{7}{2x - 8} + \dfrac{1}{3x - 12}$.

Solution   Factor each denominator completely.

$$\frac{5}{2 \cdot 2(x - 4)} + \frac{7}{2(x - 4)} + \frac{1}{3(x - 4)}$$

Each denominator contains at most two 2s, one 3, and one $(x - 4)$.

$$\frac{5}{2 \cdot 2(x - 4)} + \frac{7}{2(x - 4)} + \frac{1}{3(x - 4)}$$

    needs 3   needs 2 and 3   needs two 2s

Multiply each numerator and denominator by the factors needed.
The LCD is $2 \cdot 2(x - 4) \cdot 3$.

$$\frac{5 \cdot 3}{2 \cdot 2(x - 4) \cdot 3} + \frac{7 \cdot 2 \cdot 3}{2(x - 4) \cdot 2 \cdot 3} + \frac{1 \cdot 2 \cdot 2}{3(x - 4) \cdot 2 \cdot 2} =$$

$$\frac{15}{12(x - 4)} + \frac{42}{12(x - 4)} + \frac{4}{12(x - 4)} = \frac{61}{12(x - 4)}$$

EXAMPLE 4 Simplify $\dfrac{3y - 5}{2y - 6} - \dfrac{4y - 2}{5y - 15}$.

Solution

$$\dfrac{3y - 5}{2y - 6} - \dfrac{4y - 2}{5y - 15} = \dfrac{3y - 5}{2(y - 3)} - \dfrac{4y - 2}{5(y - 3)}$$

Express each fraction in terms of the LCD, $2 \cdot 5(y - 3)$.

$$= \dfrac{(3y - 5) \cdot 5}{2(y - 3) \cdot 5} - \dfrac{(4y - 2) \cdot 2}{5(y - 3) \cdot 2}$$

$$= \dfrac{15y - 25}{2 \cdot 5(y - 3)} - \dfrac{8y - 4}{2 \cdot 5(y - 3)}$$

Use $\dfrac{a}{b} - \dfrac{c}{b} = \dfrac{a - c}{b}$.

$$= \dfrac{15y - 25 - (8y - 4)}{2 \cdot 5(y - 3)}$$

$$= \dfrac{15y - 25 - 8y + 4}{2 \cdot 5(y - 3)}$$

Factor $7y - 21$. Divide out common factors.

$$= \dfrac{7y - 21}{2 \cdot 5(y - 3)} = \dfrac{7(y - 3)}{2 \cdot 5(y - 3)} = \dfrac{7}{10}$$

EXAMPLE 5 Simplify $\dfrac{3a - 39}{a^2 - 7a + 10} + \dfrac{3}{2 - a}$.

Solution

$$\dfrac{3a - 39}{a^2 - 7a + 10} + \dfrac{3}{2 - a} = \dfrac{3a - 39}{(a - 2)(a - 5)} + \dfrac{3}{2 - a}$$

$a - 2$ is the opposite of $2 - a$.
Multiply the numerator and denominator of the second fraction by -1.

$$= \dfrac{3a - 39}{(a - 2)(a - 5)} + \dfrac{-3}{a - 2}$$

$$= \dfrac{3a - 39}{(a - 2)(a - 5)} + \dfrac{-3(a - 5)}{(a - 2)(a - 5)}$$

$$= \dfrac{3a - 39 - 3a + 15}{(a - 2)(a - 5)}$$

$$= \dfrac{-24}{(a - 2)(a - 5)}$$

Classroom Exercises

Find the LCM of each pair of numbers or expressions.

1. 3, 13   **2.** 4, 12   **3.** $6x^2$, $x(x - 1)$ **4.** $x^2 - 1$, $x + 1$

Simplify.

5. $\dfrac{2b}{7} + \dfrac{3b}{14}$ **6.** $\dfrac{7x}{10} + \dfrac{3x}{2} - \dfrac{x}{5}$ **7.** $\dfrac{7}{x - 3} + \dfrac{4}{(x - 3)(x + 1)}$

Written Exercises

Simplify.

1. $\dfrac{3a}{25} + \dfrac{2a}{25}$

2. $\dfrac{7}{2a} + \dfrac{5}{2a}$

3. $\dfrac{11a}{15} - \dfrac{2a}{15}$

4. $\dfrac{2x + 5}{4x + 8} + \dfrac{1 + x}{4x + 8}$

5. $\dfrac{4b - 1}{b^2 - 9} - \dfrac{14 - b}{b^2 - 9}$

6. $\dfrac{7x + 15}{x^2 + 7x + 10} - \dfrac{2x + 5}{x^2 + 7x + 10}$

7. $\dfrac{7x}{10} + \dfrac{3x}{20} - \dfrac{x}{5}$

8. $\dfrac{2a}{7} + \dfrac{3a}{14} + \dfrac{7a}{2}$

9. $\dfrac{2a}{3a + 12} + \dfrac{5a}{6a + 24} + \dfrac{9}{5a + 20}$

10. $\dfrac{7}{9a - 27} - \dfrac{4}{15a - 45}$

11. $\dfrac{2}{9x + 45} + \dfrac{7}{6x + 30}$

12. $\dfrac{5}{6a + 15} - \dfrac{5}{8a + 20}$

13. $\dfrac{5}{m - 7} - \dfrac{55}{m^2 - 3m - 28}$

14. $\dfrac{3x - 10}{x^2 - 8x + 12} - \dfrac{2}{x - 6}$

15. $\dfrac{-2x - 3}{x^2 - 3x} + \dfrac{x}{x - 3}$

16. $\dfrac{-9b - 27}{b^2 - 3b - 18} - \dfrac{b}{6 - b}$

17. $\dfrac{-9y - 3}{y^2 - 11y + 18} + \dfrac{y + 3}{y - 9}$

18. $\dfrac{-24}{a^2 - 7a + 10} + \dfrac{a + 3}{a - 5}$

19. $\dfrac{2a - 3}{3a^2 - 13a - 10} + \dfrac{2a + 1}{5 - a} + \dfrac{1}{3a + 2}$

20. $\dfrac{5}{xy + 3y - 2x - 6} + \dfrac{4}{x + 3} - \dfrac{2}{2 - y}$

21. Prove the first part of Theorem 7.2.

22. Prove the second part of Theorem 7.2.

Midchapter Review

Simplify.   *7.1, 7.2, 7.4*

1. $\dfrac{4x + 2}{10x + 5}$

2. $\dfrac{a + 4}{9y^3} \cdot \dfrac{3y^4}{a + 4}$

3. $\dfrac{6x + 2}{x^2 - 2x - 8} + \dfrac{2}{8 - 2x}$

Convert each measure as indicated.   *7.3*

4. 60 mg to grams

5. 10 yd^2 to square ft

6. 20 gal to quarts

7.5 Complex Rational Expressions

Objective

To simplify complex rational expressions

When a quotient has fractions in the numerator or the denominator (see the fractions at the right), it is called a *complex rational expression.*

$$\frac{\frac{1}{10} + \frac{3x}{5}}{\frac{x}{2} - \frac{1}{5}} \qquad \frac{a^{-1} - b^{-1}}{a^{-2} - b^{-2}}$$

Definition

A **complex rational expression** is a rational expression in which the numerator, the denominator, or both contain at least one rational expression.

There are two methods for simplifying complex rational expressions.

Method 1: Express the complex rational expression as a quotient using the division symbol (÷). Then simplify the result.

Method 2: Use the property $\dfrac{a}{b} = \dfrac{a \cdot c}{b \cdot c}$, where c is the LCM of all the denominators of the individual expressions. Multiply the numerator and the denominator of the complex rational expression by the LCM.

EXAMPLE 1   Simplify $\dfrac{\frac{1}{10} + \frac{3x}{5}}{\frac{x}{2} - \frac{1}{5}}$.

Solution

The LCM of 10, 5, and 2 is 10, or $5 \cdot 2$.

Method 1

$$\left(\frac{1}{10} + \frac{3x}{5}\right) \div \left(\frac{x}{2} - \frac{1}{5}\right)$$

$$= \left(\frac{1}{10} + \frac{3x \cdot 2}{5 \cdot 2}\right) \div \left(\frac{x \cdot 5}{2 \cdot 5} - \frac{1 \cdot 2}{5 \cdot 2}\right)$$

$$= \frac{1 + 6x}{10} \div \frac{5x - 2}{10}$$

$$= \frac{1 + 6x}{10} \cdot \frac{10}{5x - 2}$$

$$= \frac{1 + 6x}{5x - 2}$$

Method 2

$$\frac{5 \cdot 2\left(\frac{1}{10} + \frac{3x}{5}\right)}{5 \cdot 2\left(\frac{x}{2} - \frac{1}{5}\right)}$$

$$= \frac{5 \cdot 2 \cdot \frac{1}{10} + 5 \cdot 2 \cdot \frac{3x}{5}}{5 \cdot 2 \cdot \frac{x}{2} - 5 \cdot 2 \cdot \frac{1}{5}}$$

$$= \frac{1 + 2 \cdot 3x}{5 \cdot x - 2 \cdot 1}, \text{ or } \frac{1 + 6x}{5x - 2}$$

EXAMPLE 2   Simplify $\dfrac{\dfrac{3x}{x^2 - 3x - 10} - \dfrac{2}{x + 2}}{\dfrac{4}{3x - 15} + \dfrac{2x}{x^2 - 3x - 10}}$.

Plan   Factor each denominator. Then multiply the numerator and denominator of the complex expression by the LCM of all the denominators.

Solution

$\dfrac{3(x - 5)(x + 2)\left[\dfrac{3x}{(x - 5)(x + 2)} - \dfrac{2}{x + 2}\right]}{3(x - 5)(x + 2)\left[\dfrac{4}{3(x - 5)} + \dfrac{2x}{(x - 5)(x + 2)}\right]}$ ←The LCM is $3(x - 5)(x + 2)$.

$= \dfrac{3(x - 5)(x + 2) \cdot \dfrac{3x}{(x - 5)(x + 2)} + 3(x - 5)(x + 2) \cdot \dfrac{-2}{x + 2}}{3(x - 5)(x + 2) \cdot \dfrac{4}{3(x - 5)} + 3(x - 5)(x + 2) \cdot \dfrac{2x}{(x - 5)(x + 2)}}$

$= \dfrac{3 \cdot 3x + 3(x - 5) \cdot (-2)}{(x + 2)4 + 3 \cdot 2x} = \dfrac{9x + (-6x + 30)}{4x + 8 + 6x}$, or $\dfrac{3x + 30}{10x + 8}$

EXAMPLE 3   Simplify $\dfrac{1 - 7x^{-1} - 18x^{-2}}{1 - 4x^{-2}}$.

Plan   Rewrite the rational expression so that it contains only positive exponents.

Solution

$\dfrac{1 - \dfrac{7}{x} - \dfrac{18}{x^2}}{1 - \dfrac{4}{x^2}} = \dfrac{x^2\left(1 - \dfrac{7}{x} - \dfrac{18}{x^2}\right)}{x^2\left(1 - \dfrac{4}{x^2}\right)} = \dfrac{x^2 \cdot 1 - x^2 \cdot \dfrac{7}{x} - x^2 \cdot \dfrac{18}{x^2}}{x^2 \cdot 1 - x^2 \cdot \dfrac{4}{x^2}}$

$= \dfrac{x^2 - 7x - 18}{x^2 - 4} = \dfrac{(x - 9)(x + 2)}{(x - 2)(x + 2)}$, or $\dfrac{x - 9}{x - 2}$

Classroom Exercises

Find the LCD for each complex rational expression.

1. $\dfrac{\dfrac{x}{3} + \dfrac{1}{2}}{\dfrac{5}{6} - \dfrac{x}{2}}$ **2.** $\dfrac{\dfrac{1}{y} + \dfrac{1}{3}}{\dfrac{1}{y} - \dfrac{1}{3}}$ **3.** $\dfrac{\dfrac{5}{a} - 2}{\dfrac{3}{a} + 1}$ **4.** $\dfrac{2a + \dfrac{1}{2}}{\dfrac{a}{3} + \dfrac{2}{5}}$

5–8. Simplify the rational expressions in Classroom Exercises 1–4 above.

Written Exercises

Simplify.

1. $\dfrac{\dfrac{7}{x^2-4}+\dfrac{2}{x-2}}{\dfrac{6}{x-2}+\dfrac{5}{x+2}}$

2. $\dfrac{\dfrac{3}{2a-10}+\dfrac{6}{a+2}}{\dfrac{9}{a^2-3a-10}+\dfrac{1}{a-5}}$

3. $\dfrac{\dfrac{4}{x^2-6x-16}-\dfrac{3}{5x-40}}{\dfrac{5}{2x-16}+\dfrac{3}{x+2}}$

4. $\dfrac{1-5m^{-1}+4m^{-2}}{1-16m^{-2}}$

5. $\dfrac{1-25a^{-2}}{1-3a^{-1}-10a^{-2}}$

6. $\dfrac{1-2x^{-1}+x^{-2}}{1+2x^{-1}-3x^{-2}}$

7. $\dfrac{\dfrac{2}{x}-\dfrac{10}{x^2+7x}}{\dfrac{5}{x+7}+\dfrac{2}{3x}}$

8. $\dfrac{\dfrac{2}{x-3}-\dfrac{1}{x-5}}{\dfrac{ax-7a-x+7}{x^2-8x+15}}$

9. $\dfrac{\dfrac{3}{a}-\dfrac{9}{a^2+3a}}{\dfrac{4}{a+3}-\dfrac{1}{a}}$

10. $\dfrac{\dfrac{10a}{a^2+6a+8}}{\dfrac{7}{a+4}+\dfrac{3}{a+2}}$

11. $\dfrac{\dfrac{1}{2}+\dfrac{1+\dfrac{1}{x}}{2}}{\dfrac{1}{x}+\dfrac{1}{2}}$

12. $1+\dfrac{3+\dfrac{1}{x}}{1+\dfrac{1}{x}}$

Simplify $\dfrac{f(3+h)-f(3)}{h}$ for each function in Exercises 13–16.

13. $f(x)=\dfrac{1}{x}$

14. $f(x)=\dfrac{1}{x-1}$

15. $f(x)=\dfrac{1}{x^2}$

16. $f(x)=\dfrac{1}{2x+1}$

Simplify.

17. $\dfrac{(x+y)y^{-1}-2x(x+y)^{-1}}{(x-y)y^{-1}+2x(x-y)^{-1}}$

Mixed Review

Graph each sentence. *3.5, 3.9*

1. $3x-2y=4$ **2.** $4x-4=8$ **3.** $2x-3y\ge 9$

4. Write an equation of the line containing the points $A(3,7)$ and $B(-2,-3)$. *3.4*

7.6 Equations Containing Rational Expressions

Objective	To solve equations containing rational expressions

The easiest way to solve equations that contain rational expressions is to multiply each side of the equation by the LCD of all the fractions.

EXAMPLE 1 Solve $\dfrac{2x - 3}{4} + 2 = \dfrac{2x + 1}{3}$.

Plan Factor each denominator to find the LCD.

Solution $\dfrac{2x - 3}{2 \cdot 2} + \dfrac{2}{1} = \dfrac{2x + 1}{3}$ ← The LCD is $2 \cdot 2 \cdot 3$.

Multiply each side by the LCD.

$$2 \cdot 2 \cdot 3 \cdot \left(\dfrac{2x - 3}{2 \cdot 2} + \dfrac{2}{1}\right) = 2 \cdot 2 \cdot 3 \cdot \left(\dfrac{2x + 1}{3}\right)$$

$$2 \cdot 2 \cdot 3 \cdot \left(\dfrac{2x - 3}{2 \cdot 2}\right) + 2 \cdot 2 \cdot 3 \cdot \dfrac{2}{1} = 2 \cdot 2 \cdot 3 \cdot \left(\dfrac{2x + 1}{3}\right)$$

$$3(2x - 3) + 2 \cdot 2 \cdot 3 \cdot 2 = 2 \cdot 2(2x + 1)$$
$$6x + 15 = 8x + 4$$
$$-2x = -11$$
$$x = 5\tfrac{1}{2}$$

Thus, the solution is $5\tfrac{1}{2}$. The check is left for you.

Possible solutions of a fractional equation may not check in the original equation. Therefore, the check is especially important.

EXAMPLE 2 Solve $\dfrac{x + 1}{x - 3} = \dfrac{3}{x} + \dfrac{12}{x^2 - 3x}$ and check.

Solution $$\dfrac{x + 1}{x - 3} = \dfrac{3}{x} + \dfrac{12}{x(x - 3)}$$

$$x(x - 3) \cdot \dfrac{x + 1}{x - 3} = x(x - 3) \cdot \dfrac{3}{x} + x(x - 3) \cdot \dfrac{12}{x(x - 3)}$$

$$x^2 + x = 3x - 9 + 12$$
$$x^2 - 2x - 3 = 0$$
$$(x - 3)(x + 1) = 0$$
$$x = 3 \quad or \quad x = -1$$

Checks

Substitute 3 for x.

$$\frac{x + 1}{x - 3} = \frac{3}{x} + \frac{12}{x^2 - 3x}$$

$$\frac{3 + 1}{3 - 3} \overset{?}{=} \frac{3}{3} + \frac{12}{3^2 - 3 \cdot 3}$$

$$\frac{4}{0} \overset{?}{=} 1 + \frac{12}{0}$$

Substitute -1 for x.

$$\frac{x + 1}{x - 3} = \frac{3}{x} + \frac{12}{x^2 - 3x}$$

$$\frac{-1 + 1}{-1 - 3} \overset{?}{=} \frac{3}{-1} + \frac{12}{(-1)^2 - 3(-1)}$$

$$\frac{0}{-4} \overset{?}{=} -3 + 3$$

$$0 = 0 \quad \text{True}$$

Because the symbols $\frac{4}{0}$ and $\frac{12}{0}$ are undefined, 3 is not a solution of the original equation.

Thus, -1 is the only solution of the original equation.

In Example 2 above, 3 is a solution of the *derived* equation, $x^2 - 2x - 3 = 0$. However, 3 is not a solution of the *original* equation. In this case, 3 is called an *extraneous solution*, or *extraneous root*.

Definition

An **extraneous root** is a solution of a derived equation that is not a solution of the original equation.

The formula $\frac{1}{R} = \frac{1}{x} + \frac{1}{y}$ is a literal equation that contains rational expressions. It can be simplified by multiplying each side by the LCD. Then it can be solved for any one of its variables.

EXAMPLE 3 Solve $\frac{1}{R} = \frac{1}{x} + \frac{1}{y}$ for x. Then find the value of x for $R = 4.5$ and $y = 6.0$.

Solution

$$\frac{1}{R} = \frac{1}{x} + \frac{1}{y}$$

$$Rxy \cdot \frac{1}{R} = Rxy \cdot \frac{1}{x} + Rxy \cdot \frac{1}{y}$$

$$xy = Ry + Rx$$

$$xy - Rx = Ry$$
$$x(y - R) = Ry$$

$$x = \frac{Ry}{y - R}$$

Next, substitute 4.5 for R and 6.0 for y.

$$x = \frac{Ry}{y - R}$$

$$= \frac{4.5(6.0)}{6.0 - 4.5}$$

$$= \frac{27.0}{1.5}$$

$$= 18$$

Thus, $x = \frac{Ry}{y - R}$, and $x = 18$ when $R = 4.5$ and $y = 6.0$.

EXAMPLE 4   The numerator of a fraction is 7 less than twice its denominator. If the numerator is doubled and the denominator is increased by 4, the resulting fraction is $\frac{2}{3}$. Find the original fraction.

What are you to find?	the original fraction
Choose two variables. What do they represent?	Let n = the numerator. Let d = the denominator. So, the original fraction is $\frac{n}{d}$.
What is given?	(1) The numerator is 7 less than twice the denominator. (2) When the numerator is doubled and the denominator increased by 4, the result is $\frac{2}{3}$.
Write a system.	(1) $n = 2d - 7$ (2) $\dfrac{2n}{d + 4} = \dfrac{2}{3}$
Solve the system.	(2) $\dfrac{2(2d - 7)}{d + 4} = \dfrac{2}{3}$ ← The substitution method is used.

Use the Proportion Property: $2(2d - 7) \cdot 3 = (d + 4) \cdot 2$
If $\frac{a}{b} = \frac{c}{d}$, then $a \cdot d = b \cdot c$.
$$12d - 42 = 2d + 8$$
$$10d = 50$$
$$d = 5 \leftarrow \text{denominator}$$
$$(1)\ n = 2d - 7$$
$$= 2(5) - 7$$
$$= 3 \leftarrow \text{numerator}$$

Check in the original problem.	The numerator is 7 less than twice the denominator. $$3 = 2 \cdot 5 - 7$$ If the numerator is doubled and the denominator is increased by 4, the resulting fraction is $\frac{2}{3}$. $$\frac{3 \cdot 2}{5 + 4} = \frac{6}{9} = \frac{2}{3}$$
State the answer.	The original fraction $\frac{n}{d}$ is $\frac{3}{5}$.

◢ *Focus on Reading*

1. A solution of a derived equation that is not a solution of the original equation is called _____.
2. One method for solving a fractional equation is to multiply each side of the equation by the _____ of the _____.

Classroom Exercises

Find the least common denominator (LCD) for each equation.

1. $\dfrac{x}{2} + \dfrac{2}{3} = \dfrac{x}{6}$

2. $\dfrac{3a}{5} + \dfrac{3}{2} = \dfrac{7a}{10}$

3. $\dfrac{1}{m} + \dfrac{2}{3} = 1$

4. $\dfrac{4}{5} + \dfrac{3}{a} = 2$

5. $\dfrac{2a - 3}{7} - \dfrac{a}{2} = \dfrac{a + 3}{14}$

6. $\dfrac{3x}{4} - \dfrac{2x - 1}{2} = \dfrac{x - 7}{6}$

7–12. Solve the equations in Classroom Exercises 1–6 above. Check for extraneous roots.

Written Exercises

Solve each equation. Check for extraneous roots.

1. $\dfrac{2}{5} + \dfrac{2}{y} = 1$

2. $\dfrac{6}{x} + \dfrac{9}{2x} = 3$

3. $\dfrac{4a + 3}{3} = \dfrac{2a + 5}{4}$

4. $\dfrac{x}{10} + \dfrac{x}{6} + \dfrac{x}{15} = 1$

5. $\dfrac{2}{3n^2} = \dfrac{1}{4n^2} + \dfrac{5}{6n}$

6. $\dfrac{2n - 3}{2} = \dfrac{3}{4} + \dfrac{n - 4}{8}$

7. $\dfrac{3x - 2}{2} + \dfrac{x - 4}{3} = \dfrac{1}{4}$

8. $\dfrac{2x - 3}{5} + 1 = \dfrac{x + 3}{3}$

9. $\dfrac{3n - 7}{n - 5} + \dfrac{n}{2} = \dfrac{8}{n - 5}$

10. $\dfrac{a - 4}{a + 3} = \dfrac{3a + 2}{a + 3} + \dfrac{a}{4}$

11. $\dfrac{3}{x - 4} + \dfrac{2}{x + 4} = \dfrac{14}{x^2 - 16}$

12. $\dfrac{-3a}{a^2 - 4a - 32} = \dfrac{2}{a - 8} + \dfrac{3}{a + 4}$

13. $\dfrac{14}{x^2 - 3x} - \dfrac{8}{x} = \dfrac{-10}{x - 3}$

14. $\dfrac{2m - 1}{m^2 - 9m + 20} = \dfrac{7}{m - 5} - \dfrac{4}{m - 4}$

15. $\dfrac{x + 5}{x - 4} = \dfrac{3}{x} + \dfrac{36}{x^2 - 4x}$

16. $\dfrac{2a - 9}{a - 7} + \dfrac{a}{2} = \dfrac{5}{a - 7}$

17. $\dfrac{5}{2y + 6} - \dfrac{3}{y - 4} = \dfrac{2y - 4}{y^2 - y - 12}$

18. $\dfrac{4}{y^2 - 8y + 12} = \dfrac{y}{y - 2} + \dfrac{1}{y - 6}$

19. $\dfrac{6}{x + 2} + \dfrac{3}{x^2 - 4} = \dfrac{2x - 7}{x - 2}$

20. $\dfrac{2}{a + 2} + \dfrac{a}{2 - a} = \dfrac{13}{4 - a^2}$

21. $\dfrac{2x - 3}{x - 5} = \dfrac{x}{x + 4} + \dfrac{20x - 37}{x^2 - x - 20}$

22. $\dfrac{2b + 3}{b - 1} - \dfrac{10}{b^2 - 1} = \dfrac{2b + 3}{b + 1}$

23. $\dfrac{x - 2}{x + 1} = \dfrac{x - 3}{x^2 - 5x - 6} - \dfrac{2x - 7}{x - 6}$

24. $\dfrac{a - 3}{3a} = \dfrac{1}{3a^2 + 9a} + \dfrac{1}{a + 3}$

25. $\dfrac{7y - 20}{y^2 - 7y + 12} = \dfrac{y}{y - 3} - \dfrac{2}{4 - y}$

26. $\dfrac{8}{12 + 4x - x^2} + \dfrac{x + 1}{6 - x} = \dfrac{5}{x + 2}$

Solve each formula for the specified variable. Then find the value of that variable for the given data.

27. Solve $\dfrac{a+b}{ab} = \dfrac{2}{x}$ for x.

    $a = 8.12; b = 8.7$

28. Solve $\dfrac{1}{R} = \dfrac{1}{x_1} + \dfrac{1}{x_2}$ for x_2.

    $R = 3.2; x_1 = 4.8$

29. Solve $\dfrac{a-b}{b} = cd$ for b.

    $a = 48.35; c = 5.8; d = 16.5$

30. Solve $\dfrac{t+mf}{m} = g$ for m.

    $t = 65.8; g = 46.3; f = 13.4$

Solve each problem.

31. The denominator of a fraction is 1 more than twice its numerator. If the numerator is increased by 1 and the denominator is decreased by 1, the resulting fraction is $\frac{2}{3}$. Find the original fraction.

32. The denominator of a fraction is 14 less than the square of its numerator. The fraction can be simplified to $\frac{1}{5}$. What could the original fraction be?

33. The numerator of a fraction is 6 less than twice its denominator. If the numerator is increased by 1 and the denominator is doubled, the resulting fraction is $\frac{1}{2}$. Find the value of the original fraction.

34. The denominator of a fraction is the same as the numerator increased by its square. If the numerator and denominator are each increased by 2, the resulting fraction, simplified, is $\frac{1}{2}$. Find the original fraction.

Solve each equation. Check for extraneous roots.

35. $\dfrac{3x^2}{x^3 - 8} = \dfrac{2x - 5}{x^2 + 2x + 4}$

36. $\dfrac{x^3 + x^2 - 7x - 1}{x - 1} = x^2 + x + 1$

Mixed Review

Given the points $A(-2, -8)$, $B(6,8)$, and $C(6, -2)$, find the following slope or equation. *3.3, 3.4*

1. slope of $\overleftrightarrow{AB}$

2. equation of $\overleftrightarrow{AB}$

3. equation of $\overleftrightarrow{BC}$

Brainteaser

Starting 3 mi away, a boy walks home at 3 mi/h. His dog, starting with him, runs home at 5 mi/h, then immediately turns around and runs back to the boy, then back home, back to the boy, and so on until the boy reaches home. How far does the dog run?

Application: *Flotation*

Two thousand years ago, Archimedes discovered that objects placed in water are buoyed up by a force working against gravity that is equal to the weight of the water the object displaces. The key is the object's density, or mass per unit volume. For example, lead is more dense than wood because a block of lead weighs more than a block of wood of the same size. Objects denser than water sink, although more slowly than they would in air. Objects less dense than water float.

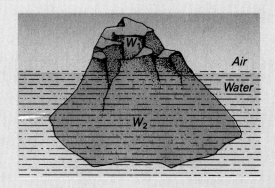

In algebraic terms, the weight w of an object *in water* is $w_1 - w_2$, where w_1 is the weight of the object and w_2 is the weight of the displaced water. For objects denser than water, $w_1 > w_2$ (the volume being equal), and w is positive. Therefore, the objects weigh more than an equal volume of water and sink. For objects less dense than water, however, w is negative, and the objects float.

To find what part of a floating object is submerged, first set $w = 0$. You can do this because floating is an equilibrium state, and therefore the pull of gravity downward is exactly offset by the buoyant force of the water acting upward. Next, divide the object into two parts, one part above the water line and the other below.

In the case of an iceberg, for example, let W_1 = the weight of the portion of the iceberg above water and W_2 = the weight of the portion submerged. The pull of gravity on the iceberg is then equal to the total weight of the iceberg ($W_1 + W_2$), and the equivalent buoyant force is the weight of the displaced water. The weight of the displaced water is 1.08 times the weight of the submerged part of the iceberg because water is 1.08 times as dense as ice. That is, a given volume of water weighs 1.08 times as much as an equal volume of ice.

To find the relationship between W_1 and W_2, use the equation below.

pull of gravity $\rightarrow W_1 + W_2 = 1.08\,W_2 \leftarrow$ buoyant force

1. How many times larger is the submerged part of an iceberg than the part above the water line?

2. A large pine log floats in water. Given that water is 2.5 times more dense than pine, what part of the log is below the water?

7.7 Problem Solving: Rates of Work

Objective

To solve work problems

Wanda can mow a lawn in 5 h. Therefore, in 1 h she can mow $\frac{1}{5}$ of the lawn. Her work rate is $\frac{1}{5}$ of the job per hour. At this rate, in 4 h she should complete $4 \cdot \frac{1}{5}$, or $\frac{4}{5}$ of the job. This suggests the following relationship:

(number of hours worked) · (rate per hour) = fractional part of job completed

$$4 \cdot \frac{1}{5} = \frac{4}{5}$$

Suppose two painters work together to paint a house. One of them can do $\frac{3}{5}$ of the work while the other is doing $\frac{2}{5}$ of the work.

Working together, they complete the job, because $\frac{3}{5} + \frac{2}{5} = 1$. The sum of the fractional parts of the job done by each worker is 1.

EXAMPLE 1

It takes Fay 5 h to paint a room. Rob can do the job in 10 h. How long will it take them to do the job if they work together? (THINK: Will it take more or less time than 5 h?)

Solution

Let x = the number of hours Fay and Rob work together.

Use a table to represent the part done by each person.

	Part of job done in 1 h	Number of hours working together	Part of job completed
Fay	$\frac{1}{5}$	x	$x \cdot \frac{1}{5} = \frac{x}{5}$
Rob	$\frac{1}{10}$	x	$x \cdot \frac{1}{10} = \frac{x}{10}$

$$\text{Fay's part} + \text{Rob's part} = \text{total job}$$
$$\frac{x}{5} + \frac{x}{10} = 1$$
$$10 \cdot \frac{x}{5} + 10 \cdot \frac{x}{10} = 10 \cdot 1$$
$$2x + 1x = 10$$
$$3x = 10$$
$$x = \frac{10}{3}, \text{ or } 3\frac{1}{3} \text{ hours working together}$$

Check	To check, add the parts done in $3\frac{1}{3}$ h.

$$x \cdot \frac{1}{5} + x \cdot \frac{1}{10} = \frac{10}{3} \cdot \frac{1}{5} + \frac{10}{3} \cdot \frac{1}{10} = \frac{2}{3} + \frac{1}{3} = 1$$

Thus, the job will take $3\frac{1}{3}$ h if Fay and Rob work together.

EXAMPLE 2 Working together, Pat, Frank, and Pam can wallpaper an apartment in 20 h. Pam can do the job alone in 35 h, and Pat can do the job alone in twice the time it would take Frank working alone. Find the time it would take Frank to do the job alone.

Solution Let x = the number of hours for Frank to do the job alone.

Then $2x$ = the number of hours for Pat to do the job alone.

	Part of job done in 1 hr	Number of hours working together	Part of job completed
Pam	$\dfrac{1}{35}$	20	$20 \cdot \dfrac{1}{35} = \dfrac{20}{35} = \dfrac{4}{7}$
Pat	$\dfrac{1}{2x}$	20	$20 \cdot \dfrac{1}{2x} = \dfrac{20}{2x} = \dfrac{10}{x}$
Frank	$\dfrac{1}{x}$	20	$20 \cdot \dfrac{1}{x} = \dfrac{20}{x}$

Pam's part + Pat's part + Frank's part = total job

$$\frac{4}{7} + \frac{10}{x} + \frac{20}{x} = 1$$

$$7x \cdot \frac{4}{7} + 7x \cdot \frac{10}{x} + 7x \cdot \frac{20}{x} = 7x \cdot 1$$

$$4x + 70 + 140 = 7x$$

$$70 = x$$

Thus, it would take Frank 70 h working alone.
The check is left for you.

Classroom Exercises

It takes Bill 18 h to build a cabinet. Represent the part built in the given period of time.

1. 5 h
2. 8 h
3. x h
4. $2x$ h

5. 10 h
6. $3x$ h
7. $(2x + 1)$ h
8. $\dfrac{x}{2}$ h

9. If it takes Ed 12 h to build Bill's cabinet (see above), how long would it take Bill and Ed to build the cabinet working together?

Written Exercises

Solve each problem.

1. It takes Don 6 h and Joyce 8 h to paint a room alone. How long would it take them to paint the room if they worked together?

2. It takes Regina 3 h to prepare dinner. Her husband, Rich, can prepare the same dinner in 2 h. How long would it take them working together?

3. Working together, it takes Dick and Gail 16 h to tile a floor. It would take Gail 40 h to do it alone. How long would it take Dick working alone?

4. It takes Lois 3 times as long as Richie to mow a lawn. How long would it take each of them alone, if together they can do it in 5 h?

5. Working together, Andy and Sal can build a fence in 7 h. Alone, it takes Andy twice the time it takes Sal. How long does it take each working alone?

6. Mary can wallpaper a room in 6 h. It takes Jackie 2 h and Kevin 3 h to do the same job alone. How long would it take to do the job if they all worked together?

7. Dick, Florence, and Betty can paint a house in 12 h if they work together. If each worked alone, Dick would take 30 h, and Betty twice as long as Florence. How long would it take each girl working alone?

8. A large pipe can fill a tank in 5 h, and a smaller one can fill the tank in 8 h. A drain pipe can empty the tank in 10 h. Find the total time needed to fill the tank when all three pipes are left open.

9. Holly can harvest a strawberry patch in 12 h, and Evelyn can do the job in 8 h. Given that Evelyn starts 2 h after Holly has begun working, find the total time needed to do the job.

10. It takes a man 6 days less than his son to build a garage. If they work together, they can do it in $\frac{1}{3}$ the time it takes the son. How long would it take each working alone?

11. If 5 men and 2 boys work together, a job can be completed in one day. If 3 men and 6 boys work together, the job can also be completed in one day. How long would it take 1 boy working alone to do the work?

Mixed Review

Simplify. Use positive exponents.   *5.2, 5.3, 5.5, 7.1, 7.2, 7.4*

1. $(5n^2)^{-3}$

2. $3x(2x + 1)(x - 6)$

3. $(8x - 3)^2$

4. $\dfrac{x^3 - 8}{2 - x}$

5. $\dfrac{5x^3}{2a^4} \cdot \dfrac{6a}{25x}$

6. $\dfrac{3}{x^2 - 4} - \dfrac{2}{x + 2}$

1. How many milliliters of water must be added to 400 ml of a 77% salt solution to dilute it to a 55% solution?

2. How many liters of a 75% sulfuric acid solution must be added to 30 liters of a 20% sulfuric acid solution to obtain a 45% acid solution?

3. Given that y varies directly as x and $y = 9.6$ when $x = 4.5$, find y when $x = 15$.

4. Convert 4 ounces per square inch to pounds per square foot.

5. Mr. Brown can carpet a floor in 3 hours. If his apprentice helps him, the job can be done in 2 hours. How long would it take the apprentice to do the job working alone?

6. The hypotenuse of a right triangle is 1 cm longer than one leg, and 4 cm longer than 3 times the length of the other leg. Find the length of the hypotenuse.

7. The height of a triangle is 7 ft less than 5 times the length of the base. Find the length of the base and the height of the triangle given that its area is 12 ft^2.

8. The length of a rectangle is 7 ft more than 3 times its width. Find the set of all possible lengths given that the perimeter is less than 102 ft.

9. How many milliliters of a 72% alcohol solution should be added to 300 ml of a 27% alcohol solution to make a 36% alcohol solution?

10. The measures of two supplementary angles have the ratio 2:3. Find the measure of the complement of the smaller angle.

11. Find three consecutive multiples of 5 such that 7 times the second number is the same as 55 increased by 4 times the third number.

12. Find three consecutive multiples of 3 given that twice the square of the first number is 18 more than the product of the other two numbers.

13. A second number is twice a first number. The product of the first number and 2 more than the second number is 60. Find all such pairs of numbers.

14. Mrs. Montero invested $10,000, part at 9% and the remainder at 10% per year. If at the end of 5 years the two investments had earned $4,800 in simple interest, what amount was invested at 9%?

15. One bus left a terminal at 6:00 A.M. and headed east at 80 km/h. A second bus left the same terminal at 7:30 A.M. and traveled west at 90 km/h. At what time were the buses 460 km apart?

16. If 90 ml of a 40% citrus solution and 20 ml of a 25% citrus solution were mixed with a 60% citrus solution to make a 50% solution, how many milliliters of the 60% solution were used?

Key Terms

Cancellation Property (p.234)
complex rational expression (p.249)
conversion fraction (p.242)
dimensional analysis (p.242)

extraneous root (p.253)
rational expression (p.233)
rational function (p.233)
zero of a function (p.234)

Key Ideas and Review Exercises

7.1 For a function such as $f(x) = \dfrac{x^2 - 9x + 8}{3x + 2}$:

1. To find the zeros, set the numerator equal to 0 and solve.
2. To find the domain, find the values of x for which the denominator is *not* 0.

Find the zeros and the domain of each function.

1. $f(x) = \dfrac{x^2 - 9x + 8}{3x + 2}$

2. $f(x) = \dfrac{x^2 - 7x + 12}{x^2 - 49}$

To simplify a rational expression such as $\dfrac{2x^3 + 2x^2 - 24x}{6x^3(15 + x - 2x^2)}$:

1. Put $15 + x - 2x^2$ into a more convenient form by factoring out -1.
2. Factor the numerator and denominator completely.
3. Divide out the common factors.

Simplify the expression that describes each function. Then find the indicated value of the function.

3. $f(x) = \dfrac{2x^3 + 2x^2 - 24x}{6x^3(15 + x - 2x^2)}; f(-2)$

4. $f(x) = \dfrac{-10 + 22x - 4x^2}{3x^2 - 16x + 5}; f(3)$

7.2 To simplify products or quotients of rational expressions, first apply $\dfrac{a}{b} \cdot \dfrac{c}{d} = \dfrac{ac}{bd}$ or $\dfrac{a}{b} \div \dfrac{c}{d} = \dfrac{a}{b} \cdot \dfrac{d}{c}$. Then simplify the result.

Simplify.

5. $\dfrac{n^2 + 3n - 18}{2n^2 - 5n - 3} \cdot \dfrac{1 - 4n^2}{n^2 - 36}$

6. $\dfrac{x}{x + 4} \div \dfrac{6x^2}{3x + 12}$

7.3 To convert from one unit of measurement to another, use pairs of equivalent measures.

7. Convert 144 ft² to square yards.

8. Convert 90 mi/h to feet per second.

7.4   To add or subtract rational expressions with unlike denominators, see Examples 3–5 in Lesson 7.4.

Simplify.

9. $\dfrac{6a - 23}{a^2 - 7a + 12} + \dfrac{a - 5}{a - 4}$

10. $\dfrac{5a}{a^2 - 9} - \dfrac{4}{a + 3} + \dfrac{2}{3 - a}$

11. Write in your own words how to add several rational expressions with unlike denominators.

7.5   To simplify a complex rational expression, use either of the two methods shown in Lesson 7.5.

Simplify.

12. $\dfrac{1 - 6x^{-1} + 5x^{-2}}{1 - 3x^{-1} - 10x^{-2}}$

13. $\dfrac{\dfrac{1}{x + 6} + \dfrac{1}{x + 2}}{\dfrac{x^2 + 11x + 28}{x^2 + 8x + 12}}$

7.6   To solve an equation containing rational expressions,
   1. factor each denominator if possible,
   2. multiply each side of the equation by the LCD of all the fractions,
   3. solve the resulting equation, and
   4. check for extraneous solutions. (No denominator can be 0.)

Solve each equation for x.

14. $\dfrac{2x + 3}{x - 1} - \dfrac{2x - 3}{x + 1} = \dfrac{10}{x^2 - 1}$

15. $\dfrac{5a}{x + b} = \dfrac{a}{x - b}$

To solve word problems involving the numerator and denominator of a fraction, use two equations as shown in Example 4 of Lesson 7.6.

16. The numerator of a fraction is 5 less than the denominator. If the numerator were increased by 3 and the denominator increased by 4, the resulting fraction would be $\frac{1}{2}$. Find the original fraction.

7.7   To solve work problems,
   (1) represent the fractional part of the work done by each worker, and
   (2) set the sum of the fractional parts done by each worker equal to 1 and solve the equation.

17. Bill can do a job in 15 h. Jane can do the same job in 10 h. If Jane were to join Bill 3 h after he had begun, what would be the total time needed to do the job?

Find the zeros and the domain of each function.

1. $f(x) = \dfrac{3x - 4}{2x + 10}$

2. $f(x) = \dfrac{x^2 - 11x + 30}{x^2 + x - 2}$

Simplify the expression that describes each function. Then find the indicated value of the function.

3. $f(x) = \dfrac{x^2 - 12x + 20}{x^2 - 11x + 10}$; $f(3)$

4. $f(x) = \dfrac{2x^2 - 4x}{x^3 - 4x}$; $f(-6)$

Simplify.

5. $\dfrac{3n^2 - 15n + 18}{45 - 5n^2}$

6. $\dfrac{3c^2 - 5c - 2}{6c^2} \cdot \dfrac{4c^2 - 8c}{c^2 - 4c + 4}$

7. $\dfrac{x}{x + 4} \div \dfrac{6x^2}{3x + 12}$

8. $\dfrac{-2x^2 - 5x}{x^2 + 7x} + \dfrac{x - 2}{x + 7} + \dfrac{2x - 3}{x}$

9. $\dfrac{2x + 7}{x^2 - 5x + 4} - \dfrac{2x - 3}{x - 4}$

10. $\dfrac{1 - 4x^{-1} - 45x^{-2}}{2 + 7x^{-1} - 15x^{-2}}$

11. Convert 288 in^2 to square feet.

12. Convert 30 T/h to pounds per minute.

Solve each equation. Check for extraneous roots.

13. $\dfrac{5}{x - 4} - \dfrac{x}{2} = \dfrac{x + 1}{x - 4}$

14. $\dfrac{n - 6}{n^2 - 2n - 8} + \dfrac{3}{n - 4} = \dfrac{2}{n + 2}$

15. Solve $A = \dfrac{h}{2}(x + y)$ for x. Then find x when $h = 10$, $A = 70$, and $y = 8$.

16. Bill can do a job alone in 8 days. After he works alone for 2 days, Marie joins him. They finish the job together in 2 more days. How long would it take Marie to do the entire job alone?

17. The denominator of a fraction is 3 less than twice the numerator. If the numerator were doubled and the denominator increased by 8, the resulting fraction would be $\frac{2}{3}$. Find the original fraction.

Simplify.

18. $\dfrac{ab + 5ac - 2bd - 10cd}{b^2 + 7bc + 10c^2} \cdot \left(\dfrac{a^2 + 2ad - 8d^2}{b^2 - 4c^2} \right)^{-1}$

Choose the *one* best answer to each question or problem.

1. Given that $y = \dfrac{1}{xz}$, what is the value of z when $x = \frac{1}{4}$ and $y = 2$?

(A) 2　(B) $\frac{1}{2}$　(C) 8　(D) 6

(E) None of these

2. The figure below represents the scale drawing of a rectangular living room. The scale used is 3 cm = 1 m. Find the actual area of the room.

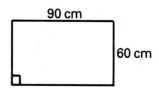

90 cm

60 cm

(A) 50 m^2　(B) 100 m^2

(C) 150 m^2　(D) 600 m^2

(E) 5,400 m^2

3. *ABCD* is a square, and the shaded region is the interior of a circle. What is the ratio of the shaded area to the area of the square?

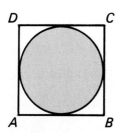

(A) $\dfrac{\pi}{4}$　(B) $\dfrac{4}{\pi}$　(C) π

(D) $\dfrac{1}{\pi}$　(E) $\dfrac{\pi}{8}$

4. If 5 more than x is 2 less than y, what is y in terms of x?

(A) $x + 3$　(B) $y - 7$　(C) $y - 3$

(D) $x - 7$　(E) $x + 7$

5. $\dfrac{1}{10^{15}} - \dfrac{1}{10^{16}} = \underline{\ ?\ }$

(A) $-\dfrac{9}{10^{16}}$　(B) $\dfrac{9}{10^{16}}$　(C) $\dfrac{1}{10}$

(D) $-\dfrac{1}{10}$　(E) $\dfrac{1}{10^{16}}$

6. If $x + y = 4$ and $x + 2y = 5$, then $x - y = \underline{\ ?\ }$

(A) 4　(B) 2　(C) -2

(D) $3\frac{1}{3}$　(E) None of these

7. If $a \neq 0$, then $\dfrac{(-4a)^3}{-4a^3} = \underline{\ ?\ }$

(A) -16　(B) 1　(C) 16

(D) -1　(E) 3

8. If $\dfrac{1}{a} + \dfrac{1}{a} = 12$, then $a = \underline{\ ?\ }$

(A) 6　(B) $\frac{1}{6}$　(C) $\frac{1}{12}$

(D) 12　(E) $\frac{1}{24}$

9. Three lines intersect as shown. What is the sum of the degree measures of the marked angles?

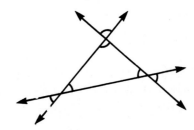

(A) 90　(B) 180　(C) 360

(D) 540　(E) It cannot be determined from the given information.

10. If $\dfrac{4}{a} = \dfrac{5}{b} + \dfrac{1}{c}$, then $ab - 4bc = \underline{\ ?\ }$

(A) $-5ac$　(B) $-5bc$　(C) $5ac$

(D) $5bc$　(E) $5abc$

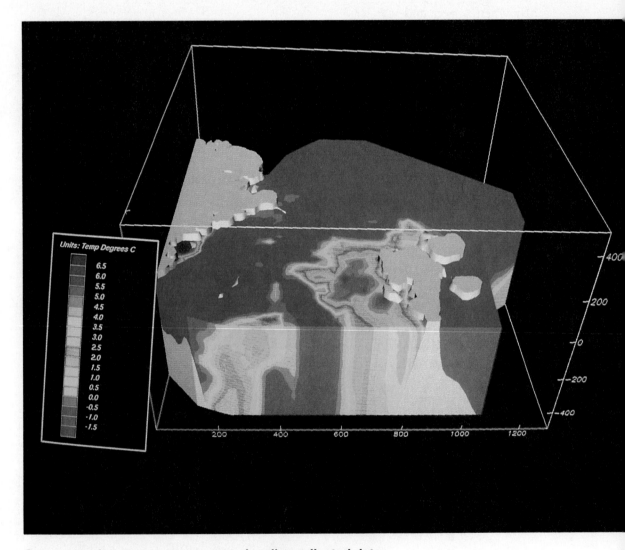

Oceanographers use computers to visualize collected data and gain a greater understanding of the physical environment they are studying. The images above show the 11-year mean temperature field in the upper 500 m of Fram Strait. The red and orange colors outline the warmer Atlantic waters.

8.1 Square Roots and Functions

Objectives

To find square roots of positive numbers
To find domains of functions involving square roots
To solve simple quadratic equations

Every positive real number has two *square roots*, one positive and the other negative. To find the two square roots of 64, find the two numbers that have 64 as their square. That is, find the values of x such that $x^2 = 64$. Because $8 \cdot 8 = 64$ and $(-8) \cdot (-8) = 64$, the square roots of 64 are 8 and -8.

Definitions

> The **square roots** of any nonnegative number b are the solutions of the equation $x^2 = b$. The **principal square root** is the positive square root of that number, written as $\sqrt{b}$. The negative square root is written $-\sqrt{b}$.

From the above definitions, it follows that

$$\sqrt{b} \cdot \sqrt{b} = (\sqrt{b})^2 = b.$$

The symbol $\sqrt{b}$ is called a **radical**, and $\sqrt{}$ is the **radical sign**. The number b under the radical sign is called the **radicand**.

Notice that the above definitions do not include negative radicands because the square root of a negative number is not a real number.

For example, $\sqrt{-9}$ does not name a real number because there is no real number whose square is -9.

Because $\sqrt{}$ indicates the principal square root, which is positive, $\sqrt{a^2} = |a|$ for any real number a. That is, the principal square root of a^2 is the absolute value of a. (Notice that when $a \geq 0$, $\sqrt{a^2} = a$.) For example, $\sqrt{64} = \sqrt{8^2} = |8| = 8$, and $\sqrt{64} = \sqrt{(-8)^2} = |-8| = 8$. On a calculator, only the principal square root is displayed.

EXAMPLE 1 Simplify.

a. $\sqrt{0.25}$ b. $\sqrt{\dfrac{9}{49}}$

Solutions

a. $\sqrt{0.25} = 0.5$ because $(0.5)^2 = 0.25$

b. $\sqrt{\dfrac{9}{49}} = \dfrac{3}{7}$ because $\left(\dfrac{3}{7}\right)^2 = \dfrac{9}{49}$

In Example 1, the radicands 0.25 and $\frac{9}{49}$ are *perfect squares* because their principal square roots, 0.5 and $\frac{3}{7}$, are rational numbers.

Definition

> A rational number is a **perfect square** if and only if its principal square root is also a rational number. That is, a is a perfect square if and only if $\sqrt{a}$ is a rational number.

Although a number such as $\sqrt{43}$ is not rational, there are some rational numbers that have squares that are very close to 43. For instance,

$$(6.557)^2 = 42.994249 \text{ and } (6.558)^2 = 43.007364.$$
$$\text{Notice that } 6.557 < \sqrt{43} < 6.558.$$

Calculators, however, cannot be used reliably to determine whether the square root of a number is or is not rational. For example, if the rational number 6.5574385 is squared on some scientific calculators, the answer will be displayed as exactly 43. If correct, this would imply that $\sqrt{43}$ is 6.5574385, which is rational. But, in fact, the displayed answer is rounded and thus is slightly too large.

EXAMPLE 2   Evaluate each of the following expressions for $x = 8$. Approximate irrational results to the nearest hundredth using a calculator or the Square Root Table on page 789.

a. $\sqrt{\dfrac{5x + 9}{16}}$

b. $\sqrt{4x + 3}$

Solutions

a. $\sqrt{\dfrac{5 \cdot 8 + 9}{16}} = \sqrt{\dfrac{49}{16}} = \dfrac{7}{4}$

b. $\sqrt{4 \cdot 8 + 3} = \sqrt{35} \approx 5.92$

Recall that the *domain* of a function f is the set of all values of x for which the function is defined. For a function involving square roots, such as

$$f(x) = \sqrt{3x - 15} \text{ or } f(x) = -\frac{x}{\sqrt{6 - x}},$$

$f(x)$ is not defined when a radicand is negative (or when a denominator is zero). Therefore, unless otherwise stated, the domain of a function that involves square roots is assumed to be the set of all real numbers for which the radicand is nonnegative, and for which the denominator is not zero.

EXAMPLE 3   Find the domain of each function.

a. $f(x) = \sqrt{3x - 15}$ b. $f(x) = -\dfrac{x}{\sqrt{6 - x}}$

Solutions   a. Find the values of x for which $3x - 15$ is nonnegative.

Solve $3x - 15 \geq 0$.
$$3x \geq 15$$
$$x \geq 5$$

The domain is
$\{x \mid x \geq 5\}$.

b. Find the values of x for which $6 - x$ is nonnegative *and* for which $\sqrt{6 - x} \neq 0$.

Solve $6 - x \geq 0$ *and* $\sqrt{6 - x} \neq 0$.
$$6 \geq x \quad and \quad 6 - x \neq 0$$
$$x \leq 6 \quad and \quad x \neq 6$$

The domain is
$\{x \mid x < 6\}$.

The definition of square root can also be applied to simple quadratic equations, as shown below.

$$x^2 = 25$$
$$x = \sqrt{25} \quad or \quad x = -\sqrt{25}$$
$$x = 5 \quad or \quad x = -5$$

In general, if $x^2 = b$, and $b \geq 0$, then $x = \sqrt{b}$ *or* $x = -\sqrt{b}$.

EXAMPLE 4   Solve each equation. Approximate irrational results to the nearest hundredth.

a. $4x^2 + 9 = 45$ b. $3x^2 - 86 = x^2 + 86$

Solutions   a. $4x^2 + 9 = 45$
$$4x^2 = 36$$
$$x^2 = 9$$
$$x = \sqrt{9} \quad or \quad x = -\sqrt{9}$$
$$x = 3 \quad or \quad x = -3$$

The roots are 3 and -3, or ± 3.

b. $3x^2 - 86 = x^2 + 86$
$$2x^2 = 172$$
$$x^2 = 86$$
$$x = \sqrt{86} \quad or \quad x = -\sqrt{86}$$
$$x \approx 9.27 \quad or \quad x \approx -9.27$$

The roots are 9.27 and -9.27, or ± 9.27 to the nearest hundredth.

Sometimes, as in Example 4, a number and its opposite are represented by a single symbol. Thus, the symbol ± 3 represents the two numbers 3 and -3 and is read as "plus or minus 3."

Classroom Exercises

Indicate if each number is rational.

1. $\sqrt{25}$ **2.** $\sqrt{53}$ **3.** $-\sqrt{36}$ **4.** $\sqrt{\frac{4}{9}}$ **5.** $\sqrt{0.04}$ **6.** $\sqrt{0.9}$

Written Exercises

Simplify if the indicated square root is a real number. Otherwise, write *not real.*

1. $\sqrt{81}$ **2.** $-\sqrt{49}$ **3.** $\sqrt{1.44}$ **4.** $\sqrt{\frac{25}{49}}$ **5.** $\sqrt{-\frac{16}{81}}$ **6.** $\sqrt{0.0121}$

Evaluate each of the following expressions for the given value of x. Approximate irrational results to the nearest hundredth using the Square Root Table on page 789 or a calculator.

7. $\sqrt{2x + 6}$; $x = 5$ **8.** $\sqrt{\frac{7 - 5x}{9}}$; $x = -8$ **9.** $\sqrt{\frac{3x + 1}{2x}}$; $x = 8$

Find the domain of each function.

10. $f(x) = \sqrt{2x - 6}$ **11.** $f(x) = \sqrt{8 - 4x}$ **12.** $f(x) = -\frac{1}{\sqrt{2x + 3}}$

Solve each equation. Approximate irrational results to the nearest hundredth.

13. $x^2 = 144$ **14.** $9a^2 = 2a^2 + 294$ **15.** $3a^2 - 4 = a^2 + 68$

16. $y^2 = 0.25$ **17.** $x^2 - 0.0009 = 0.004$ **18.** $5x^2 - 1 = 4x^2 + 0.21$

Find the domain of each function.

19. $f(x) = \sqrt{x^2 - 7x + 12}$ **20.** $f(x) = \sqrt{\frac{x^2 - 5x + 4}{6x - 48}}$

21. $f(x) = \sqrt{\frac{x^4 - 16}{2x - 6}}$ **22.** Simplify $\sqrt{x^2 + 6x + 9}$. (HINT: $\sqrt{a^2} = |a|$ for each real number a.)

Mixed Review

Simplify. **5.1, 5.2, 7.1**

1. $(3x^2y^3)^4$ **2.** $5a^0$ **3.** $\frac{4x^{-3}y^{-5}}{16x^{-5}y^{-2}}$ **4.** $\frac{x^2 - 7x + 12}{9 - x^2}$

8.2 Simplifying Square Roots

Objectives

To simplify square roots of positive numbers
To simplify algebraic expressions containing square roots

The following illustration suggests a theorem for products of square roots.

$$\sqrt{16 \cdot 9} \overset{?}{=} \sqrt{16} \cdot \sqrt{9}$$
$$\sqrt{144} \overset{?}{=} 4 \cdot 3$$
$$12 = 12 \quad \text{True}$$

So, $\sqrt{16 \cdot 9} = \sqrt{16} \cdot \sqrt{9}$. In general, the square root of the product of two nonnegative real numbers is the product of the square roots of the numbers.

Theorem 8.1

For all real numbers a and b, where $a \geq 0$ and $b \geq 0$, $\sqrt{a} \cdot \sqrt{b} = \sqrt{ab}$.

Proof

Statement	Reason
1. $(\sqrt{a} \cdot \sqrt{b})^2 = (\sqrt{a})^2 (\sqrt{b})^2$	1. Power of a product: $(ab)^n = a^n b^n$
2. $\qquad\qquad = ab$	2. $(\sqrt{a})^2 = \sqrt{a} \cdot \sqrt{a} = a$; $(\sqrt{b})^2 = \sqrt{b} \cdot \sqrt{b} = b$
3. $(\sqrt{ab})^2 = ab$	3. Def of principal square root
4. $(\sqrt{a} \cdot \sqrt{b})^2 = (\sqrt{ab})^2$	4. Trans and Sym Prop of Eq
5. $\sqrt{a} \cdot \sqrt{b} = \sqrt{ab}$	5. Take the square root of both sides.

A radical such as $\sqrt{108}$ is not in simplest form as long as its radicand contains a factor that is a perfect square. Example 1 below illustrates a method for simplifying square roots whose radicands contain perfect square factors.

EXAMPLE 1   Simplify $5\sqrt{108}$. Use a calculator to check the result.

Plan   Find the greatest perfect square factor of 108. Then use Theorem 8.1.

Solution

$$5\sqrt{108} = 5\sqrt{36 \cdot 3}$$
$$= 5\sqrt{36} \cdot \sqrt{3}$$
$$= 5 \cdot 6\sqrt{3} = 30\sqrt{3}$$

Calculator check:

5 ⊗ 108 √ = 51.961524

30 ⊗ 3 √ = 51.961524

In Example 1, the greatest perfect square factor of 108 is 36. Smaller perfect square factors (4 and 9) may also be used, as shown below.

$$5\sqrt{108} = 5\sqrt{4 \cdot 9 \cdot 3} = 5\sqrt{4} \cdot \sqrt{9} \cdot \sqrt{3} = 5 \cdot 2 \cdot 3\sqrt{3} = 30\sqrt{3}$$

By definition, $\sqrt{b}$ is a solution of $x^2 = b$. Therefore, the positive number $\sqrt{3^{10}}$ is a solution of the equation $x^2 = 3^{10}$. But 3^5 is also the positive solution of this equation because $(3^5)^2 = 3^{10}$ is also true. It therefore follows that $\sqrt{3^{10}}$ and 3^5 are the same number. In general,

$$\sqrt{b^{2n}} = b^n, \text{ if } n \text{ is a natural number and } b \geq 0.$$

If n is an *even* number, the above formula is also true for $b < 0$, as illustrated below for $b = -2$ and $n = 2$.

$$\sqrt{(-2)^{2 \cdot 2}} = \sqrt{(-2)^4} = \sqrt{16} = 4 = (-2)^2$$

If, however, n is *odd* and $b < 0$, then the formula must be modified to $\sqrt{b^{2n}} = |b^n|$ to ensure that the right side of the equation is a positive number. For example, if $b = -2$ and $n = 3$, then

$$\sqrt{(-2)^{2 \cdot 3}} = \sqrt{(-2)^6} = \sqrt{64} = 8 = |(-2)^3|.$$

So, if b is any real number, then

$$\sqrt{b^{2n}} = b^n \text{ if } n \text{ is an } \textit{even} \text{ natural number, and}$$
$$\sqrt{b^{2n}} = |b^n| \text{ if } n \text{ is an } \textit{odd} \text{ natural number.}$$

EXAMPLE 2　　Simplify each radical.

a. $\sqrt{28a^{16}}$　　　　　　　　　　　b. $2\sqrt{45b^{10}}$

Solutions　　a. $\dfrac{16}{2} = 8$, an even number.　　b. $\dfrac{10}{2} = 5$, an odd number.

So, no absolute value symbol is needed.　　　　So, the absolute value symbol is necessary.

$\sqrt{28a^{16}}$　　　　　　　　　$2\sqrt{45b^{10}}$
$= \sqrt{4 \cdot 7 \cdot a^{16}}$　　　　　$= 2\sqrt{9 \cdot 5 \cdot b^{10}}$
$= \sqrt{4} \cdot \sqrt{7} \cdot \sqrt{a^{16}}$　　　$= 2\sqrt{9} \cdot \sqrt{5} \cdot \sqrt{b^{10}}$
$= 2\sqrt{7} \cdot a^8$　　　　　　$= 2 \cdot 3\sqrt{5}|b^5|$
$= 2a^8\sqrt{7}$　　　　　　　$= 6|b^5|\sqrt{5}$

If it can be assumed that $b \geq 0$, then absolute value symbols are not necessary when simplifying $\sqrt{b^{2n}}$, regardless of whether n is even or odd. Thus, $\sqrt{b^6} = b^3$, if $b \geq 0$.

Radicals such as $\sqrt{b^1}$ and $\sqrt{b^7}$, whose radicands have exponents that are odd numbers, are not real numbers if b is negative. For example, $\sqrt{(-4)^3} = \sqrt{-64}$, and there is no real number whose square is -64. However, if $b \geq 0$, then a radical such as $\sqrt{b^7}$ can be simplified by first rewriting the radicand as the product of two factors, one of which is b^1.

Thus, $\sqrt{b^7}$, where $b \geq 0$, can be rewritten as

$$\begin{aligned} \sqrt{b^7} &= \sqrt{b^6 \cdot b^1} \\ &= \sqrt{b^6} \cdot \sqrt{b^1} = b^3\sqrt{b} \end{aligned}$$

No absolute value symbol is needed because b is given as nonnegative. In general, if $b \geq 0$ and n is a natural number, then $2n + 1$ is odd and

$$\sqrt{b^{2n+1}} = \sqrt{b^{2n} \cdot b^1} = \sqrt{b^{2n}} \cdot \sqrt{b} = b^n\sqrt{b}$$

EXAMPLE 3 Simplify $\sqrt{27a^9}$, where $a \geq 0$.

Solution
$$\begin{aligned} \sqrt{27a^9} &= \sqrt{9 \cdot 3 \cdot a^8 \cdot a^1} = \sqrt{9 \cdot a^8 \cdot 3 \cdot a} \\ &= \sqrt{9} \cdot \sqrt{a^8} \cdot \sqrt{3a} = 3a^4\sqrt{3a} \end{aligned}$$

Sometimes simplifying a radical involves radicands with both odd and even exponents, as illustrated in Example 4.

EXAMPLE 4 Simplify $4a^3b\sqrt{72a^7b^{10}}$, where $a \geq 0$ and $b \geq 0$.

Solution There is no need for the absolute value symbol because $a \geq 0$ and $b \geq 0$.

$$\begin{aligned} 4a^3b\sqrt{72a^7b^{10}} &= 4a^3b\sqrt{36 \cdot 2 \cdot a^6 \cdot a^1 \cdot b^{10}} \quad \leftarrow \text{Write } a^7 \text{ as } a^6 \cdot a^1 \\ &\qquad\qquad\qquad\qquad\qquad\qquad\qquad\quad \text{because 7 is odd.} \\ &= 4a^3b\sqrt{36 \cdot a^6 \cdot b^{10} \cdot 2 \cdot a} \\ &= 4a^3b\sqrt{36} \cdot \sqrt{a^6} \cdot \sqrt{b^{10}} \cdot \sqrt{2a} \\ &= 4a^3b \cdot 6 \cdot a^3 \cdot b^5 \cdot \sqrt{2a} = 24a^6b^6\sqrt{2a} \end{aligned}$$

When simplifying a square root, keep the following points in mind.

(1) A square root is not in simplest form if its radicand contains a factor that is a perfect square.

(2) If the exponent of b is *even*, then:

$\sqrt{b^{2n}} = |b^n|$ if n is an *odd* natural number,

$\sqrt{b^{2n}} = b^n$ if n is an *even* natural number, and

$\sqrt{b^{2n}} = b^n$ if n is a natural number and $b \geq 0$.

(3) If the exponent of b is *odd*, then:

$\sqrt{b^{2n+1}} = b^n\sqrt{b}$ if n is a natural number and $b \geq 0$.

Indicate whether each of the following statements is always true, sometimes true, or never true.

1. $\sqrt{a^n}$ is defined for all real numbers a, if n is even.
2. $\sqrt{a^3} = a\sqrt{a}$, if $a < 0$.
3. $\sqrt{a^{2n}} = |a^n|$
4. $\sqrt{a} \cdot \sqrt{b} = \sqrt{ab}$

Classroom Exercises

Find the greatest perfect square factor of each radicand.

1. $\sqrt{8}$ 2. $\sqrt{12}$ 3. $\sqrt{18}$ 4. $\sqrt{32}$ 5. $\sqrt{200}$

Simplify each radical.

6. $\sqrt{t^{32}}$ 7. $\sqrt{9a^2}$ 8. $\sqrt{c^{19}}, c \geq 0$ 9. $\sqrt{25y^8}$ 10. $\sqrt{t^{22}}, t \geq 0$

Written Exercises

Simplify.

1. $\sqrt{24}$ 2. $\sqrt{40}$ 3. $\sqrt{45}$ 4. $\sqrt{44}$

5. $\sqrt{50}$ 6. $\sqrt{48}$ 7. $\sqrt{72}$ 8. $\sqrt{98}$

9. $3\sqrt{32}$ 10. $-5\sqrt{20}$ 11. $0.75\sqrt{80}$ 12. $-\frac{1}{14}\sqrt{490}$

13. $\sqrt{16a^4}$ 14. $\sqrt{49b^6}$ 15. $\sqrt{36c^{10}}$ 16. $\sqrt{144x^{16}}$

17. $\sqrt{m^7}, m \geq 0$ 18. $\sqrt{b^{15}}, b \geq 0$ 19. $b^4\sqrt{b^9}, b \geq 0$ 20. $a\sqrt{a^5}, a \geq 0$

21. $4\sqrt{12a^{18}}$ 22. $-3\sqrt{18m^{12}}$ 23. $\sqrt{32a^3}, a \geq 0$ 24. $\sqrt{45m^9}, m \geq 0$

Simplify. For Exercises 25–32 assume that x and y are positive real numbers.

25. $\sqrt{12x^4y^5}$ 26. $\sqrt{24x^7y^{14}}$ 27. $\sqrt{80x^{16}y^{11}}$ 28. $2xy^2\sqrt{75x^7y^6}$

29. $-5x^3y\sqrt{20x^8y^9}$ 30. $3xy\sqrt{18x^5y^{10}}$ 31. $3xy^3\sqrt{32x^6y^3}$ 32. $4y^3z^5\sqrt{44y^{13}z^{14}}$

33. Write how to simplify an expression such as $\sqrt{50a^7b^6}$, where $a \geq 0$ and $b \geq 0$.

Simplify, assuming that m and n are positive integers and that x and y are positive real numbers.

34. $\dfrac{\sqrt{x^{6m}}}{x^{2m}}$

35. $\dfrac{\sqrt{x^{4m-2}}}{x^{m-1}}$

36. $x^m\sqrt{x^{4m+7}}$

37. $x^{m+3}y^{2n-5}\sqrt{x^{8m-2}y^{6n+5}}$

38. Simplify $\sqrt{8x^2 - 16x + 8}$.

39. Prove that for all nonnegative real numbers x and y, $\sqrt{x^2 + y^2} \le x + y$.
(HINT: Assume that $\sqrt{x^2 + y^2} > x + y$. Then show that this leads to a contradiction.)

Mixed Review

Simplify.   *5.2, 5.5*

1. $(3x - 4)^0$

2. $\left|\dfrac{2a^{-2}}{b^{-1}}\right|^3$

3. $(4x - 3)(4x + 3)$

Application: *Inventory Modeling*

To maximize sales, retailers must maintain adequate inventories of the items they sell. To maximize profit, however, they need to maintain those inventories at the lowest possible cost. Inventory modeling balances these objectives by considering the following variables:

k = set-up cost per order h = monthly holding cost per item
c = purchase cost per item s = monthly sales (30-day months)

If Q = the number of items ordered each cycle, then the total cost per order includes the set-up cost k, the purchasing cost $c \cdot Q$, and the holding cost. The average inventory during a cycle is $Q \div 2$ items. The length of a cycle is $Q \div s$ 30-day months. So, the holding cost per order is $hQ^2 \div 2s$.

Therefore, the total cost per month, T, for Q items ordered per cycle is

$$T(Q) = \left(k + cQ + \frac{hQ^2}{2s}\right) \div \frac{Q}{s} = \frac{sk}{Q} + sc + \frac{hQ}{2}.$$

The value of Q that minimizes T is $Q = \sqrt{(2sk) \div h}$.

Each month, a retailer sells 106 radios for which she pays $12 each. Set-up cost is $35 per order, and her holding cost is $1 per radio. Find the number of radios per cycle that minimizes her inventory cost, the minimum inventory cost, and the length of each cycle.

8.3 Sums, Differences, and Products of Square Roots

Objectives

To simplify expressions containing sums or differences of square roots

To simplify expressions containing products of square roots

The radical expressions $5\sqrt{3}$ and $-7\sqrt{3}$ are called *like radicals* because their radical factors are the same. Recall that the Distributive Property allows like terms to be added.

So, for like radicals such as $a\sqrt{b}$ and $c\sqrt{b}$, $a\sqrt{b} + c\sqrt{b} = (a + c)\sqrt{b}$.

Thus, adding like radicals resembles adding like terms:

$$8\sqrt{3} + 5\sqrt{3} = 13\sqrt{3}$$

Sometimes radicals have to be simplified before like radicals can be recognized and combined. This is illustrated below.

EXAMPLE 1 Simplify $4\sqrt{12} - 2\sqrt{18} - 2\sqrt{27} + \sqrt{50}$.

Solution

$4\sqrt{12} - 2\sqrt{18} - 2\sqrt{27} + \sqrt{50}$

$= 4\sqrt{4 \cdot 3} - 2\sqrt{9 \cdot 2} - 2\sqrt{9 \cdot 3} + \sqrt{25 \cdot 2}$

$= 4\sqrt{4} \cdot \sqrt{3} - 2\sqrt{9} \cdot \sqrt{2} - 2\sqrt{9} \cdot \sqrt{3} + \sqrt{25} \cdot \sqrt{2}$

$= 4 \cdot 2\sqrt{3} - 2 \cdot 3\sqrt{2} - 2 \cdot 3\sqrt{3} + 5\sqrt{2}$

$= 8\sqrt{3} - 6\sqrt{3} - 6\sqrt{2} + 5\sqrt{2}$ ← Group like radicals.

$= 2\sqrt{3} - 1\sqrt{2}$, or $2\sqrt{3} - \sqrt{2}$ ← Combine like radicals.

EXAMPLE 2 Simplify $3x^2\sqrt{20x^3y} - x\sqrt{45x^5y} - x^3\sqrt{5xy}$, where $x \geq 0$ and $y \geq 0$.

Solution

$3x^2\sqrt{4 \cdot 5 \cdot x^2 \cdot x \cdot y} - x\sqrt{9 \cdot 5 \cdot x^4 \cdot x \cdot y} - x^3\sqrt{5xy}$

$= 3x^2\sqrt{4 \cdot x^2 \cdot 5xy} - x\sqrt{9 \cdot x^4 \cdot 5xy} - x^3\sqrt{5xy}$

$= 3x^2 \cdot 2 \cdot x\sqrt{5xy} - x \cdot 3 \cdot x^2\sqrt{5xy} - x^3\sqrt{5xy}$

$= 6x^3\sqrt{5xy} - 3x^3\sqrt{5xy} - x^3\sqrt{5xy}$

$= 2x^3\sqrt{5xy}$

When simplifying the product of two square roots, first determine whether the radicands are the same. If they are, use $\sqrt{a} \cdot \sqrt{a} = a$. Thus, $(5\sqrt{x})^2$ becomes $5\sqrt{x} \cdot 5\sqrt{x}$, or $25x$.

If the product of two square roots results in a large integral radicand, it is often easier to factor the radicands first before multiplying them. Both methods are shown in Example 3.

EXAMPLE 3 Simplify $\sqrt{15} \cdot \sqrt{10}$.

Solution

Method 1	**Method 2**
$\sqrt{15} \cdot \sqrt{10} = \sqrt{150}$	$\sqrt{15} \cdot \sqrt{10} = \sqrt{5} \cdot \sqrt{3} \cdot \sqrt{5} \cdot \sqrt{2}$
$= \sqrt{25 \cdot 6}$	$= \sqrt{5} \cdot \sqrt{5} \cdot \sqrt{3} \cdot \sqrt{2}$
$= 5\sqrt{6}$	$= 5\sqrt{6}$

EXAMPLE 4 Simplify $4\sqrt{3x^3} \cdot 5\sqrt{15x^8}$, where $x \geq 0$.

Solution
$$4\sqrt{3x^3} \cdot 5\sqrt{15x^8} = 20\sqrt{45x^{11}}$$
$$= 20\sqrt{9 \cdot 5 \cdot x^{10} \cdot x^1}$$
$$= 20 \cdot 3 \cdot x^5\sqrt{5 \cdot x} = 60x^5\sqrt{5x}$$

The following example utilizes two extensions of the Distributive Property: $a(b + c + d) = ab + ac + ad$ and $(a + b)(c + d) = ac + ad + bc + bd$. The second of these duplicates the FOIL method (see Lesson 5.3).

EXAMPLE 5 Simplify each expression.

a. $2\sqrt{5}\,(6\sqrt{5} - \sqrt{10} + 4\sqrt{7})$ **b.** $(4\sqrt{3} + \sqrt{2})(2\sqrt{3} - 5\sqrt{2})$

Solutions

a. Apply the Distributive Property to $2\sqrt{5}\,(6\sqrt{5} - \sqrt{10} + 4\sqrt{7})$.
$$2\sqrt{5} \cdot 6\sqrt{5} - 2\sqrt{5} \cdot \sqrt{10} + 2\sqrt{5} \cdot 4\sqrt{7}$$
$$= 12\sqrt{5} \cdot \sqrt{5} - 2\sqrt{50} + 8\sqrt{35}$$
$$= 12 \cdot 5 - 2\sqrt{25 \cdot 2} + 8\sqrt{35}$$
$$= 60 - 2 \cdot 5\sqrt{2} + 8\sqrt{35}$$
$$= 60 - 10\sqrt{2} + 8\sqrt{35}$$

b. Apply the FOIL method to $(4\sqrt{3} + \sqrt{2})(2\sqrt{3} - 5\sqrt{2})$.
$$4\sqrt{3} \cdot 2\sqrt{3} - 4\sqrt{3} \cdot 5\sqrt{2} + \sqrt{2} \cdot 2\sqrt{3} - \sqrt{2} \cdot 5\sqrt{2}$$
$$= 8\sqrt{3} \cdot \sqrt{3} - 20\sqrt{6} + 2\sqrt{6} - 5\sqrt{2} \cdot \sqrt{2}$$
$$= 8 \cdot 3 - 20\sqrt{6} + 2\sqrt{6} - 5 \cdot 2$$
$$= 24 - 20\sqrt{6} + 2\sqrt{6} - 10 = 14 - 18\sqrt{6}$$

Classroom Exercises

Simplify. Assume that variables represent nonnegative real numbers.

1. $5\sqrt{3} - 7\sqrt{3}$
2. $7\sqrt{y} + \sqrt{y}$
3. $9\sqrt{5y} - 7\sqrt{5y}$
4. $9\sqrt{a} - 9\sqrt{a}$
5. $\sqrt{5} \cdot \sqrt{11}$
6. $3\sqrt{2} \cdot 4\sqrt{3}$
7. $\sqrt{7} \cdot \sqrt{7}$
8. $3\sqrt{2} \cdot 3\sqrt{2}$

Written Exercises

Simplify. Assume that variables represent nonnegative real numbers.

1. $2\sqrt{3} + \sqrt{27}$
2. $\sqrt{20} - 4\sqrt{5}$
3. $\sqrt{2} - \sqrt{8}$
4. $\sqrt{12} - 2\sqrt{3}$
5. $\sqrt{98} + \sqrt{50}$
6. $6\sqrt{18} + \sqrt{32}$
7. $3\sqrt{99} - 5\sqrt{44}$
8. $5\sqrt{45} - \sqrt{80}$
9. $7\sqrt{20} + 8\sqrt{5} - 2\sqrt{45}$
10. $\sqrt{28} - 2\sqrt{7} - 4\sqrt{63}$
11. $6\sqrt{8} - \sqrt{24} + 3\sqrt{72} - \sqrt{54}$
12. $8\sqrt{5} - 2\sqrt{48} - \sqrt{45} + 4\sqrt{75}$
13. $(\sqrt{8x})^2$
14. $(9\sqrt{x})^2$
15. $(\sqrt{4 - x})^2$
16. $(-2\sqrt{x} - 1)^2$
17. $\sqrt{14} \cdot \sqrt{32}$
18. $-\sqrt{15} \cdot \sqrt{35}$
19. $4\sqrt{5} \cdot 6\sqrt{10}$
20. $3\sqrt{6} \cdot (-4\sqrt{8})$
21. $2\sqrt{3}(7\sqrt{3} - \sqrt{8} + 2\sqrt{5})$
22. $4\sqrt{2}(\sqrt{12} - 3\sqrt{2} + 4\sqrt{8})$
23. $(4\sqrt{2} - 2\sqrt{3})(5\sqrt{2} - \sqrt{3})$
24. $(4\sqrt{5} + \sqrt{6})(3\sqrt{5} - 2\sqrt{6})$
25. $(3\sqrt{2} - 4\sqrt{6})^2$
26. $(7\sqrt{5} + 3\sqrt{11})^2$
27. $(\sqrt{3x} - \sqrt{5y})(2\sqrt{3x} + \sqrt{5y})$
28. $(3\sqrt{x} - 4\sqrt{y})^2$
29. $\sqrt{49ab^3} - \sqrt{ab^3} + 4b\sqrt{ab}$
30. $y\sqrt{12x^5y} - x^2y\sqrt{3xy} + x^2\sqrt{27xy^3}$
31. $\sqrt{4y - 12} + \sqrt{9y - 27}$
32. $\sqrt{4x - 4} - \sqrt{x^3 - x^2}$

Mixed Review

Solve for x. *1.4, 2.3, 5.4, 6.4*

1. $4x - (7 - x) = 12$
2. $|2x - 8| = 4$
3. $ax - b = cx + m$
4. $x^2 - 4 < 0$

Brainteaser

Given identical cups of cocoa and milk, a teaspoon of milk is taken from the milk cup and stirred into the cocoa cup. Then a teaspoon is taken from the cocoa cup and returned to the milk cup. Is there more milk in the cocoa or more cocoa in the milk? Explain.

8.4 Quotients of Square Roots

Objective   To simplify expressions containing quotients of square roots

In the study of trigonometry, one occasionally encounters such equations as

$$\tan x = \frac{\dfrac{\sqrt{3}}{3} + 1}{1 - \dfrac{\sqrt{3}}{3}}.$$

To solve for x in these equations, it is necessary first to simplify the complex fractions. In this lesson, you will learn how to do this.

First notice that $\sqrt{\dfrac{9}{16}} = \dfrac{3}{4}$ and $\dfrac{\sqrt{9}}{\sqrt{16}} = \dfrac{3}{4}$.

Thus, $\sqrt{\dfrac{9}{16}} = \dfrac{\sqrt{9}}{\sqrt{16}}$.

This suggests the following theorem, the proof of which is similar to that of Theorem 8.1.

Theorem 8.2   For all real numbers a and b, where $a \geq 0$ and $b > 0$, $\dfrac{\sqrt{a}}{\sqrt{b}} = \sqrt{\dfrac{a}{b}}$.

Theorem 8.2 can be used to simplify such expressions as $\dfrac{\sqrt{48}}{\sqrt{6}}$, where the radicand of the numerator is exactly divisible by the radicand of the denominator. Example 1 illustrates such a simplification.

EXAMPLE 1   Simplify each expression.

a. $\dfrac{\sqrt{48}}{\sqrt{6}}$

b. $\dfrac{\sqrt{54x^7}}{\sqrt{2x^2}}$, $x > 0$

Solutions

a. $\dfrac{\sqrt{48}}{\sqrt{6}} = \sqrt{\dfrac{48}{6}}$

$= \sqrt{8}$

$= \sqrt{4 \cdot 2}$

$= 2\sqrt{2}$

b. $\dfrac{\sqrt{54x^7}}{\sqrt{2x^2}} = \sqrt{\dfrac{54x^7}{2x^2}}$

$= \sqrt{27x^5}$

$= \sqrt{9 \cdot 3 \cdot x^4 \cdot x}$

$= 3x^2\sqrt{3x}$

The denominator of $\dfrac{2}{3\sqrt{5}}$ contains a radical.

This expression can be rewritten so that the *denominator is rational* as follows:

$$\frac{2}{3\sqrt{5}} = \frac{2 \cdot \sqrt{5}}{3\sqrt{5} \cdot \sqrt{5}} = \frac{2\sqrt{5}}{3 \cdot 5} = \frac{2\sqrt{5}}{15}$$

The above procedure, known as *rationalizing the denominator*, makes use of the property $\dfrac{a}{b} = \dfrac{a \cdot c}{b \cdot c}$, where $b \neq 0$ and $c \neq 0$.

EXAMPLE 2   Simplify. Rationalize the denominator.

a. $\dfrac{15}{\sqrt{18}}$

b. $\dfrac{6x^2y^4}{\sqrt{24x^3y^{10}}}$, $x > 0, y > 0$

Plan   Simplify each radical. Then rationalize the denominator.

a.
$$\frac{15}{\sqrt{18}} = \frac{15}{\sqrt{9 \cdot 2}}$$
$$= \frac{15}{3\sqrt{2}}$$
$$= \frac{5}{\sqrt{2}}$$
$$= \frac{5 \cdot \sqrt{2}}{\sqrt{2} \cdot \sqrt{2}}$$
$$= \frac{5\sqrt{2}}{2}$$

b.
$$\frac{6x^2y^4}{\sqrt{24x^3y^{10}}} = \frac{6x^2y^4}{\sqrt{4 \cdot 6 \cdot x^2 \cdot x \cdot y^{10}}}$$
$$= \frac{6x^2y^4}{2xy^5\sqrt{6x}}$$
$$= \frac{3x}{y\sqrt{6x}}$$
$$= \frac{3x \cdot \sqrt{6x}}{y\sqrt{6x} \cdot \sqrt{6x}}$$
$$= \frac{3x\sqrt{6x}}{6xy} = \frac{\sqrt{6x}}{2y}$$

Example 2a can be checked using a calculator, as shown below.

15 ÷ 18 √ = 3.5355339

5 × 2 √ ÷ 2 = 3.5355339

EXAMPLE 3   Simplify $\sqrt{\dfrac{27a^3}{20m^5n^2}}$, $a \geq 0$, $m > 0$, and $n > 0$.

Solution $\sqrt{\dfrac{27a^3}{20m^5n^2}} = \dfrac{\sqrt{27a^3}}{\sqrt{20m^5n^2}}$ ←Rewrite as the quotient of two radicals.

$$= \frac{3a\sqrt{3a}}{2m^2n\sqrt{5m}} = \frac{3a\sqrt{3a} \cdot \sqrt{5m}}{2m^2n\sqrt{5m} \cdot \sqrt{5m}} = \frac{3a\sqrt{15am}}{2m^2n \cdot 5m} = \frac{3a\sqrt{15am}}{10m^3n}$$

The denominator of $\dfrac{4}{\sqrt{5} - \sqrt{3}}$ contains a difference of radicals.

Multiplying both the numerator and denominator by $\sqrt{5}$ or $\sqrt{3}$ will *not* rationalize the denominator. Instead, to rationalize the denominator use the *conjugate* of $\sqrt{5} - \sqrt{3}$, which is $\sqrt{5} + \sqrt{3}$. Notice that the product of these two irrational numbers is a rational number.

$$(\sqrt{5} + \sqrt{3})(\sqrt{5} - \sqrt{3}) = (\sqrt{5})^2 - (\sqrt{3})^2 = 5 - 3, \text{ or } 2$$

The above procedure is based upon the special product $(a + b)(a - b) = a^2 - b^2$ (see Lesson 5.5). Here, the expressions $a + b$ and $a - b$ are sometimes called a pair of **conjugates**. The conjugate of $a + b$ is $a - b$, and the conjugate of $a - b$ is $a + b$.

EXAMPLE 4　Simplify the given expressions. Rationalize the denominators.

　　a. $\dfrac{14}{\sqrt{5} + \sqrt{3}}$　　　　b. $\dfrac{\sqrt{x} + \sqrt{y}}{\sqrt{x} - \sqrt{y}}$, $x > 0, y > 0, x \neq y$

Solutions　a. $\dfrac{14}{\sqrt{5} + \sqrt{3}}$　　　　b. $\dfrac{\sqrt{x} + \sqrt{y}}{\sqrt{x} - \sqrt{y}}$

$\qquad = \dfrac{14(\sqrt{5} - \sqrt{3})}{(\sqrt{5} + \sqrt{3})(\sqrt{5} - \sqrt{3})}$ 　　　$= \dfrac{(\sqrt{x} + \sqrt{y})(\sqrt{x} + \sqrt{y})}{(\sqrt{x} - \sqrt{y})(\sqrt{x} + \sqrt{y})}$

$\qquad = \dfrac{14(\sqrt{5} - \sqrt{3})}{5 - 3}$ 　　　　　$= \dfrac{(\sqrt{x})^2 + 2 \cdot \sqrt{x} \cdot \sqrt{y} + (\sqrt{y})^2}{(\sqrt{x})^2 - (\sqrt{y})^2}$

$\qquad = \dfrac{14(\sqrt{5} - \sqrt{3})}{2}$ 　　　　　$= \dfrac{x + 2\sqrt{xy} + y}{x - y}$

$\qquad = 7(\sqrt{5} - \sqrt{3})$

To simplify such complex expressions as

$$\dfrac{\dfrac{\sqrt{3}}{3} + 1}{1 - \dfrac{\sqrt{3}}{3}},$$

which contain one or more radicals, first multiply the numerator and denominator of the complex expression by their LCD. Then rationalize the denominator.

EXAMPLE 5

Simplify $\dfrac{\dfrac{\sqrt{3}}{3} + 1}{1 - \dfrac{\sqrt{3}}{3}}$.

Plan

Multiply the numerator and denominator by 3.

Solution

$$\dfrac{3\left(\dfrac{\sqrt{3}}{3} + 1\right)}{3\left(1 - \dfrac{\sqrt{3}}{3}\right)} = \dfrac{\sqrt{3} + 3}{3 - \sqrt{3}}$$

$$= \dfrac{(3 + \sqrt{3})(3 + \sqrt{3})}{(3 - \sqrt{3})(3 + \sqrt{3})}$$

$$= \dfrac{9 + 2 \cdot 3\sqrt{3} + 3}{9 - 3} \quad \leftarrow \text{From the special product}$$
$$\qquad\qquad\qquad\qquad (a + b)^2 = a^2 + 2ab + b^2$$

$$= \dfrac{12 + 6\sqrt{3}}{6} = \dfrac{6(2 + \sqrt{3})}{6} = 2 + \sqrt{3}$$

A radical expression is in simplest form if:
(1) no radicand contains a perfect-square factor other than 1,
(2) no denominator contains a radical,
(3) no radicand contains a fraction, and
(4) the expression is not a complex expression.

Classroom Exercises

Find the conjugate of each expression.

1. $4 + \sqrt{5}$ **2.** $2\sqrt{3} - 6$ **3.** $4\sqrt{2} + 6\sqrt{7}$ **4.** $-\sqrt{11} - \sqrt{3}$

Simplify. Assume that the variables represent positive real numbers.

5. $\dfrac{\sqrt{24}}{\sqrt{6}}$ **6.** $\dfrac{\sqrt{a^7}}{\sqrt{a}}$ **7.** $\dfrac{6}{\sqrt{12}}$ **8.** $\dfrac{12}{\sqrt{5} - 1}$

Written Exercises

Simplify. Assume that no denominator has a value of zero and that all variables represent positive real numbers.

1. $\dfrac{\sqrt{24}}{\sqrt{3}}$ **2.** $\dfrac{\sqrt{140}}{\sqrt{7}}$ **3.** $\dfrac{\sqrt{60}}{\sqrt{5}}$ **4.** $\dfrac{\sqrt{96}}{\sqrt{2}}$

5. $\dfrac{\sqrt{x^{11}}}{\sqrt{x^3}}$ **6.** $\dfrac{\sqrt{b^9}}{\sqrt{b^3}}$ **7.** $\dfrac{\sqrt{56a^7}}{\sqrt{2a^3}}$ **8.** $\dfrac{\sqrt{80x^3}}{\sqrt{2x}}$

9. $\dfrac{4}{\sqrt{2}}$ **10.** $\dfrac{12}{\sqrt{6}}$ **11.** $\dfrac{4}{\sqrt{8}}$ **12.** $\dfrac{8}{\sqrt{20}}$

13. $\dfrac{-20d}{\sqrt{28d^3}}$ **14.** $\dfrac{15x^2}{\sqrt{27x^5}}$ **15.** $\dfrac{6a^2}{\sqrt{18a^8}}$ **16.** $\dfrac{10y^4}{\sqrt{20y^7}}$

17. $\sqrt{\dfrac{4}{a^3}}$ **18.** $\sqrt{\dfrac{8}{3}}$ **19.** $\sqrt{\dfrac{x^5}{y^3}}$ **20.** $\sqrt{\dfrac{4a^2}{125b^3}}$

21. $\dfrac{14}{\sqrt{3}-\sqrt{2}}$ **22.** $\dfrac{10}{\sqrt{7}+\sqrt{2}}$ **23.** $\dfrac{3}{2+\sqrt{3}}$ **24.** $\dfrac{8}{\sqrt{5}-3}$

25. $\dfrac{4xy^4}{\sqrt{32x^5y^8}}$ **26.** $\dfrac{2a^3b^4}{\sqrt{48a^6b^7}}$ **27.** $\dfrac{-18x^3y}{\sqrt{27x^7y^4}}$ **28.** $\dfrac{8x^5y^4}{\sqrt{72x^2y^5}}$

29. $\dfrac{\sqrt{3}-\sqrt{2}}{\sqrt{3}+\sqrt{2}}$ **30.** $\dfrac{2\sqrt{5}-\sqrt{3}}{\sqrt{5}-\sqrt{3}}$ **31.** $\dfrac{2\sqrt{x}}{3\sqrt{x}-4\sqrt{y}}$ **32.** $\dfrac{\sqrt{a}-\sqrt{b}}{\sqrt{a}+\sqrt{b}}$

33. $\dfrac{1+\dfrac{\sqrt{2}}{5}}{1-\dfrac{\sqrt{2}}{5}}$ **34.** $\dfrac{\dfrac{\sqrt{2}}{2}+4}{1-\dfrac{\sqrt{2}}{2}}$ **35.** $\dfrac{\dfrac{\sqrt{6}-\sqrt{2}}{2}}{\dfrac{\sqrt{6}}{2}+\sqrt{2}}$ **36.** $\dfrac{\dfrac{\sqrt{5}}{2}-\dfrac{\sqrt{3}}{3}}{\dfrac{\sqrt{5}}{2}+\dfrac{\sqrt{3}}{3}}$

37. $\dfrac{\sqrt{a+b}+\sqrt{a-b}}{\sqrt{a+b}-\sqrt{a-b}}$ **38.** $\dfrac{\sqrt{4x^2+1}+2x}{\sqrt{4x^2+1}-2x}$ **39.** $\sqrt{\dfrac{\sqrt{20}+\sqrt{12}}{\sqrt{5}+\sqrt{3}}}$

40. Prove Theorem 8.2.

Midchapter Review

Simplify. Assume that variables represent nonnegative real numbers.   *8.1, 8.2, 8.3, 8.4*

1. $-\sqrt{144}$ **2.** $\sqrt{90x^6}$ **3.** $-2\sqrt{45x^4}$ **4.** $\sqrt{44a^7b^8}$

5. $\sqrt{35}\cdot\sqrt{21}$ **6.** $2\sqrt{27}+4\sqrt{75}$ **7.** $\dfrac{\sqrt{65}}{\sqrt{5}}$ **8.** $\dfrac{-2}{\sqrt{5}-\sqrt{7}}$

Solve each equation. Approximate irrational results to the nearest hundredth.   *8.1*

9. $y^2=196$ **10.** $x^2=7$ **11.** $b^2-8=12$ **12.** $a^2-0.36=0$

8.5 Simplifying Radicals With Indices Greater Than 2

Objective

To simplify expressions containing cube, fourth, or fifth roots

To solve the equation $x^2 = 36$, the definition of square root is used: $x = \sqrt{36} = 6$ or $x = -\sqrt{36} = -6$. The *principal square root* is 6.

Equations such as $x^4 = 81$ can be solved in a similar way. Because $3^4 = 81$ and $(-3)^4 = 81$, $x = \sqrt[4]{81} = 3$ or $x = -\sqrt[4]{81} = -3$. The numbers 3 and -3 are *fourth roots* of 81. Of the two real fourth roots of 81, the *principal fourth root* is defined as the positive root 3, symbolized as $\sqrt[4]{81}$.

Definition

> The **nth roots** of a real number b are the solutions of the equation $x^n = b$, where n is a positive integer.

In general, the *principal nth root* of a number b is the real-number solution of the equation $x^n = b$ and is written as $\sqrt[n]{b}$. As with square roots, the number b is the *radicand*, and the positive integer n, where $n > 1$, is the **index**. For square roots, the index 2 is understood.

The principal nth root of a real number b is *not* necessarily positive, and may not exist as a real number. Consider the following.

(1) n even, radicand positive: $\sqrt[4]{16} = \sqrt[4]{2^4} = 2$
$$\sqrt[4]{16} = \sqrt[4]{(-2)^4} = |-2| = 2$$

(2) n even, radicand negative: $\sqrt[4]{-16}$ is not a real number because there is no real number whose fourth power is negative.

(3) n odd, radicand positive: $\sqrt[3]{8} = \sqrt[3]{2^3} = 2$

(4) n odd, radicand negative: $\sqrt[3]{-8} = \sqrt[3]{(-2)^3} = -2$

Thus, if n is odd and the radicand is negative, the principal root is negative.

Definition

> The **principal nth root** of a real number b, symbolized by $\sqrt[n]{b}$, represents either
>
> (1) the nonnegative nth root of b if $b \geq 0$ and n is even, or
>
> (2) the one real nth root of b if n is odd.

Illustrations (1)–(4) and the definition just given also suggest the following generalizations:

$$\left(\sqrt[n]{b}\right)^n = b, \text{ because } \sqrt[n]{b} \text{ is a solution of } x^n = b.$$

$$\sqrt[n]{b^n} = |b| \text{ if } n \text{ is even.}$$

$$\sqrt[n]{b^n} = b \text{ if } n \text{ is odd. (Do not use absolute value symbols if } n \text{ is odd.)}$$

EXAMPLE 1 Simplify each radical.

a. $\sqrt[4]{(4a)^4}$ b. $\left(\sqrt[4]{x-1}\right)^4$ c. $\left(2\sqrt[3]{2x-1}\right)^3$

Solutions

a. $\sqrt[4]{(4a)^4} = |4a|$ b. $\left(\sqrt[4]{x-1}\right)^4 = x - 1$

$\quad\quad = 4|a|$

c. $\left(2\sqrt[3]{2x-1}\right)^3 = 2^3\left(\sqrt[3]{2x-1}\right)^3$

$\quad\quad\quad\quad = 8(2x - 1)$

$\quad\quad\quad\quad = 16x - 8$

Theorem 8.3

For all odd integers $n > 1$ and all real numbers a and b, and for all even integers $n > 1$ and all real numbers $a \geq 0$ and $b \geq 0$,

$$\sqrt[n]{a \cdot b} = \sqrt[n]{a} \cdot \sqrt[n]{b}.$$

To simplify radicals such as $\sqrt[5]{-128}$, first find the greatest fifth power that is a factor of -128.

EXAMPLE 2 Simplify.

a. $\sqrt[5]{-128}$ b. $\sqrt[4]{160}$ c. $\sqrt[3]{27x^{12}}$

Solutions

a. $\sqrt[5]{-128}$

$\quad = \sqrt[5]{(-2)^7}$

$\quad = \sqrt[5]{(-2)^5} \cdot \sqrt[5]{(-2)^2}$

$\quad = -2\sqrt[5]{4}$

b. $\sqrt[4]{160}$

$\quad = \sqrt[4]{2^4 \cdot 10}$

$\quad = \sqrt[4]{2^4} \cdot \sqrt[4]{10}$

$\quad = 2\sqrt[4]{10}$

c. $\sqrt[3]{27x^{12}}$

$\quad = \sqrt[3]{27} \cdot \sqrt[3]{x^{12}}$

$\quad = \sqrt[3]{3^3} \cdot \sqrt[3]{(x^4)^3}$

$\quad = 3x^4$

EXAMPLE 3 Simplify $\sqrt[3]{250a^{11}}$.

Solution

$\sqrt[3]{250a^{11}} = \sqrt[3]{125 \cdot 2 \cdot a^9 \cdot a^2}$ ← Use $125 = 5^3$ and $(a^3)^3 = a^9$.

$\quad\quad\quad = \sqrt[3]{5^3 \cdot a^9 \cdot 2a^2}$

$\quad\quad\quad = \sqrt[3]{5^3} \cdot \sqrt[3]{(a^3)^3} \cdot \sqrt[3]{2a^2}$

$\quad\quad\quad = 5a^3\sqrt[3]{2a^2}$

Like radicals can be added or subtracted just as square roots are. Similarly, products and quotients of nth roots can also be simplified in the same manner as square roots.

EXAMPLE 4   Simplify $5\sqrt[4]{2} + 6\sqrt[4]{32} - \sqrt[4]{162}$.

Solution
$$5\sqrt[4]{2} + 6\sqrt[4]{32} - \sqrt[4]{162}$$
$$= 5\sqrt[4]{2} + 6\sqrt[4]{2^4 \cdot 2} - \sqrt[4]{3^4 \cdot 2} \quad \leftarrow 32 = 16 \cdot 2 = 2^4 \cdot 2 \text{ and}$$
$$= 5\sqrt[4]{2} + 6 \cdot 2\sqrt[4]{2} - 3\sqrt[4]{2} \qquad 162 = 81 \cdot 2 = 3^4 \cdot 2$$
$$= 5\sqrt[4]{2} + 12\sqrt[4]{2} - 3\sqrt[4]{2} = 14\sqrt[4]{2}$$

EXAMPLE 5   Simplify $\sqrt[4]{3y^3} \cdot \sqrt[4]{6y} \cdot \sqrt[4]{9y^2}$, where $y \geq 0$.

Solution
$$\sqrt[4]{3y^3} \cdot \sqrt[4]{6y} \cdot \sqrt[4]{9y^2} = \sqrt[4]{3y^3 \cdot 6y \cdot 9y^2}$$
$$= \sqrt[4]{162y^6}$$
$$= \sqrt[4]{81 \cdot 2 \cdot y^4 \cdot y^2}$$
$$= 3y\sqrt[4]{2y^2}$$

The property of $\left(\sqrt[n]{b}\right)^n = b$ may be used to rationalize a denominator.

$$\frac{10}{\sqrt[3]{25}} = \frac{10}{\sqrt[3]{5 \cdot 5}} = \frac{10}{\sqrt[3]{5} \cdot \sqrt[3]{5}} = \frac{10 \cdot \sqrt[3]{5}}{\sqrt[3]{5} \cdot \sqrt[3]{5} \cdot \sqrt[3]{5}} = \frac{10\sqrt[3]{5}}{5} = 2\sqrt[3]{5}$$

Theorem 8.4

For all odd integers $n > 1$ and all real numbers a and b, $b \neq 0$, and for all even integers $n > 1$ and all real numbers a and b, $a \geq 0$ and $b > 0$,

$$\sqrt[n]{\frac{a}{b}} = \frac{\sqrt[n]{a}}{\sqrt[n]{b}}.$$

EXAMPLE 6   Simplify $\sqrt[4]{\dfrac{3}{8a^2}}$, $a > 0$.

Solution
$$\sqrt[4]{\frac{3}{8a^2}} = \frac{\sqrt[4]{3}}{\sqrt[4]{2^3 a^2}}$$

$$= \frac{\sqrt[4]{3}}{\sqrt[4]{2^3 a^2}} \cdot \frac{\sqrt[4]{2a^2}}{\sqrt[4]{2a^2}} \quad \leftarrow \text{Choose the radicand } 2a^2 \text{ to get the fourth root}$$
$$\text{of a fourth power: } 2^3 a^2 \cdot 2a^2 = 2^4 a^4 = (2a)^4.$$

$$= \frac{\sqrt[4]{3 \cdot 2a^2}}{\sqrt[4]{2^4 a^4}} = \frac{\sqrt[4]{6a^2}}{2a}$$

Classroom Exercises

Simplify.

1. $\sqrt[3]{8}$
2. $\sqrt[4]{16}$
3. $\left(\sqrt[4]{3}\right)^4$
4. $\left(\sqrt[5]{a}\right)^5$
5. $\sqrt[3]{5} \cdot \sqrt[3]{5} \cdot \sqrt[3]{5}$

6. $\sqrt[3]{-27}$
7. $\sqrt[3]{\dfrac{8}{27}}$
8. $\sqrt[6]{a^6}$
9. $\sqrt[5]{64}$
10. $3\sqrt[4]{64} + \sqrt[4]{4}$

Written Exercises

Simplify. Assume that no denominator is 0.

1. $\left(\sqrt[4]{7x}\right)^4$
2. $\left(\sqrt[4]{x+2}\right)^4$
3. $\left(-2\sqrt[3]{3y}\right)^3$
4. $\left(4\sqrt[3]{2x-1}\right)^3$

5. $\sqrt[4]{81}$
6. $\sqrt[3]{125}$
7. $\sqrt[5]{-32}$
8. $\sqrt[5]{243}$

9. $\sqrt[4]{64}$
10. $\sqrt[3]{-54}$
11. $\sqrt[3]{-250}$
12. $\sqrt[5]{128}$

13. $\sqrt[3]{x^{21}}$
14. $\sqrt[4]{(2x)^4}$
15. $\sqrt[6]{a^{30}}$, $a \geq 0$
16. $\sqrt[4]{b^{16}}$

17. $\sqrt[4]{16a^{23}}$, $a \geq 0$
18. $\sqrt[3]{27x^9}$
19. $\sqrt[5]{32b^{10}}$
20. $\sqrt[4]{81x^{12}}$, $x \geq 0$

21. $7\sqrt[4]{2} + 8\sqrt[4]{32}$
22. $3\sqrt[3]{54} - 5\sqrt[3]{2}$

23. $4\sqrt[4]{80} - 2\sqrt[4]{5}$
24. $\sqrt[4]{8a^5} \cdot \sqrt[4]{4a^7}$, $a \geq 0$

25. $\sqrt[3]{9b^7} \cdot \sqrt[3]{12b^5}$
26. $\sqrt[4]{8x^5} \cdot \sqrt[4]{20x^2}$, $x \geq 0$

27. $\dfrac{14}{\sqrt[3]{7}}$
28. $\dfrac{6x}{\sqrt[4]{2}}$
29. $\sqrt[4]{\dfrac{7}{8}}$
30. $\sqrt[4]{\dfrac{5}{6}}$

31. $5\sqrt[3]{3} + \sqrt[3]{24} - 7\sqrt[3]{81}$
32. $7\sqrt[4]{3} + 8\sqrt[4]{243} - \sqrt[4]{48}$

33. $\sqrt[4]{7d^5} \cdot \sqrt[4]{8d^3} \cdot \sqrt[4]{2d^3}$, $d \geq 0$
34. $\sqrt[5]{10a^7} \cdot \sqrt[5]{2a} \cdot \sqrt[5]{16a^4}$

35. $\sqrt[3]{\dfrac{7}{25y^2}}$
36. $\dfrac{4h^3}{\sqrt[5]{8h^7}}$
37. $\sqrt[5]{\dfrac{3}{16a^4b^2}}$
38. $\dfrac{6a^2}{\sqrt[4]{8a^2}}$

39. Prove Theorem 8.3.
40. Prove Theorem 8.4.

Mixed Review

Solve. *5.1, 6.4, 6.6*

1. $2^{3n-1} = 32$
2. $x^2 < 9$
3. $2x^3 + 3x^2 - 8x + 3 = 0$

8.6 Rational-Number Exponents

Objectives

To write expressions with rational-number exponents in radical form, and vice versa

To evaluate expressions containing rational-number exponents

Recall that b^0, where $b \neq 0$, is defined as 1, and b^{-n}, where $b \neq 0$ and n is a positive integer, is defined as $\frac{1}{b^n}$.

These two definitions allowed the Laws of Exponents to be extended to nonpositive integral exponents. In this lesson, a further extension will be made to cover exponents such as $\frac{1}{5}$ and $-\frac{2}{3}$ that are rational but not integral.

Recall the power-of-a-power law, $(b^m)^n = b^{mn}$, where m and n are integers and $b \neq 0$. If this law is to be extended to rational-number exponents, then the following statement must be true.

$$\left(4^{\frac{1}{5}}\right)^5 = 4^{\frac{1}{5} \cdot 5} = 4^1, \text{ or } 4.$$

Thus, the fifth power of $4^{\frac{1}{5}}$ must be 4, and the fifth power of $\sqrt[5]{4}$ is also 4, so $4^{\frac{1}{5}}$ can be *defined* as $\sqrt[5]{4}$.

This definition of the meaning of the exponent $\frac{1}{5}$ can be extended to the following general definition.

Definition

For all integers $n > 1$ and all real numbers b,

$b^{\frac{1}{n}} = \sqrt[n]{b}$, provided that $\sqrt[n]{b}$ is a real number.

The definition of negative rational exponents such as $-\frac{1}{2}$ and $-\frac{1}{4}$ is somewhat similar to the definition of negative integer exponents given in Chapter 5.

Definition

For all integers $n > 0$ and all nonzero real numbers b,

$b^{-\frac{1}{n}} = \dfrac{1}{b^{\frac{1}{n}}}$, provided that $b^{\frac{1}{n}}$ is a real number.

From this definition, $(-8)^{-\frac{1}{3}} = \dfrac{1}{(-8)^{\frac{1}{3}}} = \dfrac{1}{\sqrt[3]{-8}} = \dfrac{1}{-2}$, or $-\dfrac{1}{2}$.

EXAMPLE 1   Evaluate each of the following.

 a. $49^{\frac{1}{2}}$ b. $(-64)^{\frac{1}{3}}$ c. $81^{-0.25}$

Solutions a. $49^{\frac{1}{2}} = \sqrt{49} = 7$

 b. $(-64)^{\frac{1}{3}} = \sqrt[3]{-64} = \sqrt[3]{(-4)^3} = -4$

 c. $81^{-0.25} = 81^{-\frac{1}{4}} = \dfrac{1}{81^{\frac{1}{4}}} = \dfrac{1}{\sqrt[4]{81}} = \dfrac{1}{3}$

The Laws of Exponents can also be extended to rational exponents of the form $\frac{m}{n}$. To see how this is done, assume that the power-of-a-power law holds in the two parallel examples below.

$$8^{\frac{2}{3}} = 8^{2 \cdot \frac{1}{3}} = (8^2)^{\frac{1}{3}} \qquad \text{or} \qquad 8^{\frac{2}{3}} = 8^{\frac{1}{3} \cdot 2} = (8^{\frac{1}{3}})^2$$
$$= \sqrt[3]{8^2} \qquad\qquad\qquad\qquad = (\sqrt[3]{8})^2$$
$$= \sqrt[3]{64} = 4 \qquad\qquad\qquad\quad\; = 2^2 = 4$$

Thus, $8^{\frac{2}{3}} = \sqrt[3]{8^2}$, or $(\sqrt[3]{8})^2$.

This development suggests the following definition.

Definition

> For all integers m and n, where $n > 1$, and all real numbers $b \neq 0$,
> $$b^{\frac{m}{n}} = (\sqrt[n]{b})^m = \sqrt[n]{b^m} \text{ provided that } \sqrt[n]{b} \text{ is real.}$$

With these definitions it can be shown that all the Laws of Exponents stated in Chapter 5 do hold true for rational exponents.

EXAMPLE 2   Write in exponential form.

 a. $\sqrt[7]{x^4}$ b. $(\sqrt[3]{13})^2$

Solutions a. $\sqrt[7]{x^4} = x^{\frac{4}{7}}$ b. $(\sqrt[3]{13})^2 = 13^{\frac{2}{3}}$

EXAMPLE 3   Write in radical form and simplify.

 a. $2^{\frac{4}{5}}$ b. $(-8)^{\frac{2}{3}}$ c. $-8^{\frac{2}{3}}$

Solutions a. $2^{\frac{4}{5}} = \sqrt[5]{2^4}$ b. $(-8)^{\frac{2}{3}} = \sqrt[3]{(-8)^2}$ c. $-8^{\frac{2}{3}} = -\sqrt[3]{8^2}$
 $= \sqrt[5]{16}$ $= \sqrt[3]{64} = 4$ $= -\sqrt[3]{64} = -4$

When finding the value of a power with a rational-number exponent, such as $b^{\frac{m}{n}}$, it is easier to find the nth root first and then to find the mth power of the result.

EXAMPLE 4 Find the value of each of the following expressions.

  a. $x^{-\frac{4}{5}}$, for $x = 32$ **b.** $5y^{\frac{3}{4}}$, for $y = 16$

Solutions **a.** $x^{-\frac{4}{5}} = 32^{-\frac{4}{5}}$ **b.** $5y^{\frac{3}{4}} = 5 \cdot 16^{\frac{3}{4}}$

$\qquad\qquad = \dfrac{1}{32^{\frac{4}{5}}}$ $\qquad\qquad = 5(\sqrt[4]{16})^3$

$\qquad\qquad\qquad\qquad\qquad\qquad\qquad = 5 \cdot 2^3$

$\qquad = \dfrac{1}{(\sqrt[5]{32})^4} = \dfrac{1}{2^4}$, or $\dfrac{1}{16}$ $\qquad = 5 \cdot 8$, or 40

Classroom Exercises

Give each expression in exponential form.

1. $\sqrt[3]{43}$ **2.** $\sqrt{19}$ **3.** $\sqrt[7]{32}$ **4.** $\sqrt[5]{x}$ **5.** $\sqrt[n]{a}, a > 0$

Give each expression in radical form. Simplify, if possible.

6. $27^{\frac{1}{3}}$ **7.** $5^{\frac{1}{4}}$ **8.** $28^{\frac{1}{2}}$ **9.** $4^{\frac{2}{3}}$ **10.** $5^{-\frac{1}{2}}$

Written Exercises

Evaluate.

1. $36^{\frac{1}{2}}$ **2.** $8^{\frac{1}{3}}$ **3.** $81^{-\frac{1}{4}}$ **4.** $(-27)^{-\frac{1}{3}}$ **5.** $400^{-0.5}$

6. $4 \cdot 16^{\frac{1}{2}}$ **7.** $5 \cdot 4^{-\frac{1}{2}}$ **8.** $-7 \cdot 27^{\frac{1}{3}}$ **9.** $-4 \cdot 81^{0.25}$ **10.** $-3 \cdot 16^{0.25}$

Write each expression in exponential form.

11. $\sqrt[5]{a^3}$ **12.** $\sqrt[4]{b^3}, b \geq 0$ **13.** $(\sqrt[3]{11})^4$ **14.** $(\sqrt{13})^5$ **15.** $\sqrt[5]{23^4}$

Write each expression in radical form and simplify.

16. $3^{\frac{2}{3}}$ **17.** $5^{\frac{3}{4}}$ **18.** $(-7)^{\frac{2}{3}}$ **19.** $3^{-\frac{4}{5}}$ **20.** $a^{\frac{4}{5}}$

Evaluate.

21. $16^{\frac{3}{4}}$ **22.** $25^{-\frac{5}{2}}$ **23.** $(-27)^{\frac{2}{3}}$ **24.** $-27^{\frac{2}{3}}$ **25.** $32^{\frac{2}{5}}$

26. $6 \cdot 4^{-\frac{5}{2}}$ **27.** $-4 \cdot 343^{-\frac{2}{3}}$ **28.** $8 \cdot 16^{-\frac{3}{4}}$ **29.** $6^0 \cdot 27^{-\frac{5}{3}}$ **30.** $8^{\frac{2}{3}} \cdot 2^{-1}$

31. The price p, in dollars, of a certain type of solar heater is approximated by $p = 3x^{\frac{1}{2}} + 5x^{\frac{2}{3}}$, where x is the number of units. Find the price of 64 units.

Simplify.

32. $\sqrt[3]{x^3 - 6x^2 + 12x - 8}$

33. $[1^m - (-27)^{\frac{1}{3}}]^{-\frac{3}{2}}$

34. Given that $f(x) = (x^3 + 31)^{-\frac{3}{2}}$, find $f(\sqrt[3]{5})$.

Mixed Review

1. Graph $3x - 4y = 12$. *3.5*

2. Simplify $(3x^{-2}y^5)^{-2}$. *5.2*

3. Solve: $\begin{aligned} 3x - 2y &= 4 \\ 4x + 3y &= 11 \end{aligned}$ *4.3*

4. Write an equation of the line containing the point $(2, -1)$ and parallel to the line whose equation is $4x - y = 8$. *3.7*

Using the Calculator

Though most scientific calculators do not have special keys for roots with indices greater than 2, many do have an $\boxed{x^y}$ (or $\boxed{y^x}$) key and other special-function keys. To find $\sqrt[6]{64}$ using such a calculator, rewrite $\sqrt[6]{64}$ as $64^{\frac{1}{6}}$ because the exponent $\frac{1}{6}$ can be entered directly.

If you have the reciprocal key ($\boxed{1/x}$), enter: 64 $\boxed{x^y}$ 6 $\boxed{1/x}$ $\boxed{=}$ 2.

Without a reciprocal key, use the left and right parentheses keys to express the exponent as $(1 \div 6)$: 64 $\boxed{x^y}$ $\boxed{(}$ 1 $\boxed{\div}$ 6 $\boxed{)}$ $\boxed{=}$ 2.

If your calculator has an inverse key ($\boxed{INV}$ or $\boxed{2nd}$), $\sqrt[6]{64}$ can be found using the fact that finding the 6th root is the inverse (reverse) of finding the 6th power: 64 $\boxed{INV}$ $\boxed{x^y}$ 6 $\boxed{=}$ 2.

The change-of-sign key $\boxed{+/-}$ can be used to show that $243^{-\frac{1}{5}} = \frac{1}{3}$.

Enter: 243 $\boxed{x^y}$ 5 $\boxed{1/x}$ $\boxed{+/-}$ $\boxed{=}$ 0.33333333

To find $64^{\frac{5}{6}}$, enter $\frac{5}{6}$ as $(5 \div 6)$ or find $64^{\frac{1}{6}}$ and raise the result to the 5th power, because $64^{\frac{5}{6}} = \left(64^{\frac{1}{6}}\right)^5$.

Use a scientific calculator to check your answers to Written Exercises 1–10 and 21–25. (Many calculators do not accept a negative number as a base. If yours does not, compute the result with a positive base, and then determine the sign of the result.

8.7 Applying Rational-Number Exponents

Objectives   To simplify expressions containing rational-number exponents
To solve exponential equations

It has now been demonstrated that the Laws of Exponents hold for all rational-number exponents. Several of these laws are shown below.

$$b^r \cdot b^s = b^{r+s} \qquad\qquad \frac{b^r}{b^s} = b^{r-s} \qquad\qquad (b^r)^s = b^{rs}$$

$$(ab)^r = a^r b^r \qquad\qquad \left(\frac{a}{b}\right)^r = \frac{a^r}{b^r} \qquad\qquad b^{-r} = \frac{1}{b^r}$$

$$(\sqrt[n]{b})^n = b$$

$$b^{\frac{m}{n}} = (\sqrt[n]{b})^m = \sqrt[n]{b^m}$$

All of the above laws are true when a and b are positive real numbers, r and s are rational numbers, m is an integer, and n is a natural number. Under some conditions, the laws also hold for negative values of a and b. Review Chapter 5 and earlier lessons of this chapter for the various restrictions.

These laws can be used to simplify expressions that involve rational exponents.

EXAMPLE 1   Simplify $6^{-\frac{3}{5}}$.

Plan   First rewrite $6^{-\frac{3}{5}}$ as $\dfrac{1}{6^{\frac{3}{5}}}$. Then rationalize the denominator by obtaining a power of 6 that is a positive integer.

Solution   The least positive integer greater than $\frac{3}{5}$ is 1.

$\frac{3}{5} + \frac{2}{5} = \frac{5}{5} = 1$, so multiply numerator and denominator of

$\dfrac{1}{6^{\frac{3}{5}}}$ by $6^{\frac{2}{5}}$.

$$6^{-\frac{3}{5}} = \frac{1}{6^{\frac{3}{5}}} = \frac{1 \cdot 6^{\frac{2}{5}}}{6^{\frac{3}{5}} \cdot 6^{\frac{2}{5}}} = \frac{6^{\frac{2}{5}}}{6^{\frac{5}{5}}} = \frac{6^{\frac{2}{5}}}{6^{1}}, \text{ or } \frac{6^{\frac{2}{5}}}{6}$$

Note that an expression such as $6^{-\frac{3}{5}}$ is considered to be in the simplest form when its exponent is positive and its denominator is rational,

as in $\dfrac{6^{\frac{2}{5}}}{6}$.

EXAMPLE 2 Simplify $4x^{-1} \cdot 7x^{\frac{1}{3}} \cdot 2x^{-\frac{2}{3}}$, where $x > 0$.

Solution
$$4x^{-1} \cdot 7x^{\frac{1}{3}} \cdot 2x^{-\frac{2}{3}} = 56x^{-1+\frac{1}{3}-\frac{2}{3}}$$
$$= 56x^{-\frac{4}{3}}$$
$$= \frac{56}{x^{\frac{4}{3}}} = \frac{56 \cdot x^{\frac{2}{3}}}{x^{\frac{4}{3}} \cdot x^{\frac{2}{3}}} = \frac{56x^{\frac{2}{3}}}{x^2}$$

EXAMPLE 3 Simplify $\left(27x^{12}y^{-6}\right)^{-\frac{2}{3}}$, where $x > 0$ and $y > 0$.

Plan Use $(ab)^r = a^r b^r$ and $(b^r)^s = b^{rs}$.

Solution
$$\left(27 x^{12}y^{-6}\right)^{-\frac{2}{3}} = 27^{-\frac{2}{3}} \cdot x^{12\left(-\frac{2}{3}\right)} \cdot y^{-6\left(-\frac{2}{3}\right)}$$
$$= 27^{-\frac{2}{3}}x^{-8}y^4 = \frac{y^4}{27^{\frac{2}{3}}x^8} = \frac{y^4}{(\sqrt[3]{27})^2 x^8} = \frac{y^4}{3^2 x^8} = \frac{y^4}{9x^8}$$

The laws $\dfrac{b^r}{b^s} = b^{r-s}$ and $\left(\dfrac{a}{b}\right)^r = \dfrac{a^r}{b^r}$ are used in Examples 4 and 5.

EXAMPLE 4 Simplify $\dfrac{8^{\frac{3}{4}}x^{-\frac{1}{2}}y^{\frac{1}{4}}}{8^{\frac{1}{12}}x^{\frac{3}{4}}y^{-\frac{1}{6}}}$, where $x > 0$ and $y > 0$.

Plan Rewrite the exponents of like bases so that they have the same denominator.

Solution
$$\frac{8^{\frac{9}{12}}x^{-\frac{2}{4}}y^{\frac{3}{12}}}{8^{\frac{1}{12}}x^{\frac{3}{4}}y^{-\frac{2}{12}}} = 8^{\frac{8}{12}}x^{-\frac{5}{4}}y^{\frac{5}{12}} = \frac{8^{\frac{2}{3}}y^{\frac{5}{12}}}{x^{\frac{5}{4}}} = \frac{4y^{\frac{5}{12}} \cdot x^{\frac{3}{4}}}{x^{\frac{5}{4}} \cdot x^{\frac{3}{4}}} = \frac{4y^{\frac{5}{12}}x^{\frac{3}{4}}}{x^2}$$

EXAMPLE 5 Simplify $\left|\dfrac{4c^{-6}}{25d^{-4}}\right|^{\frac{3}{2}}$, where $c > 0$ and $d > 0$.

Solution
$$\left|\frac{4c^{-6}}{25d^{-4}}\right|^{\frac{3}{2}} = \frac{4^{\frac{3}{2}}c^{-6\left(\frac{3}{2}\right)}}{25^{\frac{3}{2}}d^{-4\left(\frac{3}{2}\right)}} = \frac{4^{\frac{3}{2}}c^{-9}}{25^{\frac{3}{2}}d^{-6}} = \frac{4^{\frac{3}{2}}d^6}{25^{\frac{3}{2}}c^9} = \frac{8d^6}{125c^9}$$

The Laws of Exponents can also be used to simplify the products and quotients of radicals with unequal indices.

To simplify an expression such as $\sqrt[3]{x^2} \cdot \sqrt[4]{x^3}$, $x > 0$, first write the expression with rational-number exponents.

$$\sqrt[3]{x^2} \cdot \sqrt[4]{x^3} = x^{\frac{2}{3}} \cdot x^{\frac{3}{4}} = x^{\frac{17}{12}} = x^{1 + \frac{5}{12}} = x \cdot x^{\frac{5}{12}} = x\sqrt[12]{x^5}.$$

To solve exponential equations involving rational-number exponents, set the exponents equal to each other and solve the resulting equation.

EXAMPLE 6 Solve $32^{3x-2} = \frac{1}{16}$ and check.

Plan First, rewrite 32 and $\frac{1}{16}$ as powers of the same positive base.

Solution

$$32^{3x-2} = \frac{1}{16}$$
$$32^{3x-2} = 16^{-1}$$
$$(2^5)^{3x-2} = (2^4)^{-1}$$
$$2^{15x-10} = 2^{-4}$$
$$15x - 10 = -4$$
$$15x = 6$$
$$x = \frac{2}{5}$$

Check:
$$32^{3x-2} = \frac{1}{16}$$
$$32^{3 \cdot \frac{2}{5} - 2} \overset{?}{=} \frac{1}{16}$$
$$32^{-\frac{4}{5}} \overset{?}{=} \frac{1}{16}$$
$$\frac{1}{32^{\frac{4}{5}}} \overset{?}{=} \frac{1}{16}$$
$$\frac{1}{16} = \frac{1}{16} \quad \text{True}$$

Classroom Exercises

Simplify each expression. Assume that $x > 0$.

1. $7^{-\frac{1}{3}}$ **2.** $x^{-\frac{1}{2}}$ **3.** $x^{-\frac{2}{5}} x^{\frac{7}{5}}$ **4.** $\dfrac{x^{\frac{4}{7}}}{x^{\frac{2}{7}}}$

Written Exercises

Simplify each expression. Assume that the values of the variables are positive.

1. $5^{-\frac{3}{4}}$ **2.** $4^{-\frac{2}{3}}$ **3.** $30x^{-\frac{4}{3}}$ **4.** $2a^{-\frac{7}{3}}$

5. $4^{\frac{2}{9}} \cdot 4^{-\frac{1}{9}} \cdot 4^{\frac{4}{9}}$ **6.** $a^{\frac{5}{4}} \cdot a^{\frac{3}{4}} \cdot a^{-\frac{7}{4}}$ **7.** $t^4 \cdot t^{-\frac{1}{4}}$ **8.** $7a^3 \cdot 8a^{-\frac{2}{5}}$

9. $(x^{\frac{2}{3}} y^{\frac{1}{4}})^{12}$ **10.** $(x^{\frac{3}{4}} \cdot y^{\frac{1}{2}})^8$ **11.** $(16x^4 y^6)^{-\frac{1}{2}}$ **12.** $(16a^8 b^4)^{\frac{5}{4}}$

13. $\dfrac{a}{a^{\frac{3}{7}}}$

14. $\dfrac{x^{\frac{2}{5}}}{x^{-\frac{1}{5}}}$

15. $\dfrac{a^{-\frac{1}{4}}}{a^{-\frac{5}{4}}}$

16. $\dfrac{b^4}{b^{\frac{2}{5}}}$

17. $\left(\dfrac{x^{\frac{2}{5}}}{y^{\frac{1}{2}}}\right)^{10}$

18. $\left(\dfrac{a^{\frac{2}{5}}}{b^{\frac{1}{3}}}\right)^{15}$

19. $\left(\dfrac{49m^8}{16m^4}\right)^{\frac{1}{2}}$

20. $\left(\dfrac{125c^9}{27z^{15}}\right)^{-\frac{2}{3}}$

Simplify. Write each expression using *one* radical. Assume that $x > 0$.

21. $\sqrt{x} \cdot \sqrt[3]{x}$

22. $\sqrt[5]{x^2} \cdot \sqrt[3]{x}$

23. $\dfrac{\sqrt[3]{x}}{\sqrt[4]{x}}$

24. $\dfrac{\sqrt[4]{x^3}}{\sqrt[6]{x}}$

Solve each equation. Check.

25. $8^x = 16$

26. $3^{6x} = \dfrac{1}{27}$

27. $3^{8x} = \dfrac{1}{81}$

28. $4^{6x} = 32$

29. $4^{2x-1} = 8$

30. $2^{2x+4} = \dfrac{1}{16}$

31. $10^{3x-2} = \dfrac{1}{1,000}$

32. $3^{1-x} = 9^{1-x}$

Simplify each expression. Assume that the values of the variables are positive.

33. $x^{\frac{3}{7}} \cdot x^{\frac{2}{7}} \cdot x^{-\frac{6}{7}}$

34. $b^{-3} \cdot b^{\frac{3}{4}}$

35. $-5a^{-\frac{3}{4}} \cdot 6a^{-\frac{5}{8}}$

36. $3x^{-2} \cdot \left(-5x^{-\frac{1}{4}}\right)$

37. $(16x^{20}y^{-8})^{-\frac{3}{4}}$

38. $\dfrac{27^{\frac{7}{12}}x^{-\frac{1}{10}}y^{\frac{1}{2}}}{27^{\frac{1}{4}}x^{\frac{1}{5}}y^{-\frac{1}{3}}}$

39. $\dfrac{t^{\frac{1}{2}}v^{-\frac{1}{5}}}{t^{\frac{2}{3}}v^{-\frac{4}{5}}}$

40. $\left(\dfrac{64m^{-3}}{27m^{-9}}\right)^{\frac{2}{3}}$

Simplify. Write each expression with the least number of radicals. Assume that the values of the variables are positive.

41. $(\sqrt[4]{x^3} + \sqrt[6]{y})(\sqrt[4]{x^3} - \sqrt[6]{y})$ **42.** $(\sqrt[4]{a} - \sqrt[4]{b})(\sqrt[4]{a} + \sqrt[4]{b})$ **43.** $(2\sqrt{x} - 3\sqrt[6]{y})^2$

44. Rationalize the denominator of $\dfrac{21}{5^{\frac{1}{3}} + 2^{\frac{1}{3}}}$. (HINT: Use the factors of $a^3 + b^3$.)

45. Solve $8x^{-\frac{2}{3}} - x^{\frac{1}{3}} = 0$.

46. Solve $8(2x - 6)^{-\frac{1}{2}} - (2x - 6)^{\frac{1}{2}} = 0$.

Mixed Review

Solve.   *2.2, 6.1, 6.6, 7.6*

1. $-6 < 2x - 4 \le 8$

2. $2x^2 - 15 = -7x$

3. $2x^3 - 13x^2 + 23x - 12 = 0$

4. $\dfrac{x}{5} = \dfrac{2}{x + 3}$

Chapter 8 Review

Key Terms

conjugate (p. 281)
index (p. 284)
nth root (p. 284)
perfect square (p. 268)
principal nth root (p. 284)

principal square root (p. 267)
radical (p. 267)
radical sign (p. 267)
radicand (p. 267)
square root (p. 267)

Key Ideas and Review Exercises

8.1 To identify the domain of a function such as $f(x) = \sqrt{4x - 12}$, find the values of x for which $4x - 12 \geq 0$.

To solve an equation such as $3x^2 - 2 = 25$, use the fact that if $x^2 = a$, then $x = \pm\sqrt{a}$.

Find the domain of each function.

1. $f(x) = \sqrt{4x - 32}$

2. $f(x) = \sqrt{21 - 7x}$

Solve each equation. Approximate irrational results to the nearest hundredth.

3. $3x^2 - 2 = 25$

4. $c^2 + 6 = 19$

8.2 To simplify the square root of a positive number, find a, the greatest perfect-square factor of the radicand, and use $\sqrt{ab} = \sqrt{a} \cdot \sqrt{b}$.

To simplify square roots when the sign of the variable is not indicated, use $\sqrt{x^{2n}} = x^n$ when n is even and $\sqrt{x^{2n}} = |x^n|$ when n is odd.

Simplify.

5. $\sqrt{50}$

6. $-2\sqrt{108}$

7. $\sqrt{16b^4}$

8. $\sqrt{25a^{10}}$

9. Write a brief paragraph describing how to simplify $\sqrt{x^n}$ if x is positive and n is an odd integer.

8.3 To simplify sums and differences of square roots, simplify each radical and combine like radicals.

To simplify products of square roots, use $\sqrt{a} \cdot \sqrt{b} = \sqrt{ab}$, $a \geq 0$, $b \geq 0$.

Simplify. Assume that the values of the variables are positive.

10. $5\sqrt{27} - 6\sqrt{32} + 3\sqrt{12} - \sqrt{50}$

11. $7\sqrt{3} \cdot 8\sqrt{6}$

12. $4\sqrt{3}(7\sqrt{2} - \sqrt{6} + 2\sqrt{3})$

13. $6\sqrt{8y^5} \cdot 2\sqrt{3y^4}$

8.4 To simplify quotients of square roots, use $\dfrac{\sqrt{a}}{\sqrt{b}} = \sqrt{\dfrac{a}{b}}$, $a \geq 0$, $b > 0$.

If a denominator contains a radical, rationalize the denominator as shown in Lesson 8.4.

To simplify a complex fraction involving square roots, multiply the numerator and denominator by the LCD of the fractions in the expression and simplify.

Simplify. Assume that the values of the variables are positive.

14. $\dfrac{\sqrt{72}}{\sqrt{6}}$

15. $\dfrac{6}{\sqrt{8}}$

16. $\dfrac{4xy^4}{\sqrt{20x^5y^8}}$

17. $\dfrac{24z^5}{\sqrt{45y^4z^3}}$

18. $\dfrac{10}{\sqrt{7} - \sqrt{2}}$

19. $\dfrac{5\sqrt{2}}{3 - \sqrt{2}}$

20. $\dfrac{1 + \dfrac{\sqrt{2}}{2}}{1 - \dfrac{\sqrt{2}}{2}}$

21. $\dfrac{1 - \dfrac{\sqrt{5}}{5}}{\dfrac{2 + \sqrt{5}}{10}}$

8.5 To simplify radicals with index n, for all integers $n > 1$, use Theorem 8.3. For sums, differences, products, and quotients of such radicals, see Examples 4, 5, and 6 in Lesson 8.5.

Simplify. Assume that the values of the variables are positive.

22. $\sqrt[3]{50y^5}$

23. $\sqrt[3]{56a^{11}}$

24. $\dfrac{21}{\sqrt[3]{7}}$

25. $\sqrt[4]{\dfrac{5}{8x}}$

26. $5\sqrt[3]{16} - 8\sqrt[3]{2}$

27. $\sqrt[4]{8x^3} \cdot \sqrt[4]{4x^9}$

8.6, 8.7 To write expressions with rational exponents in radical form, or vice versa, use $b^{\frac{m}{n}} = (\sqrt[n]{b})^m = \sqrt[n]{b^m}$.

To simplify expressions containing rational exponents, apply the Laws of Exponents. An expression is in simplest form when all its exponents are positive and any denominators are rational.

28. Write $\sqrt[7]{x^2}$ in exponential form.

29. Write $2^{\frac{4}{5}}$ in radical form. Simplify.

Simplify. Assume that the values of the variables are positive.

30. $(-32)^{\frac{1}{5}}$

31. $8^{-\frac{2}{3}}$

32. $16^{0.25}$

33. $5^{\frac{3}{7}} \cdot 5^{-\frac{1}{7}} \cdot 5^{\frac{4}{7}}$

34. $4x^{-\frac{3}{5}}$

35. $\dfrac{x}{x^{\frac{5}{8}}}$

36. $\left(\dfrac{8y^{-9}}{125x^{-6}}\right)^{\frac{2}{3}}$

37. $\dfrac{x^{\frac{1}{4}}y^{-\frac{2}{7}}}{x^{\frac{3}{2}}y^{-\frac{4}{7}}}$

38. Solve and check $5^{2x-1} = \dfrac{1}{25}$.

1. Evaluate $\sqrt{\dfrac{5x + 11}{4}}$ for $x = 5$.

2. Give the domain of the function $f(x) = \sqrt{4x - 20}$.

Solve each equation. Approximate irrational results to the nearest hundredth.

3. $a^2 - 7 = 29$ **4.** $5x^2 + 17 = 47$

5. Simplify $4\sqrt{24x^6}$ using absolute value symbols, if necessary.

Simplify. Assume that the values of the variables are positive.

6. $\sqrt{81x^4}$ **7.** $8\sqrt{44}$ **8.** $\sqrt[3]{-40b^{13}}$ **9.** $\sqrt[4]{16x^8}$

10. $\dfrac{\sqrt{48}}{\sqrt{6}}$ **11.** $\dfrac{4}{\sqrt{12}}$ **12.** $\dfrac{20x^6}{\sqrt{8x^3}}$ **13.** $\sqrt[4]{\dfrac{3}{2x}}$

14. $25^{\frac{3}{2}}$ **15.** $32^{\frac{4}{5}}$ **16.** $(-27)^{-\frac{2}{3}}$ **17.** $3y^{-\frac{2}{5}}$

18. $\dfrac{6}{\sqrt{5} - \sqrt{2}}$ **19.** $\dfrac{1 + \dfrac{\sqrt{3}}{3}}{1 - \dfrac{\sqrt{3}}{3}}$ **20.** $\dfrac{x}{x^{\frac{3}{4}}}$ **21.** $\dfrac{x^{\frac{1}{2}}y^{-\frac{2}{3}}}{x^{\frac{3}{4}}y^{-\frac{4}{3}}}$

22. $7\sqrt{12} - 5\sqrt{48} - 2\sqrt{75}$ **23.** $5\sqrt{2}(8\sqrt{2} - 3\sqrt{6} - \sqrt{8})$

24. $(3\sqrt{2} + \sqrt{5})(6\sqrt{2} - \sqrt{5})$ **25.** $6\sqrt[4]{32} - 8\sqrt[4]{2}$

26. $\sqrt[3]{4x^{10}} \cdot \sqrt[3]{40x^2}$ **27.** $8\sqrt{5a^3b^2} - 4ab\sqrt{20a}$

28. Write $\sqrt[5]{x^3}$ in exponential form.   **29.** Write $3^{\frac{3}{4}}$ in radical form. Simplify.

30. Solve and check $4^{3x-2} = \dfrac{1}{64}$.

31. Four less than the square of a number is 8. Find the number to the nearest hundredth.

Simplify.

32. $\sqrt{\dfrac{\sqrt[10]{32} \cdot \sqrt[4]{4}}{2^{-3} \cdot 3^{-2}}}$ **33.** $\dfrac{a^2 + b^2}{a^{\frac{2}{3}} + b^{\frac{2}{3}}}, a \neq 0, b \neq 0$

34. Solve $-3y^{-\frac{1}{3}} + 7y^{\frac{2}{3}} = 0, y > 0$.

Choose the *one* best answer to each question or problem.

1. Which of the following products has the greatest value less than 100?
   (A) $2 \times 4 \times 6$ (B) $2 \times 4 \times 9$
   (C) $4 \times 4 \times 9$ (D) $3 \times 3 \times 9$
   (E) $4 \times 4 \times 6$

2.

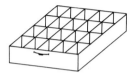

   The figure above represents a drawer with compartments in a chest of identical drawers. If each compartment in each drawer contains 30 screws, and if the chest contains 9,000 screws, how many drawers are in the chest?
   (A) 12 (B) 300 (C) 15
   (D) 60 (E) None of these

3. On a map, 1 in. represents 2 mi. A circle on the map has a circumference of 5π in. This represents a circular region with what area?
   (A) 10π mi^2 (B) 25π mi^2
   (C) 5π mi^2 (D) 100π mi^2
   (E) 50π mi^2

4. If $5x + 3y = 23$, and x and y are positive integers, then y can equal which one of the following?
   (A) 3 (B) 4 (C) 5
   (D) 6 (E) 7

5. $\frac{1}{4} \div \left(\frac{1}{4} \div \frac{1}{4}\right) = $?
   (A) $\frac{1}{4}$ (B) 4 (C) 1
   (D) $\frac{1}{16}$ (E) 16

6. In $5p$ years, Bill will be $10q + 1$ times his current age. Find Bill's current age in terms of p and q.
   (A) $\frac{5p - 1}{10q - 1}$ (B) $\frac{p}{2q}$
   (C) $\frac{5p}{10q + 1}$ (D) $10q + 1 - 5p$
   (E) It cannot be determined from the information given.

7. If $x + y = 4$ and $2x - y = 5$, then $x + 2y = $?
   (A) 5 (B) 2 (C) 1
   (D) 4 (E) 6

8.

   The interior of a rectangle is shaded inside a regular hexagon as shown above. What is the ratio of the shaded area to the area of the hexagon?
   (A) $\frac{1}{2}$ (B) $\frac{2}{3}$ (C) $\frac{3}{4}$
   (D) $\frac{4}{5}$ (E) $\frac{5}{6}$

9. $\sqrt{1 - \left(\frac{1}{2} + \frac{1}{4} + \frac{1}{8} + \frac{1}{16} + \frac{1}{32}\right)} = $
   (A) $\frac{3\sqrt{6}}{8}$ (B) $\frac{\sqrt{62}}{8}$ (C) $\frac{1}{2\sqrt{2}}$
   (D) $\frac{1}{4}$ (E) $\frac{\sqrt{2}}{8}$

10. $\left(-27x^6y^{21}\right)^{\frac{2}{3}} = $?
    (A) $-9x^4y^{14}$ (B) $9x^4y^{14}$
    (C) $18x^2y^7$ (D) $-18x^2y^7$
    (E) $-27x^4y^{14}$

Cumulative Review (Chapters 1–8)

Match each equation with one of the properties (A-E) for operations with real numbers.

1. $0 + 8x = 8x$ **1.3**

2. $5x + 5y = 5(x + y)$

3. $-xy + xy = 0$

4. $x + (y + 7) = (y + 7) + x$

5. $7 + (y + 9) = (7 + y) + 9$

   (A) Add Identity
   (B) Distr
   (C) Associate for Add
   (D) Add Inverse
   (E) Commutative for Add

Solve the equation. (Exercises 6–10)

6. $2(3x + 8) - 3(x - 2) =$ **1.4**
   $9x - (4x - 6)$

7. $|2x - 8| = 20$ **2.3**

8. $3^{2n-4} = 27$ **5.1**

9. $\dfrac{5}{2y + 6} = \dfrac{2y - 4}{y^2 - y - 12}$ **7.6**

10. $x^2 - 6 = \frac{1}{4}$ **8.1**

11. Solve $ax + b = c$ for x.   **1.4**

Find the solution set.

12. $3(4 - 2x) \geq 4x - 23$ **2.1**

13. $2x + 9 \leq 3$ or $3x - 7 > 5$ **2.2**

14. $|2x - 6| < 4$ **2.3**

15. $x^2 - 6x - 16 \geq 0$ **6.4**

Write an equation in standard form $(ax + by = c)$ for the line described.

16. Passing through $P(2, -1)$ and   **3.4**
   $Q(12,3)$

17. Passing through $A(-5,2)$ and   **3.7**
   parallel to the line $y = -x$

Draw the graph on a coordinate plane. (Exercises 18–19)

18. $5x - 5y - 15 = 0$ **3.5**

19. $4x - 2y > 6$ **3.9**

20. The voltage V of a given   **3.8**
   electric circuit varies directly
   as the current I. If V is 120
   volts when I is 9 amperes,
   what is the voltage V when
   the current I is 15 amperes?

21. Solve the system: **4.3**
   $5x - 2y = 20$
   $7x + 4y = 11$

22. Solve the system by graphing: **4.7**
   $x + y > -1$
   $3x - 2y > 4$

For Exercises 23–25, $f(x) = 2x - 7$ and $g(x) = x^2 - 4x - 2$. Find the following values.

23. $f(-4) - g(-2)$ **3.2**

24. $f(g(-1))$ **5.9**

25. $\dfrac{f(a) - f(b)}{a - b}$ **7.1**

26. Simplify $\dfrac{4a^{-3}b^5}{8a^2b^{-2}c^{-3}}$ **5.2**
   using positive exponents.

Factor completely. (Exercises 27–29)

27. $8x^3 - 1$ **5.6**

28. $x^4 - 2x^2 - 8$ **5.7**

29. $3xy + 6x - y - 2$

30. Use synthetic division and the   **6.6**
   Factor Theorem to solve
   $2x^3 - 12x^2 + 22x - 12 = 0$.

31. Find the domain of the   **7.1**
   function $f(x) = \dfrac{x - 7}{x^2 - 5x + 4}$.

Simplify. (Exercises 32–41)

32. $3x(4x - 2)^2$ *5.5*

33. $\dfrac{4a + 8}{5a - 20} \div \dfrac{10 + 3a - a^2}{a^2 - 4a}$ *7.2*

34. $\dfrac{5x}{x^2 - 9} - \dfrac{4}{x + 3} + \dfrac{2}{3 - x}$ *7.4*

35. $\dfrac{1 - \dfrac{6}{x} + \dfrac{5}{x^2}}{1 - \dfrac{3}{x} - \dfrac{10}{x^2}}$ *7.5*

36. $3y\sqrt{27y^{11}}, \; y \geq 0$ *8.2*

37. $7\sqrt{27} - 4\sqrt{3} - 8\sqrt{12}$ *8.3*

38. $(2\sqrt{3} - \sqrt{2})(4\sqrt{3} + 3\sqrt{2})$

39. $\dfrac{28}{4 - \sqrt{2}}$ *8.4*

40. $27^{-\frac{2}{3}}$ *8.6*

41. $\left(16a^{16}\right)^{\frac{3}{4}}$ *8.7*

42. The second of three numbers   *1.6*
is 3 times the first. The third
number is 2 more than the
second number. Seven less
than twice the second is the
same as 12 more than the
third. Find the three numbers.

43. The length of a rectangle is   *1.7*
4 m longer than twice its
width. The perimeter is
26 m. Find the length and
the width.

44. Carol is 6 times as old as her   *2.5*
nephew, Juan. Betty is 22
years younger than her aunt,
Carol. In 4 years, Carol's age
will be twice the sum of
Juan's and Betty's ages now.
How old is each person now?

45. If 3 less than a positive inte-   *2.4*
ger is divided by 2, the result
is less than 1. Find the set of
all such positive integers.

46. A passenger train and a   *2.6*
freight train start toward each
other at the same time from
stations 870 km apart. The
passenger train travels at 80
km/h and the freight train at
65 km/h. In how many
hours will they meet?

47. Bill has twice as many quar-   *4.4*
ters as dimes. The total face
value of the money is $9.00.
How many coins are there of
each type?

48. Find three consecutive even   *6.2*
integers such that the first
integer times the second integer is 80.

49. A square and a rectangle   *6.3*
have the same width. The
length of the rectangle is 4
times its width. If the sum of
the two areas is 45 cm^2, what
is the area of each figure?

50. The denominator of a frac-   *7.6*
tion is 3 more than twice the
numerator. If the numerator
is decreased by 1 and the
denominator is increased
by 1, the resulting fraction
is equal to $\frac{1}{4}$. Find the
original fraction.

51. Work Crew A takes 15 h to   *7.7*
do a job that Crew B can do
in 10 h. If Crew B joins
Crew A 3 h after Crew A has
begun, what is the total time
for the job?

The image shown here is an example of a type of fractal known as a Julia set. Such images are produced by repeating a simple procedure over and over. With Julia sets, the process involves squaring *complex numbers,* numbers with both real and "imaginary" components.

9.1 Complex Numbers: Addition and Absolute Value

Objectives

To solve equations of the form $x^2 = k$, where $k < 0$
To add and subtract complex numbers
To find the absolute value of complex numbers

The equation $x^2 = -1$ has no *real-number* roots because there is no real number whose square is -1. To solve this equation, mathematicians have defined the numbers i and $-i$ as follows:

$$i = \sqrt{-1} \qquad i^2 = -1 \qquad -i = -\sqrt{-1} \qquad (-i)^2 = -1$$

Having defined i as $\sqrt{-1}$, we can extend the idea to other radicals of the form $\sqrt{-k}$. First observe that $(2i)^2 = 2^2 \cdot i^2 = 4(-1) = -4$. Then

$$(2i)^2 = -4$$
$$2i = \sqrt{-4}$$

Also, $\qquad\qquad\qquad\qquad\qquad 3i = \sqrt{-9}$, and so on.

In general, $\sqrt{-y} = i\sqrt{y}$ if $y > 0$, that is, if $-y < 0$.

EXAMPLE 1

Simplify each expression below.

a. $-2 + \sqrt{-16}$ **b.** $6 - 4\sqrt{-5}$

Solutions

$= -2 + i\sqrt{16}$ $\qquad = 6 - 4i\sqrt{5}$

$= -2 + 4i$

Definition

A **complex number** is a number that can be written in the **standard form** $a + bi$, where a and b are real numbers.

The numbers in Example 1 are *complex numbers*. If $a = 0$, then the complex number $a + bi$, or bi, is called an **imaginary number**. If $b = 0$, then the complex number $a + bi$, or a, is a *real number*.

Equations such as $x^2 + 16 = 0$ and $3y^2 = -24$ have roots in the set of complex numbers. Thus, the general solution of $x^2 = k$ (see Lesson 8.1) can be extended from $k \geq 0$ to include $k < 0$ as follows:

If $x^2 = k$, then $x = \pm\sqrt{k}$, for all real numbers k.

EXAMPLE 2   Solve each equation below.

a. $x^2 + 16 = 0$ b. $3y^2 = -24$

Solutions

$$x^2 = -16$$

$$x = \pm\sqrt{-16} = \pm 4i$$

The roots are $4i$ and $-4i$.

$$y^2 = -8$$

$$y = \pm\sqrt{-8} = \pm 2i\sqrt{2}$$

The roots are $2i\sqrt{2}$ and $-2i\sqrt{2}$.

The equation $(2x - 1)(2x + 1)(x^2 + 9) = 0$ has *four complex* roots. Two $(\pm\frac{1}{2})$ are real numbers obtained from $2x - 1 = 0$ and $2x + 1 = 0$. Two $(\pm 3i)$ are imaginary numbers obtained from $x^2 + 9 = 0$.

EXAMPLE 3   Solve $y^4 + 21y^2 - 100 = 0$.

Plan   Factor completely. Then use the Zero-Product Property.

Solution

$$y^4 + 21y^2 - 100 = 0$$
$$(y^2 - 4)(y^2 + 25) = 0$$
$$(y + 2)(y - 2)(y^2 + 25) = 0$$
$$y + 2 = 0 \quad or \quad y - 2 = 0 \quad or \quad y^2 + 25 = 0$$
$$y = -2 \qquad\qquad y = 2 \qquad\qquad y^2 = -25 \text{ and } y = \pm 5i$$

The solutions are $-2, 2, 5i,$ and $-5i$.

In Example 3, the equation $y^4 + 21y^2 - 100 = 0$ is in **quadratic form** because, by substituting z for y^2, it is possible to obtain the equation $z^2 + 21z - 100 = 0$, which is a quadratic equation in z.

The eleven properties of a field listed in Lesson 1.3 also apply to the set of complex numbers. Therefore, you can add and subtract complex numbers just as you would real-number binomials.

EXAMPLE 4   Write in the form $a + bi$, where a and b are real numbers.

a. $(7 - 8i) - (5 - 3i)$ b. $(-7 + \sqrt{-3}) + (2 - \sqrt{-48})$

Solutions

$$= (7 - 8i) + (-5 + 3i)$$
$$= (7 - 5) + (-8 + 3)i$$
$$= 2 - 5i$$

$$= (-7 + i\sqrt{3}) + (2 - 4i\sqrt{3})$$
$$= (-7 + 2) + (\sqrt{3} - 4\sqrt{3})i$$
$$= -5 - 3i\sqrt{3}$$

Each complex number $a + bi$ can be associated with a unique ordered pair of real numbers (a,b), which in turn can be associated with a unique point on a coordinate plane. In this coordinate plane, the x- and y-axes are called the **real axis** and the **imaginary axis**, respectively.

Complex number	Ordered pair	Point
$-5 + 3i$	$(-5,3)$	A
$-2i$	$(0,-2)$	B
-7	$(-7,0)$	C

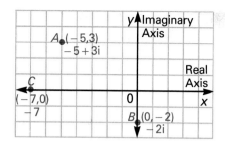

The distance PO between $P(a,b)$ and the origin is the length of the hypotenuse of the right triangle OPM (see diagram). Therefore, PO can be determined by the Pythagorean relation.

$$a^2 + b^2 = (PO)^2, \text{ or}$$
$$\sqrt{a^2 + b^2} = PO$$

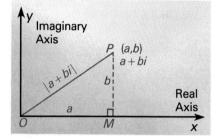

The distance between $A(-5, 3)$ and the origin is $\sqrt{(-5)^2 + 3^2}$, or $\sqrt{34}$. This distance is called the *absolute value* of $-5 + 3i$ and is written as $|-5 + 3i| = \sqrt{34}$.

For the imaginary number $-2i$, $B(0,-2)$ is 2 units from the origin and $|0 - 2i| = \sqrt{0^2 + (-2)^2} = \sqrt{4} = 2$. For the real number -7, $C(-7,0)$ is 7 units from the origin and $|-7 + 0i| = \sqrt{(-7)^2 + 0^2} = \sqrt{49} = 7$, which confirms the real-number definition of $|-7|$ as 7. In general, the **absolute value of a complex number** is given by $|a + bi| = \sqrt{a^2 + b^2}$.

EXAMPLE 5 Find the absolute value of each complex number. Simplify the answer.

  a. $3\sqrt{2} - 2i$ **b.** $3i\sqrt{2}$ **c.** -6

Solutions

  a. $|3\sqrt{2} + (-2)i|$ **b.** $|0 + (3\sqrt{2})i|$ **c.** $|-6 + 0i|$
  $= \sqrt{(3\sqrt{2})^2 + (-2)^2}$ $= \sqrt{0^2 + (3\sqrt{2})^2}$ $= \sqrt{(-6)^2 + 0^2}$
  $= \sqrt{18 + 4}$, or $\sqrt{22}$ $= \sqrt{18}$, or $3\sqrt{2}$ $= \sqrt{36}$, or 6

Classroom Exercises

Simplify.

  1. $\sqrt{-4}$ **2.** $\sqrt{-5}$ **3.** $(4 + 2i) - (2 - i)$ **4.** $|3 + 4i|$

Solve each equation.

  5. $x^2 = -1$ **6.** $y^2 = -4$ **7.** $2n^2 = -10$ **8.** $4x^2 + 36 = 0$

Written Exercises

Write in simplest radical form using *i*.

1. $\sqrt{-64}$ 2. $\sqrt{-81}$ 3. $-\sqrt{-36}$ 4. $-\sqrt{-14}$

5. $\sqrt{-21}$ 6. $3\sqrt{-4}$ 7. $-\sqrt{-3}$ 8. $-5\sqrt{-16}$

9. $\sqrt{-50}$ 10. $-3\sqrt{-20}$ 11. $-4\sqrt{-40}$ 12. $10\sqrt{-80}$

13. $8 - \sqrt{-36}$ 14. $-18 - \sqrt{-30}$ 15. $-7 + 3\sqrt{-50}$ 16. $3 - \sqrt{-45}$

Solve each equation.

17. $x^2 = -25$ 18. $2y^2 = -20$ 19. $3a^2 + 150 = 0$

20. $a^4 + 7a^2 - 18 = 0$ 21. $x^4 - 16 = 0$ 22. $y^4 + 13y^2 + 36 = 0$

Write in the form $a + bi$, where a and b are real numbers.

23. $(2 + 5i) + (6 + 3i)$ 24. $(6 + 7i) - (2 + 4i)$ 25. $(-3 - 6i) + (7 + 10i)$

26. $(6 - 8i) - (-3 + 4i)$ 27. $(-3 - i) - (5 - 2i)$ 28. $(-10 + i) + (8 - 4i)$

Find the absolute value of each complex number. Simplify the answer.

29. $-3 - 4i$ 30. $6 - 2i$ 31. $-4i$

32. $5\sqrt{2} + 3i$ 33. $4i\sqrt{2}$ 34. $-2 + 3i\sqrt{5}$

Solve each equation.

35. $7t^2 + 20 = 4t^2 - 28$ 36. $5x^2 + 21 = 3 - x^2$ 37. $7y^2 + 12 = 8 - 2y^2$

38. $y^4 = 7y^2 + 8$ 39. $n^4 + 2n^2 = 15$ 40. $x^4 + 10x^2 + 16 = 0$

Write in the form $a + bi$, where a and b are real numbers.

41. $15 - 10\sqrt{-12}$ 42. $(5 + \sqrt{-36}) + (-9 - \sqrt{-4})$

Simplify.

43. $|6 + 8i| + |6 - 8i|$ 44. $|-2 + 4i| + |-2 - 4i|$

45. $|5 - 3i| - |-5 + 3i|$ 46. $|3 - i| - |3 + i|$

Prove that each theorem is true for all complex numbers.

47. $|a + bi| = |a - bi|$ 48. $|a - bi| = |-a + bi|$

Mixed Review

Simplify. *8.3, 8.4*

1. $2\sqrt{3}(3\sqrt{6} - 4\sqrt{3})$ 2. $(3\sqrt{5} - 4\sqrt{3})(2\sqrt{5} + \sqrt{3})$ 3. $\dfrac{6}{7\sqrt{2}}$ 4. $\dfrac{4}{3\sqrt{3} - \sqrt{5}}$

9.2 Products, Quotients, Conjugates

Objectives

To simplify products, powers, and quotients of complex numbers

To factor $x^2 + y^2$

The fact that i^2 is defined as -1 can be used to simplify the products of complex numbers. Begin by writing each factor in the form $a + bi$.

EXAMPLE 1 Simplify each product.

   a. $5i \cdot 4i$ b. $-\sqrt{-10} \cdot \sqrt{-5}$ c. $5\sqrt{6} \cdot 4\sqrt{-3}$

Solutions
   a. $= 20i^2$ b. $= -i\sqrt{10} \cdot i\sqrt{5}$ c. $= 5\sqrt{6} \cdot 4i\sqrt{3}$
   $= 20(-1)$ $= -\sqrt{50} \cdot i^2$ $= 20i\sqrt{18}$
   $= -20$ $= -5\sqrt{2} \cdot (-1)$ $= 20i \cdot 3\sqrt{2} = 60i\sqrt{2}$
   $= 5\sqrt{2}$

The real-number property for multiplying square-root radicals,

$$\sqrt{x} \cdot \sqrt{y} = \sqrt{x \cdot y}, \text{ where } x \geq 0 \text{ and } y \geq 0,$$

does *not* extend to the case where $x < 0$ and $y < 0$. As a counter-example, $\sqrt{-4} \cdot \sqrt{-9}$ is *not* equal to $\sqrt{-4(-9)}$ or $\sqrt{36}$, which is 6. Instead, $\sqrt{-4} \cdot \sqrt{-9} = 2i \cdot 3i = 6i^2 = 6(-1) = -6$, and in general, $\sqrt{-x} \cdot \sqrt{-y} = i\sqrt{x} \cdot i\sqrt{y} = i^2\sqrt{xy} = -\sqrt{xy}$, where $-x < 0$ and $-y < 0$.

Simplifying a quotient such as $\frac{-3}{8i}$ is similar to rationalizing the denominator in the real-number system, as shown below.

$$\frac{-3}{8i} = \frac{-3}{8i} \cdot \frac{i}{i} = \frac{-3i}{8i^2} = \frac{-3i}{-8} = \frac{3i}{8} \quad \begin{array}{l}\text{(no imaginary}\\ \text{number in the}\\ \text{denominator)}\end{array}$$

To simplify a product of the form $(a + bi)(c + di)$, use the FOIL method.

EXAMPLE 2 Simplify $(2 + 3i)(4 - 5i)$.

Solution

$$\ \ \text{F}\quad\ \ \text{O}\quad\ \ \text{I}\quad\ \ \text{L}$$
$$(2 + 3i)(4 - 5i) = 8 - 10i + 12i - 15i^2$$
$$= 8 + 2i - 15(-1) = 23 + 2i$$

Thus, $(2 + 3i)(4 - 5i) = 23 + 2i$.

Recall that $5 + \frac{1}{2}\sqrt{2}$ and $5 - \frac{1}{2}\sqrt{2}$ are a pair of *irrational* conjugates, and that their product $24\frac{1}{2}$ is a *rational* number. Similarly, $5 + \pi i$ and $5 - \pi i$ are *complex* conjugates, and their product $25 + \pi^2$ is a *real* number. In general, two complex numbers of the form $a + bi$ and $a - bi$ are **complex conjugates**. Their product, $(a + bi)(a - bi)$, is a real number, $a^2 + b^2$.

The conjugate of $a + bi$ is sometimes written as $\overline{a + bi}$. For example, $(5 + 3i)(\overline{5 + 3i}) = (5 + 3i)(5 - 3i) = 25 - 9i^2 = 34$.

A quotient with a complex-number divisor can be simplified by using a conjugate.

EXAMPLE 3 Simplify $\dfrac{8}{5 - 3i}$ by rationalizing the denominator.

Solution

$$\frac{8}{5 - 3i} \cdot \frac{5 + 3i}{5 + 3i} = \frac{8(5 + 3i)}{25 - 9i^2} = \frac{8(5 + 3i)}{34}$$

$$= \frac{4(5 + 3i)}{17}, \text{ or } \frac{20 + 12i}{17}$$

Powers of i can be simplified using (1) $i^2 = -1$ and (2) $i^4 = i^2 \cdot i^2 = -1(-1) = 1$, or $i^4 = 1$.

For example, $i^7 = i^4 \cdot i^2 \cdot i = 1(-1)i = -i$, and
$(2i)^{10} = 2^{10} \cdot i^{10} = 2^{10} \cdot (i^4)^2 \cdot i^2 = 1{,}024 \cdot 1^2 \cdot (-1) = -1{,}024.$

EXAMPLE 4 Factor $4x^2 + 25$. Use complex numbers.

Plan Write 25 as $-(-25)$, or $-25i^2$.

Solution $4x^2 + 25 = 4x^2 - (-25) = 4x^2 - 25i^2 = (2x - 5i)(2x + 5i)$

Thus, $4x^2 + 25 = (2x - 5i)(2x + 5i)$.

Classroom Exercises

Determine whether each statement is true or false.

1. $\sqrt{-2} \cdot \sqrt{-8} = \sqrt{16}$ **2.** $\overline{5 - 2i} = -5 + 2i$ **3.** $i^{15} = i^3 = i^2 \cdot i = -i$

Simplify.

4. $-i\sqrt{8} \cdot i\sqrt{2}$ **5.** $(1 - i)^2$ **6.** $(6 - 2i) \div i$

Written Exercises

Simplify each product or power.

1. $4i \cdot 7i$

2. $-i\sqrt{5} \cdot i\sqrt{2}$

3. $3\sqrt{2} \cdot 2\sqrt{-3}$

4. $i\sqrt{3} \cdot i\sqrt{6} \cdot i\sqrt{2}$

5. $\sqrt{-10} \cdot \sqrt{-2} \cdot \sqrt{-5}$

6. $-2\sqrt{-3} \cdot 5\sqrt{-12}$

7. $\sqrt{-3}(2\sqrt{3} - \sqrt{-12})$

8. $2\sqrt{-9}(5 + \sqrt{-16})$

9. $3\sqrt{2}(5 - 2\sqrt{-24})$

10. $(3 + 4i)(6 + 5i)$

11. $(2 - 2i)(3 - 4i)$

12. $(5 - 6i)(-2 + i)$

13. $(4 + 3i)^2$

14. $(5 - 2i)^2$

15. $(-3 - i)^2$

16. $(5 + 8i)(\overline{5 + 8i})$

17. $(-3 - 5i)(\overline{-3 - 5i})$

18. $(\overline{6 - 2i})(6 - 2i)$

19. $(-5i)^3$

20. $(i\sqrt{5})^4$

21. $(2i)^5$

Simplify each quotient by rationalizing the denominator.

22. $29 \div i$

23. $5 \div (4i)$

24. $-4 \div (3i)$

25. $7 \div (-2i)$

26. $\dfrac{3 - 5i}{4i}$

27. $\dfrac{2 + 5i}{-6i}$

28. $\dfrac{10}{4 - 3i}$

29. $\dfrac{6i}{-1 - i}$

30. $\dfrac{1 - 5i}{2 + 4i}$

31. $\dfrac{1 + i}{1 - i}$

32. $\dfrac{1 - 2i}{-2 - 6i}$

33. $\dfrac{4\sqrt{3}}{2\sqrt{3} + i\sqrt{3}}$

Factor each binomial.

34. $x^2 + 16$

35. $9y^2 + 4$

36. $36x^2 + y^2$

37. $5a^2 + 3$

38. Write in your own words how you would simplify an odd power of i greater than i^4.

Simplify each power.

39. $(-10i)^6$

40. i^{14}

41. $(4 - 2i\sqrt{3})^2$

42. $(2i\sqrt{3} + i\sqrt{2})^2$

43. $(2i)^{-3}$

44. $(3 + i)^{-1}$

45. $i^{-2} + i^{-4}$

46. $(1 - 2i)^{-2}$

In Exercises 47–50, w and z are defined as the complex numbers $a + bi$ and $c + di$, respectively. Show each of the following.

47. $\overline{w + z} = \overline{w} + \overline{z}$

48. $\overline{w - z} = \overline{w} - \overline{z}$

49. $\overline{wz} = \overline{w} \cdot \overline{z}$

50. $\overline{\overline{w}} = w$

Mixed Review   5.5, 5.6, 8.1

1. Simplify $(x - 6)^2$.

2. Factor $t^2 - 8t + 16$.

$(t - 4)^2$

3. Solve $y^2 = 18$.

$\pm 3\sqrt{2}$

9.3 Completing the Square

Objective

To solve quadratic equations for their complex roots by completing the square

You have solved quadratic equations such as $y^2 = 28$ and $n^2 = -9$ by using the following property: If $x^2 = k$, then $x = \pm\sqrt{k}$.

If $y^2 = 28$, then $y = \pm\sqrt{28}$, or $y = \pm 2\sqrt{7}$; the two roots are real numbers.

If $n^2 = -9$, then $n = \pm\sqrt{-9}$, or $n = \pm 3i$; the two roots are imaginary numbers.

This property can be extended to equations of the form $(x - t)^2 = k$, where $x - t = \pm\sqrt{k}$, or $x = t \pm \sqrt{k}$. For example:

If $(y + 3)^2 = 8$,
then $y + 3 = \pm\sqrt{8} = \pm 2\sqrt{2}$,
or $y = -3 \pm 2\sqrt{2}$
The roots are $-3 + 2\sqrt{2}$ and $-3 - 2\sqrt{2}$.

If $(n - 2)^2 = -25$,
then $n - 2 = \pm\sqrt{-25} = \pm 5i$,
or $n = 2 \pm 5i$
The roots are $2 + 5i$ and $2 - 5i$.

The equation $x^2 - 5x = -7$ can be solved in a similar way after **completing the square** on the left side. Notice that the coefficient of x in $x^2 - 5x = -7$ is -5, and the square of $\frac{1}{2}$ of -5 is $-\left(\frac{5}{2}\right)^2$, or $\frac{25}{4}$.

EXAMPLE 1

Solve $x^2 - 5x = -7$ by completing the square.

Plan

Add $\left(-\frac{5}{2}\right)^2$, or $\frac{25}{4}$, to each side of the equation to complete the square.

Solution

$$x^2 - 5x = -7$$
$$x^2 - 5x + \frac{25}{4} = -7 + \frac{25}{4}$$
$$\left(x - \frac{5}{2}\right)^2 = \frac{-3}{4}$$

Use the property:
If $y^2 = k$, then $y = \pm\sqrt{k}$.

$$x - \frac{5}{2} = \pm\sqrt{\frac{-3}{4}} = \pm\frac{\sqrt{-3}}{2}$$
$$x - \frac{5}{2} = \pm\frac{i\sqrt{3}}{2}$$
$$x = \frac{5}{2} \pm \frac{i\sqrt{3}}{2}, \text{ or } \frac{5 \pm i\sqrt{3}}{2}$$

The roots are $\dfrac{5 + i\sqrt{3}}{2}$ and $\dfrac{5 - i\sqrt{3}}{2}$.

> **Solving a Quadratic Equation by Completing the Square**
> In general, to solve a quadratic equation of the form $x^2 + bx = c$, divide b by 2, square the result, and add $(\frac{b}{2})^2$ to each side. Then factor and use the property: If $y^2 = k$, then $y = \pm\sqrt{k}$.

To solve $6y^2 + 7y - 5 = 0$ by completing the square, begin by writing the equation in the form $y^2 + by = c$, as shown in Example 2.

EXAMPLE 2   Solve $6y^2 + 7y - 5 = 0$ by completing the square.

Plan   Add 5 to each side; then divide each side by 6 to obtain the form $y^2 + by = c$.

Then $b = \frac{7}{6}$, so $\frac{b}{2} = \frac{7}{6} \cdot \frac{1}{2} = \frac{7}{12}$ and $(\frac{b}{2})^2 = (\frac{7}{12})^2 = \frac{49}{144}$.

Solution

$$6y^2 + 7y = 5$$
$$y^2 + \frac{7}{6}y = \frac{5}{6}$$
$$y^2 + \frac{7}{6}y + \frac{49}{144} = \frac{5}{6} + \frac{49}{144}$$
$$\left(y + \frac{7}{12}\right)^2 = \frac{169}{144}$$
$$y + \frac{7}{12} = \pm\frac{13}{12}$$
$$y = \frac{-7 \pm 13}{12}$$
$$y = \frac{6}{12} \quad or \quad y = -\frac{20}{12}$$

The roots are $\frac{1}{2}$ and $-\frac{5}{3}$.

Classroom Exercises

Find the number to be added to each binomial to complete the square.

1. $a^2 + 12a$ **2.** $n^2 - 4n$ **3.** $x^2 + 3x$

4. $y^2 - \frac{5}{3}y$ **5.** $x^2 + 2x\sqrt{5}$ **6.** $y^2 - 10y\sqrt{3}$

Solve each equation.

7. $(x - 3)^2 = 25$ **8.** $y^2 + 4y = 3$ **9.** $t^2 - 8t + 25 = 0$

Written Exercises

Solve each equation by completing the square.

1. $x^2 - 4x = 21$ **2.** $n^2 + 12n = -20$ **3.** $y^2 - 7y = -10$

4. $a^2 + 10a + 5 = 0$ **5.** $c^2 - 8c - 2 = 0$ **6.** $x^2 + 6x - 41 = 0$

7. $2x^2 + 5x = 3$ **8.** $3y^2 - 8y + 4 = 0$ **9.** $6x^2 + 5x + 1 = 0$

10. $y^2 - 3y - 6 = 0$ **11.** $t^2 + 7t = -1$ **12.** $a^2 + 5a + 2 = 0$

13. $x^2 - 6x + 25 = 0$ **14.** $a^2 + 10a + 34 = 0$ **15.** $x^2 + 8x + 20 = 0$

16. $y^2 - 6y\sqrt{5} + 40 = 0$ **17.** $n^2 - 4n\sqrt{2} = 10$ **18.** $a^2 + 8a\sqrt{6} = 54$

19. $x^2 + 7x + 4 = 0$ **20.** $y^2 - 5y - 5 = 0$ **21.** $16x^2 + 24x = -9$

22. $2y^2 = 5y + 5$ **23.** $7x = 5x^2 + 1$ **24.** $y^2 = 7y - 15$

25. $z^2 - 4iz = -12$ **26.** $a^2 - 10ia - 16 = 0$ **27.** $x^2 + 6ix - 5 = 0$

28. $2x^2 + 6x\sqrt{2} = 3$ **29.** $a^2 - 4a\sqrt{2} = -8$ **30.** $n^2 + 6n\sqrt{5} = -45$

Mixed Review

Simplify. Use positive exponents.   *5.1, 5.2*

1. $5x^2(-2x^3y)^3$

2. $5x^{-2}y \cdot (-3)x^{-4}y^5 \cdot xy^{-1}$

3. $\dfrac{-10a^{-3}b^4c^2}{15a^{-5}b^{-1}c^{-3}}$

4. $(5m^{-2}n^4)^{-3}$

Application: *Stopping Distance*

The distance d (in ft) that a car needs to stop on a particular road surface is given by the formula

$d = \dfrac{s^2}{30F}$, where s is the car's speed (in mi/h) and F is the friction coefficient of the road.

Friction Coefficients

	Concrete	Tar
Wet	0.4	0.5
Dry	0.8	1.0

Police often measure skid marks to find the speed at which the car that made them was traveling. Use the friction coefficients above to find the speed, to the nearest tenth, at which the car that made each of these marks was traveling. (HINT: First solve the formula for s.)

1. 132 ft on a dry tar road

2. 280 ft on a wet concrete road

9.4 Problem Solving: Maximum Values

Objectives

To find the maximum value of given quadratic expressions

To solve word problems involving maximum value

The quadratic polynomial $30t - 5t^2$ represents the height h in meters at the end of t seconds of an object sent upward with an initial velocity of 30 m/s. When $h = 30t - 5t^2$, the value of h increases from 0 to some *maximum* value and then decreases again to 0 as the value of t increases from 0 to 6.

t	0	2	3	4	6
$30t - 5t^2$	$0 - 0$	$60 - 20$	$90 - 45$	$120 - 80$	$180 - 180$
h	0	40	45	40	0

It seems from the table that the maximum height is 45 m, and that this occurs at the end of 3 s. The technique of completing the square can be used to prove that this is true.

Let $h = 30t - 5t^2$.

Then $h = -5t^2 + 30t$.

Factor out -5, the coefficient of t^2.

$$h = -5(t^2 - 6t)$$

Complete the square *inside* the parentheses by adding $(\frac{-6}{2})^2$, or 9. Add $-5 \cdot 9$ to the left side to preserve the equality.

$$h - 5 \cdot 9 = -5(t^2 - 6t + 9)$$
$$h - 45 = -5(t - 3)^2$$

Now, either $t = 3$ or $t \neq 3$. Consider these two cases.

If $t = 3$, then:	$t - 3 = 0$	If $t \neq 3$, then: $\quad t - 3 \neq 0$
	$(t - 3)^2 = 0$	$(t - 3)^2 > 0$
	$-5(t - 3)^2 = 0$	$-5(t - 3)^2 < 0$
Substitute.	$h - 45 = 0$	Substitute. $\quad h - 45 < 0$
	$h = 45$	$h < 45$

Thus, if $h - 45 = -5(t - 3)^2$, then $h = 45$ when $t = 3$, and $h < 45$ when $t \neq 3$. The maximum height h is 45 m, and this occurs when the time t is 3 s. In general:

If $y - y_1 = a(x - x_1)^2$ and $a < 0$, then the maximum value of y is y_1 and this occurs when $x = x_1$.

EXAMPLE 1   Find the maximum value of P and the corresponding value of x given that $P = 500 + 80x - 10x^2$.

Solution   Subtract the constant term, 500. $P - 500 = -10x^2 + 80x$

Factor out -10, the $P - 500 = -10(x^2 - 8x + \underline{?})$
coefficient of x^2.

Complete the square and $P - 500 - 10 \cdot 16 = -10(x^2 - 8x + 16)$
preserve the equality. $P - 660 = -10(x - 4)^2$

The maximum value is 660, and it occurs when $x = 4$.

One objective in the business world is to sell your product or service at a rate that will yield the maximum income.

Suppose you can sell 80 items at $6 each, and for every 50¢ decrease in price, you can sell 10 more items. Sales will keep increasing, but there is a "best price" that will yield the maximum income, as shown

No. of 50¢ decreases	No. of items ×	Item Price	= Income
0	80	$6	$480
1	80 + 10	6 − 0.50	495
2	80 + 20	6 − 1.00	500
3	80 + 30	6 − 1.50	495
4	80 + 40	6 − 2.00	480

in the table at the right. Two 50¢ decreases from $6, or $5, is that best price, and the maximum income earned is $500 from selling 80 + 20, or 100 items.

EXAMPLE 2   If tickets cost $4 each, 800 people will attend a certain concert. For each 25¢ increase in the ticket price, attendance will decrease by 20 people. What ticket price will yield the maximum income? What is the maximum income from the concert? How many tickets need to be sold to yield the maximum income?

Solution

No. of 25¢ increases	No. of tickets ×	Ticket price
0	800	· $4
1	(800 − 20 · 1)	· [4 + 0.25(1)]
2	(800 − 20 · 2)	· [4 + 0.25(2)]
3	(800 − 20 · 3)	· [4 + 0.25(3)]
.	.	.
.	.	.
.	.	.
x	(800 − 20x)	· (4 + 0.25x) = income I for x 25¢ increases

Now,
$$I = (800 - 20x)(4 + 0.25x)$$
$$I = -5x^2 + 120x + 3{,}200$$
$$I - 3{,}200 = -5x^2 + 120x$$
$$I - 3{,}200 = -5(x^2 - 24x + \underline{\,?\,})$$
$$I - 3{,}200 - 5 \cdot 144 = -5(x^2 - 24x + 144)$$
$$I - 3{,}920 = -5(x - 12)^2$$

So, $I = 3{,}920$ when $x = 12$.

The maximum income I occurs when the number x of 25¢ increases is 12. The best ticket price, $(4 + 0.25x)$ dollars, is $4 + 0.25(12)$, or $7. The maximum income of $3,920 occurs when the number of tickets sold, $(800 - 20x)$, is $800 - 20 \cdot 12$, or 560.

Classroom Exercises

Find the maximum value of P and the corresponding value of x.

1. $P - 18 = -2(x - 3)^2$ **2.** $P - 16 - 34 = -1(x - 4)^2$

3. $P - 2{,}000 - 20 \cdot 25 = -20(x - 5)^2$ **4.** $P + 10 - 4 \cdot 4 = -4(x - 2)^2$

Fill in the blanks to complete the square in parentheses and to preserve the equality.

5. $P - \underline{\,?\,} = -4(x^2 - 6x + \underline{\,?\,})$ **6.** $A - \underline{\,?\,} = -3(w^2 - 8w + \underline{\,?\,})$

7. $y - \underline{\,?\,} = -(x^2 - 4x + \underline{\,?\,})$ **8.** $I - \underline{\,?\,} = -20(t^2 - 10t + \underline{\,?\,})$

Written Exercises

Find the maximum value of each polynomial P and the corresponding value of x.

1. $P = -2x^2 + 20x$ **2.** $P = 64x - 16x^2$ **3.** $P = 25 + 6x - x^2$

4. $P = 50 + 12x - 3x^2$ **5.** $P = 675 + 200x - 20x^2$ **6.** $P = 750 + 500x - 25x^2$

7. An arrow is shot upward at 40 m/s. Find its maximum height and the corresponding time. Use the formula $h = vt - 5t^2$.

8. If there are 20 passengers, a chartered bus ride costs $30 per ticket. The ticket price is reduced $1 for each additional passenger beyond 20. How many passengers will produce the maximum income? What is the maximum income? What is the best ticket price?

9. A second number is 12 decreased by a first number. Among all such pairs of numbers, find the pair with the greatest product.

10. One number is 24 decreased by 3 times another number. Among all such pairs, find the pair with the greatest product.

11. A rectangular field next to the straight bank of a river is to be fenced in with no fencing along the riverbank. If 240 yd of fencing are available, what is the maximum area that can be enclosed? What will the dimensions of the field be?

12. Six hundred people will buy tickets at $4 each for a district championship game. For each 20¢ decrease in price, 50 more people will buy tickets. Find the best ticket price, the maximum income, and the corresponding number of tickets.

13. An orange grove has 20 trees per acre, and the average yield is 300 oranges per tree. For each additional tree per acre, the average yield will be reduced by 10 oranges per tree. How many trees per acre will yield the maximum number of oranges per acre? What is the maximum number of oranges per acre?

14. Prove: If $y - y_1 = a(x - x_1)^2$ and $a > 0$, then the *minimum* value of y is y_1, and this occurs when $x = x_1$. (HINT: Either $x = x_1$ or $x \neq x_1$.)

Find the minimum value of P and the corresponding value of x.

15. $P = 10x + x^2$

16. $P = 3x^2 - 18x$

17. $P = 5x^2 + 40x - 20$

18. One number is 12 more than twice another number. Among all such pairs of numbers, find the pair with the minimum product. What is the product?

Midchapter Review

Simplify. *9.1, 9.2*

1. $5 - 2\sqrt{-18}$
2. $(8 - 2i) - (-3 + 4i)$
3. $4\sqrt{-6} \cdot \sqrt{-3}$
4. $(-3i)^3$
5. $(3 - 4i)(2 + 5i)$
6. $\dfrac{-7}{3 + 5i}$

7. Write $9x^2 + 4$ as the difference of two squares and factor the result. *9.2*

Solve by completing the square. *9.3*

8. $x^2 + 10x = 11$
9. $x^2 - 4x = 4$
10. $y^2 + 6y + 14 = 0$

11. Find the maximum value of P and the corresponding value of x given that $P = 100 + 48x - 6x^2$. *9.4*

9.5 The Quadratic Formula

Objectives

To solve quadratic equations for their complex roots using the quadratic formula

To find the three cube roots of nonzero integers that are perfect cubes

Every quadratic equation can be written in the form $ax^2 + bx + c = 0$, where $a > 0$. This general equation can be solved by completing the square. Begin by adding $-c$ to each side of the equation.

$$ax^2 + bx = -c$$

Divide by a and complete the square.

$$x^2 + \frac{b}{a}x + (\frac{b}{2a})^2 = (\frac{b}{2a})^2 + \frac{-c}{a}$$

$$x^2 + \frac{b}{a}x + (\frac{b}{2a})^2 = \frac{b^2}{4a^2} + \frac{-c}{a} \cdot \frac{4a}{4a}$$

Factor the left side.

$$(x + \frac{b}{2a})^2 = \frac{b^2 - 4ac}{4a^2}$$

Use: If $y = k^2$, then $y = \pm\sqrt{k}$.

$$x + \frac{b}{2a} = \frac{\pm\sqrt{b^2 - 4ac}}{2a}$$

$$x = \frac{-b \pm \sqrt{b^2 - 4ac}}{2a}$$

The Quadratic Formula
The solutions of a quadratic equation of the form $ax^2 + bx + c = 0$, where $a > 0$, are given by the formula

$$x = \frac{-b \pm \sqrt{b^2 - 4ac}}{2a}.$$

A polynomial equation with coefficients that are integers is an **integral polynomial equation**. The roots of a second-degree integral polynomial equation can be a pair of rational numbers, a pair of irrational conjugates, or, as in Example 1 below, a pair of complex conjugates.

EXAMPLE 1   Solve $3x^2 + 4x = -2$ using the quadratic formula.

Solution

Write the general form. $3x^2 + 4x + 2 = 0 \leftarrow a = 3, b = 4, c = 2$

Write the quadratic formula. $x = \frac{-b \pm \sqrt{b^2 - 4ac}}{2a}$.

$$\text{Substitute. } x = \frac{-4 \pm \sqrt{16 - 4 \cdot 3 \cdot 2}}{2 \cdot 3}$$

$$= \frac{-4 \pm \sqrt{-8}}{6} = \frac{-4 \pm 2i\sqrt{2}}{6}, \text{ or } \frac{-2 \pm i\sqrt{2}}{3}$$

The roots are $\dfrac{-2 + i\sqrt{2}}{3}$ and $\dfrac{-2 - i\sqrt{2}}{3}$.

EXAMPLE 2   Find the three cube roots of -8.

Plan   Solve a third-degree equation. The 3 cube roots of -8 are the 3 roots of the equation $x^3 = -8$. Solve $x^3 = -8$.

First rewrite the equation as $x^3 + 8 = 0$. Then write $x^3 + 8$ as the sum of two cubes, $x^3 + 2^3$. Use $a^3 + b^3 = (a + b)(a^2 - ab + b^2)$ to factor.

Solution
$$x^3 + 2^3 = 0$$
$$(x + 2)(x^2 - 2x + 4) = 0$$

Use the Zero-Product Property to solve the last equation.

$$x + 2 = 0 \quad \text{or} \quad x^2 - 2x + 4 = 0$$

$$x = -2 \qquad\qquad x = \frac{2 \pm \sqrt{4 - 16}}{2} = \frac{2 \pm \sqrt{-12}}{2}$$

$$= \frac{2 \pm 2i\sqrt{3}}{2}, \text{ or } 1 \pm i\sqrt{3}$$

The three cube roots of -8 are -2, $1 + i\sqrt{3}$, and $1 - i\sqrt{3}$.

Focus on Reading

Place the following steps in the best sequence for solving a quadratic equation using the quadratic formula.

a. Simplify the radical, if possible.

b. Simplify the radicand.

c. Identify the values of a, b, and c.

d. Simplify the quotient, if possible.

e. Write the equation in the form $ax^2 + bx + c = 0$, where $a > 0$.

f. Substitute for a, b, and c in the quadratic formula.

Classroom Exercises

Identify a, b, and c in the general form of the quadratic equation.

1. $5x^2 + 3x - 2 = 0$ **2.** $0 = x^2 - 4x + 3$ **3.** $x^2 - x - 4 = 0$

Simplify.

4–6. Solve the equations of Classroom Exercises 1–3 using the quadratic formula.

Written Exercises

Solve each equation using the quadratic formula.

1. $x^2 - 6x + 8 = 0$ **2.** $n^2 + 3n - 10 = 0$ **3.** $y^2 - 5y - 2 = 0$

4. $5x^2 + x - 2 = 0$ **5.** $2n^2 - n - 1 = 0$ **6.** $2y^2 - 5y + 4 = 0$

7. $4y^2 = 2y - 1$ **8.** $2x^2 = 6x - 3$ **9.** $4t^2 + 13 = 12t$

10. $3n^2 + 2 = 2n$ **11.** $8y^2 = 4y - 5$ **12.** $5x^2 = 1 - 2x$

13. $t^2 + 4t = 1$ **14.** $y^2 + 2y + 2 = 0$ **15.** $n^2 = 6n + 11$

Find the three cube roots of each number. Simplify the results.

16. -1 **17.** 27 **18.** 8 **19.** 64 **20.** 125 **21.** $-1{,}000$

Solve each equation using the quadratic formula. Simplify each answer.

22. $\frac{1}{4}x^2 = \frac{1}{2}x + \frac{3}{8}$ **23.** $\frac{1}{3}y^2 + \frac{3}{2} = \frac{1}{3} - y$

24. $2x^2\sqrt{2} + 3x - \sqrt{2} = 0$ **25.** $x^2\sqrt{3} - 2x - \sqrt{3} = 0$

26. $x^2 - 2ix + 3 = 0$ **27.** $2ix^2 - 5x - 2i = 0$

28. Find the 4 fourth roots of 16.   **29.** Find the 3 cube roots of $\frac{1}{8}$.

Mixed Review

Simplify each expression.   *7.1, 7.4, 8.2, 8.5*

1. $\dfrac{f(a) - f(b)}{a - b}$, given that $f(x) = 5x + 4$ **2.** $\dfrac{9x}{x^2 - 9} - \dfrac{2}{x + 3} + \dfrac{5}{3 - x}$

3. $\sqrt{48x^{10}y^9}$, $x \geq 0$, $y \geq 0$ **4.** $\sqrt[3]{-16x^4y^6}$

9.6 Problem Solving: Irrational Answers

Objective To solve word problems using the quadratic formula

The Pythagorean relation, $a^2 + b^2 = c^2$, and the quadratic formula can be used to solve some problems involving right triangles.

EXAMPLE 1 The length of a rectangle is 1 ft more than twice its width, and 1 ft less than the length of a diagonal. Find the length and the area of the rectangle. State the answers in simplest radical form.

What are you to find? the length and the area

Choose a variable. Let w = the width.
What does it represent? Then $2w + 1$ = the length of the rectangle, and $2w + 2$ = the length of the diagonal.

What do you know? $\triangle ABC$ is a right triangle. Use the Pythagorean relation, $a^2 + b^2 = c^2$.

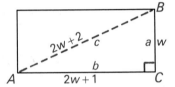

Write an equation. $w^2 + (2w + 1)^2 = (2w + 2)^2$

Solve it. $w^2 + 4w^2 + 4w + 1 = 4w^2 + 8w + 4$

$$w^2 - 4w - 3 = 0$$

$$w = \frac{4 \pm \sqrt{16 - 4 \cdot 1(-3)}}{2} \leftarrow \text{quadratic formula}$$

$$= \frac{4 \pm \sqrt{28}}{2}, \text{ or } 2 \pm \sqrt{7}$$

$w = 2 + \sqrt{7} \leftarrow w \neq 2 - \sqrt{7}$ because $2 - \sqrt{7} < 0$.

$2w + 1 = 2(2 + \sqrt{7}) + 1 = 5 + 2\sqrt{7}$

Area $= w(2w + 1)$

$= (2 + \sqrt{7})(5 + 2\sqrt{7})$

$= 10 + 9\sqrt{7} + 14$, or $24 + 9\sqrt{7}$

Check in the original problem. The check is left for you.

State your answer. The length is $5 + 2\sqrt{7}$ ft, and the area is $24 + 9\sqrt{7}$ ft^2.

EXAMPLE 2   An object is sent upward from the earth's surface with an initial velocity of 96 ft/s. When will the height of the object be 136 ft? State the answer to the nearest tenth of a second.

Solution   Use the formula $h = vt - 16t^2$. Substitute 136 for h and 96 for v.

$$h = vt - 16t^2$$
$$136 = 96t - 16t^2$$
$$17 = 12t - 2t^2 \leftarrow \text{Each side is divided by 8, the GCF of each term.}$$
$$2t^2 - 12t + 17 = 0$$
$$t = \frac{12 \pm \sqrt{144 - 136}}{4} = \frac{12 \pm 2\sqrt{2}}{4} = \frac{6 \pm \sqrt{2}}{2} \approx \frac{6 \pm 1.414}{2}$$

$$\frac{6 + 1.414}{2} = 3.707 \quad or \quad \frac{6 - 1.414}{2} = 2.293$$

The height to the nearest 0.1 s will be 136 ft at the end of 2.3 s (on the way up) and again at the end of 3.7 s (on the way down).

EXAMPLE 3   A rectangular lawn is 3 times as long as it is wide. It is surrounded by a sidewalk with a uniform width of 5.00 ft. The total area of the lawn and sidewalk is 2,600 ft^2. Find the dimensions of the lawn to the nearest 0.1 ft.

Solution   Use the given data to draw and label a figure such as that at the right.

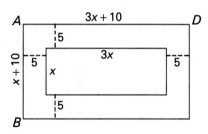

$$AB \cdot AD = \text{total area}$$
$$(x + 10)(3x + 10) = 2,600$$
$$3x^2 + 40x + 100 = 2,600$$
$$3x^2 + 40x - 2,500 = 0$$

Now use the quadratic formula to solve the equation above.

$$x = \frac{-40 \pm \sqrt{1,600 + 30,000}}{6} = \frac{-40 \pm \sqrt{31,600}}{6} = \frac{-40 \pm 20\sqrt{79}}{6}$$

$$x = \frac{-20 + 10\sqrt{79}}{3} \approx \frac{-20 + 10(8.888)}{3} \approx 23.0 \leftarrow \text{The negative value of } x \text{ is discarded.}$$

$$3x = -20 + 10\sqrt{79} \approx -20 + 10(8.888) \approx 68.9$$

The lawn measures 23.0 ft by 68.9 ft, to the nearest 0.1 ft.

Classroom Exercises

Solve the following problems.

1. A diagonal of a square is 5 cm longer than a side of the square. Find the length of the diagonal and the area of the square. Give your answer in simplest radical form.

2. An object is sent upward with an initial velocity of 144 ft/s. To the nearest 0.1 s, when will the height of the object be 96 ft? Use the formula $h = vt - 16t^2$.

Written Exercises

Solve. Write the answer in simplest form.

1. One leg of a right triangle is 2 ft longer than the other leg and 4 ft shorter than the hypotenuse. Find the length of the longer leg and the area of the triangle.

2. If a pellet is launched straight upward with an initial velocity of 200 ft/s, when will its altitude be 568 ft? Answer to the nearest 0.1 s. Use the formula $h = vt - 16t^2$.

3. The product of a pair of numbers is 1. The second number is 4 less than the first number. Find all such pairs of numbers.

4. A second number is 2 less than twice a first number. Find all such pairs of numbers whose product is 3.

5. A second complex number is 6 less than a first complex number. Find all such pairs of numbers given that their product is the real number -14.

6. If a second number is multiplied by 2 more than a first number, the product is -2. Find all such pairs of numbers given that the second is 4 more than 4 times the first.

7. A rectangular lawn measures 20 yd by 40 yd and is surrounded by a sidewalk of uniform width. The sidewalk's outer edge is a rectangle with an area of 1,000 yd^2. Find the width of a strip of the sidewalk to the nearest tenth of a yard.

Mixed Review

Solve each equation.   *1.4, 6.1, 7.6*

1. $\dfrac{5x - 3}{10} = \dfrac{6x + 2}{15}$

2. $\dfrac{x}{6} + \dfrac{x}{3} + \dfrac{x}{4} = 1$

3. $2(x - 1)^2 + 3x = (x + 3)^2 - x$

4. $\dfrac{y + 8}{3y} = \dfrac{y + 3}{2y - 1}$

9.7 Vectors: Addition and Subtraction

Objectives
To add and subtract vectors by drawing
To draw scalar multiples of vectors
To draw vectors whose resultants are zero vectors

The displacement (movement) of an object through a distance of 8 mi to the southeast is represented by the directed line segment $\overrightarrow{AB}$ shown at the right. Such directed line segments are called **vectors**. The vector shown is written as $\overrightarrow{AB}$ (with a half-arrow $\rightharpoonup$) and read as "vector AB."

Notice:

(1) $\overrightarrow{AB}$ has a **magnitude** of 8 mi and its **direction** is southeast, as shown by the arrowhead in the drawing.

(2) The initial point (*tail*) of $\overrightarrow{AB}$ is A. The terminal point (*head*) of $\overrightarrow{AB}$ is B.

A second use of vectors is to represent forces. The force vector $\overrightarrow{CD}$ represents the force exerted by a 7-newton weight (magnitude) suspended on a coiled spring, stretching it downward (direction).

A single letter can be used to name a vector. The velocity vector, $\vec{G}$, represents the velocity of a car traveling at a speed of 40 mi/h (magnitude) to the west (direction). Observe that the speed tells only how fast the object moves, while the velocity vector indicates both speed and direction.

Four vectors are shown at the right in a coordinate plane. Vectors $\vec{R}$ and $\vec{R}'$ are **equivalent vectors** because they have the same magnitude and the same direction. Vectors $\vec{T}$ and $-\vec{T}$ are **opposite vectors** because they have the same magnitude but opposite directions.

Vectors can be added, the sum of two vectors being the result of two consecutive displacements. If an object moves from A to B and then from B to C, the result is equivalent to one move directly from A to C. Thus, $\overrightarrow{AB} + \overrightarrow{BC} = \overrightarrow{AC}$. Notice that the tail of $\overrightarrow{BC}$ is the head of $\overrightarrow{AB}$ and that the sum, $\overrightarrow{AC}$, is drawn from the tail of $\overrightarrow{AB}$ to the head of $\overrightarrow{BC}$.

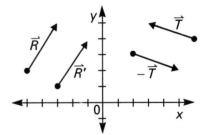

$$\overrightarrow{AB} + \overrightarrow{BC} = \overrightarrow{AC}$$

The method just shown is called the *triangle method* of addition for vectors. The sum, $\overrightarrow{AC}$, is the **resultant** of $\overrightarrow{AB}$ and $\overrightarrow{BC}$. Given the disjoint vectors $\overrightarrow{V}$ and $\overrightarrow{W}$ below, you can use the equivalent vector $\overrightarrow{W'}$ to draw $\overrightarrow{V} + \overrightarrow{W}$ as shown below at the right. Move the tail of $\overrightarrow{W'}$ to the head of $\overrightarrow{V}$, and use the triangle method to draw the resultant.

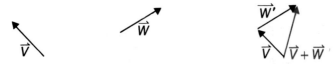

Three vectors can be added by extending the method above.

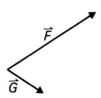 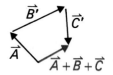

Because vectors $\overrightarrow{F}$ and $\overrightarrow{G}$ below have the same initial point, the *parallelogram method* can be used to draw $\overrightarrow{F} + \overrightarrow{G}$. Draw equivalent vectors $\overrightarrow{F'}$ and $\overrightarrow{G'}$ to form a parallelogram with $\overrightarrow{F}$ and $\overrightarrow{G}$ as shown. The diagonal from the common initial point is $\overrightarrow{F} + \overrightarrow{G}$.

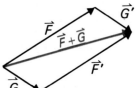 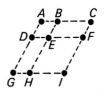

EXAMPLE 1 Use the figure at the right to find the following vectors.

a. $\overrightarrow{AG} + \overrightarrow{GH}$ b. $\overrightarrow{DB} + \overrightarrow{BF}$ c. $\overrightarrow{CH} + \overrightarrow{HD}$
d. $\overrightarrow{BE} + \overrightarrow{HI}$ e. $\overrightarrow{CE} + \overrightarrow{BA}$ f. $\overrightarrow{AI} + \overrightarrow{CB}$
g. $\overrightarrow{AB} + \overrightarrow{AD}$ h. $\overrightarrow{HB} + \overrightarrow{HG}$ i. $\overrightarrow{EF} + \overrightarrow{EH}$
j. $\overrightarrow{GB} + \overrightarrow{BC} + \overrightarrow{CF}$ k. $\overrightarrow{DB} + \overrightarrow{EF} + \overrightarrow{BH}$

Solutions Use the triangle method for parts a–c.

a. $\overrightarrow{AH}$ b. $\overrightarrow{DF}$ c. $\overrightarrow{CD}$

Use equivalent vectors for parts d–f.
d. $\overrightarrow{BE} + \overrightarrow{EF} = \overrightarrow{BF}$ e. $\overrightarrow{CE} + \overrightarrow{ED} = \overrightarrow{CD}$ f. $\overrightarrow{AI} + \overrightarrow{IH} = \overrightarrow{AH}$

Use the parallelogram method for parts g–i.
g. $\overrightarrow{AE}$ h. $\overrightarrow{HA}$ i. $\overrightarrow{EI}$

Use any method or a combination of methods for parts j and k.
j. $\overrightarrow{GF}$ k. $\overrightarrow{DB} + \overrightarrow{BC} + \overrightarrow{CI} = \overrightarrow{DI}$

The Rule of Subtraction for vectors is similar to the Rule of Subtraction for real numbers, as stated below.

$$a - b = a + (-b) \text{ for all real numbers } a \text{ and } b.$$
$$\vec{X} - \vec{Y} = \vec{X} + (-\vec{Y}) \text{ for all vectors } \vec{X} \text{ and } \vec{Y}.$$

EXAMPLE 2 Draw $\vec{P} - \vec{Q}$ for each of the cases shown.

a.

b.

Solutions

a. Draw $-\vec{Q}$. Use the parallelogram method to draw $\vec{P} + (-\vec{Q})$.

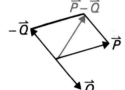

b. Draw $-\vec{Q}$ from the head of $\vec{P}$. Use the triangle method to draw $\vec{P} + (-\vec{Q})$.

A vector can also be multiplied by a real number. Consider, for example, a displacement of 4 mi to the east occurring every hour for 3 h. The result is a total displacement of 12 mi to the east, as shown below.

$\vec{D}$
4 mi $\vec{D}$
4 mi $\vec{D}$
4 mi $3 \cdot \vec{D}$
12 mi

$\vec{D}$ above was multiplied by the real number 3 to give the vector $3 \cdot \vec{D}$. In this case, the real number 3 is called a **scalar**, and the vector $3 \cdot \vec{D}$ is called a **scalar multiple** of $\vec{D}$.

EXAMPLE 3 Given $\vec{A}$ as shown, draw $-2 \cdot \vec{A}$.

$\vec{A}$
12 ft

Solution $-2 \cdot \vec{A} = 2(-\vec{A})$

$-\vec{A}$
12 ft $-\vec{A}$
12 ft

$-2 \cdot \vec{A}$
24 ft

EXAMPLE 4   Draw $\frac{1}{2} \cdot \vec{P} - 2 \cdot \vec{Q}$ given $\vec{P}$ and $\vec{Q}$.

Plan   Draw $\frac{1}{2} \cdot \vec{P}$ and $-2 \cdot \vec{Q}$.

Use the parallelogram method to draw $\frac{1}{2} \cdot \vec{P} + (-2 \cdot \vec{Q})$.

Solution

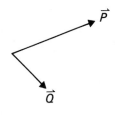

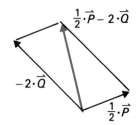

The set of vectors is closed for addition. That is, the sum of any two vectors is a vector. Consider two forces of 50 lb each acting in opposite directions on an object at C, as shown in the figure at the right. The object will remain in equilibrium because $\overrightarrow{CD} + \overrightarrow{CE} = \overrightarrow{CC}$.

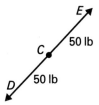

$\overrightarrow{CC}$ is a *zero vector*. Its magnitude is 0 and it has no direction. In general, the sum of a pair of *opposite vectors* is a **zero vector**.

EXAMPLE 5   Using the figure at the right, draw $\vec{R}$ so that $\vec{P} + \vec{Q} + \vec{R}$ is a zero vector ($\overrightarrow{AA}$) that leaves the object at A in equilibrium.

Solution   Draw $\vec{P} + \vec{Q}$.   Draw $\vec{R} = -(\vec{P} + \vec{Q})$.

$\vec{P} + \vec{Q} + \vec{R} = \overrightarrow{AA}$

Classroom Exercises

Match the phrase at the left with a vector in the figure.

1. velocity vector
2. force vector
3. displacement vector
4. northeast direction
5. west direction
6. upward direction
7. magnitude of 10 km
8. eastward direction

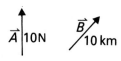

Use the figure at the right to find the following.

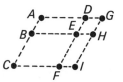

9. two vectors equivalent to $\overrightarrow{ED}$

10. three vectors that are opposites of $\overrightarrow{AD}$

11. $\overrightarrow{CE} + \overrightarrow{EA}$ **12.** $\overrightarrow{CB} + \overrightarrow{CF}$ **13.** $\overrightarrow{IE} + \overrightarrow{FC}$ **14.** $\overrightarrow{GI} + \overrightarrow{GD}$

15. $\overrightarrow{AF} + \overrightarrow{DG}$ **16.** $\overrightarrow{AB} + \overrightarrow{BF} + \overrightarrow{FI}$ **17.** $\overrightarrow{CF} + \overrightarrow{IH} + \overrightarrow{BA}$ **18.** $\overrightarrow{AD} - \overrightarrow{HG}$

19. $\overrightarrow{HI} - \overrightarrow{DG}$ **20.** $\overrightarrow{FE} - \overrightarrow{AD}$ **21.** $\overrightarrow{AE} - \overrightarrow{FE}$ **22.** $\overrightarrow{FB} - \overrightarrow{DA}$

Written Exercises

Carefully copy the figures below. Then draw and label the resultant vectors given in Exercises 1–26.

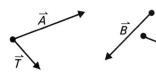 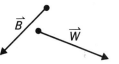

$\overleftarrow{P}$
12 cm

$\overleftarrow{Q}$
7 cm

$\overrightarrow{R}$
9 cm

1. $\vec{A} + \vec{T}$ **2.** $\vec{B} + \vec{W}$ **3.** $\vec{C} + \vec{K} + \vec{L}$ **4.** $\vec{X} + \vec{Y} + \vec{Z}$

5. $\vec{A} - \vec{T}$ **6.** $\vec{B} - \vec{W}$ **7.** $\vec{W} - \vec{B}$ **8.** $\vec{T} - \vec{A}$

9. $3 \cdot \vec{Q}$ **10.** $\frac{2}{3} \cdot \vec{R}$ **11.** $-2 \cdot \vec{Q}$ **12.** $-1.5 \cdot \vec{P}$

13. $2 \cdot \vec{A} + 3 \cdot \vec{T}$ **14.** $3 \cdot \vec{A} - 2 \cdot \vec{T}$ **15.** $3 \cdot \vec{B} - 2 \cdot \vec{W}$ **16.** $\frac{1}{2} \cdot \vec{W} - 2 \cdot \vec{B}$

17. $\vec{C} + \vec{K} - \vec{L}$ **18.** $\vec{K} + \vec{L} - \vec{C}$ **19.** $\vec{X} + \vec{Y} - \vec{Z}$ **20.** $\vec{X} - \vec{Y} + \vec{Z}$

21. $\vec{V}$, so that $\vec{A} + \vec{T} + \vec{V}$ is a zero vector   **22.** $\vec{V}$, so that $\vec{B} + \vec{W} + \vec{V}$ is a zero vector

23. $\vec{C} - (\vec{K} + \vec{L})$ **24.** $2 \cdot (\vec{C} + \vec{K} + \vec{L})$

25. $\vec{V}$, so that $\vec{C} + \vec{L} + \vec{K} + \vec{V}$ is a zero vector

26. $\vec{V}$, so that $\vec{X} + \vec{Y} + \vec{Z} + \vec{V}$ is a zero vector

Mixed Review

Identify the property of operations with real numbers illustrated below. All variables represent real numbers.   *1.3*

1. $-5 + 3.26$ is a real number.   **2.** $x + x^2 = x^2 + x$

3. $(-8 + 17) + 23 = -8 + (17 + 23)$ **4.** $7.1 + (-7.1) = 0$

5. $8(c + 12) = 8 \cdot c + 8 \cdot 12$ **6.** $1(c + d) = c + d$

9.8 Vectors and Complex Numbers

Objectives

To match vectors in standard position with corresponding complex numbers

To draw sums and differences of two vectors corresponding to the sums and differences of two complex numbers.

To draw scalar multiples of vectors corresponding to the products of real numbers and complex numbers

Recall that each complex number $a + bi$ can be matched with a unique ordered pair of real numbers (a,b), and that each ordered pair can in turn be matched with a unique point in a coordinate plane.

Now, let $\vec{V}$ be a vector with its initial point at the origin of a coordinate plane and its terminal point matched with a unique ordered pair of real numbers (a,b). The table below illustrates this matching for the four vectors shown in the diagram. The corresponding complex numbers for these vectors are also shown.

Vector	Terminal point	Ordered pair (a,b)	Complex number
$\vec{V_1}$	A	$(5,3)$	$5 + 3i$
$\vec{V_2}$	B	$(0,-2)$	$-2i$
$\vec{V_3}$	C	$(-4,0)$	-4
$\vec{V_4}$	D	$(-2,4)$	$-2 + 4i$

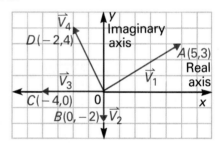

As suggested by the table, there is a one-to-one correspondence between all complex numbers $a + bi$ and all vectors $\vec{V}$ with their initial points at the origin. Such vectors are said to be in **standard position**. If $\vec{V}$ is in standard position with its terminal point at (a,b), then (a,b) can be used to represent $\vec{V}$. (a,b) is called the **rectangular form** of $\vec{V}$.

Addition and subtraction of complex numbers correspond to addition and subtraction of vectors in standard position, as shown below.

Addition of complex numbers:
$(5 + 3i) + (-3 + 2i) = 2 + 5i$

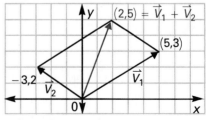

Addition of vectors:

Let $\vec{V_1} = (5,3)$ and $\vec{V_2} = (-3,2)$. Use the parallelogram method to draw $\vec{V_1} + \vec{V_2}$.

Notice that $(5 + 3i) + (-3 + 2i) = 2 + 5i$ corresponds to $(5,3) + (-3,2) = (2,5)$.

Subtraction of complex numbers:
$(3 + 4i) - (5 - 2i) =$
$= (3 + 4i) + (-5 + 2i)$
$= -2 + 6i$

Subtraction of vectors:
Let $\vec{V_1} = (3,4)$ and $\vec{V_2} = (5,-2)$.
Then $-\vec{V_2} = (-5,2)$.
Draw $\vec{V_1} - \vec{V_2} = \vec{V_1} + (-\vec{V_2})$.

Notice that $(3,4) - (5,-2)$
$\qquad = (3,4) + (-5,2)$
$\qquad = (-2,6)$

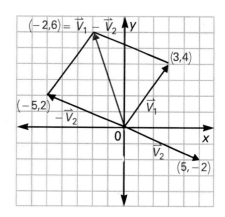

EXAMPLE

Draw the two vectors that correspond to $4 - 2i$ and $1 + 3i$. Then draw their sum and their difference.

Solutions

Draw $(4,-2)$ and $(1,3)$.

Sum: $(4,-2) + (1,3)$
$\qquad = (5,1)$

Difference: Draw $-(1,3) = (-1,-3)$.
Then, $(4,-2) - (1,3)$
$\qquad = (3,-5)$

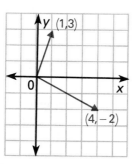

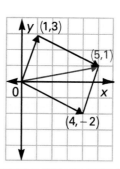

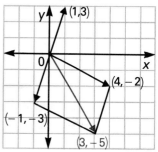

The product of a real number and a complex number corresponds to the product of a scalar and a vector. For $3(-2 + i) = -6 + 3i$, let $\vec{V} = (-2,1)$. Draw $3 \cdot \vec{V}$ as shown above at the right. Notice that

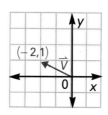

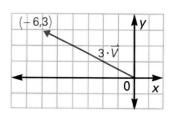

$$3 \cdot \vec{V} = 3 \cdot (-2,1) = (-6,3), \text{ and } 3(-2 + i) = -6 + 3i$$

Classroom Exercises

Match each complex number at the left with the rectangular form of the corresponding vector at the right.

1. 5

2. $2 + 3i$

a. $(5,5)$

e. $(2,-3)$

3. $5i$

4. $-2 - 3i$

b. $(-2,3)$

f. $(0,5)$

5. $\overline{2 + 3i}$

6. $-2 + 3i$

c. $(2,3)$

g. $(5,0)$

7. $\overline{5 - 5i}$

8. $\overline{5 + 5i}$

d. $(-2,-3)$

h. $(5,-5)$

9. Draw the vectors for $5 - i$ and $6 + 2i$. Then draw their sum and the difference of the second from the first.

Written Exercises

Draw the two vectors that correspond to the two given complex numbers. Then draw their sum and the difference of the second from the first.

1. $4 + 2i, -2 + 3i$ **2.** $-4 - i, 3 - 2i$ **3.** $3 + 5i, 2 - 3i$ **4.** $-2i, 5$

5. $-4 - 2i, 5i$ **6.** $2 + 4i, -5$ **7.** $5i, -5i$ **8.** $2 - 4i, 2 + 4i$

In Exercises 9–12, each expression corresponds to the product $s\vec{V}$ of a scalar s and a vector $\vec{V}$, represented as a complex number. Draw both $\vec{V}$ and $s\vec{V}$.

9. $2(3 - 4i)$ **10.** $1.5(-4 + 2i)$ **11.** $-3(2 + 2i)$ **12.** $-0.5(6i)$

13. Use a vector diagram to prove that the sum of a pair of conjugate complex numbers is a real number. (HINT: Use the general complex number $a + bi$ and its conjugate $a - bi$, and recall that the x-axis is the axis of real numbers.)

14. Use a vector diagram to prove that the difference of two conjugate complex numbers is an imaginary number.

15. Draw the vectors, in standard position, that correspond to the consecutive integral powers of i: $i, i^2, i^3, i^4, \ldots, i^8$.

Mixed Review

Identify the property of operations with real numbers illustrated below. All variables represent real numbers.   *1.3*

1. $-8 \cdot 112$ is a real number.

2. $5(3c) = (5 \cdot 3)c$

3. $y(x + 2) = (x + 2)y$

4. $m \cdot \dfrac{1}{m} = 1, m \neq 0$

Extension: *The Field of Complex Numbers*

The eleven properties of a field that are true for real numbers can be used to prove that the complex numbers form a field under the operations of addition and multiplication. The following definitions will be used.

Definitions 1, 2, 3, and 4

1. $z = x + yi$ is a complex number, where x and y are real numbers.

If $z_1 = a + bi$ and $z_2 = c + di$, then:

2. $z_1 = z_2$ if and only if $a = c$ and $b = d$,

3. $z_1 + z_2 = (a + bi) + (c + di) = (a + c) + (b + d)i$, and

4. $z_1 \cdot z_2 = (a + bi)(c + di) = (ac - bd) + (ad + bc)i$.

Proofs of several theorems about complex numbers are outlined below with $z_1 = a + bi$ and $z_2 = c + di$, where a, b, c, and d are real numbers. Try to supply a reason for each step.

Theorem 9.1
Closure for addition

$z_1 + z_2$
$= (a + bi) + (c + di)$
$= (a + c) + (b + d)i$
$a + c$ and $b + d$ are real numbers.
$z_1 + z_2$ is a complex number.

Theorem 9.2
Addition is commutative.

$z_1 + z_2$
$= (a + bi) + (c + di)$
$= (a + c) + (b + d)i$
$= (c + a) + (d + b)i$
$= (c + di) + (a + bi)$
$= z_2 + z_1$ So, $z_1 + z_2 = z_2 + z_1$.

If $z_1 + z_2 = z_1$, then z_2 is an additive identity. For example, $0 + 0i$ is an additive identity since $(a + bi) + (0 + 0i) = (a + 0) + (b + 0)i = a + bi$. Is $0 + 0i$ the only additive identity?

Theorem 9.3
$0 + 0i$ is *the* additive identity.

$(a + bi) + (x + yi) = a + bi$ if $x + yi$ is an identity element.

So, $(a + x) + (b + y)i = a + bi$. By Definition 2: $a + x = a$ and $b + y = b$.

The only solutions are $x = 0$ and $y = 0$.

Thus, $x + yi$, or $0 + 0i$ is the *only* additive identity.

If $z_1 + z_2 = 0 + 0i$, then z_2 is an additive inverse of z_1. For example, $-a - bi$ is an additive inverse of $a + bi$ since $(a + bi) + (-a - bi) = [a + (-a)] + [b + (-b)]i = 0 + 0i$. You can show that $-a - bi$ is the *only* additive inverse of $a + bi$.

Theorem 9.4

$-a - bi$ is *the* additive inverse of $a + bi$.

$(a + bi) + (x + yi) = 0 + 0i$ if $x + yi$ is an additive inverse of $a + bi$.

So, $(a + x) + (b + y)i = 0 + 0i$, or $a + x = 0$ and $b + y = 0$.

The only solutions are $x = -a$ and $y = -b$.

Thus, $x + yi$, or $-a - bi$, is the *only* additive inverse of $a + bi$.

Exercises

Use Definition 2 to solve each equation for x and y.

1. $x + yi = 5 - 2i$
2. $4 - yi = x + 6i$
3. $3x - 4yi = -6 + 12i$

Outline the proofs of the following theorems for complex numbers. Use the outlines for Theorems 9.1–9.4 as models. Also, use Definition 4 for $z_1 \cdot z_2$.

4. $z_1 \cdot z_2$ is a complex number. (Closure for multiplication)
5. Multiplication is commutative: $z_1 \cdot z_2 = z_2 \cdot z_1$.
6. Addition is associative: $(z_1 + z_2) + z_3 = z_1 + (z_2 + z_3)$.

If $z_1 \cdot z_2 = z_1$, then z_2 is a **multiplicative identity.**

7. Show that $1 + 0i$ is a multiplicative identity by showing that $(a + bi)(1 + 0i) = a + bi$.

8. Prove that $1 + 0i$ is the *only* multiplicative identity by solving $(a + bi)(x + yi) = a + bi$ for $x + yi$.

If $z_1 \cdot z_2 = 1 + 0i$, then z_2 is a **multiplicative inverse** of z_1, $z_1 \neq 0 + 0i$.

9. Show that a multiplicative inverse of $z = 3 + 4i$ is

$$\frac{3}{3^2 + 4^2} + \frac{-4}{3^2 + 4^2}i, \text{ or that } (3 + 4i)\left(\frac{3}{3^2 + 4^2} + \frac{-4}{3^2 + 4^2}i\right) = 1 + 0i.$$

10. Prove that $\dfrac{a}{a^2 + b^2} + \dfrac{-b}{a^2 + b^2}i$ is the *only* multiplicative inverse of $a + bi$, $a + bi \neq 0 + 0i$ by solving $(a + bi)(x + yi) = 1 + 0i$ for $x + yi$.

Solve each problem.

1. A rowing crew traveled 8.5 km downstream in the same time it took to travel 6.5 km upstream. Find the rate of the current given that the crew travels 30.0 km/h in still water.

2. Machine A can do a job in 18 h. If Machines A and B work together, the time is cut to 8 h. How long would it take Machine B to do the job working alone?

3. Find three consecutive multiples of 4 such that the product of the second and third numbers is 192.

4. Convert a pressure of 180 pounds per square foot to ounces per square inch.

5. How many liters of water must be added to 6 liters of a 40% iodine solution to dilute it to a 10% iodine solution?

6. How many milliliters of a 25% iodine solution should be added to 600 ml of a 65% iodine solution to obtain a 55% iodine solution?

7. The height of a triangle is 6 cm more than 3 times the length of its base. Find the height, given that the area of the triangle is 36 cm^2.

8. Ninety-six limes are packed in boxes so that the number of boxes is 2 less than 3 times the number of limes in a box. Find the number of boxes.

9. One leg of a right triangle is twice as long as the other leg. The hypotenuse is 5 in. longer than the longer leg. Find the length of the hypotenuse in simplest radical form.

10. Ralph is 3 years older than Sonya, and Teresa is 3 times as old as Ralph. In 8 years, Teresa's age will be 6 years more than the combined ages of Ralph and Sonya then. Find their present ages.

11. The stretch S in a spring balance varies directly as the applied weight w. Given that $S = 4.8$ in. when $w = 7.2$ lb, find the weight needed to cause a stretch of 6.4 in.

12. The denominator of a fraction is 2 less than the square of its numerator. If the numerator and denominator are both increased by 1, the fraction is equal to $\frac{1}{2}$. Find the original fraction.

13. A patio floor is 2 yd longer than it is wide. It is then extended by a second floor 3 yd longer and 1 yd narrower than the first floor. Find the area of the original floor given that the new total area of the patio is 31 yd^2.

14. Some red and some black pens are sold in packages of 20. The red pens are worth 50¢ each and the black pens are worth 30¢ each. The package of 20 pens is worth $8.20. How many black pens are in the package?

15. The units digit of a two-digit number is 1 more than 3 times the tens digit. If the digits are reversed, the new number is 9 less than 3 times the original number. Find the new number.

16. Six times the tens digit of a two-digit number is 4 more than the units digit. If the digits are reversed, the sum of the new number and twice the original one is 138. Find the original number.

Chapter 9 Review

Key Terms

absolute value of a complex number (p. 305)
completing the square (p. 310)
complex conjugates (p. 308)
complex number (p. 303)
direction of a vector (p. 323)
equivalent vectors (p. 323)
imaginary axis (p. 304)
imaginary number (p. 303)
integral polynomial equation (p. 317)
magnitude of a vector (p. 323)
opposite vectors (p. 323)

quadratic form (p. 304)
Quadratic Formula (p. 317)
real axis (p. 304)
rectangular form of a vector (p. 328)
resultant (p. 324)
scalar (p. 325)
scalar multiple of a vector (p. 325)
standard form of a complex number (p. 303)
standard position of a vector (p. 328)
vector (p. 323)
zero vector (p. 326)

Key Ideas and Review Exercises

9.1 To simplify the sum or difference of two complex numbers, use $\sqrt{-x} = i\sqrt{x}$, where $-x < 0$, and combine like terms. The absolute value of $a + bi$, denoted by $|a + bi|$, is $\sqrt{a^2 + b^2}$.

Write in the form $a + bi$, where a and b are real numbers.

1. $(6 - 4i) - (-3 + 2i)$

2. $(5 - 3\sqrt{-12}) + (-2 - \sqrt{-27})$

3. Simplify $|2 - 5i|$.

4. Simplify $|2 + 4i| + |2 - 4i|$.

9.1 If $x^2 = k$, then $x = \pm\sqrt{k}$, for all real numbers k.

5. Solve $3y^2 + 33 = 0$.

6. Solve $x^4 + 2x^2 = 24$.

9.2 To simplify the product of two complex numbers, use $\sqrt{-1} = i$ and $i^2 = -1$. Also use the conjugate of $a + bi$, denoted by $\overline{a + bi}$, which is $a - bi$. To simplify the quotients $\dfrac{2}{3i}$ and $\dfrac{3}{4 + 2i}$, multiply by $\dfrac{i}{i}$ and $\dfrac{4 - 2i}{4 - 2i}$, respectively.

Simplify.

7. $3\sqrt{-2} \cdot 2\sqrt{-6}$

8. i^7

9. $4 \div (-5i)$

10. $\dfrac{2}{3 - 4i}$

9.2 To factor $x^2 + y^2$, write $x^2 - (-y^2) = x^2 - y^2 i^2 = (x + yi)(x - yi)$.

11. Factor $16x^2 + 9$.

12. Factor $49a^2 + 121$.

9.3 To complete the square for $x^2 + bx$, write $x^2 + bx + (\frac{b}{2})^2 = (x + \frac{b}{2})^2$.

13. Solve $x^2 - 8x - 4 = 0$ by completing the square.

9.4   If $y - y_1 = -a(x - x_1)^2$ and $-a < 0$, then the maximum value of y is y_1 and this occurs when $x = x_1$.

14. Find the maximum value of the polynomial P and the corresponding value of x given that $P = -5x^2 + 30x + 12$.

15. Leon High School will sell 400 of its yearbooks if each book sells for $18.00. For each 50¢ decrease in the price, 20 more books will be sold. What book price will produce the maximum income?

9.5   To solve a quadratic equation $ax^2 + bx + c = 0$ using the quadratic formula, use $x = \dfrac{-b \pm \sqrt{b^2 - 4ac}}{2a}$. To find the three cube roots of 125, solve the equation $x^3 = 125$, or $x^3 - 125 = 0$, by writing $x^3 - 125$ as $x^3 - 5^3$ and factoring.

16. Solve $2x^2 + 7 = 6x$ using the quadratic formula. Simplify.

17. Find the three cube roots of 216. Simplify.

9.6   To solve a word problem involving geometric figures, first draw and label the figures.

18. The length of a rectangle is 1 ft more than twice its width, and 2 ft less than its diagonal's length. Find the length of the diagonal in simplest radical form.

9.7   If $\vec{X}$ and $\vec{Y}$ are noncollinear vectors, $\vec{X} + \vec{Y}$ is drawn using the triangle method or the parallelogram method. If $\vec{X}$ and $\vec{Y}$ are collinear, draw $\vec{X} + \vec{Y}$ using the tail-to-head method. To draw $\vec{X} - \vec{Y}$, draw $-\vec{Y}$ and then draw $\vec{X} + (-\vec{Y})$.

Carefully copy the figures at the right. Then draw and label the following.

19. $\vec{A} + \vec{B}$

20. $\vec{V}$, so that $\vec{A} + \vec{B} + \vec{V}$ is a zero vector.

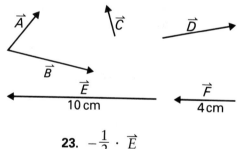

21. $\vec{C} - \vec{D}$ **22.** $\vec{E} - \vec{F}$ **23.** $-\dfrac{1}{2} \cdot \vec{E}$

9.8   The vector with its initial point at the origin of a coordinate plane and its terminal point at $P(a,b)$ corresponds to the complex number $a + bi$.

24. Draw the two vectors that correspond to $-3 + 2i$ and $2 + 3i$ and draw their sum.

25. Draw the vector $\vec{V}$ that corresponds to $5 + 2i$ and the scalar multiple $s\vec{V}$ that corresponds to $-2(5 + 2i)$.

Write in the form $a + bi$, where a and b are real numbers.

1. $(5 - 2i) - (-7 + 4i)$

2. $(7 - 3\sqrt{-8}) + (-2 - \sqrt{-50})$

3. Simplify $|5 - 3i|$.

4. Simplify $|7 - i| - |7 + i|$.

5. Solve $4y^2 + 40 = 0$.

6. Solve $x^4 + 4x^2 - 32 = 0$.

Simplify.

7. $2\sqrt{-3} \cdot 3\sqrt{-6}$

8. i^9

9. $(3 + 5i)(2 - 4i)$

10. $5 \div (-6i)$

11. $(3 + 4i)^2$

12. $(-i\sqrt{3})^5$

13. $(3 - 6i)(\overline{3 - 6i})$

14. $\dfrac{3}{5 + 2i}$

15. Factor $25x^2 + 36$.

16. Solve $x^2 - 8x - 5 = 0$ by completing the square.

17. Three hundred people will enter a bowling tournament if the entry fee is $10 each. The number of entries will decrease by 20 for each $2 increase in the fee. What entry fee will yield the maximum income from all the fees?

18. Solve $2x^2 + 5 = 4x$ using the quadratic formula. Simplify.

19. Find the three cube roots of 64. Simplify.

20. The length of a rectangle is 3 ft more than twice its width, and 2 ft less than a diagonal's length. Find the length of the diagonal in simplest radical form.

Copy the figures below. Then draw and label the following for Exercises 21–25.

21. $\vec{A} + \vec{B}$

22. $\vec{C} - \vec{B}$

23. $\vec{E} - \vec{D}$

24. $-2 \cdot \vec{D}$

25. $\vec{V}$, so that $\vec{A} + \vec{B} + \vec{V}$ is a zero vector.

26. Draw the two vectors that correspond to $-3 + 4i$ and $4 + 2i$ and draw their sum.

27. Draw the vector $\vec{V}$ that corresponds to $2 - i$ and the scalar multiple $s\vec{V}$ that corresponds to $-3(2 - i)$.

28. Prove that $|a + bi| = |a|\sqrt{2}$, if $a = b$.

29. Solve $x^2 + 10ix - 16 = 0$ by completing the square.

Choose the *one* best answer to each question or problem.

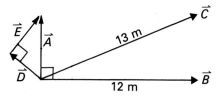

1. Use the drawing above to find the magnitude of $\vec{E}$, given that $\vec{A} + \vec{B} = \vec{C}$, $\vec{D} + \vec{E} = \vec{A}$, and the magnitude of $\vec{D}$ is 3 m.
   (A) 5 m   (B) 4 m   (C) 3 m
   (D) 2 m   (E) None of these

2. $a + bi = c + di$ if and only if $a = c$ and $b = d$. Find x and y given that $3x + y + 9i = 1 + (2x - y)i$.
   (A) $x = \frac{8}{5}$ and $y = \frac{11}{5}$
   (B) $x = 4$ and $y = -1$
   (C) $x = 3$ and $y = -8$
   (D) $x = 2$ and $y = -5$
   (E) None of these

3. For every complex number z, z^* is defined to be z^{-2}. Find the value of $[(-3i)^*]^*$.
   (A) $81i$ (B) 81 (C) $-81i$
   (D) -81 (E) None of these

4. The formula $h = 5t(40 - t)$ gives the height h of an object at t seconds. The greatest height is reached at which time?
   (A) 0 s (B) 10 s (C) 20 s (D) 40 s
   (E) All heights are the same.

5. Choose the number that is not equal to the other three.
   (A) $4 + 2\sqrt{5}$ (B) $\dfrac{8 + \sqrt{80}}{2}$
   (C) $6 + \sqrt{80} - 2 - \sqrt{20}$
   (D) $\sqrt{2}(\sqrt{8} + \sqrt{10})$
   (E) They are all equal.

6. The complex number $x + yi$ corresponds to the point $P(x,y)$. The sum of $5 - 4i$, the conjugate of $2 + 3i$, and the opposite of $6 - i$ corresponds to a point in which quadrant?
   (A) Quadrant I
   (B) Quadrant II
   (C) Quadrant III
   (D) Quadrant IV
   (E) None of these

7. If $7(3 - 2i)^2 = n(2 + 3i)^2$, then $n = \underline{\quad?\quad}$.
   (A) 7   (B) -7 (C) $7i$
   (D) $-7i$ (E) None of these

8. For which equation is the sum of the roots the greatest?
   (A) $(x - 6)^2 = 4$
   (B) $(x - 2)^2 = 9$
   (C) $(x + 5)^2 = 16$
   (D) $(x + 8)^2 = 25$
   (E) $x^2 = 36$

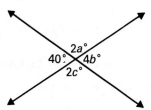

9. In the figure above, what is the value of $a + b + c$?
   (A) 40   (B) 140   (C) 150
   (D) 160   (E) 320

10. Each of the numbers below is the product of two consecutive positive integers. For which of these is the greater of the two consecutive integers an even integer?
    (A) 20   (B) 42   (C) 72
    (D) 90   (E) None of these

10 POLYNOMIAL EQUATIONS AND FUNCTIONS

Mathematical equations are used to study population growth. The equations are often simple, but the repetition of them reveals fascinating patterns. The "science-fiction landscape" shown here is actually a computer study of what happens when different arrangements of numbers are used in a simple population-growth equation.

10.1 Sums and Products of Roots

Objectives

To find the sums and products of roots of given quadratic equations without solving the equations

To write quadratic equations in the general form $ax^2 + bx + c = 0$ given their roots

The polynomial equation $6x^5 - 17x^4 - 29x^3 + 74x^2 + 20x - 24 = 0$ has roots, but cannot be solved directly for x. Even so, some information about its roots *can* be obtained by carefully examining the coefficients. You will study such higher-degree equations later in this chapter, but first, some simpler cases. One such case is the quadratic equation.

An interesting property of the general quadratic equation is that the sum and product of its solutions, or roots, can be found without solving the equation for the roots themselves. For example, if x_1 and x_2 are the roots of $3x^2 + 7x - 20 = 0$, where $a = 3$, $b = 7$, and $c = -20$, then $x_1 + x_2 = -\frac{7}{3} (= -\frac{b}{a})$ and $x_1 \cdot x_2 = \frac{-20}{3} (= \frac{c}{a})$. This can be

shown using the quadratic formula, $x = \dfrac{-b \pm \sqrt{b^2 - 4ac}}{2a}$.

If $3x^2 + 7x - 20 = 0$,

$$\text{then } x_1 = \frac{-7 + \sqrt{289}}{6} \text{ and } x_2 = \frac{-7 - \sqrt{289}}{6}.$$

$$x_1 + x_2 = \frac{-7 + \sqrt{289}}{6} + \frac{-7 - \sqrt{289}}{6} = \frac{-14}{6} = -\frac{7}{3} (= -\frac{b}{a}).$$

$$x_1 \cdot x_2 = \frac{-7 + \sqrt{289}}{6} \cdot \frac{-7 - \sqrt{289}}{6}$$

$$= \frac{49 - 289}{36} = \frac{-240}{36} = \frac{-20}{3} (= \frac{c}{a})$$

This suggests Theorem 10.1 below.

Theorem 10.1   If $ax^2 + bx + c = 0$, or $x^2 + \frac{b}{a}x + \frac{c}{a} = 0$, where $a \neq 0$ and x_1 and x_2 are the roots, then

$$x_1 + x_2 = -\frac{b}{a} \text{ and } x_1 \cdot x_2 = \frac{c}{a}.$$

EXAMPLE 1   Find the sum and product of the roots of $5x^2 + 2 = 6x$ without solving the equation.

Solution $5x^2 - 6x + 2 = 0$ $a = 5, b = -6, c = 2$

$$x_1 + x_2 = -\frac{b}{a} = -\left(\frac{-6}{5}\right) = \frac{6}{5} \qquad x_1 \cdot x_2 = \frac{c}{a} = \frac{2}{5}$$

The sum of the roots is $\frac{6}{5}$, and their product is $\frac{2}{5}$.

The converse of Theorem 10.1 is also true.

Theorem 10.2   If $x_1 + x_2 = -\frac{b}{a}$ and $x_1 \cdot x_2 = \frac{c}{a}$, then x_1 and x_2 are the roots of the equation $x^2 + \frac{b}{a}x + \frac{c}{a} = 0$, or $ax^2 + bx + c = 0$.

Therefore, if $\frac{6}{5}$ and $\frac{2}{5}$ are the sum and product, respectively, of the two roots of a quadratic equation, then

$$\frac{6}{5} = -\frac{b}{a}, \text{ or } -\frac{6}{5} = \frac{b}{a}, \text{ and } \frac{2}{5} = \frac{c}{a},$$

and the equation is

$$x^2 - \frac{6}{5}x + \frac{2}{5} = 0, \text{ or } 5x^2 - 6x + 2 = 0.$$

EXAMPLE 2   Write a quadratic equation, in general form, that has $\frac{4}{5}$ and $-\frac{1}{2}$ as its roots.

Plan   Find the sum and product of the roots. Then use Theorem 10.2.

Solution   Let $x_1 = \frac{4}{5}$ and $x_2 = -\frac{1}{2}$.

$$x_1 + x_2 = \frac{4}{5} + \frac{-1}{2} = \frac{8}{10} + \frac{-5}{10} = \frac{3}{10}$$

So, $-\frac{b}{a} = \frac{3}{10}$ and $\frac{b}{a} = -\frac{3}{10}$

$$x_1 \cdot x_2 = \frac{4}{5} \cdot \frac{-1}{2} = -\frac{4}{10}$$

So, $\frac{c}{a} = -\frac{4}{10}$

$x^2 + \frac{b}{a}x + \frac{c}{a} = 0$ becomes $x^2 - \frac{3}{10}x - \frac{4}{10} = 0.$

The equation, in general form, is $10x^2 - 3x - 4 = 0.$

A quadratic equation may have exactly one root. For example, if $x^2 - 6x + 9 = 0$, then $(x - 3)(x - 3) = 0$ and $x - 3 = 0$ or $x - 3 = 0$. The only root is 3. For this case, let x_1 and x_2 represent the roots of the quadratic equation, and let $x_1 = x_2$.

EXAMPLE 3 Write a quadratic equation, in general form, that has $\frac{3}{5}$ as its only root.

Solution Let $x_1 = x_2 = \frac{3}{5}$.

$x_1 + x_2 = \frac{3}{5} + \frac{3}{5} = \frac{6}{5} = -\frac{b}{a}$ and $x_1 \cdot x_2 = \frac{3}{5} \cdot \frac{3}{5} = \frac{9}{25} = \frac{c}{a}$

Because $-\frac{b}{a} = \frac{6}{5}, \frac{b}{a} = -\frac{6}{5}$.

Thus, $x^2 + \frac{b}{a}x + \frac{c}{a} = x^2 - \frac{6}{5}x + \frac{9}{25} = 0$, or $25x^2 - 30x + 9 = 0$.

The procedure shown above can also be used with roots that are pairs of conjugate irrational numbers or conjugate complex numbers, as illustrated in Example 4.

EXAMPLE 4 Write a quadratic equation, in general form, that has $-5 \pm 3\sqrt{2}$ as its roots.

Solution Let $x_1 = -5 + 3\sqrt{2}$ and $x_2 = -5 - 3\sqrt{2}$.

Then $x_1 + x_2 = (-5 + 3\sqrt{2}) + (-5 - 3\sqrt{2}) = -10 \leftarrow -\frac{b}{a} = -10$

$$\frac{b}{a} = 10$$

and $x_1 \cdot x_2 = (-5 + 3\sqrt{2})(-5 - 3\sqrt{2}) = 25 - 18 = 7 \leftarrow \frac{c}{a} = 7$

Thus, the equation with the given roots is $x^2 + 10x + 7 = 0$.

Classroom Exercises

Match each equation at the left with one of the items a–j in the column at the right.

1. $x^2 + 11x + 24 = 0$
2. $x^2 + 5x - 24 = 0$
3. $x^2 - 11x + 24 = 0$
4. $x^2 - 5x - 24 = 0$
5. $x^2 + 6x - 10 = 0$
6. $x^2 - 6x + 10 = 0$
7. $x^2 - 6x - 10 = 0$
8. $2y^2 - 5y + 8 = 0$
9. $2y^2 + 5y - 8 = 0$
10. $2y^2 + 5y + 8 = 0$

a. roots: 3, 8
b. roots: -3, 8
c. roots: 3, -8
d. roots: -3, -8
e. sum and product of roots: 6, -10
f. sum and product of roots: $-\frac{5}{2}$, 4
g. sum and product of roots: -6, -10
h. sum and product of roots: $-\frac{5}{2}$, -4
i. sum and product of roots: 6, 10
j. sum and product of roots: $\frac{5}{2}$, 4

Written Exercises

Without solving each equation, find the sum and product of its roots.

1. $x^2 + 9x - 22 = 0$ **2.** $y^2 - 7y - 18 = 0$ **3.** $3n^2 - 12n = 0$

4. $5y^2 - 6y - 1 = 0$ **5.** $6n^2 - 12n + 3 = 0$ **6.** $t^2 = 6 - 5t$

7. $-2y^2 + y + 6 = 0$ **8.** $8x^2 + 6x = 0$ **9.** $2y^2 - 5 = 0$

Write a quadratic equation, in general form, that has the given root or roots.

10. 4 and 6   **11.** -5 and 3   **12.** 0 and 2   **13.** $\frac{1}{6}$ and $-\frac{1}{3}$

14. $-\frac{3}{8}$ and $-\frac{1}{4}$ **15.** $-\frac{4}{5}$ and 3   **16.** -5 **17.** $\frac{2}{3}$

18. $\pm 2\sqrt{3}$ **19.** $\pm 4i$ **20.** $3 \pm \sqrt{7}$ **21.** $-6 \pm 2\sqrt{5}$

22. $4 \pm 2i$ **23.** $\dfrac{5 \pm \sqrt{10}}{2}$ **24.** $\dfrac{-1 \pm 2\sqrt{6}}{3}$ **25.** $\dfrac{-6 \pm 5i\sqrt{2}}{4}$

Without solving each equation, find the sum and product of its roots.

26. $3x^2 + \frac{1}{2}x - \frac{2}{3} = 0$ **27.** $\frac{1}{2}y^2 - 6y + 3 = 0$ **28.** $\frac{2}{3}t^2 + 4 = \frac{1}{6}t$

29. $0.4x^2 + 1.2x = 4.8$ **30.** $12y^2 = 1.44$ **31.** $y^2 - 3y\sqrt{2} = 20$

32. For what values of k will the sum of the roots of $x^2 - (k^2 - 3k)x + 21 = 0$ be 10?

33. For what values of k will the product of the roots of $x^2 - 8x + (k^2 - 10k + 21) = 0$ be 12?

Mixed Review

Find the solution set.   *6.1, 6.4*

1. $x^2 - 4x - 12 = 0$ **2.** $-12k^2 + 48k + 144 = 0$

3. $x^2 - 4x - 12 > 0$ **4.** $x^2 - 4x - 12 < 0$

◢ Brainteaser

Let $f(x) = 10x^2 - 29x + 10$. For what real values of x does $f(x + \frac{1}{x}) = 0$?

10.2 The Discriminant

Objectives

To describe the nature of the roots of quadratic equations by finding their discriminants

To find coefficients of quadratic equations, such that the equations will have exactly one real root, two real roots, or two nonreal roots

It is not necessary to solve a quadratic equation in order to know the number of its roots and whether they are rational, irrational, or complex. This information can be determined instead by using the value of $b^2 - 4ac$ in $x = \dfrac{-b \pm \sqrt{b^2 - 4ac}}{2a}$.

Equation	$b^2 - 4ac$	Nature of roots	Roots
$8x^2 - 2x - 3 = 0$	100 (positive, a perfect square)	2 rational roots	$\dfrac{3}{4}, -\dfrac{1}{2}$
$x^2 + 5x - 2 = 0$	33 (positive, not a perfect square)	2 irrational roots	$\dfrac{-5 \pm \sqrt{33}}{2}$
$4x^2 - 12x + 9 = 0$	0 (perfect square)	1 rational root	$\dfrac{3}{2}$
$2x^2 + 5x + 4 = 0$	-7 (negative)	2 nonreal roots	$\dfrac{-5 \pm i\sqrt{7}}{4}$

Nature of the Roots of a Quadratic Equation

If $ax^2 + bx + c = 0$, where a, b, and c are real numbers ($a \neq 0$),

then $x = \dfrac{-b \pm \sqrt{b^2 - 4ac}}{2a}$, and $b^2 - 4ac$ is the **discriminant** of the equation.

(1) If $b^2 - 4ac > 0$, there are two real roots.
(2) If $b^2 - 4ac = 0$, there is exactly one real root.
(3) If $b^2 - 4ac < 0$, there are two nonreal roots.

Also, if a, b, and c are rational and if $b^2 - 4ac > 0$, then the roots are *rational* when $b^2 - 4ac$ is a perfect square, and the roots are *irrational* when $b^2 - 4ac$ is not a perfect square.

EXAMPLE 1 Use the discriminant to determine the nature of the roots.

      a. $16x^2 + 24x + 9 = 0$ **b.** $x^2 = 6x - 25$

      c. $2x^2 + 3x - 7 = 0$ **d.** $2x^2 = 9x + 5$

Plan   Write each equation in general form in order to identify a, b, and c.

Solutions

a. $16x^2 + 24x + 9 = 0$

$a = 16, b = 24, c = 9$

$b^2 - 4ac$

$\quad = 24^2 - 4 \cdot 16 \cdot 9$

$\quad = 576 - 576,$ or 0

Because $b^2 - 4ac = 0$, there is exactly one root, and it is real and rational

b. $x^2 = 6x - 25$

$x^2 - 6x + 25 = 0$

$a = 1, b = -6, c = 25$

$b^2 - 4ac$

$\quad = (-6)^2 - 4 \cdot 1 \cdot 25$

$\quad = 36 - 100,$ or -64

Because $b^2 - 4ac < 0$, there are two nonreal roots

c. $2x^2 + 3x - 7 = 0$

$b^2 - 4ac$

$\quad = 9 - 4 \cdot 2(-7)$

$\quad = 9 + 56,$ or 65

Because 65 is positive, there are two real roots. Because 65 is not a perfect square, the roots are irrational

d. $2x^2 = 9x + 5$

$2x^2 - 9x - 5 = 0$

$b^2 - 4ac$

$\quad = (-9)^2 - 4 \cdot 2(-5)$

$\quad = 81 + 40,$ or 121

Because 121 is a perfect square, there are two real, rational roots

---

In an equation such as $4x^2 + 2kx + 9 = 0$, the letter k can be treated as a temporary variable. If k is replaced by 6, the equation formed is $4x^2 + 12x + 9 = 0$. Its discriminant, $b^2 - 4ac$, is $12^2 - 4 \cdot 4 \cdot 9$, or 0, and $4x^2 + 12x + 9 = 0$ has exactly one root.

If $k = 7$, then $4x^2 + 2kx + 9 = 0$ becomes $4x^2 + 14x + 9 = 0$, where $b^2 - 4ac = 14^2 - 4 \cdot 4 \cdot 9 = 52$.

If $k = 5$, then $4x^2 + 2kx + 9 = 0$ becomes $4x^2 + 10x + 9 = 0$, where $b^2 - 4ac = 10^2 - 4 \cdot 4 \cdot 9 = -44$.

Thus, $b^2 - 4ac > 0$ and $4x^2 + 14x + 9 = 0$ has two real roots.

Thus, $b^2 - 4ac < 0$ and $4x^2 + 10x + 9 = 0$ has two nonreal roots.

EXAMPLE 2   For what values of k will $(k - 4)x^2 - 12x + 3k = 0$ have exactly one root?

Plan   Express the discriminant $b^2 - 4ac$ in terms of k.

Solution   The equation $(k - 4)x^2 - 12x + 3k = 0$ is in the general form $ax^2 + bx + c = 0$. Thus, $a = k - 4$, $b = -12$, and $c = 3k$.

$$b^2 - 4ac = 144 - 4(k - 4) \cdot 3k$$
$$= 144 - 12k^2 + 48k = -12k^2 + 48k + 144$$

To have exactly one root, $b^2 - 4ac = 0$.
$$-12k^2 + 48k + 144 = 0$$

Divide by -12. $k^2 - 4k - 12 = 0$
$$(k + 2)(k - 6) = 0$$
$$k = -2 \quad \text{or} \quad k = 6$$

The equation will have exactly one root if $k = -2$ or $k = 6$

Focus on Reading

Given the quadratic equation $ax^2 + bx + c = 0$, where a, b, and c are integers, match each expression or statement at the left with exactly one phrase at the right.

1. $b^2 - 4ac$ **a.** two rational roots
2. $b^2 - 4ac < 0$ **b.** two nonreal roots
3. $b^2 - 4ac = 0$ **c.** two real roots
4. $b^2 - 4ac > 0$ **d.** the discriminant
5. $b^2 - 4ac$ is a positive perfect square.   **e.** exactly one root

Classroom Exercises

Match each value of the discriminant of a quadratic equation with the nature of the roots of the equation. The numbers a, b, and c are rational.

1. $b^2 - 4ac = 0$ **a.** two nonreal
2. $b^2 - 4ac = 39$ **b.** two rational
3. $b^2 - 4ac = -16$ **c.** one rational
4. $b^2 - 4ac = 25$ **d.** two irrational

Determine whether the quadratic equation has exactly one rational root, two rational roots, two irrational roots, or two nonreal roots.

5. $(x - 4)(x - 4) = 0$

6. $(x - 2i)(x + 2i) = 0$

7. $(2x + 5)(3x - 2) = 0$

8. $(x + 3\sqrt{2})(x - 3\sqrt{2}) = 0$

9. $(x + 3)^2 = 0$

10. $x^2 - 5x - 1 = 0$

Written Exercises

Find the discriminant, $b^2 - 4ac$, and determine the nature of the roots without solving the equation.

1. $4x^2 - 20x + 25 = 0$ **2.** $9y^2 + 12y + 2 = 0$ **3.** $4n^2 + 4n - 3 = 0$

4. $9t^2 - 6t + 4 = 0$ **5.** $x^2 = 21x - 110$ **6.** $49y^2 + 25 = 70y$

7. $1 = 8n - 4n^2$ **8.** $-t^2 + 6t - 10 = 0$ **9.** $7x^2 + 10x = 0$

For what values of k will the quadratic equation have exactly one root?

10. $kx^2 - 12x + k = 0$ **11.** $y^2 + 8y + k^2 = 0$

12. $2y^2 + 5ky + 50 = 0$ **13.** $x^2 - (k + 6)x + 16 = 0$

For what values of k will the quadratic equation have two *real* roots?

14. $kx^2 + 10x + k = 0$ **15.** $y^2 - 6y + k^2 = 0$ **16.** $y^2 - 2ky + 16 = 0$

For what values of k will the quadratic equation have two *nonreal* roots?

17. $x^2 + 5kx + 25 = 0$ **18.** $y^2 - (k + 6)y + 64 = 0$

Determine the nature of the roots without solving the equation.

19. $1.5x^2 - 2.5x - 2 = 0$ **20.** $5y^2 + 3.5y + 2.2 = 0$

21. $1.8x^2 - 7.2x + 7.2 = 0$ **22.** $(2x - 7)(3x + 8) = -50$

Determine the nature of the roots without solving the equation. Note that at least one coefficient in each equation is an irrational number.

23. $x^2 - 4x\sqrt{3} + 12 = 0$ **24.** $y^2\sqrt{2} - 6y - 20\sqrt{2} = 0$

25. $4z^2 + 12iz - 9 = 0$ **26.** $6ix^2 + 11x - 3i = 0$

Mixed Review

Simplify each expression.   *8.3, 8.5*

1. $(\sqrt{x} + 5)^2$ **2.** $(\sqrt[3]{2x} - 1)^3$ **3.** $(3\sqrt{x} - 2)^2$ **4.** $(3 + \sqrt{x})^2$

10.3 Radical Equations

Objective

To solve radical equations

Equations such as $x - \sqrt{2x + 1} = 7$, $\sqrt[3]{y^2 - 6y} = -2$, and $2\sqrt[4]{3x} = \sqrt[4]{3x + 15}$ are called **radical equations** because variables appear in the radicands. Such equations can be solved by (1) isolating the radicals each on one side, (2) raising each side of the equation to the power indicated by the index of the radicals, (3) solving for the variable, and (4) checking the apparent solutions. Although radical equations are not polynomial equations, polynomial equations often arise in the steps of the solution.

In Example 1 below, each radical can be isolated on one side of the equation.

EXAMPLE 1

Solve $2\sqrt[4]{3x} - \sqrt[4]{3x + 15} = 0$. Check.

Plan

The indices are both 4, so isolate the radicals and raise each side to the fourth power.

Solution

$$2\sqrt[4]{3x} - \sqrt[4]{3x + 15} = 0 \qquad\qquad 2\sqrt[4]{3x} - \sqrt[4]{3x + 15} = 0$$

$$2\sqrt[4]{3x} = \sqrt[4]{3x + 15} \qquad 2\sqrt[4]{3 \cdot \frac{1}{3}} - \sqrt[4]{3 \cdot \frac{1}{3} + 15} \overset{?}{=} 0$$

$$(2\sqrt[4]{3x})^4 = (\sqrt[4]{3x + 15})^4 \qquad\qquad 2\sqrt[4]{1} - \sqrt[4]{16} \overset{?}{=} 0$$

$$16 \cdot 3x = 3x + 15 \qquad\qquad\qquad 0 = 0$$

$$45x = 15 \qquad\qquad\qquad\qquad\qquad \text{True}$$

$$x = \frac{1}{3}$$

The solution is $\frac{1}{3}$.

The apparent solutions of a radical equation must always be checked in the original equation because raising each side to the same power may introduce an *extraneous* solution, one that does not satisfy the original equation (see Lesson 7.6). Notice that it is possible to square each side of an equation that is false for all values of x, such as $3 = -\sqrt{x}$, and obtain an equation, such as $3^2 = x$, that is true for some values of x ($x = 9$). In this case, 9 is an extraneous solution because it does not satisfy the original equation, $3 = -\sqrt{x}$, even though it satisfies the squared equation, $9 = x$.

EXAMPLE 2   Solve $x - \sqrt{2x + 1} = 7$. Check for extraneous solutions.

Plan   Isolate the radical and square each side.

Solution
$$x - 7 = \sqrt{2x + 1}$$
$$(x - 7)^2 = (\sqrt{2x + 1})^2$$
$$x^2 - 14x + 49 = 2x + 1$$
$$x^2 - 16x + 48 = 0$$
$$(x - 4)(x - 12) = 0$$
$$x = 4 \quad or \quad x = 12$$

Checks

$x = 4$:

$x - \sqrt{2x + 1} = 7$

$4 - \sqrt{9} \stackrel{?}{=} 7$

$1 = 7$

False

$x = 12$:

$x - \sqrt{2x + 1} = 7$

$12 - \sqrt{25} \stackrel{?}{=} 7$

$7 = 7$

True

The only solution is 12. (4 is extraneous.)

An equation such as $\sqrt{3x + 1} - \sqrt{x - 1} = 2$ has two radicals and a third term. To solve it, it is necessary to isolate radicals twice.

EXAMPLE 3   Solve $\sqrt{3x + 1} - \sqrt{x - 1} = 2$. Check.

Solution   Isolate one of the radicals and square each side.
$$\sqrt{3x + 1} = 2 + \sqrt{x - 1}$$

Square each side. $3x + 1 = 4 + 4\sqrt{x - 1} + (x - 1)$

Isolate the radical. $2x - 2 = 4\sqrt{x - 1}$

Divide each side by 2. $x - 1 = 2\sqrt{x - 1}$

Square each side. $x^2 - 2x + 1 = 4(x - 1)$
$$x^2 - 6x + 5 = 0$$
$$(x - 1)(x - 5) = 0$$
$$x = 1 \quad or \quad x = 5$$

The solutions are 1 and 5. The check is left for you.

Classroom Exercises

For each equation, the apparent solutions are -2 and/or 5. Check -2 and 5 in each equation to find the correct root, or roots, of the equation. If there are no roots, write *no roots*.

1. $\sqrt{x + 11} = 3$ **2.** $\sqrt{x + 4} = -3$ **3.** $\sqrt{x + 11} = x - 1$

Solve the equation. Check.

4. $\sqrt{x + 7} = 2$

5. $\sqrt{x + 12} = x$

6. $\sqrt[3]{y - 5} = -2$

Written Exercises

Solve the equation. Check.

1. $\sqrt{3x - 5} = 4$

2. $\sqrt[3]{4a} = 4$

3. $\sqrt{6n - 2} = \sqrt{4n + 4}$

4. $y + 1 = \sqrt{y + 7}$

5. $x = 4 + \sqrt{2x}$

6. $\sqrt{2w - 1} = w - 2$

7. $n = 5 + \sqrt{2n + 5}$

8. $\sqrt[3]{c^2 - 8} = 2$

9. $\sqrt[3]{2a^2 - 9} = 3$

10. $\sqrt[4]{x^3 + 8} = 2$

11. $\sqrt[4]{x^2 + 9} = 3$

12. $2\sqrt{y} = \sqrt{y} + 2$

13. $\sqrt{3x} - \sqrt{x - 2} = 2$

14. $\sqrt{x + 11} = \sqrt{5x} - 1$

15. $\sqrt{n + 7} - \sqrt{n} = 1$

16. $4 + \sqrt{5t + 1} = 0$

17. $-3 - \sqrt{4t + 1} = 0$

18. $2 + \sqrt{3x + 7} = 2x$

19. $2\sqrt{3c - 2} - c - 2 = 0$

20. $\sqrt{x^2 - x + 5} = 5$

21. $\sqrt{y^2 + 8y} = 4\sqrt{3}$

22. $\sqrt[4]{x^2 + 7} = \sqrt[4]{5x + 1}$

23. $\sqrt{4x - 3} - \sqrt{2x - 5} = 2$

24. The radius r of a circle with area A is given by the formula

$r = \sqrt{\dfrac{A}{\pi}}$. Solve this formula for A in terms of r.

25. The radius of the base of a cylinder with height h and volume V is

given by $r = \sqrt{\dfrac{V}{\pi h}}$. Solve for V in terms of r and h.

Solve each equation. Check.

26. $\sqrt{x + 1} = \sqrt[4]{10x + 1}$

27. $\sqrt[3]{x + 2} = \sqrt[6]{9x + 10}$

Midchapter Review

1. Without solving $4x^2 - 3x - 8 = 0$, find the sum and the product of its roots.   **10.1**

Find the discriminant and determine the nature of the roots without solving the equation.   **10.2**

2. $2x^2 - 3x - 4 = 0$

3. $9y^2 = 12y - 4$

Solve the equation. Check.   **10.3**

4. $n = 3 + \sqrt{2n + 2}$

5. $\sqrt[4]{2x^2 - 19} = 3$

Application: *Escape Velocity*

When an object is launched upward from the earth, the force of gravity acts to pull it back down. Whether or not it actually falls back, however, depends on the velocity with which it is launched. With a strong enough boost, even a 6,200,000-lb rocket like the Saturn V can overcome the earth's gravitational field, never to return.

An object launched vertically at an initial speed of 11.2 km/s, for instance, will never come down because 11.2 km/s is the earth's escape velocity. The **escape velocity** for a particular planet is the minimum speed at which an object must be launched in order to escape that planet's gravitational attraction permanently.

Escape velocity can be calculated using the following formula,
$$v_e = \sqrt{2gr},$$
where g is the surface gravity of the planet and r is its radius.

Example

The radius of the earth is 6,380,000 m. Its surface gravity is 9.8 m/s^2. Find its escape velocity.

Solution

$$v_e = \sqrt{2gr}$$
$$= \sqrt{2(9.8)(6,380,000)}$$
$$\approx 11,200 \text{ m/s, or } 11.2 \text{ km/s}$$

The earth's escape velocity is about 11.2 km/s.

---

1. The surface gravity of the moon is 1.6 m/s^2. Its radius is 1,740,000 m. Find its escape velocity.

2. If the surface gravity of Mars is 3.6 m/s^2, and its escape velocity is 5.1 km/s, what is the radius of Mars?

3. The formula for escape velocity can be derived from another formula, shown below, for the velocity needed to reach height h above a planet.

$$v = \sqrt{2gr}\left(\sqrt{\frac{h}{h + r}}\right)$$

Explain how the derivation is done. (HINT: Let $h =$ an infinite distance, or infinity.)

10.4 Higher-Degree Equations; Complex Roots

Objectives

To find all the roots of integral polynomial equations

To find upper and lower bounds for the real zeros of integral polynomials

Let $P(x) = x^4 - 2x^3 - 11x^2 + 16x + 24$. Recall that according to the Integral Zero Theorem (see Lesson 6.6), the integral roots, *if any*, of the equation $P(x) = 0$ must be among the integral factors of the constant term—in this case, 24.

$$\pm 1, \pm 2, \pm 3, \pm 4, \pm 6, \pm 8, \pm 12, \pm 24$$

After finding the integral zeros, if any, of $P(x)$, you can find its *irrational* zeros as shown below. Synthetic substitution is used.

$$x^4 - 2x^3 - 11x^2 + 16x + 24 \leftarrow P(x)$$

	1	-2	-11	16	24
		(-1)	(3)	(8)	(-24)
-1	1	-3	-8	24	⓪ $\leftarrow (x + 1)(x^3 - 3x^2 - 8x + 24)$
		(3)	(0)	(-24)	
3	1	0	-8		⓪ $\leftarrow (x + 1)(x - 3)(x^2 - 8)$

Solve $x^2 - 8 = 0$.

$$x^2 = 8$$

$$x = 2\sqrt{2} \text{ or } x = -2\sqrt{2} \leftarrow (x + 1)(x - 3)(x - 2\sqrt{2})(x + 2\sqrt{2})$$

So, the zeros of $P(x)$ and the roots of $P(x) = 0$ are $-1, 3, 2\sqrt{2},$ and $-2\sqrt{2}$.

The *Fundamental Theorem of Algebra* states that each polynomial of degree n, where $n > 0$, has *at least one* zero.

Theorem 10.3

The Fundamental Theorem of Algebra
If $P(x)$ is a polynomial of degree n, where $n > 0$, then there is a zero r among the complex numbers for which $P(r) = 0$.

EXAMPLE 1

Let $Q(x) = x^3 + 2x^2 + 4x + 8$. Find all the roots of $Q(x) = 0$.

Solution

Find the zeros of $Q(x)$. Try the integral factors of 8: $\pm 1, \pm 2, \pm 4, \pm 8$.

The *compact form* of synthetic substitution is used.

$$
\begin{array}{c|cccc}
 & 1 & 2 & 4 & 8 & \leftarrow Q(x) = x^3 + 2x^2 + 4x + 8 \\
\hline
1 & 1 & 3 & 7 & 15 & \leftarrow 1 \text{ is not a root. Try another.} \\
-2 & 1 & 0 & 4 & \boxed{0} & \leftarrow -2 \text{ is a root.}
\end{array}
$$

$Q(x) = (x + 2)(x^2 + 4)$ Solve. The roots are $-2, 2i,$ and $-2i$.

Notice in Example 1 that synthetic substitution is not used after a second-degree factor is found. Instead, a quadratic equation is solved.

The potential integral zeros of $P(x) = 12x^4 + x^3 - 42x^2 - 3x + 18$ are $\pm1, \pm2, \pm3, \pm6, \pm9, \pm18$. It is possible to reduce this list to $\pm1, \pm2$ by using the **Upper- and Lower-Bound Theorem** below.

Theorem 10.4

Let $P(x)$ be an integral polynomial, and U and L be real numbers such that $U \geq 0$ and $L \leq 0$.

(1) Let $P(U)$ be found by synthetic substitution. If the bottom row of numbers is *all nonnegative* or *all nonpositive*, then U is an **upper bound** for the real zeros of $P(x)$.

(2) Let $P(L)$ be found by synthetic substitution. If the bottom row of numbers *alternates in sign*, then L is a **lower bound** for the real zeros of $P(x)$.

This theorem can be used to find upper and lower bounds for the real zeros of $P(x) = 12x^4 + x^3 - 42x^2 - 3x + 18$, as shown below.

1. Find $P(x)$ for $x = 0, 1, 2, \ldots$ by synthetic substitution.

x	12	1	-42	-3	18	$P(x)$
0	12	1	-42	-3	18	18
1	12	13	-29	-32	-14	-14
2	12	25	8	13	44	44

All nonnegative

3	12	37	69	204	630	630

All nonnegative

These numbers and $P(x)$ will continue to increase. So, $P(x)$ cannot be zero for $x > 2$. Thus, 2 is an upper bound for the real zeros of $P(x)$.

2. Find $P(x)$ for $x = -1, -2, -3, \ldots$ by synthetic substitution.

x	12	1	-42	-3	18	$P(x)$
-1	12	-11	-31	28	-10	-10
-2	12	-23	4	-11	40	40

Signs alternate

-3	12	-35	63	-192	594	594

Signs alternate

These signs will continue to alternate, and $P(x)$ will continue to be positive. So, $P(x)$ cannot be zero for $x < -2$. Thus, -2 is a lower bound for the real zeros of $P(x)$.

Because 2 and -2 are bounds for the real zeros of $P(x)$, the list of potential integral zeros has been reduced from $\pm 1, \pm 2, \pm 3, \pm 6, \pm 9, \pm 18$ to $\pm 1, \pm 2$.

The zeros of $P(x) = 12x^4 + x^3 - 42x^2 - 3x + 18$, or
$$P(x) = (4x + 3)(3x - 2)(x - \sqrt{3})(x + \sqrt{3}),$$
are $-\frac{3}{4}, \frac{2}{3}, \sqrt{3}$, and $-\sqrt{3}$. Each zero z is in the interval $-2 \le z \le 2$, where 2 is an upper bound and -2 is a lower bound for the four real zeros.

EXAMPLE 2 Find the least integral upper bound U and the greatest integral lower bound L for the real zeros of $P(x) = -2x^4 + 3x^3 + 7x^2 + 3x + 9$.

Solution

1. For U, try $x = 0, 1, 2, \ldots$

x	-2	3	7	3	9
0	-2	3	7	3	9
1	-2	1	8	11	20
2	-2	-1	5	13	35
3	-2	-3	-2	-3	0

All nonpositive: $U = 3$

2. For L, try $x = -1, -2, -3, \ldots$

x	-2	3	7	3	9
-1	-2	5	2	1	8
-2	-2	7	-7	17	-25

Signs alternate: $L = -2$

Thus, 3 is the least integral upper bound and -2 is the greatest integral lower bound.

The Integral Zero Theorem and the Upper- and Lower-Bound Theorem can often be used together to solve higher-degree equations.

EXAMPLE 3   The polynomial $P(x) = 6x^5 - 17x^4 - 29x^3 + 74x^2 + 20x - 24$ has five distinct zeros. Solve $P(x) = 0$.

Plan   The integral roots of $P(x) = 0$ are among the integral factors of -24: $\pm 1, \pm 2, \pm 3, \pm 4, \pm 6, \pm 8, \pm 12, \pm 24$. Use synthetic substitution.

Solution   Use $x = 0, 1, 2, \ldots$ in synthetic substitution. Look for zeros and an upper bound.

x	6	-17	-29	74	20	-24
0	6	-17	-29	74	20	-24
1	6	-11	-40	34	54	30
2	6	-5	-39	-4	12	$\boxed{0}$

Thus, 2 is a zero of $P(x)$.

$$P(x) = (x - 2)(6x^4 - 5x^3 - 39x^2 - 4x + 12)$$

Next, find the zeros of the *depressed* polynomial $Q(x) = 6x^4 - 5x^3 - 39x^2 - 4x + 12$.

x	6	-5	-39	-4	12
3	6	13	0	-4	$\boxed{0}$

Thus, 3 is a zero of $Q(x)$ and also of $P(x)$.

$$P(x) = (x - 2)(x - 3)(6x^3 + 13x^2 - 4)$$

Next, find the zeros of the depressed polynomial $R(x) = 6x^3 + 13x^2 - 4$.

x	6	13	0	-4
4	6	37	148	588

All nonnegative: 4 is not a zero, but it *is* an upper bound.

Use $x = -1, -2, -3, \cdots$ in synthetic substitution with $R(x) = 6x^3 + 13x^2 - 4$. Look for negative zeros or a lower bound.

x	6	13	0	-4
-1	6	7	-7	3
-2	6	1	-2	$\boxed{0}$

$$P(x) = (x - 2)(x - 3)(x + 2)(6x^2 + x - 2)$$

Factor. $6x^2 + x - 2 = (3x + 2)(2x - 1)$

Thus, $P(x) = (x - 2)(x - 3)(x + 2)(3x + 2)(2x - 1)$, and the roots of $P(x) = 0$ are $2, 3, -2, -\frac{2}{3}$, and $\frac{1}{2}$.

Classroom Exercises

Find all the zeros of each polynomial.

1. $5x + 2$ **2.** $x^2 + 25$ **3.** $x^2 - 18$ **4.** $(x - 12)(x^2 - 12)$
5. $(x + 1)(x - 4)(x^2 - 7)$ **6.** $(x - 1)(x + 2)(x^2 + 9)$ **7.** $x^3 - 5x^2 + x - 5$

Written Exercises

Find all the roots of each equation.

1. $x^3 - 2x^2 - x + 2 = 0$ **2.** $x^3 - x^2 - 14x + 24 = 0$
3. $a^3 + 3a^2 + a + 3 = 0$ **4.** $x^3 - 5x^2 + 9x - 45 = 0$
5. $y^3 - 2y^2 - 18y + 36 = 0$ **6.** $c^3 + 4c^2 - 20c - 80 = 0$
7. $x^3 + 5x^2 + 2x + 10 = 0$ **8.** $a^3 - 3a^2 + 12a - 36 = 0$

Find the least integral upper bound U and the greatest integral lower bound L for the real zeros of each polynomial.

9. $y^3 - 2y^2 + 6y - 12$ **10.** $6x^4 - x^3 + 58x^2 - 10x - 20$
11. $x^4 + x^3 - 10x^2 - 6x + 36$ **12.** $t^4 - 7t^2 - 15$
13. $4y^4 - 8y^3 - 16y^2 + 40y - 20$ **14.** $x^3 + 2x^2 - 10x - 20$

Find all the roots of each equation.

15. $x^4 + 5x^3 + 16x^2 + 60x + 48 = 0$ **16.** $a^4 - a^3 - 14a^2 + 8a + 48 = 0$
17. $6y^5 - 13y^4 - 6y^3 + 17y^2 - 4 = 0$ (HINT: Insert a missing term.)
18. $9y^4 - 146y^2 + 32 = 0$
19. $x^4 - x^3 - 21x^2 + 43x - 10 = 0$ **20.** $x^4 + 6x^3 + 27x^2 + 32x - 66 = 0$

Mixed Review

1. Write an equation, in standard form, of the line through $P(-3,2)$ and $Q(7,4)$. *3.4*
2. Factor $5x^3 - 40y^3$ completely. *5.7*
3. If $f(x) = 2x^2$ and $g(x) = -3x$, what is $f(g(2))$? *5.9*
4. Solve $\dfrac{3x}{x^2 - 5x - 14} - \dfrac{5}{x - 7} = \dfrac{4}{x + 2}$. *7.6*

10.5 The Rational Zero Theorem

Objective

To solve integral polynomial equations using the Rational Zero Theorem

The polynomial $P(x) = 12x^3 - 4x^2 - 3x + 1$ has no integral zeros. Instead, it has three *rational zeros*, $\frac{1}{3}$, $\frac{1}{2}$, and $-\frac{1}{2}$, which cannot be found by the Integral Zero Theorem. It can be proved, however, that if $\frac{1}{r}$ is a zero of $P(x)$, and r is an integer, then r is a factor of the first coefficient of $P(x)$, which in this instance is 12.

Let $\frac{1}{r}$ be a zero of $12x^3 - 4x^2 - 3x + 1$, where r is an integer.

$$12 \cdot \frac{1}{r^3} - 4 \cdot \frac{1}{r^2} - 3 \cdot \frac{1}{r} + 1 = 0$$

Next, multiply each side by r^3.

$$12 - 4r - 3r^2 + r^3 = 0$$

$$r^3 - 3r^2 - 4r = -12$$

Factor.

$$r(r^2 - 3r - 4) = -12$$

So, since r is a factor of -12, it is also a factor of 12, and is thus among the integers

$$\pm 1, \ \pm 2, \ \pm 3, \ \pm 4, \ \pm 6, \ \pm 12.$$

Thus, $\frac{1}{r}$ is among $\pm\frac{1}{1}$, $\pm\frac{1}{2}$, $\pm\frac{1}{3}$, $\pm\frac{1}{4}$, $\pm\frac{1}{6}$, $\pm\frac{1}{12}$.

This line of reasoning can be used to extend the Integral Zero Theorem to rational numbers in the *Rational Zero Theorem* below.

Theorem 10.5

Rational Zero Theorem

If $\frac{c}{d}$, where c and d are relatively prime integers, is a zero of an integral polynomial $P(x)$, then c is a factor of the constant term in $P(x)$ and d is a factor of the coefficient of the term of highest degree. That is, if $\frac{c}{d}$ is a zero of

$$a_0x^n + a_1x^{n-1} + a_2x^{n-2} + \cdots + a_{n-1}x^1 + a_n,$$

then c is a factor of a_n and d is a factor of a_0.

NOTE: The Rational Zero Theorem does not assure the existence of rational zeros. In fact, an integral polynomial may have no rational zeros.

For $P(x) = 6x^4 + 7x^3 - 21x^2 - 21x + 9$, you can list the *potential* rational zeros that are not integers as shown below.

c is a factor of 9: 1, 3, 9. d is a factor of 6: 1, 2, 3, 6.

Thus, any nonintegral rational zeros $\frac{c}{d}$, if they exist, are among
$\pm\frac{1}{2}, \pm\frac{1}{3}, \pm\frac{1}{6}, \pm\frac{3}{2}, \pm\frac{9}{2}.$

EXAMPLE 1 $P(x) = 6x^4 + 7x^3 - 21x^2 - 21x + 9$ has four distinct zeros. Find the zeros.

Solution   **1.** Test the potential *integral* zeros first: $\pm1, \pm3, \pm9$.

	6	7	-21	-21	9
1	6	13	-8	-29	-20
3	6	25	54	141	432

	6	7	-21	-21	9
-1	6	1	-22	1	8
-3	6	-11	12	-57	180

All nonnegative:
3 is an upper bound.

Signs alternate:
-3 is a lower bound.

It is unnecessary to test ±9. $P(x)$ has no integral zeros.

2. Test the remaining potential rational zeros between the bounds
-3 and 3: $\pm\frac{1}{2}, \pm\frac{1}{3}, \pm\frac{1}{6}, \pm\frac{3}{2}$.

	6	7	-21	-21	9
$\frac{1}{3}$	6	9	-18	-27	0
$-\frac{3}{2}$	6	0	-18	0	

Successive Factorizations:

$(x - \frac{1}{3})(6x^3 + 9x^2 - 18x - 27)$

$(x - \frac{1}{3})(x + \frac{3}{2})(6x^2 - 18)$

After two rational zeros are found, identify a second-degree factor.

3. Solve $6x^2 - 18 = 0$.
$$6(x^2 - 3) = 0 \qquad \leftarrow \quad (x - \tfrac{1}{3})(x + \tfrac{3}{2}) \cdot 6(x^2 - 3)$$
$$x^2 = 3$$
$$x = \pm\sqrt{3} \qquad \leftarrow \quad 6(x - \tfrac{1}{3})(x + \tfrac{3}{2})(x - \sqrt{3})(x + \sqrt{3})$$

The zeros of $P(x)$ are $\frac{1}{3}$, $-\frac{3}{2}$, $\sqrt{3}$, and $-\sqrt{3}$.

The fourth-degree integral polynomial in Example 1 has two rational (but nonintegral) zeros, $\frac{1}{3}$ and $-\frac{3}{2}$, and two irrational zeros that are conjugates, $\sqrt{3}$ and $-\sqrt{3}$. A fourth-degree integral polynomial may also have one integral zero, one rational zero, and two complex zeros that are conjugates, as in $P(x) = 2x^4 - x^3 + 3x^2 - 2x - 2$.

Recall that a polynomial may have zeros with multiplicities greater than 1. If $P(x) = (x - 5)(x - 5)(x - 5)(2x + 3)$, then the zero 5 has a multiplicity of 3. The zero $-\frac{3}{2}$ has a multiplicity of 1. Notice that the sum of the multiplicities is $3 + 1$, or 4, which is the same as the degree of $P(x)$.

The *Factorization Theorem* stated below guarantees that a polynomial $P(x)$ of degree n, where $n > 0$, has exactly n first-degree factors, and thus that the sum of the multiplicities of the zeros of $P(x)$ is equal to n.

Theorem 10.6

Factorization Theorem
Each polynomial $P(x)$ of degree n, where $n > 0$, can be factored into exactly n first-degree factors and a constant factor.

EXAMPLE 2 $P(x) = 16x^4 - 16x^3 - 32x^2 + 36x - 9$. Solve $P(x) = 0$. State the multiplicity m of a root when $m > 1$.

Solution   1. Test the potential integral zeros: $\pm1, \pm3, \pm9$.

	16	-16	-32	36	-9	
1	16	0	-32	4	-5	
3	16	32	64	228	675	$\leftarrow$ 3 is an upper bound.
-1	16	-32	0	36	-45	
-3	16	-64	160	-444	1,323	$\leftarrow -3$ is a lower bound.

Thus, there are no integral roots.

2. Try to "narrow" the bounds.

	16	-16	-32	36	-9	
2	16	16	0	36	63	$\leftarrow$ 2 is an upper bound.
-2	16	-48	64	-92	175	$\leftarrow -2$ is a lower bound.

3. List the remaining potential rational zeros between the bounds -2 and 2:

$$\pm\tfrac{1}{2}, \ \pm\tfrac{1}{4}, \ \pm\tfrac{1}{8}, \ \pm\tfrac{1}{16}, \ \pm\tfrac{3}{2}, \ \pm\tfrac{3}{4}, \ \pm\tfrac{3}{8}, \ \pm\tfrac{3}{16}, \ \pm\tfrac{9}{8}, \ \pm\tfrac{9}{16}$$

4. Test these zeros until a second-degree factor is found.

$$\frac{1}{2} \, \bigg| \, \frac{16 \;\; -16 \;\; -32 \;\; 36 \;\; -9}{16 \;\;\;\; -8 \;\; -36 \;\; 18 \;\;\;\; 0}$$

Try $\frac{1}{2}$ again. $\quad \dfrac{1}{2} \, \bigg| \; 16 \;\;\;\; 0 \;\; -36 \;\;\;\; 0$

$$P(x) = (x - \tfrac{1}{2})(x - \tfrac{1}{2})(16x^2 - 36)$$

5. Solve $16x^2 - 36 = 0$.
$$4(4x^2 - 9) = 0$$
$$4(2x - 3)(2x + 3) = 0 \quad P(x) = 4(x - \tfrac{1}{2})(x - \tfrac{1}{2})(2x - 3)(2x + 3)$$

The roots are $\frac{1}{2}$, $\frac{3}{2}$, and $-\frac{3}{2}$. The multiplicity of $\frac{1}{2}$ is 2.

Classroom Exercises

State the potential integral zeros of each polynomial. Then give the potential rational, but nonintegral, zeros of the polynomial.

1. $8x^3 + 6x^2 - 5x + 1$

2. $y^4 - 8y^3 + y - 10$

3. $3z^4 - z^3 + 2z - 6$

4. $6x^3 + x^2 - x + 4$

Solve each equation. State the multiplicity m of a root when $m > 1$.

5. $(x - 3)(x + 4)(x - 3) = 0$

6. $5(n + 2)(n + 2)(n - 4)(n + 2) = 0$

Written Exercises

Solve each equation. State the multiplicity m of a root when $m > 1$.

1. $3y^3 - y^2 - 24y + 8 = 0$

2. $4x^3 + 3x^2 + 36x + 27 = 0$

3. $18c^3 + 9c^2 - 23c + 6 = 0$

4. $18y^3 + 3y^2 - 4y - 1 = 0$

5. $3x^4 - 4x^3 - 19x^2 + 20x + 20 = 0$

6. $4n^4 + n^3 + 29n^2 + 8n - 24 = 0$

7. $3y^4 - 13y^3 - 4y^2 - 26y - 20 = 0$

8. $4x^4 + 4x^3 - 11x^2 - 12x - 3 = 0$

9. $24x^4 - 20x^3 - 6x^2 + 9x - 2 = 0$

10. $4c^4 - 12c^3 + 25c^2 - 48c + 36 = 0$

11. $9c^4 + 23c^2 - 12 = 0$

12. $4y^4 - 17y^2 + 18 = 0$

13. $16x^5 - 12x^4 - 68x^3 + 51x^2 + 16x - 12 = 0$

14. $2y^5 + y^4 - 16y^3 + 7y^2 + 24y - 18 = 0$

15. Write in your own words how you would form a list of the potential rational zeros of an integral polynomial $P(x)$.

16. If A, B, and C are odd integers, then $Ax^2 + Bx + C = 0$ has no integral roots. Prove that this is true using the following analysis:
   a. A, B, and C are odd and zero is even.   b. Let O mean odd and E mean even.
   c. If $Ax^2 + Bx + C = 0$, then $O \cdot x \cdot x + O \cdot x + O = E$.
   d. Show that x can be neither odd nor even, and thus, not an integer. Consider two cases. (Case I) x is even. (Case II) x is odd.

Irrational zeros of integral polynomials occur in conjugate pairs. Thus, if $4 - 3\sqrt{2}$ is a zero of $P(x)$, then $4 + 3\sqrt{2}$ is another zero of $P(x)$.

Example: Solve $x^4 - 8x^3 - 5x^2 + 24x + 6 = 0$, given that $x_1 = 4 - 3\sqrt{2}$ is one of its roots.

$$
\begin{array}{c|ccccc}
 & 1 & -8 & -5 & 24 & 6 \\
\hline
4 - 3\sqrt{2} \ | & 1 & -4 - 3\sqrt{2} & -3 & 12 + 9\sqrt{2} & \boxed{0} \\
\hline
4 + 3\sqrt{2} \ | & 1 & 0 & -3 & \boxed{0} &
\end{array}
$$

Solve $x^2 - 3 = 0$. $x^2 = 3$, or $x = \pm\sqrt{3}$

The roots are $4 - 3\sqrt{2}$, $4 + 3\sqrt{2}$, $\sqrt{3}$, and $-\sqrt{3}$.

Solve each equation, given one of its roots.

17. $x^4 - 14x^2 + 24 = 0$, $x_1 = 2\sqrt{3}$
18. $n^4 - 2n^3 - 19n^2 + 36n + 18 = 0$, $n_1 = -3\sqrt{2}$
19. $x^4 - 4x^3 - 9x^2 + 32x + 8 = 0$, $x_1 = 2 + \sqrt{5}$
20. $c^4 + 2c^3 - 20c^2 - 36c + 36 = 0$, $c_1 = -1 - \sqrt{3}$

Mixed Review

Graph each equation in a coordinate plane.   **3.5**

   1. $y = -2x + 3$ 2. $2y - 6 = 0$ 3. $5x + 10 = 0$ 4. $2x - 3y = 6$

Brainteaser

The mathematics department of Kenyon High School is planning to hold a departmental meeting next month.

- The date of the meeting is an even number.
- The date is sometime after the 12th.
- The date is sometime before the 19th.
- The date is not a perfect cube.
- The date is a perfect square.

Only one of these statements is true. What is the date of the meeting?

10.6 Graphing Polynomial Functions

Objectives

To graph integral polynomial functions of degree n with n distinct real zeros

To bound (locate) between consecutive integers the real zeros of given polynomials

The graph of $y = x^3 - x^2 - 14x + 14$ can be sketched using a table of ordered pairs, as demonstrated below. Note that all the integral values of x fall between upper and lower bounds for the real zeros.

x	1	-1	-14	14	(x,y)
-4	1	-5	6	-10	$(-4,-10)$

Signs alternate:
-4 is a lower bound.

-3	1	-4	-2	20	$(-3,20)$
-2	1	-3	-8	30	$(-2,30)$
-1	1	-2	-12	26	$(-1,26)$
0	1	-1	-14	14	$(0,14)$
1	1	0	-14	0	$(1,0)$
2	1	1	-12	-10	$(2,-10)$
3	1	2	-8	-10	$(3,-10)$
4	1	3	-2	6	$(4,6)$
5	1	4	6	44	$(5,44)$

All nonnegative:
5 is an upper bound.

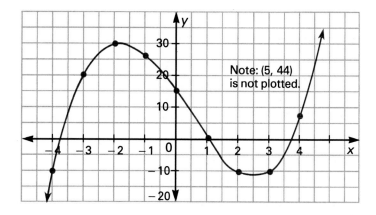

Note: (5, 44) is not plotted.

The graph of $y = x^3 - x^2 - 14x + 14$ reveals the following facts.

(1) The curve crosses the x-axis between -4 and -3, at 1, and between 3 and 4. So, this third-degree function has three real zeros: 1 is a zero. There is a zero between -4 and -3, and a third zero between 3 and 4.

(2) The "*turning points*" (where the graph changes direction) in Quadrants II and IV cannot be determined precisely at this time. However, approximate turning points can be located.

(3) The graph of this third-degree function has this characteristic shape.

The table can be used to locate each real zero as follows.

(1) The ordered pair $(1,0)$ indicates that 1 is a zero.

(2) The ordered pairs $(-4,-10)$ and $(-3,20)$ have y-values that differ in sign (-10 and 20). So, there is a zero between the x-values -4 and -3.

(3) The pairs $(3,-10)$ and $(4,6)$ have y-values with different signs. So, there is a zero between 3 and 4.

If we were to use a table of ordered pairs to graph the second-degree equation $y = 4x^2 - 20x - 20$, and locate each real zero, we could find that:

(1) there are no integral zeros for the function;

(2) there is a zero between -1 and 0, because the y-values of $(-1,4)$ and $(0,-20)$ change sign; and

(3) there is a zero between 5 and 6 for the same reason.

The graph would show that:

(1) this second-degree function with two real zeros has this characteristic shape, _/ and

(2) the turning point cannot be determined precisely at this time. However, an approximate turning point can be located.

EXAMPLE   Use a table of ordered pairs to locate each real zero of the function described by $y = 6x^4 - 5x^3 - 58x^2 + 45x + 36$. Graph the function.

Solution   Construct a table of ordered pairs (x,y) for all integral values of x between an upper bound and a lower bound.

x	6	-5	-58	45	36	(x,y)	Zeros
-4	6	-29	58	-187	784	$(-4,784)$	There are no zeros for $x \leq -4$.
			-4 is a lower bound.				
-3	6	-23	11	12	0	$(-3,0)$	$\leftarrow$ -3 is a zero.
-2	6	-17	-24	93	-150	$(-2,-150)$	
-1	6	-11	-47	92	-56	$(-1,-56)$	There is a zero between
0	6	-5	-58	45	36	$(0,36)$	-1 and 0: $-56 < 0 < 36$.
1	6	1	-57	-12	24	$(1,24)$	There is a zero between
2	6	7	-44	-43	-50	$(2,-50)$	1 and 2: $-50 < 0 < 24$.
3	6	13	-19	-12	0	$(3,0)$	$\leftarrow$ 3 is a zero.
4	6	19	18	117	504	$(4,504)$	There are no zeros for $x \geq 4$.
			4 is an upper bound.				

A scientific calculator can be quite helpful in constructing these tables of ordered pairs. Review the *Using the Calculator* feature in Lesson 6.5. Remember to record the display each time you press $\boxed{=}$.

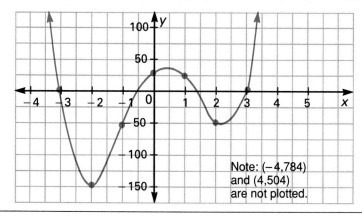

Note: $(-4,784)$ and $(4,504)$ are not plotted.

In the example, observe that (1) the graph of this fourth-degree function with four real zeros has this characteristic shape, and that (2) the turning points cannot be determined precisely at this time.

Classroom Exercises

Match the degree of a polynomial function at the left with its characteristic graph at the right.

1. degree 4
2. degree 3
3. degree 2
4. degree 1
5. degree 0

Use the graph of each function to locate its real zeros, either as integers or between consecutive integers.

6. 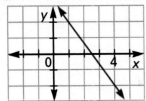   **7.** 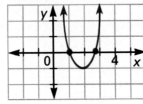   **8.**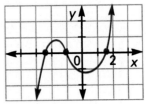

Written Exercises

Graph each function. Locate each real zero as an integer or between consecutive integers.

1. $y = -2x + 6$

2. $y = 3x - 4$

3. $y = x^2 + 4x - 4$

4. $y = x^3 - 4x^2 - 2x + 8$

5. $y = x^4 + x^3 - 14x^2 - 2x + 24$

6. $y = x^4 - 10x^2 + 9$

7. $y = 4x^2 + 20x + 19$

8. $y = -3x^2 - 6x + 4$

9. $y = 4x^3 - 4x^2 - 19x + 10$

10. $y = -2x^3 + x^2 + 18x - 9$

11. $y = 2x^4 - x^3 - 35x^2 + 16x + 48$

12. $y = -x^4 + 10x^2 - 9$

The turning points for the third-degree function $y = x^3 + bx^2 + cx + d$ occur where the values of x are the solutions of the equation

$$3x^2 + 2bx + c = 0, \text{ or at } x = \frac{-2b \pm \sqrt{4b^2 - 12c}}{6}.$$

Find the values of x, both in simplest radical form and to the nearest tenth, for the two turning points of each graph described below.

13. $y = x^3 - x^2 - 14x + 14$ (the first graph of this lesson)

14. $y = x^3 - 4x^2 - 2x + 8$ (the graph in Exercise 4)

15. Determine the range of values of c for which the function $f(x) = 2x^3 - x^2 - 6x + c$ has a zero between $x = 0$ and $x = 1$. (HINT: Find values of c so that $f(x)$ is positive at one of the points 0 and 1 and negative at the other.)

Mixed Review

Determine whether the lines are parallel, perpendicular, or neither.   *3.7*

1. $3x - 2y = 7$
$4y = 6x + 9$

2. $5x - 2y = 10$
$2y + 5x = 10$

3. $4x - 3y = 12$
$4y + 3x = 12$

4. $2 - 3x = 14$
$4y - 5 = 23$

Statistics: *Correlation*

Researchers often assert that there is a "high correlation" between two things, such as smoking and cancer or certain pollutants and ozone depletion. That is, they seem to be *connected* or *related*.

Statisticians may use a line to show the relation between 2 sets of data plotted on a *scatter diagram* (see page 123). A number between −1 and 1 called the **correlation coefficient**, denoted by the letter r, tells how closely the points cluster about a line. If r is positive, there is a **positive correlation** between the data sets, and a line drawn to fit the points has positive slope. If r is negative, there is a **negative correlation**, and a line drawn to fit the points has negative slope.

If r is −1 or 1, the points all lie on a line, as shown at the right. If r is zero, they appear to be randomly distributed, and show no relationship. The closer r is to −1 or 1, the *stronger* the correlation, and the flatter an ellipse drawn to contain the points becomes. Some illustrations of values of r are shown below.

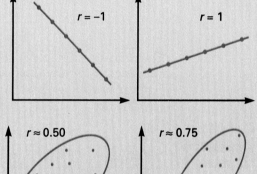

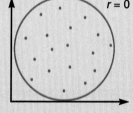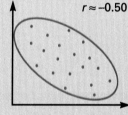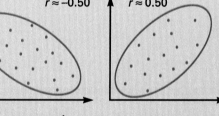

The Dow Jones Industrial Average and Standard and Poor's 500 Index are two of the primary measures of the stock market. The table at the right gives each measure for 15 consecutive trading days in a recent summer, with the Dow Average listed first.

(2895.30, 359.10) (2862.38, 358.71)
(2893.56, 358.47) (2897.33, 363.15)
(2882.18, 356.88) (2911.65, 364.96)
(2935.89, 362.91) (2925.00, 366.64)
(2928.22, 362.90) (2935.19, 367.40)
(2929.95, 364.90) (2900.97, 363.16)
(2933.42, 366.25) (2876.66, 361.23)
(2892.57, 361.63)

1. Construct a scatter diagram of the points. Is the correlation positive or negative? Does one measure cause the other to go up or down or might other factors, called *hidden variables*, control both?

2. These ordered pairs group the average monthly rainfall in inches with the average monthly percent of sunshine blocked by clouds (listed first) in Amarillo, Texas. Construct a scatter diagram. Describe the correlation. Are the results surprising?
(32, 0.49), (32, 0.46), (31, 0.57), (29, 0.87), (27, 1.08), (28, 2.79)
(23, 3.50), (22, 2.70), (23, 2.95), (26, 1.72), (25, 1.39), (28, 0.58)

10.7 The Conjugate Zero Theorem and Descartes' Rule of Signs

Objectives

To find the remaining zeros of integral polynomials given one nonreal zero

To predict the sums of the multiplicities of the positive and negative real zeros of integral polynomials

---

The zeros of $P(x) = x^2 - 8x + 20$ are a pair of conjugates, as shown below.

Let $P(x) = 0$. Then by the quadratic formula,

$$x = \frac{8 \pm \sqrt{(-8)^2 - 80}}{2} = \frac{8 \pm \sqrt{-16}}{2} = \frac{8 \pm 4i}{2}.$$

The zeros of $P(x)$ are $4 + 2i$ and $4 - 2i$, a pair of complex conjugates. In general, for any integral polynomial $P(x)$, its nonreal zeros, if any, appear in *conjugate pairs*. This is stated in the *Conjugate Zero Theorem* below.

Theorem 10.7

Conjugate Zero Theorem
If $a + bi$ is a zero of an integral polynomial $P(x)$, then $a - bi$ is also a zero of $P(x)$.

EXAMPLE 1

Given that $P(x) = x^4 - 5x^3 + x^2 + 49x - 78$ and that $3 - 2i$ is a zero of $P(x)$, find the remaining zeros of $P(x)$.

Plan

If $3 - 2i$ is a zero, then $3 + 2i$ is also a zero. Use synthetic substitution to find a second-degree factor of $P(x)$.

Solution

$$
\begin{array}{r|ccccc}
 & 1 & -5 & 1 & 49 & -78 \\
 & & (3 - 2i) & (-10 - 2i) & (-31 + 12i) & (78) \\
\hline
3 - 2i & 1 & -2 - 2i & -9 - 2i & 18 + 12i & \boxed{0} \\
\end{array}
$$

$$
\begin{array}{r|cccc}
 & & (3 + 2i) & (3 + 2i) & (-18 - 12i) \\
\hline
3 + 2i & 1 & 1 & -6 & \boxed{0} \\
\end{array}
$$

$$x^2 + x - 6 = (x + 3)(x - 2)$$

The zeros of $P(x)$ are $3 - 2i$, $3 + 2i$, -3, and 2.

The polynomial $P(x)$ in Example 1 has one positive real zero, one negative real zero and two nonreal zeros, each with a multiplicity of 1. The sum of these multiplicities, or the total number of first-degree factors, is equal to 4, which is also the degree of $P(x)$.

Consider $P(x) = 2x^5 - 11x^4 + x^3 + 65x^2 - 39x - 90$.
$$= (x - 3)^2(2x - 5)(x + 1)(x + 2)$$

$P(x)$ has two positive real zeros: 3 (with a multiplicity of 2) and $\frac{5}{2}$ (with a multiplicity of 1). The sum of these multiplicities is $2 + 1$, or 3.

$P(x)$ has two negative real zeros, -1 and -2, each with a multiplicity of 1. The sum of these multiplicities is $1 + 1$, or 2.

There are three *changes in sign* in the coefficients of $P(x)$, as shown below.

$$P(x) = \underbrace{+2x^5 \quad -11x^4}_{1} \underbrace{\quad +1x^3}_{2} \quad \underbrace{+65x^2 \quad -39x}_{3} \quad -90$$

Notice that the number of sign changes in $P(x)$ is the same as the sum of the multiplicities of the positive real zeros.

There are two sign changes in $P(-x)$, as shown below.

$$P(-x) = 2(-x)^5 - 11(-x)^4 + 1(-x)^3 + 65(-x)^2 - 39(-x) - 90$$
$$= -2x^5 \quad -11x^4 \underbrace{\quad -1x^3}_{1} \quad +65x^2 \quad \underbrace{+39x}_{2} \quad -90$$

Notice that the number of sign changes in $P(-x)$ is the same as the sum of the multiplicities of the negative real zeros. Also notice that the sum of all the multiplicities is 5, which is the degree of $P(x)$.

This example suggests *Descartes' Rule of Signs*, stated below. It allows us to predict both the number of first-degree factors of $P(x)$ that give *positive* real roots and the number that give *negative* real roots, that is, the sums of the multiplicities of the positive and negative real zeros.

Descartes' Rule of Signs
The sum of the multiplicities of the positive real zeros of an integral polynomial $P(x)$ is either equal to the number of sign changes in $P(x)$ or is less than that number by a multiple of two.

The sum of the multiplicities of the negative real zeros of $P(x)$ is either equal to the number of sign changes in $P(-x)$ or is less than that number by a multiple of two.

EXAMPLE 2 Find the possible combinations of the sums of the multiplicities of the positive and negative real zeros of $P(x)$, where

$$P(x) = x^5 - 9x^3 - x^2 + 9.$$

Solution Find the number of sign changes in $P(x)$ and in $P(-x)$.

$$P(x) \quad = \quad \underbrace{+x^5}_{1} \underbrace{-9x^3 \quad -x^2}_{2} +9 \quad \leftarrow 2 \text{ sign changes}$$

So, the sum of the multiplicities of the positive real zeros of $P(x)$ is either 2 or 0.

$$P(-x) = (-x)^5 \quad -9(-x)^3 \quad -(-x)^2 \quad +9$$
$$= \underbrace{-x^5}_{1} \underbrace{+9x^3}_{2} \underbrace{-x^2}_{3} +9 \leftarrow 3 \text{ sign changes}$$

So, the sum of the multiplicities of the negative real zeros of $P(x)$ is either 3 or 1.

Thus, there are four possible combinations for the sums of the multiplicities of the real zeros, as shown in the table at the right.

positive	2	2	0	0
negative	3	1	3	1

Classroom Exercises

Determine another zero of the integral polynomial $P(x)$ given the non-real zero of $P(x)$ below.

1. $-3i$ **2.** $5i$ **3.** $2 + 4i$ **4.** $-1 - 6i$

Find the number of sign changes in the coefficients of each polynomial.

5. $P(x) = 7x^4 + 6x^3 - 5x^2 + 4x - 8$ **6.** $Q(x) = 6x^5 + 3x^4 - 4x^2 - 1$
7. $P(-x) = 7(-x)^4 + 6(-x)^3 - 5(-x)^2 + 4(-x) - 8$
8. Find the remaining zeros of $P(x) = 2x^3 - x^2 + 32x - 16$ given that $x_1 = 4i$.

Written Exercises

Find the remaining zeros of $P(x)$ given that x_1 is a zero of $P(x)$.

1. $P(x) = 3x^3 - x^2 + 75x - 25, x_1 = 5i$
2. $P(x) = x^3 - 4x^2 + 4x - 16, x_1 = -2i$
3. $P(x) = x^4 - 5x^3 + 15x^2 - 5x - 26, x_1 = 2 - 3i$
4. $P(x) = 3x^4 - 22x^3 + 50x^2 - 16x - 40, x_1 = 3 + i$

Find the possible combinations of the sums of the multiplicities of the positive and negative real zeros of each polynomial.

5. $x^5 - x^4 + 12$ **6.** $8x^3 + 4x^2 + x + 2$ **7.** $x^4 - 4x^3 + 6x^2 - 5$

8. $x^6 - x^5 + x^2 + x + 4$ **9.** $2x^5 - x^3 + 3x^2 - x - 8$

10. $x^7 - x^6 + x^4 + x^2 + 1$ **11.** $2x^8 + x^7 - 3x^6 + x + 2$

12. Find all the zeros of $P(x) = 2x^5 - 7x^4 + 8x^3 - 2x^2 + 6x + 5$ given that i and $2 - i$ are two of the zeros.

13. Find all the zeros of $P(x) = x^5 - 5x^4 + 8x^3 - 40x^2 + 16x - 80$ given that $2i$ is one of the zeros and that it has a multiplicity of two.

Mixed Review 5.4, 7.3

1. Solve $a(bx - c) - (3x + ac) = d$ for x.

2. Convert 75 mi/h to feet per second.

Application: *Rational Powers in Music*

On a piano, a **half-step** is the *musical interval* from one key to the next. For example, the interval from B to C is a half-step. The interval from C to C-sharp (a black key) is also a half-step. The interval from one C to the next C, which is called an octave, has 12 half-steps.

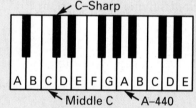

In the **well-tempered system of tuning,** invented by J. S. Bach, there is a constant ratio of *musical frequencies* of any note to the next higher note. That ratio is $1 : 2^{\frac{1}{12}}$. If the frequency of a note is f, then the frequency f' of the note n half-steps *higher* is

$$f' = f \times 2^{\frac{n}{12}}, \text{ or } f \times \sqrt[12]{2^n}.$$

The frequency f'' of a note n half-steps *lower* is

$$f'' = f \times 2^{-\frac{n}{12}}, \text{ or } f \div \sqrt[12]{2^n}.$$

The standard tuning note is the A above Middle C. This note, which is called A-440, has a frequency of 440 cycles per second.

1. From A-440 to the C above there are 3 half-steps. Find the frequency of the C.

2. From A-440 to the C below (Middle C) there are 9 half-steps. Find the frequency of Middle C.

3. What is the ratio of Middle C to the C above? How could you show this using a frequency formula?

Key Terms

Conjugate Zero Theorem (p. 366)
Descartes' Rule of Signs (p. 367)
discriminant (p. 343)
Factorization Theorem (p. 358)
Fundamental Theorem of Algebra
  (p. 351)
lower bound (p. 352)

radical equation (p. 347)
rational zero (p. 356)
Rational Zero Theorem (p. 356)
Upper- and Lower-Bound Theorem
  (p. 352)
upper bound (p. 352)

Key Ideas and Review Exercises

10.1   If $ax^2 + bx + c = 0$, or $x^2 + \frac{b}{a}x + \frac{c}{a} = 0$, has roots x_1 and x_2, then

$x_1 + x_2 = -\frac{b}{a}$ and $x_1 \cdot x_2 = \frac{c}{a}$.

If $x_1 + x_2 = -\frac{b}{a}$ and $x_1 \cdot x_2 = \frac{c}{a}$, then x_1 and x_2 are the roots of

$x^2 + \frac{b}{a}x + \frac{c}{a} = 0$, or $ax^2 + bx + c = 0$.

Without solving the equation, find the sum and product of its roots.

1. $3w^2 - 5w + 6 = 0$ **2.** $\frac{1}{2}y^2 + 4y = 3$ **3.** $-x^2 + 6 = 4x$

Write a quadratic equation, in general form, that has the given root or roots.

4. $-\frac{1}{3}$ and 4   **5.** $2 \pm 3\sqrt{2}$ **6.** 7

10.2   The discriminant $b^2 - 4ac$ determines the nature of the roots of any
quadratic equation $ax^2 + bx + c = 0$. It determines the number
of roots (1 or 2), and whether they are real or nonreal. If the roots are real,
and a, b, and c are rational, then the discriminant also determines whether
the roots are rational or irrational.

**Find the discriminant and determine the nature of the roots without
solving the equation.**

7. $4x^2 - 5x + 2 = 0$ **8.** $6y^2 = 4 - 5y$ **9.** $20x = 25x^2 + 4$

10. For what values of k will $y^2 + (k + 2)y + 9 = 0$ have exactly one
  root? (HINT: $b^2 - 4ac = 0$ when there is exactly one root)

11. For what values of k will $kx^2 - 8x + (k + 6) = 0$ have two real
  roots? (HINT: $b^2 - 4ac > 0$ when there are two real roots)

10.3   To solve a radical equation such as $\sqrt{5y - 16} - y = 2$, (1) isolate the
radical, (2) raise each side to the appropriate power, (3) solve for the
variable, and (4) check the apparent solutions.

Solve the equation. Check.

12. $3\sqrt{y - 5} = \sqrt{3y - 3}$

13. $\sqrt{2x + 7} - \sqrt{x + 4} = 2$

14. Solve $y = \sqrt{\dfrac{3x}{z}}$ for x, in terms of y and z.

10.4   The Upper- and Lower-Bound Theorem is used to identify intervals within which all the real zeros of a polynomial lie. It can be used to reduce the number of potential zeros that must be tested.

15. Find the least integral upper bound U and the greatest integral lower bound L for the real zeros of $3x^4 - 5x^3 - 25x^2 - 5x - 28$.

Use the Integral Zero Theorem to solve the equation.

16. $x^3 - 3x^2 + 4x - 12 = 0$

17. $x^4 - 2x^3 - 8x^2 + 10x + 15 = 0$

18. $4y^5 + 8y^4 - 29y^3 - 42y^2 + 45y + 54 = 0$

10.5   The Rational Zero Theorem can be used to list the potential rational zeros of a polynomial.

Use the Rational Zero Theorem to solve the equation.

19. $3x^4 - 2x^3 + 8x^2 - 6x - 3 = 0$

20. $8y^4 - 6y^3 + 17y^2 - 12y + 2 = 0$

10.6   To graph a function such as $y = 2x^3 - x^2 - 10x + 5$, use synthetic substitution to form a table of ordered pairs (x,y) for all integral values of x between an upper bound and a lower bound for the real zeros of the function.

Graph each function. Locate each real zero as an integer or between consecutive integers.

21. $y = 2x^3 - x^2 - 10x + 5$

22. $y = x^4 - 2x^3 - 13x^2 + 14x + 24$

10.7   Review the Conjugate Zero Theorem and Descartes' Rule of Signs.

23. Write in your own words how you would find all the zeros of a fourth-degree integral polynomial given one of its nonreal zeros.

24. Find all the remaining zeros of $P(x) = x^4 - 5x^3 - 3x^2 + 43x - 60$ given that $2 + i$ is one of the zeros.

25. Use Descartes' Rule of Signs to find all the possible combinations of the sums of the multiplicities of the positive and negative real zeros of $3x^5 - 2x^4 + 5x^3 - 4x - 2$.

Without solving the equation, find the sum and product of its roots.

1. $2x^2 + 7x - 8 = 0$ **2.** $\frac{1}{3}y^2 = 5y - 2$ **3.** $14 = -\frac{1}{2}y^2$

Write a quadratic equation, in general form, that has the given root or roots.

4. $\frac{1}{2}$ and -3 **5.** $3 \pm i$ **6.** -6

Find the discriminant, $b^2 - 4ac$, and determine the nature of the roots without solving the equation.

7. $3x^2 - 2x + 4 = 0$ **8.** $16y^2 = 8y - 1$ **9.** $5z^2 + 2z - 4 = 0$

10. For what values of k will $2x^2 + (k + 2)x + 18 = 0$ have exactly one solution?

Solve the equation. Check.

11. $2\sqrt[4]{x} = \sqrt[4]{14x + 32}$ **12.** $3 + \sqrt{3x + 1} = x$ **13.** $\sqrt{x + 16} - \sqrt{x} = 2$

14. Find the least integral upper bound U and the greatest integral lower bound L for the real zeros of $2x^4 + x^3 - 11x^2 - 5x + 5$.

Find all the roots of the equation.

15. $2x^4 + x^3 - 11x^2 - 5x + 5 = 0$

16. $x^4 + 4x^3 - x^2 + 16x - 20 = 0$

17. Graph $y = x^4 - 7x^2 + 6$. Locate each real zero as an integer or between consecutive integers.

18. Find all the remaining zeros of $2x^4 - 7x^3 + 14x^2 - 63x - 36$ given that $3i$ is one of the zeros.

19. Use Descartes' Rule of Signs to find all the possible combinations of the sums of the multiplicities of the positive and negative real zeros of $5x^4 - 2x^3 + 7x^2 - 10x - 4$.

20. For what values of k will the sum of the roots of $x^2 - (k^2 - 2k)x + 12 = 0$ be 8?

21. Solve $\sqrt{x + 2} = \sqrt[4]{10x + 11}$.

22. Solve $x^4 - 6x^3 + 5x^2 + 12x - 14 = 0$ given that $3 - \sqrt{2}$ is one of the roots.

College Prep Test

Choose the *one* best answer to each question or problem.

1. A polynomial with *exactly* one integral zero is __?__.
   (A) $x^2 - 1$ (B) $4z^2 + 1$
   (C) $a^3 + a^2 + a - 1$
   (D) $15y^3 - 10y^2 - 4y - 1$
   (E) None of these

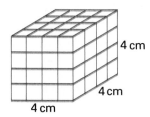

2. A solid cube is 4 cm long on each edge. How many straight cuts, *through* the cube, are needed to produce 64 smaller cubes that are 1 cm long on each edge?
   (A) 3   (B) 6   (C) 9
   (D) 12   (E) 16

3. If $\dfrac{2(\sqrt{2x} - \sqrt{2x - 5})}{2\sqrt{x + 5} - 2\sqrt{x}} = 1$,

   then $x =$ __?__
   (A) 3   (B) 4   (C) 5
   (D) 6   (E) None of these

4. If $y = \sqrt[3]{\dfrac{x}{z}}$, and y is tripled while z is doubled, by what number is x multiplied?
   (A) 2   (B) 3   (C) 8
   (D) 27   (E) 54

5. One diagonal of a rectangle is $\sqrt{15}$ cm long. The rectangle's length is $\sqrt{12}$ cm. Find the rectangle's area.
   (A) $3\sqrt{2}$ cm^2 (B) 6 cm^2
   (C) 9 cm^2 (D) $6\sqrt{5}$ cm^2
   (E) None of these

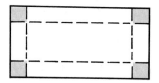

6. A rectangular piece of cardboard is 80 cm by 70 cm. A 10-cm square is cut from each corner, and the four flaps are folded up to form an open box. Find the volume of the box.
   (A) 3,000 cm^3 (B) 30,000 cm^3
   (C) 42,000 cm^3 (D) 56,000 cm^3
   (E) None of these

7. If -2, with a multiplicity of 3, is a zero of $P(x) = x^4 + 4x^3 - 16x - 16$, what is another zero of $P(x)$?
   (A) -1 (B) 1   (C) 2
   (D) 3   (E) 8

8. $\sqrt[4]{(\sqrt{8})^3} =$
   (A) $\sqrt[4]{8}$ (B) $\sqrt[7]{8}$ (C) $2\sqrt[8]{2}$
   (D) $4\sqrt[4]{4}$ (E) $\sqrt[12]{8}$

9. If $(a,b)^* = \sqrt{a^2 + b^2}$, which of the following is the greatest?
   (A) $(3,4)^*$ (B) $(-4,-3)^*$
   (C) $(5,0)^*$ (D) $(-1,-5)^*$
   (E) $(\sqrt{7},\sqrt{10})^*$

10. $4 + 3\sqrt{2}$ is a root of which equation?
    (A) $x^2 - 8x - 2 = 0$
    (B) $x^2 - 8x - 18 = 0$
    (C) $x^2 - 16x - 2 = 0$
    (D) $x^2 - 16x + 18 = 0$
    (E) None of these

11. If $kx^2 = k^2$, for what values of k will there be exactly two real values of x?
    (A) All values of k
    (B) All values of $k \neq 0$
    (C) All values of $k > 0$
    (D) All values of $k < 0$
    (E) None of these

Cumulative Review (Chapters 1–10)

Identify the property of operations **1.3** with real numbers. All variables represent real numbers.

1. $3a \cdot \frac{1}{3a} = 1$

2. $3a \cdot 1 = 3a$

3. $(c + 4)d = cd + 4d$

4. $cd + 4d = 4d + cd$

5. $(c + 4)d = d(c + 4)$

6. $(c + 4) + d = c + (4 + d)$

Solve.

7. $5a - (6 - 3a) - 2(a - 4) = 14$ **1.4**

8. $\frac{x}{6} + \frac{x}{4} + \frac{x}{3} = 1$

9. $a + 2.65 = 7.45 - 2.2a$

10. $|4y - 6| = 14$ **2.3**

11. $2^{3n-1} = 32$ **5.1**

12. $\frac{2x - 1}{x - 1} = \frac{3}{x + 3}$ **7.6**

13. $\frac{24}{x + 2} - \frac{x - 4}{x^2 - 5x - 14} = \frac{2}{x - 7}$

14. $\frac{5}{n} - \frac{3}{n + 8} = 1$

15. $x^2 + 14 = 0$ **9.1**

16. $2t^2 - 5t + 4 = 0$ **9.5**

17. $3\sqrt{2x - 1} = \sqrt{x + 25}$ **10.3**

18. $\sqrt[3]{3x^2 - 11} = 4$

19. $y + \sqrt{5y + 1} = 7$

20. Solve $a(2x - 4) = 3a - bx$ **7.6**
    for x. Use the result to find x
    given that $a = 1.5$ and $b = 0.5$.

Find the solution set.

21. $5x - 4(2x - 3) \geq x - 8$ **2.1**

22. $3 - 5x < x - 9$ and   **2.2**
    $8 > 2x - 6$

23. $-14 < 3t + 4 < 22$

24. $|4y - 6| < 10$ **2.3**

25. $a^2 - a - 12 > 0$ **6.4**

26. $y^3 - 36y > 0$

Find the domain of each function.

27. $\{(3, -1), (4, 2), (5, 0), (6, 7)\}$ **3.1**

28. $y = 4$ **3.5**

29. $y = \dfrac{x + 5}{x^2 - 3x - 10}$ **7.1**

Given $f(x) = 4x - 3$ and $g(x) = x^2 + 3$, find the following values.

30. $f(5) - g(-2)$ **3.2**

31. $f(g(c + 4))$ **5.9**

32. $\dfrac{f(a) - f(b)}{a - b}$ **7.1**

33. Use the slope formula and   **3.3**
    the point-slope form to write   **3.4**
    an equation, in standard
    form, for the line $\overleftrightarrow{PQ}$
    through $P(2, -4)$ and
    $Q(-1, 5)$.

Draw the graph in a coordinate plane.

34. $4x - 3y - 6 = 0$ **3.5**

35. $3x - 4y \leq 8$ **3.9**

36. If p varies directly as v, and   **3.8**
    $p = 314$ when $v = 50$, what
    is v when $p = 785$?

37. Solve the system.   **4.3**
    $2x - 5y = 16$
    $3x + 4y = 1$

38. Solve the system by graphing.   **4.7**
    $y + 3x > 5$
    $y < \frac{1}{3}x - 5$

Simplify.

39. $\left(\dfrac{2a^{-2}}{b^{-1}}\right)^3$ *5.2*

40. $2a(3a - 4)^2$ *5.3*

41. $\dfrac{a^5x^2}{5x + 20} \div \dfrac{8ax^6}{x^2 + 5x + 4}$ *7.2*

42. $\dfrac{3y}{y^2 - 2y - 8} + \dfrac{4}{y - 4} - \dfrac{5}{y + 2}$ *7.4*

43. $2c\sqrt{12c^7},\ c \geq 0$ *8.2*

44. $5\sqrt{8} + 6\sqrt{32} - 4\sqrt{2}$ *8.3*

45. $(5\sqrt{3} - \sqrt{5})(5\sqrt{3} + \sqrt{5})$

46. $\dfrac{5}{4 - 2\sqrt{2}}$ *8.4*

47. $5a\sqrt[3]{16a^6b^4}$ *8.5*

48. $(7 + \sqrt{-8}) - (2 - 3\sqrt{-2})$ *9.1*

49. $3\sqrt{-2} \cdot 5\sqrt{-12}$ *9.2*

50. $(6 + 2i)^2$

51. $\dfrac{2}{4 - 3i}$

Factor completely. (Exercises 52–53)   *5.6*

52. $12y^3 - 27y$

53. $4x^3 + 32$

54. Use synthetic substitution to   *6.6*
solve.
$2x^4 - 5x^3 - 5x^2 + 5x + 3 = 0.$

55. Convert 8 oz/in^2 to pounds   *7.3*
per square foot.

Evaluate. (Exercises 56–59)

56. $\dfrac{3 + 5^2 - (-3)^2}{(4 - 10)^2 + 9 \cdot 4}$ *1.2*

57. $(1.5 \times 10^{-11}) \times (4 \times 10^9)$ *5.2*

58. $2 \cdot 81^{\frac{3}{4}}$ *8.6*

59. $16^{-\frac{1}{4}}$

60. Find the maximum value of   *9.4*
P given that
$P = -3x^2 + 12x + 10.$

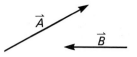

Copy the vectors above and draw the
following. (Exercises 61–64)

61. $\vec{A} + \vec{B}$ *9.7*

62. $\vec{B} - \vec{A}$

63. $-2 \cdot \vec{B}$

64. $\vec{V}$, so that $\vec{A} + \vec{B} + \vec{V}$ is a
zero vector.

65. Use the Rational Zero Theo-   *10.5*
rem to find all the roots of
$4x^4 + 3x^2 - 1 = 0.$

66. A van left a terminal at 8:00   *2.6*
A.M. and traveled east at 25
mi/h. At 10:00 A.M., an
auto left the same terminal
and averaged 55 mi/h along
the same route. At what time
did the auto overtake the
van?

67. Raul has three more dimes   *4.4*
than quarters. The total face
value of his coins is $1.35.
How many coins of each
kind does Raul have?

68. An airplane traveled 840 mi   *4.5*
in 3 h aided by a tailwind. It
would have taken 4 h to
travel the same distance
against the wind. Find the
wind speed.

69. Machine A can produce 300   *7.7*
bolts in 6 h, while Machine
B produces 300 bolts in 4 h.
How long will it take Ma-
chines A and B, operating
together, to make 1,500 bolts?

70. The length of a rectangle is 2   *9.6*
ft more than its width, and 1
ft less than the length of a
diagonal. Find the length of
the diagonal in simplest radical form.

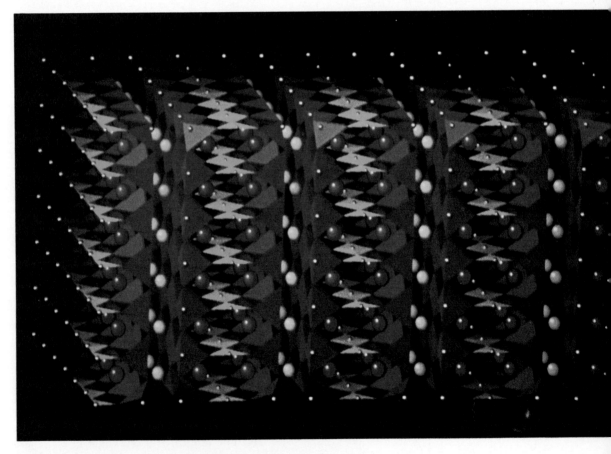

Scientists use computer graphics to construct images of objects that they cannot see. The pattern above is a computer-generated picture showing the crystal structure of a high temperature superconductor.

11.1 Applying Coordinate Geometry

To find distances between points
To find midpoints of segments
To apply the distance and midpoint formulas

The distance between points A and B on a number line is the absolute value of the difference between their coordinates. AB is read as "the *distance* between A and B."

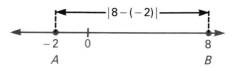

$$AB = |8 - (-2)| = |10| = 10, \text{ or } AB = |-2 - 8| = |-10| = 10$$

The concept of distance between points on a number line can be extended to the distance between points on a horizontal or vertical line in a plane.

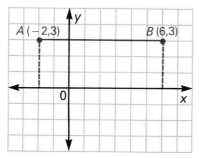

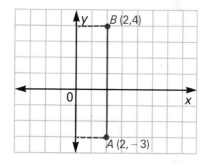

$$AB = |6 - (-2)| = 8 \qquad\qquad AB = |4 - (-3)| = 7$$

In general, $AB = |x_2 - x_1|$ \qquad or \qquad $AB = |y_2 - y_1|$.

The Pythagorean relation, $a^2 + b^2 = c^2$, can be used to find the distance between points on a nonvertical, nonhorizontal line. To find the distance between $A(7,5)$ and $B(3,2)$, first draw a right triangle with $\overline{AB}$ (read "segment AB") as its hypotenuse. Then apply the Pythagorean relation to find c, the length of the hypotenuse.

$$
\begin{aligned}
c^2 &= a^2 + b^2 \\
c &= \sqrt{a^2 + b^2} \\
&= \sqrt{|7 - 3|^2 + |5 - 2|^2} \\
&= \sqrt{4^2 + 3^2} \\
&= \sqrt{25} = 5
\end{aligned}
$$

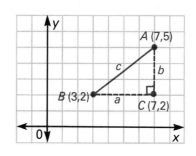

In the triangle at the right, the lengths of the legs are $|x_2 - x_1|$ and $|y_2 - y_1|$.

Use the Pythagorean relation.

$c^2 = a^2 + b^2$

$c = \sqrt{a^2 + b^2}$

$ = \sqrt{|x_2 - x_1|^2 + |y_2 - y_1|^2}$

$ = \sqrt{(x_2 - x_1)^2 + (y_2 - y_1)^2} \leftarrow$ for all real numbers a, $|a|^2 = a^2$

This generalized result is known as the **distance formula**.

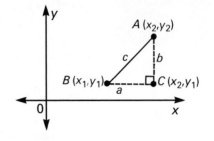

Theorem 11.1

The distance between any two points $A(x_1,y_1)$ and $B(x_2,y_2)$ is

$d = \sqrt{(x_2 - x_1)^2 + (y_2 - y_1)^2}.$

EXAMPLE 1   Find the distance between $R(3, -5)$ and $S(5, -9)$.

Solution
$d = \sqrt{(x_2 - x_1)^2 + (y_2 - y_1)^2}$

$ = \sqrt{(5 - 3)^2 + [-9 - (-5)]^2} = \sqrt{2^2 + (-4)^2} = \sqrt{20} = 2\sqrt{5}$

Calculator check:

(5 − 3) x^2 + (9 +/− − 5 +/−) x^2 = √

Also, 2 × 5 √ =   Display: 4.472136

Thus, the distance between R and S is $2\sqrt{5}$.

EXAMPLE 2   Use the distance formula to determine whether $\triangle ABC$ has at least two congruent sides (is *isosceles*).

Solution
$AB = \sqrt{(8 - 2)^2 + (2 - 4)^2}$

$ = \sqrt{36 + 4} = \sqrt{40} = 2\sqrt{10}$

$AC = \sqrt{(7 - 2)^2 + (9 - 4)^2}$

$ = \sqrt{25 + 25} = \sqrt{50} = 5\sqrt{2}$

$BC = \sqrt{(7 - 8)^2 + (9 - 2)^2}$

$ = \sqrt{1 + 49} = \sqrt{50} = 5\sqrt{2}$

Thus, $\triangle ABC$ is isosceles, because $AC = BC = 5\sqrt{2}$.

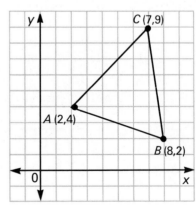

A point is the **midpoint of a segment** if it divides the segment into two segments of equal length. In the figure at the right, the distance formula can be used to show that M is the midpoint of $\overline{AB}$ by demonstrating that $AM = MB$.

$$AM = \sqrt{(3-2)^2 + (5-3)^2}$$
$$= \sqrt{1+4} = \sqrt{5}$$

$$MB = \sqrt{(4-3)^2 + (7-5)^2}$$
$$= \sqrt{1+4} = \sqrt{5}$$

Thus, M is the midpoint of $\overline{AB}$ because $AM = MB$.

Notice that the coordinates of M, the midpoint of $\overline{AB}$, are the arithmetic means (averages) of the coordinates of A and B.

$$\text{Coordinates of } M = \left| \frac{2+4}{2}, \frac{3+7}{2} \right| = (3,5)$$

This suggests the *Midpoint Formula*.

Theorem 11.2

The Midpoint Formula: The coordinates of the midpoint M of the segment joining $A(x_1, y_1)$ and $B(x_2, y_2)$ are $M\left(\dfrac{x_1 + x_2}{2}, \dfrac{y_1 + y_2}{2} \right)$.

EXAMPLE 3 Determine the coordinates of Q, an endpoint of $\overline{PQ}$, given that the other endpoint is $P(-6, -4)$ and the midpoint is $M(-2, 1)$.

Solution $P(-6, -4)$, $M(-2, 1)$, and $Q(x_2, y_2)$.
Use the midpoint formula.

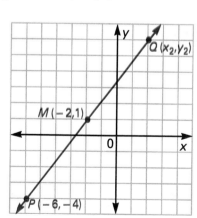

$$x_M = \frac{x_1 + x_2}{2} \qquad y_M = \frac{y_1 + y_2}{2}$$

$$-2 = \frac{-6 + x_2}{2} \qquad 1 = \frac{-4 + y_2}{2}$$

$$-4 = -6 + x_2 \qquad 2 = -4 + y_2$$

$$x_2 = 2 \qquad y_2 = 6$$

Thus, the coordinates of Q are $(2,6)$.

Classroom Exercises

Determine the length of $\overline{AB}$.

1.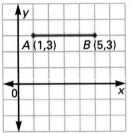

2.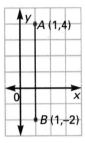

3.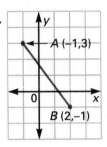

4.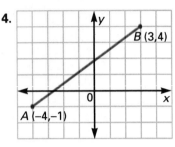

5–8. Using the diagrams for Exercises 1–4, find the coordinates of M, the midpoint of $\overline{AB}$.

Written Exercises

Find the distance between the given points. Give the result in simplest radical form.

1. $A(3,0)$, $B(7,0)$
2. $C(0,5)$, $D(0,-2)$
3. $P(2,4)$, $Q(2,-1)$

4. $A(4,7)$, $B(3,10)$
5. $P(2,4)$, $Q(6,6)$
6. $M(2,1)$, $N(6,5)$

7. $S(5,6)$, $T(9,10)$
8. $G(3,6)$, $H(5,12)$
9. $F(2,9)$, $G(8,13)$

10. $A(-2,5)$, $B(6,7)$
11. $T(-4,-3)$, $U(0,2)$
12. $Z(-3,-6)$, $J(4,-4)$

13. $A(2,-4)$, $B(4,6)$
14. $K(1,5)$, $L(-3,7)$
15. $U\left(-\frac{3}{2},2\right)$, $V\left(\frac{1}{2},\frac{1}{4}\right)$

Use the distance formula to determine whether the points with the given coordinates are the vertices of an isosceles triangle.

16. $A(8,4)$, $B(4,2)$, $C(6,0)$
17. $A(0,0)$, $B(2,4)$, $C(4,2)$

18. $P(-6,0)$, $Q(1,-3)$, $R(0,4)$
19. $G(0,4)$, $H(10,10)$, $K(4,0)$

Determine the coordinates of the midpoint M of the segment joining points P and Q.

20. $P(8,6)$, $Q(4,10)$
21. $P(6,5)$, $Q(8,3)$
22. $P(5,4)$, $Q(7,-10)$

23. $P(-5,-7)$, $Q(-1,-3)$
24. $P(-1,-5)$, $Q(-4,-6)$
25. $P(-8,-6)$, $Q(-3,1)$

Determine the coordinates of Q, an endpoint of $\overline{PQ}$, given that the other endpoint is P and the midpoint is M.

26. $P(2,3)$, $M(7,-7)$
27. $P(7,-1)$, $M(-3,-2)$
28. $P(9,-2)$, $M(5,7)$

29. $P(-3,-2)$, $M(-5,-4)$
30. $P(8,-2)$, $M(-4,-3)$
31. $P(0,-5)$, $M(-2,-6)$

Use the figure at the right for Exercises 32–34.

32. A *rhombus* is a four-sided figure with all sides equal in length. Is *ABCD* a rhombus?

33. Determine whether the diagonals $\overline{AC}$ and $\overline{BD}$ bisect each other. (HINT: Determine whether they have the same midpoint.)

34. Find the perimeter of the quadrilateral formed by joining the midpoints of the four sides of *ABCD*.

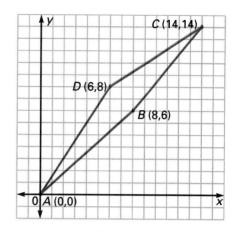

Use the figure at the right for Exercises 35–37.

35. Find the length of $\overline{CM}$, where *M* is the midpoint of $\overline{AB}$.

36. Show that the length of $\overline{PQ}$, the segment joining the midpoints of $\overline{AC}$ and $\overline{BC}$, is half the length of $\overline{AB}$.

37. Find the perimeter of the triangle *PQM* formed by joining the midpoints of the three sides.

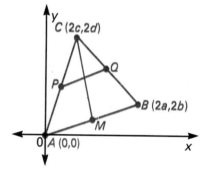

Mixed Review

Write an equation of the line containing the given points. *3.4*

1. $A(5,7)$ and $B(0,-3)$

2. $P(-4,7)$ and $Q(5,7)$

Find the slope of the line described. *3.7*

3. perpendicular to the line containing $P(3,7)$, and $Q(8,9)$

4. parallel to the line $3x - 4y = 12$

▰▰▰ *Brainteaser*

In the figure at the right, point *C* is the center of the square and also a vertex of the triangle. The lengths of the sides of the square and the triangle are as shown. Find the area of the shaded region.

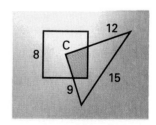

11.2 Other Coordinate Geometry Applications

Objective

To apply the formulas of coordinate geometry

Recall from Chapter 3 how to find the slope of a line given two of its points, and how to write the equation of a line using the point-slope form. Consider segment $\overline{AB}$ with endpoints $A(3,1)$ and $B(9,5)$.

(1) Slope of $\overline{AB}$: $\quad m = \dfrac{y_2 - y_1}{x_2 - x_1} = \dfrac{5 - 1}{9 - 3} = \dfrac{4}{6}$, or $\dfrac{2}{3}$

(2) Equation of $\overleftrightarrow{AB}$:

$$\text{Point-slope form: } y - y_1 = m(x - x_1)$$

$$
\begin{array}{ccc}
y - 1 = \tfrac{2}{3}(x - 3) & \text{or} & y - 5 = \tfrac{2}{3}(x - 9) \\
3y - 3 = 2x - 6 & & 3y - 15 = 2x - 18 \\
2x - 3y = 3 & & 2x - 3y = 3
\end{array}
$$

Standard form: $ax + by = c$

A **median** of a triangle is the segment joining a vertex to the midpoint of the opposite side.

EXAMPLE 1

Determine whether the median $\overline{CM}$ of $\triangle ABC$ is also an altitude of the triangle.

Plan

Because $\overline{CM}$ is a median, M is the midpoint of $\overline{AB}$. Find the coordinates of M.
Then determine whether $\overline{CM}$ is perpendicular to $\overline{AB}$. Remember that the product of the slopes of two non-vertical, nonhorizontal perpendicular lines is -1.

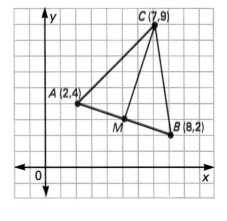

Solution

Coordinates of M: $\left(\dfrac{2 + 8}{2}, \dfrac{4 + 2}{2}\right)$
$= (5,3)$

Slope of $\overline{AB}$: $\dfrac{2 - 4}{8 - 2} = \dfrac{-2}{6} = -\dfrac{1}{3}$ Slope of $\overline{CM}$: $\dfrac{9 - 3}{7 - 5} = \dfrac{6}{2} = 3$

$\overline{CM}$ is perpendicular to $\overline{AB}$ because the product of their slopes is -1.

Thus, $\overline{CM}$ is an altitude of $\triangle ABC$.

In the figure at the right, M is the mid-point of $\overline{AB}$. This makes $\overleftrightarrow{CM}$ a *bisector* of $\overline{AB}$. $\overleftrightarrow{CM}$ is also *perpendicular* to $\overline{AB}$. Therefore, $\overleftrightarrow{CM}$ is the **perpendicular bisector** of $\overline{AB}$.

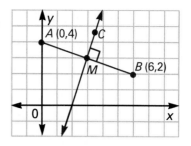

To find an equation of $\overleftrightarrow{CM}$, it suffices to know the coordinates of a point on the line and the slope of the line. Both can be found as follows.

(1) Point M is a point on both $\overline{AB}$ and $\overleftrightarrow{CM}$. The midpoint formula can be used to find the coordinates of M.

(2) Because $\overleftrightarrow{CM}$ is perpendicular to $\overline{AB}$, its slope is the opposite of the reciprocal of the slope of $\overline{AB}$.

EXAMPLE 2 Write an equation in standard form of the perpendicular bisector $\overleftrightarrow{CM}$ of $\overline{AB}$ for $A(0,4)$ and $B(6,2)$.

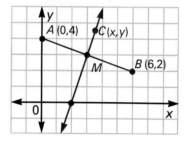

Plan Use the two given facts about $\overleftrightarrow{CM}$:
(1) that it bisects $\overline{AB}$, and
(2) that it is perpendicular to $\overline{AB}$.
Then use the point-slope form to find an equation of $\overleftrightarrow{CM}$.

Solution Find the coordinates of M, the midpoint of $\overline{AB}$.

$$\left(\frac{0+6}{2}, \frac{4+2}{2}\right) = (3,3)$$

Find the slope of $\overleftrightarrow{CM}$, the opposite of the reciprocal of the slope of $\overline{AB}$.

Slope of $\overline{AB} = \dfrac{2-4}{6-0} = -\dfrac{1}{3}$

Slope of $\overleftrightarrow{CM} = 3$

Write an equation of $\overleftrightarrow{CM}$ using the point-slope form.

$$y - y_1 = m(x - x_1)$$
$$y - 3 = 3(x - 3)$$
$$y - 3 = 3x - 9$$

Write an equation in standard form.

$$6 = 3x - y, \text{ or } 3x - y = 6$$

Thus, $3x - y = 6$ is an equation in standard form of the perpendicular bisector of $\overline{AB}$.

Classroom Exercises

Tell whether $\overline{CM}$ is a median of $\triangle ABC$.

1.

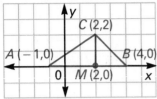

2.

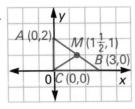

3.

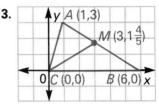

Written Exercises

Determine whether the median $\overline{CM}$ is also an altitude of the given triangle. M is the midpoint of $\overline{AB}$.

1.

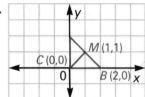

2.

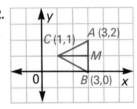

3. $A(2,-4)$, $B(8,-6)$, $C(6,-8)$

4. $A(4,-3)$, $B(6,3)$, $C(-4,3)$

Write an equation in standard form of the perpendicular bisector of $\overline{PQ}$.

5. $P(2,4)$, $Q(6,6)$

6. $P(0,4)$, $Q(-4,0)$

7. $P(2,-1)$, $Q(3,7)$

8. $P(5,8)$, $Q(-8,-5)$

Use the coordinates of the following triangles for Exercises 9–14.

$\triangle ABC$: $A(0,0)$, $B(8,4)$, $C(1,8)$ $\triangle PQR$: $P(0,0)$, $Q(8,4)$, $R(6,8)$

9. Is $\triangle ABC$ isosceles?

10. Is $\triangle PQR$ isosceles?

Find an equation of the line containing the following.

11. median from C to $\overline{AB}$

12. perpendicular bisector of $\overline{AB}$

13. median from R to $\overline{PQ}$

14. perpendicular bisector of $\overline{PQ}$

A *rectangle* is a quadrilateral in which adjacent sides form right angles. Determine whether quadrilateral $ABCD$ is a rectangle.

15. $A(0,0)$, $B(2,8)$, $C(-2,9)$, $D(-4,1)$

16. $A(-2,0)$, $B(8,-1)$, $C(13,3)$, $D(3,4)$

17. The vertices of quadrilateral $ABCD$ are $A(1,-1)$, $B(9,5)$, $C(15,13)$, and $D(7,7)$. Determine whether $ABCD$ is a parallelogram (a quadrilateral with two pairs of parallel sides); a rectangle (see Exercises 15–16); a rhombus (a parallelogram with four sides of equal length).

18. Using quadrilateral $ABCD$ of Exercise 17, determine whether the diagonals are perpendicular bisectors of each other.

19. Any point on the perpendicular bisector of a segment is equidistant from the endpoints of that segment. Verify this statement for the segment with endpoints $A(2,4)$ and $B(8,6)$. (HINT: Find a random point whose coordinates satisfy the equation of the perpendicular bisector of $\overline{AB}$, and show that this point is the same distance from A as it is from B.)

20. The coordinates of the vertices of $\triangle ABC$ are $A(0,0)$, $B(4a,0)$, and $C(2a, 2a\sqrt{3})$. Show that $\triangle ABC$ is equilateral.

21. Use coordinate geometry to prove that any median of an equilateral triangle is also an altitude.

22. The coordinates of the vertices of $\triangle ABC$ are $A(2,1)$, $B(5,5)$, and $C(6,3)$. Show that $\triangle ABC$ is a right triangle.

Mixed Review

1. Graph the equation $y = x^2 - 5$. Use the following values for x to make a table of ordered pairs: $-3, -2, -1, 0, 1, 2, 3$.   *3.2*

2. Solve $x^2 - 4 = 0$.   *6.1*   3. Solve $\sqrt{2x - 1} = 2$.   *10.3*

Application: *Highway Curves*

When designing curves, highway planners take the following into account: the degree D of the curve, which measures the number of degrees turned while traveling 100 feet; the superelevation e, which indicates how steeply the curve is banked; the coefficient of side friction f, which describes the friction between tire and pavement; and the speed v (in miles per hour) for which the road was designed. They use the following formula.

$$D_{max} = \frac{85{,}950(e + f)}{v^2}$$

Find the maximum degree of curve for a road with a superelevation of 0.06, a coefficient of side friction of 0.13, and a design speed of 60 mi/h. Give your answer to the nearest tenth.

11.3 Absolute Value Functions

Objective   To graph equations of the form $y - k = a|x - h|$

Recall that $|x| = x$ when $x \geq 0$, and $|x| = -x$ when $x < 0$.

EXAMPLE 1   Graph the function whose equation is $y = |x|$.

Solution   First, prepare a table of values.

x	-6	-4	-2	0	2	4	6
y	6	4	2	0	2	4	6

The resulting graph is V-shaped.

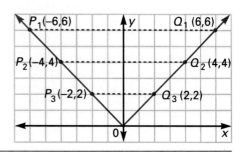

The function in Example 1 is called the basic **absolute value function**. Notice that the y-axis is the perpendicular bisector of P_1Q_1. Also, if the graph of $y = |x|$ is folded along the y-axis, P_1 and Q_1 coincide. So, P_1 and Q_1 are a pair of **mirror-image points** with respect to the y-axis.

In fact, every point on the graph of $y = |x|$ (except the origin) can be paired with another point that is its mirror image. Thus, the graph is *symmetric* with respect to the y-axis, and the y-axis is the **axis of symmetry** of the graph. The point $(0,0)$ is the **vertex** of the graph.

Compare the graphs of $y = \frac{1}{2}|x|$ and $y = 2|x|$ with the graph of the basic absolute value function $y = |x|$.

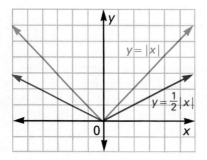

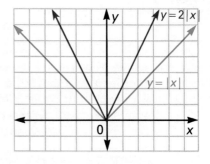

The graph of $y = \frac{1}{2}|x|$ is a *widening* of the graph of $y = |x|$. The slopes of the rays are $\frac{1}{2}$ and $-\frac{1}{2}$.

The graph of $y = 2|x|$ is a *narrowing* of the graph of $y = |x|$. The slopes of the rays are 2 and -2.

EXAMPLE 2   Graph each function below on the same set of axes as a graph of $y = |x|$, and compare its graph with the graph of $y = |x|$.

a. $y = -|x|$

b. $y = -\frac{1}{2}|x|$

Solutions

x	-4	-2	0	2	4
y	-4	-2	0	-2	-4

x	-4	-2	0	2	4
y	-2	-1	0	-1	-2

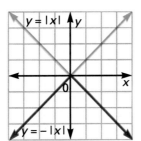

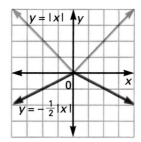

The graph of $y = -|x|$ has the *same shape* as that of $y = |x|$. The vertex is at the origin, and the graph opens *downward*.

The graph of $y = -\frac{1}{2}|x|$ is a *widening* of the graph of $y = |x|$. The vertex is at the origin, and the graph opens *downward*.

In Example 2, the graphs are inverted Vs because the coefficients of $|x|$ are negative. Another way in which the basic absolute value function $y = |x|$ can be modified is by adding a constant to $|x|$.

EXAMPLE 3   Graph the functions below on the same set of axes as a graph of $y = |x|$, and compare each graph with the graph of $y = |x|$.

a. $y = |x| + 3$

b. $y = |x| - 2$

Solutions

x	-4	-2	0	2	4
y	7	5	3	5	7

x	-4	-2	0	2	4
y	2	0	-2	0	2

a. When $y = |x|$ is changed to $y = |x| + 3$, the graph of $y = |x|$ moves up 3 units. The shape remains unchanged.

b. When $y = |x|$ is changed to $y = |x| - 2$, the graph of $y = |x|$ moves down 2 units. The shape remains unchanged.

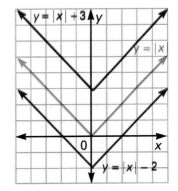

A move, such as one of those in Example 3, that neither changes the shape of a graph nor rotates it, is called a **translation** of the graph. The graph of $y = |x|$ can also be translated horizontally, left or right.

Notice that when the equation $y = |x|$ is changed to $y = |x - 5|$, the vertex moves 5 units to the *right*. The coordinates of the new vertex are (5,0). The equation of the new axis of symmetry is $x = 5$.

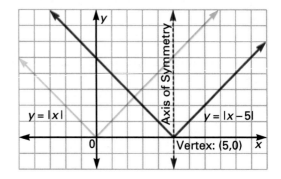

The shape of the graph, however, is unchanged, as is the *direction* of the graph, in the sense that it is not inverted.

A function such as $y = -\frac{1}{3}|x - 2| + 4$ involves a horizontal translation, a vertical translation, a widening, and an inversion. It can be graphed by first rewriting the equation in the **standard form** $y - k = a|x - h|$.

EXAMPLE 4   Graph $y = -\frac{1}{3}|x - 2| + 4$ without using a table of values.

Plan   Rewrite the equation in standard form to find the vertex.

Solution
$y - 4 = -\frac{1}{3}|x - 2|$
The vertex is at (h,k), or (2,4).

Draw the axis of symmetry, $x = h$, or $x = 2$.
Draw the V *downward* because the coefficient, $a = -\frac{1}{3}$, is negative.
The slopes of the two rays of the V are $\frac{1}{3}$ and $-\frac{1}{3}$ ($\pm a$).

Thus, the V is wider in appearance.

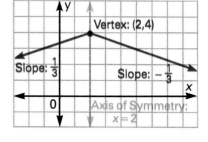

Here are the important characteristics of the absolute value function:

(1) Equation (standard form):
   $y - k = a|x - h|$
(2) Vertex: (h,k)
(3) Equation of the axis of symmetry: $x = h$

(4) Graph opens upward for $a > 0$.
   Graph opens downward for $a < 0$.
(5) V widens for $|a| < 1$.
   V narrows for $|a| > 1$.
(6) Slopes of the rays: a and $-a$

Classroom Exercises

Compare the shape and the direction of the graph of each function with the graph of the function $y = |x|$.

1. $y = 3|x|$ **2.** $y = -\frac{1}{4}|x|$ **3.** $y = \frac{2}{3}|x|$ **4.** $y = -5|x|$

Give the coordinates of the vertex of each graph and the equation of its axis of symmetry.

5. $y = \frac{1}{2}|x|$ **6.** $y - 3 = |x - 5|$

7. $y + 4 = |x + 6|$ **8.** $y = |x + 1| + 2$

9–12. Graph each function in Classroom Exercises 5–8.

Written Exercises

Graph each function. Use a table of values for x and y. Give the coordinates of the vertex.

1. $y = 4|x|$ **2.** $y = \frac{1}{4}|x|$ **3.** $y = -3|x|$ **4.** $y = |x| + 6$

5. $y = |x| - 7$ **6.** $y = |x| - 5$ **7.** $y = |x - 8|$ **8.** $y = |x + 3|$

Graph each function without using a table of values. Give the coordinates of the vertex and the equation of the axis of symmetry.

9. $y - 1 = |x - 1|$ **10.** $y + 1 = 2|x + 1|$ **11.** $y - 2 = -|x + 2|$

12. $y = 2|x - 4| + 6$ **13.** $y = -|x + 5| - 4$ **14.** $y = \frac{1}{2}|x - 1| + 7$

15. $y = 4|x - 2| - 3$ **16.** $y = -\frac{1}{4}|x + 5| + 2$ **17.** $y = -2|x + 6| - 8$

Give the coordinates of the vertex of each graph and the equation of its axis of symmetry. (HINT: $|a(x + b)| = |a||x + b|$)

18. $y = |4x - 8| - 10$ **19.** $y + 5 = |3x - 12|$ **20.** $y = -\frac{1}{2}|7x + 21| - 10$

21. $y + 6 = |\frac{1}{2}x - 2|$ **22.** $y = 3|\frac{1}{3}x + 1| - 1$ **23.** $6 - y = 2|-2x + 3|$

Graph the following relations. (HINT: Consider separate cases, one for $y \geq 0$ and one for $y < 0$.)

24. $x = |y|$ **25.** $|y| = |x| + 4$ **26.** $|x| - 2|y| = 2$

Mixed Review

Simplify. Use positive exponents.   *5.2, 8.2, 8.6, 9.2*

1. $(3x^{-2})^2 x^{-3}$ **2.** $\sqrt{28a^3b^4}$; $a \geq 0$ **3.** $(-27)^{-\frac{2}{3}}$ **4.** $(3 - 4i)^2$

Problem Solving Strategies

Using a Graph

Using a graph can provide the key to solving many new problems. For example, by examining a graph it is possible to derive the distance formula for points in space, much as was done for points in a plane.

In space, there are three mutually perpendicular axes: x, y, and z. So, three planes are represented: xy, yz, and xz. By convention, the xy-plane is represented as a horizontal plane. A point in space is represented by an **ordered triple**, (x, y, z).

Plotting a point in space resembles doing so in a plane. For example, to plot the point $(4, -4, 3)$, move 4 units in the positive x-direction, 4 units in the negative y-direction, and 2 units in the positive z-direction, as shown at the right.

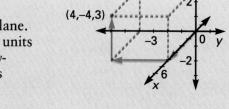

To find the distance between two points $P(x_1, y_1, z_1)$ and $Q(x_2, y_2, z_2)$, refer to the graph at the right. As in the plane, $e^2 = a^2 + b^2$. The Pythagorean relation also gives the following.

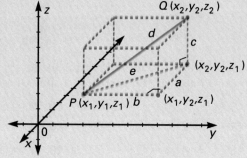

$$d^2 = e^2 + c^2$$
$$d^2 = a^2 + b^2 + c^2 \text{ (Subst for } e^2)$$
$$d = \sqrt{a^2 + b^2 + c^2}$$

Now since a, b, and c are the distances between P and Q along the x-, y-, and z-axes, the formula above becomes the following.

$$d = \sqrt{|x_2 - x_1|^2 + |y_2 - y_1|^2 + |z_2 - z_1|^2}, \text{ or}$$
$$d = \sqrt{(x_2 - x_1)^2 + (y_2 - y_1)^2 + (z_2 - z_1)^2}$$

Exercises

Use the graph at the right for Exercises 1–3.

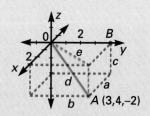

1. Find the distances a, b, c, and e.
2. Find the distance d.
3. Find the distance between A and B.
4. Find the distance between $P(1,2,2)$ and $Q(4,6,14)$.
5. Which is farther from the origin, $(0,0,15)$ or $(9,9,9)$?

11.4 Quadratic Functions

Objectives

Given parabolas of the form $y - k = a(x - h)^2$:
To determine the coordinates of their vertices
To write equations for their axes of symmetry
To graph the parabolas

The graph of the quadratic function $y = x^2$ behaves much like the graph of the absolute value function $y = |x|$.

$y = x^2$:

x	-3	-2	-1	0	1	2	3
y	9	4	1	0	1	4	9

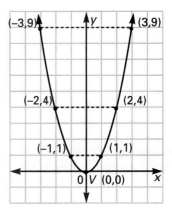

Notice the pairs of mirror-image points with respect to the y-axis: $(-3,9)$, $(3,9)$; $(-2,4)$, $(2,4)$; $(-1,1)$, $(1,1)$

This is the basic graph of a **parabola**. The point $V(0,0)$ is called the *turning point*, or *vertex*, of the parabola. The y-axis is the *axis of symmetry* of the parabola.

EXAMPLE 1

Graph each of the following on the same set of axes.

a. $y = x^2$ b. $y = 2x^2$ c. $y = -\frac{1}{2}x^2$

Solutions

Make a table of values for each graph.

a. The table of values for $y = x^2$ is given above.

b. $y = 2x^2$

x	-2	-1	0	1	2
y	8	2	0	2	8

c. $y = -\frac{1}{2}x^2$

x	-3	-2	-1	0	1	2	3
y	$-4\frac{1}{2}$	-2	$-\frac{1}{2}$	0	$-\frac{1}{2}$	-2	$-4\frac{1}{2}$

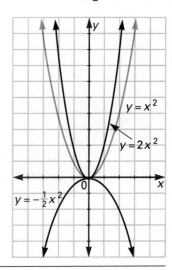

Compare the graphs of $y = 2x^2$ and $y = -\frac{1}{2}x^2$ with that of $y = x^2$.

(1) $y = 2x^2$: The graph is *narrower* and opens *upward*.

(2) $y = -\frac{1}{2}x^2$: The graph is *wider* and opens *downward*.

So, the coefficient a of x^2 in $y = ax^2$ determines the shape of the parabola and the direction in which it opens.

Next, look at the graph of $y = (x + 3)^2 - 6$ below.

x	-6	-5	-4	-3	-2	-1	0
y	3	-2	-5	-6	-5	-2	3

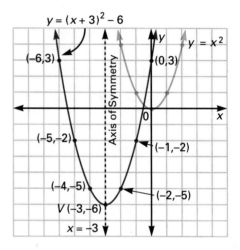

Rewrite $y = (x + 3)^2 - 6$ in *standard form*:
$$y - k = a(x - h)^2$$

Notice the pattern below.

$$y - (-6) = [x - (-3)]^2$$

Vertex: $(-3, -6)$

Axis of symmetry: $x = -3$

Thus, the equation $y + 6 = (x + 3)^2$ is a *translation* of the basic parabola $y = x^2$ to a new position 3 units to the left and 6 units down.

EXAMPLE 2   Graph $y = -2(x - 7)^2 + 4$.

Solution   Rewrite the equation in standard form: $y - 4 = -2(x - 7)^2$

Find and plot the vertex, $(7,4)$.

Draw the axis of symmetry from its equation, $x = 7$.

Find and plot four more points.
Let $x = 7 \pm 1$ (that is, 6 or 8).
Let $x = 7 \pm 2$ (that is, 5 or 9).

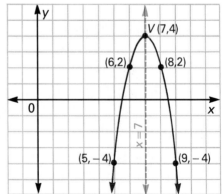

x	5	6	7	8	9
y	-4	2	4	2	-4

Sketch the graph.

The graph is a parabola with vertex $(7,4)$, opening downward.

In summary, the graph of a quadratic function is a parabola with the following characteristics:

(1) Equation (standard form):
$$y - k = a(x - h)^2$$

(2) Vertex: (h,k)

(3) Equation of the axis of symmetry: $x = h$

(4) Graph opens upward for $a > 0$.
Graph opens downward for $a < 0$.

(5) Graph widens for $|a| < 1$.
Graph narrows for $|a| > 1$.

The standard form $y - k = a(x - h)^2$ can also be used to find the equation of a quadratic function when one point of its graph is given along with the vertex of the parabola.

EXAMPLE 3   Find an equation of the function whose graph is given below.

Plan   Because the graph is a parabola, the function is quadratic. Use the vertex V to find the values of h and k in $y - k = a(x - h)^2$. Then find the value of a using point $P(0,1)$ of the parabola.

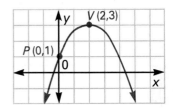

Solution

Substitute 2 for h and 3 for k.

Substitute 0 for x and 1 for y.

Solve for a.

$$y - k = a(x - h)^2$$
$$y - 3 = a(x - 2)^2$$
$$1 - 3 = a(0 - 2)^2$$
$$-2 = a(-2)^2$$
$$-2 = 4a, \text{ or } a = -\frac{1}{2}$$

The equation of the quadratic function is $y - 3 = -\frac{1}{2}(x - 2)^2$.

Classroom Exercises

Give the vertex of each parabola.

1. $y - 1 = 2(x - 2)^2$ **2.** $y = (x + 3)^2 - 9$ **3.** $y = (x - 3)^2 + 1$

4–6. Give the equation of the axis of symmetry of each parabola in Exercises 1–3.

7. Graph the parabola of Classroom Exercise 1.

8. What is the equation of the quadratic function which passes through $(4,10)$ and whose vertex is $(2,2)$?

9. How do the shape and direction of the graph of $y = -\frac{1}{3}(x-2)^2 + 7$ compare with those of the graph of $y = x^2$?

Written Exercises

Graph each function and $y = x^2$ in the same coordinate plane. Use a table of values.

1. $y = 3x^2$ **2.** $y = \frac{1}{3}x^2$ **3.** $y = -3x^2$ **4.** $y = 2x^2$ **5.** $y = -\frac{1}{4}x^2$ **6.** $y = -4x^2$

Graph each quadratic function. Give the vertex of the parabola and write the equation of the axis of symmetry.

7. $y = (x - 1)^2$
8. $y - 2 = x^2$
9. $y - 1 = (x - 2)^2$

10. $y = -(x + 4)^2 + 2$
11. $y = 3(x - 2)^2 - 5$
12. $y = -2(x + 4)^2 + 3$

13. $y = (x - 8)^2$
14. $y = -(x + 5)^2 - 1$
15. $y = x^2 - 7$

16. $y = \frac{1}{2}(x - 7)^2 - 4$
17. $y = -\frac{1}{2}(x + 2)^2 + 3$
18. $y = \frac{2}{3}(x + 1)^2 - 5$

Find an equation of the quadratic function whose graph is given.

19.

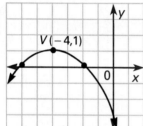

20.

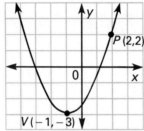

21.

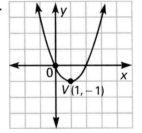

22. Write, in your own words, the steps in graphing $y - 1 = 2(x - 3)^2$.

23. For what values of p will the graph of the parabola $y = (-2p + 5)(x - 5)^2 + 4$ open downward?

24. Determine the values of a and b for which the vertex of the parabola $y = -3(x + 2a + b)^2 + a - b$ will be $(-5, 8)$.

25. For what values of h and k will the vertex of the graph of $y - 6 = a(x + h)^2 + k$ be in Quadrant I?

Midchapter Review

Determine the coordinates of the midpoint M of the segment $\overline{AB}$. Then determine whether the median $\overline{CM}$ is also an altitude of $\triangle ABC$.
11.1, 11.2

1. $A(-3, -4)$, $B(2,5)$, $C(-5,3)$ **2.** $A(2,2)$, $B(4, -1)$, $C(6,1)$

Graph each function. Give the vertex and the equation of the axis of symmetry. **11.3, 11.4**

3. $y - 1 = |x + 3|$ **4.** $y = -|x - 5| - 2$ **5.** $y - 1 = (x + 3)^2$

11.5 Changing to Standard Form

Objectives

To change equations of parabolas from general form to standard form

To determine the vertices and axes of symmetry of parabolas described by equations in general form

Sometimes the equation of a parabola is given in the **general form**, $y = ax^2 + bx + c$, rather than the **standard form**, $y - k = a(x - h)^2$. Because certain characteristics of the parabola are not apparent in the general form of the equation, it can be helpful to change the equation to standard form by completing the square.

EXAMPLE 1   Rewrite $y = x^2 - 8x + 15$ in standard form. Identify the vertex of the parabola and write the equation of its axis of symmetry.

Solution

$$y = x^2 - 8x + 15$$

Isolate the x- and x^2-terms.

$$y - 15 = x^2 - 8x$$

Complete the square.

$$y - 15 + 16 = x^2 - 8x + 16$$

Factor the right side.

$$y + 1 = (x - 4)^2$$

Write in the form $y - k = a(x - h)^2$.

$$y - (-1) = 1(x - 4)^2$$

Vertex: $(4, -1)$ Equation of the axis of symmetry: $x = 4$

EXAMPLE 2   Graph the parabola described by $y = -2x^2 + 12x - 13$.

Solution

Write the equation in standard form by completing the square.

$$y = -2x^2 + 12x - 13$$
$$y + 13 = -2(x^2 - 6x)$$
$$y + 13 - 2 \cdot 9 = -2(x^2 - 6x + 9)$$
$$y - 5 = -2(x - 3)^2$$

Graph the parabola $y - 5 = -2(x - 3)^2$.

Vertex: $(3,5)$

Equation of the axis of symmetry: $x = 3$

Make a table of values.

x	1	2	3	4	5
y	-3	3	5	3	-3

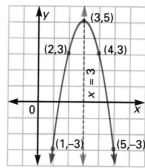

Classroom Exercises

Tell what number must be added to each side of the equation to put it in the standard form of the equation of a parabola.

1. $y - 2 = x^2 - 10x$ **2.** $y - 1 = x^2 + 8x$ **3.** $y = x^2 - 4x$

4. $y + 3 = x^2 - 12x$ **5.** $y = x^2 - 6x$ **6.** $y + 2 = x^2 + 14x$

7. $y - 7 = x^2 + 3x$ **8.** $y + 4 = x^2 + x$ **9.** $y - 2 = x^2 + 5x$

10–12. Rewrite the equations of Classroom Exercises 1–3 in standard form. Give the vertex of the parabola and the equation of its axis of symmetry. Then graph the parabola.

Written Exercises

Rewrite each equation in standard form. Give the vertex and the equation of the axis of symmetry. Then graph the parabola.

1. $y = x^2 + 4x - 1$ **2.** $y = x^2 - 6x + 2$ **3.** $y = x^2 - 10x + 15$

4. $y = x^2 - 6x + 12$ **5.** $y = x^2 + 8x + 20$ **6.** $y = x^2 + 4x - 5$

7. $y = x^2 + 10x + 27$ **8.** $y = x^2 - 6x$ **9.** $y = x^2 - 2x$

10. $y = x^2 - 4x$ **11.** $y = x^2 - 8x$ **12.** $y = x^2 + 6x$

13. $y = 2x^2 - 8x + 6$ **14.** $y = 3x^2 - 6x - 5$ **15.** $y = 4x^2 - 8x + 1$

16. $y = -5x^2 - 10x + 7$ **17.** $y = -2x^2 - 4x + 6$ **18.** $y = -2x^2 - 16x + 3$

19. $y = -x^2 - 6x - 4$ **20.** $y = -3x^2 + 18x$ **21.** $y = -x^2 - 2x$

22. Graph $y = \left| \frac{1}{4}x^2 + \frac{3}{2}x \right|$ for $-9 \le x \le 3$.

23. Graph $y = \frac{1}{4}x^2 + \frac{1}{2}|x| + 2$ for $-4 \le x \le 4$.

24. The graph of a parabola has the same shape as the graph of $y = -2x^2 - 8x + 3$, but is shifted (translated) 2 units to the right and 4 units up. Also, this parabola opens in the opposite direction. Write an equation of this parabola in standard form.

25. Find b and c so that the graph of $y = x^2 + bx + c$ has its vertex at $(3,4)$.

Mixed Review

For the points $A(-2,4)$ and $B(6,8)$, determine each of the following.
3.4, 11.1, 11.2

1. the equation in standard form of the line containing the points A and B

2. the distance between A and B

3. the coordinates of the midpoint of $\overline{AB}$

4. the equation of the perpendicular bisector of $\overline{AB}$

11.6 Problem Solving: Maximum and Minimum Problems

Objective

To solve maximum and minimum problems

In the last two lessons, it was shown that the graph of a quadratic (second-degree) function is a parabola.

The parabola $y = -x^2 + 6x + 1$ can be expressed in standard form as $y - 10 = -(x - 3)^2$. Its graph is shown at the right.

Because $a < 0$, the parabola opens *downward*.

The coordinates of the vertex are (3,10).

Because the vertex is the highest point of the parabola, the y-coordinate of the vertex, 10, is the greatest value, or **maximum** value, of the function.

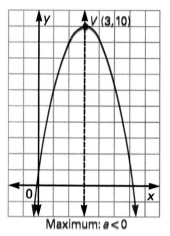
Maximum: $a < 0$

Similarly, for a parabola that opens *upward*, (with $a > 0$), the vertex is the lowest point of the parabola. In the graph of $y - 5 = (x - 4)^2$ at the right, the y-coordinate of the vertex, 5, is the least, or **minimum**, value of the function. The general form of its equation is $y = x^2 - 8x + 21$.

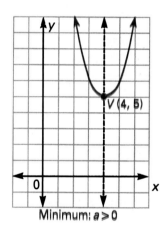
Minimum: $a > 0$

The general equation $y = ax^2 + bx + c$ can be written in standard form by completing the square. The result follows.

$$y - \left(\frac{4ac - b^2}{4a} \right) = a\left(x + \frac{b}{2a} \right)^2$$

The coordinates of the vertex follow from the equation above.

$$(h,k) = \left(-\frac{b}{2a}, \frac{4ac - b^2}{4a} \right)$$

This result is expressed in the theorem at the top of the next page.

Theorem 11.3	The x-coordinate of the vertex of the parabola defined by $y = ax^2 + bx + c$ is $x = -\dfrac{b}{2a}$.

EXAMPLE 1 Find the maximum value of the function $y = -x^2 + 8x - 4$.

Plan Since the coefficient of x^2 is negative, the function has a *maximum* at the vertex of the parabola. Use Theorem 11.3 to find the x-coordinate of the vertex.

Solution From the general equation $y = ax^2 + bx + c$, identify a and b.

$$a = -1, \ b = 8$$

Next, use Theorem 11.3 to find the x-coordinate of the vertex.

$$x = -\frac{b}{2a} = -\frac{8}{2(-1)}, \text{ or } 4$$

Now substitute 4 for x to find the y-coordinate of the vertex, the function's maximum value.

$$y = -x^2 + 8x - 4$$
$$y = -(4)^2 + 8(4) - 4 = 12$$

The maximum value of the function $y = -x^2 + 8x - 4$ is 12.

EXAMPLE 2 Determine whether $y = 2x^2 + 10x - 4$ has a maximum or a minimum value. Then compute that value.

Solution Because 2, the coefficient of x^2, is positive, y has a *minimum*. Use $x = -\frac{b}{2a}$ to find the x-coordinate of the vertex of the parabola.

Substitute 2 for a and 10 for b.

$$x = -\frac{b}{2a}$$
$$= -\frac{10}{2(2)} = -\frac{10}{4}, \text{ or } -\frac{5}{2}$$

Substitute $-\frac{5}{2}$ for x.

$$y = 2x^2 + 10x - 4$$
$$y = 2\left(-\frac{5}{2}\right)^2 + 10\left(-\frac{5}{2}\right) - 4$$
$$y = 2\left(\frac{25}{4}\right) - 25 - 4$$
$$y = 12\frac{1}{2} - 25 - 4 = -16\frac{1}{2}$$

Thus, the minimum value of $y = 2x^2 + 10x - 4$ is $-16\frac{1}{2}$.

EXAMPLE 3   Plans call for a rectangular berry patch next to a summer cottage to be enclosed on three sides by a total of 100 ft of fencing and on the fourth side by a cottage wall. Find the maximum area that can be enclosed in this way. Also, find the corresponding dimensions of the berry patch.

Plan   Distribute the 100 ft of fencing among the three sides as shown in the diagram. Let x = the length of each of the two shorter sides. Then the length of the third side is $100 - x - x$, or $100 - 2x$.

Solution   Use the formula for the area of a rectangle.

$A = lw$

$A = (100 - 2x)x$

$\quad = 100x - 2x^2$, or $-2x^2 + 100x$

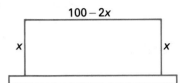

Use Theorem 11.3 to find the value of x for which A is a maximum.

Substitute -2 for a and 100 for b. $x = -\dfrac{b}{2a} = -\dfrac{100}{2(-2)} = 25$

When $x = 25$, $A = -2(25)^2 + 100(25)$, or 1,250. The maximum value of A is 1,250 and this occurs when $x = 25$.

Thus, the maximum area of the berry patch is $1{,}250$ ft^2 when the width x is 25 ft and the length, $100 - 2x$, is 50 ft.

Focus on Reading

Indicate whether each of the following is true or false. If false, tell why.

1. The equation $y = x^2 - 5x + 2$ is in standard form.
2. The function $y = 4x^2 - 7x + 8$ has a minimum value.
3. The minimum value of the function $y = 4x^2 - 8x - 10$ occurs when $x = 1$.
4. The function $y = x^2 + 1$ has a maximum value when $x = 0$.

Classroom Exercises

Determine whether each function has a maximum value or a minimum value.

1. $y = 3x^2 - 12x + 2$ 2. $y = -x^2 - 8x - 10$ 3. $y = 4x^2 + 8x - 4$
4. $y = -3x^2 - 6x - 12$ 5. $y = -6x^2 - x - 7$ 6. $y = x^2 - x + 1$

7. Determine whether $y = -2x^2 - 16x + 5$ has a maximum or a minimum value. Then compute that value.

Written Exercises

1–6. For Classroom Exercises 1–6, find the maximum or minimum value of each function.

7. For a typical day, the net income I, in hundreds of dollars, for a small pizza stand is given by the formula $I = -\frac{1}{2}p^2 + 7p - \frac{45}{2}$, where p is the asking price in dollars per pizza. What price maximizes net income? What is this income?

8. The formula for the height h reached by a rocket fired straight up from a 50-ft platform with an initial velocity of 96 ft/s is $h = -16t^2 + 96t + 50$. Find the number of seconds t required to reach the maximum height. Find the maximum height.

9. A biologist's formula for predicting the number of impulses fired after stimulation of a nerve is $i = -x^2 + 40x - 45$, where i is the number of impulses per millisecond after stimulation, and x is a measure of the strength of the stimulus. Find the maximum possible number of impulses.

10. In a 220-volt electric circuit with a resistance of 15 ohms, the available power in watts W is given by the formula $W = 220I - 15I^2$, where I is the amount of current in amperes. Find the maximum power that the electric circuit can deliver.

11. A rectangular piece of ground is to be enclosed on three sides by 160 ft of fencing. The fourth side is the side of a barn. Find the dimensions of the enclosure that maximizes the area.

12. A rectangular field is to be enclosed by 300 ft of fencing. Find the dimensions of the enclosure that maximizes the area.

13. Find two numbers whose sum is 20 and the sum of whose squares is a minimum.

14. Find two numbers summing to 16.8 whose product is maximal.

15. A rain gutter is made by bending up the edges of a piece of metal of width 36 in. so that the cross-sectional area is a maximum. Find the maximal cross-sectional area.

16. Find the dimensions of the rectangle of maximum area that has a perimeter of p units.

17. From the general form of the equation of a parabola, $y = ax^2 + bx + c$, derive the standard form, $y - k = a(x - h)^2$.

Mixed Review

Solve for x. *2.3, 6.1, 6.6, 9.5*

1. $|2x - 8| = 12$

2. $x^2 - 8x + 12 = 0$

3. $x^3 - 7x^2 + 14x - 8 = 0$

4. $x^3 - 8 = 0$

11.7 Odd and Even Functions

Objective

To determine whether a function is odd or even

Recall that the graph of $y = x^2$, or $f(x) = x^2$, is symmetric with respect to the y-axis. Notice that $f(-2) = f(2) = 4$, and $f(-3) = f(3) = 9$. In general, $f(-x) = f(x)$. Such a function is called an *even function*.

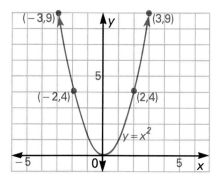

Definition

A function is an **even function** if $f(-x) = f(x)$ for all x in the domain of f.

EXAMPLE 1

Determine whether or not each function is even.

a. $f(x) = x^5$ b. $f(x) = x^4$

Solution

Replace x by $-x$.

a. $f(-x) = (-x)^5 = -x^5$
$$x^5 \neq -x^5$$
$f(-x) \neq f(x) \leftarrow f$ is *not* even.

b. $f(-x) = (-x)^4 = x^4$
$$x^4 = x^4$$
$f(-x) = f(x) \leftarrow f$ is even.

Consider the following definition.

Definition

A curve is said to be **symmetric with respect to the origin** if for any segment through the origin with its endpoints on the curve, the segment is bisected by the origin.

The graph of $f(x) = \frac{1}{3}x^3$ is symmetric with respect to the origin, as can be seen at the right. The origin bisects the segment $\overline{PP'}$. So, $P(3,9)$ and $P'(-3,-9)$ are a pair of symmetric points on the curve. Notice that $f(-x) = -f(x)$; that is, $f(-x)$ and $f(x)$ are *opposites*. So, $f(-3) = -9$, while $f(3) = 9$, the opposite of -9.

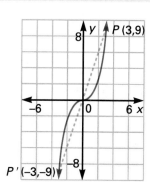

When the graph of f is symmetric with respect to the origin, then f is called an *odd* function.

A function f is an **odd function** if $f(-x) = -f(x)$ for all x in the domain of f.

EXAMPLE 2   Determine whether or not $f(x) = -2x^3$ is odd.

Solution

$$f(x) = -2x^3$$

Replace x by $-x$. $f(-x) = -2(-x)^3$ opposites
$$= -2(-x^3)$$
$$= 2x^3$$

Thus, $f(-x) = -f(x)$, and $f(x) = -2x^3$ is an *odd* function.

Classroom Exercises

Determine whether or not each function is symmetric with respect to the origin, and state the reason for your answer.

1. $f(x) = x^4$ **2.** $f(x) = -x^5$ **3.** $f(x) = -3x^2$ **4.** $f(x) = x^3 + 8$

Determine whether each function is odd, even, or neither.

5. $f(x) = -4x^2$ **6.** $f(x) = 6x^3$ **7.** $f(x) = 3x^2$ **8.** $f(x) = x^3 + 8$

Written Exercises

Determine whether each function is odd, even, or neither.

1. $f(x) = 6x^2$ **2.** $f(x) = 2x^5$ **3.** $f(x) = 8 - x^2$ **4.** $f(x) = -6x$
5. $f(x) = -3x$ **6.** $y = 3x$ **7.** $f(x) = -2x^3$ **8.** $f(x) = 8 - x^4$
9. $f(x) = x^4 + x^2$ **10.** $y - 4 = x^3 + x^2$ **11.** $y + 2 = x^3$ **12.** $y - x^6 = 4x^2$

Points A and B are symmetric with respect to the origin. Find the possible values of x.

13. $A(x^2 - 2x, 5)$, $B(2x - 9, -5)$ **14.** $A(x^3 + 2x^2, 1)$, $B(-7x - 2, -1)$

Mixed Review   5.6, 2.1, 3.5, 4.2, 8.2

1. Factor $9x^2 - 25$.   **2.** Solve $3x - 4 < 5x + 8$.   **3.** Graph $3x - 2y = 8$.

4. Solve the system $\begin{aligned} 2x - y &= 4 \\ 3x + 4y &= 17. \end{aligned}$ **5.** Simplify $\sqrt{28x^7}$ for $x \geq 0$.

1. A number y varies directly as the square of x. If $y = 100$ when $x = 5$, what is y when $x = 3$?

2. Find the set of positive integers for which 3 more than 8 times the sum of a number and 9 is less than 115.

3. Convert 30 mi/h to ft/s.

4. Find the positive number whose square exceeds 3 times the number by 88.

5. Find the maximum height reached by an object shot upward from the earth's surface with an initial velocity of 160 ft/s. Use the formula $h = vt - 16t^2$.

6. One diagonal of a square is 2 cm longer than the length of a side. Find the length of the side in simplest radical form.

7. The perimeter of a rectangle is 24 ft, and its area is 32 ft^2. Find its dimensions.

8. Find two numbers whose product is 20 and the sum of whose reciprocals is $\frac{9}{20}$.

9. How many liters of a 20% alcohol solution must be added to 12 liters of a 60% alcohol solution to obtain a 35% solution?

10. The product of two numbers is 2, and the second number is 4 less than the first number. Find two pairs of such numbers.

11. Working together, Terry and Kim can sand a set of chairs in 6 h. Working alone, Terry takes 5 h longer than Kim. How long does Kim take to sand the chairs working alone?

12. Ted drove a distance of 80 mi in 2 h. Because of construction work, his average speed for the first 20 mi was half of his speed for the rest of his trip. What was his average speed for the first 20 mi?

13. A racing crew rowed 1,120 yd downstream in 3.5 min, and rowing at the same pace, returned the same distance upstream in 4 min. Find the rate of the current in yd/min.

14. Teresa mixed Brand A dog food worth 55¢/lb with Brand B dog food worth 77¢/lb to make a 20-lb package worth $12.76. Find the amount of Brand A that was used.

15. If the numerator of a fraction is increased by 3 and the denominator is decreased by 9, the result is a fraction equal to $\frac{5}{3}$. Find the original fraction, given that the numerator is 8 less than the denominator.

16. A rectangular garden is 3 yd longer than it is wide. A second rectangular garden is planned so that it will be 6 yd wider and twice as long as the first garden. Find the area of the first garden if the sum of the areas of both gardens will be 216 yd^2.

17. The sum of the digits of a two-digit number is 15. If the digits are reversed, the new number is 27 less than the original number. Find the original number.

18. The tens digit of a two-digit number is twice the units digit. If the digits are reversed, the new number is 36 less than the original number. Find the original number.

Key Terms

absolute value function (p. 386)
axis of symmetry (p. 386)
distance formula (p. 378)
even function (p. 401)
general form of a parabola (p. 395)
maximum (p. 397)
median (p. 382)
midpoint formula (p. 379)
midpoint of a segment (p. 379)
minimum (p. 397)

mirror-image points (p. 386)
odd function (p. 402)
ordered triple (p. 390)
parabola (p. 391)
perpendicular bisector (p. 383)
standard form (pp. 388, 395)
symmetry with respect to
  the origin (p. 401)
translation (p. 388)
vertex (p. 386)

Key Ideas and Review Exercises

11.1 Given two points $P_1(x_1,y_1)$ and $P_2(x_2,y_2)$, the following are true.

(1) The distance d between P_1 and P_2 is $\sqrt{(x_2 - x_1)^2 + (y_2 - y_1)^2}$.

(2) The midpoint M of $\overline{P_1P_2}$ is $M\left(\dfrac{x_1 + x_2}{2}, \dfrac{y_1 + y_2}{2}\right)$.

Give the distance between the following points in simplest radical form.

1. $A(2,-4)$; $B(4,6)$

2. $Z(-3,-6)$; $J(4,-4)$

3. $D(-2,0)$; $E(10,\sqrt{5})$

4. $W\left(-\dfrac{2}{3},-\dfrac{1}{3}\right)$; $Q\left(\dfrac{1}{6},-1\right)$

Use $\triangle ABC$ with vertices $A(-3,2)$, $B(1,-2)$, and $C(3,4)$ for Exercises 5–7.

5. Find the distance between A and B.

6. Find the midpoint of $\overline{AB}$.

7. Determine whether $\triangle ABC$ is isosceles.

8. Determine the coordinates of Q, an endpoint of $\overline{PQ}$, given that the other endpoint is $P(-2,4)$ and the midpoint is $M(1,5)$.

11.2 To illustrate geometric relationships involving either
(1) lengths and midpoints of line segments, or
(2) parallelism and perpendicularity, use the properties in Lesson 11.2.

For Exercises 9 and 10, use $\triangle ABC$ of Exercises 5–7.

9. Determine whether the median from C to $\overline{AB}$ is also an altitude of the triangle.

10. Write an equation in standard form of the perpendicular bisector of $\overline{BC}$.

11.3   To graph a function of the form $y - k = a|x - h|$, use the properties summarized in Lesson 11.3.

Graph each function without using a table of values.

11. $y = |x - 1|$ **12.** $y = 3|x - 4| + 7$ **13.** $y = -2|x + 5| - 1$

11.4   To graph a quadratic function of the form $y - k = a(x - h)^2$, use the properties summarized in Lesson 11.4.

Graph each quadratic function. Give the vertex of the parabola and the equation of its axis of symmetry.

14. $y = -(x + 5)^2 - 4$ **15.** $y = \frac{1}{2}(x - 1)^2 + 4$ **16.** $y = (x - 3)^2 + 2$

11.5   To graph a parabola of the form $y = ax^2 + bx + c$, first use the technique of completing the square to change the equation into standard form.

For each parabola, give the vertex and the equation of its axis of symmetry. Then graph the parabola.

17. $y = x^2 - 2x + 5$ **18.** $y = -4x^2 - 8x - 1$ **19.** $y = 2 + 2x - 2x^2$
20. Write, in your own words, the effect of the coefficient a on the shape and the direction of the parabola $y = ax^2 + bx + c$.

11.6   To find the maximum or minimum value of the quadratic function $y = ax^2 + bx + c$, use $x = -\dfrac{b}{2a}$ to find the x-coordinate of the vertex of the parabola. Then use this x-value to find the corresponding y-value.

Determine whether the given function has a maximum value or a minimum value. Then compute that value.

21. $y = -8x^2 + 16x - 9$ **22.** $y = 3x^2 - 9x + 1$ **23.** $y = 2x - x^2 + 3$
24. The sum of two numbers is 18. Their product is a maximum. Find the two numbers.
25. Sarita intends to use 180 yards of fencing to build two pens of equal size to separate two incompatible pets. To make the best use of the fencing, she plans to build the pens adjacent to each other, as shown at the right. What are the dimensions for each pen that will maximize the area enclosed?

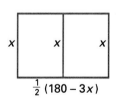

11.7   A function $y = f(x)$ is even if $f(-x) = f(x)$. A function $y = f(x)$ is odd if $f(-x) = -f(x)$.

Determine whether each of the following functions is *even*, *odd*, or *neither.*

26. $f(x) = -2x^2$ **27.** $f(x) = \frac{1}{2}x^3$ **28.** $f(x) = -|x + 1|$

Use $\triangle ABC$ with vertices $A(2,4)$, $B(6,0)$, and $C(10,8)$ for Exercises 1–5.

1. Determine whether or not $\triangle ABC$ is isosceles.
2. Write an equation in standard form of the perpendicular bisector of $\overline{AC}$.
3. Determine whether or not the median from A to $\overline{BC}$ is also an altitude of the triangle.
4. Find the length of the median from C to $\overline{AB}$.
5. Find the length of $\overline{AB}$.

Use the following points for Exercises 6 and 7: $A(0,0)$, $B(5,12)$, $C(-19,22)$, $D(-24,10)$.

6. Determine whether or not $ABCD$ is a rectangle.
7. Determine whether or not the diagonals of $ABCD$ are perpendicular bisectors of each other.
8. Determine the coordinates of Q, an endpoint of $\overline{PQ}$, given that the other endpoint is $P(-6,3)$ and the midpoint is $M(4,5)$.

Graph each function without using a table of values.

9. $y = |x| + 3$ 10. $y + 5 = 2|x - 6|$ 11. $y = -3|x + 1| + 4$

For each parabola, give the vertex and the equation of its axis of symmetry. Then graph the parabola.

12. $y = -\frac{1}{8}(x + 5)^2 + 2$ 13. $y = -(x - 1)^2 + 6$ 14. $y = x^2 - 12x + 11$

Determine whether the given function has a maximum value or a minimum value. Then compute that value.

15. $y = -x^2$ 16. $y = 10x^2 - x + 4$ 17. $y = 10x - x^2 + 4$
18. Find two numbers whose sum is 30 and the sum of whose squares is a minimum.

Determine whether the functions in Exercises 19–21 are *even*, *odd*, or *neither*.

19. $f(x) = 55x - 3$ 20. $f(x) = 4$ 21. $f(x) = \sqrt[3]{x^4}$
22. Graph $y = |36 - x^2|$.   23. Graph $x = -|y - 6| + 5$.
24. What is the equation in standard form of the parabola formed by translating the graph of the parabola $y = x^2 - 4x + 2$ four units to the left and four units upward and then inverting it so that it opens downward?

Choose the *one* best answer to each question or problem.

1. If $y - 3 = 7$, then $y + 6 =$ __?__ .

(A) 10   (B) 16   (C) 24
(D) 4   (E) 27

2. If $k = \frac{n}{18} + \frac{n}{18} + \frac{n}{18}$, then the least positive integer n for which k is an integer is __?__ .

(A) 18   (B) 4   (C) 12
(D) 6   (E) 18

3.

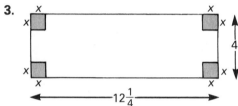

In the figure above, the area of the unshaded region is __?__ .

(A) $(7 - 2x)(7 + 2x)$

(B) $(12\frac{1}{4} - x)(4 - x)$

(C) $(12\frac{1}{4} - 4x)(4 - 4x)$

(D) $16\frac{1}{4}x$ (E) $49 - x^2$

4. Which of the following values for N will maximize $\left(-\frac{1}{3}\right)^N$?

(A) 2   (B) 3   (C) 7
(D) 5   (E) 0

5. $P = 2 - \frac{9}{10}$, $Q = 2 - 0.099$,
$R = 2 \div 9$
In which list below are P, Q, and R in order from greatest to least?

(A) P, Q, R (B) Q, P, R
(C) R, P, Q (D) P, R, Q
(E) R, Q, P

6. Fifty percent of 50% of 2 is __?__ .

(A) 1   (B) $\frac{1}{2}$ (C) $\frac{1}{4}$
(D) 4   (E) 2

7.

A 9-hour clock is shown. If at 12 noon today the pointer is at 0, where will the pointer be at 12 noon tomorrow?

(A) 0   (B) 2   (C) 6
(D) 7   (E) 8

8. After giving her $5, Ian had $9 more than Gail. How much more money than Gail did Ian have originally?

(A) $14   (B) $4   (C) $19
(D) $9   (E) None of these

9. If $a^3 b^2 c < 0$, which of the following must be true?
I. $abc < 0$ II. $ac < 0$ III. $bc < 0$

(A) II only   (B) I only
(C) II and III   (D) I and III
(E) None of these

10. If $y = \frac{1}{xz}$, what is the value of x when $y = 12^{-1}$ and $z = 3$?

(A) -36 (B) $-\frac{1}{36}$ (C) -4
(D) 4   (E) None of these

11.

If the area of the triangle above is 30, then the length of $\overline{DB}$ is __?__ .

(A) 2   (B) 8   (C) 10
(D) 3   (E) 5

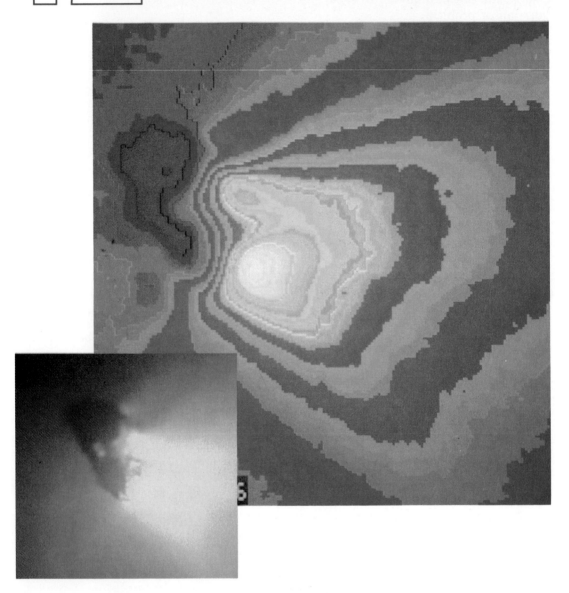

Comets that reappear on a regular basis, like Halley's Comet, have elliptical orbits. These images of Halley's Comet were produced from data gathered from the European Giotto space mission. Giotto is the name of an Italian pre-Renaissance artist who represented the Christmas star as a comet.

12.1 The Circle

Objectives

To write equations of circles given certain conditions

To graph circles given their equations

People have been fascinated by circles for thousands of years. Around 2000 B.C., an ancient people erected huge standing stones in a large circular ditch at Stonehenge in southern England. Most archaeologists believe that Stonehenge served a religious function, but some scientists suspect that it also served as an astronomical instrument for tracing the movements of the sun and the moon and for observing eclipses.

Definitions

A **circle** is the set of all points in a plane that are the same distance from a given point, called the **center**. This distance is equal to the length of any segment, called a **radius**, which joins the center to a point on the circle.

The word *radius* can also denote the *length* of the radius. Thus, *radius 5* means "radius of length 5." In a coordinate plane, the distance formula, $d = \sqrt{(x_2 - x_1)^2 + (y_2 - y_1)^2}$, can be used to write an equation of the circle of radius 5 shown below, which has its center at the origin. Let $P(x,y)$ be any point on the circle. Then use the distance formula to express the distance between $C(0,0)$ and $P(x,y)$, which is already known to be 5.

$$\sqrt{(x - 0)^2 + (y - 0)^2} = 5$$
$$\sqrt{x^2 + y^2} = 5$$
$$(\sqrt{x^2 + y^2})^2 = 5^2$$
$$x^2 + y^2 = 25$$

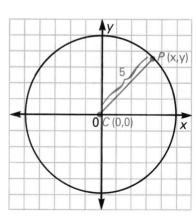

Thus, an equation of the circle of radius 5 having its center at the origin is $x^2 + y^2 = 5^2$, or $x^2 + y^2 = 25$.

Theorem 12.1   The standard form of the equation of a circle of radius r with its center at the origin is $x^2 + y^2 = r^2$.

EXAMPLE 1   Write the equation of the circle centered at the origin and passing through the point $A(12,5)$. Graph the circle.

Plan   Substitute 12 for x and 5 for y in the equation above to find r^2.

Solution $x^2 + y^2 = r^2$

$12^2 + 5^2 = r^2$

$144 + 25 = 169 = r^2; r = \sqrt{169}$, or 13

Thus, the equation is $x^2 + y^2 = 169$. Plot convenient points 13 units from the origin and draw the circle.

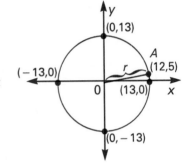

Circles can be *translated* (moved without rotation) so that their centers are not at the origin. In the figure at the right, circle S of radius 3 is centered at the origin. Circle T is a translation of circle S to $T(4,-2)$. The length of the radius, however, remains the same.

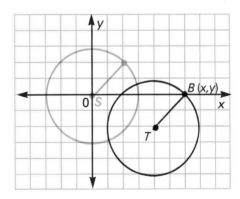

An equation of circle T can be found using the distance formula. Let $B(x,y)$ be any point on circle T. The distance between $T(4,-2)$ and $B(x,y)$ is 3. Therefore,

$$\sqrt{(x - 4)^2 + [y - (-2)]^2} = 3$$
$$\sqrt{(x - 4)^2 + (y + 2)^2} = 3$$
$$(x - 4)^2 + (y + 2)^2 = 9 \qquad \text{Square each side.}$$

Notice the following relationship: $(x - 4)^2 + (y + 2)^2 = 9$

center: $(4, -2)$ radius: $\sqrt{9} = 3$

Theorem 12.2   The standard form of the equation of a circle of radius r with its center at $C(h,k)$ is $(x - h)^2 + (y - k)^2 = r^2$.

EXAMPLE 2 Write, in standard form, the equation of the circle centered at $C(-4,5)$ and passing through the point $P(2,10)$. Graph the circle.

Solution Substitute 2 for x, 10 for y, -4 for h, and 5 for k in the standard form.

$$(x - h)^2 + (y - k)^2 = r^2$$
$$[2 - (-4)]^2 + (10 - 5)^2 = r^2$$
$$6^2 + 5^2 = r^2$$
$$61 = r^2$$
$$r = \sqrt{61} \approx 7.8$$

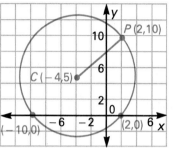

Thus, the standard form of the equation of this circle is $(x + 4)^2 + (y - 5)^2 = 61$.

Plot several points 7.8 units from $C(-4,5)$. Draw the graph as shown.

EXAMPLE 3 Rewrite $x^2 - 8x + y^2 + 4y + 16 = 0$ in standard form. Identify the circle's center and radius.

Plan Isolate the x- and y-terms. Then complete the squares in both x and y.

Solution
$$x^2 - 8x + \underline{} + y^2 + 4y + \underline{} = -16$$
$$x^2 - 8x + \left(\tfrac{1}{2} \cdot 8\right)^2 + y^2 + 4y + \left(\tfrac{1}{2} \cdot 4\right)^2 = -16 + 16 + 4$$
$$(x - 4)^2 + (y + 2)^2 = 4 \leftarrow r^2 = 4; r = 2$$

Standard form: $(x - 4)^2 + (y + 2)^2 = 4$; center: $(4, -2)$; radius: 2

Classroom Exercises

Give the radius and the coordinates of the center of each circle. Graph each circle.

1. $x^2 + y^2 = 49$
3. $(x - 4)^2 + (y - 1)^2 = 100$

2. $x^2 + y^2 = 19$
4. $(x - 1)^2 + y^2 = 144$

Written Exercises

Write, in standard form, the equation of the circle with center C and radius r given, or with center C and a point P on the circle given.

1. $C(0,0)$, $r = 13$ **2.** $C(0,0)$, $r = 3.1$ **3.** $C(2,3)$, $P(8,0)$ **4.** $C(4,2)$, $P(0,8)$
5. $C(-2,8)$, $r = 9$ **6.** $C(1,1)$, $r = \sqrt{3}$ **7.** $C(1,-1)$, $P\left(\tfrac{1}{2},1\right)$ **8.** $C\left(\tfrac{1}{4},\tfrac{1}{4}\right)$, $P\left(1,-1\right)$
9–16. Graph each circle in Written Exercises 1–8.

Graph, rewriting each equation in standard form if necessary. Identify the circle's center and radius.

17. $x^2 + y^2 = 4$ **18.** $x^2 + y^2 = 25$ **19.** $x^2 + y^2 = 64$ **20.** $x^2 + y^2 = 81$

21. $x^2 + y^2 = 8$ **22.** $x^2 + y^2 = 12$ **23.** $x^2 + y^2 = 22$ **24.** $x^2 + y^2 = 15$

25. $x^2 - 6x + y^2 - 10y - 2 = 0$ **26.** $x^2 + 8x + y^2 - 4y - 5 = 0$

27. $x^2 + 6x + y^2 - 14y - 42 = 0$ **28.** $x^2 - 12x + y^2 - 2y - 8 = 0$

29. A point $P(x,y)$ moves in a coordinate plane about a fixed point $C(-1,2)$ so that its distance from C is always $2\sqrt{3}$ units. Use the distance formula, $d = \sqrt{(x_2 - x_1)^2 + (y_2 - y_1)^2}$, to write an equation describing the path of point P.

Use the distance formula for Exercises 30 and 31.

30. Prove Theorem 12.1. **31.** Prove Theorem 12.2.

32. The circle with equation $(x - 1)^2 + (y + 1)^2 = 9$ is translated 4 units to the right and 2 units downward. Find the equation of the resulting circle.

33. The circle with equation $x^2 - 8x + y^2 + 4y + 9 = 0$ is translated 2 units to the left and 5 units upward. Find the equation of the resulting circle.

Write, in standard form, the equation of the circle having the given properties.

34. center on the line $x = 5$, tangent to the y-axis at $P(0,9)$

35. tangent to both rays of the graph of $y = |x|$, radius 4
(HINT: The slope of $y = x$ is one, so the triangle formed by the origin, the circle's center, and the point of tangency is isosceles.)

Use the circle $x^2 + y^2 = 100$ and triangle ABC for Exercises 36–38.

36. Show that the vertices $A(-6,8)$, $B(0,10)$, and $C(6,-8)$ of triangle ABC are points on the circle, and that the triangle is a right triangle.

37. Write an equation of the line tangent to the circle at $A(-6,8)$.

38. Show that the perpendicular bisector of $\overline{AB}$ passes through the center of the circle.

Mixed Review

1. Graph $y = |x - 4| + 5$. *11.3* **2.** Graph $y = x^2 - 8x + 2$. *11.5*

Simplify. *5.2, 8.2, 8.6, 9.2*

3. $(2x - 4)^0, x \neq 2$ **4.** $\sqrt{32x^5}$ **5.** $27^{-\frac{2}{3}}$ **6.** $(3 + 4i)^2$

12.2 The Ellipse

To graph ellipses given their equations
To write equations of ellipses given certain conditions

The figure at the right shows a 20-cm piece of
string fastened to two points, F_1 and F_2, 16 cm
apart. A pencil holds the string taut. As it moves,
it generates the points P_1, P_2, P_3, and so on.
These points form an "egg-shaped" curve called
an *ellipse*. At each point on this curve, the sum
of the distance to F_1 and the distance to F_2 is
always 20 cm, the length of the string. For example, $P_3F_1 + P_3F_2 = 20$. This experiment suggests the following definitions.

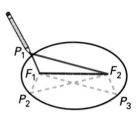

Definitions

An **ellipse** is the set of all points in a plane such that for each point,
the sum of the distances from two fixed points is constant. Each
fixed point is called a **focus** (plural: *foci*). The distance from any
point on the ellipse to a focus is called a **focal radius**.

In the ellipse at the right, the foci are
$F_1(-8,0)$ and $F_2(8,0)$. The midpoint of
$\overline{F_1F_2}$ is the *center*, $C(0,0)$, of the ellipse.
The sum of the distances from any
point $P(x,y)$ on the ellipse to the foci
is 20. The equation of the ellipse
can be written using the definition
above and the distance formula.

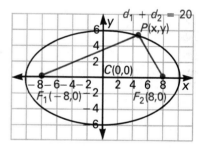

Use the definition of ellipse. $d_1 + d_2 = 20$

Substitute. $\sqrt{(x+8)^2 + y^2} + \sqrt{(x-8)^2 + y^2} = 20$

Isolate a radical. $\sqrt{x^2 + 16x + 64 + y^2} = 20 - \sqrt{x^2 - 16x + 64 + y^2}$

Square each side. $x^2 + 16x + 64 + y^2 = 400 -$
$$40\sqrt{x^2 - 16x + 64 + y^2} + x^2 - 16x + 64 + y^2$$

Isolate a radical. $32x - 400 = -40\sqrt{x^2 - 16x + 64 + y^2}$

Divide each side
by 8. $4x - 50 = -5\sqrt{x^2 - 16x + 64 + y^2}$

Square each side. $16x^2 - 400x + 2{,}500 = 25x^2 -$
$$400x + 1{,}600 + 25y^2$$

The resulting equation, $9x^2 + 25y^2 = 900$, can be analyzed to reveal facts about the ellipse it describes.

Divide each side by 900 and simplify. $\dfrac{x^2}{100} + \dfrac{y^2}{36} = 1$

Rewrite the denominators as squares. $\dfrac{x^2}{10^2} + \dfrac{y^2}{6^2} = 1$

If $y = 0$, then $x = \pm 10$. The x-intercepts of the ellipse are ± 10.

If $x = 0$, then $y = \pm 6$. The y-intercepts are ± 6.

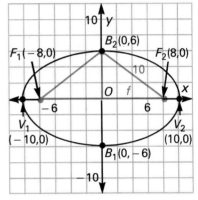

In the diagram at the right, $\overline{V_1 V_2}$ is the **major axis** of the ellipse, and $\overline{B_1 B_2}$ is the **minor axis**. The points B_1, B_2, V_1, and V_2 are the **vertices** of the ellipse. Notice that the numbers 10 and 6 in the equation of the ellipse are, respectively, the lengths of *half* the major axis $\overline{V_1 V_2}$ and *half* the minor axis $\overline{B_1 B_2}$.

Another relationship can be found by calculating $B_2 F_1$ and $B_2 F_2$.

By the distance formula,
$$B_2 F_1 = \sqrt{[0 - (-8)]^2 + (6 - 0)^2} = \sqrt{100} = 10,$$
and $B_2 F_2 = \sqrt{(0 - 8)^2 + (6 - 0)^2} = \sqrt{100} = 10.$

Therefore, both $B_2 F_1$ and $B_2 F_2$ are equal to OV_2, half the length of the major axis.

In the figure, triangle $OB_2 F_2$ is a right triangle. The Pythagorean relation can be used to relate the lengths 6, 10, and the distance f from the origin to the focus F_2. That relationship is $f^2 + 6^2 = 10^2$, so $f = 8$.

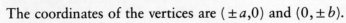

Theorem 12.3

The equation, in standard form, of an ellipse with its center at the origin and a horizontal major axis is

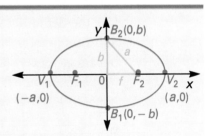

$\dfrac{x^2}{a^2} + \dfrac{y^2}{b^2} = 1$, where $0 < b < a$.

The length of the *semimajor axis* is a.

The length of the *semiminor axis* is b.

The coordinates of the vertices are $(\pm a, 0)$ and $(0, \pm b)$.

The coordinates of the foci are $(\pm f, 0)$, where $f^2 + b^2 = a^2$, or $f = \sqrt{a^2 - b^2}$.

EXAMPLE 1   Graph $16x^2 + 25y^2 = 400$. Find the coordinates of the ellipse's vertices and foci.

Solution   Write in standard form.

$$\frac{x^2}{25} + \frac{y^2}{16} = 1$$

Plot the vertices: $(\pm 5, 0)$, $(0, \pm 4)$
Draw the graph.

$$f^2 + 4^2 = 5^2$$

$$f = \sqrt{25 - 16} = \sqrt{9} = 3$$

The coordinates of the foci are $(\pm 3, 0)$.

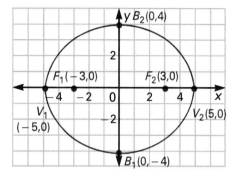

The major axis of an origin-centered ellipse can also lie along the y-axis. In this case, the foci will be points on the y-axis. The resulting changes in the standard form of the equation of an ellipse are outlined in the following theorem.

Theorem 12.4   The equation of an ellipse with its center at the origin and a vertical major axis is

$$\frac{y^2}{a^2} + \frac{x^2}{b^2} = 1, \text{ where } 0 < b < a.$$

The length of the semimajor axis is a.

The length of the semiminor axis is b.

The coordinates of the vertices are $(\pm b, 0)$ and $(0, \pm a)$.

The coordinates of the foci are $(0, \pm f)$, where $f = \sqrt{a^2 - b^2}$.

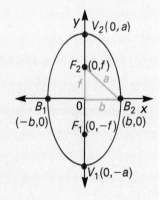

Both Theorems 12.3 and 12.4 can be used to write equations for origin-centered ellipses given the coordinates of their vertices, or given those of a vertex and a focus. Sketching each ellipse can help determine whether its major axis (length a) is horizontal or vertical, and thus whether a^2 is the denominator of x^2 or y^2 in the equation.

EXAMPLE 2   Given the following information about each ellipse, write its equation in standard form.

a. Vertices: $(\pm 8, 0)$, $(0, \pm 5)$ **b.** Vertices: $(0, \pm 4)$; Foci: $(0, \pm 2)$

Solutions   First, sketch each ellipse.

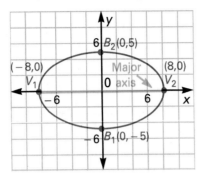

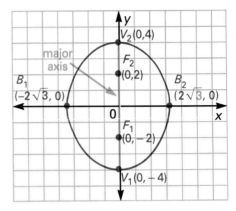

Major axis is horizontal: $8 > 5$

Use $\dfrac{x^2}{a^2} + \dfrac{y^2}{b^2} = 1$, with
$a = 8$ and $b = 5$.

$$\frac{x^2}{8^2} + \frac{y^2}{5^2} = 1, \text{ or } \frac{x^2}{64} + \frac{y^2}{25} = 1$$

The equation is $\dfrac{x^2}{64} + \dfrac{y^2}{25} = 1$.

Foci lie on the y-axis so
major axis is vertical.

Use $\dfrac{y^2}{a^2} + \dfrac{x^2}{b^2} = 1$, with $a = 4$.

$f = \sqrt{a^2 - b^2}$

$2 = \sqrt{4^2 - b^2}$, so $4 = 16 - b^2$,
or $b^2 = 12$

The equation is $\dfrac{y^2}{16} + \dfrac{x^2}{12} = 1$.

Classroom Exercises

Indicate whether the major axis is horizontal or vertical. Find
the lengths of the semimajor axis a and the semiminor axis b.

1. $\dfrac{y^2}{16} + \dfrac{x^2}{4} = 1$ **2.** $\dfrac{x^2}{49} + \dfrac{y^2}{25} = 1$ **3.** $\dfrac{y^2}{100} + \dfrac{x^2}{16} = 1$ **4.** $\dfrac{x^2}{81} + \dfrac{y^2}{9} = 1$

5. $\dfrac{x^2}{49} + \dfrac{y^2}{4} = 1$ **6.** $\dfrac{y^2}{144} + \dfrac{x^2}{64} = 1$ **7.** $\dfrac{x^2}{25} + \dfrac{y^2}{4} = 1$ **8.** $\dfrac{x^2}{5} + \dfrac{y^2}{3} = 1$

Written Exercises

Graph each ellipse. Give the coordinates of the vertices and the foci.

1. $\dfrac{x^2}{25} + \dfrac{y^2}{9} = 1$ **2.** $\dfrac{y^2}{25} + \dfrac{x^2}{16} = 1$ **3.** $\dfrac{y^2}{169} + \dfrac{x^2}{25} = 1$

4. $4x^2 + 25y^2 = 100$ **5.** $4x^2 + 49y^2 = 196$ **6.** $4x^2 + y^2 = 4$

7. $x^2 + 25y^2 = 25$ **8.** $9x^2 + y^2 = 36$ **9.** $16x^2 + 9y^2 = 144$

10. $x^2 + 4y^2 = 4$ **11.** $25x^2 + 4y^2 = 100$ **12.** $100x^2 + 36y^2 = 3,600$

Given the following information, write an equation in standard form of each ellipse.

13.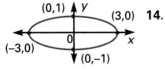

14.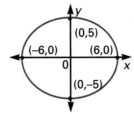

15.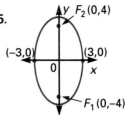

16. vertices: $(\pm 5, 0)$, $(0, \pm 8)$

17. vertices: $(\pm \sqrt{5}, 0)$, $(0, \pm \sqrt{3})$

18. vertices: $(\pm \sqrt{7}, 0)$, $(0, \pm \sqrt{6})$

19. vertices: $(\pm 8, 0)$ foci: $(\pm 3, 0)$

20. vertices: $(\pm 6, 0)$ foci: $(\pm 4, 0)$

21. vertices: $(0, \pm 8)$ foci: $(0, \pm 3)$

22. vertices: $(0, \pm 5)$ foci: $(0, \pm 2)$

23. vertices: $(\pm 4, 0)$ foci: $(0, \pm 3)$

24. vertices: $(0, \pm 8)$ foci: $(\pm 5, 0)$

25. Find the standard form of the equation of the ellipse with foci $(0, \pm 3)$ and the sum of whose focal radii is 10.

26. Find the standard form of the equation of the ellipse with vertices $(\pm 2, 0)$ and that contains the point $\left(1, -\dfrac{\sqrt{3}}{2}\right)$.

27. Given the ellipse with equation $9x^2 + 36y^2 = 324$ and the circle with equation $x^2 - 2x + y^2 - 2y + 1 = 0$ in the same coordinate plane, find the area of the region inside the ellipse, but not inside the circle. (HINT: Area of the ellipse $= \pi ab$)

28. Prove Theorem 12.3 using the definition of an ellipse and the figure at the right. (HINT: Let the two fixed points, the foci, be $(\pm f, 0)$. Let $P(x, y)$ be any point on the ellipse. Let the sum of the distances d_1 and d_2 be $2a$, $a > 0$. At the final stage of development, group the terms that contain $a^2 - f^2$ as a factor. Then substitute b^2 for $a^2 - f^2$ so that b becomes the length of the semiminor axis.)

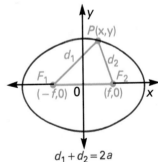

$d_1 + d_2 = 2a$

29. Prove Theorem 12.4. (HINT: Let the foci be $(0, \pm f)$. Then refer to the hint for Exercise 28.)

Mixed Review

Graph each equation.   *3.5, 11.3, 11.5, 12.1*

1. $4x - 3y + 12 = 0$

2. $y = 2|x - 5| - 4$

3. $x^2 + 6x - 8y + 25 = 0$

4. $x^2 - 6x + y^2 + 4y - 23 = 0$

12.3 The Hyperbola

Objectives

To graph hyperbolas given their equations

To write equations of hyperbolas given certain conditions

To find the vertices, foci, and equations of the asymptotes of hyperbolas given their equations

Consider the following definitions.

Definition

A **hyperbola** is the set of all points in a plane such that from each point, the difference of the distances to two fixed points is constant. The fixed points are called the **foci**, and the distances are called **focal radii**.

The hyperbola at the right has foci $F_1(-5,0)$ and $F_2(5,0)$. Let $P(x,y)$ represent any point on the hyperbola. Given that the difference of the distances from $P(x,y)$ to the foci is 6, an equation for the hyperbola can be written using the definition above and the distance formula.

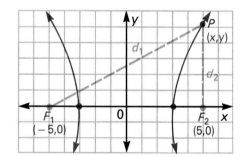

$$d_1 = \sqrt{[x - (-5)]^2 + (y - 0)^2} \qquad d_2 = \sqrt{(x - 5)^2 + (y - 0)^2}$$

$$d_1 - d_2 = \sqrt{x^2 + 10x + 25 + y^2} \quad - \quad \sqrt{x^2 - 10x + 25 + y^2} = 6$$

Simplify this equation in the same manner as that of the ellipse in the previous lesson. The result is $16x^2 - 9y^2 = 144$.

The equation above can be solved for y to give the equation $y = \pm\frac{4}{3}\sqrt{x^2 - 9}$. Because $x^2 - 9$ must be nonnegative for y to be real, x cannot assume values between -3 and 3. Thus, the domain of the relation $16x^2 - 9y^2 = 144$ is $\{x \mid x \le -3 \text{ or } x \ge 3\}$. Note that the two extreme values, $x = \pm 3$, occur when the two *branches* of the hyperbola intersect the x-axis at the two *vertices*, $V_1(-3,0)$ and $V_2(3,0)$.

The equation $y = \pm\frac{4}{3}\sqrt{x^2 - 9}$ can also be rewritten as follows.

$$y = \pm\frac{4}{3}\sqrt{x^2 - 9} = \pm\frac{4}{3}\sqrt{x^2\left(1 - \frac{9}{x^2}\right)} = \pm\frac{4x}{3}\sqrt{1 - \frac{9}{x^2}}$$

When $|x|$ is very large, $\dfrac{9}{x^2}$ is very close to zero and $\sqrt{1 - \dfrac{9}{x^2}}$ is very close to 1. Therefore, $y = \pm\dfrac{4x}{3}\sqrt{1 - \dfrac{9}{x^2}} \approx \pm\dfrac{4x}{3}$.

In the figure below, as $|x|$ gets very large, each branch of the hyperbola gets very close to the two lines $y = \pm\dfrac{4}{3}x$. These lines are called the **asymptotes**. Their slopes can be found as follows.

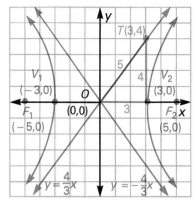

Original equation: $16x^2 - 9y^2 = 144$

Divide each side by 144.

$$\dfrac{16x^2}{144} - \dfrac{9y^2}{144} = \dfrac{144}{144}$$

$$\dfrac{x^2}{9} - \dfrac{y^2}{16} = 1$$

$$\dfrac{x^2}{3^2} - \dfrac{y^2}{4^2} = 1$$

Slopes of the asymptotes: $\pm\dfrac{4}{3}$

Also in the figure, $\overline{TV_2}$ is a vertical line segment drawn to connect the vertex V_2 with a point T on the asymptote of slope $\dfrac{4}{3}$. Because $\overline{TV_2}$ is vertical, T and V_2 have the same x-coordinate, 3. The y-coordinate of T is $y = \dfrac{4}{3}(3)$, or 4.

Note that, by the distance formula, $OT = \sqrt{3^2 + 4^2} = 5$. Thus, OT is equal to f, the distance from the center of the hyperbola, $(0,0)$, to either of the foci, F_1 or F_2.

Theorem 12.5

The equation, in standard form, of a hyperbola centered at the origin with its foci on the x-axis is

$$\dfrac{x^2}{a^2} - \dfrac{y^2}{b^2} = 1, \text{ where } a > 0 \text{ and } b > 0.$$

The coordinates of the vertices are $(\pm a, 0)$.

The coordinates of the foci are $(\pm f, 0)$, where $f = \sqrt{a^2 + b^2}$.

The slopes of the asymptotes are $\pm\dfrac{b}{a}$.

The equations of the asymptotes are $y = \pm\dfrac{b}{a}x$.

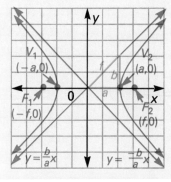

EXAMPLE 1   Graph $25x^2 - 4y^2 = 100$. Find the coordinates of the vertices and the foci. Write the equations of the asymptotes.

Solution   Divide each side by 100 to obtain the standard form.

$$\frac{25x^2}{100} - \frac{4y^2}{100} = \frac{100}{100}$$

$$\frac{x^2}{4} - \frac{y^2}{25} = 1 \leftarrow a = 2 \text{ and } b = 5$$

The vertices are $V(\pm 2, 0)$.

Because $f = \sqrt{a^2 + b^2} = \sqrt{4 + 25} = \sqrt{29}$, the foci are $F(\pm\sqrt{29}, 0)$.

The equations of the asymptotes are $y = \pm\frac{5}{2}x$.

Draw the asymptotes and plot the vertices and foci. It is possible to find a few other points on the hyperbola by solving for y and making a table as shown below.

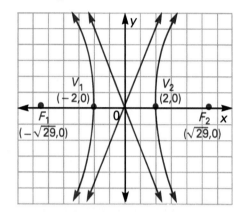

x	$y = \pm\frac{5}{2}\sqrt{x^2 - 4}$
± 2	0
± 2.5	± 3.8
± 3	± 5.6

Draw the graph so that as $|x|$ increases, the hyperbola approaches the asymptotes without touching them.

If the foci of a hyperbola lie on the y-axis, the following is true.

Theorem 12.6

The equation, in standard form, of a hyperbola centered at the origin with its foci on the y-axis is $\dfrac{y^2}{a^2} - \dfrac{x^2}{b^2} = 1$, where $a > 0$ and $b > 0$.

The coordinates of the vertices are $(0, \pm a)$. The coordinates of the foci are $(0, \pm f)$, where $f = \sqrt{a^2 + b^2}$. The slopes of the asymptotes are $\pm\dfrac{a}{b}$. The equations of the asymptotes are $y = \pm\dfrac{a}{b}x$.

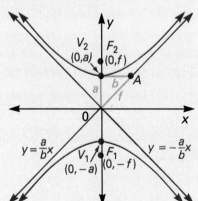

EXAMPLE 2 A hyperbola has vertices $(0, \pm 4)$ and foci $(0, \pm 5)$.

Write the equation of the hyperbola in standard form. Write the equations of the asymptotes.

Solution Plot the vertices and foci. Sketch the hyperbola. The foci are points on the y-axis. So, $a = 4$ and $a^2 = 16$. $OF_2 = OA = f = 5$

Use $f = \sqrt{a^2 + b^2}$ to find b^2 and b.

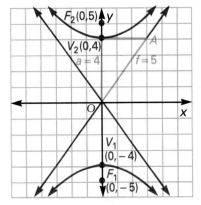

$$f = \sqrt{a^2 + b^2}$$
$$5 = \sqrt{4^2 + b^2}$$
$$25 = 16 + b^2$$
$$9 = b^2, \text{ or } b^2 = 9 \quad \text{Thus, } b = 3.$$

Write the equation using the standard form. The equation of the hyperbola is
$$\frac{y^2}{16} - \frac{x^2}{9} = 1.$$

The equations of the asymptotes are $y = \pm \frac{4}{3} x$.

Classroom Exercises

For each hyperbola, indicate whether the foci are points on the x-axis or the y-axis.

1. $\frac{x^2}{4} - \frac{y^2}{9} = 1$ **2.** $\frac{y^2}{36} - \frac{x^2}{16} = 1$ **3.** $\frac{x^2}{25} - \frac{y^2}{1} = 1$ **4.** $y^2 - \frac{x^2}{64} = 1$

5–8. Give the slopes of the asymptotes of each hyperbola in Classroom Exercises 1–4.

9–12. Graph each hyperbola in Classroom Exercises 1–4. Give the coordinates of the vertices and the foci. Write the equations of the asymptotes.

Written Exercises

Graph each hyperbola using the properties of a hyperbola stated in Theorems 12.5 and 12.6. Give the coordinates of the vertices and the foci. Write the equations of the asymptotes.

1. $\frac{x^2}{4} - \frac{y^2}{16} = 1$ **2.** $\frac{x^2}{49} - \frac{y^2}{9} = 1$ **3.** $\frac{x^2}{36} - \frac{y^2}{16} = 1$

4. $\frac{x^2}{25} - \frac{y^2}{100} = 1$ **5.** $\frac{y^2}{64} - \frac{x^2}{49} = 1$ **6.** $\frac{y^2}{121} - \frac{x^2}{16} = 1$

7. $x^2 - 16y^2 = 144$ **8.** $9x^2 - 25y^2 = 225$ **9.** $4x^2 - 16y^2 = 64$

10. $y^2 - x^2 = 144$ **11.** $25x^2 - 36y^2 = 900$ **12.** $36x^2 - y^2 = 144$

Use the information in Exercises 13–20 to write the equation of each
hyperbola in standard form.

13.

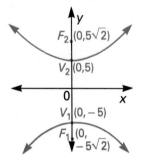

14.

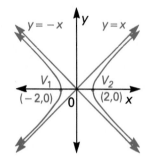

15. vertices: $(\pm 4, 0)$
   foci: $(\pm 5, 0)$

16. vertices: $(0, \pm 2)$
   foci: $(0, \pm 8)$

17. vertices: $(0, \pm 5)$
   foci: $(0, \pm 7)$

18. vertices: $(0, \pm 3)$
   foci: $(0, \pm 5)$

19. equations of asymptotes: $5y + 2x = 0$, $5y - 2x = 0$; vertices: $(\pm 10, 0)$

20. equations of asymptotes: $2y + 7x = 0$, $2y - 7x = 0$; vertices: $(0, \pm 14)$

21. The foci of a hyperbola are $F_1(13, 0)$ and $F_2(-13, 0)$. Write its
   equation in standard form given that the difference between its
   focal radii is 10.

22. Use the definition of hyperbola to prove Theorem 12.5. (HINT: Let the coordinates of the
   foci be $F(\pm f, 0)$, and let the difference between the focal radii be $2a$. At the final stage
   of development, group terms that contain
   $f^2 - a^2$ as a factor. Then substitute b^2 for
   $f^2 - a^2$.)

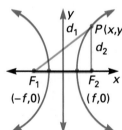

23. Prove Theorem 12.6. (HINT: Let the foci be $F(0, \pm f)$.)
   Then refer to the hint for Exercise 22.)

Mixed Review

1. Graph $4x^2 + 9y^2 = 36$.   *12.2*

2. Graph $25x^2 + 4y^2 = 100$.   *12.2*

3. Solve by completing the square.
   $x^2 - 4x = -3$ *9.3*

4. Write an equation of the ellipse with
   vertices $(\pm 10, 0)$ and foci $(\pm 8, 0)$.   *12.2*

12.4 Translations of Ellipses and Hyperbolas

Objectives

To find the centers, vertices, and foci of ellipses and hyperbolas given their equations

To graph ellipses and hyperbolas given their equations

Just as a circle, an ellipse can be translated so that its center is no longer at the origin. Shown below are the standard forms of the equation of an ellipse with center (h,k). For both forms, the distance from the center to either focus is $f = \sqrt{a^2 - b^2}$.

Horizontal major axis

$$\frac{(x - h)^2}{a^2} + \frac{(y - k)^2}{b^2} = 1,$$

$$0 < b < a$$

Vertical major axis

$$\frac{(y - k)^2}{a^2} + \frac{(x - h)^2}{b^2} = 1,$$

$$0 < b < a$$

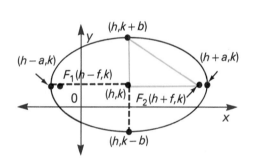

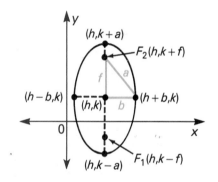

EXAMPLE 1 For the ellipse $\dfrac{(y - 2)^2}{9} + \dfrac{(x + 1)^2}{4} = 1$, find the coordinates of the center, the vertices, and the foci. Graph the ellipse.

Solution Center: $(-1,2)$

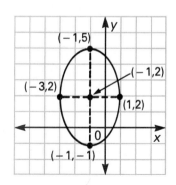

Because $9 > 4$, and 9 is the denominator for $(y - 2)^2$, $a = \sqrt{9} = 3$, $b = \sqrt{4} = 2$, and the major axis is vertical.

Vertices $(h, k \pm a)$, $(h \pm b, k)$:

$(-1,5)$, $(-1,-1)$, $(1,2)$, $(-3,2)$

Use $f = \sqrt{a^2 - b^2}$ to find the foci.

$$f = \sqrt{9 - 4} = \sqrt{5}$$

Foci $(h, k \pm f)$: $F_1(-1, 2 - \sqrt{5})$, $F_2(-1, 2 + \sqrt{5})$

Use the vertices to graph the ellipse as shown.

By translating a hyperbola so that its center moves from the origin to $C(h,k)$, the standard forms change as follows. The distance from the center to the foci is $f = \sqrt{a^2 + b^2}$, where $a > 0$ and $b > 0$. The foci and vertices lie on lines parallel to the x- and y-axes respectively.

$$\frac{(x - h)^2}{a^2} - \frac{(y - k)^2}{b^2} = 1 \qquad\qquad \frac{(y - k)^2}{a^2} - \frac{(x - h)^2}{b^2} = 1$$

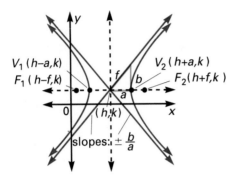
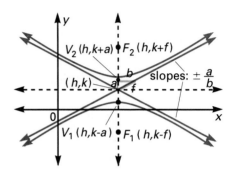

EXAMPLE 2

For the hyperbola $9x^2 + 54x - 4y^2 + 8y + 41 = 0$, find the coordinates of the center, the vertices, and the foci. Write the equations of the asymptotes. Sketch the hyperbola.

Plan

Complete the squares of both the x- and y-terms.

Solution

$9(x^2 + 6x + \underline{}) - 4(y^2 - 2y + \underline{}) = -41 + \underline{}$

$9(x^2 + 6x + 9) - 4(y^2 - 2y + 1) = -41 + 9 \cdot 9 - 4 \cdot 1$

$9(x + 3)^2 - 4(y - 1)^2 = 36$

$\dfrac{9(x + 3)^2}{36} - \dfrac{4(y - 1)}{36} = 1$

$\dfrac{(x + 3)^2}{4} - \dfrac{(y - 1)^2}{9} = 1 \leftarrow a^2 = 4, b^2 = 9$

Center: $(-3,1)$

Use $a = 2$ to find the vertices.

Vertices: $(-3 \pm 2, 1)$, or $(-1,1)$, $(-5,1)$

Use $f = \sqrt{3^2 + 2^2} = \sqrt{13}$ to find the foci.

Foci: $F_1(-3 - \sqrt{13},1)$, $F_2(-3 + \sqrt{13},1)$

Use $C(-3,1)$ and $m = \pm\dfrac{b}{a} = \pm\dfrac{3}{2}$ to find equations of the asymptotes.

$y - y_1 = m(x - x_1)$

$y - 1 = \pm\dfrac{3}{2}[x - (-3)]$, or $y = \dfrac{3}{2}x + \dfrac{11}{2}$ and $y = -\dfrac{3}{2}x - \dfrac{7}{2}$

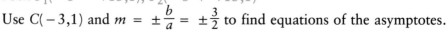

Use the vertices and the asymptotes to sketch the hyperbola.

Classroom Exercises

Give the coordinates of the center of each ellipse or hyperbola.

1. $\dfrac{(x-4)^2}{25} + \dfrac{(y+1)^2}{9} = 1$

2. $\dfrac{(y-5)^2}{16} + \dfrac{(x+7)^2}{4} = 1$

3. $\dfrac{(x-8)^2}{9} - \dfrac{(y-5)^2}{4} = 1$

4. $\dfrac{(y+1)^2}{9} + \dfrac{(x-6)^2}{49} = 1$

Written Exercises

For each ellipse or hyperbola, find the coordinates of the center, the vertices, and the foci. Then graph the ellipse or hyperbola.

1. $\dfrac{(x-3)^2}{100} + \dfrac{(y-2)^2}{36} = 1$

2. $\dfrac{(x+6)^2}{25} + \dfrac{(y+5)^2}{16} = 1$

3. $\dfrac{(x-3)^2}{16} - \dfrac{(y-2)^2}{9} = 1$

4. $\dfrac{(x-6)^2}{64} + \dfrac{(y+1)^2}{100} = 1$

5. $\dfrac{(x+5)^2}{36} - \dfrac{(y-1)^2}{64} = 1$

6. $\dfrac{(y+6)^2}{25} - \dfrac{(x-1)^2}{144} = 1$

7. $16x^2 + 4y^2 - 96x + 8y + 84 = 0$ **8.** $25x^2 - 4y^2 - 150x - 16y + 109 = 0$

9. $x^2 - y^2 - 2x - 4y - 4 = 0$ **10.** $x^2 + 4y^2 - 2x + 40y + 100 = 0$

11. $4x^2 - y^2 + 24x + 4y + 28 = 0$ **12.** $4x^2 + 25y^2 + 16x + 50y - 59 = 0$

13. $4x^2 + 9y^2 - 24x + 18y + 9 = 0$ **14.** $9x^2 - 16y^2 - 90x + 32y + 65 = 0$

15. Find the equation of the ellipse with foci $(2, -3)$ and $(2,5)$, and the sum of whose focal radii equals 10.

16. The hyperbola with equation $4x^2 - 9y^2 + 32x + 18y + 19 = 0$ is translated 4 units to the right and 1 unit downward. Give the standard form of the equation of the hyperbola in its new position.

17. Write an equation of the hyperbola with foci at $F_1(-5,2)$ and $F_2(3,2)$ and asymptotes $y - 2 = \pm\sqrt{3}(x + 1)$.

18. Write an equation for the set of all points, the sum of whose distances from $A(2,3)$ and $B(-4,3)$ is always 12. Draw the resulting graph.

Mixed Review

Solve.   *7.6, 9.5, 6.6*

1. $\dfrac{3}{x-5} = \dfrac{x}{2}$

2. $x^2 - 2x + 4 = 0$

3. $x^3 - 7x^2 + 14x - 8 = 0$

12.5 The Rectangular Hyperbola and Inverse Variation

Objectives   To graph rectangular hyperbolas
To determine whether equations express inverse variations
To solve word problems involving inverse variation

In 1662, Robert Boyle noted that if a fixed amount of gas is held at a constant temperature, then its volume goes *down* when it pressure goes *up*. This observation, known as *Boyle's Law*, is represented more precisely by the equation $pv = k$, where k is a real nonzero constant. A similar equation is shown in Example 1.

EXAMPLE 1   Graph $xy = 4$.

Plan   Solve for y in terms of x.

Then make a table of values.

Solution

$xy = 4$

$y = \dfrac{4}{x}$

x	-4	-2	-1	1	2	4
y	-1	-2	-4	4	2	1

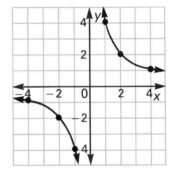

Plot the points. Draw the two branches of the resulting hyperbola.

Example 1 suggests that the graph of an equation of the form $xy = k$, where k is a real nonzero constant, is a hyperbola with the x- and y-axes as its asymptotes. Such a hyperbola is called a **rectangular hyperbola**.

If $k > 0$, then the branches of the hyperbola lie in quadrants I and III.

If $k < 0$, then the branches lie in quadrants II and IV.

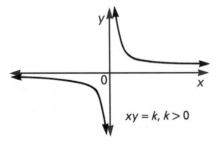

$xy = k,\ k > 0$

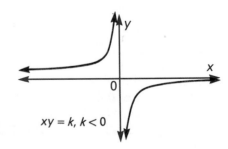

$xy = k,\ k < 0$

If the product xy is constant, then as one factor increases, the other factor decreases, and y is said to "vary inversely as x."

Definitions

A function defined by an equation of the form $xy = k$, where k is a real nonzero constant, is called an **inverse variation**. The constant k is called the **constant of variation**.

EXAMPLE 2

Determine whether the equation or table represents an inverse variation. If so, give the constant of variation.

a. $\dfrac{4}{r} = t$

b. $y = \dfrac{x}{4}$

c.

x	1	3	4	12
y	12	4	3	1

Solutions

a. Rewrite as $rt = 4$, which is of the form $xy = k$.

Inverse variation; constant of variation: 4

b. Rewrite as $\dfrac{y}{x} = \dfrac{1}{4}$, which is not of the form $xy = k$.

Not an inverse variation

c. This is a table for $xy = 12$, which is of the form $xy = k$.

Inverse variation; constant of variation: 12

Note that the equation in Example 2b is an example of *direct* variation.

Suppose (x_1, y_1) and (x_2, y_2) represent two ordered pairs of an inverse variation. By definition, $x_1 y_1 = k$ and $x_2 y_2 = k$, since the product of x and y is *constant*.

By substitution, $x_1 y_1 = x_2 y_2$. This equation is applied below.

EXAMPLE 3

A number y varies inversely as x. If $y = 27$ when $x = 2$, what is y when $x = 9$?

Plan

Use $x_1 y_1 = x_2 y_2$.

Solution

Let $(x_1, y_1) = (2, 27)$ and $(x_2, y_2) = (9, y)$.

$2 \cdot 27 = 9 \cdot y$

$54 = 9y$

$y = 6$

Thus, $y = 6$ when $x = 9$.

EXAMPLE 4

The illumination I from a lamp varies inversely as the square of the distance d between the lamp and the illuminated object. If $I = 8$ ft-candles when $d = 3$ ft, what is d when $I = 2$ ft-candles? (THINK: Will the distance be less than 3 ft or greater than 3 ft? How do you know?)

Because I varies inversely as the square of d, use $x_1y_1 = x_2y_2$.

Solution

$I_1 \cdot d_1^2 = I_2 \cdot d_2^2 \leftarrow$ Let $x_1 = I_1$, $y_1 = d_1^2$, $x_2 = I_2$, and $y_2 = d_2^2$.

$8 \cdot 3^2 = 2 \cdot d_2^2$

$72 = 2d_2^2$

$36 = d_2^2$

$d_2 = \sqrt{36} = 6$

Thus, $d = 6$ ft when $I = 2$ ft-candles.

Classroom Exercises

Complete each table so that an inverse variation will be represented.

1.

x	2	?	16
y	24	8	3

2.

a	-8	40	?
b	5	-1	-20

3.

x	?	25	10
y	50	4	10

Determine whether each equation expresses an inverse variation (c is a nonzero constant).

4. $4x = y$ **5.** $x = \dfrac{c}{y}$ **6.** $\dfrac{c}{x} = \dfrac{y}{1}$ **7.** $1 = \dfrac{-x}{y}$ **8.** $y = \dfrac{1}{cx + x}$, $c \neq -1$

Written Exercises

Graph each rectangular hyperbola.

1. $xy = 6$ **2.** $xy = -24$ **3.** $y = -\dfrac{1}{x}$ **4.** $y = \dfrac{20}{x}$ **5.** $4xy = -48$

Determine whether the equation or table expresses an inverse variation. If so, find the constant of variation.

6. $y = 8x - 7$ **7.** $\dfrac{a}{3} = b$ **8.** $A = 7.3M$ **9.** $p = \dfrac{5}{q}$ **10.** $7b = \dfrac{3}{c}$

11.

x	2	3	5	6	12
y	1.5	1	0.6	0.5	0.25

12.

x	1	2	3	6	10
y	5	10	15	30	50

In Exercises 13–16, y varies inversely as x.

13. If $y = 24$ when $x = 3$, what is y when $x = 8$?

14. If $y = -32$ when $x = 2$, what is x when $y = 4$?

15. If $y = 4$ when $x = 9$, what is x when $y = 72$?

16. If $y = -3$ when $x = 25$, what is y when $x = 15$?

17. The current in an electric circuit of constant voltage varies inversely as the resistance. When the current is 30 amps, the resistance is 20 ohms. Find the current when the resistance is 25 ohms.

18. The length of a rectangle with constant area varies inversely as the width. When the length is 18, the width is 8. Find the length when the width is 9.

19. A number y varies inversely as the square of x. If $y = 2$ when $x = 6$, what is y when $x = 3$?

20. A number y varies inversely as $\sqrt{x}$. If $y = 3$ when $x = 64$, what is x when $y = 6$?

21. The illumination I from a light varies inversely as the square of its distance d from the illuminated object. If $I = 8.00$ ft-candles when $d = 5.00$ ft, what is I when $d = 4.00$ ft?

22. The height of a cylinder with constant volume is inversely proportional to the square of its radius. If $h = 8$ when $r = 4$, what is r when $h = 2$?

23. The volume of a fixed amount of an ideal gas at constant temperature is inversely proportional to the pressure. What happens to the volume when the pressure is doubled?

24. The graph of $xy - 3y + 4x - 12 = 8$ represents a translation of the rectangular hyperbola $xy = 8$. Find the center of the translated hyperbola.

25. The weight of an object varies inversely as the square of its distance from the center of the earth. If Daniel weighs 170.0 lb on the surface of the earth 3960 mi from its center, what will he weigh on top of a nearby mountain rising 3 mi above him?

26. Prove that if y is inversely proportional to x, then $\dfrac{1}{y}$ is directly proportional to x.

Midchapter Review

Write an equation of each circle, ellipse, or hyperbola given the following conditions. *12.1, 12.2, 12.3*

1. Circle: center $(-4, 2)$, radius 5

2. Ellipse: vertices $(\pm 6, 0)$, foci $(0, \pm 8)$

3. Hyperbola: vertices $(\pm 5, 0)$, foci $(\pm 13, 0)$

Graph each circle, ellipse, or hyperbola. If the graph is a circle, label the center and radius. If the graph is an ellipse, label the vertices, center, and foci. If the graph is a hyperbola, label the vertices, center, foci, and equations of the asymptotes. *12.1, 12.2, 12.3, 12.4*

4. $x^2 + y^2 = 100$

5. $4x^2 - 25y^2 = 100$

6. $16x^2 + 25y^2 = 400$

7. $9x^2 + 4y^2 - 18x + 16y - 11 = 0$

8. $y^2 - 4x^2 - 6y - 16x - 23 = 0$

9. A number y varies inversely as the square root of x. If $y = 8$ when $x = 9$, what is x when $y = 12$? *12.5*

Application: *Eccentricity of an Ellipse*

When undisturbed by outside forces, a satellite of the earth—whether a communications satellite, a space station, or an astronaut—follows an elliptical orbit with the center of the earth at one focus. The **eccentricity**, denoted by e, tells how round or flat the ellipse is. The eccentricity is the ratio of the distance between the foci of any ellipse to the length of its major axis. So, $e = \dfrac{2f}{2a} = \dfrac{f}{a}$.

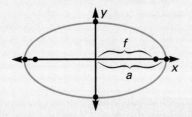

If $e = 0$, then $f = 0$ and the ellipse is actually a circle. For values of e closer to 1, the ellipse becomes flatter.

The eccentricity of a satellite can be found from its maximum and minimum distances from the surface of the earth, often called its **apogee** and **perigee**. The satellite $OV_{1\ 13}$ had an apogee of 9,320 km and a perigee of 560 km. Using 6,370 km as the radius of the earth,

$$2a = 9,320 + 2(6,370) + 560$$
$$= 22,620, \text{ so } a = 11,310.$$

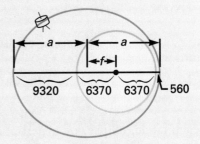

$$f = a - (6370 + 560)$$
$$= 11,310 - 6,930 = 4,380$$

So, $e = \dfrac{f}{a} = \dfrac{4,380}{11,310} \approx 0.387$.

1. Vanguard I, launched in 1958, was the second U.S. satellite. Its apogee was 3,950 km and its perigee 660 km. Find its eccentricity.

2. 10th Molniya$_1$, a Soviet satellite launched in 1968, had an apogee of 39,600 km and a perigee of 430 km. Find its eccentricity.

3. Explorer 41, launched in 1969 to explore interplanetary fields, had an apogee of 213,850 km (over half the distance to the moon) and an eccentricity of 0.9405. Find its closest pass to the earth.

12.6 Problem Solving: Joint and Combined Variation

Objective

To solve word problems involving joint or combined variation

Recall the two types of variation studied so far: direct variation (see Lesson 3.8), in which the *quotient* of y and x is constant ($\frac{y}{x} = k$, $k \neq 0$); and inverse variation, in which the *product* of y and x is constant ($yx = k$, $k \neq 0$).

Direct variation can be extended to several variables. For example, if y varies directly as v, w, and x, then the equation of variation is $\frac{y}{vwx} = k$.

The statement "y varies directly as v, w, and x" can also be written as "y varies *jointly* as v, w, and x."

EXAMPLE 1

A number y varies jointly as x and z. If $y = 2$ when $x = 4$ and $z = 6$, what is y when $x = 3$ and $z = 8$?

Plan

Write "y varies jointly as x and z" as an equation with subscripted variables.

Solution

$$\frac{y_1}{x_1 z_1} = \frac{y_2}{x_2 z_2} \quad \leftarrow \quad \frac{y_1}{x_1 z_1} = k = \frac{y_2}{x_2 z_2}$$

$$\frac{2}{4 \cdot 6} = \frac{y_2}{3 \cdot 8}$$

$$\frac{1}{12} = \frac{y_2}{24}$$

$$12 y_2 = 24$$

$$y_2 = 2$$

You can use a calculator to find $y_2 = \dfrac{2 \cdot 3 \cdot 8}{4 \cdot 6}$ directly.
2 ⊗ 3 ⊗ 8 ⊘ 4 ⊘ 6 ⊜ 2.

Thus, $y = 2$ when $x = 3$ and $z = 8$.

EXAMPLE 2

The safe load for a beam varies directly as the width of the beam and the square of its depth. A particular beam 1.6 cm wide and 3.5 cm deep can safely support 6,300 lb. Find the safe load for a similar beam 1.4 cm wide and 4.0 cm deep.

Solution

Let s = the safe load, w = the width, and d = the depth. So, s varies jointly as w and the *square* of d.

$$\frac{s_1}{w_1 d_1^2} = \frac{s_2}{w_2 d_2^2} \quad \leftarrow \frac{s}{wd^2} \text{ is constant.}$$

$$\frac{6{,}300}{(1.6)(3.5)^2} = \frac{s_2}{(1.4)(4.0)^2}$$

$$\frac{6{,}300}{19.6} = \frac{s_2}{22.4} \quad \leftarrow \text{Use a calculator to find } s_2 = \frac{6{,}300(1.4)(4.0)^2}{(1.6)(3.5)^2}.$$

$$6{,}300 \;\boxed{\times}\; 1.4 \;\boxed{\times}\; 4.0 \;\boxed{x^2}\; \boxed{\div}\; 1.6 \;\boxed{\div}\; 3.5 \;\boxed{x^2}\; \boxed{=}$$

$$19.6 s_2 = 141{,}120 \qquad\qquad\qquad\qquad\qquad\qquad\qquad 7{,}200$$

$$s_2 = 7{,}200$$

Thus, the safe load for a similar beam 1.4 cm wide and 4.0 cm deep is 7,200 lb.

Both direct and inverse variation may occur at the same time. Such instances are examples of **combined variation**. It is helpful to keep in mind that the *quotient* is constant for direct variation, and the *product* is constant for inverse variation.

EXAMPLE 3

The force of gravitational attraction, F, between two objects varies directly as the product of their masses m and M and inversely as the square of the distance r between them. The force F_1 is 134 dynes when $m_1 = 600$ kg, $M_1 = 750$ kg, and $r_1 = 15$ cm. Find the value of F_2 when $m_2 = 1{,}080$ kg, $M_2 = 6{,}700$ kg, and $r_2 = 201$ cm.

Plan

First, write an equation that defines the variation.
F varies directly as $m \cdot M$ and inversely as r^2.

This means that $\dfrac{Fr^2}{m \cdot M}$ is constant.

Solution

$$\frac{F_1 r_1^2}{m_1 M_1} = \frac{F_2 r_2^2}{m_2 M_2}$$

$$\frac{134 \cdot 15^2}{600 \cdot 750} = \frac{F_2 \cdot 201^2}{1{,}080 \cdot 6{,}700}$$

Simplify the fractions. $\quad \dfrac{67}{1{,}000} = \dfrac{67 F_2}{12{,}000} \qquad$ Solve for F_2. $\quad F_2 = 12$

Thus, the value of F_2 is 12 dynes.

Classroom Exercises

Write an equation to describe each variation.

1. a varies jointly as b and c.
2. t varies directly as m and g.
3. y varies directly as x and inversely as z.
4. p varies directly as q and the square of r.
5. y varies directly as x and the square root of z, and inversely as m.
6. y varies jointly as the square root of x and the cube of r.
7. A number y varies directly as x and z. If $y = 6$ when $x = 4$ and $z = 5$, what is x when $y = 21$ and $z = 7$?

Written Exercises

1. A number y varies directly as x and z. If $y = 6$ when $x = 8$ and $z = 3$, what is y when $x = 12$ and $z = 4$?

2. A number y varies jointly as x and z. If $y = 5$ when $x = 3$ and $z = 10$, what is y when $x = 15$ and $z = 4$?

3. A number a varies jointly as b and the square of c. If $a = 8$ when $b = 9$ and $c = 2$, what is c when $a = 4$ and $b = 2$?

4. A number y varies directly as x and the square root of z, and inversely as t. If $y = 4$ when $x = 2$, $z = 9$, and $t = 5$, what is t when $y = 5$, $x = 12$, and $z = 4$?

5. A number z varies directly as m and n, and inversely as the square of p. If $z = 30$ when $m = 5$, $n = 4$, and $p = 2$, what is p when $z = 15$, $m = 5$, and $n = 32$?

6. A number y varies directly as the cube root of x and inversely as the square of z. If $y = 3$ when $x = 8$ and $z = 4$, what is y when $x = 27$ and $z = 6$?

7. The volume V of a given mass of gas varies directly as the absolute temperature T and inversely as the pressure P. If $V = 462$ cm^3 when $T = 42°$ and $P = 40$ kg/cm^2, what is the volume when $T = 30°$ and $P = 30$ kg/cm^2?

8. The electrical resistance in a wire varies directly as its length and inversely as the area of its cross section. A silver wire 100 m long and 0.0125 cm in cross-sectional radius has a resistance of 30 ohms. Find the length of silver wire of radius 0.075 cm needed to achieve a resistance of 20 ohms. (Assume the wires are circular in cross section, and use $A = \pi r^2$.)

9. If y varies directly as x and the cube of z, what is the effect on y of doubling z?

10. If a varies jointly as b and the square root of c, what is the effect on a of multiplying c by 4?

11. If y varies directly as x and inversely as the square of z, what is the effect on y of doubling x and doubling z?

12. If t varies directly as s and inversely as the square of r, what is the effect on t when the value of r is increased by 50%?

13. Under certain conditions of artificial light, the exposure time to photograph an object is directly proportional to the square of its distance from the source of light and inversely proportional to the candlepower of the light. If the exposure time is 0.01 s when the light is 6.00 ft from the object, how far from the object should the light be placed when both the candlepower and the exposure time are doubled?

14. The lifting force exerted by the atmosphere on the wings of an airplane in flight varies directly as the surface area of the wings and the square of the plane's airspeed. A small private plane has a cruising airspeed of 250 mi/h. In order to obtain three times the lifting force, a new plane is designed with a wing surface area twice that of the older model. What cruising speed (to the nearest mile per hour) is planned for the new model?

15. As a consequence of Newton's Law of Universal Gravitation, the magnitude of the attraction between two spheres varies jointly as the product of their masses and inversely as the square of the distance between their centers. Write this relationship as a formula and use it to find the point between the centers of the earth and the moon at which a space vehicle would experience equal attraction from both the earth and the moon. Use the following data: earth's mass: 5.98×10^{24} kg; moon's mass: 7.35×10^{22} kg; mean distance between the centers of the earth and the moon: 4.60×10^{8} m.

16. The maximum safe load (in pounds) for a horizontal beam varies jointly as its width and the square of its depth, and inversely as the distance between its supports. For the construction of a certain house, a beam is positioned between two supports 16 ft apart. The width and depth of the beam are 2.0 in. and 6.0 in., respectively. The beam can safely support 1,200 lbs. A second beam of similar material has a width of 4.0 in. and a depth of 8.0 in. The supports for this beam are 18 ft apart. Find the safe load for the second beam to the nearest hundred pounds.

Mixed Review

Simplify. *7.1, 7.4, 8.4, 9.2*

1. $\dfrac{x^3 - 8}{x^2 - 4x + 4}$

2. $\dfrac{5}{x^2 - 7x + 12} - \dfrac{2x - 1}{x - 4}$

3. $\dfrac{28}{4 - \sqrt{2}}$

4. $\dfrac{13}{3 - 2i}$

5. Find all the roots of $x^4 + 3x^3 - 3x^2 + 3x - 4 = 0$. *10.4*

12.7 The Parabola: Focus and Directrix

Objectives

To write equations of parabolas given certain conditions

To write an equation for the directrix and to find the coordinates of the focus of a parabola given its equation

To graph parabolas

In the graph at the right, the distance from $P_1(4,4)$ to $F(0,1)$ is
$\sqrt{(4-0)^2 + (4-1)^2} = \sqrt{16 + 9} = \sqrt{25} = 5.$

The distance from P_1 to H on the line d is also 5. Thus, P_1 is equidistant from the point F and the line d.

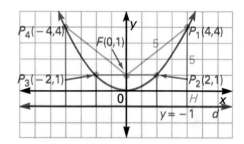

The graph also shows several other points that are equidistant from the point F and the line d. These points form a *parabola*.

Definitions

A **parabola** is the set of all points in a coordinate plane equidistant from a given point, the **focus**, and a given line, the **directrix**. The **vertex** of the parabola is the midpoint of the perpendicular segment from the focus to the directrix.

Theorem 12.7

If a vertical parabola has focus $(0,p)$, directrix $y = -p$, and its vertex at the origin, then its equation in standard form is $x^2 = 4py$.

There are two alternate forms of the equation of a parabola with its vertex at the origin and the y-axis as its axis of symmetry.

$$x^2 = 4py, \text{ or } y = \frac{1}{4p}x^2, \text{ and } y = ax^2 \text{ (from Lesson 11.4)}$$

For both forms, a *positive* coefficient indicates that the parabola opens *upward*, and a *negative* coefficient indicates that the parabola opens *downward*. For example, the parabola

$y = \frac{1}{4}x^2$ opens *upward* because $\frac{1}{4} > 0$.

Or, $x^2 = 4y$ *opens upward* because the coefficient of y is *positive*.

$y = -5x^2$ opens *downward* because $-5 < 0$.

Or, $x^2 = -\frac{1}{5}y$ opens *downward* because the coefficient of y is *negative*.

The figures below show that p represents the *directed* (signed) distance from the vertex to the focus.

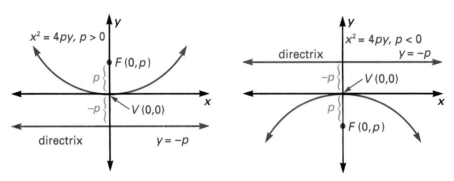

These relationships can also be extended to parabolas whose vertices are not at the origin.

The standard form of the equation of a vertical parabola with vertex (h,k) (see Lesson 11.4), $y - k = a(x - h)^2$, can be written as $y - k = \frac{1}{4p}(x - h)^2$, where $a = \frac{1}{4p}$. Then $(x - h)^2 = 4p(y - k)$, where p is the directed distance from the vertex to the focus.

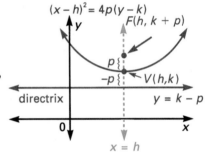

EXAMPLE 1   Write an equation of the parabola with focus $F(3, -6)$ and directrix $y = 4$. Graph the parabola.

Plan   Find the vertex $V(h,k)$. Then use points F and V to find p, the directed distance from V to F. Finally, substitute in $(x - h)^2 = 4p(y - k)$.

Solution   (1) Plot the focus $F(3, -6)$ and draw the directrix $y = 4$.

(2) Draw a line through F perpendicular to the directrix at D. This is the axis of symmetry. Points D and F have the same x-coordinate, 3.

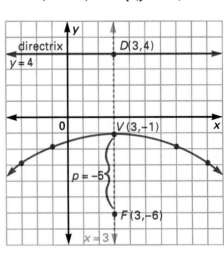

(3) Find V, the midpoint of $\overline{FD}$.
$$V = \left(\frac{3 + 3}{2}, \frac{-6 + 4}{2} \right) = (3, -1)$$

(4) Find p, the directed distance from V to F.
$$p = -|-1 - (-6)| = -5$$

(5) Write the equation using $p = -5$ and $V(3, -1)$.

$$(x - h)^2 = 4p(y - k)$$
$$(x - 3)^2 = 4(-5)[y - (-1)]$$
$$(x - 3)^2 = -20(y + 1)$$

Thus, an equation of the parabola is $(x - 3)^2 = -20(y + 1)$.

Plot the vertex. Also, find a few other points on the parabola by solving for y and making a table. Draw the graph as shown.

Each figure below shows a horizontal parabola with vertex $V(0,0)$, focus $(p,0)$, and directrix $x = -p$.

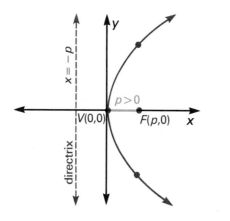

If $p > 0$, the parabola opens to the *right*.

If $p < 0$, the parabola opens to the *left*.

The general equation for a horizontal parabola with its vertex at the origin is given in Theorem 12.8.

Theorem 12.8

If a horizontal parabola has focus $(p,0)$, directrix $x = -p$, and its vertex at the origin, then its equation is $y^2 = 4px$.

The standard form of the equation of a horizontal parabola with vertex (h, k), $x - h = a(y - k)^2$, can be written as $(y - k)^2 = 4p(x - h)$, where p is the directed distance from the vertex to the focus.

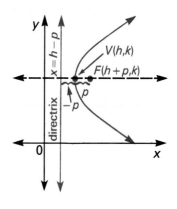

EXAMPLE 2 For the parabola $y^2 - 2y - 2x + 5 = 0$, find the coordinates of the vertex and the equation of the axis of symmetry. Then find the coordinates of the focus and the equation of the directrix. Graph the parabola.

Plan The graph is a horizontal parabola because there is a y^2-term.

Solution

$$y^2 - 2y - 2x + 5 = 0$$

Complete the square.

$$y^2 - 2y + \underline{\quad} = 2x - 5 + \underline{\quad}$$

$$y^2 - 2y + 1 = 2x - 5 + 1$$

$$(y - 1)^2 = 2x - 4$$

Factor out 2 and then rewrite it as $4(\frac{1}{2})$ to obtain standard form.

$$(y - 1)^2 = 2(x - 2)$$

$$(y - 1)^2 = 4 \cdot \frac{1}{2} \cdot (x - 2)$$

Vertex: $(2,1)$; $p = \frac{1}{2}$

The axis of symmetry is *horizontal*.

Axis of symmetry: $y = 1$

Use $p = \frac{1}{2}$ to find the focus.

Focus: $(2 + p, 1)$, or $(2\frac{1}{2}, 1)$

The directrix is *vertical*.

Directrix: $x = 2 - p$, or $x = 1\frac{1}{2}$

Plot the vertex and draw the axis of symmetry. Because $p > 0$, the parabola opens to the right.

Find several other points on the parabola by solving the equation for y and making a table of values.

Draw the horizontal parabola.

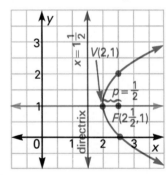

![] **Focus on Reading**

Complete each of the following to make a true statement.

1. The graph of $y^2 = 4px$ is a _____ parabola with its vertex at $V(\underline{\quad})$.

2. The graph of $x^2 = 4py$ is a _____ parabola opening _____ when $p > 0$.

3. The graph of $(x - h)^2 = 4p(y - k)$ is a _____ parabola with its vertex at $V(\underline{\quad})$.

Classroom Exercises

Determine whether each parabola is vertical or horizontal.

1. $x^2 = 4y$ **2.** $y^2 = -8x$ **3.** $x^2 = -20y$

4–6. For each parabola in Classroom Exercises 1–3, find the coordinates of the vertex and the focus, and the equations of the directrix and the axis of symmetry. Graph the parabola.

Written Exercises

Use the information provided to write an equation of each parabola.

1. Focus: $(-8,0)$
Directrix: $x = 8$

2. Focus: $(3,1)$
Directrix: $y = 10$

3. Focus: $(3,8)$
Directrix: $x = 6$

4. Focus: $(5,-1)$
Directrix: $x = 6$

5. Vertex: $(-2,3)$
Directrix: $y = -3$

6. Vertex: $(-4,3)$
Directrix: $x = 5$

7. Focus: $(3,8)$
Vertex: $(3,2)$

8. Focus: $(-8,-5)$
Vertex: $(2,-5)$

For each parabola, find the coordinates of the vertex and the focus, and the equations of the directrix and the axis of symmetry. Graph the parabola.

9. $(x - 7)^2 = 8(y - 2)$ **10.** $(x - 4)^2 = -12(y + 5)$ **11.** $(x - 6)^2 = 10(y + 1)$

12. $(x + 3)^2 = -2(y - 4)$ **13.** $(y - 6)^2 = -8(x - 5)$ **14.** $(y - 1)^2 = 12(x + 7)$

15. $x^2 - 6x - 8y + 25 = 0$ **16.** $x^2 + 10x + 12y + 31 = 0$

17. $x^2 - 2x - 2y - 9 = 0$ **18.** $y^2 - 8y - 4x + 28 = 0$

19. $y^2 - 10y + 2x + 27 = 0$ **20.** $x^2 - 4x + 8y + 12 = 0$

21. A point P moves in a coordinate plane in such a way that it is always equidistant from the line $x = -2$ and the point $Q(2,0)$. Use the distance formula, $d = \sqrt{(x_2 - x_1)^2 + (y_2 - y_1)^2}$, to write an equation describing the path of point P.

22. Prove Theorem 12.8.

23. Write an equation for the set of all points equidistant from the point $F(h, p + k)$ and the line $y = k - p$. Verify that this equation is equivalent to the standard form of the equation of a vertical parabola.

Mixed Review

Graph each equation.   *3.5, 11.4, 12.1, 12.3*

1. $3x - 4y = 8$ **2.** $x^2 + y^2 = 25$ **3.** $4x^2 - 9y^2 = 36$ **4.** $y = -x^2$

12.8 Identifying Conic Sections

Objective

To identify conic sections

The graphs of circles, parabolas, ellipses, and hyperbolas can be represented by passing a plane through a hollow double cone. Therefore, each of these curves is called a **conic section**.

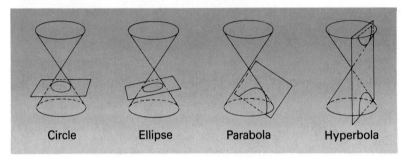

Circle   Ellipse   Parabola   Hyperbola

The general equation of a conic section can be written in the form
$$Ax^2 + Bxy + Cy^2 + Dx + Ey + F = 0,$$
where A, B, C, D, E, and F are not all zero.

To identify conic sections described by equations of the form above, change the equations to one of the standard forms shown below.

Conic Section	Center at (0,0)	Center at (h,k)	
Circle	$x^2 + y^2 = r^2$	$(x - h)^2 + (y - k)^2 = r^2$	$r > 0$
Ellipse	$\dfrac{x^2}{a^2} + \dfrac{y^2}{b^2} = 1$	$\dfrac{(x - h)^2}{a^2} + \dfrac{(y - k)^2}{b^2} = 1$	$0 < b < a$
	$\dfrac{y^2}{a^2} + \dfrac{x^2}{b^2} = 1$	$\dfrac{(y - k)^2}{a^2} + \dfrac{(x - h)^2}{b^2} = 1$	
Hyperbola	$\dfrac{x^2}{a^2} - \dfrac{y^2}{b^2} = 1$	$\dfrac{(x - h)^2}{a^2} - \dfrac{(y - k)^2}{b^2} = 1$	$a > 0,$ $b > 0$
	$\dfrac{y^2}{a^2} - \dfrac{x^2}{b^2} = 1$	$\dfrac{(y - k)^2}{a^2} - \dfrac{(x - h)^2}{b^2} = 1$	
Rectangular Hyperbola	$xy = k$		$k \neq 0$
Parabola	$x^2 = 4py$ $y^2 = 4px$	$(x - h)^2 = 4p(y - k)$ $(y - k)^2 = 4p(x - h)$	$p \neq 0$

EXAMPLE 1 Identify each conic section whose equation is given.

  a. $4x^2 + 4y^2 = 36$ **b.** $4x^2 + 9y^2 = 36$ **c.** $4x^2 - 9y^2 = 36$
  d. $-3xy = 24$ **e.** $12 + 2x^2 = y + 7x$

Plan If there is only one squared term, and no other second-degree variable, then the conic section is a parabola. Otherwise, change the equation to a standard form.

Solutions

Equation	Standard form	Conic section
a. $4x^2 + 4y^2 = 36$	$x^2 + y^2 = 3^2$	circle
b. $4x^2 + 9y^2 = 36$	$\dfrac{x^2}{3^2} + \dfrac{y^2}{2^2} = 1$	ellipse
c. $4x^2 - 9y^2 = 36$	$\dfrac{x^2}{3^2} - \dfrac{y^2}{2^2} = 1$	hyperbola
d. $-3xy = 24$	$xy = -8$	rectangular hyperbola
e. $12 + 2x^2 = y + 7x$		parabola

EXAMPLE 2 Identify the conic section defined by
$4x^2 + 9y^2 + 32x - 90y + 253 = 0$.

Plan Complete both squares. Then write in a standard form.

Solution
$$4x^2 + 32x + \underline{\quad} + 9y^2 - 90y + \underline{\quad} = -253 + \underline{\quad} + \underline{\quad}$$
$$4(x^2 + 8x + \underline{\quad}) + 9(y^2 - 10y + \underline{\quad}) = -253 + \underline{\quad} + \underline{\quad}$$
$$4(x^2 + 8x + 16) + 9(y^2 - 10y + 25) = -253 + 4 \cdot 16 + 9 \cdot 25$$
$$4(x + 4)^2 + 9(y - 5)^2 = 36$$
$$\frac{(x + 4)^2}{9} + \frac{(y - 5)^2}{4} = 1 \quad \leftarrow \text{standard form of an ellipse}$$

Thus, the conic section is an ellipse.

Classroom Exercises

Identify the conic section whose equation is given.

  1. $\dfrac{x^2}{25} - \dfrac{y^2}{9} = 1$ **2.** $\dfrac{x^2}{25} + \dfrac{y^2}{9} = 1$ **3.** $\dfrac{x^2}{9} + \dfrac{y^2}{9} = 1$ **4.** $y = x^2$

Written Exercises

Identify each conic section whose equation is given.

1. $4x^2 - 8y^2 = 32$ **2.** $x^2 + 4 = y$ **3.** $5x^2 - y = 20$ **4.** $5x^2 + 5y^2 = 20$

5. $6x^2 + 4y^2 = 24$ **6.** $7xy = -28$ **7.** $4x^2 - y^2 = 4$ **8.** $8x^2 + 5y^2 = 40$

9. $\dfrac{(x-4)^2}{16} + \dfrac{(y+3)^2}{4} = 1$ **10.** $\dfrac{(x+8)^2}{25} - \dfrac{(y-1)^2}{4} = 1$

11. $\dfrac{(x+3)^2}{4} + (y-5)^2 = 1$ **12.** $4(x-2)^2 + 4(y+6)^2 = 100$

13. $6x^2 + 6y^2 + 36y + 18x - 2 = 0$ **14.** $25x^2 - 9y^2 + 200x - 175 = 0$

15. $y^2 + x^2 - 8y + 2x - 8 = 0$ **16.** $16x^2 - 9y^2 - 72y - 288 = 0$

17. $x^2 - 12x - y + 30 = 0$ **18.** $x^2 + y^2 - 4x + 6y - 1 = 0$

19. $(3x - 2y)(3x + 2y) = 36$ **20.** $(x-6)(x+8) = (y-2)^2$

21. Write, in your own words, why an equation of the form $Ax^2 + By^2 = C$, where A, B, and C are real numbers, cannot describe a parabola.

22. For what values of k will $(k^2 - 2k)x^2 - 4x + 8y^2 + 2y + 3 = 0$ define a circle?

23. For what values of k will $x^2 + (k^2 - 7k + 12)y^2 - 5x - y + 6 = 0$ define a parabola?

24. The graph of $4x^2 - 9y^2 = F$ is a hyperbola if $F = 36$. What is the graph if $F = 0$?

Mixed Review

For Exercises 1–4, use $f(x) = x^2 - 5x + 4$ and $g(x) = 2x - 6$. *3.2, 5.9*

1. Find $f(3)$. **2.** Find $f(-3) + g(-3)$. **3.** Give the domain of f. **4.** Simplify $f(g(x))$.

Application: LORAN

LORAN is a long-range navigation system that uses a pair of radio transmitters placed several hundred miles apart. LORAN receivers measure the difference in microseconds between the reception of the two signals. Each pair of signals is synchronized so that, knowing the speed of the signal, a navigator can calculate the difference between the distances from the ship to the transmitters.

The locus of all possible ship locations is what conic section? What are its foci?

12.9 Linear-Quadratic Systems

Objectives

To solve systems consisting of a linear and a quadratic equation

To solve word problems involving linear-quadratic systems

The graphs of the line $y = 2x + 1$ and the parabola $y = x^2 - 2$ are shown at the right. They have two points in common, $A(-1, -1)$ and $B(3, 7)$. The coordinates of these points satisfy both equations, which can be shown by substituting these values for x and y in each equation.

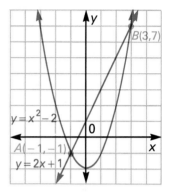

Thus, $(-1, -1)$ and $(3, 7)$ are the solutions of the linear-quadratic system consisting of $y = 2x + 1$ and $y = x^2 - 2$. As illustrated below, a system of one quadratic and one linear equation may have 0, 1, or 2 real solutions.

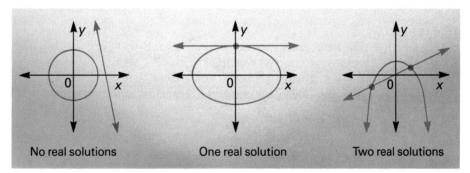

No real solutions   One real solution   Two real solutions

EXAMPLE 1 Solve this system graphically: $\dfrac{x^2}{4} + \dfrac{y^2}{25} = 1$
$$5x + 2y = 10$$

Solution Ellipse: $\dfrac{x^2}{4} + \dfrac{y^2}{25} = 1$

x-intercepts; ± 2; y-intercepts: ± 5

Line: $5x + 2y = 10$

x-intercept: 2; y-intercept: 5

The solutions are $(2, 0)$ and $(0, 5)$ because the graphs intersect as shown at the right.

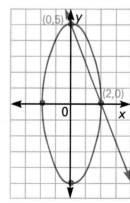

Algebraic methods are usually more accurate than graphical methods for solving linear-quadratic systems. An algebraic solution by substitution is illustrated in Example 2.

EXAMPLE 2 Solve $\begin{array}{l} 2x - 3y = 3 \\ x^2 - 2y^2 = 7 \end{array}$ algebraically.

Solution

Solve the linear equation for one of its variables.

$$2x - 3y = 3$$

$$x = \frac{3y + 3}{2}$$

Substitute this expression for x in the quadratic equation.

$$x^2 - 2y^2 = 7$$

$$\left(\frac{3y + 3}{2}\right)^2 - 2y^2 = 7$$

$$\frac{9y^2 + 18y + 9}{4} - 2y^2 = 7$$

Multiply each side by 4.

$$9y^2 + 18y + 9 - 8y^2 = 28$$

Solve the quadratic equation.

$$y^2 + 18y - 19 = 0$$
$$(y + 19)(y - 1) = 0$$
$$y = -19 \qquad or \qquad y = 1$$

Find the corresponding values of x by substituting for y in the linear equation already solved for x.

$$x = \frac{3y + 3}{2} \qquad\qquad x = \frac{3y + 3}{2}$$

$$x = \frac{3(-19) + 3}{2} \qquad x = \frac{3(1) + 3}{2}$$

$$x = -27 \qquad\qquad x = 3$$

Check by substituting the ordered pairs $(-27, -19)$ and $(3,1)$ in both equations. This is left for you.

Thus, $(-27, -19)$ and $(3,1)$ are the solutions of the system.

EXAMPLE 3 Find the length of a side of each of two squares given that the sum of their perimeters is 48 cm and the sum of their areas is 74 cm^2.

Plan

Let x = the length of a side of the larger square and y = the length of a side of the smaller square. Then the respective perimeters are $4x$ and $4y$, and the respective areas are x^2 and y^2.

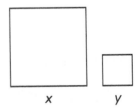

Solution   Write a system and solve by substitution.

The sum of the perimeters is 48 cm. (1) $4x + 4y = 48$

The sum of the areas is 74 cm^2. (2) $x^2 + y^2 = 74$

Solve (1) for x.

$$4x = 48 - 4y$$
$$x = 12 - y$$

Substitute for x in (2).

$$(12 - y)^2 + y^2 = 74$$
$$144 - 24y + y^2 + y^2 = 74$$
$$2y^2 - 24y + 70 = 0$$
$$y^2 - 12y + 35 = 0$$
$$(y - 5)(y - 7) = 0$$
$$y = 5 \text{ or } y = 7$$

Use $x = 12 - y$ to find x. If $y = 5$, then $x = 12 - 5 = 7$.

If $y = 7$, then $x = 12 - 7 = 5$.

Because $x > y$, discard $x = 5$, $y = 7$.

Check $x = 7$ and $y = 5$. Sum of perimeters $= 28 + 20 = 48$

Sum of areas $= 49 + 25 = 74$

Thus, 7 cm and 5 cm are the lengths of the sides of the two squares.

Classroom Exercises

Tell what substitution might be made as a first step in the solution of each system.

1. $x^2 + 4y^2 = 16$ $\quad$ **2.** $x^2 + y^2 = 13$ $\quad$ **3.** $4x^2 + 9y^2 = 36$ $\quad$ **4.** $x^2 + 4y^2 = 25$
 $x = 2y - 4$ $y = 2x - 1$ $x = y - 3$ $x + 2y = 1$

5–8. Solve the systems of Classroom Exercises 1–4 algebraically.

Written Exercises

Solve each system of equations graphically.

1. $x^2 + y^2 = 4$ $\quad$ **2.** $y = x^2 - 5$ $\quad$ **3.** $xy = 9$ $\quad$ **4.** $\dfrac{x^2}{16} + \dfrac{y^2}{9} = 1$
 $x + y = 2$ $2x - y = -3$ $y = x$ $3x - 4y = 12$

Solve each system of equations algebraically.

5. $y = x^2 + 3x - 1$ **6.** $x^2 + y^2 = 25$ **7.** $x + y = 4$ $\quad$ **8.** $2a + b = 7$
 $4x - y = -1$ $2y = x + 5$ $y = x^2 + 4x - 20$ $a^2 - b^2 = 8$

9. $4x^2 + 9y^2 = 36$
$3y = 12 - 4x$

10. $c^2 + d^2 = 25$
$3c - 4d = 25$

11. $y^2 - x^2 = 16$
$3x + 5y = 16$

12. $x^2 + y^2 = 50$
$9x + 7y = 70$

13. $y = x^2 + 4$
$y = x + 1$

14. $x^2 - y^2 = 1$
$x - y = 3$

15. $x^2 + 4y^2 = 4$
$2x - 2y = -6$

16. $l = w + 6$
$l = -w^2 + 6w$

Use a linear-quadratic system in two variables to solve each problem.

17. Find the length of a side of each of two squares given that the sum of their perimeters is 44 ft and the sum of their areas is 73 ft^2.

18. The perimeter of a rectangular lot is 88 m, and its area is 480 m^2. Find the width and the length.

19. If the numerator of a fraction $\frac{n}{d}$ is increased by 3, and the denominator is decreased by 3, the resulting fraction is the reciprocal of the original fraction. The numerator of the original fraction is 1 more than one-half its denominator. What is the original fraction?

20. A positive two-digit number is represented by $10t + u$. The sum of the squares of the digits is 26. If the number is decreased by the number with its digits reversed, the result is 36. Find the two-digit number.

Solve each system of equations. Give irrational solutions in simplest radical form.

21. $2x + 3y = 7$
$\dfrac{x^2}{2y} + 2y = \dfrac{5}{y} + 1$

22. $x - 2y = 2$
$\dfrac{y^2}{x + 2} = x - 2$

23. $\dfrac{1}{x} - \dfrac{1}{y} = 1$
$3x + 2y = 2$

24. For what values of m will the line $y = mx + 8$ and the circle $x^2 + y^2 = 25$ have exactly one point in common?

Mixed Review

Simplify.   *7.1, 8.3, 8.4, 8.6, 9.2*

1. $\dfrac{x^3 - 27}{x - 3}$

2. $\sqrt{32} - 6\sqrt{8}$

3. $\dfrac{26}{4 - \sqrt{3}}$

4. $(-27)^{-\frac{2}{3}}$

5. $\dfrac{-4}{1 - i}$

Brainteaser

A mule and a donkey were carrying bundles of wheat to a market. The mule said, "If you give me one bundle of wheat, then I will have twice as many as you. However, if I give you one bundle of wheat, then we will each be carrying the same number." How many bundles of wheat was each carrying?

12.10 Quadratic-Quadratic Systems

Objectives

To solve systems of two quadratic equations
To solve word problems involving systems of two quadratic equations
To graph quadratic inequalities

A system of two quadratic equations may have 0, 1, 2, 3, or 4 real solutions, as illustrated by the systems of a parabola and a circle shown below.

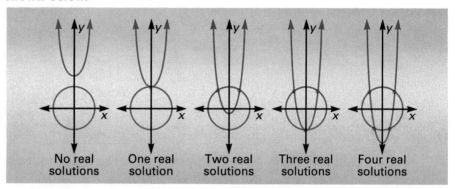

| No real solutions | One real solution | Two real solutions | Three real solutions | Four real solutions |

The graphical solution of a system of quadratic equations usually yields only approximate results, as illustrated in Example 1.

EXAMPLE 1

Solve the system $x^2 + y^2 = 25$ graphically.
$\quad\quad\quad\quad\quad\quad\quad\quad\quad\quad y = x^2 + 1$

Estimate the solutions to the nearest half unit.

Solution

$x^2 + y^2 = 25$: circle of radius 5 with its center at $(0,0)$

Write $y = x^2 + 1$ in the standard form of a parabola.

$(x - 0)^2 = (y - 1)$: parabola opening upward with its vertex at $(0,1)$

The graphs intersect at two points whose approximate coordinates are $(2,4.5)$ and $(-2,4.5)$.

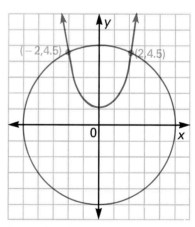

Checks

Check by substituting $(2,4.5)$ and $(-2,4.5)$ in both equations.

The approximate solutions are $(2,4.5)$ and $(-2,4.5)$, or $(\pm 2,4.5)$.

Systems of two quadratic equations can also be solved by the substitution and linear-combination methods. The system in the next example is solved algebraically by the linear-combination method.

EXAMPLE 2   Solve $\dfrac{x^2}{4} + \dfrac{y^2}{10} = 5$ algebraically.
$2x^2 + 3y^2 = 84$

Give irrational solutions in simplest radical form, and then find each to the nearest tenth using a square root table or a calculator.

Solution

Multiply each side of the first equation by the LCD, 20, to eliminate the fractions.

$$\frac{x^2}{4} + \frac{y^2}{10} = 5$$

$$5x^2 + 2y^2 = 100$$

Solve the system.

(1) $5x^2 + 2y^2 = 100$
(2) $2x^2 + 3y^2 = 84$

Multiply each side of (1) by 3.
$15x^2 + 6y^2 = 300$

Multiply each side of (2) by -2.
$\underline{-4x^2 - 6y^2 = -168}$

Add the resulting equations.
$11x^2 \qquad\quad = 132$

$$x^2 = 12$$

$$x = \pm\sqrt{12} = \pm 2\sqrt{3}$$

Find y by substituting $\pm 2\sqrt{3}$ for x in (1).

$$5x^2 + 2y^2 = 100$$
$$5(\pm 2\sqrt{3})^2 + 2y^2 = 100$$
$$5(12) + 2y^2 = 100$$
$$2y^2 = 40$$
$$y^2 = 20$$
$$y = \pm\sqrt{20} = \pm 2\sqrt{5}$$

The four solutions are $(\pm 2\sqrt{3}, \pm 2\sqrt{5})$.

$2\sqrt{3}$ to the nearest tenth is 3.5. $2\sqrt{5}$ to the nearest tenth is 4.5.

Thus, the solutions to the nearest tenth are $(\pm 3.5, \pm 4.5)$.

The check is left for you.

EXAMPLE 3   Find the dimensions of the rectangle whose area is 12 ft^2 and whose diagonal is 5 ft long.

Plan

What are you to find?   the length and width

Choose two variables.   Let $l =$ the length.
What do they represent?   Let $w =$ the width.

Solution	What is given?	The area is 12, so $lw = 12$. The diagonal is the hypotenuse of a right triangle, so $l^2 + w^2 = 25$.

Write a system.	(1) $lw = 12$ (2) $l^2 + w^2 = 25$
Solve the system.	Solve (1) for either variable. (3) $l = \frac{12}{w}$

Substitute (3) in (2).

(2) $\left(\frac{12}{w}\right)^2 + w^2 = 25$

Solve for w.

$$\frac{144}{w^2} + w^2 = 25$$

$$144 + w^4 = 25w^2$$

$$w^4 - 25w^2 + 144 = 0$$

$$(w^2 - 16)(w^2 - 9) = 0$$

$$w^2 - 16 = 0 \qquad or \qquad w^2 - 9 = 0$$

$$w^2 = 16 \qquad or \qquad w^2 = 9$$

$$w = \pm 4 \qquad or \qquad w = \pm 3$$

$$w = 4 \qquad or \qquad w = 3$$

Lengths are positive; discard the negative solutions.

Substitute 4 and 3 for w in (3).

$$l = \frac{12}{w} \qquad\qquad l = \frac{12}{w}$$

$$l = \frac{12}{4} = 3 \qquad l = \frac{12}{3} = 4$$

So, $l = 3$ and $w = 4$, or $l = 4$ and $w = 3$.

The two solutions describe the same rectangle, so choose $l = 4$ and $w = 3$.

Check in the original problem.	The area is 12: $3 \cdot 4 = 12$ True The diagonal's length is 5: $\sqrt{3^2 + 4^2} = 5$ True
State the answer.	The dimensions are 3 ft by 4 ft.

The graph of $x^2 + y^2 = 25$ at the right separates the plane into three sets of points: a circle, points outside the circle, and points inside the circle. When graphing such inequalities as $x^2 + y^2 < 25$ or $x^2 + y^2 > 25$, you can decide whether to shade the *interior* or the *exterior* by choosing sample points.

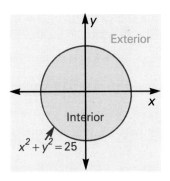

EXAMPLE 4   Graph $x^2 + y^2 > 25$.

Solution   First, graph the circle $x^2 + y^2 = 25$. Use a dashed curve because the inequality symbol $>$ indicates that the circle is not part of the graph. Test sample points.

Checks   Choose $(0,0)$ in the interior.

$$x^2 + y^2 > 25$$
$$0^2 + 0^2 \overset{?}{>} 25$$
$$0 + 0 \overset{?}{>} 25$$
$$0 > 25 \text{ False}$$

Choose $(7,6)$ in the exterior.

$$x^2 + y^2 > 25$$
$$7^2 + 6^2 \overset{?}{>} 25$$
$$49 + 36 \overset{?}{>} 25$$
$$85 > 25 \text{ True}$$

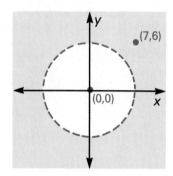

So, the graph consists of all points exterior to the circle.

Classroom Exercises

Tell what multipliers can be used to solve each system algebraically.

1. $x^2 + y^2 = 34$
   $2x^2 - 3y^2 = 23$

2. $x^2 - y = 5$
   $x^2 + y^2 = 25$

3. $2x^2 - y^2 = -1$
   $x^2 + 2y^2 = 22$

4. $5x^2 - y^2 = 1$
   $x^2 - y^2 = -3$

5. $2x^2 + y^2 = 8$
   $x^2 - 2y^2 = 4$

6. $3x^2 + 2y^2 = 13$
   $4x^2 + 5y^2 = 22$

7–9. Solve the systems of Classroom Exercises 1–3 algebraically.

10. Determine whether $\dfrac{x^2}{9} - \dfrac{y^2}{16} > 1$ describes the points in the interior or the exterior of the branches of the hyperbola $\dfrac{x^2}{9} - \dfrac{y^2}{16} = 1$. Is the point $(6,8)$ in the interior or the exterior?

Written Exercises

Solve each system of equations graphically.

1. $x^2 + y^2 = 9$
   $4x^2 + 9y^2 = 36$

2. $x^2 - 4y^2 = 1$
   $x^2 - 1 = y^2$

3. $x^2 + y^2 = 4$
   $4x^2 - 9y^2 = 36$

Solve each system algebraically. Give irrational solutions in simplest radical form.

4–6. Solve the systems of Classroom Exercises 4–6.

7. $x^2 - y^2 = 5$
$xy - 6 = 0$

8. $xy = 6$
$x^2 + y^2 = 12$

9. $9x^2 - 4y^2 = 32$
$xy = 2$

10–12. Algebraically solve the systems of Written Exercises 1–3.

Solve each system algebraically. Use a calculator to approximate irrational solutions to the nearest tenth.

13. $y^2 - x^2 = -10$
$\dfrac{y^2}{6} - \dfrac{x^2}{9} = 1$

14. $\dfrac{y^2}{6} - \dfrac{x^2}{3} = 1$
$xy = 6$

15. $\dfrac{x^2}{5} - \dfrac{4y^2}{5} = 1$
$y = \dfrac{3}{x}$

16. $\dfrac{y^2 - x^2}{5} = 1$
$xy + 6 = 0$

Write a system of two quadratic equations and solve the problem.

17. Find the dimensions of a rectangle with an area of 60 ft^2 and a diagonal of 13 ft.

18. The area of a rectangle is 48 m^2. The length of a diagonal is 10 m. Find the perimeter of the rectangle.

19. The area of a rectangle is 12 in^2. The perimeter is 24 in. Find the dimensions of the rectangle.

20. Find two negative numbers such that the sum of their squares is 170, and twice the square of the first minus three times the square of the second is 95.

Graph each inequality.

21. $x^2 + y^2 > 64$

22. $\dfrac{x^2}{100} - \dfrac{y^2}{25} \le 1$

23. $x^2 + y^2 \le 1$

24. $\dfrac{x^2}{36} + \dfrac{y^2}{4} > 1$

Solve each system algebraically. Answer in simplest radical form.

25. $x^2 - 3xy + 2y^2 = 0$
$x^2 - xy + y^2 = 3$

26. $x^2 + 3xy = 28$
$xy + 4y^2 = 8$

27. Find an equation of the ellipse centered at the origin, with its foci on the x-axis, containing the points $(\sqrt{2},\sqrt{3})$, $(\sqrt{6},1)$.

Mixed Review

Simplify. *6.5, 7.1, 8.6, 9.2*

1. $\dfrac{x^3 - 27}{x^2 - 9}$

2. $32^{-\frac{4}{5}}$

3. $\dfrac{30}{2 + i}$

4. $(x^3 + x^2 - 2x + 8) \div (x + 3)$

Key Terms

asymptote (p. 419)
center (p. 409)
circle (p. 409)
combined variation (p. 432)
constant of variation (p. 427)
directrix (p. 435)
ellipse (p. 413)
focal radius (pp. 413, 418)

focus (pp. 413, 418, 435)
hyperbola (p. 418)
inverse variation (p. 427)
joint variation (p. 431)
parabola (p. 435)
radius (p. 409)
rectangular hyperbola (p. 426)
vertex (pp. 414, 418, 435)

Key Ideas and Review Exercises

12.1 To identify the center (h,k) and radius r of a circle, write the equation in standard form: $(x - h)^2 + (y - k)^2 = r^2$, where $r > 0$.

Graph each circle whose equation is given.

1. $x^2 + y^2 = 4$

2. $x^2 - 2x + y^2 + 8y - 8 = 0$

Write the equation in standard form of the circle with given center C and given radius r, or with given center C and a given point P on the circle.

3. $C(0,0)$, $r = \sqrt{2}$

4. $C(-3,1)$, $P(3,9)$

12.2, To graph an origin-centered ellipse or hyperbola with foci on the x- or
12.3 y-axis, first write its equation in standard form.
For an ellipse, use Theorems 12.3 and 12.4 in Lesson 12.2.
For a hyperbola, use Theorems 12.5 and 12.6 in Lesson 12.3.

Graph each ellipse. Find the coordinates of its vertices and foci.

5. $4x^2 + y^2 = 4$

6. $36x^2 + 9y^2 = 144$

Given the following, write an equation in standard form of each ellipse.

7. Vertices: $(\pm 6,0)$, $(0,\pm 2)$

8. Vertices: $(0,\pm 7)$, Foci: $(0,\pm 4)$

Graph each hyperbola. Find the coordinates of its vertices and foci. Write the equations of its asymptotes.

9. $x^2 - 25y^2 = 25$

10. $100y^2 - 25x^2 = 2,500$

Given the following, write an equation in standard form of each hyperbola.

11. Vertices: $(0,\pm 8)$; foci: $(0,\pm 10)$

12. Vertices: $(\pm 4,0)$; foci: $(\pm 5,0)$

12.4   To graph an ellipse or hyperbola not centered at the origin, use the appropriate standard form (see Lesson 12.4).

Graph each ellipse or hyperbola. Label the vertices, center, and foci.

13. $\dfrac{(x + 7)^2}{36} + \dfrac{(y - 6)^2}{9} = 1$

14. $4x^2 - 24x - 9y^2 - 18y - 9 = 0$

12.5,   To graph a rectangular hyperbola, $xy = k$, $k \neq 0$, first solve for y.
12.6   To solve problems involving variation, use the following: (1) $xy = k$, when y varies inversely as x, (2) $\dfrac{y}{ab} = k$, when y varies directly (jointly) as a and b, or (3) $\dfrac{yd}{bc} = k$, when y varies directly (jointly) as b and c and inversely as d.

15. Graph $xy = -8$.

16. Graph $xy = 48$.

17. If y varies inversely as x, and $y = 12$ when $x = 5$, what is y when $x = 18$?

18. A number y varies jointly as x and z. If $y = -24$ when $x = 4$ and $z = 3$, what is y when $x = -6$ and $z = -2$?

19. A number y varies jointly as x and z and inversely as $\sqrt{w}$. If $y = 12$ when $x = 2$, $z = 6$, and $w = 9$, what is y when $x = 5$, $z = 7$, and $w = 25$?

12.7   To graph a parabola, or to write its equation, use Theorems 12.7 and 12.8 in Lesson 12.7.

Use the information provided to write an equation of each parabola.

20. focus: $F(6,8)$; directrix: $y = 12$

21. focus: $F(-2,4)$; vertex: $V(6,4)$

Graph each parabola. Label the vertex, focus, directrix, and axis of symmetry.

22. $x^2 - 6x + 8y - 7 = 0$

23. $y^2 - 8y + 4x - 8 = 0$

12.8   To identify a conic section, change its equation to a standard form.

Identify each conic section whose equation is given.

24. $4(x - 3)^2 + 9(y + 1)^2 = 36$

25. $4x^2 + 8x - 9y^2 + 36y - 68 = 0$

12.9,   Solve linear-quadratic and quadratic-quadratic systems by graphing, or
12.10   use one of the algebraic methods: substitution or linear-combination.

26. Solve by graphing. Estimate answers to the nearest half unit.
$2x - 3y = 6$
$9x^2 + 16y^2 = 144$

27. Solve algebraically. Give irrational answers to the nearest tenth.
$4x^2 + 2y^2 = 20$
$3x^2 - 4y^2 = 4$

Given the following data, write the equation of each conic section in standard form.

1. circle; center: $(-6,4)$
  radius: 7

2. ellipse; vertices: $(\pm 6,0)$, $(0,\pm 4)$

3. ellipse; vertices: $(\pm 3,0)$
  foci: $(0,\pm 4)$

4. hyperbola; vertices: $(0,\pm 5)$
  foci: $(0,\pm 8)$

5. parabola; focus: $(4,6)$
  directrix: $y = -6$

6. parabola; focus: $(8,2)$
  vertex: $(3,2)$

7. Graph the rectangular hyperbola $xy = -72$.

Graph each equation. For a circle, label the center and the radius. For a parabola, label the vertex, the focus, and the directrix. For an ellipse, label the vertices, the center, and the foci. For a hyperbola, label the vertices, the center, and the foci, and write the equations of the asymptotes.

8. $x^2 + y^2 = 49$

9. $16x^2 + y^2 = 16$

10. $\dfrac{(x-1)^2}{4} - \dfrac{(y+2)^2}{25} = 1$

11. $\dfrac{(x-4)^2}{4} + \dfrac{(y+1)^2}{9} = 1$

12. $y^2 - 4y - 6x - 14 = 0$

13. $4x^2 - y^2 - 4y - 8x - 4 = 0$

14. A number y varies directly as x and z and inversely as the square of r. If $y = 6$ when $x = 3$, $z = 4$, and $r = 7$, what is y when $x = 6$, $z = 8$, and $r = 4$?

15. A number y varies jointly as x and the cube of z. If $y = 160$ when $x = 4$ and $z = 2$, what is y when $x = -5$ and $z = 3$?

16. The natural frequency of a string under constant tension varies inversely as its length. If a string 40 cm long vibrates 680 times per second, what length must the string be to vibrate 850 times per second under the same tension?

17. Solve $\begin{aligned} x^2 + y^2 &= 25 \\ x^2 - y^2 &= 25 \end{aligned}$ graphically.

Solve each system algebraically. Give irrational solutions in simplest radical form.

18. $2x^2 + y^2 = 22$
  $x^2 + 3y^2 = 21$

19. $x^2 + y^2 = 15$
  $xy = 6$

20. Solve algebraically.
  $x^2 - 5xy + 6y^2 = 0$
  $x + y^2 - 7y = 0$

21. Find the square roots of $3 - 4i$. (HINT: Let $x + yi$ be a square root. Then $(x + yi)^2 = 3 - 4i$.)

Choose the *one* best answer to each question or problem.

Sample Question

*(x, y, z) is defined as follows:

$$*(x, y, z) = \frac{\frac{x}{y}}{z}$$

Then *$(6,10,12) = $ ___?___

(A) $\frac{1}{20}$ (B) $\frac{1}{5}$ (C) $\frac{1}{6}$

(D) 5   (E) $\frac{36}{5}$

The answer is A: $\frac{6}{10} \div 12 = \frac{1}{20}$.

1. Use the definition in the sample question.

*$(3,9,3) = $ ___?___

(A) *$(1,3,1)$ (B) *$(9,3,9)$
(C) *$(3,1,3)$ (D) *$(6,18,6)$
(E) *$(12,9,12)$

2. If $\star p$ means $p(p + 1)(p + 2)$, then $\frac{\star 8}{\star 2} = $ ___?___

(A) 4   (B) 10
(C) 20   (D) 30
(E) 36

3. $(a,b) * (c,d)$ is defined as $(ac + bd, ad + bc)$. If $(a,b) * (x,y) = (a,b)$, then $(x,y) = $ ___?___

(A) $(0,0)$ (B) $(1,0)$
(C) $(0,1)$ (D) $(1,1)$
(E) None of these

4. The lines with equations $x = 2$, $x = 5$, $y = 7$, and $y = 3$ form a rectangle. What is the area of this rectangle?

(A) 12   (B) 14   (C) 15
(D) 17   (E) 28

5. The following are the dimensions of five rectangular solids. All have the same volume *except* ___?___

(A) 10 by 15 by 2
(B) 6 by 50 by 3
(C) 4 by 75 by 1
(D) 8 by 75 by $\frac{1}{2}$
(E) $\frac{1}{4}$ by 25 by 48

6.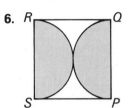

PQRS is a square, and the two shaded regions are semicircles. What is the ratio of the total area of the shaded regions to the area of the square?

(A) $\frac{\pi}{4}$ (B) $\frac{\pi}{2}$ (C) π

(D) $\frac{2\pi}{3}$ (E) $\frac{3\pi}{4}$

7. What is one-third of the perimeter of the square whose area is 36 square units?

(A) 4   (B) 6   (C) 8
(D) 12   (E) 48

8. Which of the following conditions guarantee that $a - b$ is a negative number?

(A) $b > 0$ (B) $a > 0$
(C) $b > a$ (D) $b = a$
(E) $a > b$

9. How many points do the graphs of $y = x^2$ and $xy = 27$ have in common?

(A) 0   (B) 1   (C) 2
(D) 3   (E) 4

1. What property of operations with real numbers is illustrated by $(3a + 2b) + c = 3a + (2b + c)$?   *1.3*

Solve for x.

2. $2 - 3[4 - 2(3 - x)] = -16$ *1.4*

3. $6x^3 - 27x^2 + 30x = 0$ *6.1*

4. $2x^4 - 5x^3 - 6x^2 + 20x = 8$ *6.6*

5. $M = \frac{P}{3}(x + y)$ *7.6*

6. $x^2 + 2x + 4 = 0$ *9.5*

7. $\sqrt{2x - 5} + 4 = x$ *10.3*

Find the solution set of each inequality. Graph the solutions.

8. $2x - 4 \le 4 - (2 - 3x)$ *2.1*

9. $|2n + 4| < 8$ *2.3*

10. $x^2 + x - 30 > 0$ *6.4*

For Exercises 11–13, use the relations described by $f(x) = 3x + 1$ and $g(x) = x^2 - 2$.

11. Determine the range of $f(x)$ for $D = \{-6, 0, 3, 9\}$. Is $f(x)$ a function?   *3.1*

12. Find $f(g(x))$.   *5.9*

13. Find $g(f(-2))$.

Use points $A(3,8)$ and $B(9,16)$ for Exercises 14–17.

14. Find the slope of $\overleftrightarrow{AB}$, and describe its slant.   *3.3*

15. Write an equation of $\overleftrightarrow{AB}$.   *3.4*

16. Find AB.   *11.1*

17. Write an equation of the perpendicular bisector of $\overline{AB}$.   *11.2*

Solve each system algebraically.

18. $2x + y = 8$
  $x = y - 4$ *4.2*

19. $3x - 2y = -2$
  $5x + 3y = 22$ *4.3*

20. $2x + y - 3z = -5$
  $3x - y + z = 4$
  $x + 3y - z = 4$ *4.6*

Solve by graphing.

21. $2x - y \le -4$ and $x \ge 3$ *4.7*

Simplify. Use positive exponents.

22. $(4a^{-3})^2$ *5.2*

23. $\dfrac{28x^{-4}y^3}{21x^{-2}y^{-1}}$ *5.5*

Factor completely.

24. $27y^3 - 125$ *5.6*

25. $xa - ay - 2x + 2y$ *5.7*

Divide.

26. $(3x^3 - 18x + 12) \div (x - 3)$ *5.8*

Simplify.

27. $\dfrac{x^2 - 5x}{3x^3} \div \dfrac{25 - x^2}{9x^2}$ *7.2*

28. $\dfrac{1 - 5x^{-1} - 14x^{-2}}{2 - 3x^{-1} - 14x^{-2}}$ *7.5*

29. $8x^2y\sqrt{28x^5y^4},\ x > 0,\ y > 0$ *8.2*

30. $\dfrac{6}{\sqrt{5} - \sqrt{3}}$ *8.4*

31. $\dfrac{26}{3 + 2i}$ *9.2*

32. Find the maximum value of P given that $P = -2x^2 + 8x + 4$.   *9.4*

33. Draw the vectors corresponding to $-5 + 3i$ and $3 + 2i$. Then draw their sum.   *9.8*

34. Determine the nature of the roots of $3x^2 + 5x - 2 = 0$ without solving the equation.   *10.2*

Graph each conic section. Label whatever applies: center, radius, vertices, foci, directrix, or asymptotes.

35. $x^2 + 6x + y^2 - 4y - 12 = 0$ *12.1*
36. $4x^2 - 9y^2 = 36$ *12.3*
37. $9x^2 + 25y^2 + 36x - 150y + 36 = 0$ *12.4*
38. $y = -\dfrac{24}{x}$ *12.5*
39. $y^2 + 8x - 6y + 1 = 0$ *12.7*

Write the equation in standard form of the given conic section. Graph the equation. (Exercises 40–41)

40. Circle; center: $(-5,4)$ radius: 6   *12.1*

41. Parabola; directrix: $y = -4$ focus: $(2,6)$ *12.7*

42. A number y varies jointly as r and s^2, and inversely as t. If $y = 120$ when $r = 5$, $s = 3$, and $t = 2$, what is r when $y = 80$, $s = 1$, and $t = 6$?   *12.6*

Solve each system. Give irrational answers to the nearest tenth. (Exercises 43–44)

43. $y = x^2 - 7x - 3$ $2x - y = 3$ *12.9*
44. $4x^2 + 5y^2 = 21$ $xy = 1$ *12.10*

45. A rectangle is 3 cm longer than it is wide. Find the set of all possible widths given that the perimeter is less than 34 cm and greater than 22 cm.   *2.4*

46. A passenger train leaves a station 2 h after a freight train, and, traveling at an average speed of 60 mi/h, overtakes it in 7 h. What is the average speed of the freight train?   *2.6*

47. How many ounces of water must be added to 30 oz of a 20% salt solution to dilute it to a 10% salt solution?   *4.4*

48. Find three consecutive multiples of 4 such that the first times the third is 240.   *6.2*

49. A painter working alone can paint a room in 6 h, while her helper needs 9 h to do it. The painter works alone for 2 h, and then they finish the room together. What is the total time needed to paint the room?   *7.7*

50. One leg of a right triangle is 3 cm longer than twice the length of the other leg, and 1 cm shorter than the hypotenuse. Find the length of the hypotenuse in simplest radical form.   *9.6*

51. The mass of a metal cylinder varies jointly as its height and the square of the radius of its base. One cylinder has a mass of 120 g. Find the mass of a second cylinder made of the same metal, 3 times as high, and having one-half the base radius of the first.   *12.6*

13 EXPONENTIAL AND LOGARITHMIC FUNCTIONS

The subtle patterns in this computer graphic of the Monterey coast were generated by repeated functions. The rock formations, the spray of the water, and other features have fractal-like structures that readily lend themselves to computer representation.

13.1 Inverse Relations and Functions

To describe the inverses of relations
To determine whether the inverses of functions are functions

In earlier chapters, you solved exponential equations such as $125 = 5^{2x}$, where the exponent either was a variable or contained a variable. In this chapter, you will study *exponential functions*, such as $y = 5^{2x}$, and related functions known as *logarithmic functions*. These functions have applications both in science and everyday life. In order to study them properly, however, you must first understand what is meant by the *inverse* of a function.

Adding 2 and *subtracting* 2 are *inverse operations* because one "reverses" the effect of the other. For example, if you begin with 7 and use both operations in succession, you will return to 7, as shown below.

$$7 + 2 = 9 \text{ and } 9 - 2 = 7 \qquad 7 - 2 = 5 \text{ and } 5 + 2 = 7$$

These two operations can be represented by the relations $A: y = x + 2$ and $S: y = x - 2$. If their domain (see Lesson 3.1) is the set of integers, then the following are true.

$$A = \{\ldots, (-2,0), (-1,1), (0,2), (1,3), (2,4), \ldots\}$$
$$S = \{\ldots, (0,-2), (1,-1), (2,0), (3,1), (4,2), \ldots\}$$

Notice that reversing each ordered pair (a,b) in A gives the ordered pairs (b,a) in S. Such relations are called *inverse relations*.

Definition

> The inverse of a relation A is the relation, denoted A^{-1}, obtained by reversing the order of the elements of each ordered pair in A. A and A^{-1} are called **inverse relations.**

Note that here the symbol -1 is *not* an exponent. For the inverse relations A and S above, S is the inverse of A, and A is the inverse of S.

$$S = A^{-1} \quad \text{and} \quad A = S^{-1}$$

If the domain of A is the set of all real numbers, then for each real number x, A is described by the equation $y = x + 2$. You can *trade x and y* as follows to obtain the equation for A^{-1}, or S.

	$A: y = x + 2$
Trade x and y.	S (or A^{-1}): $x = y + 2$
Solve for y.	$S: y = x - 2$

The relation A is a function because no two ordered pairs have the same first coordinate. Its inverse S is also a function. However, the inverse of a function is not always a function.

EXAMPLE 1 Describe the inverse of each function. Is the inverse a function?

a. Function f, described by the equation $y = \frac{3}{2}x - 6$.

$$f: y = \frac{3}{2}x - 6$$

Trade x and y. $\qquad\qquad f^{-1}: x = \frac{3}{2}y - 6$

Solve for y. $\qquad\qquad\quad f^{-1}: y = \frac{2}{3}x + 4$

f^{-1} is described by $y = \frac{2}{3}x + 4$. There is exactly one value of y for each value of x. Thus, f^{-1} is a function.

b. Constant function G, with equation $y = 3$.

Replace y with x in $y = 3$.

G^{-1}, with equation $x = 3$, is *not a function* because it contains many pairs, such as $(3,5)$ and $(3,6)$, with the same first coordinate of 3.

The diagram below shows the graphs of the inverse relations C and C^{-1}. The vertical line test (see Lesson 3.1) will determine whether they are functions.

$C = \{(1,3), (0,2), (-2,2), (-3,1)\}$
$C^{-1} = \{(3,1), (2,0), (2,-2), (1,-3)\}$

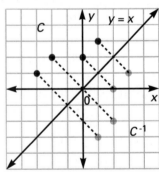

The graphs of C and C^{-1} are mirror images about the line $y = x$; that is, they are symmetric to each other with respect to the line $y = x$.

C is a function, but C^{-1} is not a function because a vertical line intersects its graph at two points when $x = 2$.

EXAMPLE 2 The function H is described by its graph at the right. Draw the graph of H^{-1}, the inverse of H. Is H^{-1} a function?

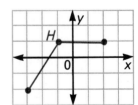

Plan The graph of H consists of two segments with endpoints $(2,1)$, $(-1,1)$ and $(-1,1)$, $(-3,-2)$. Plot $(1,2)$, $(1,-1)$, and $(-3,-2)$. Then draw the two segments that form the mirror image of H with respect to the line $y = x$.

Solution H^{-1} is represented by the graph at the right.

Although H is a function, H^{-1} is *not a function* because its graph does not pass the vertical line test at $x = 1$.

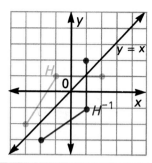

EXAMPLE 3   Write the slope-intercept form of the equation that describes g^{-1} if g is described by $3x - 5y = 15$.

Solution

$$g: \quad 3x - 5y = 15$$
$$g^{-1}: \quad 3y - 5x = 15$$
$$3y = 5x + 15; \; y = \tfrac{5}{3}x + 5$$

Thus, $y = \tfrac{5}{3}x + 5$ describes g^{-1}.

In Example 3, both g and g^{-1} are *linear nonconstant functions* because their equations are of the form $ax + by = c$, where $a \neq 0$ and $b \neq 0$.

Theorem 13.1   The inverse of any linear nonconstant function is also a linear non-constant function.

The inverse of a quadratic function is described in Example 4.

EXAMPLE 4   Function Q is described by $y = 2x^2 - 6$. Write the equation describing Q^{-1}, solving for y. Are Q and Q^{-1} a pair of inverse functions?

Solution

$$Q: \quad y = 2x^2 - 6$$
$$Q^{-1}: \quad x = 2y^2 - 6$$
$$2y^2 = x + 6$$
$$y = \pm\sqrt{\frac{x + 6}{2}}$$

Q^{-1} is described by $y = \pm\sqrt{\dfrac{x + 6}{2}}$. Choose a value for x, such as 8. Then Q^{-1} contains $(8,\sqrt{7})$ and $(8,-\sqrt{7})$, so Q^{-1} is not a function.

So, Q and Q^{-1} are *not a pair of inverse functions*.

Is the statement always, sometimes, or never true?

1. The inverse of a function is a function.
2. The inverse of a relation is a relation.
3. If A is a relation, then $(A^{-1})^{-1} = A$.
4. The inverse of a constant function $y = c$ is a function.
5. The inverse of a linear nonconstant function is a function.
6. If f is a quadratic function whose domain is the set of real numbers, then the inverse of f is a function.

Classroom Exercises

Describe the inverse of each.

1. $\{(6,3), (4,4), (2,5)\}$
2. subtracting 5
3. dividing by 2
4. multiplying by -4
5. $y = -2$
6. $y = x$
7. $y = x + 6$
8. $y = 2x$

Written Exercises

Describe the inverse of each function. Is the inverse a function?

1. $\{(1,4), (-3,4), (2,0)\}$
2. $\{(7,8), (6,4), (5,0)\}$
3. $y = 4x$
4. $y = -3x$
5. $y = 5$
6. $y = x$
7. $3y = 2x$
8. $y = 2x - 4$
9. $y = 2x - 6$
10. $y = -3x + 6$

Graph the inverse of each relation.

11.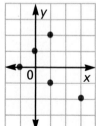
12.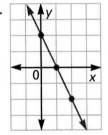
13.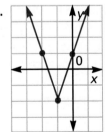
14.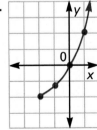

15–18. For the graphs of Exercises 11–14, is the inverse a function? Are the relation and its inverse a pair of inverse functions?

Write the equation of the inverse of each function, solving for *y*.
Are the function and its inverse a pair of inverse functions?

19. $y = \frac{3}{4}x - 3$ **20.** $y = -\frac{5}{3}x + 10$ **21.** $y = -\frac{2}{3}x + 2$ **22.** $y = \frac{1}{2}x^2$

23. $y = 2x^2$ **24.** $y = -x^2$ **25.** $2x - 3y = 6$ **26.** $3x + 4y = 12$

27. $2x + 5y = 10$ **28.** $y = x^2 + 2$ **29.** $y = -2x^2 + 3$ **30.** $y = \frac{1}{2}x^2 - 3$

For each function *f*, find f^{-1} and the values $f(12)$, $f^{-1}(12)$, $f(f^{-1}(12))$,
and $f^{-1}(f(12))$.

31. $f(x) = 3x - 3$ \hspace{3cm} **32.** $f(x) = \frac{3}{4}x + 6$

33. Given $g(x) = \frac{1}{2}x + 5$, choose the function *t* such that $g(t(x)) = x$
and $t(g(x)) = x$.

    a. $t(x) = 2x + 10$ **b.** $t(x) = 2x - 10$ **c.** $t(x) = 2x - 5$

Mixed Review

1. Find the solution set of $5 - 3x < 17$ *and* $9 > 2x - 3$. **2.2**

2. Factor $x^3 - 8$. **5.6** \hspace{4cm} **3.** Simplify $\dfrac{x^2 - x - 2}{x^2 - 1}$. **7.1**

Application: *Paraboloidal Reflectors*

Parabolas have a very interesting property. Take any parabola, where *F* is the focus, *P* is any point on the curve, and *l* is the line passing through *P* parallel to the parabola's axis of symmetry. It can be proved that any beam of light from a source at *F* striking a parabolic mirror at *P* will reflect along the line *l* parallel to the parabola's axis of symmetry.

Now imagine a parabola spun around its axis of symmetry until it becomes three-dimensional, a shape much like a cup. This shape, called a **paraboloid**, is used to make the reflectors for automobile headlights, searchlights, flashlights, and so on, because all light emanating from its focus will be concentrated ahead, rather than dispersed by the bulb. Paraboloidal reflectors are also used to collect sunlight for solar energy and to receive radio waves. Explain why. (THINK: Where would the receivers be placed?)

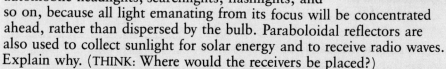

13.2 Exponential Functions

Objectives

To graph exponential functions
To interpret graphs of exponential functions
To approximate irrational powers of positive numbers

The equation $y = 5 \cdot 2^x$ describes an **exponential function** with a base of 2. Notice that both x and y can have rational and irrational values.

x	-2	-1	0	1	1.5	2	3	$\sqrt{10}$
2^x	0.25	0.5	1	2	2.83	4	8	8.95
$y = 5 \cdot 2^x$	1.25	2.5	5	10	14.1	20	40	44.8

1. If $x = -2$, then $2^x = 2^{-2} = \frac{1}{4}$. So, $y = 5 \cdot \frac{1}{4} = 5(0.25) = 1.25$.

2. If $x = 1.5$, then $2^x = 2^{1.5} = 2^{\frac{3}{2}} = \sqrt{2^3} = \sqrt{8}$.
   $\sqrt{8} = 2.828427. . .$, an irrational number, so $y \approx 5(2.828) \approx 14.1$.

3. If $x = \sqrt{10}$, an *irrational exponent*, then $2^x = 2^{\sqrt{10}} = 2^{3.1622776}. . .$.

Since the rational numbers 3, 3.1, 3.16, 3.162, 3.1622, 3.16227, . . .
are getting closer and closer to $\sqrt{10}$, it follows that the rational powers
$2^3, 2^{3.1}, 2^{3.16}, 2^{3.162}, 2^{3.1622}, 2^{3.16227}$, . . . are getting closer and closer
to $2^{\sqrt{10}}$. These powers can be found using a calculator's $\boxed{x^y}$ key:

8, 8.5741 . . . , 8.9382 . . . , 8.9506 . . . , 8.9519 . . . , 8.9523 . . .

Notice that the list appears to be rounding off near 8.95.

So, $2^{\sqrt{10}} \approx 8.95$, and $y = 5 \cdot 2^{\sqrt{10}} \approx 44.8$.

Also, a scientific calculator can be used to approximate $5 \cdot 2^{\sqrt{10}}$
directly: $5 \boxed{\times} 2 \boxed{x^y} 10 \boxed{\sqrt{}} \boxed{=} 44.762098 \approx 44.8$.

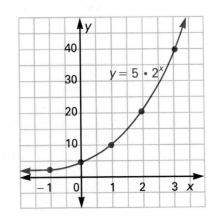

The graph of $y = 5 \cdot 2^x$ is drawn
using the table. Note the scales
on the axes. The graph reveals the
following properties of $f: y = 5 \cdot 2^x$.

(1) The domain is the set of real
numbers.

(2) The range is $\{y \mid y > 0\}$.
(3) The function is *always increasing*;
that is, if $x_2 > x_1$, then $y_2 > y_1$.

(4) The ordered pair $(0,5)$ belongs to
the function because the graph
has 5 as its y-intercept.

The base of an exponential function must be a positive number not equal to 1. It may, however, be a fraction between 0 and 1, as in $y = \left(\frac{1}{3}\right)^x$. Recall that $\left(\frac{1}{3}\right)^x = \frac{1}{3^x} = 3^{-x}$ (see Lesson 5.2).

EXAMPLE

Draw the graphs of $A: y = 3^x$ and $B: y = \left(\frac{1}{3}\right)^x$ in the same coordinate plane. Choose convenient scales for the axes.

Plan

Construct tables of ordered pairs for A and for B. Then plot the points and draw two smooth curves.

Solution

$A: y = 3^x$ $\qquad$ $B: y = \left(\frac{1}{3}\right)^x$

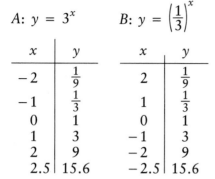

x	y
-2	$\frac{1}{9}$
-1	$\frac{1}{3}$
0	1
1	3
2	9
2.5	15.6

x	y
2	$\frac{1}{9}$
1	$\frac{1}{3}$
0	1
-1	3
-2	9
-2.5	15.6

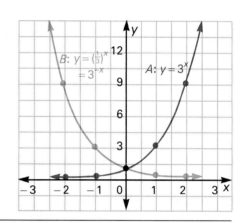

The graphs above show the following four properties of A and B. These properties hold for all exponential functions of the form $y = b^x$, where $b > 0$, $b \neq 1$.

(1) The domains of A and B are the set of real numbers.

(2) The ranges of A and B are $\{y \mid y > 0\}$.

(3) A is always increasing: $y_2 > y_1$ if $x_2 > x_1$.
$\quad$ B is always decreasing: $y_2 < y_1$ if $x_2 > x_1$.

(4) The ordered pair $(0,1)$ belongs to both A and B.

Note also that the combined graph of A and B is symmetric with respect to the y-axis. But A and B are *not* inverses of each other because their graphs are not symmetric to each other with respect to the line $y = x$.

Classroom Exercises

For each equation, give another equation such that the combined graph of the equations is symmetric with respect to the y-axis.

1. $y = 2^x$ $\qquad$ **2.** $y = \left(\frac{1}{5}\right)^x$ $\qquad$ **3.** $y = \left(\frac{1}{4}\right)^x$ $\qquad$ **4.** $y = 6^x$

Determine whether the function is always increasing or always decreasing.

5. $y = 4^x$ **6.** $y = \left(\frac{1}{2}\right)^x$ **7.** $y = \left(\frac{1}{6}\right)^x$ **8.** $y = 5^x$

9. Graph $y = 4^x$. Choose convenient scales for the axes.

10. Graph $y = \left(\frac{1}{4}\right)^x$. Choose convenient scales for the axes.

11–12. Give the domain and range for Classroom Exercises 9 and 10.

Written Exercises

Graph each function. Choose convenient scales for the axes. Find the domain and range. Determine whether the function is always increasing or always decreasing. Find the y-intercept of the graph.

1. $y = 2^x$ **2.** $y = 5^x$ **3.** $y = \left(\frac{1}{2}\right)^x$ **4.** $y = \left(\frac{1}{5}\right)^x$

5. $y = 3 \cdot 2^x$ **6.** $y = 3\left(\frac{1}{2}\right)^x$ **7.** $y = 3 \cdot 4^x$ **8.** $y = 2\left(\frac{1}{3}\right)^x$

9. $y = -2 \cdot 3^x$ **10.** $y = -\left(\frac{1}{4}\right)^x$ **11.** $y = 2^x + 6$ **12.** $y = 3^x - 7$

Use the $\boxed{x^y}$ key on a scientific calculator to approximate each power to the nearest tenth.

13. $3^{4.2}$ **14.** $6.5^{3.4}$ **15.** $5^{\sqrt{2}}$ **16.** $14^{\sqrt{3}}$

Sketch the graph of $y = 2^x$ and the given function in the same coordinate plane. Describe the relationship between the two graphs.

17. $y = \left(\frac{1}{2}\right)^x$ **18.** $y = 4 + 2^x$ **19.** $y = 4 \cdot 2^x$

20. $y = 2^{x+3}$ **21.** $y = 2^{x-1}$ **22.** $y = -2^x$

Graph each function for the given domain.

23. $y = 3 \cdot 2^{x+1} - 4$ for $-3 \le x \le 3$ **24.** $y = 9^{\sqrt{x}}$ for $0 \le x \le 4$

25. Why is $y = a^x$ not an exponential function when $a = 1$?

26. Will any exponential function of the form $y = a^x$, $a > 0$, $a \ne 1$ intersect the x-axis? Why or why not?

27. Find the asymptote for $y = a^x - c$, where $a > 0$, $a \ne 1$, and c is a constant.

Mixed Review

Find the value of each expression.   *5.2, 8.6*

1. 8^0 **2.** 3^{-2} **3.** $25^{\frac{1}{2}}$ **4.** $16^{\frac{3}{4}}$

5. Write $\sqrt[3]{x^2}$ in exponential form.   *8.6*

13.3 Base-*b* Logarithms

Objectives

To find base-*b* logarithms of positive numbers
To write base-*b* logarithmic equations as exponential equations
To solve base-*b* logarithmic equations

The base-2 exponential function f: $y = 2^x$ is a mapping from the exponent x to the power 2^x.

The inverse of f is the mapping from the power 2^x to x. The exponent x is called a **logarithm**.

Exponent	Power
4	2^4, or 16
3	2^3, or 8
0	2^0, or 1
-3	2^{-3}, or $\frac{1}{8}$

Power	Exponent
16, or 2^4	4
8, or 2^3	3
1, or 2^0	0
$\frac{1}{8}$, or 2^{-3}	-3

The *base-2 logarithm* of a positive number, denoted by $\log_2$, is found by changing an exponential equation into a **logarithmic equation** as shown below. $\log_2 8$ is read as "the base-two log of eight."

Base-2 exponential equation	Base-2 logarithmic equation
$16 = 2^4$	$\log_2 16 = \log_2 2^4 = 4$
$8 = 2^3$	$\log_2 8 = \log_2 2^3 = 3$
$1 = 2^0$	$\log_2 1 = \log_2 2^0 = 0$
$\frac{1}{8} = 2^{-3}$	$\log_2 \frac{1}{8} = \log_2 2^{-3} = -3$
$\vdots$	$\vdots$
$y = 2^x$	$\log_2 y = \log_2 2^x = x$

In general, if $y = 2^x$, then $\log_2 y = \log_2 2^x = x$. Any positive number b not equal to 1 can be the base of the **base-*b* logarithm**.
For example, $\log_5 125$ is a base-5 logarithm: $\log_5 125 = \log_5 5^3 = 3$.

Definition

For each positive number y and each positive number b where $b \neq 1$, if $y = b^x$, then $\log_b y = \log_b b^x = x$.

EXAMPLE 1

Find each logarithm.
a. $\log_5 625$ **b.** $\log_3 \frac{1}{9}$ **c.** $\log_6 \sqrt[3]{6}$ **d.** $\log_{10} \sqrt[4]{1,000}$

Solutions

a. $\log_5 5^4 = 4$ **b.** $\log_3 3^{-2} = -2$ **c.** $\log_6 6^{\frac{1}{3}} = \frac{1}{3}$ **d.** $\log_{10} 10^{\frac{3}{4}} = \frac{3}{4}$

EXAMPLE 2 Find the value of the variable in each logarithmic equation.

a. $\log_b 25 = 2$ b. $\log_2 128 = x$ c. $\log_3 y = 4$ d. $\log_b \frac{1}{16} = -4$

Plan   Rewrite each logarithmic equation as an exponential equation.

Solutions

a. $b^2 = 25$
$b = 5$
$(b > 0, \text{ so}$
$b \neq -5)$

b. $2^x = 128$
$2^x = 2^7$
$x = 7$

c. $3^4 = y$
$y = 81$

d. $b^{-4} = \frac{1}{16}$
$b^{-4} = 2^{-4}$
$b = 2$
$(b > 0, \text{ so}$
$b \neq -2)$

Classroom Exercises

Write each logarithmic equation as an exponential equation.

1. $\log_4 16 = c$ **2.** $\log_2 16 = d$ **3.** $\log_2 c = 4$ **4.** $\log_3 9 = x$

5. $\log_5 p = -2$ **6.** $\log_4 m = 0$ **7.** $\log_b \sqrt[5]{3} = \frac{1}{5}$ **8.** $\log_n p = m$

Written Exercises

Find each logarithm.

1. $\log_8 64$ **2.** $\log_5 1$ **3.** $\log_6 6$ **4.** $\log_6 \frac{1}{6}$ **5.** $\log_2 \frac{1}{16}$

6. $\log_3 81$ **7.** $\log_4 1$ **8.** $\log_2 \sqrt[5]{2}$ **9.** $\log_5 \sqrt{5}$ **10.** $\log_3 \sqrt[4]{3}$

Find the value of the variable in each logarithmic equation.

11. $\log_b 100 = 2$ **12.** $\log_2 y = 6$ **13.** $\log_3 243 = x$ **14.** $\log_4 y = 3$

15. $\log_5 \sqrt[3]{5} = x$ **16.** $\log_{\frac{1}{3}} y = 2$ **17.** $\log_b \frac{1}{8} = -3$ **18.** $\log_{\frac{2}{3}} \frac{8}{27} = x$

Find each logarithm.

19. $\log_2 \sqrt[5]{8}$ **20.** $\log_{10} \sqrt[3]{100}$ **21.** $\log_5 \sqrt[4]{125}$ **22.** $\log_{\frac{1}{4}} 16$

23. $\log_{\frac{1}{2}} 4^3$ **24.** $\log_c (c^2)^d$ **25.** $\log_{4n} 16n^2$ **26.** $\log_{2n} \frac{1}{8n^3}$

Mixed Review

1. Solve $3^{4x-2} = 81$. **5.1**

3. Factor $8a^3 - b^3$. **5.6**

2. Find $f(2)$ for $f(x) = 8x^{-3}$. **5.2**

4. Solve $x^2 + 8x + 25 = 0$. **9.5**

13.4 Graphing $y = \log_b x$

Objectives
To describe the inverses of exponential functions as logarithmic functions

To graph logarithmic functions

To interpret the graphs of logarithmic functions

To describe the inverse of an exponential function such as $A: y = 3^x$, trade x and y in $y = 3^x$. Then write the new equation, solving for y.

$$A: \qquad y = 3^x$$
$$A^{-1}: \qquad x = 3^y \quad \leftarrow \text{ Trade } x \text{ and } y.$$
$$\log_3 x = y, \quad \leftarrow \text{ Change the exponential equation}$$
$$\text{or } y = \log_3 x \qquad \text{to a logarithmic equation.}$$

Thus, the inverse of $A: y = 3^x$ is described by $y = \log_3 x$. In general, the inverse of an exponential function is a **logarithmic function**.

Theorem 13.2
The inverse of the exponential function
$$y = b^x \ (y > 0, b > 0, b \neq 1)$$
is the logarithmic function described by
$$y = \log_b x \ (x > 0, b > 0, b \neq 1).$$
The converse is also true.

You will be asked to prove Theorem 13.2 in Exercises 16 and 17.

EXAMPLE 1
Write the equation for the inverse of each function, solving for y.
a. $B: y = \left(\frac{1}{3}\right)^x$
b. $C: y = \log_5 x$

Solutions
a. The inverse of $B: y = \left(\frac{1}{3}\right)^x$ is described by $y = \log_{\frac{1}{3}} x$.

b. The inverse of $C: y = \log_5 x$ is described by $y = 5^x$.

Because $y = \log_3 x$ and $y = 3^x$ describe inverse functions, their graphs are symmetric to each other with respect to the line $y = x$. To draw the graph of $y = \log_3 x$, make a table of ordered pairs (x,y) for $y = 3^x$. Then reverse the order in each pair to form a table for $y = \log_3 x$. Plot these points and draw a smooth curve through them, as shown at the top of the next page.

$$y = 3^x \qquad y = \log_3 x$$

x	y		x	y
-2	$\frac{1}{9}$		$\frac{1}{9}$	-2
-1	$\frac{1}{3}$		$\frac{1}{3}$	-1
0	1		1	0
1	3		3	1
2	9		9	2

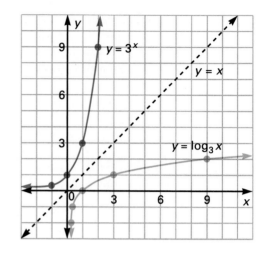

The graph of $y = \log_3 x$ shows four properties of the $\log_3$ function.

(1) The domain is $\{x \mid x > 0\}$.

(2) The range is the set of real numbers.

(3) The function is always increasing: $y_2 > y_1$ if $x_2 > x_1$.

(4) The ordered pair $(1,0)$ belongs to the function because 1 is the x-intercept of the graph. That is, $\log_3 1 = 0$.

EXAMPLE 2 Draw the graphs of $C\colon y = \log_5 x$ and $D\colon y = \log_{\frac{1}{5}} x$ in the same coordinate plane. Choose convenient scales for the axes.

Plan Because it is not yet convenient to use a calculator to find such values as $\log_5 11.2$ directly, make tables of ordered pairs for C^{-1} and D^{-1}. Then reverse the order to make tables for C and D.

Solution

C^{-1}: C: D^{-1}: D:

$y = 5^x \qquad y = \log_5 x \qquad y = \left(\frac{1}{5}\right)^x = 5^{-x} \qquad y = \log_{\frac{1}{5}} x$

x	y	x	y	x	y	x	y
-2	$\frac{1}{25}$	$\frac{1}{25}$	-2	-2	25	25	-2
-1	$\frac{1}{5}$	$\frac{1}{5}$	-1	-1.5	11.2	11.2	-1.5
0	1	1	0	-1	5	5	-1
1	5	5	1	0	1	1	0
1.5	11.2	11.2	1.5	1	$\frac{1}{5}$	$\frac{1}{5}$	1
2	25	25	2	2	$\frac{1}{25}$	$\frac{1}{25}$	2

Using the tables on the previous page yields the graphs of $C: y = \log_5 x$ and $D: y = \log_{\frac{1}{5}} x$ as shown at the right.

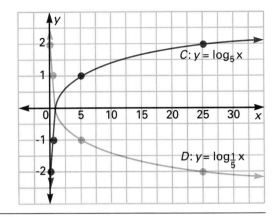

These graphs show the following for $C: y = \log_5 x$ and $D: y = \log_{\frac{1}{5}} x$.

(1) The domains of C and D are $\{x \mid x > 0\}$.

(2) The ranges of C and D are the set of real numbers.

(3) The function $C: y = \log_5 x$ is always increasing.

   The function $D: y = \log_{\frac{1}{5}} x$ is always decreasing.

(4) The ordered pair $(1,0)$ belongs to both C and D.

Notice that the graphs of the logarithmic functions $C: y = \log_5 x$ and $D: y = \log_{\frac{1}{5}} x$ are symmetric to each other with respect to the x-axis.

Pairs of Functions $(b > 1)$	Inverses?	Line of symmetry
$y = b^x$ and $y = \log_b x$	yes	$y = x$
$y = \left(\frac{1}{b}\right)^x$ and $y = \log_{\frac{1}{b}} x$	yes	$y = x$
$y = \log_b x$ and $y = \log_{\frac{1}{b}} x$	no	x-axis

Classroom Exercises

What is the line of symmetry for the graph of each pair of functions? Are the functions a pair of inverse functions?

1. $y = \log_4 x$, $y = \log_{\frac{1}{4}} x$

2. $y = 4^x$, $y = \log_4 x$

3. $y = \log_{\frac{1}{6}} x$, $y = \left(\frac{1}{6}\right)^x$

4. $y = \log_{\frac{1}{6}} x$, $y = \log_6 x$

5. Use the first table to complete the second.

x	4	0	1	-3	2
$y = 2^x$	16	1	2	$\frac{1}{8}$	4

x	16	1	2	$\frac{1}{8}$	4
$y = \log_2 x$					

Determine whether the function is always increasing or always decreasing. What two quadrants contain the graph of the function?

6. $y = \log_4 x$ **7.** $y = 4^x$ **8.** $y = \log_{\frac{1}{4}} x$ **9.** $y = \left(\frac{1}{4}\right)^x$, or $y = 4^{-x}$

Written Exercises

Write the equation of the inverse of each function, solving for y.

1. $y = 2^x$ **2.** $y = \log_{\frac{1}{2}} x$ **3.** $y = \log_8 x$

Graph each function. Choose convenient scales for the axes. Give the domain and range. Determine whether the function is always increasing or always decreasing. Find the x-intercept of the graph.

4. $y = \log_2 x$ **5.** $y = \log_4 x$ **6.** $y = \log_{\frac{1}{2}} x$

7. $y = \log_{\frac{1}{4}} x$ **8.** $y = \log_6 x$ **9.** $y = \log_{\frac{1}{6}} x$

10. $y = 3 \cdot \log_2 x$ **11.** $y = 2 \cdot \log_3 x$ **12.** $y = 4 \cdot \log_{\frac{1}{2}} x$

13. $y = 3 + \log_2 x$ **14.** $y = 2 + \log_3 x$ **15.** $y = 4 + \log_{\frac{1}{2}} x$

16. Prove that the inverse of f: $y = b^x$ is $y = \log_b x$.

17. Prove that the inverse of g: $y = \log_b x$ is $y = b^x$.

Graph each function. Choose convenient scales for the axes.

18. $y = \log_2 (x - 4)$ **19.** $y = \log_3 (x + 2)$ **20.** $y = \log_4 (x - 4)$

Midchapter Review

Write the equation that describes the inverse of the function, solving for y.   *13.1, 13.4*

1. $2x - 5y = 10$ **2.** $y = x^2 + 4$ **3.** $y = 6^x$ **4.** $y = \log_4 x$

Graph each function. Choose convenient scales for the axes.   *13.2, 13.4*

5. $y = \left(\frac{1}{2}\right)^x$ **6.** $y = \log_2 x$

Find each logarithm.   *13.3*

7. $\log_6 36$ **8.** $\log_2 \frac{1}{8}$ **9.** $\log_5 \sqrt[3]{5}$ **10.** $\log_3 1$

Find the value of the variable.   *13.3*

11. $\log_b 81 = 4$ **12.** $\log_2 y = -3$

13.5 $\text{Log}_b \ (x \cdot y)$ and $\text{Log}_b \ \dfrac{x}{y}$

Objectives

To expand logarithms of products and quotients
To simplify sums and differences of logarithms
To solve logarithmic equations

In the late sixteenth-century, Scottish nobleman John Napier devised tables of numbers he called *logarithms* that could be used to convert multiplications and divisions into additions and subtractions, respectively. Contemporaries soon adopted Napier's tables to help reduce the extraordinary drudgery of their arithmetic calculations. The methods for calculating with logarithms follow.

The logarithm of the *product* of two positive numbers can be written as the *sum* of two logarithms. For example, consider $\log_2 \ (16 \cdot 8)$.

$$\log_2 \ (16 \cdot 8) = \log_2 \ (2^4 \cdot 2^3) = \log_2 \ (2^7) = 7$$
$$\log_2 \ 16 + \log_2 \ 8 = \log_2 \ 2^4 + \log_2 \ 2^3 = 4 + 3 = 7$$

So, $\log_2 \ (16 \cdot 8) = \log_2 \ 16 + \log_2 \ 8$. This suggests Theorem 13.3.

Theorem 13.3

$\log_b \ xy = \log_b \ x + \log_b \ y$, where $x > 0$, $y > 0$, $b > 0$, and $b \neq 1$.

Proof

(1) Let $\qquad \log_b \ x = p \qquad$ and $\qquad \log_b \ y = q$.

(2) Then $\qquad x = b^p \qquad$ and $\qquad y = b^q$.

(3) So, $\qquad xy = b^p \cdot b^q = b^{p+q}$

(4) and $\qquad \log_b \ xy = \log_b \ b^{p+q} = p + q$.

(5) Thus, $\qquad \log_b \ xy = \log_b \ x + \log_b \ y$ (Steps 1 and 4).

The logarithm of the *quotient* of two positive numbers can be written as the *difference* of two logarithms. Consider $\log_2 \ \dfrac{128}{16}$.

$$\log_2 \ \frac{128}{16} = \log_2 \frac{2^7}{2^4} = \log_2 \ 2^3 = 3$$
$$\log_2 \ 128 - \log_2 \ 16 = \log_2 \ 2^7 - \log_2 \ 2^4 = 7 - 4 = 3$$

So, $\log_2 \ \dfrac{128}{16} = \log_2 \ 128 - \log_2 \ 16$.

This suggests Theorem 13.4, which you will be asked to prove in Exercise 30.

$\log_b \frac{x}{y} = \log_b x - \log_b y$, where $x > 0$, $y > 0$, $b > 0$, and $b \neq 1$.

Theorems 13.3 and 13.4 can be used to *expand the logarithm* of a product or a quotient.

EXAMPLE 1 Write each expression in expanded form.

a. $\log_3 7rt$ b. $\log_5 \frac{4c}{d}$ c. $\log_b \frac{mn}{5p}$

Solutions

a. $\log_3 7rt = \log_3 (7r \cdot t) = \log_3 7r + \log_3 t$
$$= \log_3 7 + \log_3 r + \log_3 t$$

b. $\log_5 \frac{4c}{d} = \log_5 4c - \log_5 d = \log_5 4 + \log_5 c - \log_5 d$

c. $\log_b \frac{mn}{5p} = \log_b mn - \log_b 5p$
$$= \log_b m + \log_b n - (\log_b 5 + \log_b p)$$
$$= \log_b m + \log_b n - \log_b 5 - \log_b p$$

Similarly, the sum or difference of two logarithms can be written as the logarithm of one term. To do this, use the *converse* (see page 103) of Theorems 13.3 and 13.4 as shown in Example 2.

EXAMPLE 2 Simplify $\log_3 8 - \log_3 m + \log_3 t^2 - \log_3 2t$.

Solution

$$\log_3 8 - \log_3 m + \log_3 t^2 - \log_3 2t$$

Rearrange the terms. $= \log_3 8 + \log_3 t^2 - \log_3 m - \log_3 2t$

Factor out -1. $= \log_3 8 + \log_3 t^2 - (\log_3 m + \log_3 2t)$

$\log_b x + \log_b y = \log_b xy$ $= \log_3 8t^2 - \log_3 2mt$

$\log_b x - \log_b y = \log_b \frac{x}{y}$ $= \log_3 \frac{8t^2}{2mt}$

Simplify. $= \log_3 \frac{4t}{m}$

Logarithmic equations can be solved by using the following property.

Property of Equality for Logarithms: If $\log_b x = \log_b y$, then $x = y$.

Thus, if $\log_2 5c = \log_2 (c + 12)$, then $5c = c + 12$, and $c = 3$.

Each apparent solution of a logarithmic equation must be checked, however, because $\log_b x$ is defined only for $x > 0$. For example, if $\log_3 2y = \log_3 (y - 5)$, then $2y = y - 5$, and $y = -5$. However, if $y = -5$, then $\log_3 2y = \log_3 (-10)$, but $\log_3 (-10)$ is undefined because $-10 < 0$. Thus, $\log_3 2y = \log_3 (y - 5)$ has no solution.

EXAMPLE 3   Solve $\log_b c + \log_b (c + 2) = \log_b (6c + 40) - \log_b 2$ for c. Check.

Solution   Simplify each side.

$$\log_b c(c + 2) = \log_b \frac{6c + 40}{2}$$

$$c(c + 2) = \frac{6c + 40}{2} \leftarrow \text{If } \log_b x = \log_b y, \text{ then } x = y.$$

$$c^2 + 2c = 3c + 20$$

$$c^2 - c - 20 = 0$$

$$(c - 5)(c + 4) = 0$$

$$c = 5 \quad or \quad c = -4$$

Check $c = 5$: $\log_b c + \log_b (c + 2) = \log_b 5 + \log_b 7$
$= \log_b (5 \cdot 7) = \log_b 35$, and $\log_b (6c + 40) -$
$\log_b 2 = \log_b 70 - \log_b 2 = \log_b \frac{70}{2} = \log_b 35$

Check $c = -4$: $\log_b c = \log_b (-4)$, and $\log_b (-4)$ is undefined
because $-4 < 0$.

Thus, 5 is the only solution.

Classroom Exercises

Determine whether each statement is true or false.

1. $\log_2 (5 \cdot 6) = \log_2 5 + \log_2 6$ **2.** $\log_5 18 = \log_5 6 + \log_5 3$

3. $\log_b 10 = \log_b 4 + \log_b 6$ **4.** $\log_6 7 + \log_6 4 = \log_6 28$

5. $\log_b \frac{23}{5} = \log_b 23 - \log_b 5$ **6.** $\log_2 \frac{27}{3} = \log_2 27 \div \log_2 3$

7. $\log_4 \left(\frac{5 \cdot 9}{11} \right) = \log_4 5 + \log_4 9 - \log_4 11$

8. $\log_2 3 - \log_2 5 + \log_2 6 - \log_2 7 = \log_2 3 + \log_2 6 - (\log_2 5 + \log_2 7)$

9. If $\log_3 2x + \log_3 4y = \log_3 z$, then $2x \cdot 4y = z$.

10. If $\log_2 (c + 7) - \log_2 5 = \log_2 3c$, then $\frac{c + 7}{5} = 3c$.

Match each expression at the left with one expression at the right.

11. $\log_3 2 + \log_3 a - \log_3 c - \log_3 d$ **a.** $\log_3 \frac{2c}{ad}$

12. $\log_3 2 - \log_3 a + \log_3 c - \log_3 d$ **b.** $\log_3 \frac{ad}{2c}$

13. $\log_3 a + \log_3 c - (\log_3 2 + \log_3 d)$ **c.** $\log_3 \frac{ac}{2d}$

14. $\log_3 a - \log_3 2 + \log_3 d - \log_3 c$ **d.** $\log_3 \frac{2a}{cd}$

Written Exercises

Write each expression in expanded form.

1. $\log_2 6c$ **2.** $\log_3 5tv$ **3.** $\log_4 \frac{n}{9}$ **4.** $\log_5 \frac{7}{m}$

5. $\log_b 3twz$ **6.** $\log_b \frac{4k}{p}$ **7.** $\log_2 \frac{5m}{cd}$ **8.** $\log_3 \frac{10a}{\pi r}$

Simplify each expression.

9. $\log_2 10 + \log_2 3c$ **10.** $\log_3 12a - \log_3 4$

11. $\log_4 9 + \log_4 5 - \log_4 3a$ **12.** $\log_b 8t^2 - \log_b 2 - \log_b t$

13. $\log_5 2c - \log_5 3d + \log_5 6c$ **14.** $\log_b (c^2 - 9) - \log_b (c + 3)$

Solve each equation. Check.

15. $\log_2 3k + \log_2 5 = \log_2 45$ **16.** $\log_b (6c - 12) - \log_b 3 = \log_b 6$

17. $\log_b 5t + \log_b 2t = \log_b 40$ **18.** $\log_5 t + \log_5 (t - 6) = \log_5 16$

19. $\log_b 2c + \log_b 4 = \log_b (3c + 10)$ **20.** $\log_3 32 - \log_3 x = \log_3 2x$

21. $\log_4 (3x + 6) - \log_4 (x + 2) = \log_4 (x - 2)$

22. $\log_b (2x + 5) - \log_b (x + 6) = \log_b (2x - 2) - \log_b x$

23. $\log_3 y^2 + \log_3 2y^2 = \log_3 10 + \log_3 5y^2$

24. $\log_2 (x + 4) + \log_2 (x - 1) = \log_2 (x - 2) + \log_2 (2x + 2)$

25. $\log_5 4 + \log_5 t + \log_5 (t + 4) = \log_5 84$

Simplify each expression.

26. $\log_b 3t - \log_b 2v - \log_b 9t^3 + \log_b 12v^2$

27. $\log_4 5c^2 - \log_4 (c^2 - 4c) + \log_4 (c^2 - 9) - \log_4 (2c + 6)$

Solve each equation. Check. (HINT: If $\log_b x = y$, then $x = b^y$.)

28. $\log_2 t + \log_2 (t - 4) = 5$ **29.** $\log_8 (3c + 1) + \log_8 (c - 1) = 2$

30. Prove Theorem 13.4.

Mixed Review

Write each expression in exponential form. *8.6*

1. $\sqrt{5}$ **2.** $\sqrt[5]{4}$ **3.** $\sqrt[3]{x}$ **4.** $\sqrt[4]{c^3}$

Write each expression in radical form. *8.6*

5. $x^{\frac{1}{4}}$ **6.** $10^{\frac{2}{3}}$ **7.** $5x^{\frac{1}{2}}$ **8.** $3^{-\frac{1}{2}}$

13.6 $\operatorname{Log}_b x^r$ and $\operatorname{Log}_b \sqrt[n]{x}$

Objectives

To expand logarithms of powers and radicals
To simplify multiples of logarithms
To solve logarithmic equations

Because $\log_2 7^3 = \log_2 (7 \cdot 7 \cdot 7) = \log_2 7 + \log_2 7 + \log_2 7$, it follows that $\log_2 7^3 = 3 \log_2 7$. That is, the logarithm of a *power* is a *multiple* of a logarithm. This suggests Theorem 13.5.

Theorem 13.5

$\log_b x^r = r \cdot \log_b x$, where $x > 0$, $b > 0$, $b \neq 1$, and r is a rational number.

Proof

1. Let $\log_b x = p$.
2. Then $x = b^p$.
3. So, $x^r = (b^p)^r = b^{r \cdot p}$
4. and $\log_b x^r = \log_b b^{r \cdot p} = r \cdot p$.
5. Thus, $\log_b x^r = r \cdot \log_b x$ (Steps 1 and 4).

If $b = 2$ and $r = \frac{1}{3}$, then $\log_b x^r = \log_2 x^{\frac{1}{3}} = \frac{1}{3} \cdot \log_2 x$, or $\log_2 \sqrt[3]{x} = \frac{1}{3} \cdot \log_2 x$. This illustrates a *corollary* to Theorem 13.5.

A **corollary** is a statement that can be immediately inferred from another statement that has already been proved.

Corollary

For each positive number x and each positive integer n,
$\log_b \sqrt[n]{x} = \frac{1}{n} \cdot \log_b x$, where $b > 0$ and $b \neq 1$.

This corollary is used to expand the logarithms of *radicals*, while Theorem 13.5 is used to expand the logarithms of *powers*.

EXAMPLE 1

Write in expanded form.

a. $\log_4 (tv^2)^3$

b. $\log_b \sqrt[4]{\dfrac{c^3}{d}}$

Plan

a. Use $\log_b x^r = r \cdot \log_b x$.

b. Use $\log_b \sqrt[n]{x} = \frac{1}{n} \cdot \log_b x$.

Solutions

a. $\log_4 (tv^2)^3 = 3 \log_4 tv^2$
$= 3(\log_4 t + \log_4 v^2)$
$= 3(\log_4 t + 2 \log_4 v)$

b. $\log_b \sqrt[4]{\dfrac{c^3}{d}} = \frac{1}{4} \log_b \dfrac{c^3}{d}$
$= \frac{1}{4}(\log_b c^3 - \log_b d)$
$= \frac{1}{4}(3 \log_b c - \log_b d)$

In Examples 2–4, multiples of logarithms are simplified using the converse forms of Theorem 13.5 and its corollary.

EXAMPLE 2 Simplify each expression.

a. $4 \log_2 m + \frac{1}{3} \log_2 n - 5 \log_2 p$ b. $\frac{1}{4}(\log_5 c + 3 \log_5 d - \log_5 f)$

Solutions

a. $\log_2 m^4 + \log_2 \sqrt[3]{n} - \log_2 p^5$

$= \log_2 (m^4 \sqrt[3]{n}) - \log_2 p^5$

$= \log_2 \dfrac{m^4 \sqrt[3]{n}}{p^5}$

b. $\frac{1}{4}(\log_5 c + \log_5 d^3 - \log_5 f)$

$= \frac{1}{4} \log_5 \dfrac{cd^3}{f}$

$= \log_5 \sqrt[4]{\dfrac{cd^3}{f}}$

EXAMPLE 3 Solve $2 \log_b t = \log_b 3 + \log_b (t + 6)$ for t. Check.

Solution

$2 \log_b t = \log_b 3 + \log_b (t + 6)$

$\log_b t^2 = \log_b 3(t + 6)$

$t^2 = 3(t + 6)$

$t^2 - 3t - 18 = 0$

$(t - 6)(t + 3) = 0$

$t = 6$ or $t = -3$

Check $t = 6$:

$2 \log_b t = \log_b 3 + \log_b (t + 6)$

$2 \log_b 6 \overset{?}{=} \log_b 3 + \log_b 12$

$\log_b 6^2 \overset{?}{=} \log_b (3 \cdot 12)$

$\log_b 36 = \log_b 36$ True

If $t = -3$, $\log_b t$ is undefined.

Thus, the only root is 6.

EXAMPLE 4 Solve $\frac{1}{2} \log_3 t + \frac{1}{2} \log_3 (t - 5) = \log_3 6$ for t. Check.

Solution

$\frac{1}{2} \log_3 t + \frac{1}{2} \log_3 (t - 5) = \log_3 6$

$\frac{1}{2} \log_3 (t^2 - 5t) = \log_3 6$

$\log_3 \sqrt{t^2 - 5t} = \log_3 6$

$\sqrt{t^2 - 5t} = 6$

$t^2 - 5t = 36$

$t^2 - 5t - 36 = 0$

$(t - 9)(t + 4) = 0$

$t = 9$ or $t = -4$

Check $t = 9$: $\frac{1}{2} \log_3 t + \frac{1}{2} \log_3 (t - 5) = \frac{1}{2} \log_3 9 + \frac{1}{2} \log_3 4 =$

$\frac{1}{2} \log_3 (9 \cdot 4) = \log_3 36^{\frac{1}{2}} = \log_3 6$ True

If $t = -4$, $\log_3 t$ is undefined. Thus, the only root is 9.

Find the expression that does *not* belong in each list.

1. a. $3 \log_5 2x$

   b. $\log_5 2x^3$

   c. $\log_5 8x^3$

   d. $\log_5 (2x)^3$

2. a. $\log_3 \sqrt[4]{2x}$

   b. $\frac{1}{4} \log_3 2x$

   c. $4 \log_3 2x$

   d. $\log_3 (2x)^{\frac{1}{4}}$

3. a. $\frac{2}{3} \log_b 5$

   b. $\frac{1}{3} \log_b 25$

   c. $\log_b \sqrt[3]{5^2}$

   d. $2 \log_b \frac{5}{3}$

Classroom Exercises

Match each expression at the left with one expression at the right.

1. $\log_5 \sqrt[3]{x}$

3. $\log_5 x^2$

5. $4 \log_5 x$

7. $5 \log_4 x$

9. $4(\log_3 c - \log_3 d)$

11. $4(\log_3 c + \log_3 d)$

2. $\log_5 x^3$

4. $\log_5 \sqrt{x}$

6. $\frac{1}{4} \log_5 x$

8. $\frac{1}{5} \log_4 x$

10. $\frac{1}{4}(\log_3 c - \log_3 d)$

12. $\log_3 c + \frac{1}{4} \log_3 d$

a. $\log_4 \sqrt[5]{x}$

b. $\frac{1}{2} \log_5 x$

c. $3 \log_5 x$

d. $\log_5 x^4$

e. $\log_3 c \sqrt[4]{d}$

f. $\log_3 \sqrt[4]{\frac{c}{d}}$

g. $\log_5 \sqrt[4]{x}$

h. $2 \log_5 x$

i. $\frac{1}{3} \log_5 x$

j. $\log_4 x^5$

k. $\log_3 \left(\frac{c}{d}\right)^4$

l. $\log_3 (cd)^4$

Written Exercises

Write each expression in expanded form.

1. $\log_3 (mn)^2$

2. $\log_b \sqrt{5g}$

3. $\log_5 \sqrt[3]{\frac{4}{h}}$

4. $\log_b \left(\frac{t}{y}\right)^3$

Simplify each expression.

5. $3 \log_2 x + 4 \log_2 x$

6. $\frac{1}{2} \log_5 3c + \frac{1}{2} \log_5 2d$

7. $\frac{1}{3}(\log_4 7 + \log_4 t)$

8. $4(\log_b 2 + \log_b c)$

9. $5 \log_b m - 2 \log_b n$

10. $\frac{1}{4} \log_3 ax - \frac{1}{4} \log_3 cy$

11. $\frac{1}{2} \log_5 c + \frac{1}{2} \log_5 (c - 2)$

12. $2 \log_b x - 2 \log_b (x + 3)$

Solve each equation. Check.

13. $2 \log_3 t = \log_3 2 + \log_3 (t + 12)$

14. $2 \log_b x - \log_b 2 = \log_b (3x + 8)$

15. $\frac{1}{2} \log_4 (3c - 5) = \log_4 5$

16. $\log_5 2 + \frac{1}{3} \log_5 (x - 1) = \frac{1}{3} \log_5 4x$

17. $\log_b 3 + 2 \log_b y = \log_b (8y + 3)$

18. $\log_5 2 + 2 \log_5 t = \log_5 (3 - t)$

19. $\frac{1}{2} \log_3 a + \frac{1}{2} \log_3 (a - 6) = \log_3 4$

20. $\frac{1}{3} \log_b y + \frac{1}{3} \log_b (y - 2) = \log_b 2$

21. $2 \log_b (2y + 4) = \log_b 9 + 2 \log_b (y - 1)$

22. $2 \log_5 (3y - 1) = \log_5 4 + 2 \log_5 (y + 2)$

23. $\frac{1}{2} \log_3 (c + 2) + \frac{1}{2} \log_3 (c - 3) = \log_3 6$

24. $\frac{1}{2} \log_b (2x - 1) - \frac{1}{2} \log_b (3x + 1) = \log_b 3 - \log_b 4$

Write each expression in expanded form.

25. $\log_5 \left(\dfrac{xy^3}{z^2} \right)^2$

26. $\log_b \sqrt[4]{\dfrac{6m}{np^3}}$

Simplify each expression.

27. $\log_3 6 + \frac{1}{2} \log_3 m - 2 \log_3 n$

28. $\frac{1}{3} \log_4 5a + \frac{1}{3} \log_4 2b - 3 \log_4 a - \log_4 b$

29. $3(\log_b 2m + 2 \log_b n - \log_b p)$

30. $\frac{1}{4} (\log_b 4 + 3 \log_b x - \log_b 3 - \log_b 5y)$

Solve each equation. Check. (HINT: If $\log_b x = y$, then $x = b^y$.)

31. $\log_5 (a^2 + 2a + 5) - \log_5 (a - 5) = 2$ **32.** $2 \log_3 (y - 3) - \log_3 (y + 1) = 2$

Simplify.

33. $\log_b \sqrt[3]{y^2} - \log_b \sqrt[4]{y^3} + 4 \log_b \sqrt[3]{y} - 5 \log_b \sqrt[4]{y}$

34. $\frac{2}{5} \log_b 3 - \frac{1}{5} \log_b 2 - \frac{2}{5} \log_b 4 + \frac{1}{5} \log_b 27$

Mixed Review

Rationalize the denominator.   *8.4, 9.2*

1. $\dfrac{4}{2\sqrt{7} - 5}$

2. $\dfrac{11}{5 + 3i}$

3. Solve $\sqrt{6x - 2} = x + 1$. Check.   *10.3*

4. Draw the vectors $(2,5)$ and $(-3,2)$ on a coordinate plane. Then draw their sum and difference.   *9.8*

Brainteaser

Find the error in the sequence of steps below.

$$3 > 2$$
$$3 \log_8 \left(\tfrac{1}{2} \right) > 2 \log_8 \left(\tfrac{1}{2} \right)$$
$$\log_8 \left(\tfrac{1}{2} \right)^3 > \log_8 \left(\tfrac{1}{2} \right)^2$$
$$\left(\tfrac{1}{2} \right)^3 > \left(\tfrac{1}{2} \right)^2$$
$$\tfrac{1}{8} > \tfrac{1}{4}$$

13.7 Common Logarithms and Antilogarithms

Objectives

To find base-10 logarithms of positive rational numbers
To find base-10 antilogarithms of rational numbers
To solve word problems involving pH

Three rows in a table of **common logarithms** (base-10) are displayed below. From the table, $\log_{10} 5.26 = 0.7210$ to four decimal places.

n	0	1	2	3	4	5	6	7	8	9
5.1	7076	7084	7093	7101	7110	7118	7126	7135	7143	7152
5.2	7160	7168	7177	7185	7193	7202	7210	7218	7226	7235
5.3	7243	7251	7259	7267	7275	7284	7292	7300	7308	7316

If $\log_b q = r$, then q is the base-b **antilogarithm** (antilog) of r. To find antilog$_{10}$ 0.7210, reverse the steps above. Look for 7210 in the table. It is in row 5.2, column 6. So, antilog$_{10}$ 0.7210 = 5.26.

When approximating common logarithms, it is customary to use the *equal* sign (=) rather than the *approximately equal* sign (≈).

EXAMPLE 1   **a.** Find $\log_{10} 5.31$.   **b.** Find antilog$_{10}$ 0.7093.

Solutions   **a.** $\log_{10} 5.31 = 0.7251$ **b.** antilog$_{10}$ 0.7093 = 5.12

The table on pages 790–791 lists the common logarithms of numbers between 1 and 10. However, using scientific notation (see Lesson 5.2), you can also use the table to estimate the common logarithms of positive numbers less than 1 or greater than 10.

Using $\log_{10} 5.29 = 0.7235$ and scientific notation, for example, you can estimate $\log_{10} 5{,}290$ as shown below.

$$\log_{10} 5{,}290 = \log_{10} (5.29 \times 10^3)$$
$$= \log_{10} 5.29 + \log_{10} 10^3$$
$$= 0.7235 + 3$$

So, $\log_{10} 5{,}290 = 3 + 0.7235$, or 3.7235.

The integer 3 is the *characteristic* of 3.7235, and the decimal 0.7235 is the *mantissa*.

If $y = n \cdot 10^c$, where $1 \leq n < 10$ and c is an integer, then $\log_{10} y = c + \log_{10} n$. The integer c is the **characteristic** and $\log_{10} n$ is the **mantissa** of $\log_{10} y$.

Notice that the mantissa, $\log_{10} n$, is a *positive* decimal less than 1. The subscript 10 can be omitted from expressions such as $\log_{10} 5{,}290$ when working with common logarithms. Thus, $\log 5{,}290$ (without a written base) means $\log_{10} 5{,}290$.

EXAMPLE 2   **a.** Find $\log 534$.   **b.** Find $\log 0.00534$

Solutions

a. $\log 534 = \log (5.34 \times 10^2)$
$\qquad\qquad = 2 + 0.7275$
$\qquad\qquad = 2.7275$

b. $\log 0.00534$
$\qquad = \log (5.34 \times 10^{-3})$
$\qquad = -3 + 0.7275$
$\qquad = -2.2725$

The log key on a calculator can be used to find the logarithms in Example 2 as shown below.

a. $\log 534$: 534 $\boxed{\log}$ $= 2.7275413 \approx 2.7275$

b. $\log 0.00534$: 0.00534 $\boxed{\log}$ $= -2.2724587 \approx -2.2725$

If you are using a table of logarithms to find the antilogarithm of a *negative* number, remember to write the number as the sum of an integer (the characteristic) and a *positive* decimal (the mantissa).

Thus, from Example 2b above comes the following.

$\qquad$ antilog (-2.2725)
$\qquad = $ antilog $(-3 + 0.7275)$ $\longleftarrow$ The mantissa must be between 0 and 1.
$\qquad = 5.34 \times 10^{-3}$ $\longleftarrow$ Take the antilog of 0.7275.
$\qquad = 0.00534$

EXAMPLE 3   **a.** Find antilog $(6.7193 - 10)$.   **b.** Find antilog 4.7135.

Solutions

a. antilog $(6.7193 - 10)$
$\qquad$ characteristic: $6 - 10$, or -4
$\qquad$ mantissa: 0.7193
$\qquad$ antilog $0.7193 = 5.24$
antilog $(6.7193 - 10)$
$\qquad = $ antilog $(-4 + 0.7193)$
$\qquad = 5.24 \times 10^{-4} = 0.000524$

b. antilog 4.7135
$\qquad$ characteristic: 4
$\qquad$ mantissa: 0.7135
$\qquad$ antilog $0.7135 = 5.17$
antilog 4.7135
$\qquad = $ antilog $(4 + 0.7135)$
$\qquad = 5.17 \times 10^4 = 51{,}700$

The antilogarithms in Example 3 can be found using the inverse and log keys on a calculator, as shown below.

a. antilog $(6.7193 - 10)$: 6.7193 (−) 10 (=) (INV) (log) = 0.000523962

b. antilog 4.7135: 4.7135 (INV) (log) = 51701.126

On some calculators, the inverse key is shown as (2nd) or (2nd f).

Also, a calculator display of 5.23962   − 04 in Example a above means 5.23962×10^{-4}.

In the study of chemistry, common logs are used to calculate the pH of a solution. pH is a measure of the solution's relative acidity or alkalinity, and it is defined as follows: $pH = -\log [H^+]$, where $[H^+]$ is the concentration of hydrogen ions in moles per liter.

EXAMPLE 4

Calculate the pH of a solution containing the given concentration of H^+ per liter.

a. 0.001 mole   **b.** 0.000375 mole

Plan

Use the formula $pH = -\log [H^+]$.

Solutions

a. $[H^+] = 0.001 = 10^{-3}$
$pH = -\log [H^+] = -\log_{10} 10^{-3} = -(-3) = 3$

The pH of the solution is 3.

b. $[H^+] = 0.000375 = 3.75 \times 10^{-4}$
$pH = -\log [H^+] = -\log (3.75 \times 10^{-4}) = -(-4 + 0.5740)$
$= -(-3.4260) \approx 3.4$

Calculator steps: 0.000375 (log) (+/−) = 3.4259687

The pH of the solution is 3.4.

EXAMPLE 5

Calculate the concentration of H^+ in a solution with a pH of 8.2.

Plan

Use the formula $pH = -\log [H^+]$, or $\log [H^+] = -pH$.

Solution

$pH = 8.2$
$8.2 = -\log [H^+]$, or $\log [H^+] = -8.2$
antilog $-8.2 =$ antilog $[-8 + (-0.2)]$ ← −0.2 is negative.
The mantissas in the table are positive. Rewrite −8.2 as −9 + 0.8.

$= $ antilog $[-9 + (0.8)]$
$= 6.3 \times 10^{-9}$

The concentration is 6.3×10^{-9} moles of H^+ per liter.

Classroom Exercises

Given log 536 = 2.7292, match each expression at the left with its equivalent at the right.

1. log 536
2. log 5.36
3. log 0.0536
4. characteristic of log 536
5. characteristic of log 53,600
6. characteristic of log 0.00536
7. mantissa of log 536
8. mantissa of log 5.36
9. antilog 2.7292
10. antilog 0.7292
11. antilog (9.7292 − 10)

a. − 3
b. 8.7292 − 10
c. 0.536
d. 0.7292
e. 2
f. 2.7292
g. 4
h. 5.36
i. 536

Express each number in scientific notation.

12. 234,000
13. 78.9
14. 0.000321

Express each number in ordinary decimal notation.

15. 5.43×10^{-1}
16. 4.56×10^{0}
17. 1.25×10^{3}

Find the characteristic of each logarithm.

18. log 3,450
19. log 6.54
20. log 0.000321

Written Exercises

Find each logarithm to four decimal places or antilogarithm to three significant digits. Where possible, give an exact answer.

1. log 18,700
2. log 0.305
3. antilog 2.4548
4. antilog (7.5490 − 10)
5. log 0.00594
6. antilog (− 2 + 0.9320)
7. log 23,500,000
8. antilog (6.7582 − 10)
9. antilog 4.8457
10. log 1,000
11. log 0.001
12. antilog 3
13. antilog − 1
14. log 1
15. log 0.1
16. antilog 1
17. log 10
18. antilog − 4
19. $\log \sqrt{10}$
20. antilog $\frac{1}{3}$
21. $\log \sqrt[3]{100}$

Calculate the pH of the solution containing the given concentration of H^{+} per liter.

22. 0.0001 mole
23. 0.0029 mole
24. $7.5 \cdot 10^{-12}$ mole

Calculate the concentration of H^+ in a solution with the given pH.

25. pH = 5   **26.** pH = 10.2   **27.** pH = 4.8

28. Find log (antilog (log (antilog 3.9350))).

29. Simplify antilog 3 + antilog 2 + antilog 1 + antilog 0 + antilog (− 1).

Use the information below for Exercises 30 and 31.

$$6^3 = 216 \qquad\qquad 6^4 = 1{,}296 \qquad\qquad 6^8 = 1{,}679{,}616$$
$$3 \cdot \log 6 \approx 2 + 0.33 \qquad 4 \cdot \log 6 \approx 3 + 0.11 \qquad 8 \cdot \log 6 \approx 6 + 0.23$$

30. Find the number of digits in the numeral for 6^{20}, the 20th power of 6.

31. For two positive integers x, the numeral for 4^x has exactly 17 digits. Find the greater value of x.

Mixed Review

Solve each equation.   *5.1, 7.6, 8.7*

1. $5^{2x-1} = 125$

2. $\dfrac{8}{x + 5} + \dfrac{3}{x} = 2$

3. $6^{x+2} = \dfrac{1}{36}$

4. $8^{2x} = 16$

Application: *Carbon Dating*

Radioactive elements decay at fixed exponential rates, called half-lives. The half-life of radioactive carbon-14, for example, is 5,730 years. In that time, half of any particular mass of carbon-14 will decay to form stable carbon-12 atoms. Because carbon-14 is found in all organic matter, it is often used to date fossils and other ancient artifacts.

The formula for radioactive decay used in carbon dating is $\dfrac{r}{r_0} = 2^{-\frac{t}{h}}$, where r is the amount of the radioactive element currently in the sample, r_0 is the amount in the sample when it was alive, t is the age of the sample, and h is the half-life of the radioactive element. The amount of the radioactive element present during life is generally assumed to be approximately that present in living matter today.

Find the age of a fossil having the given ratio of r to r_0 and h = 5,730 yr.

1. $\dfrac{1}{2}$ **2.** $\dfrac{1}{4}$ **3.** $\dfrac{1}{\sqrt{2}}$ **4.** $\dfrac{\sqrt{2}}{32}$

13.8 Exponential Equations

To solve exponential equations using common logarithms
To find base-b logarithms of positive numbers using common logarithms
To solve word problems using exponential formulas

In the *exponential equation* $3^{2x+1} = 27$, 27 is an integral power of three. So, $3^{2x+1} = 3^3$, $2x + 1 = 3$, and $x = 1$. In the equation $3^{2x+1} = 212$, however, 212 is not an integral power of three. Instead, it lies in an interval between two integral powers of three. Because $81 < 212 < 243$, it follows that $3^4 < 3^{2x+1} < 3^5$. So, $4 < 2x + 1 < 5$, or $3 < 2x < 4$, and thus $1.5 < x < 2$.

To find x to three significant digits, use base-10 logarithms.

EXAMPLE 1 Solve $3^{2x+1} = 212$ for x to three significant digits.

Solution $3^{2x+1} = 212$, so $\log 3^{2x+1} = \log 212$.

$$(2x + 1)\log 3 = \log 212 \leftarrow \log a^r = r \log a$$
$$2x + 1 = \frac{\log 212}{\log 3} = \frac{2.3263}{0.4771}$$
$$2x + 1 = 4.876 \leftarrow \text{Divide to four significant digits.}$$
$$2x = 3.876$$
$$x = 1.938$$

Thus, $x = 1.94$ to three significant digits.

A base-b logarithm such as $\log_4 6$ can be found to three significant digits by solving an exponential equation.

EXAMPLE 2 Find $\log_4 6$ to three significant digits.

Solution Let $\log_4 6 = x$.   Then $4^x = 6$. $\leftarrow$ Change to an exponential equation.

$$\log 4^x = \log 6$$
$$x \cdot \log 4 = \log 6$$
$$x = \frac{\log 6}{\log 4} = \frac{0.7782}{0.6021} = 1.292$$

Calculator steps: 6 $\boxed{\text{log}}$ $\boxed{\div}$ 4 $\boxed{\text{log}}$ $\boxed{=}$ 1.2924813

So, $\log_4 6 = 1.29$ to three significant digits.

Notice in Example 2 that $\log_4 6 = \dfrac{\log 6}{\log 4}$. In general, $\log_b y = \dfrac{\log y}{\log b}$, where $b > 0$, $b \neq 1$, and $y > 0$.

Exponential formulas are used in business and science, and may lead to large powers of rational numbers.

If \$8,000 is invested at 14% per year and compounded semiannually, then every 6 months the interest earned for the half-year is added to the previous principal to form a new principal. The total value of this investment over several 6-month periods is found and generalized below. Observe that 14% per year is 7% per 6 months.

Number of 6-month periods	Total value (principal plus interest)
0	8,000
1	$8,000 + 8,000(0.07)$ $= 8,000(1 + 0.07) = 8,000(1.07)^1$
2	$8,000(1.07) + 8,000(1.07)(0.07)$ $= 8,000(1.07)(1 + 0.07) = 8,000(1.07)^2$
3	$8,000(1.07)^2 + 8,000(1.07)^2(0.07)$ $= 8,000(1.07)^2(1 + 0.07) = 8,000(1.07)^3$
$\vdots$	$\vdots$
x	$8,000(1.07)^x$

This suggests the formula for *compound interest*: $A = p\left(1 + \dfrac{r}{n}\right)^{nt}$, where A is the total amount at the end of t years when p dollars are invested at r% per year, compounded n times each year.

EXAMPLE 3 An investment of \$7,400 at 12% per year is compounded quarterly (4 times each year). How much will the investment be worth in 15 years? Use logarithms or a calculator with the x^y function.

Solution
$$A = p\left(1 + \frac{r}{n}\right)^{nt} = 7,400\left(1 + \frac{0.12}{4}\right)^{4\cdot15} = 7,400(1.03)^{60}.$$
$$\log A = \log 7,400(1.03)^{60} = \log 7,400 + 60 \cdot \log 1.03$$
$$= 3.8692 + 60(0.0128)$$
$$= 4.6372$$

$$A = \text{antilog }(4 + 0.6372) = 4.34 \times 10^4 = \$43,400$$

Using logarithms as shown above, you will find that the \$7,400 investment will be worth \$43,400 in 15 years. Using a scientific calculator, you will obtain a value of \$43,597.86.

By a process called *simple fission*, some bacteria reproduce by splitting themselves into two new bacteria at the end of each growth period. If 600 bacteria are present at the start of an experiment, and simple fission takes place every 20 minutes, the number of bacteria increases rapidly.

Periods	0	1	2	3	4	5
Number of bacteria	600	1,200	2,400	4,800	9,600	19,200
	600	$600 \cdot 2$	$600 \cdot 2^2$	$600 \cdot 2^3$	$600 \cdot 2^4$	$600 \cdot 2^5$

At the end of 5 hours, the bacteria have gone through 15 growth periods, and the number of bacteria is $600 \cdot 2^{15}$, or 19,660,800.

This suggests the formula $N = N_0 \cdot 2^k$, where N is the number of bacteria present at the end of k periods of simple fission and N_0 is the number of bacteria at the start.

EXAMPLE 4 Find the number of bacteria produced from 800 bacteria at the end of 6 h if simple fission occurs every 12 min.

Plan There are 5 twelve-minute periods in 1 hour, and 30 such periods in 6 h. Use $N = N_0 \cdot 2^k$, where $N_0 = 800$ and $k = 30$.

Solution
$$N = N_0 \cdot 2^k = 800 \cdot 2^{30} \leftarrow \text{Calculator steps: } 800 \ \boxed{\times}\ 2 \ \boxed{x^y}\ 30 \ \boxed{=}$$
$$8.58993 \times 10^{11}$$

$$\log N = \log (800 \cdot 2^{30})$$
$$= \log 800 + 30 \cdot \log 2$$
$$= 2.9031 + 30(0.3010) = 11.9331$$
$$N = \text{antilog } 11.9331 = \text{antilog } (11 + 0.9331) = 8.57 \times 10^{11}$$

There will be 8.57×10^{11} bacteria at the end of 6 h.
If a scientific calculator is used, this value will be 8.59×10^{11} bacteria.

Classroom Exercises

The exponent x lies between what two consecutive integers?

1. $3^x = 22$ **2.** $2^x = 22$ **3.** $5^x = 22$ **4.** $10^x = 754$

Solve each equation for x to three significant digits.

5. $3^x = 22$ **6.** $5^{x-1} = 45$ **7.** $\log_2 7 = x$ **8.** $x = \log_4 5.2$

Written Exercises

Solve each equation for x to three significant digits.

1. $4^{2x} = 30$ **2.** $8^{2x+1} = 600$ **3.** $7^{3x-2} = 750$ **4.** $9^{2x-4} = 5$

Find each logarithm to three significant digits.

5. $\log_7 4$ **6.** $\log_5 35$ **7.** $\log_3 45.3$ **8.** $\log_6 850$

Find the value to the nearest cent of each investment at the given annual interest rate. Use the formula $A = p\left(1 + \dfrac{r}{n}\right)^{nt}$ and a calculator.

9. $600, paying 10%, compounded semiannually for 4 years

10. $850, paying 8%, compounded quarterly for 3 years

11. $8,000, paying 12%, compounded quarterly for 11 years

12. $7,200, paying 14%, compounded semiannually for 20 years

Find N, the number of bacteria present, for the given data to three significant digits. Use $N = N_0 \cdot 2^k$ and a calculator.

13. $N_0 = 500$, and simple fission occurs every 30 min for 4 h.

14. $N_0 = 4,000$, and simple fission occurs every 15 min for 10 h.

15. $N_0 = 7,500$, and simple fission occurs every 20 min for 15 h.

16. $N_0 = 35,000$, and simple fission occurs every 90 min for 24 h.

Solve each equation for x to three significant digits.

17. $5^x = 3 \cdot 4^x$ **18.** $3 \cdot 5^{x+1} = 5 \cdot 3^{x+2}$ **19.** $2^{4x+3} = 5 \cdot 2^{3x}$

20. $x = 2^{\sqrt{6}}$ **21.** $3x = 4^{\sqrt{5}}$ **22.** $2x - 1 = 12^{\sqrt{2}}$

Graph $y = \log_2 x$ and the given function in the same coordinate plane. Describe the relationship between the two graphs.

23. $y = 3 + \log_2 x$ **24.** $y = \log_2 (x - 4)$ **25.** $y = \log_2 (x + 3)$

Mixed Review

1. Solve the system: $\begin{array}{l} 3x - 2y = 18 \\ 10x + 3y = 2 \end{array}$ **4.3**

2. Convert 18 lb/ft^2 to ounces per square inch.   **7.3**

3. Use synthetic substitution and the Integral Zero Theorem to solve $x^4 + x^3 + 7x^2 + 9x - 18 = 0$.   **10.4**

4. Find the maximum value of y given that $y = -2x^2 - 12x + 35$.   **9.4**

5. A number b varies inversely as the square of a. If $b = 4$ when $a = 3$, what is b when $a = 2$?   **12.5**

Extension: Linear Interpolation

The table below shows values of x to one decimal place and values of $f(x)$ to three significant digits.

x	2.0	2.1	2.2	2.3	2.4	2.5	2.6	2.7	2.8	2.9
$f(x)$	5.40	5.47	5.54	5.61	5.68	5.75	5.82	5.89	5.96	6.03

The equation for the function is $f(x) = 0.7x + 4$, but even when the equation is not known, the value of $f(x)$ for $x = 2.76$ (two decimal places) can still be found using only the table and a process called **linear interpolation**. Study the graph for the interval $2.7 < x < 2.8$.

In the graph, notice the following.
(1) $f(2.7) = 5.89$, $f(2.8) = 5.96$, and
$f(2.76) = 5.89 + BD$

(2) Triangle ABD is similar to triangle ACE.

Hence, $\dfrac{AD}{AE} = \dfrac{BD}{CE}$, or $\dfrac{0.06}{0.1} = \dfrac{BD}{0.07}$.

$BD = \dfrac{0.06 \times 0.07}{0.1} = 0.042 \approx 0.04$

Then $5.89 + BD = 5.89 + 0.04 = 5.93$.
So, $f(2.76) = 5.93$.

Linear interpolation can be performed as shown below.

Example 1   Find $f(2.53)$ to three significant digits. Use the table above.

Solution

$$
0.1 \left\{ 0.03 \left\{ \begin{array}{c|c} x & f(x) \\ \hline 2.5 & 5.75 \\ 2.53 & ? \\ 2.6 & 5.82 \end{array} \right\} n \right\} 0.07
$$

$$\frac{0.03}{0.1} = \frac{n}{0.07}$$

$$n = \frac{0.03 \times 0.07}{0.1}, \text{ or } n = 0.021 \approx 0.02$$

Thus, $f(2.53) = 5.75 + n = 5.75 + 0.02 = 5.77$.

Example 2

Find x to two decimal places given that $f(x) = 5.58$. Use the table on the previous page.

Solution

$$0.1 \left\{ n \left\{ \begin{array}{c|c} x & f(x) \\ \hline 2.2 & 5.54 \\ ? & 5.58 \\ 2.3 & 5.61 \end{array} \right\} 0.04 \right\} 0.07$$

$$\frac{n}{0.1} = \frac{0.04}{0.07}$$

$$n = \frac{0.1 \times 0.04}{0.07}$$

$$n \approx 0.06$$

Thus, $x = 2.2 + n = 2.2 + 0.06 = 2.26$.

Over narrow intervals, the graphs of logarithmic functions are close to straight lines. Hence, linear interpolation can be used to find $\log x$ when x is given to four significant digits. This is shown in Example 3.

Example 3

Find $\log 2{,}346$ to four decimal places. Use linear interpolation and the logarithm table.

Plan

2,346 is between 2,340 and 2,350. From the table, $\log 2340 = 3.3692$ and $\log 2350 = 3.3711$. Omit all decimal points and commas while interpolating.

Solution

$$10 \left\{ 6 \left\{ \begin{array}{c|c} x & \log x \\ \hline 2340 & 33692 \\ 2346 & ? \\ 2350 & 33711 \end{array} \right\} n \right\} 19$$

$$\frac{6}{10} = \frac{n}{19}, \text{ or } n = \frac{6 \times 19}{10} \approx 11$$

So, $33692 + n = 33692 + 11 = 33703$.

Thus, $\log 2{,}346 = 3.3703$.

Example 4

Example 4   Find antilog 1.5123 to four significant digits. Use linear interpolation and the logarithm table.

Plan   If antilog 1.5123 = x, then log x = 1.5123. From the table, 5123 is between 5119 and 5132. So, antilog 1.5123 is between antilog 1.5119 = 32.5 and antilog 1.5132 = 32.6. While interpolating, omit all decimal points and commas.

Solution

$$
10 \left\{ \begin{array}{c} n \left\{ \begin{array}{c|c} x & \log x \\ \hline 3250 & 15119 \\ ? & 15123 \\ 3260 & 15132 \end{array} \right. \end{array} \right.
$$

with brackets: $\left. 15119 \atop 15123 \right\} 4 \Big\} 13$

$$\frac{n}{10} = \frac{4}{13}$$

$$n = \frac{10 \times 4}{13} \approx 3$$

So, $3250 + n = 3250 + 3 = 3253$.

Thus, antilog 1.5123 = 32.53.

Exercises

Use linear interpolation and the table below for Exercises 1–8.

x	3.0	3.1	3.2	3.3	3.4	3.5	3.6	3.7
$f(x)$	10.52	10.77	11.02	11.27	11.52	11.77	12.02	12.27

Find $f(x)$ to two decimal places.

1. $x = 3.42$ **2.** $x = 3.04$ **3.** $x = 3.52$ **4.** $x = 3.36$

Find x to two decimal places.

5. $f(x) = 11.08$ **6.** $f(x) = 10.67$ **7.** $f(x) = 10.93$ **8.** $f(x) = 11.09$

Find each logarithm to four decimal places and each antilogarithm to four significant digits. Use linear interpolation.

9. log 3,582   **10.** log 23.74   **11.** antilog 2.5735   **12.** antilog 4.3454

Mixed Problem Solving

1. Find the value after 5 years of $7,500 earning 12% per year, compounded quarterly, to the nearest cent. Use the formula $A = p\left(1 + \frac{r}{n}\right)^{nt}$ and a calculator.

2. The pressure P of a gas at constant temperature varies inversely as the volume V. If $V = 600$ in^3 when $P = 45$ lb/in^2, what is P when $V = 54$ in^3?

3. If y varies jointly as x and $\sqrt{z}$, and $y = 14$ when $x = 4$ and $z = 36$, what is z when $x = 12$ and $y = 63$?

4. Find the maximum area A of a rectangle whose dimensions are x and $800 - 2x$ units.

5. Corn worth $2.10/kg and oats worth $2.80/kg are mixed to make 100 kg of feed worth a total of $227.50. How many kilograms of oats are used?

6. How many milliliters of a 40% acid solution must be added to 150 milliliters of a 65% acid solution to obtain a 50% acid solution?

7. Two cars took the same trip. One averaged 40 mi/h and the other averaged 45 mi/h. If the slower car took 30 min longer, what distance did the cars drive?

8. Alice is 4 years younger than Betty and Carl is 3 times as old as Alice. Five years ago, Carl's age was 3 years less than twice the sum of the girls' ages then. How old is Carl now?

9. The length of a rectangle is 1 ft more than its width and 3 ft less than the length of a diagonal. Find the length of the diagonal in simplest radical form.

10. An object is shot upward from the earth's surface with an initial velocity of 80 m/s. To the nearest 0.1 second, when will its height be 200 m? Use the formula $h = vt - 5t^2$.

11. The Drama Workshop raised $615 by selling 280 tickets to its annual play. Adult tickets cost $3.00 each and children's tickets cost $1.50 each. How many $3.00 tickets were sold?

12. The perimeter of a rectangle is 72 in. If the width and length are each increased by 6 inches, their ratio will be 7:9. What are the dimensions of the original rectangle?

13. A certain boat can travel 50 km/h in still water. One day it travels 10.4 km downstream in the same time that it travels 9.6 km upstream. Find the rate of the current.

14. Machine A can do a certain job in 18 hours. Machines B and C can do the same job in 6 h and 4.5 h, respectively. How long will the job take if all three machines operate at the same time?

15. Find three consecutive odd integers such that the sum of the squares of the second and third integers is 130.

16. The product of two numbers is 8 and the sum of their squares is 20. Find all such pairs of numbers.

17. The tens digit of a two-digit number is one more than 4 times the units digit. The sum of the number and the number with its digits reversed is 121. Find the number.

18. Working together, Jake and Felicia can clean a house in 3 hours. It takes Jake 4 times longer than Felicia to do it alone. How long would it take each alone?

Key Terms

antilogarithm (p. 481)
base (p. 467)
base-b logarithm (p. 467)
characteristic (p. 482)
common logarithm (p. 481)
corollary (p. 477)
exponential function (p. 464)
inverse relations (p. 459)

linear interpolation (p. 490)
logarithm (p. 467)
logarithmic equation (p. 467)
logarithmic function (p. 469)
mantissa (p. 482)
Property of Equality for
   Logarithms (p. 474)

Key Ideas and Review Exercises

13.1   For the inverse relations A and A^{-1}, each ordered pair (a,b) in A is reversed to (b,a) in A^{-1}. The graphs of A and A^{-1} are symmetric with respect to the line $y = x$. The equation for A^{-1} can be obtained from the equation for A by trading x and y.

The graph of relation A is shown at the right.

1. Draw the graph of A^{-1}.

2. Is A^{-1} a function?

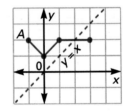

The function B is described by $y = x^2 + 6$.

3. Write the equation for B^{-1}, solving for y.

4. Is B^{-1} a function?

13.2   Exponential functions are described by the equations $y = b^x$ and $y = \left(\frac{1}{b}\right)^x$ (or $y = b^{-x}$), where $b > 1$. Their domains are the set of real numbers. Their ranges are $\{y \mid y > 0\}$. $y = b^x$ is always increasing; $y = \left(\frac{1}{b}\right)^x$ is always decreasing. Their y-intercepts are 1, and their combined graph is symmetric with respect to the y-axis.

5. Draw the graph of g: $y = 3 \cdot 2^x$. Find the domain, range, and y-intercept. Is the function always increasing or always decreasing?

13.3   The equation $\log_b y = x$ can be written as $y = b^x$. Then $\log_b y = \log_b b^x = x$.

Find each logarithm.

6. $\log_4 64$ **7.** $\log_8 1$ **8.** $\log_7 \sqrt{7}$ **9.** $\log_2 2^6$

10. Solve $\log_3 y = 4$ for y.   **11.** Solve $\log_b 16 = 4$ for b.

13.4   Logarithmic functions are described by the equations $y = \log_b x$ and $y = \log_{\frac{1}{b}} x$, where $b > 1$.

Their domains are $\{x \mid x > 0\}$. Their ranges are the set of real numbers. $y = \log_b x$ is always increasing; $y = \log_{\frac{1}{b}} x$ is always decreasing.

Their x-intercepts are 1, and their combined graph is symmetric with respect to the x-axis.

Note that $f: y = b^x$ where $y > 0$, $b > 0$, $b \neq 1$, and $f^{-1}: y = \log_b x$ where $x > 0$, $b > 0$, $b \neq 1$ are a pair of inverse functions.

12. Draw the graph of $f: y = \log_2 x$. Find the domain, range, and x-intercept. Determine whether the function is always increasing or always decreasing.

13.5, 13.6   To expand or simplify a logarithmic expression, use the following.

$$\log_b xy = \log_b x + \log_b y \qquad \log_b \frac{x}{y} = \log_b x - \log_b y$$

$$\log_b x^r = r \cdot \log_b x \qquad \log_b \sqrt[n]{x} = \frac{1}{n} \cdot \log_b x$$

To solve a logarithmic equation, use: If $\log_b x = \log_b y$, then $x = y$.

Write each expression in expanded form.

13. $\log_3 \dfrac{6a}{m}$ **14.** $\log_b c^2 d$ **15.** $\log_5 \sqrt[4]{tv}$ **16.** $\log_4 \sqrt{\dfrac{x}{y}}$

Simplify each expression.

17. $\log_b 7 + \log_b a - \log_b d$ **18.** $4 \log_2 c + \frac{1}{3} \log_2 d$

19. Write in your own words how to solve $\frac{1}{4} \log_6 (x - 3) = \log_6 3$ for x.

20. Solve $2 \log_5 y = \log_5 2 + \log_5 (3y + 8)$ for y. Check.

13.7   Common logarithms ($\log x$) are base-10 logarithms. Log x means $\log_{10} x$.

21. Find log 0.00453.

22. Find antilog 4.8007.

23. Find the pH of a solution that has 0.00060 mole of H^+ per liter. Use $\text{pH} = -\log [H^+]$.

24. Find the concentration of H^+ in a solution with a pH of 3.8. Use $\log [H^+] = -\text{pH}$.

13.8   To solve $a^x = b$ for x, use the fact that $\log a^x = \log b$.

Then $x \log a = \log b$, and $x = \dfrac{\log b}{\log a}$.

To find $\log_b y$, let $\log_b y = x$, rewrite as $b^x = y$, and solve for x.

25. Solve $6^{3x-2} = 476$ for x to three significant digits.

26. Find $\log_3 78$ to three significant digits.

The graph of relation A is shown at the right.

1. Draw the graph of A^{-1}.
2. Is A^{-1} a function?
3. Are A and A^{-1} a pair of inverse functions?
4. Write the equation for the inverse of
   $B: y = x^2 - 3$, solving for y.

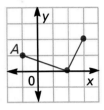

Graph each function. Find the domain, range, and x- or y-intercept.

5. $y = 2 \cdot 3^x$
6. $y = \log_4 x$

Find each logarithm.

7. $\log_2 64$
8. $\log_2 \sqrt{32}$

Find the value of the variable in each logarithmic equation.

9. $\log_5 y = 4$
10. $\log_b 64 = 2$
11. Write the equation for the inverse of $f: y = 4^x$, solving for y.

Write each expression in expanded form.

12. $\log_2 \dfrac{7a}{c}$
13. $\log_5 \sqrt[3]{pc}$

Simplify each expression.

14. $\log_b 5 + \log_b 3v - \log_b c$
15. $3 \log_b c + \frac{1}{5} \log_b d$

Solve each equation. Check.

16. $\log_4 a + \log_4 (a - 5) = \log_4 24$
17. $\frac{1}{3} \log_b (x - 2) = \log_b 2$
18. Find $\log 0.0532$.
19. Find antilog 3.4942.
20. Calculate the pH of a solution containing 0.000025 mole of H^+ per liter. Use $pH = -\log [H^+]$.
21. Solve $3^x = 6 \cdot 2^x$ for x to three significant digits.
22. Find $\log_4 25$ to three significant digits.
23. Use $A = p\left(1 + \dfrac{r}{n}\right)^{nt}$ and a calculator to find the value of a $750 investment, paying 10% per year, compounded semiannually for 8 years, to the nearest cent.
24. Draw the graph of $y = \log_2 (x - 3)$. Find the domain, range, and x-intercept.
25. Solve $\log_2 (3c - 4) + \log_2 c = 5$. Check.

Choose the *one* best answer to each question or problem.

1. Let A be any constant function $y = c$. Which statement(s) is (are) true?

   I. A^{-1} is a function.
   II. A^{-1} is a relation.
   III. A and A^{-1} are a pair of inverse relations.

(A) I only   (B) II only
(C) I and II only
(D) II and III only
(E) I, II, and III

2.

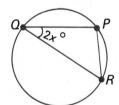

$\overline{QR}$ is a diameter of the circle above. Point P is on the circle so that m $\angle PQR = 2x$. Find m $\angle PRQ$.

(A) $6x$

(B) $\dfrac{180 - 2x}{2}$

(C) $2x - 90$
(D) $2(45 - x)$
(E) There is not enough information.

3. $\log \dfrac{1}{mn} =$ ___?___
(A) $1 - \log m - \log n$
(B) $\log m + \log n$
(C) $-\log m - \log n$
(D) $\log m - \log n$
(E) None of these

4. If x and y are real numbers, and $y^2 = 6 - 2x$, then ___?___
(A) $x \geq 6$ (B) $x \geq 3$
(C) $x \leq 6$ (D) $x \leq 3$
(E) None of these

5. The only solution to the equation $\log_2 x - \log_3 (x + 1) = \log_5 (x - 3)$ is ___?___
(A) 8   (B) 6
(C) 4   (D) 2
(E) None of these

6. If $\log_2 y = x$, then ___?___
(A) $y < x$ (B) $y > x$
(C) $y = x$ (D) $y = 2$
(E) None of these

7. If $\log_b (4x - 8) < \log_b (2x + 6)$, where $b > 1$, then ___?___
(A) $-3 < x < 7$ (B) $0 < x < 7$
(C) $-3 < x < 2$ (D) $2 < x < 7$
(E) None of these

8. If $2^y = 50$ and $y = 2x - 1$, then ___?___
(A) $x = 13$
(B) $16.5 < x < 32.5$
(C) $2.0 < x < 2.5$
(D) $3.0 < x < 3.5$
(E) None of these

9. Find the false statement, if any.
(A) If $y = \log_{\frac{1}{2}} x$, then $x = 2^{-y}$.

(B) $\log_b (x^2 - 9) = \log_b (x + 3) + \log_b (x - 3)$

(C) $\log_5 (x^2 + y^2) = 2 \log_5 x + 2 \log_5 y$

(D) $\log_{\frac{1}{2}} 8 = \log_2 \frac{1}{8}$

(E) None of these

10. The graphs of which pair of functions do *not* intersect?
(A) $y = \log_2 x, \; y = \log_{\frac{1}{2}} x$

(B) $y = 3x - 2, \; y = -3x + 2$
(C) $y = 2^x, \; y = \log_2 x$

(D) $y = 3^x, \; y = \left(\frac{1}{3}\right)^x$

(E) Each pair of graphs intersects.

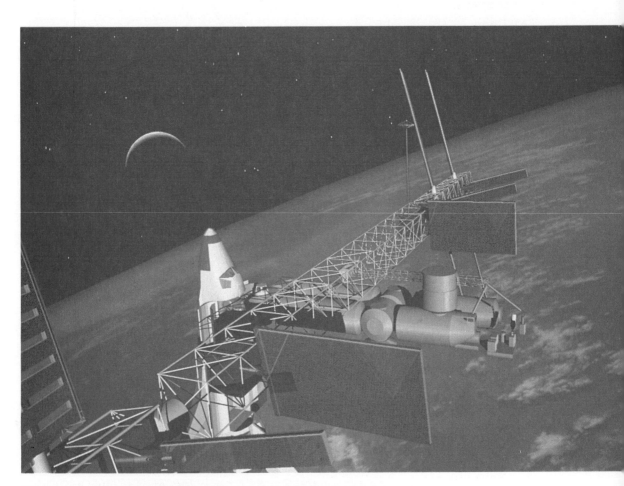

Computers are used by engineers in the aerospace industry
to design and test complex machinery. The image above is
a computer-generated model of the space station
Freedom—a project that is currently being developed at
NASA's Marshall Space Flight Center.

14.1 Fundamental Counting Principle

Objective

To find the number of possible arrangements of objects

There are 3 different paths that joggers can choose to run from the west side of Hudson Park to the center. From the center of the park, a jogger has a choice of 4 paths leading to the east side of the park.

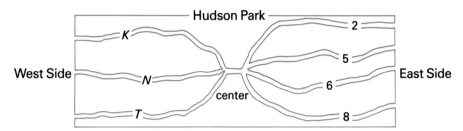

One example of a west-to-east route is to take Path K and then take Path 2. This route can be called $K2$. All of the possible routes are listed below. By counting you find that there are 12.

$K2$ $K5$ $K6$ $K8$ $N2$ $N5$ $N6$ $N8$ $T2$ $T5$ $T6$ $T8$

A west-to-east jogger has to make 2 *decisions*.

1st decision: Choose a west-to-center path.
There are 3 *choices*: K, N, or T.

2nd decision: Choose a center-to-east path.
There are 4 *choices*: 2, 5, 6, or 8.

$$\underset{\text{(letter-choices)}}{3} \quad \times \quad \underset{\text{(number-choices)}}{4} \quad = \quad 12 \text{ possible routes}$$

This illustrates a short way for computing the total number of possible choices. It is called the *Fundamental Counting Principle*.

Fundamental Counting Principle
If a choice can be made in a ways, and for each of these a second choice can be made in b ways, then the choices can be made in $a \times b$ ways.

The Fundamental Counting Principle can be extended to 3 or more choices. If the jogger above wants to make a round trip, west-center-east-center-west, that jogger has to make 4 decisions. The number of possible routes for the round trip is as follows.

$$\underline{\ 3\ } \times \underline{\ 4\ } \times \underline{\ 4\ } \times \underline{\ 3\ } = 144$$

EXAMPLE 1 How many 3-digit numbers can be formed from the 5 digits 1, 2, 4, 7, 9, if no digit may be repeated in a number?

Plan Three decisions must be made, one for each digit. Form a number from the given digits, say 427, to study the available choices.

Solution There are 5 choices for a 100s digit: 1, 2, 4, 7, 9. Choose the 4.
This leaves 4 choices for a 10s digit: 1, 2, 7, 9. Choose the 2.
Three choices remain for a 1s digit: 1, 7, 9. Choose the 7.

Now use the Fundamental Counting Principle.
$$\underline{\ 5\ } \times \underline{\ 4\ } \times \underline{\ 3\ } = 60$$

Thus, 60 3-digit numbers can be formed.

At times, digits may be repeated in a number. For example, at least one digit is repeated in each of the numbers 5,383, 5,355, and 3,838.

EXAMPLE 2 How many 4-digit numbers can be formed from the digits 2, 4, 6, 7, 8, if a digit may be repeated in a number?

Solution There are 5 choices for each of the 4 digits. $\underline{\ 5\ } \times \underline{\ 5\ } \times \underline{\ 5\ } \times \underline{\ 5\ } = 625$

Thus, 625 4-digit numbers can be formed.

EXAMPLE 3 A manufacturer makes school jackets in 4 different colors. Each jacket is available in 5 fabrics, 3 kinds of collars, and a choice of buttons or a zipper. If the jackets are also available with or without the school name, how many different types of jackets does the manufacturer make?

Solution There are 5 decisions to be made.

$$\underset{\text{color}}{\underline{\ 4\ }} \times \underset{\text{fabric}}{\underline{\ 5\ }} \times \underset{\text{collar}}{\underline{\ 3\ }} \times \underset{\substack{\text{button/}\\\text{zipper}}}{\underline{\ 2\ }} \times \underset{\substack{\text{name/}\\\text{no name}}}{\underline{\ 2\ }} = 240$$

Thus, 240 types of jackets are made.

There is a short way to indicate a product of consecutive positive integers starting with 1. For example, $1 \times 2 \times 3 \times 4 \times 5$ is abbreviated as 5!, which is read "5 *factorial*." Other examples using **factorial** notation are shown below.

$$3! = 1 \times 2 \times 3 = 6 \qquad 6! = 1 \times 2 \times 3 \times 4 \times 5 \times 6 = 720$$

For each positive integer n, $n!$ is defined as follows.

$$n! = 1 \cdot 2 \cdot 3 \cdot \ldots \cdot n$$

EXAMPLE 4 In how many different ways can 7 waiters be assigned to 7 tables if each waiter serves one table?

Plan The first waiter has a choice of 7 tables, the second waiter then has a choice of 6 tables, and so on.

Solution $\underline{\ 7\ } \times \underline{\ 6\ } \times \underline{\ 5\ } \times \underline{\ 4\ } \times \underline{\ 3\ } \times \underline{\ 2\ } \times \underline{\ 1\ } = 7!$, or 5,040

A calculator with an $\boxed{x!}$ key can be used to find 7!.
Press 7 $\boxed{x!}$ and read 5,040.
The waiters can be assigned in 5,040 different ways.

Classroom Exercises

In Exercises 1–6, match each item at the left with one item at the right.

1. 3!   **2.** 5!   **3.** 9!

4. the number of digits in our numeration system

5. a 4-digit number in which a digit is repeated

6. a 4-digit number in which no digit is repeated

a. 9   **e.** 1,236
b. 10   **f.** 6
c. 120   **g.** 9 × 8!
d. 2,880   **h.** Infinitely many

7. How many 2-digit numbers can be formed from the 4 digits 2, 3, 5, 8, if no digit may be repeated in a number?

Written Exercises

1. How many 5-digit numbers can be formed from the digits 1, 2, 3, 4, 7, 8, if no digit may be repeated in a number?

2. If a digit may be repeated in a number, how many 5-digit numbers can be formed from the digits 2, 3, 4, 7, 8, 9?

3. Find the number of ways that 8 classes can be assigned to 8 rooms for the first hour of school.

4. In how many ways can 6 different books be placed side by side on a shelf?

5. Find the number of 3-letter arrangements of the letters in the word MONTH if no letter may be repeated in an arrangement.

6. How many 4-letter code words can be formed from the letters in NUMBER if each letter may be repeated in a word?

7. How many different automobile license plates can be made if each plate has a letter followed by three digits?

8. Find the number of 7-digit phone numbers that can be dialed if the first digit cannot be zero.

9. A school cafeteria offers each student 3 choices of meat, 4 choices of vegetable, 5 choices of drink, and 2 choices of fruit. How many different meat-vegetable-drink-fruit lunch trays are available?

10. A car-rental service offers 2-door and 4-door models. Each model is available in 5 different exterior colors, 2 different interior colors, 3 different types of interior upholstery, with or without a radio, and with or without air-conditioning. How many different types of cars are available for rent?

A sports arena has 4 west gates and 6 east gates. Solve Exercises 11–12 using this information.

11. In how many ways can you enter the arena through a west gate and later leave through an east gate?

12. In how many different ways can a person enter and then leave the arena?

13. How many different signals can be shown by arranging 5 flags in a row if 8 different flags are available?

14. In how many different ways can 6 multiple-choice questions be answered if each question has 4 choices for the answers?

15. A baseball coach must present a batting order for his 9 starting hitters before a game can begin. How many different batting orders are possible?

16. In 1988, Jefferson High School had 4 valedictorians. In how many different orders could they have given their graduation speeches?

17. A secretary typed 4 letters and addressed 4 envelopes. The letters were placed in the envelopes at random with one letter per envelope. In how many ways could this be done?

Write each of the following in factorial notation.

18. $1 \times 2 \times 3 \times \cdots (n - 4)(n - 3)(n - 2)$
19. $(n + 3)(n + 2)(n + 1) \cdots \times 3 \times 2 \times 1$

Mixed Review

Write an equation in standard form for the line described. *3.4*

1. slope of $\frac{3}{4}$ and y-intercept at 2

2. through $P(2, -1)$ and $Q(-3, 1)$

3. vertical and through $P(2, -1)$

4. horizontal and through $Q(-3, 1)$

14.2 Permutations

Objective	To find the number of permutations of objects under given conditions

A **permutation** is an arrangement of objects in a definite order.
For example, HDNA and ANDH are two permutations of letters in the word HAND. Altogether, there are 4 × 3 × 2 × 1, or 24 permutations. If two letters are taken at a time, there are only 12 permutations: HA, HN, HD, AN, AD, AH, NA, NH, ND, DH, DA, DN.

EXAMPLE 1   Find the number of permutations of letters in the word UNTIL under the following conditions.
   a. The permutation ends with the letter U or the letter N.
   b. The permutation begins with the prefix UN-.

Solutions   **a.** The 1st decision consists of 2 choices for the last letter: U or N. The number of choices for the remaining decisions follow.

$$\underset{\text{5th}}{1} \times \underset{\text{4th}}{2} \times \underset{\text{3rd}}{3} \times \underset{\text{2nd}}{4} \times \underset{\text{1st}}{2} = 48$$

Thus, 48 permutations of U, N, T, I, L end with U or N.
   b. There is 1 choice for each of the first two letters.

$$\underline{1} \times \underline{1} \times \underline{3} \times \underline{2} \times \underline{1} = 6$$

Thus, 6 permutations of U, N, T, I, L begin with the prefix UN-.

EXAMPLE 2   How many even numbers with one, two, *or* three digits can be formed from the digits 6, 7, 8, if no digit is repeated in a number?

Plan   Since the units digit must be even for the number to be even, there are 2 choices (6 and 8) for the units digit. Add the numbers of permutations for each type of number.

Solution

	Number of permutations	The even numbers
one-digit numbers	2	6 8
two-digit numbers	2 × 2 = 4	76 78 68 86
three-digit numbers	1 × 2 × 2 = 4	876 678 768 786

Thus, there are 2 + 4 + 4, or 10 even numbers.

Example 2 demonstrates that in permutations involving the connector *or*, numbers of permutations are *added* to find the total number.

The conjunction *and* suggests the use of *multiplication*. For example, the list below shows 12 permutations of the letters A, B, C *and* the digits 1 and 2, with the letters together and at the left of the digits.

ABC12   ACB12   BAC12   BCA12   CAB12   CBA12
ABC21   ACB21   BAC21   BCA21   CAB21   CBA21

There are 6 ways to arrange the letters and 2 ways to arrange the digits. The letters *and* digits are arranged in 6×2, or 12 ways.

EXAMPLE 3   Four different biology books and 3 different chemistry books are to be placed on a shelf with the biology books together and at the right of the chemistry books. In how many ways can this be done?

Solution

| Chemistry books (left) | Biology books (right) |

$$(\underbrace{3}_{1st} \times \underbrace{2}_{2nd} \times \underbrace{1}_{3rd}) \times (\underbrace{4}_{1st} \times \underbrace{3}_{2nd} \times \underbrace{2}_{3rd} \times \underbrace{1}_{4th}) = 144$$

There are $3! \times 4!$, or 144 possible arrangements.

Classroom Exercises

Match each description with one or more examples at the right.

1. three even numbers   2. three odd numbers   a. 80, 4, 56, 770

3. numbers with two or more digits   b. 324, 16, 4,000

4. numbers less than 1,000   c. 7, 59, 995

Match the given condition with the number of permutations of all the letters in the word NUMERAL indicated at the right.

5. end with the suffix -ER   d. $1 \cdot 2 \cdot 3 \cdot 4 \cdot 5 \cdot 6 \cdot 6$

6. begin with E or R   e. $2 \cdot 6 \cdot 5 \cdot 4 \cdot 3 \cdot 2 \cdot 1$

7. do not end with R   f. $6 \cdot 5 \cdot 4 \cdot 1 \cdot 3 \cdot 2 \cdot 1$

8. have N for the middle letter   g. $1 \cdot 2 \cdot 3 \cdot 4 \cdot 5 \cdot 1 \cdot 1$

Find the number of permutations of all the letters in the word PINCER under the following conditions. (Exercises 9 and 10)

9. The permutation begins with the letter N, C, or E.

10. The permutation ends with the suffix -NCE.

11. How many even 4-digit numbers can be formed from the digits 1, 4, 8, 9, if a digit may be repeated?

Written Exercises

1. Find the number of permutations of the 6 letters in the word JUNIOR that begin with I and end with O.

2. How many odd 4-digit numbers can be formed from the digits 1, 2, 3, 5, 6, 7, if a digit may be repeated in a number?

3. How many even 5-digit numbers can be formed from the digits 1, 2, 4, 7, 8, if no digit may be repeated in a number?

4. How many numbers of one or more digits can be formed from the digits 1, 2, 3, 4, if no digit may be repeated in a number?

5. If digits may be repeated in a number, how many numbers less than 500 can be formed from the digits 2, 3, 4, 5, 6?

6. In how many ways can 6 different algebra books be placed together on a shelf at the right of 4 different geometry books?

7. Five female cheerleaders and 4 male cheerleaders want to line up for a cheer with the males at the left of the females. In how many ways can this be done?

8. How many 3-digit numbers less than 600 can be formed from 0, 1, 5, 7, 9, without repeating digits? Consider 075 as a 2-digit number.

9. How many numbers less than 350 can be formed from 1, 2, 3, 4, 5, 6, if a digit may be repeated in a number?

10. Find the number of odd numbers less than 1,600 that can be formed from 1, 2, 5, 7, 8, if repetition of digits is allowed.

11. How many 2-, 3-, or 4-digit numbers can be formed from 0, 3, 6, 9, if no digit may be repeated in a number? Consider 093 the same as 93.

12. Five novels and 3 short stories are displayed on a shelf with the novels together and the short stories together. In how many ways can this be done? (NOTE: The novels can be at the left *or* the right.)

13. Find the number of ways that a family of 5 can stand in a line, if the mother and father are not to be separated.

14. In how many ways can 7 students be seated in a row of 7 chairs if 2 of the students are a brother and sister who do not want to sit together?

15. A 2-volume dictionary, a 4-volume atlas, and a 5-volume collection of plays are placed on a shelf with volumes of the same type together. In how many ways can this be done?

16. In how many ways can a tennis game of mixed doubles (male and female versus male and female) be arranged from a group of 6 males and 4 females?

17. The number of 4-flag signals in a row that can be formed when 10 different flags are available is the number of permutations of 10 things taken 4 at a time, or $_{10}P_4$, where $_{10}P_4 = 10 \cdot 9 \cdot 8 \cdot 7 = 5{,}040$.

In general, $_nP_r = n(n - 1)(n - 2)\cdots(n - r + 1)$.

Prove that $_nP_r = \dfrac{n!}{(n - r)!}$.

Use the formula $_nP_r = \dfrac{n!}{(n - r)!}$ from Exercise 17 for Exercises 18–21.

Find the value of the expression, or the factored form in terms of n.

18. $_8P_3$ **19.** $_{10}P_6$ **20.** $_nP_2$ **21.** $_nP_4$

Mixed Review

Given $f(x) = 3x - 4$ and $g(x) = x^2 + 2$, find the following.   *3.2, 5.5, 5.9, 7.1*

1. $f(-2)$ **2.** $g(5c)$

3. $g(f(a + 1))$ **4.** $\dfrac{f(x + h) - f(x)}{h}$

Using the Calculator

Use a calculator with the ⌈x!⌉ function to compute the following. Answers greater than 1,000,000 may be left in scientific notation.

1. $3! \cdot 4!$ **2.** $6! \cdot 5! \cdot 3!$ **3.** $4! + 7!$

4. $8! + 7! + 6! + 5!$ **5.** $15! \cdot 14!$ **6.** $15! + 14!$

Brainteaser

The first fifty natural numbers are multiplied together. How many zeros appear in the right part of the product?

Application: *The Decibel Scale*

The decibel (dB) scale is used to measure sound intensity. Since the human ear responds to changes in sound pressure, the decibel scale is based on the ratio of a measured sound pressure P to a reference value P_0. The most commonly used value for P_0 is 0.0002 dynes/cm^2, the lowest sound pressure that can be detected through the air by the average human ear. Decibel values can be computed using the formula

$dB = 20 \log_{10} \dfrac{P}{P_0}$, where P and P_0 are measured in dynes/cm^2.

1. A rock band produces a sound pressure P of 1,000 dynes/cm^2. To the nearest decibel, what is the sound intensity? Use $P_0 = 0.0002$ dynes/cm^2.

2. In underwater acoustics, a different reference value P_0 is used. Suppose an underwater sound has a sound pressure of 10,000 dynes/cm^2. The decibel level is calculated to be 80 dB. What reference value was used?

14.3 Distinguishable Permutations

Objectives
To find the value of an expression with the factorial notation
To find the number of distinguishable permutations of objects

The five letters in the word DAILY can be arranged in 5!, or 120 different ways. In the word DADDY, there are fewer than 5! *distinguishable* arrangements because the 3 Ds are alike. Consider the arrangement ADDYD. The 3 Ds can be arranged in 3! different ways.

$$AD_1D_2YD_3 \quad AD_1D_3YD_2 \quad AD_2D_1YD_3 \quad AD_2D_3YD_1 \quad AD_3D_1YD_2 \quad AD_3D_2YD_1$$

Without the subscripts, these 6 arrangements are not distinguishable. Thus, there are $\frac{5!}{3!}$ *distinguishable permutations* of the letters in the word DADDY. This example illustrates the following rule.

The number of **distinguishable permutations** of n objects of which r of the objects are alike is $\frac{n!}{r!}$.

You can extend this rule to find the number of distinguishable permutations of n objects of which more than one set of the objects are alike.

EXAMPLE 1
How many distinguishable 7-digit numbers can be formed from the digits of 6,563,656?

Plan
Find the number of permutations of 7 digits of which four are 6s and two are 5s. Use the fact that $7! = 7 \cdot 6 \cdot 5 \cdot (4!)$.

Solution

$$\frac{7!}{4!2!} = \frac{7 \cdot \overset{3}{\cancel{6}} \cdot 5 \cdot \overset{1}{\cancel{(4!)}}}{\underset{1}{\cancel{(4!)}} \cdot \underset{1}{\cancel{2}} \cdot 1} = 7 \cdot 3 \cdot 5 = 105$$

There are 105 distinguishable 7-digit numbers.

EXAMPLE 2
Find the number of distinguishable signals that can be formed by displaying 12 flags in a row if 4 flags are green, 5 are red, 2 are blue, and 1 is yellow.

Solution

$$\frac{12!}{4!5!2!} = \frac{\overset{1}{\cancel{12}} \cdot 11 \cdot 10 \cdot 9 \cdot \overset{4}{\cancel{8}} \cdot 7 \cdot \overset{3}{\cancel{6}} \cdot \overset{1}{\cancel{(5!)}}}{\underset{1}{\cancel{4}} \cdot \cancel{3} \cdot \underset{1}{\cancel{2}} \cdot 1 \cdot \underset{1}{\cancel{(5!)}} \cdot \underset{1}{\cancel{2}} \cdot 1} = 83,160$$

Thus, 83,160 distinguishable signals can be formed.

Which number does not belong in the list?

1. $\dfrac{12!}{8!}$; $99(5!)$; $\dfrac{8!4!}{8!}$; $\dfrac{12 \cdot 11 \cdot 10 \cdot 9 \cdot (8!)}{8!}$

2. $9!$; $5! + 4!$; 12^2; $4! \cdot 3!$

3. $\dfrac{6!}{3!2!}$; $\dfrac{5!}{2!}$; $\dfrac{5 \cdot 4!}{2}$; $\dfrac{4 \cdot 5!}{3!}$

Classroom Exercises

Find the number of distinguishable permutations of all the letters in the word.

1. ace   **2.** add   **3.** toot   **4.** poor   **5.** deemed

Written Exercises

Find the value of each expression.

1. $\dfrac{8!}{4!}$ **2.** $\dfrac{11!}{8!}$ **3.** $\dfrac{14!}{11!3!}$ **4.** $\dfrac{10!}{3!6!2!}$ **5.** $\dfrac{12!}{7!5!3!}$

Find the number of distinguishable permutations of all the letters in the word.

6. beef   **7.** puppy   **8.** pepper   **9.** scissors
10. parallel   **11.** murmur   **12.** nonsense   **13.** Tennessee

14. How many distinguishable signals can be formed with 10 flags in a line if 3 flags are black, 4 are orange, and the remainder are green?

15. In how many ways can 4 nickels, 2 dimes, 3 quarters, 1 half-dollar, and 1 penny be distributed among 11 children if each child is to receive 1 coin?

Simplify.

16. $\dfrac{n!}{(n-2)!}$ **17.** $\dfrac{(n+3)!}{(n+1)!}$ **18.** $\dfrac{(n+1)!}{(n-1)!}$ **19.** $\dfrac{(n-1)!}{(n-3)!}$

Mixed Review

1. Simplify $\dfrac{3}{5-i}$.   *9.2*

2. Solve $x^4 - 2x^3 + 6x^2 - 18x = 27$ *10.4*

3. Solve $\log_5 t + \log_5 (t-2) = \log_5 24$. Check.   *13.5*

4. Simplify $\log_b 4c + 2\log_b 3 - \log_b 6d$.   *13.6*

14.4 Circular Permutations

Objective	To find the number of circular permutations

Three objects A, B, and C can be arranged in a row in 3!, or 6 ways. However, they can be placed in a circle in only 2!, or 2 ways.

Consider a lazy Susan (revolving platter) with trays A, B, and C. As shown at the right, the two arrangements on top are different. However, each arrangement below the first row can be found from the one above it by rotating the tray. Therefore, they are not different arrangements.

In a circular arrangement of n objects, there is *no* first position. One object is selected as a "base" and the remaining objects are arranged *relative to the first* in $(n - 1)!$ ways.

The number of **circular permutations** of n objects is $(n - 1)!$.

EXAMPLE 1	In how many ways can 8 different types of bushes be planted in a circle around a flagpole?
Plan	Arrange 7 of the bushes in relation to 1 "base" bush.
Solution	$(n - 1)! = (8 - 1)! = 7! = 5,040$
	There are 5,040 different arrangements.

EXAMPLE 2	In how many ways can 3 males and 3 females be seated around a circular table if no 2 females may be seated next to each other?
Plan	Seat one person as a base. The remaining two of the *same sex* can be seated in 2! ways. The 3 members of the opposite sex can be seated in 3! ways.
Solution	The 6 people can be seated alternately in $2! \cdot 3!$, or 12 ways.

If keys are placed on a ring, the ring can be turned around so that the two views, front and back, represent just one arrangement. So, the number of permutations of n objects on a *ring* is $\dfrac{(n - 1)!}{2}$.

Classroom Exercises

Describe each arrangement as linear, circular, or either of these.

1. books on a table   **2.** keys on a key ring   **3.** digits in a 3-digit number

4. charms on a bracelet   **5.** trees in a park   **6.** letters in the word LETTERS

Tell which number of permutations is the greater for each pair below.

7. six people in a *line* or 6 people in a *circle*

8. all the letters in POLO or all the letters in POLE

Written Exercises

1. Find the number of ways that a football team of 11 players can form a circular huddle.

2. Find the number of different ways that a landscaper can plant 10 different types of bushes around a circular flower bed.

3. In how many ways can the principal and 7 department heads at Jackson High School be seated around a circular conference table?

4. A mother and daughter each invite two friends to lunch. In how many ways can all of them be seated around a circular table?

5. Find the number of ways that 4 teachers and 4 students can be seated alternately in a circular discussion group.

6. In how many ways can 5 couples be seated around a circular banquet table if no 2 men may be seated by each other?

7. In how many ways can 4 keys be arranged on a key ring?

8. In how many ways can 6 charms be arranged on a bracelet?

9. In how many ways can a family of 6 be seated around a circular table if the mother and father must not be seated next to each other?

Solve each equation for n.

10. $(n - 1)! = 720$ **11.** $\dfrac{(n - 1)!}{2} = 60$

Mixed Review

Find the slope and describe the slant of $\overleftrightarrow{AB}$.   *3.3*

1. $A(-2,3)$, $B(4,-2)$ **2.** $A(1,3)$, $B(7,3)$ **3.** $A(-1,-2)$, $B(3,4)$ **4.** $A(8,2)$, $B(8,7)$

14.5 Combinations

Objectives

To find the number of possible selections of n objects taken r at a time without regard to order

To find the value of $\binom{n}{r}$ for $n \geq r$

A selection of objects *without regard to order* is called a **combination**.

How many combinations of 4 letters can be selected from the 6 letters A, B, C, D, E, F? The list below shows all 15 combinations.

ABCD	ABDE	ACDE	BCDE	DEFA
ABCE	ABDF	ACDF	BCDF	DEFB
ABCF	ABEF	ACEF	BCEF	DEFC

The symbol $\binom{6}{4}$ is used to denote *the number of combinations of 6 things taken 4 at a time*. As shown above, $\binom{6}{4} = 15$.

To evaluate $\binom{6}{4}$ without listing combinations, you can reason as follows.

There are $6 \cdot 5 \cdot 4 \cdot 3$, or 360, different *arrangements* of 4 letters from the 6. Each *combination* of 4 letters has 4! arrangements. Thus,

$$\binom{6}{4} = \frac{6 \cdot 5 \cdot 4 \cdot 3}{4 \cdot 3 \cdot 2 \cdot 1} = 15.$$

This fraction can also be expressed using factorial notation.

$$\binom{6}{4} = \frac{6 \cdot 5 \cdot 4 \cdot 3}{4 \cdot 3 \cdot 2 \cdot 1} = \frac{6 \cdot 5 \cdot 4 \cdot 3 \cdot 2 \cdot 1}{4!2!} = \frac{6!}{4!2!}$$

In general, $\binom{n}{r}$ is the number of combinations of n things taken r at a time. For positive integers n and r, where $n > r$,

$$\binom{n}{r} = \frac{n!}{r!(n-r)!}.$$

EXAMPLE 1

How many different 4-member committees can be formed from 9 people.

Plan

Find $\binom{9}{4}$, the number of combinations of 9 people taken 4 at a time.

Solution

$$\binom{9}{4} = \frac{9!}{4!(9-4)!} = \frac{9!}{4!5!} = \frac{9 \cdot 8 \cdot 7 \cdot 6}{4 \cdot 3 \cdot 2 \cdot 1} = 126$$

Thus, 126 different 4-member committees can be formed.

In Example 1, notice that for each group of 4 people selected, a group of 5 people is *not* selected. Therefore $\binom{9}{4} = \binom{9}{5}$, or $\dfrac{9!}{4!5!} = \dfrac{9!}{5!4!}$.

This suggests that $\binom{n}{r} = \binom{n}{n-r}$.

EXAMPLE 2   Eight points, A through H, are on a circle.
a. How many lines are determined by the 8 points?
b. How many triangles are determined by the 8 points?

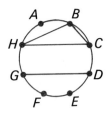

Solutions   a. A line is determined by any 2 points.

$$\binom{8}{2} = \frac{8!}{2!6!} = \frac{8 \cdot 7}{2 \cdot 1} = 28$$

Thus, 28 lines are determined by the 8 points.

b. A triangle is determined by any 3 noncollinear points.

$$\binom{8}{3} = \frac{8!}{3!5!} = \frac{8 \cdot 7 \cdot 6}{3 \cdot 2 \cdot 1} = 56$$

Calculator steps: 8 $\boxed{x!}$ $\boxed{\div}$ 3 $\boxed{x!}$ $\boxed{\div}$ 5 $\boxed{x!}$ $\boxed{=}$ 56

Thus, 56 triangles are determined by the 8 points.

With the *definition*, $0! = 1$, the meaning of $\binom{n}{r} = \binom{n}{n-r}$ can be extended to include $r = n$ and $r = 0$.

For example, consider the number of combinations of 9 objects taken 9 at a time: $\binom{9}{9} = \dfrac{9!}{9!(9-9)!} = \dfrac{9!}{9!0!} = \dfrac{1}{0!} = 1$.

Similarly, $\binom{9}{0} = \dfrac{9!}{0!(9-0)!} = \dfrac{9!}{0!9!} = \dfrac{1}{0!} = 1$.

Classroom Exercises

Tell whether a permutation or a combination is indicated.
(Exercises 1–6)

1. a 3-letter name of a triangle

2. a 5-digit number

3. a 6-person committee

4. a 5-card hand in a card game

5. a 4-letter code word

6. a 5-flag signal

7. Find the 4 different amounts of money determined by selecting 3 coins from 1 penny, 1 nickel, 1 dime, and 1 quarter.

8. How many 3-member committees can be formed from 7 people?

9. How many 5-member committees can be formed from 7 people?

Written Exercises

Find the value of each expression.

1. $\binom{9}{2}$ **2.** $\binom{9}{7}$ **3.** $\binom{15}{3}$ **4.** $\binom{8}{8}$ **5.** $\binom{7}{0}$ **6.** $\binom{50}{1}$

7. Find the number of lines determined by 12 points, A through L, on a circle.

8. Find the number of triangles determined by 14 points, A through N, on a circle.

9. How many different 12-member juries can be formed if 18 people are available for jury duty?

10. In how many ways can a team of 15 soccer players select 3 captains for its next game?

11. Find the number of different 5-card hands that can be drawn from a deck of 52 cards.

12. A student must select 8 of 10 functions to be graphed. How many selections are possible?

13. How many baseball games will be played by 12 teams if each team plays each other team once?

14. Each of 6 teams in a league will play each other team 3 times. How many games will be played?

Solve each equation for n.

15. $\binom{n}{4} = \binom{12}{8}$ **16.** $\binom{n}{3} = \binom{n}{7}$ **17.** $\binom{n}{5} = \binom{8}{0}$ **18.** $\binom{n}{n-1} = \binom{9}{1}$

Find 4 pairs of integers (n, r) for each equation.

19. $\binom{n}{6} = \binom{9}{r}$ **20.** $\binom{n}{20} = \binom{30}{r}$ **21.** $\binom{8}{r} = \binom{n}{2}$ **22.** $\binom{n}{10} = \binom{12}{r}$

23. Prove: $\binom{n}{r} = \binom{n}{n-r}$ for all integers r and n, where $0 \le r \le n$.

Mixed Review 1.3, 14.2

1. Evaluate $5(3x - 2y) - (12x + 10y)$ for $x = -3$ and $y = -4$.

2. Evaluate $3x^2y - xy - 5xy^2$ for $x = 5$ and $y = -2$.

3. In how many ways can 5 girls and 3 boys stand in a line with the girls at the left and the boys at the right?

4. How many numbers of 1 or more digits can be formed from the digits 4, 5, 6, if no digit is repeated in a number?

Using the Calculator

Use a scientific calculator to find the value of each expression.

1. $\binom{3}{1} + \binom{3}{2} + \binom{3}{3}$ **2.** $\binom{4}{1} + \binom{4}{2} + \binom{4}{3} + \binom{4}{4}$ **3.** $\binom{12}{4} \cdot \binom{8}{3} \cdot \binom{4}{2}$

4. $\binom{8}{5} \cdot \binom{10}{4} + \binom{8}{4} \cdot \binom{10}{5}$ **5.** $\binom{7}{3} \cdot \binom{5}{2} + \binom{7}{4} \cdot \binom{5}{1} + \binom{7}{5}$

14.6 Conditional Combinations

To find the number of combinations of objects given certain conditions.

Recall that in Lesson 14.2, multiplication is used with the conjunction *and*, while addition is used with the disjunction *or*.

If a box contains 7 yellow cards and 5 white cards, you can select 2 yellow cards in $\binom{7}{2}$ ways and 3 white cards in $\binom{5}{3}$ ways. So, 2 yellow cards *and* 3 white cards can be selected in

$$\binom{7}{2} \times \binom{5}{3} \text{ ways (multiplication)},$$

2 yellow cards *or* 3 white cards can be selected in

$$\binom{7}{2} + \binom{5}{3} \text{ ways (addition)},$$

and 1 *or* more white cards can be selected in

$$\binom{5}{1} + \binom{5}{2} + \binom{5}{3} + \binom{5}{4} + \binom{5}{5} \text{ ways}.$$

EXAMPLE 1 In how many ways can at least 2 girls be selected from among 4 girls?

Solution "At least 2 girls" means 2 girls, *or* 3 girls, *or* 4 girls.

$$\binom{4}{2} + \binom{4}{3} + \binom{4}{4} = \frac{4!}{2!2!} + \frac{4!}{3!1!} + \frac{4!}{4!0!} = 6 + 4 + 1 = 11$$

There are 11 ways to select at least 2 girls from among 4 girls.

EXAMPLE 2 Six boys and 3 girls are eligible for a 5-member team.
  a. In how many ways can the team be formed with exactly 3 boys?
  b. In how many ways can the team be formed with at least 2 girls?

Solutions **a.** "Exactly 3 boys" implies 3 boys *and* 2 girls. Use multiplication.

$$\binom{6}{3} \cdot \binom{3}{2} = \frac{6!}{3!3!} \cdot \frac{3!}{2!1!} = 20 \cdot 3 = 60$$

There are 60 ways that the team of 5 can have exactly 3 boys.

b. "At least 2 girls" on a team of 5 implies either 2 girls *and* 3 boys or 3 girls *and* 2 boys. Use multiplication coupled with addition.

$$\binom{3}{2} \cdot \binom{6}{3} + \binom{3}{3} \cdot \binom{6}{2} = \frac{3!}{2!1!} \cdot \frac{6!}{3!3!} + \frac{3!}{3!0!} \cdot \frac{6!}{2!4!} = 75$$

There are 75 ways that the team can have at least 2 girls.

Classroom Exercises

Five males and 8 females are eligible to begin training as astronauts. Match each selection described with one expression at the right.

1. 3 males or 4 females
2. 3 males and 4 females
3. 3 or more males
4. at least 6 females

a. $\binom{8}{6} + \binom{8}{7} + \binom{8}{8}$ d. $\binom{5}{3} + \binom{5}{4} + \binom{5}{5}$

b. $\binom{8}{6}$ e. $\binom{5}{3}$

c. $\binom{5}{3} \cdot \binom{8}{4}$ f. $\binom{5}{3} + \binom{8}{4}$

A 4-member shuttle crew will be selected from the 5 males and 8 females. Match the crew description with one expression at the right. (Exercises 5–8)

5. exactly 2 males
6. exactly 3 females
7. at least 3 females
8. no females

g. $\binom{5}{2}$ j. $\binom{8}{3} \cdot \binom{5}{1}$

h. $\binom{8}{3} + \binom{8}{4}$ k. $\binom{5}{2} \cdot \binom{8}{2}$

i. $\binom{5}{4} \cdot \binom{8}{0}$ l. $\binom{8}{3} \cdot \binom{5}{1} + \binom{8}{4} \cdot \binom{5}{0}$

9. Find the number of different ways to select exactly 3 boys from 6 boys and 4 girls for a scholarship if just 4 scholarships are given.

Written Exercises

A box contains 8 blue chips and 10 yellow chips. Find the number of different ways to select the following.

1. 5 blue chips and 3 yellow chips
2. exactly 3 blue chips
3. 2 blue chips or 8 yellow chips
4. 6 or more blue chips
5. exactly 9 yellow chips
6. at least 8 yellow chips

From a group of 15 females and 10 males, a 6-member team will be selected to represent their school at a science fair.

7. In how many ways can the team of 6 be selected if it is to have exactly 4 females?

8. Find the number of different 6-member teams that have at least 4 males.

9. How many teams of 6 could have at least 5 females?

10. In how many ways can a team of 6 have exactly 3 males?

11. How many different amounts of money can be obtained by selecting 2 or 3 coins from 1 penny, 1 nickel, 1 dime, and 1 quarter?

12. A box contains 1 silver dollar, 1 half-dollar, 1 quarter, 1 dime, 1 nickel, and 1 penny. How many different amounts of money can be obtained by selecting one or more coins from the box?

Three red marbles, 4 white marbles, and 5 blue marbles are numbered
1 through 12. Use this information in Exercises 13–16.

13. In how many ways can 2 red, or 2 white, or 3 blue marbles be selected?

14. Find the number of ways to select 3 red, 3 white, and 2 blue marbles.

15. If 5 marbles are selected, find the number of ways that exactly 3 of them can be blue.

16. If 6 marbles are selected, in how many ways can at least 4 of them be blue?

Five sophomores, 8 juniors, and 10 seniors are eligible for selection to a
5-member group that will tour Europe during the summer.

17. In how many ways can the group of 5 be selected and have 1 sophomore, 2 juniors, and 2 seniors?

18. Find the number of ways to select the 5 members so that there are exactly 2 juniors in the group.

19. How many 5-member groups can have at least 4 seniors?

The number of combinations of n things taken r at a time, or $\binom{n}{r}$, can
be denoted by $_nC_r$ where $_nC_r = \dfrac{n!}{r!(n-r)!}$.

Use the formula above to evaluate the following. (Exercises 20–24)

20. $_{20}C_{17}$ 21. $_{53}C_{53}$ 22. $_{16}C_1$ 23. $_8C_3 \cdot {_7C_4}$

24. $_4C_0 + {_4C_1} + {_4C_2} + {_4C_3} + {_4C_4}$

25. Show that $_5C_0 + {_5C_1} + {_5C_2} + {_5C_3} + {_5C_4} + {_5C_5} = 2^5$.

26. Show that $_6C_6 + {_6C_5} + {_6C_4} + {_6C_3} + {_6C_2} + {_6C_1} = 2^6 - 1$.

27. Prove: $\binom{n}{r} + \binom{n}{r+1} = \binom{n+1}{r+1}$ for all integers r and n, $0 \le r \le n$.

Midchapter Review

For Exercises 1 and 2, the digits 1, 2, 3, 5, 6, 7 are available. How many
numbers, described below, can be formed from these 6 digits?   14.1, 14.2

1. four-digit numbers if no digit is repeated in a number

2. three-digit odd numbers if a digit may be repeated in a number

3. Find the number of distinguishable permutations of all the letters in the word NINETEEN.   14.3

4. In how many ways can 6 people be seated around a circular table?   14.4

5. Each of 8 teams in a baseball league will play the other teams 5 times each. How many games will be played?   14.5

6. From a group of 6 seniors and 4 juniors, in how many ways can a 5-member committee be selected with at least 4 seniors on the committee?   14.6

14.7 Probability

To find the probability of an event

When a die (plural: dice) is tossed, the upper face will show 1, 2, 3, 4, 5, or 6 dots. These are equally likely outcomes if the die is "fair." If the die is tossed many times, the 4-dot face should appear on top about 1 out of 6 times, or $\frac{1}{6}$ of the time. Thus, the *probability* of rolling a 4 in one toss is $\frac{1}{6}$. In the ratio $\frac{1}{6}$, 1 is the number of successful outcomes s (ways the desired event can occur), and 6 is the total number of possible outcomes t. This probability is written as follows.

$$P(4) = \frac{s}{t} = \frac{1}{6}$$

The set of outcomes {1, 2, 3, 4, 5, 6} is called a *sample space* for rolling a die. A **sample space** is the set of *all* possible outcomes; a *subset* of the sample space is called an **event**.

If a sample space consists of t equally likely outcomes, then the **probability** of an event E is given by the following formula.

$$P(E) = \frac{s}{t} = \frac{\text{number of successful outcomes}}{\text{total number of outcomes}}$$

EXAMPLE 1

In one toss of a die, what is the probability of each event?
a. rolling a 5
b. rolling an odd number
c. rolling a number less than 7
d. rolling a 7

Plan

The sample space is {1, 2, 3, 4, 5, 6}. There are 6 possible outcomes in this set, so $t = 6$.

Solutions

a. rolling a 5
   1 successful outcome: $s = 1$
   $P(5) = \frac{s}{t} = \frac{1}{6}$
c. rolling a number less than 7
   6 successful outcomes: $s = 6$
   $P(\text{number} < 7) = \frac{s}{t} = \frac{6}{6} = 1$
   This event *is certain to occur.*

b. rolling an odd number: 1, 3, 5
   3 successful outcomes: $s = 3$
   $P(\text{odd number}) = \frac{s}{t} = \frac{3}{6}$, or $\frac{1}{2}$
d. rolling a 7
   no successful outcome: $s = 0$
   $P(7) = \frac{s}{t} = \frac{0}{6}$, or 0
   This event *cannot occur.*

Example 1 illustrates the *range* for the probability of an event E. If E cannot occur, $P(E) = 0$. If E is certain to occur, $P(E) = 1$. Thus:

$$0 \le P(E) \le 1$$

For a family with 3 children, there are 8 possibilities for the order in which boys B and girls G can be born. These outcomes are shown in the sample space below, where BGG means the first is a boy and the next two are girls.

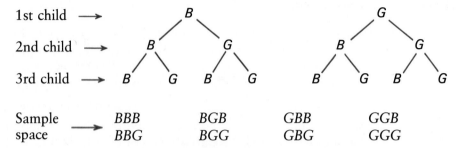

1st child ⟶

2nd child ⟶

3rd child ⟶

| Sample space ⟶ | BBB | BGB | GBB | GGB |
| | BBG | BGG | GBG | GGG |

Use this sample space for Example 2. Assume that the 8 outcomes are equally likely.

EXAMPLE 2   For a family of 3 children, find the probability that the children can be described as follows.

  a. boy, girl, boy   **b.** exactly 2 girls   **c.** at least 1 boy

Solutions   **a.** boy, girl, boy   There is 1 successful outcome: BGB.

  $$P(BGB) = \frac{s}{t} = \frac{1}{8}$$

  b. exactly 2 girls   There are 3 successful outcomes: BGG, GBG, GGB.

  $$P(\text{exactly 2 girls}) = \frac{s}{t} = \frac{3}{8}$$

  c. at least 1 boy   There are 7 successful outcomes: all except GGG.

  $$P(\text{at least 1 boy}) = \frac{s}{t} = \frac{7}{8}$$

In Example 3, cards are drawn at random from a deck of 52 playing cards. "At random" means that each card has an equal chance of being drawn. The 52 cards provide the sample space. They appear in 2 colors (red, black), 4 suits (clubs, diamonds, hearts, spades), 3 types of face cards (kings, queens, jacks), aces, and 9 numbers (2–10), as shown below.

faces

clubs	−	2 3 4 5 6 7 8 9 10 J Q K A	−	black
diamonds	−	2 3 4 5 6 7 8 9 10 J Q K A	−	red
hearts	−	2 3 4 5 6 7 8 9 10 J Q K A	−	red
spades	−	2 3 4 5 6 7 8 9 10 J Q K A	−	black

EXAMPLE 3   One card is drawn at random from an ordinary deck of 52 cards. Find the probability of each event.

a. a red face card   **b.** the 6 of clubs   **c.** a king   **d.** not a king

Solutions

a. red face card
6 face cards are red: $s = 6$

$P(\text{red face card}) = \frac{6}{52}$, or $\frac{3}{26}$

b. 6 of clubs
$s = 1$

$P(\text{6 of clubs}) = \frac{1}{52}$

c. a king
4 cards are kings: $s = 4$

$P(\text{king}) = \frac{4}{52}$, or $\frac{1}{13}$

d. not a king
$s = 52 - 4 = 48$

$P(\text{not a king}) = \frac{48}{52}$, or $\frac{12}{13}$

In Example 3, notice that $P(\text{king}) = \frac{1}{13}$, $P(\text{not a king}) = \frac{12}{13}$, and $\frac{1}{13} + \frac{12}{13} = 1$. This fact can be generalized as follows.

If $P(E)$ is the probability of an event E occurring, and $P(\text{not } E)$ is the probability of E not occurring, then the following is true.

$$P(E) + P(\text{not } E) = 1 \quad \text{and} \quad P(\text{not } E) = 1 - P(E)$$

To understand the meaning of the word *odds* in an expression such as "The odds are 3 to 2" that an event will occur, consider drawing a red marble from a bag filled with 8 red and 6 blue marbles.

In this case, $t = 14$, the total number of possible outcomes,
$s = 8$, the number of successful outcomes, and
$u = 6$, the number of unsuccessful outcomes.

The odds *for* drawing a red marble are 8 to 6, or 4 to 3, the *ratio of s to u*. The odds *against* drawing a red marble are 3 to 4, the *ratio of u to s*.

In general, if an event E has s successful outcomes and u unsuccessful outcomes, the **odds** for E occurring are s to u, and the odds against E occurring are u to s.

Classroom Exercises

One die is tossed. Find the probability that the top face will show the following number of dots.

1. an even number   **2.** a prime number   **3.** a number greater than 6

Use the sample space for the 3-child family, shown on the previous page, to find the number of successful outcomes for each event.

4. two girls and a boy
5. boy, boy, girl (in that order)
6. exactly 1 girl
7. at least 1 girl
8–11. Find the probability of each event for Classroom Exercises 4–7.

Written Exercises

For a 3-child family, find the indicated probability of each event.

1. P(girl, then 2 boys)   **2.** P(exactly 2 boys)   **3.** P(at least 2 boys)

4. P(oldest child is a boy)   **5.** P(at least 1 boy and at least 1 girl)

One card is drawn at random from an ordinary deck of 52 playing cards. Find the indicated probability of each event.

6. P(black)   **7.** P(face card)   **8.** P(not a 2)

9. P(heart)   **10.** P(7 of spades)   **11.** P(not a diamond)

12. What are the odds for drawing a red marble from a bag containing 12 red and 18 white marbles?

13. What are the odds against drawing a club at random from a deck of 52 playing cards?

In Exercises 14–16, $P(E) = \frac{3}{4}$. Find each of the following.

14. P(not E)   **15.** the odds for E   **16.** the odds against E

Prepare a sample space showing all outcomes for tossing 3 coins. Use T for tails and H for heads. Find the probability of each event.

17. 3 tails   **18.** exactly 1 tail   **19.** 2 heads and a tail

20. at least 2 heads   **21.** THT **22.** exactly 2 heads

Prepare a sample space showing all possible outcomes for boys and girls in a 4-child family. Find the probability of each event.

23. 3 girls and a boy   **24.** 2 girls and 2 boys   **25.** exactly 2 girls

26. at least 2 boys   **27.** at least 1 girl and at least 1 boy

28. The probability that it will rain on any given day in April in Pensacola is $\frac{7}{10}$. About how many days in April is it likely not to rain?

29. If the probability that it will rain on a given day in Denver is $\frac{2}{9}$, what are the odds that it will not rain on a given day?

Mixed Review

Find the solution set.   2.2, 2.3, 6.4

1. $5 - x < 7 \text{ or } 9 < 2x - 3$

2. $8 - x < 2 \text{ and } 4 > 2x - 14$

3. $|2x - 7| < 5$

4. $x^2 - 5x - 14 > 0$

14.8 Adding and Multiplying Probabilities

Objectives

To find the probability of inclusive and mutually exclusive events
To find the probability of independent events

To determine $P(A \text{ or } B)$, begin by deciding whether the two events A and B are *inclusive* or *mutually exclusive events*. If two events can occur at the same time, they are called **inclusive events**. If they cannot occur at the same time, they are called **mutually exclusive events**.

Both types of events can be illustrated by tossing a pair of dice. Suppose that the first die is red and the second die is green. Then the ordered pair (3,5) represents "3 on the red and 5 on the green." The 36-pair sample space for all such outcomes is shown at the right.

Green						
6	(1,6)	(2,6)	(3,6)	(4,6)	(5,6)	(6,6)
5	(1,5)	(2,5)	(3,5)	(4,5)	(5,5)	(6,5)
4	(1,4)	(2,4)	(3,4)	(4,4)	(5,4)	(6,4)
3	(1,3)	(2,3)	(3,3)	(4,3)	(5,3)	(6,3)
2	(1,2)	(2,2)	(3,2)	(4,2)	(5,2)	(6,2)
1	(1,1)	(2,1)	(3,1)	(4,1)	(5,1)	(6,1)
	1	2	3	4	5	6

Red

Let r and g represent outcomes on the red die and green die, respectively. There are 12 pairs where $r \geq 5$ and 18 pairs where $g \leq 3$. Since $(r \geq 5)$ and $(g \leq 3)$ have 6 pairs in common, they are *inclusive events*. The 6 pairs cannot be counted twice for $(r \geq 5 \text{ or } g \leq 3)$.

Green						
6					(5,6)	(6,6)
5					(5,5)	(6,5)
4					(5,4)	(6,4)
3	(1,3)	(2,3)	(3,3)	(4,3)	(5,3)	(6,3)
2	(1,2)	(2,2)	(3,2)	(4,2)	(5,2)	(6,2)
1	(1,1)	(2,1)	(3,1)	(4,1)	(5,1)	(6,1)
	1	2	3	4	5	6

Red

The number of pairs (outcomes) for this event is $12 + 18 - 6$, or 24.

Therefore, $P(r \geq 5 \text{ or } g \leq 3) = \frac{s}{t} = \frac{12 + 18 - 6}{36} = \frac{24}{36} = \frac{2}{3}$.

This result can also be found by the alternate method below.

$$P(r \geq 5) = \frac{12}{36} = \frac{1}{3} \qquad P(g \leq 3) = \frac{18}{36} = \frac{1}{2}$$
$$P(r \geq 5 \text{ and } g \leq 3) = \frac{6}{36} = \frac{1}{6}$$

Then, in terms of the above probabilities,

$$P(r \geq 5 \text{ or } g \leq 3) = P(r \geq 5) + P(g \leq 3) - P(r \geq 5 \text{ and } g \leq 3)$$
$$= \frac{1}{3} + \frac{1}{2} - \frac{1}{6} = \frac{2}{3}$$

Now consider the events $(g \le 2)$ and $(g \ge 6)$. These are mutually exclusive events because they cannot occur at the same time (see the sample space). There are 12 pairs where $g \le 2$ and 6 pairs where $g \ge 6$, with *no* pairs common to both sets. Therefore:

$$P(g \le 2 \ or \ g \ge 6) = \frac{s}{t} = \frac{12 + 6}{36} = \frac{1}{2}$$

This result can also be computed as follows.

$$P(g \le 2) = \frac{12}{36} = \frac{1}{3} \qquad P(g \ge 6) = \frac{6}{36} = \frac{1}{6}$$

Then $P(g \le 2 \ or \ g \ge 6) = P(g \le 2) + P(g \ge 6) = \frac{1}{3} + \frac{1}{6} = \frac{1}{2}$.

Probability of (A or B)
If A and B are inclusive events, then
 $P(A \ or \ B) = P(A) + P(B) - P(A \ and \ B)$.
If A and B are mutually exclusive events, then
 $P(A \ or \ B) = P(A) + P(B)$.

EXAMPLE 1   Use the 36-pair sample space on the previous page for tossing a red die and a green die to find the probabilities or odds below.

Solutions

a. $P(r \le 4 \ or \ g = 1)$

The two events are inclusive. There are 24 pairs where $r \le 4$, 6 pairs where $g = 1$, and 4 pairs where $r \le 4 \ and \ g = 1$.

$$P(r \le 4 \ or \ g = 1) = \frac{24 + 6 - 4}{36} = \frac{13}{18}$$

b. $P(r = 2 \ or \ r > 4)$

The two events are mutually exclusive. There are 6 pairs for $r = 2$ and 12 pairs for $r > 4$.

$$P(r = 2 \ or \ r > 4) = \frac{6 + 12}{36} = \frac{1}{2}$$

c. $P(\text{sum of } 9)$

Find $P(r + g = 9)$. The sum 9 occurs in 4 ways: $(3,6)$, $(4,5)$, $(5,4)$, $(6,3)$.

$$P(\text{sum of } 9) = \frac{4}{36} = \frac{1}{9}$$

d. odds for a sum of 9

There are 4 successful ways and $36 - 4$, or 32, unsuccessful ways to have a sum of 9. The odds for a sum of 9 are 4 to 32, or 1 to 8.

For the event (A *and* B), described by the conjunction *and*, count the outcomes that are in *both* Event A *and* Event B. Consider again the tossing of a red die and a green die. As noted before, there are 6 outcomes out of the 36 for which $r \geq 5$ *and* $g \leq 3$. Thus:

$$P(r \geq 5 \text{ and } g \leq 3) = \frac{6}{36} = \frac{1}{6}$$

You can also multiply probabilities to find $P(r \geq 5 \text{ and } g \leq 3)$. Recall that $P(r \geq 5) = \frac{12}{36} = \frac{1}{3}$ and $P(g \leq 3) = \frac{18}{36} = \frac{1}{2}$.

$$P(r \geq 5 \text{ and } g \leq 3) = P(r \geq 5) \cdot P(g \leq 3) = \frac{1}{3} \cdot \frac{1}{2} = \frac{1}{6}$$

In this case, events $(r \geq 5)$ and $(g \leq 3)$ are called **independent events** because the outcome of throwing a red die does not affect the outcome of throwing a green die, and vice versa. This example illustrates the following rule.

Probability of (A *and* B)
If A and B are independent events, then $P(A \text{ and } B) = P(A) \cdot P(B)$.

The rule for finding $P(A \text{ and } B)$ is modified when A and B are *dependent* events. This will be shown in the next lesson.

EXAMPLE 2 Use the 36-pair sample space for tossing a red die and a green die to find each probability.

a. $P(r = 3 \text{ and } g > 4)$ b. $P(r < 3 \text{ and } g \geq 4)$

Solutions a. $P(r = 3 \text{ and } g > 4) = P(r = 3) \cdot P(g > 4) = \frac{6}{36} \cdot \frac{12}{36} = \frac{1}{18}$

b. $P(r < 3 \text{ and } g \geq 4) = P(r < 3) \cdot P(g \geq 4) = \frac{12}{36} \cdot \frac{18}{36} = \frac{1}{6}$

Classroom Exercises

Use the 36-pair sample space for tossing a red die and a green die to match each event at the left with the number of its successful outcomes at the right.

1. $r = 3$ *or* $g = 4$ **2.** $r = 3$ *and* $g = 4$ a. 1   e. 2

3. $r \leq 2$ *or* $g > 5$ **4.** $r \leq 2$ *and* $g > 5$ b. 3   f. 4

5. $r + g = 11$ **6.** $r + g \leq 5$ c. 10   g. 11

7. $r + g = 3$ *or* $r + g = 12$ **8.** $r \cdot g = 6$ d. 12   h. 16

9–16. Find the probability of each event for Classroom Exercises 1–8.

Written Exercises

Use the 36-pair sample space for tossing a red die and a green die to find the following probabilities or odds.

1. $P(r = 3 \text{ or } r = 5)$
2. $P(r \le 2 \text{ or } r > 3)$
3. $P(g = 2 \text{ and } g = 4)$
4. $P(r \ge 1 \text{ and } g \le 6)$
5. $P(r < 3 \text{ or } g > 3)$
6. $P(r \ge 3 \text{ and } g \ge 5)$
7. $P(r + g = 7)$
8. $P(r + g \ge 8)$
9. $P(r + g = 6 \text{ or } r + g = 8)$
10. $P(r + g = 5 \text{ or } r + g = 9)$
11. $P(r + g < 6)$
12. $P(r \cdot g = 12)$
13. odds for $r = 2$
14. odds against $g = 5$
15. odds for sum of 7
16. odds against sum of 11

One card is drawn at random from an ordinary deck of 52 playing cards. Find the probability of drawing the indicated card. The sample space appears after Example 2 of Lesson 14.7.

17. a red card *or* an ace
18. a black card *or* a face card
19. a club *or* a 10
20. an odd number *or* a diamond
21. a red queen *or* a heart
22. a black 7 *or* a spade
23. a black card *or* a heart
24. a 2 *or* a face card
25. an ace *or* a 2
26. a club *or* a red card
27. a red card *and* a face card
28. an even number *and* a black card
29. an odd number *and* a club
30. a diamond *and* a face card

Fifteen slips of paper are numbered from 1 through 15 and placed in a box. One slip is drawn at random from the box. Find the probability that the slip drawn has the number described.

31. divisible by 3 *or* 5
32. divisible by 2 *or* 3
33. prime *or* divisible by 7
34. factor of 15 *or* 24

Mixed Review

Simplify. *5.2, 8.5*

1. $\dfrac{8.8 \times 10^{-12}}{4 \times 10^{-9}}$

2. $\sqrt[3]{16x^6y^8}$

Rationalize the denominator. *8.4, 9.2*

3. $\dfrac{5}{3\sqrt{3} - 5}$

4. $\dfrac{-2}{4 + 3i}$

14.9 Selecting More than One Object at Random

Objective

To find the probability of selecting at random more than one object from a set of objects

Given a box containing 6 white marbles and 4 red marbles, two marbles can be drawn at random in either of the following two ways.

(1) Draw one marble, record its color, and replace it before making the second draw. The sample space is the same for each draw.

(2) Draw one marble, record its color, do *not* replace it, and draw a second marble. The sample space for the second draw is reduced by 1 marble from the sample space for the first draw. In this case, the two events are **dependent** because the first event affects the outcome of the second event.

If events A and B are dependent and B follows A, then the probability of B is written as $P(B$ given $A)$, or $P(B|A)$. So, the probability of A occurring, *followed* by B, is $P(A$, then $B) = P(A) \cdot P(B|A)$.

EXAMPLE 1

A box contains 6 white marbles and 4 red marbles. Find the probability of each event when two marbles are drawn at random.
a. a white and then a red *without* replacement after the first draw
b. a red and then a white *with* replacement after the first draw
c. a white and a red *without* replacement after the first draw

Plan

The sample space for the first draw is the set of 10 marbles. For the second draw, the sample space has 10 marbles if there is replacement and 9 marbles if there is no replacement. Let W represent a *white* marble and R represent a *red* marble.

Solutions

a. $P(W$, then $R) = P(W) \cdot P(R|W) = \frac{6}{10} \cdot \frac{4}{9} = \frac{4}{15}$ ← For R, the sample
   $P(W$, then $R)$ without replacement is $\frac{4}{15}$.   space is reduced
   by 1, from 10 to 9.

b. $P(R$, then $W) = P(R) \cdot P(W) = \frac{4}{10} \cdot \frac{6}{10} = \frac{6}{25}$ ← The second sample
   $P(R$, then $W)$ with replacement is $\frac{6}{25}$.   space is unchanged.

c. $(W$ *and* $R)$ is equivalent to $[(W$, then $R)$ *or* $(R$, then $W)]$.
   $P(W$ *and* $R) = P[(W$, then $R)$ *or* $(R$, then $W)]$

   $$= P(W) \cdot P(R|W) \overset{\downarrow}{+} P(R) \cdot P(W|R)$$

   $$= \frac{6}{10} \cdot \frac{4}{9} + \frac{4}{10} \cdot \frac{6}{9} = \frac{4}{15} + \frac{4}{15} = \frac{8}{15}$$

   Therefore, $P(W$ *and* $R)$ without replacement is $\frac{8}{15}$.

When you select objects *without replacement and without regard to order*, you can use combinations to compute s and t and then $P(E)$. For example, a box of 10 microchips has 7 functional (good) chips and 3 defective (bad) chips. If 4 chips are drawn at random, what is the probability that all 4 are functional?

Number of successful outcomes: $\binom{7}{4}$ Total number of outcomes: $\binom{10}{4}$

$$P(E) = \frac{s}{t} = \frac{\binom{7}{4}}{\binom{10}{4}} = \frac{\frac{7!}{4!3!}}{\frac{10!}{4!6!}} = \frac{\frac{7 \cdot 6 \cdot 5}{3 \cdot 2 \cdot 1}}{\frac{10 \cdot 9 \cdot 8 \cdot 7}{4 \cdot 3 \cdot 2 \cdot 1}} = \frac{35}{210} = \frac{1}{6}$$

To find $P(E)$ above by calculator, calculate the numerator and denominator separately. Then simplify the fraction.

EXAMPLE 2   A 5-member committee is to be formed by selecting 5 people at random from a group of 4 males and 6 females. Find the probability that the committee has the following.

   a. all females   **b.** exactly 3 males   **c.** at least 2 males

Plan   The number t of all possible outcomes is $\binom{10}{5} = \frac{10!}{5!5!}$, or 252. Let M represent *male* and F *female*.

Solutions   **a.** $P(\text{all } F) = P(5F \text{ and no } M)$

   To find s, select 5 from among 6 and 0 from among 4. Multiply.

$$P(\text{all } F) = \frac{s}{t} = \frac{\binom{6}{5} \cdot \binom{4}{0}}{\binom{10}{5}} = \frac{6 \cdot 1}{252} = \frac{1}{42}$$

   b. $P(\text{exactly } 3M) = P(3M \text{ and } 2F)$

   To find s, select 3 from among 4 and 2 from among 6. Multiply.

$$P(3M \text{ and } 2M) = \frac{s}{t} = \frac{\binom{4}{3} \cdot \binom{6}{2}}{\binom{10}{5}} = \frac{4 \cdot 15}{252} = \frac{60}{252} = \frac{5}{21}$$

   c. $P(\text{at least } 2M) = P[(2M \text{ and } 3F) \text{ or } (3M \text{ and } 2F) \text{ or } (4M \text{ and } 1F)]$

   Find s as follows: $(2M,3F) \text{ or } (3M,2F) \text{ or } (4M,1F)$

$$s = \binom{4}{2} \cdot \binom{6}{3} + \binom{4}{3} \cdot \binom{6}{2} + \binom{4}{4} \cdot \binom{6}{1}$$
$$= 6 \cdot 20 + 4 \cdot 15 + 1 \cdot 6 = 186$$

   $P(\text{at least } 2M) = \frac{s}{t} = \frac{186}{252} = \frac{31}{42}$

Classroom Exercises

A bag contains 5 white balls and 3 black balls. Two balls are drawn at random *without* replacement. Tell whether the given expression represents the probability of (white, then black), (both black), (white *and* black), or (at least one white).

1. $\frac{3}{8} \cdot \frac{2}{7}$ **2.** $\frac{5}{8} \cdot \frac{3}{7} + \frac{3}{8} \cdot \frac{5}{7}$ **3.** $\dfrac{\binom{5}{1} \cdot \binom{3}{1} + \binom{5}{2} \cdot \binom{3}{0}}{\binom{8}{2}}$ **4.** $\frac{5}{8} \cdot \frac{3}{7}$

Two cards are drawn *with* replacement from a deck of 52 cards. Match the probability of the given event with one expression at the right.

5. First is a 7 and second is a club.

6. One is a 7 and the other is a club.

7. Both are 7s.

8. Exactly one is a club.

a. $\frac{4}{52} \cdot \frac{4}{52}$ c. $\frac{13}{52} \cdot \frac{39}{52} + \frac{39}{52} \cdot \frac{13}{52}$

b. $\frac{4}{52} \cdot \frac{13}{52}$ d. $\frac{4}{52} \cdot \frac{13}{52} + \frac{13}{52} \cdot \frac{4}{52}$

A bag contains 8 green tokens and 9 red tokens. Two tokens are drawn. Find the probability of each event.

9. red, then green *with* replacement

10. red, then green *without* replacement

Written Exercises

A bag contains 3 green marbles and 6 blue marbles. Two marbles are drawn at random *with* replacement. Find the probability of each event.

1. green, then blue   **2.** green and blue   **3.** neither one green

A bag has 5 yellow buttons, 3 red buttons, and 1 white button. Two buttons are drawn *without* replacement. Find the probability of each event.

4. red, then yellow   **5.** yellow, then white   **6.** red *and* white

From a group of 4 boys and 6 girls, three piano students are to be selected at random to represent their school at a music recital. Determine the probability of each selection.

7. all girls   **8.** exactly 1 boy   **9.** at least 2 girls

Two cards are drawn at random *with* replacement from a deck of 52 playing cards. Find the probability that the cards are as follows.

10. club, then heart   **11.** black *and* red   **12.** exactly 1 face card

Two cards are drawn at random from a deck of 52 playing cards *without* replacing the first card. Find the probability that the cards are as follows.

13. both red

14. both clubs

15. spade, then heart

16. red, then black

17. club *and* diamond

18. face card *and* ace

19. exactly 1 ace

20. exactly 1 red

21. no face card

Six members of a jury are selected at random from a group of 8 males and 7 females. Find the probability of each indicated jury selection.

22. all males

23. all females

24. exactly 3 females

25. exactly 2 males

26. at least 4 males

27. at least 4 females

28. A box contains 5 white marbles and 10 black marbles. Two marbles are drawn *without* replacement. What is the probability that both are the same color?

29. If 5 cards are drawn at random from an ordinary deck of 52 cards, what is the probability that all 5 cards are hearts?

30. The letters of the alphabet are printed on separate cards and placed in a box. Four cards are drawn at random. Find the probability that the cards in the order drawn spell the word FOUR.

31. In a carton of 12 eggs, exactly two are cracked. If 2 eggs are selected at random, what is the probability that (a) both eggs are not cracked and (b) both eggs are cracked.

Mixed Review

1. y varies directly as x and $y = 15$ when $x = 2.7$. Find y when $x = 1.8$. **3.8**

2. Solve the system for (x,y).
$3x + 2y = 5$
$5x = 7 - 4y$ **4.3**

3. Factor completely.
$4x^3 - 32$ **5.7**

4. Solve and check.
$2 \log_b x = \log_b 3 + \log_b (x + 6)$ **13.6**

Brainteaser

A conference is held at a circular table that seats 24 people, equally spaced. Place cards that have been put on the table list 24 different names. Suppose the conference participants arrive and take seats at random. They then find that no one is in the correct seat. Regardless of the seats taken by the participants, is it always possible to rotate the table until at least 2 people are seated in front of their place cards? Explain.

14.10 Frequency Distributions

Objective

To find the mean, median, mode(s), variance, and standard deviation for a set of data

The table at the right shows the scores and their frequency for 20 students who took an algebra test. The same data can be displayed in a bar graph called a **histogram** as shown below.

Test score	Number of students
100	3
97	4
95	2
92	1
86	5
85	2
82	1
76	1
72	1

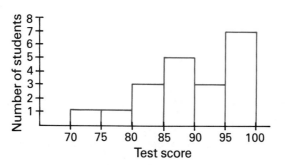

In the histogram, scores have been assigned to intervals of 5 units, with each boundary score, 75, 80, 85, 90, 95, 100, placed in the interval at the left. For example, 85 is assigned to the interval 80–85.

The table and the histogram provide information about the scores and their frequency. Such displays are called **frequency distributions**. Three numbers that measure the central tendency of a frequency distribution are the *mean*, the *median*, and the *mode*. The mean is often referred to as the "average."

Definitions

The **mean** is the number found by adding the scores and dividing the sum by the number of scores.

When the scores are arranged in order, the **median** is the middle score in an odd number of scores, or the mean (average) of the two middle scores in an even number of scores.

The **mode** is the score that occurs most frequently. There may be no mode if no score occurs more frequently than another, or there may be more than one mode if two or more scores have the same highest frequency.

For the 20 test scores above, the mean is 90, the median is $(92 + 86) \div 2$, or 89, and the mode is 86.

EXAMPLE 1 Find the mean, median, and mode(s) for the following scores.
2, 7, 5, 6, 5, 7, 2, 8, 5, 9, 2   Make a frequency table.

Solution

			score	frequency	score × frequency
2	2	2	2	3	6
5	5	5	5	3	15
6			6	1	6
7	7		7	2	14
8			8	1	8
9			9	1	9
				11 ←Totals→	58

The mean (M) is $\frac{58}{11}$, or 5.27 to the nearest 0.01.

The median is the 6th score (5) in the ordered list of 11 scores.
There are two modes, 2 and 5, each with a frequency of 3.
So, the mean is 5.27, the median is 5, and the modes are 2 and 5.

The scores in a list may cluster around the mean or may be spread from the mean. The simplest measure of the spread, or dispersion, is the **range**, which is the difference between the highest and lowest scores.

The two most common measures of dispersion are the *variance* and the *standard deviation*. Both of these measures involve the deviation from the mean for each score in a list of scores. Consider the list with scores of 92, 89, 83, 77, 77, 68 and a mean (M) of 81.

score	92	89	83	77	77	68
score minus 81	92–81	89–81	83–81	77–81	77–81	68–81
deviation from M	+11	+8	+2	−4	−4	−13

The **variance** of a set of scores is the sum of the squares of the deviations, divided by the number of entries. For the 6 entries above, the variance is 65, as shown below.

$$\text{variance} = \frac{11^2 + 8^2 + 2^2 + 2(-4)^2 + (-13)^2}{6} = \frac{390}{6} = 65$$

The **standard deviation** of a set of scores is the positive square root of the variance for the set. For the 6 entries above with a variance of 65, the standard deviation is $\sqrt{65}$, or 8.06 to 2 decimal places. The Greek letter sigma (σ) is used to denote the standard deviation. For the 6 scores above,

$$\sigma = \sqrt{\text{variance}} = \sqrt{65} \approx 8.06.$$

In general, the variance and standard deviation can be found as follows.

Formulas for Variance and Standard Deviation

For a set of n scores $(x_1, x_2, x_3, \ldots, x_n)$ with a mean of M, the variance is equal to the following expression.

$$\frac{(x_1 - M)^2 + (x_2 - M)^2 + (x_3 - M)^2 + \cdots + (x_n - M)^2}{n}$$

Standard deviation, σ: $\sigma = \sqrt{\text{variance}}$

EXAMPLE 2 Find the variance and standard deviation for the following list of scores: 35, 56, 23, 47, 68, 47, 23, 50, 56, 23, 78.

Plan Organize the data in a frequency table and find the mean (M).

Solution No. = score   Freq. = frequency   Dev. = deviation from mean (M)

No.	Freq.	No. × Freq.	Dev. (No. − 46)	Dev.^{2}	Freq. × Dev.^{2}
78	1	78	+32	1,024	1,024
68	1	68	+22	484	484
56	2	112	+10	100	200
50	1	50	+4	16	16
47	2	94	+1	1	2
35	1	35	−11	121	121
23	3	69	−23	529	1,587
	11	506			3,434

$$\text{mean}(M) = \frac{506}{11} = 46$$

$$\text{variance} = \frac{3,434}{11} \approx 312.18$$

$$\sigma = \sqrt{\frac{3,434}{11}} \approx 17.7$$

The variance is 312.2 and the standard deviation is 17.7.

Classroom Exercises

Find the mean, median, and mode(s) for each set of data.

1. 2, 3, 5, 5, 5

2. 1, 3, 3, 7, 7, 9

3. 2, 2, 2, 2, 4, 6

The mean for each list below is 20. Determine the deviation from the mean for each score in the list.

4. 28, 23, 15, 14

5. 17, 18, 20, 22, 23

6. 30, 25, 20, 20, 15, 10

Written Exercises

Find the mean, median, and mode(s) for each set of data.

1. 37, 44, 54, 64, 23, 33, 43, 53, 54
2. 75, 85, 45, 75, 35, 85, 25, 65
3. 16, 22, 19, 31, 15, 22, 16, 18, 40, 16
4. 70, 80, 40, 10, 20, 40, 80, 70, 10, 30, 80

5. score	frequency	6. score	frequency	7. score	frequency
95	2	8	7	80	2
91	3	11	3	70	2
87	4	15	4	60	3
85	6	20	4	50	1
82	3	23	1	40	1
77	2	35	1	30	3
65	1				

Find the variance and standard deviation for the following data.
First organize the data in a frequency table and find the mean
(see Example 2).

8. 80, 80, 75, 70, 70, 65, 65, 65, 60
9. 14, 9, 14, 16, 20, 7, 10, 14, 16, 10
10. 100, 97, 94, 93, 95, 93, 88, 76, 95, 89
11. 35, 65, 42, 35, 55, 25, 65, 30, 35
12. The mean for 8 scores is 94. Seven of the scores are 99, 98, 90, 95, 98, 92, 96. Find the missing score.
13. The mean for 10 scores is 48. Nine of the scores are 53, 53, 41, 41, 47, 47, 47, 45, 45. What is the missing score?
14. Fifty seniors from Northside High School had a mean score of 980 on the SAT. Thirty seniors from Southside High School had a mean score of 940 on the same test. What was the mean score for the 80 seniors?
15. A herd of 27 dairy cows produces a monthly mean of 70 lb of milk per cow. Another herd of 36 dairy cows produces a monthly mean of 84 lb of milk per cow. Find the monthly mean of milk per cow produced by both herds.

Mixed Review

1. Find the slope and describe the slant of the line whose equation is $3x - 4y = 8$. **3.3**
2. Convert 21.6 lb/ft^3 to ounces per cubic inch (oz/in^3). **7.3**
3. Simplify $\dfrac{8x}{x^2 - 25} + \dfrac{2}{5 - x} - \dfrac{4}{x + 5}$ **7.4**
4. Solve $\dfrac{2x}{3} + \dfrac{5x}{8} - \dfrac{x}{4} = 2$. **7.6**

14.11 Normal Distribution

Objective   To solve problems involving a normal distribution

If a large number of families with 4 children were surveyed to find the sexes and order of birth for their 4 children, a "representative" sample of 16 families would yield the following results.

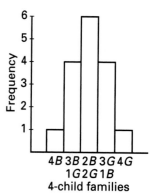

4-child families

BBBB	BBBG BBGB BGBB GBBB	BBGG GGBB BGBG GBGB BGGB GBBG	GGGB GGBG GBGG BGGG	GGGG
4B	3B, 1G	2B, 2G	3G, 1B	4G
$n = 1$	$n = 4$	$n = 6$	$n = 4$	$n = 1$

The histogram at the right shows the frequency distribution for the data without regard to order of birth.

EXAMPLE 1   Mr. and Mrs. Mason want to have 4 children. Use the histogram above to find the probability of each event for a 4-child family.

    a. 4 girls   **b.** 2 boys and 2 girls   **c.** at least 2 boys

Plan   Let the area of the bar for 4 boys be one square unit. Then the total area of the 5 bars is $1 + 4 + 6 + 4 + 1$, or 16 square units.
$$P(E) = \frac{s}{t} = \frac{s}{16}$$

Solutions   **a.** $\frac{1}{16}$ **b.** $\frac{6}{16}$, or $\frac{3}{8}$ **c.** $\frac{6 + 4 + 1}{16}$, or $\frac{11}{16}$

The bell-shaped curve, drawn through the midpoints of the tops of the bars in this histogram, is called a **normal curve**. It represents a **normal distribution** of data, such as the weights of many children, the foot sizes of many females, and so on, all selected at random.

A normal curve reflects the mean M and the dispersion from the mean in terms of the standard deviation.

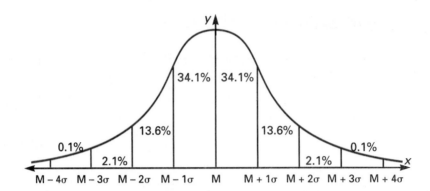

The normal curve above has the following properties.

(1) The curve is symmetric with respect to the y-axis.

(2) The area between the curve and the x-axis is 1.

(3) 50% of the area is at the right (or left) of $x = M$.

(4) 68.2% of the area is between $M - 1\sigma$ and $M + 1\sigma$.

(5) 95.4% of the area is between $M - 2\sigma$ and $M + 2\sigma$.

(6) 99.6% of the area is between $M - 3\sigma$ and $M + 3\sigma$.

EXAMPLE 2

The life expectancy of a certain type of automobile tire approximates a normal distribution. The mean life of a tire is 35,000 mi, with one standard deviation equal to 4,000 mi.

a. What is the probability of a tire lasting from 31,000 to 39,000 mi?

b. What is the probability that a tire will last more than 31,000 mi?

c. If a car rental agency buys 500 tires, how many of them can be expected to last between 27,000 and 43,000 mi?

Plan

Use the normal curve with $M = 35,000$ and $\sigma = 4,000$.

$M - 2\sigma$	$M - \sigma$	M	$M + \sigma$	$M + 2\sigma$
27,000	31,000	35,000	39,000	43,000

Solutions

a. The interval from 31,000 to 39,000 is between $M - \sigma$ and $M + \sigma$.
34.1% + 34.1%, or 68.2%, of the area is in this interval.
The probability is 0.682.

b. $31,000 = M - \sigma$
The area at the right of $M - \sigma$ is 34.1% + 50%, or 84.1%.
The probability is 0.841.

c. The interval from 27,000 to 43,000 is between $M - 2\sigma$ and $M + 2\sigma$.
2(13.6% + 34.1%), or 95.4%, of the area is in this interval.
95.4% of 500 = 0.954(500) = 477

So, 477 tires can be expected to last between 27,000 and 43,000 mi.

Classroom Exercises

The histogram shows the frequency distribution of the children in one type of family.

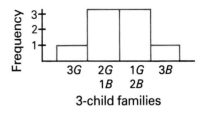

3-child families

1. How many children are in this type of family?
2. The area of the bar for 3 girls is 1 square unit. What is the total area of the 4 bars?

Find the probability of each event for a 3-child family.

3. 3 boys 4. 2 girls *and* 1 boy 5. at least 2 girls 6. at least 1 boy
7. fewer than 3 girls 8. exactly 2 boys *or* exactly 2 girls

Written Exercises

The histogram at the right shows the frequency distribution for tossing 5 coins. Use the histogram to find the probability of each event.

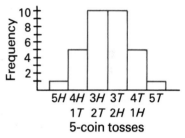

5-coin tosses

1. 4 tails *and* 1 head 2. 3 heads *and* 2 tails
3. 2 heads *or* 2 tails 4. 4 tails *or* 1 tail
5. at least 3 heads 6. more than 2 tails

The life expectancy of a certain tire approximates a normal distribution. The mean life of a tire is 45,000 mi, with one standard deviation equal to 5,000 mi. Find the probability that a tire will last the given mileage.

7. more than 45,000 mi
8. between 40,000 and 50,000 mi
9. more than 40,000 mi
10. between 40,000 and 55,000 mi
11. Out of 25 tires, how many should last between 40,000 and 50,000 mi?
12. Out of 200 tires, how many should last between 35,000 and 55,000 mi?
13. The heights, to the nearest inch, for 200 randomly selected adults approximate a normal distribution and are tallied in the table below. Which heights are between $M - \sigma$ and $M + \sigma$?

Height (in.)	78	76	73	71	68	66	63	61	59
Frequency	1	2	27	32	69	34	30	3	2

Mixed Review

Find the following, given $f(x) = 2x - 3$ and $g(x) = x^2 + 2$. *3.2, 5.9*

1. $f(4.5)$ 2. $g(-5)$ 3. $f(g(a))$ 4. $g(f(a))$

Extension

Sampling a Population

Several weeks prior to an election, a straw poll can be taken to predict the percent of the votes that each candidate for mayor will receive on election day. Since contacting every eligible voter would be impractical, a subset, or *sample*, of all eligible voters is polled. This sample should represent the total population of voters as accurately as possible.

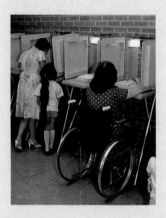

Example

In a certain city with 80,000 eligible voters, the population lives in four areas that roughly correspond to the annual family incomes shown below.

Area	No. of voters	Family income
Rolling Hills	400	over $80,000
South Park	40,600	$40,000–$60,000
West Side	27,000	$15,000–$35,000
North End	12,000	under $12,000

Explain why each of the following samples is not a good representation of the total population of voters in this city.

a. every home in Rolling Hills

b. 100 homes in each of the four areas

c. every 10th person entering the terminal building at the local commercial airport that serves this city

Solutions

a. The Rolling Hills area has families with the highest income level. Voters from this area might tend to vote for a candidate who will represent their interests by promising low taxes, wanting to provide services based on property and home values, and so on.

b. The 80,000 voters are not equally divided among the four areas. South Park and West Side would be *under*represented while Rolling Hills and North End would be *over*represented.

c. The frequency of airline trips is not a characteristic that is equally distributed among the total population. Businesspeople tend to fly more often than many other people.

In contrast to the faulty sampling procedures shown above, there are efficient procedures that provide a more representative sample of a given population. Two examples follow.

(1) Telephone each nth person listed in a telephone book.
(2) Mail a questionnaire with a stamped return envelope to each nth person in the telephone book.

Notice that even these methods are not truly representative. For example, some people don't have telephones, some have unlisted numbers, some might not return their questionnaires, and so on. Still, for a specific city or area, they often give a fairly representative sample.

Exercises

Tell whether each sampling procedure below provides a poor or a reasonable representation of the total population. Justify your answer.

1. To determine the mean (average) weight of all 20-year-old males, find the weights of all 20-year-old males at the U.S. Naval Academy.

2. To determine the majority opinion on whether a city should fluoridate its water supply, publish a ballot in the city's newspaper and ask the public to vote and mail it to the newspaper's office.

3. To determine what percent of the students in a high school drink milk at breakfast, survey all the students in the following classes or organizations.
   a. trigonometry classes
   b. English classes
   c. drama club

4. To determine the majority opinion on a very controversial topic, have television viewers telephone a specific number for a "yes" response or another specific number for a "no" response.

5. To determine the overall accuracy of a 700-page novel, randomly choose 70 of the pages to examine for errors.

6. To determine the average number of chocolate chips per cookie in a bag of cookies, find the average number in every fifth cookie.

7. To determine the quality of a given model of automobile, test the last one assembled every Friday of the model year.

Chapter 14 Review

Key Terms

circular permutation (p. 509)
combination (p. 511)
dependent events (p. 525)
distinguishable permutation (p. 507)
event (p. 517)
factorial (p. 500)
frequency distribution (p. 529)
Fundamental Counting Principle (p. 499)
histogram (p. 529)
inclusive events (p. 521)
independent events (p. 523)
mean (p. 529)

median (p. 529)
mode (p. 529)
mutually exclusive events (p. 521)
normal curve (p. 533)
normal distribution (p. 533)
odds (p. 519)
permutation (p. 503)
probability (p. 517)
range (p. 530)
sample space (p. 517)
standard deviation (p. 530)
variance (p. 530)

Key Ideas and Review Exercises

14.1 The Fundamental Counting Principle is stated on page 499.

1. How many different types of jackets can be made using 6 colors, 3 choices of collar, 2 choices of cuffs, and 4 choices of fabric?

14.2 In finding the number of permutations of objects, determine the decisions to be made and the number of choices for each decision.

2. How many odd 5-digit numbers can be formed from the digits 1, 2, 3, 4, 6, 8, 9, if a digit may be repeated in a number?

3. How many numbers of one or more digits can be formed from the digits 2, 4, 6, 8, if no digit may be repeated in a number?

4. Find the number of ways that 6 males and 4 females can be arranged in a line with the females together and at the left of the males.

14.3 The number of distinguishable permutations of n objects with r of them alike is $\frac{n!}{r!}$.

5. How many distinguishable permutations can be made using all the letters in the word SCISSORS?

14.4 There are $(n - 1)!$ ways to arrange n objects in a circle, and $\frac{(n - 1)!}{2}$ ways to arrange n objects on a ring with two views.

6. In how many ways can 8 different desserts be arranged in a circle around a floral centerpiece?

7. In how many ways can 8 different charms be arranged on a bracelet?

14.5, The number of combinations of n things taken r at a time is given by
14.6 $\binom{n}{r} = \dfrac{n!}{r!(n-r)!}$.

8. Find the number of lines determined by 7 points, A–G, on a circle.

9. From a group of 7 females and 6 males, in how many ways can you select a 5-member committee having exactly 3 females?

14.7 $P(E) = \dfrac{s}{t} = \dfrac{\text{number of successful outcomes}}{\text{total number of outcomes}}$; $P(\text{not } E) = 1 - P(E)$

The odds for E are s to u and the odds against E are u to s, where u = the number of unsuccessful outcomes.

A playing card is drawn from a regular 52-card deck. Find the probability or odds.

10. $P(\text{red})$ **11.** $P(\text{not a club})$ **12.** odds for an ace   **13.** odds against a 10

14.8 $P(A \text{ or } B) = P(A) + P(B) - P(A \text{ and } B)$ for inclusive events A and B.
$P(A \text{ or } B) = P(A) + P(B)$ for mutually exclusive events A and B.
$P(A \text{ and } B) = P(A) \cdot P(B)$ for independent events A and B.

If a red die and a green die are tossed, find the probability of each event.

14. $g = 4 \text{ or } g \le 2$ **15.** $r > 4 \text{ or } g > 3$ **16.** $r \ge 4 \text{ and } g \le 3$

14.9 $P(A, \text{ then } B) = P(A) \cdot P(B|A)$, where the sample space for B is reduced if A and B are dependent events.

A box contains 3 red tags and 6 green tags. A tag is drawn, replaced, and a second tag is drawn. Find the probability that the tags are as follows.

17. green, then red   **18.** both red   **19.** red and green

14.10 The definitions of mean, median, mode, variance, and standard deviation are given on pages 529–531.

20. Find the mean, median, mode, variance, and standard deviation for the frequency distribution shown in the table at the right.

score	32	26	24	20	18
frequency	4	3	2	3	2

21. Describe in your own words how to find the mean of a set of scores.

14.11 For a normal distribution, 68.2% of the scores are between $M - \sigma$ and $M + \sigma$, and 95.4% of the scores are between $M - 2\sigma$ and $M + 2\sigma$.

22. The scores on a college entrance test are normally distributed with a mean M of 500 and a standard deviation σ of 100. What is the probability that a student selected at random will score from 500 to 700?

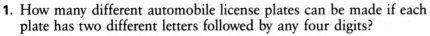

1. How many different automobile license plates can be made if each plate has two different letters followed by any four digits?

2. How many odd 4-digit numbers can be formed from the digits 1, 3, 4, 7, 8, if a digit may be repeated in a number?

3. How many numbers of one or more digits can be formed from the digits 4, 6, 7, if a digit may not be repeated in a number?

4. In how many ways can 3 different novels be placed together on a shelf at the right of 5 different biographies?

5. How many distinguishable 6-digit numbers can be formed by using all the digits in 747,475?

6. In how many ways can 6 people be seated around a circular table?

7. In how many ways can 5 different keys be arranged on a key ring?

8. Find the number of triangles determined by 6 points, A–F, on a circle.

9. How many tennis matches will be played by 6 people if each person plays each other person 3 times?

10. From a group of 9 males and 7 females, in how many ways can you select a 6-member committee having exactly 3 males?

11. How many different amounts of money are determined by selecting one or more bills from a box that contains one $1 bill, one $5 bill, one $10 bill, one $20 bill, and one $50 bill?

A playing card is drawn from a regular 52-card deck. Find the probability or odds.

12. P(black face card)   13. P(not a diamond)   14. odds for a club

A red die and a green die are tossed. Find the probability of each event.

15. $r > 4$ or $g < 3$ 16. $r > 4$ and $g < 3$ 17. $r + g = 11$

A bag contains 5 red tags and 6 green tags. A tag is drawn and replaced before a second tag is drawn. Find the probability of each event.

18. first is red, second is green

19. one is green, the other is red

20. Two cards are drawn *without* replacement from a deck of 52 playing cards. Find the probability that the two cards are a 9 and a 10.

21. A 5-member committee is to be selected at random from a group of 3 boys and 7 girls. Find the probability that at least 4 of the 5 will be girls.

In each item, compare a quantity in Column 1 with a quantity in Column 2. Write the letter of the correct answer from these choices.

A—The quantity in Column 1 is greater than the quantity in Column 2.
B—The quantity in Column 2 is greater than the quantity in Column 1.
C—The quantity in Column 1 is equal to the quantity in Column 2.
D—The relationship cannot be determined from the given information.

NOTE: Information centered over both columns refers to one or both of the quantities to be compared.

Sample Items and Answers

	Column 1	Column 2
S1.	$(7 - 7)!$	$7! - 7!$

$$n = 12 \text{ and } r = 6$$

S2.	$\binom{n}{r-1}$	$\binom{n}{r+1}$

S1: The answer is A: $(7 - 7)! = 0!$, or 1; $7! - 7! = 0$, and $1 > 0$.

S2: The answer is C: If $n = 12$ and $r = 6$, then $\binom{n}{r-1} = \dfrac{12!}{5!7!}$, $\binom{n}{r+1} = \dfrac{12!}{7!5!}$, and $\dfrac{12!}{5!7!} = \dfrac{12!}{7!5!}$.

	Column 1	Column 2
1.	$5! + 4!$	$5! \cdot 4!$
2.	$\binom{50}{0}$	$\binom{30}{30}$
3.	$\binom{7}{0} + \binom{7}{1} + \binom{7}{2}$	$\binom{7}{7} + \binom{7}{6} + \binom{7}{5}$

$$n = 8 \text{ and } r = 3$$

4.	$\binom{n}{r-1}$	$\binom{n-1}{r}$
5.	number of triangles determined by 6 points on a circle	number of lines determined by 6 points on a circle
6.	$50! \times 30!$	$51! \times 29!$

	Column 1	Column 2

$$n \text{ is an integer, } n > 4$$

7.	$\binom{n}{2}$	$\binom{n}{5}$
8.	the number of ways to select 2 males or 2 females from a group of 6 males and 4 females	the number of ways to select 1 male and 1 female from a group of 6 males and 4 females

$ABCD$ is a rectangle.

9. area of $\triangle DBC$ — area of $\triangle ABE$

Cumulative Review (Chapters 1–14)

Solve the equation.

1. $3^{2n+2} = 243$ **5.1**

2. $\dfrac{2x - 3}{x + 1} = \dfrac{x}{6}$ **7.6**

3. $3t^2 - 2t + 2 = 0$ **9.5**

4. $3\sqrt{5x + 4} = 2\sqrt{10x + 14}$ **10.3**

5. $\log_b 64 = 3$ **13.3**

6. $\log_5 x + \log_5 (x - 1) = \log_5 2$ **13.5**

Find the solution set.

7. $-4x + 5 < x - 10$ or   **2.2**
   $7 < 2x - 3$

8. $a^2 + 2a - 15 < 0$ **6.4**

Simplify.

9. $\left(\dfrac{5a^{-3}}{-2b^{-2}}\right)^2$ **5.2**

10. $\dfrac{7y}{y^2 - 36} - \dfrac{2}{y + 6} + \dfrac{3}{6 - y}$ **7.4**

11. $2a\sqrt[3]{24a^{12}b^5}$, $b \geq 0$ **8.5**

12. $2\sqrt{-3} \cdot 4\sqrt{-8}$ **9.2**

13. $\dfrac{4}{3 + 4i}$

14. $\log_2 32$ **13.3**

15. $\log_2 x + 3 \log_2 y$ **13.6**

For Exercises 16–20, use the relations
$C = \{(2,6), (3,5), (4,4), (5,3)\}$ **and**
$D = \{(-2,6), (6,-2), (3,3), (6,6)\}$.

16. Find the domain of C.   **3.1**

17. What is the range of D?

18. Is C a function?

19. Is D a function?

20. Are C and D a pair of inverse functions?   **13.1**

21. Solve the system for (x,y).   **4.3**
    $5x - 2y = 24$
    $2x + 3y = 2$

22. Solve $a(3x + 1) = 5a - 2bx$ **5.4**
    for x.

23. Factor $ax^3 - 27a$ completely.   **5.7**

For Exercises 24–26, $f(x) = 3x + 2$ and
$g(x) = x^2 - 4$. **Find the following.**

24. $g(f(2))$ **5.9**

25. $f(g(c + 1))$

26. $\dfrac{f(a) - f(b)}{a - b}$

27. Solve $2x^4 + x^3 - 7x^2 - 3x +$ **10.5**
    $3 = 0$, using the Integral Zero
    and Rational Zero theorems.

28. Convert 4 oz/in^2 to pounds per   **7.3**
    square foot.

29. Find the maximum value of P **11.6**
    if $P = -5x^2 + 20x + 7$.

30. On a coordinate plane, draw the **9.7**
    vectors $\vec{A} = (5,2)$, $\vec{B} = (-3,4)$,
    and $\vec{C} = -\vec{B}$. Use the
    parallelogram method to draw
    $\vec{A} + \vec{B}$ and $\vec{A} - \vec{B}$.

For Exercises 31–35, use $A(-2,3)$,
$B(6,-1)$, **and** $C(4,5)$. **Find the**
following.

31. distance between A and B, in   **11.1**
    simplest radical form

32. coordinates of the midpoint
    of $\overline{BC}$

33. slope and slant of $\overleftrightarrow{AB}$ **3.3**

34. equation of the horizontal line through A *3.4*

35. equation of the line parallel to $\overleftrightarrow{AB}$ and passing through C *3.7*

For points A, B, and C in Exercises 31–35, prove the following.

36. $\triangle ABC$ is an isosceles triangle.   *11.2*

37. $\overline{AC}$ is perpendicular to $\overline{BC}$.   *3.7*

Name the conic section.   *12.8*

38. $4x^2 - 9y^2 = 36$

39. $9x^2 + 4y^2 = 36$

40. $4x^2 + 4y^2 = 36$

41. $9y = 9x^2 + 36$

42. $4xy = 36$

43. If y varies directly as the square of x, and $y = 36$ when $x = 8$, find y when $x = 4$.   *3.8*

44. M varies inversely as d. Find M when $d = 1.8$, if $M = 42$ when $d = 2.4$.   *12.5*

45. If y varies jointly as x and $\sqrt{z}$, and $y = 44$ when $x = 6$ and $z = 6.25$, find z when $y = 22$ and $x = 5$.   *12.6*

For Exercises 46–48, write the equation for A^{-1}, the inverse of A, solving for y, and determine whether A^{-1} is a function.

46. $A: y = 4x - 8$ *13.1*

47. $A: y = x^2 + 5$

48. $A: y = 3^x$ *13.3*

49. Rico and José entered a non-stop, cross-country cycling race. Rico began the race at 6:30 A.M. and averaged 34 mi/h. José began the race at 7:00 A.M. and averaged 36 mi/h. At what time did José pass Rico on the race course?   *2.6*

50. Find three consecutive multiples of 4 such that the square of the second number is 32 more than 7 times the third number.   *6.2*

51. A triangle's height is 4 cm less than 3 times the length of the base. Find the height if the area of the triangle is 16 cm^2.   *6.3*

52. The product of two numbers is 4, and the second number is 4 less than the first number. Find two such pairs of numbers.   *9.6*

53. The perimeter of a rectangle is 28 ft and the area is 48 ft^2. Find the length and the width of the rectangle.   *12.10*

54. How many odd numbers of one or more digits can be formed from the digits 1, 2, 3, 4, 5, if no digit may be repeated in a number?   *14.2*

55. How many different 7-digit numbers can be formed using all the digits in 3,838,835?   *14.3*

56. How many different 4-member committees can be selected if 12 people are available for selection?   *14.5*

57. Find the probability of drawing a red face card in one draw from a deck of 52 playing cards.   *14.7*

This "surreal" image of a flower was created using the drawing capability of the computer. Artists can sketch their designs, then *paint* the figure with any color imaginable. Today's computers have color capacities of more than a million hues.

15.1 Arithmetic Sequences

Objectives

To write several consecutive terms in an arithmetic sequence

To find a specified term in an arithmetic sequence

The list of numbers 1, 1, 2, 3, 5, 8, 13, 21, . . . is called a *Fibonacci sequence*. This sequence can be used to describe certain structures in nature, such as the distribution of leaves around a stem. A **sequence** (or *progression*) is an ordered list of numbers. The numbers in the list are called *terms* of the sequence. Often, as with the Fibonacci sequence, a pattern can be observed in the sequence. Add any two consecutive terms of the Fibonacci sequence. What do you observe?

Notice the pattern in the infinite sequences below.

(1) 23, 28, 33, 38, . . .   (2) 4, −2, −8, −14, . . .
  Add 5 to any term to obtain   Add −6 to any term to obtain
  the next term.   the next term.

Each sequence above is an *arithmetic sequence*. Notice that the same number is added to each term to obtain the next term. The number added is called a *common difference*. To find the common difference, subtract any term from the one that follows it.

$$\text{In (1), } 5 = 28 - 23 = 33 - 28 = 38 - 33 = \ldots$$

$$\text{In (2), } -6 = -2 - 4 = -8 - (-2) = -14 - (-8) = \ldots$$

A sequence such as $\frac{1}{2}$, 1, 2, 4, 8, . . . is *not* an arithmetic sequence because there is no common difference. For example, $1 - \frac{1}{2} \neq 2 - 1$.

Definitions

The sequence $a_1, a_1 + d, a_1 + 2d, a_1 + 3d, a_1 + 4d, \ldots$ is an **arithmetic sequence** with first term a_1 and **common difference** d. If $x_1, x_2, x_3, x_4, \ldots$ is an arithmetic sequence with common difference d, then $d = x_2 - x_1 = x_3 - x_2 = \ldots$.

EXAMPLE 1

For each sequence that is arithmetic, find the common difference d.

a. 6, 8, 11, 15, 20, . . .   b. −7, −2, 3, 8, 13, . . .

Solutions

a. Since $8 - 6 \neq 11 - 8$, the sequence is not arithmetic.

b. Since $-2 - (-7) = 3 - (-2) = 8 - 3 = 13 - 8 = 5$, the sequence is arithmetic, with $d = 5$.

EXAMPLE 2 Write the next three terms of the arithmetic sequence 11, 3, -5, $-13, \ldots$.

Solution The common difference $d = 3 - 11$, or -8. When -8 is added to each term to obtain the next term, the next three terms are found to be $-21, -29, -37$.

EXAMPLE 3 Write the first four terms of the arithmetic sequence whose first term a_1 is 4 and whose common difference d is -5.

Solution

2nd term: 3rd term: 4th term:
$4 + (-5) = -1$ $-1 + (-5) = -6$ $-6 + (-5) = -11$

Thus, the first four terms are $4, -1, -6, -11$.

In an arithmetic sequence, you may need to find a specific term. For example, find the 71st term in the sequence 23, 28, 33, 38, ... To find the term, it is helpful to rewrite the sequence as follows.

23	28	33	38	. . .	?
$23 + 0 \cdot 5$	$23 + 1 \cdot 5$	$23 + 2 \cdot 5$	$23 + 3 \cdot 5$	. . .	$23 + 70 \cdot 5$
1st term	2nd term	3rd term	4th term	. . .	71st term

The 71st term is $23 + (71 - 1) \cdot 5$, or 373. This suggests the following formula for the nth term of an arithmetic sequence.

---

nth term of an arithmetic sequence

The nth term of an arithmetic sequence is given by

$$a_n = a_1 + (n - 1)d,$$

where a_n is the nth term ($n \geq 1$), a_1 is the first term, and d is the common difference.

EXAMPLE 4 Find the specified term of each arithmetic sequence.

a. 26th term of $-5, -2, 1, 4, \ldots$

b. 41st term of $7, 4.4, 1.8, -0.8, \ldots$

Solutions

a. $a_1 = -5, n = 26, d = 3$
$a_n = a_1 + (n - 1)d$
$\quad = -5 + (26 - 1)3$
$\quad = 70$
The 26th term is 70.

b. $a_1 = 7, n = 41, d = -2.6$
$a_n = a_1 + (n - 1)d$
$\quad = 7 + (41 - 1)(-2.6)$
$\quad = -97$
The 41st term is -97.

You can also use the formula $a_n = a_1 + (n - 1)d$ to find which term a given number is in an arithmetic sequence. For example, which term is 189 in the sequence 9, 15, 21, 27, . . . ? In this case, $a_n = 189$, $a_1 = 9$, and $d = 6$.

Solve for n in the formula $a_n = a_1 + (n - 1)d$.

$189 = 9 + (n - 1)6; \quad 180 = (n - 1)6; \quad 30 = n - 1; \quad n = 31$

So, 189 is the 31st term of the sequence.

EXAMPLE 5 A computer programmer's starting salary is $25,500 with a guaranteed minimum annual increase of $1,100. What minimum annual salary should be expected for the twelfth year?

Plan Note that the salaries form an arithmetic sequence, 25,500, 26,600, 27,700, . . . , with 1,100 as the common difference. Use the formula $a_n = a_1 + (n - 1)d$ to find the 12th term.

Solution $a_1 = 25,500 \qquad n = 12 \qquad d = 1,100$

$a_n = a_1 + (n - 1)d = 25,500 + (12 - 1)1,100 = 37,600$

The minimum salary is $37,600 for the twelfth year.

Classroom Exercises

Match each arithmetic sequence with exactly one statement at the right.

1. 6, 4, 2, . . .   **2.** 4, 5, 6, . . .   **a.** $d = 6$ **c.** Sixth term is 12.

3. 3, 9, 15, . . .   **4.** $-8, -4, 0, . . .$ **b.** First term is 6.   **d.** Third term is 6.

For each sequence that is arithmetic, find the common difference and the next two terms.

5. 3, 8, 13, 18, . . .   **6.** $-4, -2, 0, 2, . . .$ **7.** 1, 5, 25, 125, . . .

Written Exercises

For each sequence that is arithmetic, find the common difference d.

1. 33, 45, 57, 69, . . .   **2.** 35, 22, 9, -4, . . .   **3.** $-11, -9, -7, -5, . . .$

4. 5, 10, 20, 40, . . .   **5.** $8\frac{1}{2}, 10\frac{1}{4}, 12, . . .$ **6.** $-3, 6, -12, . . .$

Find the next three terms of each arithmetic sequence.

7. 8, 5, 2, -1, . . .   **8.** 21, 22.3, 23.6, . . .   **9.** $\frac{1}{4}, \frac{1}{3}, \frac{5}{12}, \frac{1}{2}, . . .$

Write the first four terms of the arithmetic sequence, given its first term a_1 and common difference d.

10. $a_1 = 23, d = 7$
11. $a_1 = 11, d = -5$
12. $a_1 = -8, d = -3$
13. $a_1 = 4, d = 2.6$
14. $a_1 = 7°C, d = 5°C$
15. $a_1 = 14$ oz, $d = -3\frac{1}{2}$ oz

Find the specified term of each arithmetic sequence.

16. 26th term of 12, 16, 20, . . .
17. 21st term of 9, 3, -3, -9, . . .
18. 31st term of 9.3, 9, 8.7, . . .
19. 36th term of -8, -6.6, -5.2, . . .
20. 83rd term of 6, $6\frac{1}{2}$, 7, . . .
21. 77th term of $11\frac{4}{5}$, $10\frac{2}{5}$, 9, . . .

22. Some cartons are stacked with 4 in the top row, 7 in the 2nd row, 10 in the 3rd row, and so on. How many cartons are in the 16th row?

23. A parachutist in a free fall travels 5 m in the 1st second, 15 m in the 2nd second, 25 m in the 3rd second, and so on. How far will he travel in the 9th second?

24. Mrs. Cooper deposited $300 in a bank on January 2 and each month after that she deposited $45 more than the preceding month. What was her deposit on December 2?

25. A landscaper's starting salary is $17,500. If he is guaranteed a minimum annual increase of $800, what minimum salary should he expect for his 10th year?

For each sequence that is arithmetic, find the indicated term.
(Exercises 26–31)

26. $2 + i$, 5, $8 - i$, . . . (fourth)
27. $\sqrt{11}$, $\sqrt{13}$, $\sqrt{15}$, $\sqrt{17}$, . . . (fifth)
28. 5, $4 + \sqrt{2}$, $3 + 2\sqrt{2}$, . . . (eleventh)
29. $6 - \sqrt{3}$, 7, $8 + \sqrt{3}$, . . . (twentieth)
30. Find n if 128 is the nth term of 4, 8, 12, 16, ...
31. Which term of -5, -2, 1, 4, . . . is 61?
32. Find the value of c so that $c + 4$, $4c + 1$, $8c - 4$ is an arithmetic sequence.
33. Find the first term of the arithmetic sequence whose 8th term is 29 and 20th term is 65.

Mixed Review

1. Simplify $5a^2(3a^2bc^3)^4$.   **5.1**

2. Express the value of $\dfrac{2.3 \times 10^{-12}}{4.6 \times 10^{-14}}$ in ordinary notation.   **5.2**

3. Find the maximum value of y if $y = -2x^2 + 12x + 9$.   **9.4, 11.6**

4. Graph the circle described by $4x^2 + 4y^2 = 48$.   **12.1**

15.2 Arithmetic Means and Harmonic Sequences

To find arithmetic means between two given numbers
To find *the* arithmetic mean of two numbers
To find a given term and harmonic means in a harmonic sequence

A sequence with a specific number of terms is a *finite* sequence. Thus, the arithmetic sequence 2, 5, 8, 11, 14, is a *finite* sequence of five terms. The terms between the first and last terms of an arithmetic sequence are called the **arithmetic means** between those two terms, as illustrated below.

$$2, \underline{5, 8, 11}, 14$$

arithmetic means

You can find any number of arithmetic means between two numbers.

EXAMPLE 1   Find the four arithmetic means between 36 and 32.

Plan   Draw 4 blanks for the means: 36, ____, ____, ____, ____, 32.

The 4 means and the given numbers will form a 6-term arithmetic sequence with $a_1 = 36$, $n = 6$, and $a_6 = 32$.

Use the formula $a_n = a_1 + (n - 1)d$ to find d, the common difference.

Solution
$$a_n = a_1 + (n - 1)d$$
$$32 = 36 + (6 - 1)d$$
$$-4 = 5d$$
$$-0.8 = d$$

Add -0.8 to each term to obtain the next term in the sequence.

$$36, \underline{35.2}, \underline{34.4}, \underline{33.6}, \underline{32.8}, 32$$

The four means are 35.2, 34.4, 33.6, and 32.8.

The procedure above can be used to find the *one* arithmetic mean between two numbers. This expression is the mean, or the average, of two numbers x and y as defined in Lesson 14.10.

The **arithmetic mean** of x and y is $\dfrac{x + y}{2}$.

EXAMPLE 2   The arithmetic mean of two numbers is 15. If the greater number is decreased by 6 times the smaller number, the result is 2. Find the two numbers.

Plan   Write and solve a system of two equations.

Solution   Let x = the smaller number and y = the greater number.

The arithmetic mean of x and y is 15.   (1) $\dfrac{x + y}{2} = 15$

If y is decreased by 6 times x, the result is 2.   (2) $y - 6x = 2$

(1) $x + y = 30$ (1) $x + (6x + 2) = 30$

(2) $y = 6x + 2$ $7x = 28$

                                 $x = 4$

                  (2) $y = 6 \cdot 4 + 2$

                       $y = 26$

So, the two numbers are 4 and 26. The check is left for you.

A sequence is *harmonic* if the reciprocals of its terms form an arithmetic sequence. Harmonic sequences have applications in music theory. Two examples are shown below.

Harmonic sequence   **Corresponding arithmetic sequence**

(1) $\frac{1}{2}, \frac{1}{3}, \frac{1}{4}, \frac{1}{5}$ 2, 3, 4, 5

(2) $4, \frac{8}{3}, 2$ $\frac{1}{4}, \frac{3}{8}, \frac{1}{2}$, or $\frac{2}{8}, \frac{3}{8}, \frac{4}{8}$

In the second example above, $\frac{8}{3}$ is the **harmonic mean** of 4 and 2, while its reciprocal, $\frac{3}{8}$, is the arithmetic mean of $\frac{1}{4}$ and $\frac{1}{2}$.

Definition   If $x_1, x_2, x_3, \ldots$ is an arithmetic sequence with no term equal to zero, then $\dfrac{1}{x_1}, \dfrac{1}{x_2}, \dfrac{1}{x_3}, \ldots$ is a **harmonic sequence**.

EXAMPLE 3   Find the 12th term in the harmonic sequence $\frac{1}{3}, \frac{1}{9}, \frac{1}{15}, \ldots$.

Plan   First, find the 12th term in the corresponding arithmetic sequence 3, 9, 15, ... Then find the reciprocal of that term.

Solution   For the 12th term in 3, 9, 15, . . . , $a_1 = 3$, $n = 12$, and $d = 6$. So, $a_n = a_1 + (n - 1)d = 3 + (12 - 1)6$, or 69.

Thus, the 12th term in the harmonic sequence is $\frac{1}{69}$.

EXAMPLE 4 Find the three harmonic means between 2 and $\frac{1}{3}$.

Plan Find three arithmetic means between $\frac{1}{2}$ and 3: $\frac{1}{2}$, ___, ___, ___, 3.

Solution Use $a_n = a_1 + (n-1)d$: $3 = \frac{1}{2} + (5-1)d$, $\frac{5}{2} = 4d$, and $d = \frac{5}{8}$.

Arithmetic sequence

$\frac{4}{8}, \frac{9}{8}, \frac{14}{8}, \frac{19}{8}, \frac{24}{8}$ or $\frac{1}{2}, \underbrace{\frac{9}{8}, \frac{7}{4}, \frac{19}{8}}_{\text{means}}, 3$

Harmonic Sequence

$2, \underbrace{\frac{8}{9}, \frac{4}{7}, \frac{8}{19}}_{\text{means}}, \frac{1}{3}$

$\underbrace{\qquad}_{\text{means}}$

The three harmonic means are $\frac{8}{9}$, $\frac{4}{7}$, and $\frac{8}{19}$.

EXAMPLE 5 Find *the* harmonic mean of 4 and $\frac{3}{2}$.

Plan Find the arithmetic mean of $\frac{1}{4}$ and $\frac{2}{3}$.

Solution $\dfrac{x+y}{2} = \dfrac{\frac{1}{4} + \frac{2}{3}}{2} = \dfrac{\frac{11}{12}}{2} = \dfrac{11}{24}$

Thus, the harmonic mean of 4 and $\frac{3}{2}$ is $\frac{24}{11}$.

Focus on Reading

Identify each sequence as arithmetic or harmonic.

1. 5, 8, 11, 14

2. $\frac{1}{5}, \frac{1}{8}, \frac{1}{11}, \frac{1}{14}$

3. $-6, 2, 10$

4. $\frac{2}{3}, 1, \frac{4}{3}, \frac{5}{3}$

5. $\frac{1}{4}, \frac{1}{7}, \frac{1}{10}$

6. $\frac{1}{2}, \frac{5}{16}, \frac{1}{8}$

7. $\sqrt{1}, \sqrt{4}, \sqrt{9}, \sqrt{16}$

8. 0.2, 0.9, 1.6

9. 0.2, 0.5, -1, -0.25

Classroom Exercises

In each arithmetic sequence, find the missing arithmetic means between the first and last terms.

1. 11, ___, 5, 2

2. 4, ___, 8, ___, ___, 14

3. 5, ___, ___, -7

Find *the* arithmetic mean of each pair of numbers.

4. 5 and 9

5. 30 and 20

6. 9 and -5

Find the arithmetic sequence related to the given harmonic sequence.

7. $\frac{1}{8}, \frac{1}{10}, \frac{1}{12}, \ldots$

8. $2, \frac{12}{5}, 3, \ldots$

9. $\frac{2}{5}, \frac{4}{11}, \frac{1}{3}, \ldots$

Written Exercises

Find the indicated number of arithmetic means between each pair of numbers.

1. two, between 14 and 35

2. three, between 40 and 12

3. three, between 9 and 7

4. four, between 8 and 16

5. four, between -3 and 14.5

6. five, between 24 and -7.2

Find _the_ arithmetic mean of each pair of numbers.

7. 36 and 82

8. 57 and -18

9. 14.8 and 6.2

10. $\sqrt{8}$ and $\sqrt{32}$

11. The arithmetic mean of two numbers is 16. Three more than the greater number is 6 times the smaller number. Find the two numbers.

12. The arithmetic mean of two numbers is 25. If one number is increased by twice the other number, the result is 68. Find both numbers.

Find the 16th term in each harmonic sequence.

13. $\frac{1}{6}, \frac{1}{7}, \frac{1}{8}, \frac{1}{9}, \ldots$

14. $\frac{2}{7}, \frac{1}{3}, \frac{2}{5}, \ldots$

15. 12, 6, 4, 3, $\ldots$

16. 16, 8, $\frac{16}{3}, \ldots$

17. Find two harmonic means between $\frac{4}{21}$ and $\frac{4}{3}$.

18. Find three harmonic means between 2 and 1.

19. Find the harmonic mean of $\frac{1}{8}$ and $\frac{1}{2}$.

20. Find the harmonic mean of 5 and $2\frac{1}{2}$.

21. A chemist's annual salary for 8 consecutive years was in an arithmetic sequence from $18,600 to $24,900. Find the salary for each year.

Find the indicated number of arithmetic means between each pair of numbers.

22. two, between $3\sqrt{6}$ and $15\sqrt{6}$

23. three, between 4 and $20 + 4\sqrt{2}$

24. four, between $-5i$ and 10

25. two, between $\sqrt{5}$ and $\sqrt{80}$

26. Prove: The arithmetic mean of x and y is $\dfrac{x + y}{2}$. (HINT: Use $a_n = a_1 + (n - 1)d$.)

27. Find the harmonic mean of x and y ($x \neq 0$, $y \neq 0$, $x \neq -y$).

Mixed Review

Factor completely. *5.7*

1. $2x^3 - 250$

2. $ax - ab - cx + bc$

3. $4x^4 - 17x^2 + 4$

4. $mx^2 - my^2$

15.3 Arithmetic Series and Sigma-Notation

Objectives

To find the sum of the first n terms of an arithmetic series
To write an arithmetic series using sigma-notation

When the terms of the arithmetic sequence 7, 2, -3, -8, . . . are also the terms of the indicated sum $7 + 2 + (-3) + (-8) + . . .$, or $7 + 2 - 3 - 8 - . . .$, the result is called an *arithmetic series*.

Definition

An **arithmetic series** is the indicated sum of the terms of an arithmetic sequence.

The sum S_{20} of the first 20 terms of the arithmetic series $6 + 11 + 16 + 21 + . . .$ can be found by writing the series in its given order and then in reverse order. The series is added to itself to give twice the sum of the first 20 terms.

$$S_{20} = 6 + 11 + 16 + . . . + 96 + 101 \leftarrow \text{20th term}$$
$$S_{20} = 101 + 96 + 91 + . . . + 11 + 6 \qquad \text{is } 6 + 19 \cdot 5,$$
$$\text{or } 101.$$
$$\overline{2S_{20} = 107 + 107 + 107 + . . . + 107 + 107}$$
$$2S_{20} = 20 \cdot 107 \text{ and } S_{20} = 1{,}070$$

Notice that $S_{20} = \frac{20}{2}(6 + 101)$, or 1,070.

The steps above suggest a method for deriving a formula for the sum S_n of the first n terms of an arithmetic series with a_1 as the first term and a_n as the nth term.

$$S_n = a_1 + (a_1 + d) + (a_1 + 2d) + . . . \dotplus (a_n - 2d) + (a_n - d) + a_n$$
$$S_n = a_n + (a_n - d) + (a_n - 2d) + . . . + (a_1 + 2d) + (a_1 + d) + a_1$$

$$2S_n = (a_1 + a_n) + (a_1 + a_n) + (a_1 + a_n) + . . . + (a_1 + a_n) + (a_1 + a_n) + (a_1 + a_n)$$

$$2S_n = n(a_1 + a_n)$$

$$S_n = \frac{n}{2}(a_1 + a_n)$$

EXAMPLE 1   Find the sum of the arithmetic series in which the number of terms n is 45, the first term a_1 is 14.3, and the last term a_{45} is 80.3.

Plan   Use the formula $S_n = \frac{n}{2}(a_1 + a_n)$ and the given values of n, a_1, and a_n.

Solution $S_n = \frac{n}{2}(a_1 + a_n) = \frac{45}{2}(14.3 + 80.3) = \frac{45}{2}(94.6) = 2{,}128.5$

The sum of the series is $2{,}128.5$.

The two formulas $S_n = \frac{n}{2}(a_1 + a_n)$ and $a_n = a_1 + (n - 1)d$ can be combined to provide a formula for the sum S_n that does not involve the last term a_n.

If $S_n = \frac{n}{2}(a_1 + a_n)$ and $a_n = a_1 + (n - 1)d$, then

$S_n = \frac{n}{2}[a_1 + a_1 + (n - 1)d]$, or $S_n = \frac{n}{2}[2a_1 + (n - 1)d]$.

Sum of the first n terms of an arithmetic series
The sum S_n of the first n terms of an arithmetic series is given by

(1) $S_n = \frac{n}{2}(a_1 + a_n)$ or

(2) $S_n = \frac{n}{2}[2a_1 + (n - 1)d]$,

where a_1 is the first term, a_n is the nth term, and d is the common difference.

EXAMPLE 2   Find the sum of the first 35 terms of the arithmetic series
$27.5 + 26 + 24.5 + 23 + \ldots$.

Plan   The 35th term is not given. Use the formula $S_n = \frac{n}{2}[2a_1 + (n - 1)d]$.

Solution $d = 26 - 27.5$, or -1.5 $n = 35$ $a_1 = 27.5$

$S_n = \frac{n}{2}[2a_1 + (n - 1)d]$

$= \frac{35}{2}[2(27.5) + (35 - 1)(-1.5)]$

$= \frac{35}{2}[55 + (-51)]$, or 70

The sum of the first 35 terms is 70.

The indicated sum $12 + 15 + 18 + 21$ can be written in the compact form $\sum\limits_{k=4}^{7} 3k$, where the terms of the series have the general form $3k$, and k takes on all integral values from 4 to 7.

To see how this compact form works, see the example below.

expanded form

$$\sum_{k=4}^{7} 3k = 3 \cdot 4 + 3 \cdot 5 + 3 \cdot 6 + 3 \cdot 7$$

$$= \underbrace{12 + 15 + 18 + 21}_{} = 66$$

series form

The Greek letter Σ (read: *sigma*) is a **summation symbol**, and k is called the **index of summation**. $\sum_{k=4}^{7} 3k$ is read "the summation of $3k$ from 4 to 7." Notice that the number of terms of the series is one more than the difference between the upper and lower values of the index. That is, the series has $(7 - 4) + 1$, or 4 terms.

To write $\sum_{j=1}^{3} (5j - 11)$ in expanded form, replace j by 1, 2, and 3.

$$\sum_{j=1}^{3} (5j - 11) = (5 \cdot 1 - 11) + (5 \cdot 2 - 11) + (5 \cdot 3 - 11)$$

$$= -6 - 1 + 4 = -3$$

Notice that 5, the coefficient of j in $5j - 11$, is the common difference of the series.

EXAMPLE 3   Write the expanded form of $\sum_{p=5}^{40} (70 - 3p)$.

Then find the sum of the series.

Solution $\sum_{p=5}^{40} (70 - 3p) = (70 - 3 \cdot 5) + (70 - 3 \cdot 6) + (70 - 3 \cdot 7) +$

$\dots + (70 - 3 \cdot 40) \leftarrow$ expanded form

$= 55 + 52 + 49 + \dots - 50 \leftarrow$ series form

Next, use $S_n = \frac{n}{2}(a_1 + a_n)$, where $n = (40 - 5) + 1$, or 36, $a_1 = 55$, and $a_{36} = -50$.

$$55 + 52 + 49 + \dots - 50 = \frac{36}{2}(55 - 50) = 90$$

The sum of the series is 90.

EXAMPLE 4 **EXAMPLE 4**   Write $18 + 25 + 32 + 39 + \ldots + 158$ using sigma-notation.

Plan   Since $d = 7$, let the general term be $7j + c$, where $j = 1, 2, 3, \ldots$.
Then find the constant c and the greatest value of j.

Solution   To find c, use the first term 18 and solve $7 \cdot 1 + c = 18$: $c = 11$.
To find j when $a_n = 158$, solve the equation $7j + 11 = 158$: $j = 21$.

So, the required notation is $\displaystyle\sum_{j=1}^{21}(7j + 11)$.

Note that the answer in Example 4 is not unique. The series can also
be expressed with the notation $\displaystyle\sum_{j=2}^{22}(7j + 4)$.

Classroom Exercises

State each summation in words. Predict the number of terms in the
series. Then write the series form.

1. $\displaystyle\sum_{k=1}^{5} 10k$ **2.** $\displaystyle\sum_{j=7}^{10} -j$ **3.** $\displaystyle\sum_{p=2}^{6}(p + 7)$ **4.** $\displaystyle\sum_{j=3}^{8} -2j$ **5.** $\displaystyle\sum_{k=0}^{4}(5 - k)$

Find the sum of each arithmetic series for the given data.

6. $n = 4$, $a_1 = 2$,   **7.** $n = 25$, $a_1 = -1$,   **8.** $\displaystyle\sum_{k=1}^{10}(8 + 3k)$
   $a_4 = -16$ $a_{25} = 10$

Written Exercises

Find the sum of each arithmetic series for the given data.

1. $n = 30$, $a_1 = 1$, $a_{30} = 134$ **2.** $n = 20$, $a_1 = 4$, $a_{20} = -115$

3. $n = 35$, $a_1 = -7.2$, $a_{35} = 58$ **4.** $n = 44$, $a_1 = -2.5$, $a_{44} = -56.7$

5. $n = 40$; $6 + 11 + 16 + 21 + \ldots$ **6.** $n = 30$; $10 + 6 + 2 - 2 - \ldots$

7. $n = 55$; $6.7 + 7 + 7.3 + 7.6 + \ldots$ **8.** $n = 65$; $-3 - 1.9 - 0.8 + 0.3 + \ldots$

9. $\displaystyle\sum_{k=1}^{5} 6k$ **10.** $\displaystyle\sum_{p=0}^{39}(p + 2)$ **11.** $\displaystyle\sum_{j=3}^{7}(2j - 5)$ **12.** $\displaystyle\sum_{p=30}^{33}(7 - p)$

Write each arithmetic series using sigma-notation.

13. $8 + 16 + 24 + 32$

14. $-6 - 12 - 18 - 24 - 30$

15. $14 + 16 + 18 + 20 + 22$

16. $15 + 12 + 9 + 6$

17. $7 + 9 + 11 + 13 + 15$

18. $14 + 11 + 8 + 5$

19. $15 + 11 + 7 + 3 - \ldots - 17$

20. $-7 - 5 - 3 - 1 + \ldots + 23$

21. $19 + 27 + 35 + 43 + \ldots + 107$

22. $75 + 66 + 57 + 48 + \ldots - 15$

Find the sum of each arithmetic series for the given data.

23. $n = 31$, $a_1 = 3\sqrt{2}$, $a_{31} = 153\sqrt{2}$

24. $n = 18$, $a_1 = 5x - 2y$, $a_{18} = 32y - 12x$

25. $n = 36$; $-0.2 + 0.3 + 0.8 + 1.3 + \ldots$

26. $n = 27$; $35.5 + 34.3 + 33.1 + 31.9 + \ldots$

27. Cartons are stacked in 16 rows with 4 in the top row, 7 in the second row, 10 in the third row, and so on. Find the total number of cartons in the stack.

28. In the month of June, Kevin saved 1 quarter the 1st day, 2 quarters the 2nd day, 3 quarters the 3rd day, and so on. How much money did he save in June?

29. The cost for printing 1000 copies of a 128-page book is $2500. The cost for printing each additional eight pages is $72. How much will it cost to print 1000 copies of a 208-page book?

30. A traveling salesperson receives a 5% commission on each of the first 5 vacuum cleaners sold, a 6% commission on each of the next 3 sold, a 7% commission on each of the next 3, and so on. If the vacuum cleaners are selling for $400 each and Mary Ings sells 23, what is her commission on the last three sold?

31. Solve $\displaystyle\sum_{k=5}^{8} (kx - 4) = 36$ for x.

32. Solve $\displaystyle\sum_{j=1}^{4} \frac{j}{x} = 75$ for x.

33. Prove $\displaystyle\sum_{k=1}^{20} 5kx = 5x \cdot \sum_{k=1}^{20} k$.

34. Prove $\displaystyle\sum_{j=1}^{15} (j + c) = 15c + \sum_{j=1}^{15} j$.

35. Prove $\displaystyle\sum_{n=1}^{k} (2n - 1) = k^2$.

36. Prove $\displaystyle\sum_{n=1}^{k} 2n = k(k + 1)$.

Mixed Review

1. Factor $x^2 - 2xy + y^2 - 16$ as the difference of two squares. **5.6**

2. Find $f(g(3a))$ if $f(x) = x^2 + 4$ and $g(x) = x - 2$. **5.9**

3. Find the solution set for $x^2 - 2x - 8 < 0$. **6.4**

4. What is the harmonic mean of $\frac{1}{2}$ and $\frac{1}{8}$? **15.2**

15.4 Geometric Sequences

Objectives

To write several consecutive terms in a geometric sequence
To find a specified term in a geometric sequence

In the sequence below, each consecutive term is obtained by multiplying the preceding term by 2.

$$3, 6, 12, 24, \ldots$$

A sequence formed in this manner is called a *geometric sequence.* Therefore, the ratio of any two successive terms is the same. In the sequence above, the *common ratio* is 2, since $2 = \frac{6}{3} = \frac{12}{6} = \frac{24}{12}$.

Definitions

$a_1, a_1r, a_1r^2, a_1r^3, \ldots$ is a **geometric sequence** with first term a_1 and common ratio r $(a_1 \neq 0, r \neq 0)$.

If $x_1, x_2, x_3, x_4, \ldots$ is a **geometric sequence** with common ratio r,
then $r = \dfrac{x_2}{x_1} = \dfrac{x_3}{x_2} = \dfrac{x_4}{x_3} = \ldots$.

EXAMPLE 1

For each sequence that is geometric, find the common ratio r.

a. $-2, -10, -50, -250, \ldots$ b. $2, 4, 6, 8, \ldots$

Solutions

a. Since $\dfrac{-10}{-2} = \dfrac{-50}{-10} = \dfrac{-250}{-50} = 5,$ b. Since $\frac{4}{2} \neq \frac{6}{4}$, the sequence
is not geometric.

the sequence is geometric with
$r = 5$.

EXAMPLE 2

Write the next three terms of the geometric sequence 64, -32, 16, $-8, \ldots$.

Solution

The common ratio $r = \dfrac{-32}{64}$, or $-\frac{1}{2}$. Multiply each term by $-\frac{1}{2}$ to obtain the next three terms.

Thus, the next three terms are 4, -2, and 1.

EXAMPLE 3

Write the first four terms of the geometric sequence whose first term a_1 is -3 and whose common ratio r is -4.

Solution

$-3(-4) = 12; \quad 12(-4) = -48; \quad -48(-4) = 192$

Thus, the first four terms are $-3, 12, -48$, and 192.

A geometric sequence in the form $a_1, a_1r, a_1r^2, \ldots$ can be used to find a specific term, such as the 20th term of $3, 6, 12, 24, \ldots$.

3	6	12	24	48	$\ldots$	?
$\downarrow$	$\downarrow$	$\downarrow$	$\downarrow$	$\downarrow$		$\downarrow$
3	$3 \cdot 2^1$	$3 \cdot 2^2$	$3 \cdot 2^3$	$3 \cdot 2^4$	$\ldots$	$3 \cdot 2^{19}$
1st term	2nd term	3rd term	4th term	5th term		20th term

The 20th term is $3 \cdot 2^{20-1}$, or 1,572,864. This suggests the following formula for the nth term in a geometric sequence.

nth term of a geometric sequence
The nth term of a geometric sequence is given by
$a_n = a_1 \cdot r^{n-1}$, where a_n is the nth term ($n \geq 1$), a_1 is the first term, and r is the common ratio.

EXAMPLE 4 Find the specified term of each geometric sequence.

  a. 5th term: $a_1 = 7$ and $r = 0.2$ **b.** 10th term of $-8, 4, -2, \ldots$

Solutions **a.** $a_1 = 7, n = 5, r = 0.2$

$a_n = a_1 \cdot r^{n-1} = 7(0.2)^{5-1} = 7(0.2)^4 = 7(0.0016) = 0.0112$

Calculator steps: 7 $\boxed{\times}$ 0.2 $\boxed{x^y}$ 4 $\boxed{=}$ 0.0112

The 5th term is 0.0112.

b. $a_1 = -8, n = 10, r = -\dfrac{1}{2}$

$a_n = a_1 \cdot r^{n-1} = -8\left(-\dfrac{1}{2}\right)^{10-1} = \dfrac{-8}{1}\left(\dfrac{1}{-2}\right)^9 = \dfrac{(-2)^3}{(-2)^9} = \dfrac{1}{(-2)^6}$

$= \dfrac{1}{64}$

The 10th term is $\dfrac{1}{64}$.

EXAMPLE 5 At the end of each hour, one-fourth of the gas in a tank is released. If the tank contains 2,048 ft^3 of gas at the beginning, how much gas will be in the tank at the end of 7 h?

Plan	Find the amount remaining at the end of each of the first 3 h. 1st hour: $\frac{3}{4}(2{,}048)$, or 1,536 ft³ remain; 2nd hour: $\frac{3}{4}(1{,}536)$, or 1,152 ft³ remain; 3rd hour: $\frac{3}{4}(1{,}152)$, or 864 ft³ remain. Then find the 7th term of the geometric sequence 1,536; 1,152; 864; . . . .
Solution	$a_1 = 1{,}536, r = \frac{3}{4}, n = 7$
	$a_7 = 1{,}536\left(\frac{3}{4}\right)^6 = \dfrac{1{,}536 \cdot 729}{4{,}096}$
	$ = 273.375$
	At the end of 7 h, 273.375, or $273\frac{3}{8}$ ft³ of gas will be in the tank.

Classroom Exercises

For each sequence that is geometric, find the common ratio and the next two terms.

1. 81, -27, 9, -3, . . .   **2.** 4, 8, 12, 16, . . .   **3.** 8, -8, 8, -8, . . .

4. $\frac{1}{12}, \frac{1}{4}, \frac{3}{4}, \frac{9}{4}, \ldots$ **5.** 18, 12, 8, $\frac{16}{3}$, . . .   **6.** 0.3, 0.06, 0.012, . . .

7. $\sqrt{2}, 2, 2\sqrt{2}, 4, \ldots$ **8.** $\sqrt{3}, 3, 2\sqrt{3}, 6, \ldots$ **9.** -4, 8, -4, 8, . . .

10. Write the first four terms of a geometric sequence in which $a_1 = -3$ and $r = -2$.

11. Find the 10th term of a geometric sequence in which $a_1 = 15$ and $r = 3$.

Written Exercises

Write the next three terms of each geometric sequence.

1. 15, 30, 60, . . .   **2.** -64, -16, -4, . . .   **3.** $-\frac{2}{9}, \frac{2}{3}, -2, \ldots$

Write the first four terms of each geometric sequence given its first term a_1 and common ratio r.

4. $a_1 = -\frac{1}{8}, r = 4$ **5.** $a_1 = 0.75, r = 2$ **6.** $a_1 = 64, r = \frac{1}{2}$

7. $a_1 = -8, r = -\frac{1}{4}$ **8.** $a_1 = 7, r = 0.1$ **9.** $a_1 = 128, r = -\frac{3}{8}$

Find the specified term of each geometric sequence.

10. 6th term: $a_1 = 23, r = 10$ **11.** 5th term: $a_1 = -10, r = 4$

12. 7th term: $a_1 = -500, r = -0.1$ **13.** 6th term: $a_1 = 15, r = -0.2$

14. 11th term: $-\frac{1}{8}, -\frac{1}{4}, -\frac{1}{2}, \ldots$ **15.** 10th term: $\frac{1}{4}, \frac{1}{2}, 1, \ldots$

16. 11th term: -96, 48, -24, . . .   **17.** 8th term: -0.36, -3.6, -36, . . .

18. 6th term: $a_1 = 25$, $r = \frac{2}{5}$

19. 7th term: $a_1 = -27$, $r = \frac{1}{3}$

20. 5th term: $a_1 = \sqrt{2}$, $r = 3\sqrt{2}$

21. 6th term: $a_1 = -4$, $r = \sqrt{3}$

Find the next two terms of each geometric sequence.

22. $1, i, -1, -i, \ldots$

23. $-2x^2, 8x^5, -32x^8, \ldots$

24. $\sqrt{3}y^3, -3y^5, 3\sqrt{3}y^7, \ldots$

25. $2i, 2, -2i, -2, \ldots$

26. A tank contains 729 liters of oil. One-third of the oil is released each time that a valve is operated. How much oil will be in the tank after the valve is operated 7 times?

27. On each bounce, a certain golf ball rebounds one-half of the distance from which it falls. If it is dropped from a height of 128 ft, how high does it go on its 10th rebound (upward)?

28. If a rubber ball dropped from a height of 81 yd rebounds on each bounce two-thirds of the distance from which it fell, how far does it fall on its 7th descent (downward)?

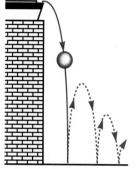

29. Find two values of y so that $\frac{1}{4}$, $y - 2$, $4y$ is a geometric sequence.

30. Find two values of t for which $4t$, $t - 2$, $\frac{2}{3}$ is a geometric sequence.

31. The 7th term of a geometric sequence is 512 and the 8th term is 1,024. Find the first term.

32. A geometric sequence has a 9th term of $\frac{1}{32}$ and a 10th term of $\frac{1}{64}$. What is the first term?

Midchapter Review

For Exercises 1–3, use the arithmetic sequence 8, 6.5, 5, 3.5, ...
Find each of the following. *15.1*

1. the common difference d

2. the next three terms

3. a_{31}, the 31st term

4. Find the four arithmetic means between 22 and 26. *15.2*

5. Find the arithmetic mean of -3.6 and 7.2. *15.2*

6. Find the two harmonic means between 4 and $\frac{8}{5}$. *15.2*

For Exercises 7 and 8, find the sum of the arithmetic series for the given number of terms. *15.3*

7. $n = 40$; $a_1 = 5$, $a_n = -144$

8. $n = 45$; $5.7 + 6 + 6.3 + 6.6 + \ldots$

9. Write $9 + 14 + 19 + 24 + 29 + 34$ using sigma-notation. *15.3*

For Exercises 10–12, use the geometric sequence $-\frac{1}{4}$, $\frac{1}{2}$, -1, ...
Find each of the following. *15.4*

10. the common ratio r

11. the next three terms

12. a_{10}, the tenth term

15.5 Geometric Means

Objectives

To find geometric means between two given numbers
To find *the* geometric mean of two numbers

The terms between two given terms in a geometric sequence are called the **geometric means** between the two terms. For example, 10, 20, 40, and 80 are the four geometric means between 5 and 160 in the geometric sequence below.

$$5, \underbrace{10, 20, 40, 80,}_{\text{geometric means}} 160$$

EXAMPLE 1 Find three real geometric means between $-\frac{2}{3}$ and -54.

Plan The three means and the given numbers will form a 5-term geometric sequence with $a_1 = -\frac{2}{3}$, $a_5 = -54$, and $n = 5$:

$$-\frac{2}{3}, \underline{\quad}, \underline{\quad}, \underline{\quad}, -54.$$

To find r, use the formula $a_n = a_1 \cdot r^{n-1}$.

Solution

$$-54 = -\frac{2}{3} \cdot r^{5-1} \leftarrow \text{Substitute for } a_1, a_n, \text{ and } n.$$

$$81 = r^4$$

$$r = \sqrt[4]{81} \qquad or \qquad r = -\sqrt[4]{81}$$

$$r = 3 \qquad or \qquad r = -3 \leftarrow \text{two real numbers for } r$$

Multiply each term by r to obtain the next term in the sequence.

$$r = 3: -\frac{2}{3}, \underline{-2}, \underline{-6}, \underline{-18}, -54 \qquad r = -3: -\frac{2}{3}, \underline{2}, \underline{-6}, \underline{18}, -54$$

The three real means are $-2, -6, -18$ or $2, -6, 18$.

In Example 1, $x^4 = 81$ has two other roots, $3i$ and $-3i$. These values of r were not used since the problem asked for *real* number means.

When finding n real geometric means between two numbers, you will obtain two sets of means if n is odd, as in Example 1, and one set of means if n is even, as in Example 2.

EXAMPLE 2 Find the two real geometric means between 18 and $-\frac{16}{3}$.

Solution To find r, use $a_n = a_1 \cdot r^{n-1}$ with $a_1 = 18$, $n = 4$, and $a_4 = -\frac{16}{3}$.

Substitute for a_1, a_n, $\rightarrow$ $-\frac{16}{3} = 18 \cdot r^{4-1}$
and n.

$$-\frac{16}{3} \cdot \frac{1}{18} = \frac{1}{18} \cdot 18 \cdot r^3$$

$$-\frac{8}{27} = r^3$$

$$r = \sqrt[3]{-\frac{8}{27}} = -\frac{2}{3}$$

Multiply each term by $-\frac{2}{3}$. $18, \underline{-12}, \underline{8}, -\frac{16}{3}$

The two real geometric means are -12 and 8.

EXAMPLE 3 Find the geometric mean between 2 and 6.

Solution $2, \underline{\quad}, 6;\ a_n = a_1 \cdot r^{n-1}$ $6 = 2 \cdot r^{3-1}$ $3 = r^2$ $r = \pm\sqrt{3}$

Use $r = \sqrt{3}$. The mean is $2\sqrt{3}$: $2, \underline{2\sqrt{3}}, 6$.

Use $r = -\sqrt{3}$. The mean is $-2\sqrt{3}$: $2, \underline{-2\sqrt{3}}, 6$.

The geometric mean is $2\sqrt{3}$ or $-2\sqrt{3}$.

The one geometric mean between two numbers is called *the geometric mean* of the two numbers. In Example 3, notice that the geometric mean between 2 and 6 is $\sqrt{2 \cdot 6}$, which equals $2\sqrt{3}$, or it is $-\sqrt{2 \cdot 6}$, which equals $-2\sqrt{3}$. This suggests the following formula.

The **geometric mean** *of x and y is* $\sqrt{xy}$ *or* $-\sqrt{xy}$ $(x \neq 0, y \neq 0)$.

EXAMPLE 4 The positive geometric mean of two numbers is 6 and one number is 9 more than the other number. Find the two numbers.

Plan Let x and y be the two numbers. Write and solve a system of equations.

Solution (1) $\sqrt{xy} = 6$ (2) $y = x + 9$

(1) $\sqrt{x(x + 9)} = 6$ $x(x + 9) = 36$ $x^2 + 9x - 36 = 0$

$(x - 3)(x + 12) = 0$

$x = 3 \ or \ x = -12$

(2) $y = x + 9$ If $x = 3$, then $y = 12$.
If $x = -12$, then $y = -3$.

The numbers are 3 and 12 or -12 and -3.

Classroom Exercises

In each geometric sequence, find the missing geometric means between the first and last terms.

1. 3, _____ , 48, 192 **2.** -2, _____, $-\frac{1}{2}$, _____, $-\frac{1}{8}$ **3.** 32, _____, _____, _____, 2

Find *the* positive geometric mean of each pair of numbers.

4. 20 and 5   **5.** 1 and 4   **6.** 4 and $\frac{1}{4}$ **7.** -2 and -8 **8.** 5 and 3

Written Exercises

Find the indicated number of real geometric means between the two numbers.

1. three, between 2 and 162

2. three, between 64 and 4

3. two, between 9 and $\frac{1}{3}$

4. two, between $\frac{3}{4}$ and 48

5. three, between -4 and -324

6. three, between $-\frac{3}{8}$ and -6

7. two, between $-\frac{1}{4}$ and 16

8. two, between 75 and $-\frac{3}{5}$

9. four, between 9 and 288

10. four, between 300,000 and 3

11. three, between $\sqrt{2}$ and $4\sqrt{2}$

12. three, between 6 and 54

Find *the* positive geometric mean of each pair of numbers.

13. 3 and 12   **14.** 36 and 4   **15.** -5 and -80 **16.** -49 and -2

17. One number is 9 times another number. The positive geometric mean of the two numbers is 6. Find the two numbers.

18. Find two numbers whose positive geometric mean is 4, if one number is 6 less than the other number.

19. Find two numbers whose difference is 12 and whose negative geometric mean is -8.

20. Nine different weights are in a geometric sequence. Find each weight if the lightest is 0.25 g and the median (middle) weight is 4 g.

21. Wheat is removed from a storage silo so that the amounts remaining at the end of successive hours are in a geometric sequence. At the end of 6 h, 270 tons remain. How much wheat was in the silo at the end of 4 h if there were 7,290 tons at the end of 3 h?

Find *the* negative geometric mean of each pair of numbers.

22. 6, 3   **23.** $2\sqrt{3}$, $8\sqrt{3}$ **24.** $-\sqrt{2}$, $-\sqrt{8}$ **25.** -5, -10

26. Write in your own words how the positive geometric mean of two positive numbers is found.

Find the positive geometric mean of each pair of numbers.

27. $0.2, 0.008$ **28.** $-1.8, -0.2$ **29.** $-2.7, -0.3$ **30.** $6, 0.24$

For the right triangle PQR at the right, the altitude to the hypotenuse separates the hypotenuse into two segments. It is proved in geometry that $\frac{a}{m} = \frac{m}{b}$, where m is called the *mean proportional* between a and b.

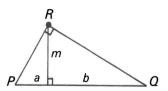

Observe that m is the geometric mean of a and b since

$$\text{if } \frac{a}{m} = \frac{m}{b}, \text{ then } m^2 = ab \text{ and } m = \sqrt{ab}.$$

Use the information and the figure above with Exercises 31–36.

31. Find m if $a = 10$ and $b = 6.4$.   **32.** Find m if $a = 2$ and $b = 6$.

33. Find b if $a = 4$ and $m = 2\sqrt{6}$.   **34.** Find m if $PR = 18$ and $a = 6$.

35. Find PR if $a = 3$ and $b = 6$.   **36.** Find PQ if $a = 3$ and $m = 6$.

37. Prove: The one positive geometric mean between x and y is $\sqrt{xy}$.
(HINT: Use the formula $a_n = a_1 \cdot r^{n-1}$ to find r and fill the blank in x, _____, y to form a geometric sequence.)

38. Prove: If a, g, and h are the arithmetic, geometric, and harmonic means, respectively, of two numbers x and y, then $g^2 = ah$.

Mixed Review

Simplify.   *5.2, 5.3, 7.4, 8.3*

1. $\dfrac{-10x^{-3}y^{-2}z^3}{15x^5y^{-6}z^{-1}}$ **2.** $8x - (7 - 2x) + 3x(4x - 5)$

3. $\dfrac{5x}{x^2 - 7x + 12} - \dfrac{4}{6 - 2x}$ **4.** $(3\sqrt{2} + \sqrt{6})(4\sqrt{2} - \sqrt{6})$

Brainteaser

Using an 11-minute hourglass and a 7-minute hourglass, what is the easiest way to time the boiling of an egg for 15 minutes?

15.6 Geometric Series

To find the sum of the first n terms of a geometric series
To solve problems involving the sum of a geometric series

The terms of the geometric sequence $3, -6, 12, -24, \ldots$ can be written as an indicated sum. The result is the *geometric series* $3 - 6 + 12 - 24 + \ldots$ with a common ratio of -2.

Definition

A **geometric series** is the indicated sum of the terms of a geometric sequence.

The sum S of the geometric series $2 + 6 + 18 + 54 + \ldots + 39{,}366$, with $r = 3$, can be found as shown below. The terms are multiplied by $-r$, or -3, and then added to the original terms.

$$
\begin{aligned}
S &= 2 + 6 + 18 + 54 + \ldots + 39{,}366 \\
-3 \cdot S &= -6 - 18 - 54 - \ldots - 39{,}366 - 118{,}098 \\
\hline
-2 \cdot S &= 2 + 0 + 0 + 0 + \ldots + 0 - 118{,}098
\end{aligned}
$$

$$
S = \frac{2 - 118{,}098}{-2}
$$

$$
= \frac{-118{,}096}{-2} = 59{,}048
$$

The steps above suggest a formula for the sum S_n of the terms of the geometric series with first term a_1, common ratio r, and last term a_n.

$$
(1 - r)S_n = a_1 - ra_n, \quad \text{or} \quad S_n = \frac{a_1 - ra_n}{1 - r}
$$

EXAMPLE 1

Find the sum of the terms in the geometric series with first term $a_1 = -9$, common ratio $r = -4$, and last term $a_n = 9{,}216$.

Solution

$$
S_n = \frac{a_1 - ra_n}{1 - r} = \frac{-9 - (-4)(9{,}216)}{1 - (-4)} = \frac{36{,}855}{5} = 7{,}371
$$

The sum of the series is 7,371.

The two formulas $S_n = \dfrac{a_1 - ra_n}{1 - r}$ and $a_n = a_1 r^{n-1}$ can be combined to provide a formula for the sum S_n that does not involve the last term a_n.

Sum of the terms of a finite geometric series
The sum S_n of the terms of a finite geometric series is given by

$$(1) \ S_n = \frac{a_1 - r a_n}{1 - r} \quad \text{or} \quad (2) \ S_n = \frac{a_1(1 - r^n)}{1 - r},$$

where a_1 is the first term, a_n is the last term, r is the common ratio, and n is the number of terms.

EXAMPLE 2   Find the sum of the terms of the geometric series
$16 - 8 + 4 - 2 + \ldots - \frac{1}{8}$.

Plan   The last term, $-\frac{1}{8}$, is given. Use the formula $S_n = \frac{a_1 - r a_n}{1 - r}$.

Solution $a_1 = 16, r = \frac{-8}{16} = -\frac{1}{2}, a_n = -\frac{1}{8}$

$$S_n = \frac{a_1 - r a_n}{1 - r} = \frac{16 - \left(-\frac{1}{2}\right)\left(-\frac{1}{8}\right)}{1 - \left(-\frac{1}{2}\right)} = \frac{16 - \frac{1}{16}}{\frac{3}{2}} = \frac{255}{16} \cdot \frac{2}{3} = \frac{85}{8}$$

The sum is $10\frac{5}{8}$, or 10.625.

EXAMPLE 3   Find the sum of the first 9 terms of the geometric series
$-\frac{1}{8} + \frac{1}{4} - \frac{1}{2} + 1 - \ldots$.

Plan   The last term is not given. Use the formula $S_n = \frac{a_1(1 - r^n)}{1 - r}$.

Solution $a_1 = -\frac{1}{8}, \quad r = \frac{1}{4} \div \left(-\frac{1}{8}\right) = -2, \quad n = 9$

$$S_n = \frac{a_1(1 - r^n)}{1 - r} = \frac{-\frac{1}{8}[1 - (-2)^9]}{1 - (-2)} = \frac{-\frac{1}{8}(1 + 512)}{3} = -21\frac{3}{8}$$

The sum is $-21\frac{3}{8}$, or -21.375.

Recall that $\displaystyle\sum_{k=2}^{5} 3k$ symbolizes $3 \cdot 2 + 3 \cdot 3 + 3 \cdot 4 + 3 \cdot 5$, or
the arithmetic series $6 + 9 + 12 + 15$ with a common difference of 3.

Similarly, $\displaystyle\sum_{k=2}^{5} 3^k$ represents $3^2 + 3^3 + 3^4 + 3^5$, or the geometric series
$9 + 27 + 81 + 243$ with a common ratio of 3.

EXAMPLE 4 Write the expanded form for $\sum\limits_{j=1}^{4} 6(-3)^j$. Find the sum of the terms.

Solution
$$\sum_{j=1}^{4} 6(-3)^j = 6(-3)^1 + 6(-3)^2 + 6(-3)^3 + 6(-3)^4 \leftarrow \text{expanded form}$$
$$= -18 + 54 - 162 + 486 = 360$$

The sum of the terms is 360.

EXAMPLE 5 On each bounce, a ball dropped from a height of 800 ft rebounds one-half of the distance from which it falls. How far will it have traveled when it hits the ground the 8th time? See the figure on page 561.

Plan The ball descends 8 times and rebounds 7 times.

The descent and rebound distances form a series,
$800 + 400 + 400 + 200 + 200 + \ldots$, for 15 terms.

The series of descents is the geometric series given by
$D = 800 + 400 + 200 + \ldots$, for 8 terms with $r = \frac{1}{2}$.

The series of rebounds is the geometric series given by
$U = 400 + 200 + 100 + \ldots$, for 7 terms with $r = \frac{1}{2}$.

Solution
$$D = \frac{a_1(1 - r^n)}{1 - r} = \frac{800\left[1 - \left(\frac{1}{2}\right)^8\right]}{1 - \frac{1}{2}} = \frac{800\left(1 - \frac{1}{256}\right)}{\frac{1}{2}} = 1{,}593.75$$

$$U = \frac{a_1(1 - r^n)}{1 - r} = \frac{400\left[1 - \left(\frac{1}{2}\right)^7\right]}{1 - \frac{1}{2}} = \frac{400\left(1 - \frac{1}{128}\right)}{\frac{1}{2}} = 793.75$$

$D + U = 1{,}593.75 + 793.75 = 2{,}387.5$

The ball will have traveled 2,387.5 ft when it hits the ground the 8th time.

Classroom Exercises

Find the common ratio r for each geometric series.

1. $18 + 0.18 + 0.0018 + \ldots$

2. $0.007 + 0.07 + 0.7 + 7 + \ldots$

3. $\sum\limits_{k=1}^{6} (-2)^k$

4. $\sum\limits_{j=1}^{6} 3(5^j)$

5. $\sum\limits_{j=0}^{2} \left(\frac{1}{3}\right)^{j-1}$

6. $\sum\limits_{p=3}^{10} 5(2^{p-3})$

7-10. Find the sum of the terms of each series in Exercises 3–6 above.

Written Exercises

Find the sum of the terms for each geometric series described below.

1. $a_1 = 5, r = 10, a_n = 50{,}000$

2. $a_1 = -8, r = -4, a_n = -32{,}768$

3. $a_1 = -486, r = \frac{1}{3}, a_n = -6$

4. $a_1 = 2{,}187, r = -\frac{1}{3}, a_n = 27$

5. $32 + 16 + 8 + \ldots + \frac{1}{4}$

6. $\frac{1}{3} + 1 + 3 + \ldots + 243$

7. $0.4 + 4 + 40 + \ldots + 40{,}000$

8. $800 - 80 + 8 - \ldots + 0.0008$

9. $n = 10; \; 3 - 6 + 12 - 24 + \ldots$

10. $n = 9; \; -5 - 10 - 20 - 40 - \ldots$

11. $n = 7; \; 0.03 + 0.3 + 3 + 30 + \ldots$

12. $n = 7; \; -1 + 10 - 100 + 1{,}000 - \ldots$

13. $n = 8; \; -4 + 8 - 16 + 32 - \ldots$

14. $n = 8; \; 64 - 32 + 16 - 8 + \ldots$

15. $a_1 = 11, r = 3, n = 6$

16. $a_1 = -6, r = -3, n = 5$

17. $a_1 = \frac{1}{5}, r = 10, n = 6$

18. $a_1 = 0.6, r = -10, n = 7$

19. $\displaystyle\sum_{k=1}^{7} 2^k$

20. $\displaystyle\sum_{j=4}^{7} (-3)^j$

21. $\displaystyle\sum_{p=0}^{6} \left(\frac{1}{2}\right)^p$

22. $\displaystyle\sum_{k=5}^{8} (0.2)^{k-4}$

23. $\displaystyle\sum_{j=1}^{4} 5(-2)^{j-1}$

24. $\displaystyle\sum_{p=1}^{4} \frac{1}{6}(3^p)$

25. $\displaystyle\sum_{k=1}^{5} -32\left(\frac{1}{4}\right)^{k-1}$

26. $\displaystyle\sum_{j=2}^{7} 27\left(-\frac{1}{3}\right)^{j-2}$

27. Smaller and smaller squares are formed consecutively, as shown at the right. Find the sum of the perimeters of the first 9 squares if the first square is 40 in. wide.

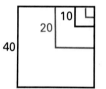

28. Find the sum of the areas of the first 6 squares at the right if the first square is 40 in. wide.

29. On each bounce, a certain golf ball rebounds two-thirds of the distance from which it falls. If the ball falls from a height of 81 yd, how far will it have traveled when it hits the ground the 7th time?

Find the sum of the terms of each series.

30. $\displaystyle\sum_{k=1}^{5} (k-1)!$

31. $\displaystyle\sum_{n=1}^{5} \frac{1}{(n-1)!}$

32. $\displaystyle\sum_{j=1}^{5} \left(\substack{\;4\\j-1}\right)$

33. $\displaystyle\sum_{p=1}^{5} \frac{p}{(p-1)!}$

34. $\displaystyle\sum_{k=1}^{5} \frac{2^{k-1}}{(k-1)!}$

35. $\displaystyle\sum_{j=1}^{3} \frac{(-1)^{j-1}(0.2)^{2j-2}}{(2j-2)!}$

Mixed Review

Let $f(x) = 3x - 2$ and $g(x) = x^2 + 4$. Find the following. *3.2, 5.5, 5.9*

1. $f(2.8)$

2. $g(c-5)$

3. $f(g(-4))$

4. $g(f(-1))$

Extension

Mathematical Induction

Suppose you want to prove the following statement, S_1: For all natural numbers n, $5 + 11 + 17 + \ldots + (6n - 1) = n(3n + 2)$. Trials using values such as $n = 1, 2, 3$ may prove S_1 true for these values, but a finite number of trials will not show that S_1 is true for *all* natural numbers n. To prove the statement, the following principle must be used.

Principle of Mathematical Induction
A statement S is true for all natural numbers n if both (1) and (2) are true: (1) S is true for $n = 1$.
(2) If S is true for $n = k$, *then* S is true for $n = k + 1$.

Since (1) guarantees that S is true for $n = 1$ and (2) guarantees that S is true for $n = 1 + 1$, or 2, and so on forever, the principle is reasonable.

Example Prove S_1: For all natural numbers n, $5 + 11 + 17 + \ldots + (6n - 1)$
$= n(3n + 2)$.

Solution (1) Show that S_1 is true for $n = 1$. $5 = 1(3 \cdot 1 + 2)$, or 5

(2) Show that if S_1 is true for $n = k$, then S_1 is true for $n = k + 1$, or
$5 + 11 + 17 + \ldots + [6(k + 1) - 1] = (k + 1)[3(k + 1) + 2]$.

Now if we assume that S_1 is true for $n = k$, then
$5 + 11 + 17 + \ldots + (6k - 1) = k(3k + 2)$. Adding the
$(k + 1)$st term, $6(k + 1) - 1$, to each side gives the following.

$5 + 11 + 17 + \ldots + (6k - 1) + [6(k + 1) - 1]$
$= k(3k + 2) + [6(k + 1) - 1]$
$= 3k^2 + 8k + 5 = (k + 1)(3k + 5) = (k + 1)[3(k + 1) + 2]$

So, if S_1 is true for $n = k$, then S_1 is true for $n = k + 1$. Thus, (1) and (2) are true, and by mathematical induction, S_1 is true.

Exercises

Use mathematical induction to prove that each statement is true for all natural numbers n.

1. $2 + 4 + 6 + \ldots + 2n = n(n + 1)$
2. $7 + 9 + 11 + \ldots + (2n + 5) = n(n + 6)$
3. $1 + 3 + 5 + \ldots + (2n - 1) = n^2$

Statistics: *Expected Value*

An insurance company must calculate the expected value of claims in order to cover their costs and set rates. A sociologist may want to know the average time-sentence for felons convicted of a particular crime. A person thinking about entering a contest may wish to compare the expected return with the contest's entry fee.

Suppose a random variable x assumes values $x_1, x_2, \ldots, x_n$, and that these values have equal probabilities of occurring. Then, the expected value is simply the numerical average of the values. However, in many instances, some values are more likely to occur than others. To reflect this, expected value is calculated by multiplying each possible value of the random variable by its probability and adding the results.

For a random variable x that assumes values $x_1, x_2, \ldots, x_i, \ldots, x_n$ with probabilities $P(x = x_1)$, $P(x = x_2)$, $\ldots$, $P(x = x_i)$, $\ldots$, $P(x = x_n)$, the expected value of x, or $E(x)$, is defined by

$$E(x) = \sum_{i=1}^{n} x_i \cdot P(x = x_i).$$

You can use the above summation to solve the following problem.

Data is collected by a company that writes hospitalization-insurance policies. They estimate the average claim for a hospital stay is $600 per year, with a probability of 0.42 for this type of claim. The average out-patient claim is $45 per year, with a probability of 0.03. Special-care claims average $520 per year, with a 0.15 probability of occurring. The probability of no claim is 0.4. What must the insurance company charge per policy, just to cover expected claims?

The random variable x is the amount paid on a claim, with the probabilities as given. So, using the formula above gives the following.

$$E(x) = 600(0.42) + 45(0.03) + 520(0.15) + 0(0.4) = 331.35$$

The company needs to charge $331.35 per year to cover costs.

Exercises

1. If a first offender is convicted of a certain crime, the probabilities of various jail sentences are as follows: $P(3 \text{ years}) = 0.1$, $P(2 \text{ years}) = 0.2$, $P(1 \text{ year}) = 0.3$, $P(\text{probation}) = 0.4$. What is the expected length of time a first offender will spend in jail?

2. One million entries are accepted for a state contest that offers the following prizes: one $50,000 prize, 2 prizes of $25,000, 6 prizes of $5,000, 100 prizes of $100, and 1,000 prizes of $10. What is the expected value of an entry? If the entry fee is $1 per entry, is it to your advantage to enter?

15.7 Infinite Geometric Series

To find the sum, if it exists, of an infinite geometric series
To solve problems involving the sum of an infinite geometric series
To write a repeating decimal as an infinite geometric series
To write a repeating decimal in the form $\frac{x}{y}$, where x and y are integers

An **infinite geometric series**, such as $2 + 1 + \frac{1}{2} + \frac{1}{4} + \ldots$, has no last term. However, the "sum" of its terms appears to approach 4 when you add the first n terms for $n = 1, 2, 3, 4, \ldots$ as follows.

$$2, \quad 2 + 1 = 3, \quad 2 + 1 + \tfrac{1}{2} = 3\tfrac{1}{2}, \quad 2 + 1 + \tfrac{1}{2} + \tfrac{1}{4} = 3\tfrac{3}{4}$$

The sequence $2, 3, 3\frac{1}{2}, 3\frac{3}{4}, \ldots$ is called the sequence of **partial sums** for the given series. The table below shows how the partial sums get closer and closer to 4, or *converge* to 4.

n	1	2	3	4	5	6	7	8	...
Partial sums	2	3	$3\frac{1}{2}$	$3\frac{3}{4}$	$3\frac{7}{8}$	$3\frac{15}{16}$	$3\frac{31}{32}$	$3\frac{63}{64}$	...

If $r = \frac{1}{2}$ in the sum formula $S_n = \dfrac{a_1(1 - r^n)}{1 - r}$, notice how rapidly r^n decreases as n increases.

$$\left(\tfrac{1}{2}\right)^1 = \tfrac{1}{2}, \ \left(\tfrac{1}{2}\right)^5 = \tfrac{1}{32}, \ \left(\tfrac{1}{2}\right)^{10} = \tfrac{1}{1{,}024}, \ \left(\tfrac{1}{2}\right)^{15} = \tfrac{1}{32{,}768}, \ldots$$

As n increases without end, $\left(\frac{1}{2}\right)^n$ approaches 0 as a *limit*. In general, when n increases without end and $-1 < r < 1$, $r \neq 0$, then

$$S_n = \frac{a_1(1 - r^n)}{1 - r} \text{ approaches } \frac{a_1(1 - 0)}{1 - r}, \text{ or } \frac{a_1}{1 - r}, \text{ as a } \textit{limit.}$$

This limit is called the *sum* of the infinite geometric series

$$a_1 + a_1 r + a_1 r^2 + a_1 r^3 + \ldots$$

Sum of an infinite geometric series
The sum S of an infinite geometric series is given by

$$S = \frac{a_1}{1 - r} \quad (-1 < r < 1, r \neq 0),$$

where a_1 is the first term and r is the common ratio.

EXAMPLE 1 Find the sum, if it exists, of each infinite geometric series.

   a. $2 + 1 + \frac{1}{2} + \frac{1}{4} + \ldots$ b. $16 - 4 + 1 - \frac{1}{4} + \ldots$

   c. $\frac{1}{8} + \frac{1}{4} + \frac{1}{2} + \ldots$

Solutions   a. $r = \frac{1}{2}$; $S = \dfrac{a_1}{1 - r} = \dfrac{2}{1 - \frac{1}{2}} = 2 \div \frac{1}{2} = 4$

   b. $r = -\frac{1}{4}$; $S = \dfrac{a_1}{1 - r} = \dfrac{16}{1 - \left(-\frac{1}{4}\right)} = 16 \div \frac{5}{4} = \frac{64}{5} = 12\frac{4}{5}$

   c. $r = 2$; Since 2 is not between -1 and 1, the sum in **c** does not exist.

The repeating decimal 7.77. . . can be written as the infinite geometric series $7 + 0.7 + 0.07 + \ldots$, where $a_1 = 7$ and $r = 0.1$.

The sum $S = \dfrac{a_1}{1 - r} = \dfrac{7}{1 - 0.1} = \dfrac{7(10)}{0.9(10)} = \dfrac{70}{9}$. Thus, 7.77. . . $= \dfrac{70}{9}$.

EXAMPLE 2 Write the repeating decimal 0.535353 . . ., or $0.53\overline{53}$, as indicated.

   a. as an infinite geometric series
   b. in the form $\dfrac{x}{y}$, where x and y are integers

Solutions   a. $0.53\overline{53} = 0.53 + 0.0053 + 0.000053 + \ldots$

   b. $a_1 = 0.53$ $r = 0.01$ $S = \dfrac{0.53}{1 - 0.01} = \dfrac{0.53}{0.99} \cdot \dfrac{100}{100} = \dfrac{53}{99}$

   So, $0.53\overline{53} = \dfrac{53}{99}$. ← THINK: Can you use a calculator to check the answer?

Classroom Exercises

For each sequence of partial sums that converges, find the limiting value.

1. $\frac{1}{2}, \frac{3}{4}, \frac{7}{8}, \frac{15}{16}, \frac{31}{32}, \frac{63}{64}, \ldots$ 2. $\frac{1}{8}, \frac{3}{16}, \frac{9}{32}, \frac{27}{64}, \frac{81}{128}, \frac{243}{256}, \frac{729}{512}, \frac{2{,}187}{1{,}024}, \ldots$

Find the sum, if it exists, of each infinite geometric series.

3. $\frac{2}{3} + \frac{1}{3} + \frac{1}{6} + \frac{1}{12} + \ldots$ 4. $0.1 + 0.01 + 0.001 + 0.0001 + \ldots$

Find the first three terms when the repeating decimal is written as an infinite geometric series.

5. 0.222 . . .   6. $4.4\overline{4}$ 7. 0.333 . . .   8. $0.34\overline{34}$ 9. $37.\overline{37}$ 10. $3.9\overline{39}$

Written Exercises

Find the sum, if it exists, of each infinite geometric series.

1. $\frac{1}{8} + \frac{1}{4} + \frac{1}{2} + \dots$ **2.** $-1 + \frac{1}{3} - \frac{1}{9} + \dots$ **3.** $-\frac{1}{4} - \frac{1}{8} - \frac{1}{16} - \dots$

4. $27 - 9 + 3 - \dots$ **5.** $4 + 0.4 + 0.04 + \dots$ **6.** $60 + 6 + 0.6 + \dots$

7. $16 + 8 + 4 + \dots$ **8.** $\frac{1}{2} - \frac{1}{4} + \frac{1}{8} - \dots$ **9.** $12 + 8 + \frac{16}{3} + \dots$

Write each repeating decimal as an infinite geometric series and then in the form $\frac{x}{y}$, where x and y are integers.

10. $0.888\dots$ **11.** $5.55\dots$ **12** $0.75\overline{75}$

13. $8.1\overline{81}$ **14.** $0.064\overline{64}$ **15.** $3.12\overline{312}$

16. A golf ball, dropped from a height of 81 yd, rebounds on each bounce two-thirds of the distance from which it falls. How far does it travel before coming to rest?

17. The midpoints of the sides of an equilateral triangle are connected to create a new triangle. This procedure is repeated on each new triangle. Find the sum of the perimeters of an infinite sequence of the triangles if one side of the original triangle is 80 cm long.

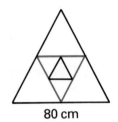

80 cm

18. The midpoints of the sides of a square are joined to create a new square. This process is performed on each new square. Find the sum of the areas of an infinite sequence of such squares if one side of the first square is 10 ft long.

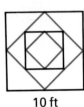

10 ft

Find the sequence of partial sums of each series for $n = 1, 2, 3, \dots, 8$. To what value does the sequence of partial sums converge?

19. $4 + 2 + 1 + \dots$ **20.** $27 + 9 + 3 + \dots$ **21.** $27 - 9 + 3 - \dots$

22. Find the sum of the infinite geometric series
$6\sqrt{2} + 6 + 3\sqrt{2} + 3 + \dots$

For what values of x does each infinite geometric series have a sum?

23. $(x - 2)^1 + (x - 2)^2 + (x - 2)^3 + \dots$
24. $(x - 1)^0 + (x - 1)^1 + (x - 1)^2 + \dots$

Mixed Review

Evaluate each expression using $\binom{n}{r} = \dfrac{n!}{r!(n - r)!}$.   *14.5*

1. $\binom{7}{4}$ **2.** $\binom{5}{0}$ **3.** $\binom{6}{6}$ **4.** $\binom{3}{0} + \binom{3}{1} + \binom{3}{2} + \binom{3}{3}$

15.8 The Binomial Theorem

Objectives

To expand a positive integral power of a binomial
To find a specified term of a binomial expansion

Lesson 5.5 presented the special product $(a + b)^2 = a^2 + 2ab + b^2$. Is there also a "special product" for $(a + b)^3$ or $(a + b)^4$? In this lesson you will learn how to expand such products.

The expansions of $(a + b)^n$ for $n = 0, 1, 2, 3, 4$ show patterns that can be used to expand $(a + b)^n$ for any positive integer n.

$$(a + b)^0 = 1$$
$$(a + b)^1 = 1a^1 + 1b^1$$
$$(a + b)^2 = 1a^2 + 2a^1b^1 + 1b^2$$
$$(a + b)^3 = 1a^3 + 3a^2b^1 + 3a^1b^2 + 1b^3$$
$$(a + b)^4 = 1a^4 + 4a^3b^1 + 6a^2b^2 + 4a^1b^3 + 1b^4$$

Notice the following in the expansions of $(a + b)^n$ above.

(1) The number of terms is always $n + 1$.
(2) The first term is a^n and the last term is b^n.
(3) The exponent of a decreases by 1 from term to term.
(4) The exponent of b increases by 1 from term to term.
(5) For each term, the sum of the exponents of a and b is n.
(6) The coefficients are symmetrical. That is, they read the same from left to right as from right to left.
(7) There is a pattern for finding the coefficients in any expansion.

Note that $(a + b)^4 = 1 \cdot a^4 + 4 \cdot a^3b + 6 \cdot a^2b^2 + 4 \cdot ab^3 + 1 \cdot b^4$

$$= \binom{4}{0}a^4 + \binom{4}{1}a^3b + \binom{4}{2}a^2b^2 + \binom{4}{3}ab^3 + \binom{4}{4}b^4$$

To expand $(a + b)^5$: (1) The coefficients will be $\binom{5}{k}$ for $k = 0, 1, 2, 3, 4, 5$.

(2) The exponents of a will decrease from 5 to 0.
(3) The exponents of b will increase from 0 to 5.

So, $(a + b)^5 = \binom{5}{0}a^5 + \binom{5}{1}a^4b^1 + \binom{5}{2}a^3b^2 + \binom{5}{3}a^2b^3 + \binom{5}{4}ab^4 + \binom{5}{5}b^5$

$$= 1a^5 + 5a^4b + 10a^3b^2 + 10a^2b^3 + 5ab^4 + 1b^5$$

Theorem 15.1

Binomial Theorem: If n is a positive integer, then

$$(a + b)^n = \binom{n}{0}a^n + \binom{n}{1}a^{n-1}b + \binom{n}{2}a^{n-2}b^2 + \binom{n}{3}a^{n-3}b^3 + \ldots + \binom{n}{n}b^n.$$

EXAMPLE 1   Expand $(2x - y^2)^3$. Simplify each term.

Plan   Rewrite $(2x - y^2)^3$ as $[2x + (-y^2)]^3$ and use the binomial theorem.

Solution $a = 2x, b = -y^2, n = 3$

$(a + b)^n = [2x + (-y^2)]^3$

$= \binom{3}{0}(2x)^3 + \binom{3}{1}(2x)^2(-y^2)^1 + \binom{3}{2}(2x)^1(-y^2)^2 + \binom{3}{3}(-y^2)^3$

$= 1 \cdot 8x^3 + 3 \cdot 4x^2(-y^2) + 3 \cdot 2x \cdot y^4 + 1(-y^6)$

$= 8x^3 - 12x^2y^2 + 6xy^4 - y^6$

Notice the pattern of signs in a binomial expansion.
(1) In the expansion of $(a + b)^n$, the signs are all $+$.
(2) In the expansion of $(a - b)^n$, the signs alternate $+$ and $-$.

The rth term in the expansion of $(a + b)^n$
The rth term in the expansion of $(a + b)^n$ is

$$\binom{n}{r-1}a^{n-(r-1)}b^{r-1}, \quad \text{or} \quad \binom{n}{r-1}a^{n-r+1}b^{r-1}.$$

EXAMPLE 2   Find the 7th term in the expansion of $(a + b)^9$.

Solution   **Method 1:**   The factors are $\binom{9}{6}$, b^6, and a^3, since the sum of the exponents is 9. Thus, the 7th term is $\binom{9}{6}a^3b^6$, or $84a^3b^6$.

Method 2:   Use the formula for the rth term with $n = 9$ and $r = 7$. The 7th term is $\binom{9}{7-1}a^{9-7+1}b^{7-1} = \binom{9}{6}a^3b^6$, or $84a^3b^6$.

EXAMPLE 3   Find the 6th term in the expansion of $(3x - y^2)^7$.

Solution $(3x - y^2)^7 = [3x + (-y^2)]^7$ Use $\binom{7}{5}a^2b^5$ with $a = 3x$ and $b = -y^2$.

The 6th term is $\binom{7}{5}(3x)^2(-y^2)^5 = 21 \cdot 9x^2(-y^{10}) = -189x^2y^{10}$.

Classroom Exercises

For each term of a binomial expansion, find the value of k.

1. $\binom{k}{3}y^3$ is the 4th term.
2. $\binom{2}{0}c^k$ is the first term.
3. $\binom{4}{k}x^{4-k}y^k$ is the middle term.

Simplify.

4. $\binom{7}{0}c^7$

5. $\binom{8}{7}c(-d)^7$

6. $\binom{8}{8}(-d)^8$

7. $\binom{6}{5}(2c)(-1)^5$

8. Expand and simplify the terms. $(x + 1)^8$

Written Exercises

Expand each power of a binomial. Simplify the terms.

1. $(a + b)^6$
2. $(a - b)^7$
3. $(x - 3)^4$
4. $(2m + 1)^5$
5. $(3c + 2d)^5$
6. $(2t - 3v)^4$
7. $(x^2 - 2y)^6$
8. $(3p + r^3)^4$

Find the specified term of each expansion. Simplify the term.

9. 4th term of $(3c + d)^5$
10. 4th term of $(x - y^2)^6$
11. middle term of $(x^2 - 2y)^4$
12. 5th term of $(4x + 2y)^7$
13. 3rd term of $(2c^3 - 3d)^5$
14. 5th term of $(3x^2 + \sqrt{y})^6$
15. middle term of $\left(4x^2 + \frac{x}{2}\right)^6$
16. 6th term of $\left(27y^4 - \frac{y^2}{3}\right)^7$

Expand each power of a binomial. Simplify the terms.

17. $(3c^3 + 2d^2)^5$
18. $(p - \sqrt{3})^6$
19. $(2p - \sqrt{p})^4$
20. $(1 + \sqrt[5]{4})^5$

Find the first four terms of each expansion. Simplify the terms.

21. $(x + 1)^{14}$
22. $(x - 3y)^{11}$
23. $(x^2 - y^3)^{10}$
24. $(1 + 0.02)^8$

Write each power in sigma-notation. Then find the sum of the terms.

25. $(2 + 0.1)^4$
26. $(1 + 0.1)^4$
27. $(1.5)^3$
28. $(2.2)^3$

Mixed Review

Solve. *1.4, 2.3, 8.1, 9.1*

1. $\frac{x}{3} = \frac{4}{x - 4}$
2. $|3x + 2| = 17$
3. $2x^2 = 20$
4. $x^2 + 4 = 0$

The concepts of probability, standard deviation, and the normal distribution can be used to create a quality-control chart that monitors the number of defective parts produced by a factory in a day. Such a chart includes the number of defective parts produced each day, the mean number of defective parts M, and upper and lower limits that represent three standard deviations from that mean, $M \pm 3\sigma$. Recall that in a normal distribution (see Lesson 14.11), 99.6% of the values will fall between $M - 3\sigma$ and $M + 3\sigma$. So, on days when the number of defective parts falls above or below the upper or lower limits, the production process should be checked to determine the cause of the unusually poor or excellent results.

Consider a process that produces 2,000 parts per day. Data shows that the number of defective parts is normally distributed and that the probability p of a defective part is 0.05. Thus, the mean number of defective parts per day is 0.05(2,000), or 100. Standard deviation from the mean can be calculated using the formula $\sigma = \sqrt{npq}$, where n = number of parts, p = probability of a defective part, and $q = 1 - p$. Thus, upper and lower limits are found as follows.

$$\sigma = \sqrt{npq}$$
$$\sigma = \sqrt{2,000(0.05)(0.95)}$$
$$\sigma = \sqrt{95}$$
$$\sigma \approx 9.75$$
$$3\sigma \approx 29.25$$

$$M = 100$$
Upper limit: $M + 3\sigma = 100 + 29.25$
$$= 129.25$$
Lower limit: $M - 3\sigma = 100 - 29.25$
$$= 70.75$$

Suppose, over a seven-day period, the plant described above produces the following numbers of defective parts: 100, 130, 115, 95, 105, 110, 65. This data is shown on the quality control chart at the right.

1. In the example above, on which day(s) should the process be checked to find the cause of the unusual performance?

2. Suppose a plant produces 5,000 parts per day. The probability of defective parts is 0.04. Over a five-day period, the plant produces the following daily totals of defective parts: 240, 155, 225, 220, 230. Use the data to make a quality-control chart.

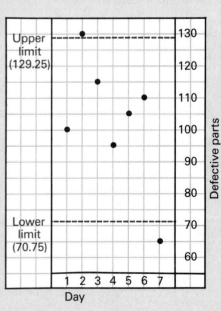

Solve each problem.

1. How many games will be played if each of 8 baseball teams plays the other 7 teams 4 times each?

2. How many signals can be formed by displaying 5 flags in a row if 10 different flags are available?

3. Book *A* has 125 fewer pages than Book *B*, and Book *C* has twice as many pages as Book *A*. Find the number of pages in Book *C* if the three books have a total of 1,225 pages.

4. The length of a rectangle is 3 m less than 4 times the width. The perimeter is 54 m. Find the following.
   a. the width   b. the length
   c. the area
   d. the length of a diagonal in simplest radical form

5. Convert 18 oz/pt to pounds per gallon.

6. If *x* is inversely proportional to $\sqrt[3]{y}$, and $x = 3.6$ when $y = 64$, find *x* when $y = 27$.

7. How many even numbers of one or more digits can be formed from the digits 1, 2, 4, 5, 8, if no digit may be repeated in a number?

8. How many liters of water must be evaporated from 75 liters of a 12% salt solution to obtain an 18% salt solution?

9. Find the number of quarters in a coin collection with 8 more dimes than nickels and twice as many quarters as dimes if the collection is worth $9.35.

10. Corn worth $2/kg is mixed with oats worth $3/kg. There are 40 kg more oats than corn. How many kilograms of oats are used if the mixture is worth $320?

11. A triangle's height is 7 cm less than 5 times the length of its base. Find the height if the area is 12 cm^2.

12. If 7 times a number is decreased by 9, the square root of the result is the same as twice the square root of the number. Find the number.

13. Ellen drove 4 h at an average speed of 42 mi/h. Her car averages 28 mi/gal. If gas costs $1.30/gal, how much did she pay for gas?

14. In a shipment of 20 lamps, two are broken. If 2 of the lamps are selected at random, what is the probability that exactly 1 of them is broken?

15. Marie bought a new car for $9,600. Each year its value will depreciate by 30% of its value at the start of the year. What will its value be at the end of 5 years? Give your answer to the nearest dollar.

16. The numerator of a fraction is one-half the sum of the denominator and 1. If the numerator is increased by 2 and the denominator is decreased by 2, the resulting fraction is the reciprocal of the original fraction. What is the original fraction?

Chapter 15 Review

Key Terms

arithmetic means (p. 549)
arithmetic sequence (p. 545)
arithmetic series (p. 553)
Binomial Theorem (p. 576)
common difference (p. 545)
common ratio (p. 558)
geometric means (p. 562)
geometric sequence (p. 558)

geometric series (pp. 566, 572)
harmonic means (p. 550)
harmonic sequence (p. 550)
index of summation (p. 555)
infinite geometric series (p. 572)
partial sums (p. 572)
sequence (p. 545)
summation symbol (p. 555)

Key Ideas and Review Exercises

15.1 For the arithmetic sequence $a_1, a_1 + d, a_1 + 2d, \ldots, a_1 + (n - 1)d$, the nth term is given by $a_n = a_1 + (n - 1)d$.

1. Write the next 3 terms of the arithmetic sequence $-3.8, -1.5, 0.8, \ldots$.

2. Find the 26th term of the arithmetic sequence $-4, -2.5, -1, \ldots$.

15.2 For the arithmetic sequence x_1, x_2, x_3, x_4, x_5, the following are true.
(1) x_2, x_3, and x_4 are the three arithmetic means between x_1 and x_5.
(2) $\dfrac{1}{x_1}, \dfrac{1}{x_2}, \dfrac{1}{x_3}, \dfrac{1}{x_4}, \dfrac{1}{x_5}$, is a harmonic sequence.
(3) The arithmetic mean of x and y is $\dfrac{x + y}{2}$.

3. Find the four arithmetic means between 8 and $28 + 5\sqrt{6}$.

4. The arithmetic mean of two numbers is 16 and one number is 4 more than 3 times the other number. Find both numbers.

5. Find the 31st term in the harmonic sequence $\dfrac{1}{3}, \dfrac{2}{9}, \dfrac{1}{6}, \dfrac{2}{15}, \ldots$.

6. Write in your own words how to find the harmonic mean of two nonzero numbers.

15.3 The sum of the first n terms of an arithmetic series is given by
(1) $S_n = \dfrac{n}{2}(a_1 + a_n)$ *or* (2) $S_n = \dfrac{n}{2}[2a_1 + (n - 1)d]$.

Find the sum of each arithmetic series for the given data.

7. $n = 34$; $a_1 = -2.5$, $a_n = 33.8$

8. $n = 41$; $14 + 9 + 4 - 1 - 6 - \ldots$

9. $\displaystyle\sum_{k=1}^{31} (2k - 7)$

10. $\displaystyle\sum_{j=8}^{10} \dfrac{j + 2}{3}$

11. Use sigma-notation to represent the series $6 + 10 + 14 + \ldots + 42$.

15.4   The nth term, a_n, of the geometric sequence $a_1, a_1r, a_1r^2, \ldots$ is a_1r^{n-1}.

Find the specified term of each geometric sequence.

12. 7th term: $a_1 = 0.03, r = 10$ **13.** 9th term: 256, 128, 64, . . .

14. A ball rebounds one-half of the distance that it falls. If the ball is dropped from a height of 256 ft, how high does it go on its 8th rebound?

15.5   For the geometric sequence x_1, x_2, x_3, x_4, x_5, the three geometric means between x_1 and x_5 are x_2, x_3, and x_4.
   The geometric mean of x and y is $\sqrt{xy}$ or $-\sqrt{xy}$.

15. Find three positive geometric means between $\frac{2}{3}$ and 54.

16. What is the negative geometric mean of 0.4 and 250?

17. The positive geometric mean of two numbers is 8. Find the two numbers if one number is 12 less than the other number.

15.6   The sum of a finite geometric series of n terms is given by

$$(1)\ S_n = \frac{a_1 - ra_n}{1 - r} \quad \text{or} \quad (2)\ S_n = \frac{a_1(1 - r^n)}{1 - r}.$$

Find the sum of the terms of the geometric series described.

18. $a_1 = -2, r = 5, a_n = -6{,}250$ **19.** $\sum\limits_{k=4}^{6} \frac{1}{6}(3^k - 2)$

20. $n = 10;\ 3 + 6 + 12 + 24 + \ldots$

21. Smaller and smaller squares are formed, one after the other, as shown at the right. Find the sum of the perimeters of the first 7 squares if the first square is 12 ft wide.

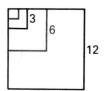

15.7   The sum of an infinite geometric series is
   $\frac{a_1}{1 - r},\ -1 < r < 1, r \neq 0.$

22. Find the sum, if it exists, of the infinite series
   $64 + 48 + 36 + 27 + \ldots$.

23. Write $0.17\overline{17}$ as an infinite geometric series and then in the form $\frac{x}{y}$, where x and y are integers.

24. Refer to the squares in Exercise 21. Find the sum of the areas of an infinite sequence of the squares if the first square is 12 ft wide.

15.8 $(a + b)^n = \binom{n}{0}a^n + \binom{n}{1}a^{n-1}b + \binom{n}{2}a^{n-2}b^2 + \binom{n}{3}a^{n-3}b^3 + \ldots + \binom{n}{n}b^n$

   In the expansion of $(a + b)^n$, the rth term is $\binom{n}{r-1}a^{n-r+1}b^{r-1}$.

25. Expand $(x^2 - 3y)^4$. Simplify the terms.

26. Find the 6th term in the expansion of $(c^2 + 2d)^8$. Simplify the term.

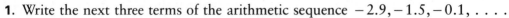

Chapter 15 Test

1. Write the next three terms of the arithmetic sequence $-2.9, -1.5, -0.1, \ldots.$
2. Find the 36th term of the arithmetic sequence $6, 2, -2, -6, \ldots.$
3. Find the three arithmetic means between 5 and $9 + 12\sqrt{5}$.
4. What is the arithmetic mean of -5.6 and 8?
5. The arithmetic mean of two numbers is 45 and one number is 6 more than 5 times the other number. Find both numbers.
6. Find the 21st term in the harmonic sequence $\frac{1}{2}, \frac{4}{9}, \frac{2}{5}, \frac{4}{11}, \ldots.$

Find the sum of each arithmetic series for the given data.

7. $n = 42;\ a_1 = -3.4,\ a_{42} = 29.4$
8. $n = 31;\ 12 + 7 + 2 - 3 - \ldots$
9. $\displaystyle\sum_{k=1}^{21} (3k - 5)$
10. $\displaystyle\sum_{j=7}^{10} \frac{j + 1}{2}$
11. Use sigma-notation to represent the series $7 + 11 + 15 + \ldots + 55$.
12. Boxes are stacked in 21 rows with 2 in the top row, 5 in the 2nd row, 8 in the 3rd row, and so on. How many boxes are in the stack?
13. Find the 10th term of the geometric sequence $128, -64, 32, \ldots.$
14. A cylinder contains 32,000 cm^3 of oxygen. One-half of the oxygen is released each time that a valve is opened. How much oxygen will be in the tank after the valve is opened 8 times?
15. A certain steel ball rebounds on each bounce one-third of the distance from which it fell. If it is dropped from a height of 729 cm, how high does it go on its 7th rebound (upward)?
16. Find the three real, positive geometric means between $\frac{3}{4}$ and 12.
17. What is the positive geometric mean of 0.9 and 40?

Find the sum of the terms of the geometric series described.

18. $a_1 = 3,\ r = 4,\ a_n = 12{,}288$
19. $n = 9;\ 5 + 10 + 20 + 40 + \ldots$
20. Find the sum, if it exists, of the infinite series $27 + 18 + 12 + 8 + \ldots.$
21. Write $0.07\overline{373}$ as an infinite geometric series, and then in the form $\frac{x}{y}$, where x and y are integers.
22. Expand $(3x - y^2)^5$. Simplify the terms.
23. Find the 5th term in the expansion of $(2c + d^2)^7$. Simplify the term.
24. Find the 1st term of the arithmetic sequence whose 6th term is 31 and 20th term is 52.
25. Find two values of c so that $\frac{1}{5}, c - 3, 4c$ is a geometric sequence.
26. Find the sum of the first n terms of the geometric series
$1 + 2x^2 + 4x^4 + 8x^6 \cdots + 2^{(n-1)}x^{(2n-2)}$.

College Prep Test

In each item, compare a quantity in Column 1 with a quantity in Column 2. Write the letter of the correct answer from these choices.

A—The quantity in Column 1 is greater than the quantity in Column 2.
B—The quantity in Column 2 is greater than the quantity in Column 1.
C—The quantity in Column 1 is equal to the quantity in Column 2.
D—The relationship cannot be determined from the given information.

NOTE: Information centered over both columns refers to one or both of the quantities to be compared.

Sample Items and Answers	
Column 1　　　　**Column 2**	
$x = 3$ and $y = -3$	S1: The answer is A: $5x > y^5$, since $5(3) > (-3)^5$, or $15 > -243$.
S1. Next term of the sequence: $2x, 3x, 4x, \ldots$　　Next term of the sequence: $y^2, y^3, y^4, \ldots$	
S2. $\sum_{k=1}^{6} 3^{k-1}$　　　$\sum_{j=4}^{9} 3^{j-4}$	S2: The answer is C: Each symbol represents the same sum, $3^0 + 3^1 + 3^2 + 3^3 + 3^4 + 3^5$.

Column 1　　**Column 2**　　　　**Column 1**　　**Column 2**

$x = 4$ and $y = -5$

1. Next term of the sequence: $\frac{1}{x}, \sqrt{\frac{1}{x}}, \frac{1}{x}, \sqrt{\frac{1}{x}} \ldots$　Next term of the sequence: $y^3, y^2, y^1, \ldots$

$x = y$ and $x > 0$

2. Arithmetic mean of x and y　　Positive geometric mean of x and y

3. Arithmetic mean of $\frac{1}{3}$ and $\frac{1}{6}$　　Harmonic mean of $\frac{1}{3}$ and $\frac{1}{6}$

4. Coefficient of the 5th term in the expansion of $(x - 4)^{10}$　Coefficient of the 5th term in the expansion of $(y + 4)^{10}$

5. $(x + 2)^4$　　$(x + 2)^5$

Midpoints of sides of a triangle are connected to form a new triangle. This procedure is continued in each new triangle.

6. Sum of the perimeters of an infinite sequence of the triangles　Twice the perimeter of the first triangle

Artists use the graphic and color capabilities of computers as a medium of artistic expression. The image above is called "After Picasso." It is a still print taken from an animated film created by computer artist Lillian Schwartz.

16.1 Addition of Matrices

To determine whether two matrices are equal

To add matrices

To prove properties of matrix addition and to use these properties

A **matrix** (plural: **matrices**) is a rectangular array of numbers arranged in rows and columns and enclosed by brackets. A matrix that contains only one row is called a *row* matrix. A matrix that contains only one column is called a *column* matrix. A matrix that has the same number of rows and columns is called a *square* matrix.

Following are examples of matrices and their classifications.

Row Matrix	Column Matrix	Square Matrix	General Rectangular Matrix (3 by 2)
$\begin{bmatrix} 1 & 4 & -3 \end{bmatrix}$	$\begin{bmatrix} 8 \\ \sqrt{3} \\ 12 \end{bmatrix}$	$\begin{bmatrix} 1 & -2 \\ 0 & 3 \end{bmatrix}$	$\begin{bmatrix} a_{11} & a_{12} \\ a_{21} & a_{22} \\ a_{31} & a_{32} \end{bmatrix}$
1 row 3 columns	3 rows 1 column	2 rows 2 columns	3 rows 2 columns

The individual numbers, such as a_{31}, are called **elements** of the matrix and are real numbers. The first number of the subscript indicates the row of the element, and the second number indicates the column of the element. So, a_{31} means the element in the third row and first column. The number of rows and columns determines the **dimensions** of a matrix. The dimensions of the matrices above in order are 1 by 3 (1 × 3), 3 by 1 (3 × 1), 2 by 2 (2 × 2), and 3 by 2 (3 × 2).

Two matrices are *equal* if and only if they have the same dimensions and the elements in corresponding positions are equal.

Two matrices can be added if they have the same dimensions. The **sum of two matrices** with the same dimensions is the matrix obtained by adding the corresponding elements of the two matrices.

For example: $\begin{bmatrix} a_{11} & a_{12} \\ a_{21} & a_{22} \\ a_{31} & a_{32} \end{bmatrix} + \begin{bmatrix} b_{11} & b_{12} \\ b_{21} & b_{22} \\ b_{31} & b_{32} \end{bmatrix} = \begin{bmatrix} a_{11} + b_{11} & a_{12} + b_{12} \\ a_{21} + b_{21} & a_{22} + b_{22} \\ a_{31} + b_{31} & a_{32} + b_{32} \end{bmatrix}$

Notice that the sum of two matrices is a unique matrix with elements that are real numbers. Thus, the **Closure Property** holds for the sum of two matrices of the same dimensions.

EXAMPLE 1   Matrices A, B, and C are given as follows.

$$A = \begin{bmatrix} 3 & -1 & 2 \\ 4 & 5 & -4 \end{bmatrix}, \; B = \begin{bmatrix} 1 & -2 & -2 \\ -3 & 4 & -1 \end{bmatrix}, \; C = \begin{bmatrix} -1 & 5 \\ 3 & -2 \\ 5 & -4 \end{bmatrix}$$

Find the following sums if they exist. If a sum does not exist, give a reason.

a. $A + B$ b. $A + C$

Solutions   a. $A + B =$

$$\begin{bmatrix} 3 + 1 & -1 + (-2) & 2 + (-2) \\ 4 + (-3) & 5 + 4 & -4 + (-1) \end{bmatrix} = \begin{bmatrix} 4 & -3 & 0 \\ 1 & 9 & -5 \end{bmatrix}$$

b. Since A and C do not have the same dimensions, the sum does not exist.

Recall that in the set of real numbers there exists a unique additive identity, the number 0, such that for each real number a, $a + 0 = a$. Also, each real number a has an opposite, or additive inverse, $-a$, such that $a + (-a) = 0$. Similarly, additive identities and inverses exist for matrix addition.

Definitions

If matrices A and O have the same dimensions and $A + O = A$, then O is called an **additive identity matrix** (zero matrix) for A. If matrices A and $-A$ have the same dimensions and $A + (-A) = O$, then $-A$ is called an **additive inverse matrix** (opposite matrix) of A.

EXAMPLE 2   Matrices A, $-A$, and O are given as follows.

$$A = \begin{bmatrix} 3 & -2 \\ -4 & 5 \end{bmatrix}, \quad -A = \begin{bmatrix} -3 & 2 \\ 4 & -5 \end{bmatrix}, \quad \text{and} \quad O = \begin{bmatrix} 0 & 0 \\ 0 & 0 \end{bmatrix}$$

a. Prove that $A + O = A$.   b. Prove that $A + (-A) = O$.

Solutions   a. $A + O = \begin{bmatrix} 3 + 0 & -2 + 0 \\ -4 + 0 & 5 + 0 \end{bmatrix} = \begin{bmatrix} 3 & -2 \\ -4 & 5 \end{bmatrix} = A$

Thus, $A + O = A$.

b. $A + (-A) = \begin{bmatrix} 3 + (-3) & -2 + 2 \\ -4 + 4 & 5 + (-5) \end{bmatrix} = \begin{bmatrix} 0 & 0 \\ 0 & 0 \end{bmatrix} = O$

Thus, $A + (-A) = O$.

The results of Example 2 can be generalized in two ways. First, all matrices of the same dimensions have the same additive identity matrix, with elements that consist entirely of zeros. Second, each element of an additive inverse matrix, $-A$, is equal to the opposite of its corresponding element in matrix A. These two points are further illustrated below.

$$A = \begin{bmatrix} a_{11} & a_{12} & a_{13} \\ a_{21} & a_{22} & a_{23} \end{bmatrix}$$

The additive identity (zero) matrix for A and the additive inverse (opposite) matrix for A are, respectively,

$$O_{2 \times 3} = \begin{bmatrix} 0 & 0 & 0 \\ 0 & 0 & 0 \end{bmatrix} \quad \text{and} \quad -A_{2 \times 3} = \begin{bmatrix} -a_{11} & -a_{12} & -a_{13} \\ -a_{21} & -a_{22} & -a_{23} \end{bmatrix},$$

where the subscript 2×3 is used to indicate the dimensions of the matrix. In general, an m-by-n matrix A may be written as $A_{m \times n}$ whenever it is desired to emphasize the dimensions of the matrix.

EXAMPLE 3   Matrices A, B, and C are given as follows.

$$A = \begin{bmatrix} 5 & -3 \\ -4 & 6 \end{bmatrix}, B = \begin{bmatrix} 2 & 4 \\ 5 & -2 \end{bmatrix}, \text{ and } C = \begin{bmatrix} -8 & -3 \\ 4 & -1 \end{bmatrix}$$

a. Prove that $A + B = B + A$.   **b.** Prove that $(A + B) + C = A + (B + C)$.

Solutions   **a.** $A + B = \begin{bmatrix} 5 + 2 & -3 + 4 \\ -4 + 5 & 6 + (-2) \end{bmatrix} = \begin{bmatrix} 7 & 1 \\ 1 & 4 \end{bmatrix}$

$$B + A = \begin{bmatrix} 2 + 5 & 4 + (-3) \\ 5 + (-4) & -2 + 6 \end{bmatrix} = \begin{bmatrix} 7 & 1 \\ 1 & 4 \end{bmatrix}$$

So, $A + B = B + A$. ← Matrix addition is commutative.

b. $(A + B) + C = \begin{bmatrix} 5 + 2 & -3 + 4 \\ -4 + 5 & 6 + (-2) \end{bmatrix} + \begin{bmatrix} -8 & -3 \\ 4 & -1 \end{bmatrix}$

$$= \begin{bmatrix} 7 + (-8) & 1 + (-3) \\ 1 + 4 & 4 + (-1) \end{bmatrix} = \begin{bmatrix} -1 & -2 \\ 5 & 3 \end{bmatrix}$$

$$A + (B + C) = \begin{bmatrix} 5 & -3 \\ -4 & 6 \end{bmatrix} + \begin{bmatrix} 2 + (-8) & 4 + (-3) \\ 5 + 4 & -2 + (-1) \end{bmatrix}$$

$$= \begin{bmatrix} 5 + (-6) & -3 + 1 \\ -4 + 9 & 6 + (-3) \end{bmatrix} = \begin{bmatrix} -1 & -2 \\ 5 & 3 \end{bmatrix}$$

So, $(A + B) + C = A + (B + C)$. ← Matrix addition is associative.

Classroom Exercises

Matrices $A_{2 \times 2}$ and $B_{2 \times 2}$ are equal, as shown at the right. Find the value of each of the following.

$$\begin{matrix} A \\ \begin{bmatrix} a_{11} & a_{12} \\ -3 & 2p \end{bmatrix} \end{matrix} = \begin{matrix} B \\ \begin{bmatrix} 1 & 6p \\ b_{21} & 0 \end{bmatrix} \end{matrix}$$

1. a_{11}

2. b_{21}

3. b_{11}

4. a_{21}

5. b_{22}

6. p

7. a_{22}

8. a_{12}

Written Exercises

Find each sum, if it exists. If it does not exist, give a reason.

1.
$$\begin{bmatrix} -3 & 5 & -6 \\ -2 & 1 & 3 \\ 5 & -2 & -9 \end{bmatrix} + \begin{bmatrix} 9 & 6 & -3 \\ 8 & -7 & 10 \\ -1 & 8 & -1 \end{bmatrix}$$

2.
$$\begin{bmatrix} -10 & 6 \\ -12 & -18 \\ 14 & 9 \end{bmatrix} + \begin{bmatrix} 1 & 3 & -8 \\ 4 & -6 & 5 \end{bmatrix}$$

3.
$$\begin{bmatrix} 12 & -6 \\ -14 & 9 \end{bmatrix} + \begin{bmatrix} -14 & 9 \\ -8 & 4 \end{bmatrix}$$

In Exercises 4–7, $A = \begin{bmatrix} 5 & -3 \\ 7 & 2 \end{bmatrix}$, $B = \begin{bmatrix} -9 & 8 \\ -6 & 4 \end{bmatrix}$, and $C = \begin{bmatrix} -1 & -5 \\ -7 & -3 \end{bmatrix}$.

Use the definition of matrix addition to prove each of the following.

4. $A + O_{2 \times 2} = A$

5. $A + (-A) = O_{2 \times 2}$

6. $-(A + B) = -A + (-B)$

7. $A + (C + B) = (A + C) + B$

In Exercises 8–10, $A = \begin{bmatrix} a_{11} & a_{12} \\ a_{21} & a_{22} \end{bmatrix}$, $B = \begin{bmatrix} b_{11} & b_{12} \\ b_{21} & b_{22} \end{bmatrix}$, and $C = \begin{bmatrix} c_{11} & c_{12} \\ c_{21} & c_{22} \end{bmatrix}$,

where all elements are real numbers. Prove each of the following.

8. $A + O_{2 \times 2} = A$

9. $B + (-B) = O_{2 \times 2}$

10. $C = -(-C)$

Mixed Review

Solve by the linear combination method. *4.3*

1. $3a + b = 6$
$4a - b = 8$

2. $5a + 6b = 4$
$-2a - 3b = -5$

3. $8a - 3b = 27$
$-4a + 5b = -17$

4. $6a - 5b = 20$
$4a - 2b = 8$

16.2 Matrices and Multiplication

To multiply a scalar and a matrix
To multiply matrices
To prove properties of matrix multiplication and to use these properties

Any matrix can be multiplied by a real number. This real number is called a **scalar**.

The product of a scalar k and a matrix A is the matrix that is represented by kA. It is obtained by multiplying each element of A by k.

EXAMPLE 1 Find kA if $k = -5$ and $A = \begin{bmatrix} 3 & -1 & 5 \\ 2 & 3 & -4 \\ 5 & -2 & -7 \end{bmatrix}$.

Plan Multiply each element of A by -5.

Solution $-5A = \begin{bmatrix} -5(3) & -5(-1) & -5(5) \\ -5(2) & -5(3) & -5(-4) \\ -5(5) & -5(-2) & -5(-7) \end{bmatrix} = \begin{bmatrix} -15 & 5 & -25 \\ -10 & -15 & 20 \\ -25 & 10 & 35 \end{bmatrix}$

One matrix can be multiplied by another if the number of *columns* of the first matrix equals the number of *rows* of the second matrix. The method is illustrated below for matrices C and D, whose product can be represented as either $C \cdot D$ or CD.

$$C = \begin{bmatrix} -3 & 5 & 2 \\ 1 & -2 & 4 \end{bmatrix}, D = \begin{bmatrix} 5 & -6 \\ -4 & 3 \\ 1 & -2 \end{bmatrix}$$

1. Multiply the elements of the *first* row of C by the corresponding elements of the *first column* of D. Then add the three products.

    $-3(5) + 5(-4) + 2(1)$
    $= -15 - 20 + 2$
    $= -33$

2. The result is the element of the *first row* and *first column* of CD.

    $\begin{bmatrix} -33 & \underline{\quad} \\ \underline{\quad} & \underline{\quad} \end{bmatrix}$

3. Multiply the elements of the *first* row of C by the corresponding elements of the *second column* of D. Then add the three products.

    $-3(-6) + 5(3) + 2(-2)$
    $= 18 + 15 - 4$
    $= 29$

4. The result is the element of the *first row* and *second column* of CD.

    $\begin{bmatrix} -33 & 29 \\ \underline{\quad} & \underline{\quad} \end{bmatrix}$

5. Use the elements of the second row of C to calculate the remaining two elements of CD in a similar manner. The entire process is summarized below.

$$CD = \begin{bmatrix} -3(5) + 5(-4) + 2(1) & -3(-6) + 5(3) + 2(-2) \\ 1(5) + (-2)(-4) + 4(1) & 1(-6) + (-2)(3) + 4(-2) \end{bmatrix}$$

$$= \begin{bmatrix} -33 & 29 \\ 17 & -20 \end{bmatrix}$$

The above example is generalized below for the case of a matrix A with 2 rows and 3 columns and a matrix B with 3 rows and 2 columns.

If $A = \begin{bmatrix} a_{11} & a_{12} & a_{13} \\ a_{21} & a_{22} & a_{23} \end{bmatrix}$ and $B = \begin{bmatrix} b_{11} & b_{12} \\ b_{21} & b_{22} \\ b_{31} & b_{32} \end{bmatrix}$, then

$$AB = \begin{bmatrix} a_{11}b_{11} + a_{12}b_{21} + a_{13}b_{31} & a_{11}b_{12} + a_{12}b_{22} + a_{13}b_{32} \\ a_{21}b_{11} + a_{22}b_{21} + a_{23}b_{31} & a_{21}b_{12} + a_{22}b_{22} + a_{23}b_{32} \end{bmatrix}.$$

In general, $A_{m \times n} \cdot B_{n \times r} = (AB)_{m \times r}$, where $m \times n$, $n \times r$, and $m \times r$ are the dimensions of the matrices.

Note that $A_{3 \times 1} \cdot B_{1 \times 2}$ exists and equals the 3-by-2 matrix $(AB)_{3 \times 2}$. However, $B_{1 \times 2} \cdot A_{3 \times 1}$ does not exist since the number of columns of B does not equal the number of rows of A.

EXAMPLE 2　Use the matrices C and D given earlier to find the matrix DC.

Plan　Since the number of columns of D is 2 and the number of rows of C is 2, DC exists; that is, $D_{3 \times 2} \cdot C_{2 \times 3} = (DC)_{3 \times 3}$.

Solution　$$DC = \begin{bmatrix} 5(-3) + (-6)(1) & 5(5) - 6(-2) & 5(2) - 6(4) \\ -4(-3) + 3(1) & -4(5) + 3(-2) & -4(2) + 3(4) \\ 1(-3) - 2(1) & 1(5) - 2(-2) & 1(2) - 2(4) \end{bmatrix}$$

$$= \begin{bmatrix} -21 & 37 & -14 \\ 15 & -26 & 4 \\ -5 & 9 & -6 \end{bmatrix}$$

Compare matrix DC of Example 2 with matrix CD that was determined earlier. Notice that $CD \neq DC$. Thus, the multiplication of two matrices *is not* a commutative operation. However, multiplication of matrices *is* an associative operation. That is, for matrices A, B, and C, if all products exist, then $(A \cdot B) \cdot C = A \cdot (B \cdot C)$.

Recall that in the set of real numbers, there exists a unique multiplicative identity, the number 1, such that for each real number a, $a \cdot 1 = a$. Also, for each nonzero real number a, there exists a multiplicative inverse $\frac{1}{a}$ such that $a \cdot \frac{1}{a} = 1$.

For *square* matrices and the multiplication of two square matrices, the situation is similar.

For each square matrix A, if $A \cdot I = I \cdot A = A$, then I is called a **multiplicative identity matrix** for A.

For each square matrix A, if $A \cdot A^{-1} = A^{-1} \cdot A = I$, then A^{-1} is called a **multiplicative inverse matrix** of A.

All square matrices of the same dimensions have the same unique multiplicative identity matrix. For example, the unique multiplicative identity matrices for 2-by-2 and 3-by-3 matrices are as follows.

$$I_{2\times 2} = \begin{bmatrix} 1 & 0 \\ 0 & 1 \end{bmatrix} \text{ and } I_{3\times 3} = \begin{bmatrix} 1 & 0 & 0 \\ 0 & 1 & 0 \\ 0 & 0 & 1 \end{bmatrix}$$

EXAMPLE 3   If $A = \begin{bmatrix} -2 & 1 & -3 \\ 4 & 2 & -1 \\ 3 & -2 & 4 \end{bmatrix}$ and $I = \begin{bmatrix} 1 & 0 & 0 \\ 0 & 1 & 0 \\ 0 & 0 & 1 \end{bmatrix}$, prove that $A \cdot I = A$.

Solution $A \cdot I =$

$$\begin{bmatrix} -2(1) + 1(0) + (-3)(0) & -2(0) + 1(1) + (-3)(0) & -2(0) + 1(0) + (-3)(1) \\ 4(1) + 2(0) + (-1)(0) & 4(0) + 2(1) + (-1)(0) & 4(0) + 2(0) + (-1)(1) \\ 3(1) + (-2)(0) + 4(0) & 3(0) + (-2)(1) + 4(0) & 3(0) + (-2)(0) + 4(1) \end{bmatrix}$$

$$= \begin{bmatrix} -2 & 1 & -3 \\ 4 & 2 & -1 \\ 3 & -2 & 4 \end{bmatrix} = A \quad \text{Thus, } A \cdot I = A.$$

The next example illustrates how to find the multiplicative inverse matrix of a given square matrix.

EXAMPLE 4   If $A = \begin{bmatrix} -2 & 1 \\ 3 & -1 \end{bmatrix}$, find A^{-1}.

Plan   Let $A^{-1} = \begin{bmatrix} a_1 & b_1 \\ a_2 & b_2 \end{bmatrix}$. Then use $A \cdot A^{-1} = I$.

Solution $$\begin{bmatrix} -2 & 1 \\ 3 & -1 \end{bmatrix} \cdot \begin{bmatrix} a_1 & b_1 \\ a_2 & b_2 \end{bmatrix} = \begin{bmatrix} 1 & 0 \\ 0 & 1 \end{bmatrix}$$

$$\begin{bmatrix} -2a_1 + a_2 & -2b_1 + b_2 \\ 3a_1 + (-1)a_2 & 3b_1 + (-1)b_2 \end{bmatrix} = \begin{bmatrix} 1 & 0 \\ 0 & 1 \end{bmatrix}$$

Since the two matrices are equal, their corresponding elements are equal.

$$-2a_1 + a_2 = 1 \qquad\qquad -2b_1 + b_2 = 0$$
$$3a_1 - a_2 = 0 \qquad\qquad 3b_1 - b_2 = 1$$

Solve each system of equations by the linear combination method (see Lesson 4.3). The result is $a_1 = 1$, $a_2 = 3$, $b_1 = 1$, and $b_2 = 2$.

Thus, $A^{-1} = \begin{bmatrix} 1 & 1 \\ 3 & 2 \end{bmatrix}$.

Focus on Reading

Let $X = \begin{bmatrix} 3 & 5 \\ 4 & -1 \end{bmatrix}$, $Y = \begin{bmatrix} -2 & -7 \\ 6 & 5 \end{bmatrix}$, and $k = 3$. Match each product in the left column with exactly one matrix in the two right columns.

1. $X \cdot Y$
2. kX
3. kY
4. $Y \cdot X$
5. $X \cdot X^{-1}$

a. $\begin{bmatrix} -6 & -21 \\ 18 & 15 \end{bmatrix}$

b. $\begin{bmatrix} 1 & 0 \\ 0 & 1 \end{bmatrix}$

c. $\begin{bmatrix} 24 & 4 \\ -14 & -33 \end{bmatrix}$

d. $\begin{bmatrix} -34 & -3 \\ 38 & 25 \end{bmatrix}$

e. $\begin{bmatrix} 9 & 15 \\ 12 & -3 \end{bmatrix}$

Classroom Exercises

Find each product, if it exists. If it does not, give a reason.

1. $-2 \begin{bmatrix} 1 \\ -3 \end{bmatrix}$

2. $\begin{bmatrix} 3 & 5 \end{bmatrix} \cdot \begin{bmatrix} 2 \\ 3 \end{bmatrix}$

3. $\begin{bmatrix} 1 \\ 2 \end{bmatrix} \cdot \begin{bmatrix} -1 & 5 \end{bmatrix}$

4. $\begin{bmatrix} -3 \\ -2 \\ 1 \end{bmatrix} \cdot \begin{bmatrix} -1 & 2 & 5 \end{bmatrix}$

5. $\begin{bmatrix} 8 & 5 \\ -1 & 6 \end{bmatrix} \cdot \begin{bmatrix} -1 & 3 \\ 2 & 4 \end{bmatrix}$

6. $\begin{bmatrix} -3 & 1 & 4 \end{bmatrix} \cdot \begin{bmatrix} 1 & 2 \\ -4 & 5 \end{bmatrix}$

Written Exercises

Find each product, if it exists. If it does not exist, give a reason.

1. $6 \begin{bmatrix} 3 \\ 0 \\ 1 \end{bmatrix}$

2. $4 \begin{bmatrix} 5 & -3 \\ 7 & -8 \\ 4 & 6 \end{bmatrix}$

3. $\begin{bmatrix} -1 & 4 & -3 \\ 5 & -2 & 2 \end{bmatrix} \cdot \begin{bmatrix} 3 & 1 & -2 \\ 4 & 5 & 3 \\ -1 & 2 & 4 \end{bmatrix}$

4. $\begin{bmatrix} 4 & -6 & -3 \\ 5 & -1 & 7 \\ 8 & 6 & -5 \end{bmatrix} \cdot \begin{bmatrix} 3 & -1 & 2 \\ 5 & 6 & -3 \end{bmatrix}$ **5.** $\begin{bmatrix} 1 & 2 & -1 \\ 3 & 4 & -2 \\ -1 & 5 & 6 \end{bmatrix} \cdot \begin{bmatrix} 1 & -1 & 3 \\ 4 & 8 & 6 \\ -5 & 7 & -3 \end{bmatrix}$

Find the multiplicative inverse of each matrix if possible. If not possible, indicate this.

6. $\begin{bmatrix} 1 & 4 \\ 2 & 9 \end{bmatrix}$ **7.** $\begin{bmatrix} -2 & \frac{1}{2} \\ 1 & -1 \end{bmatrix}$ **8.** $\begin{bmatrix} 3 & -5 \\ -2 & 6 \end{bmatrix}$ **9.** $\begin{bmatrix} -5 & 7 \\ 8 & -2 \end{bmatrix}$

10. $\begin{bmatrix} 1 & 1 \\ 1 & 1 \end{bmatrix}$ **11.** $\begin{bmatrix} 1 & 2 & -3 \\ 4 & 6 & 4 \\ -5 & -1 & 2 \end{bmatrix}$ **12.** $\begin{bmatrix} 7 & -3 & 2 \\ -1 & -2 & 5 \\ 6 & -4 & 8 \end{bmatrix}$ **13.** $\begin{bmatrix} 3 & -2 \\ -6 & 4 \end{bmatrix}$

In Exercises 14–16, $A = \begin{bmatrix} 2 & 3 \\ -1 & 4 \end{bmatrix}$, $B = \begin{bmatrix} -1 & -3 \\ 2 & 1 \end{bmatrix}$, and $C = \begin{bmatrix} -2 & 3 \\ 4 & 2 \end{bmatrix}$.

Prove that each of the following statements is true.

14. $B \cdot I = I \cdot B$ **15.** $A \cdot C \neq C \cdot A$ **16.** $A \cdot (B \cdot C) = (A \cdot B) \cdot C$

For Exercises 17–22,

$A = \begin{bmatrix} a_{11} & a_{12} \\ a_{21} & a_{22} \end{bmatrix}$, $B = \begin{bmatrix} b_{11} & b_{12} \\ b_{21} & b_{22} \end{bmatrix}$, $C = \begin{bmatrix} c_{11} & c_{12} \\ c_{21} & c_{22} \end{bmatrix}$,

and each element represents a real number. Determine which of the following statements are true. Justify your answer by performing the computations.

17. $0B = O_{2 \times 2}$ **18.** $O_{2 \times 2} \cdot B = O_{2 \times 2}$ **19.** $A \cdot B = B \cdot A$

20. $I \cdot C = C \cdot I = C$ **21.** $(A \cdot B)^2 = A^2 \cdot B^2$ **22.** $A \cdot (B + C) = A \cdot B + A \cdot C$

Mixed Review

Simplify each expression.   **8.5**

1. $\sqrt[3]{-8}$ **2.** $\sqrt[3]{-125}$ **3.** $\sqrt[3]{250a^9}$ **4.** $\sqrt[4]{81a^{16}}$

▨ *Brainteaser*

A boat is carrying cobblestones on a small, shallow lake. The boat capsizes and the cobblestones sink. The boat, now empty, displaces less water than it did before it capsized. Does the lake's water level rise or drop because of the stones that sink to the lake's bottom?

16.3 Two-by-Two Determinants

Objectives

To find the values of 2-by-2 determinants

To solve systems of two linear equations in two variables using determinants

For each square matrix there is a corresponding real number associated with it called a *determinant*. For the 2-by-2 square matrix $\begin{bmatrix} 3 & 4 \\ 2 & -1 \end{bmatrix}$, the value of the determinant is found by multiplying the elements of one diagonal, 3 and -1, and subtracting from this product the product of the elements of the other diagonal, 2 and 4. So, the determinant of the 2-by-2 square matrix is $3(-1) - 2(4) = -3 - 8$, or -11. The determinant of a matrix is written in the same form as the matrix, but with vertical line segments instead of brackets.

Definition

2-by-2 determinant The determinant of the 2-by-2 square matrix $\begin{bmatrix} a & b \\ c & d \end{bmatrix}$ is symbolized by $\begin{vmatrix} a & b \\ c & d \end{vmatrix}$ and is equal to $ad - cb$.

EXAMPLE 1 Find the value of each determinant.

a. $\begin{vmatrix} -3 & 1 \\ 4 & 5 \end{vmatrix}$

b. $\begin{vmatrix} \sqrt{3} & \sqrt{6} \\ -1 & \sqrt{2} \end{vmatrix}$

Solutions

a. $ad - cb = -3(5) - 4(1)$
$= -15 - 4 = -19$

b. $ad - ab = \sqrt{3} \cdot \sqrt{2} - (-1)(\sqrt{6})$
$= \sqrt{6} + \sqrt{6} = 2\sqrt{6}$

In Chapter 4, systems of linear equations were solved using either the *substitution method* or *linear combination method*. Such systems can also be solved by using determinants in a method known as **Cramer's Rule**. This method is used to solve linear systems in the standard form $a_1x + b_1y = c_1$, where $a_1, a_2, b_1, b_2, c_1,$ and c_2 are real numbers.
$a_2x + b_2y = c_2$

To solve this linear system, use the following formulas.

$$x = \frac{\begin{vmatrix} c_1 & b_1 \\ c_2 & b_2 \end{vmatrix}}{\begin{vmatrix} a_1 & b_1 \\ a_2 & b_2 \end{vmatrix}} = \frac{c_1b_2 - c_2b_1}{a_1b_2 - a_2b_1} \text{ and } y = \frac{\begin{vmatrix} a_1 & c_1 \\ a_2 & c_2 \end{vmatrix}}{\begin{vmatrix} a_1 & b_1 \\ a_2 & b_2 \end{vmatrix}} = \frac{a_1c_2 - a_2c_1}{a_1b_2 - a_2b_1},$$

where $a_1b_2 - a_2b_1 \neq 0$.

In the formulas for Cramer's Rule, the denominator is the determinant of the matrix of coefficients, $\begin{bmatrix} a_1 & b_1 \\ a_2 & b_2 \end{bmatrix}$. Notice that the determinant for each numerator differs in exactly one of its columns from the determinant of the denominator.

EXAMPLE 2   Solve the system $\begin{aligned} 3x - 2y &= 6 \\ y &= 4x + 4 \end{aligned}$ using Cramer's Rule.

Solution   Write each equation in the standard form $ax + by = c$. Then determine the coefficients of the variables and find the constant terms.

$$3x + (-2)y = 6 \qquad\qquad 4x + (-1)y = -4$$
$$a_1x + b_1y = c_1 \qquad\qquad a_2x + b_2y = c_2$$
$$a_1 = 3,\ b_1 = -2,\ c_1 = 6 \qquad a_2 = 4,\ b_2 = -1,\ c_2 = -4$$

$$x = \frac{\begin{vmatrix} 6 & -2 \\ -4 & -1 \end{vmatrix}}{\begin{vmatrix} 3 & -2 \\ 4 & -1 \end{vmatrix}} = \frac{6(-1) - (-4)(-2)}{3(-1) - 4(-2)} \qquad y = \frac{\begin{vmatrix} 3 & 6 \\ 4 & -4 \end{vmatrix}}{\begin{vmatrix} 3 & -2 \\ 4 & -1 \end{vmatrix}} = \frac{3(-4) - 4(6)}{3(-1) - 4(-2)}$$

$$= \frac{-6 - 8}{-3 + 8},\ \text{or}\ -\frac{14}{5} \qquad\qquad = \frac{-12 - 24}{-3 + 8},\ \text{or}\ -\frac{36}{5}$$

Thus, $\left(-\frac{14}{5},\ -\frac{36}{5}\right)$ is the solution of the system.

Classroom Exercises

State the value of each of the following determinants.

1. $\begin{vmatrix} 1 & 1 \\ 1 & 1 \end{vmatrix}$
2. $\begin{vmatrix} 1 & -1 \\ 1 & 1 \end{vmatrix}$
3. $\begin{vmatrix} 0 & -1 \\ -1 & 0 \end{vmatrix}$
4. $\begin{vmatrix} -2 & 4 \\ 6 & 5 \end{vmatrix}$

To solve $\begin{aligned} -3x + 5y &= 6 \\ 2x - 3y &= 9 \end{aligned}$, what determinant would you use for each of the following?

5. the denominator for x
6. the denominator for y
7. the numerator for x
8. the numerator for y

9. Solve the system of Classroom Exercises 5–8 using Cramer's Rule.

Written Exercises

Find the value of each determinant.

1. $\begin{vmatrix} 3 & -1 \\ 2 & 5 \end{vmatrix}$
2. $\begin{vmatrix} -4 & 3 \\ -2 & 6 \end{vmatrix}$
3. $\begin{vmatrix} 9 & 4 \\ 8 & -6 \end{vmatrix}$
4. $\begin{vmatrix} -10 & -3 \\ 4 & -8 \end{vmatrix}$

5. $\begin{vmatrix} 1 & 0 \\ 0 & 1 \end{vmatrix}$

6. $\begin{vmatrix} -1 & -6 \\ -7 & -10 \end{vmatrix}$

7. $\begin{vmatrix} 0 & 5 \\ 10 & 0 \end{vmatrix}$

8. $\begin{vmatrix} 7 & 9 \\ 11 & 12 \end{vmatrix}$

9. $\begin{vmatrix} \frac{1}{2} & \frac{1}{3} \\ \frac{1}{5} & -\frac{2}{3} \end{vmatrix}$

10. $\begin{vmatrix} \frac{3}{4} & \frac{1}{2} \\ -\frac{1}{4} & \frac{1}{3} \end{vmatrix}$

11. $\begin{vmatrix} -\frac{2}{3} & -\frac{1}{4} \\ \frac{3}{5} & -\frac{1}{2} \end{vmatrix}$

12. $\begin{vmatrix} 4\sqrt{3} & \sqrt{2} \\ -2\sqrt{3} & -3\sqrt{3} \end{vmatrix}$

Solve each system of linear equations using Cramer's Rule.

13. $3x - 2y = -7$
$-x + y = 3$

14. $-4x + 3y = 14$
$-2x + 2y = 8$

15. $5x - 3y = 18$
$2x + 4y = 2$

16. $-x + 4y = -8$
$6x - 25 = y$

17. $6x - 5y = 15$
$2y = 3x - 6$

18. $2x - 7 = y$
$3y = x - 16$

19. $x + 2 = y$
$2y = 3x + 2$

20. $y = 3x - 1$
$y = 2x + 5$

21. $4x + 3y = 11.5$
$2x - y = 0.5$

22. $a_1x + b_1y = 3$
$a_2x + b_2y = -4$

23. $\sqrt{2}x + \sqrt{3}y = 10$
$\sqrt{3}x - \sqrt{2}y = 0$

24. $m_1x - n_1y = p_1$
$m_2x + n_2y = p_2$

Suppose that $A = \begin{bmatrix} a_1 & b_1 \\ c_1 & d_1 \end{bmatrix}$ and $B = \begin{bmatrix} a_2 & b_2 \\ c_2 & d_2 \end{bmatrix}$ are 2-by-2 matrices and k is a scalar. Define $|A|$, $|B|$, and $|A \cdot B|$ as the determinants formed from the corresponding elements of A, B, and $A \cdot B$; that is,

$|A| = \begin{vmatrix} a_1 & b_1 \\ c_1 & d_1 \end{vmatrix}$ and so on. Prove that each of the following is true.

25. $|-A| = |A|$

26. $|kA| = k^2|A|$

27. $|A \cdot B| = |A| \cdot |B|$

Midchapter Review

In Exercises 1–12, $A = \begin{bmatrix} 2 & 0 \\ -1 & 8 \end{bmatrix}$, $B = \begin{bmatrix} -4 & 4 \\ 2 & 2 \end{bmatrix}$,

$C = \begin{bmatrix} -6 & 3 \\ 0 & 8 \end{bmatrix}$, $D = \begin{bmatrix} -7 \\ 1 \end{bmatrix}$, $E = \begin{bmatrix} 1 & -1 \end{bmatrix}$, $O_{2\times2} = \begin{bmatrix} 0 & 0 \\ 0 & 0 \end{bmatrix}$,

and $I_{2\times2} = \begin{bmatrix} 1 & 0 \\ 0 & 1 \end{bmatrix}$. Find each of the following. If the indicated sum or product does not exist, give a reason. **16.1, 16.2**

1. $A + B$

2. $B + D$

3. $-C$

4. $A + (-C)$

5. $A + O_{2\times2}$

6. $-2B$

7. AB

8. BA

9. CE

10. $CI_{2\times2}$

11. DE

12. AD

13. Solve $\begin{matrix} 2x - 3y = 11 \\ 4x + 7y = -30 \end{matrix}$ using Cramer's Rule. **16.3**

16.4 Three-by-Three Determinants

Objectives

To find the values of 3-by-3 determinants

To solve systems of three linear equations in three variables using determinants

For the 3-by-3 square matrix $\begin{bmatrix} a_1 & b_1 & c_1 \\ a_2 & b_2 & c_2 \\ a_3 & b_3 & c_3 \end{bmatrix}$, the value of

the corresponding determinant can be found by the *method of repeated columns*. First, repeat columns 1 and 2 as shown below. Then multiply on the down diagonals, multiply on the up diagonals, and subtract the sum of the up-diagonal products from the sum of the down-diagonal products.

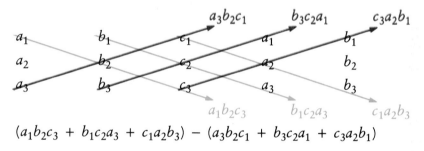

$$(a_1 b_2 c_3 + b_1 c_2 a_3 + c_1 a_2 b_3) - (a_3 b_2 c_1 + b_3 c_2 a_1 + c_3 a_2 b_1)$$

Definition

3-by-3 Determinant   The determinant of a 3-by-3 square matrix is

$$\begin{vmatrix} a_1 & b_1 & c_1 \\ a_2 & b_2 & c_2 \\ a_3 & b_3 & c_3 \end{vmatrix} = (a_1 b_2 c_3 + b_1 c_2 a_3 + c_1 a_2 b_3) - (a_3 b_2 c_1 + b_3 c_2 a_1 + c_3 a_2 b_1).$$

When evaluating 3-by-3 determinants, you will often find it easier to use the method of repeated columns than to refer to the definition above.

EXAMPLE 1   Find the value of $\begin{vmatrix} -3 & 1 & -2 \\ -4 & 3 & 2 \\ -1 & -2 & -1 \end{vmatrix}$.

Use the method of repeated columns to obtain the following.

Solution $[9 + (-2) + (-16)] - [6 + 12 + 4] = -9 - 22 = -31$

Thus, the value of the determinant is -31.

Cramer's Rule can also be used to solve systems of three linear equations in three variables which are of the form

$$\begin{aligned} a_1x + b_1y + c_1z &= d_1 \\ a_2x + b_2y + c_2z &= d_2, \text{ where all coefficients are real numbers.} \\ a_3x + b_3y + c_3z &= d_3 \end{aligned}$$

To solve this linear system, use the following formulas.

$$x = \frac{\begin{vmatrix} d_1 & b_1 & c_1 \\ d_2 & b_2 & c_2 \\ d_3 & b_3 & c_3 \end{vmatrix}}{D}, \quad y = \frac{\begin{vmatrix} a_1 & d_1 & c_1 \\ a_2 & d_2 & c_2 \\ a_3 & d_3 & c_3 \end{vmatrix}}{D}, \text{ and } z = \frac{\begin{vmatrix} a_1 & b_1 & d_1 \\ a_2 & b_2 & d_2 \\ a_3 & b_3 & d_3 \end{vmatrix}}{D},$$

where $D = \begin{vmatrix} a_1 & b_1 & c_1 \\ a_2 & b_2 & c_2 \\ a_3 & b_3 & c_3 \end{vmatrix}$ is the determinant of the matrix of coefficients of the variables x, y, and z.

EXAMPLE 2   Solve the system $\begin{aligned} -3x + 2y - z &= -8 \\ 4x + 3y + 2z &= 7 \\ -x - 4y + 3z &= 12 \end{aligned}$ using Cramer's Rule.

Solution

$$x = \frac{\begin{vmatrix} -8 & 2 & -1 \\ 7 & 3 & 2 \\ 12 & -4 & 3 \end{vmatrix}}{D}, \quad y = \frac{\begin{vmatrix} -3 & -8 & -1 \\ 4 & 7 & 2 \\ -1 & 12 & 3 \end{vmatrix}}{D}, \quad z = \frac{\begin{vmatrix} -3 & 2 & -8 \\ 4 & 3 & 7 \\ -1 & -4 & 12 \end{vmatrix}}{D}$$

D:

$$D = [-27 + (-4) + 16] - [3 + 24 + 24] = -15 - 51 = -66$$

$$x = \frac{(-72 + 48 + 28) - (-36 + 64 + 42)}{-66} = \frac{4 - 70}{-66}, \text{ or } 1$$

$$y = \frac{(-63 + 16 - 48) - (7 - 72 - 96)}{-66} = \frac{-95 + 161}{-66}, \text{ or } -1$$

$$z = \frac{(-108 - 14 + 128) - (24 + 84 + 96)}{-66} = \frac{6 - 204}{-66}, \text{ or } 3$$

Thus, $(1, -1, 3)$ is the solution of the system. Check in the original equations.

Another way to find the determinant of a matrix and to solve a system of equations is to expand a determinant by *minors*. The **minor** of an element of a determinant is the determinant found by deleting the column and row in which the element lies.

Thus, for $\begin{vmatrix} a_1 & b_1 & c_1 \\ a_2 & b_2 & c_2 \\ a_3 & b_3 & c_3 \end{vmatrix}$, the minor of a_1 is $\begin{vmatrix} b_2 & c_2 \\ b_3 & c_3 \end{vmatrix}$, the minor of

a_2 is $\begin{vmatrix} b_1 & c_1 \\ b_3 & c_3 \end{vmatrix}$, and the minor of a_3 is $\begin{vmatrix} b_1 & c_1 \\ b_2 & c_2 \end{vmatrix}$.

In general, the value of the determinant $\begin{vmatrix} a_1 & b_1 & c_1 \\ a_2 & b_2 & c_2 \\ a_3 & b_3 & c_3 \end{vmatrix}$ can be obtained

by finding the minors of all the elements of any row or column, multiplying each minor by its element, and forming an *expansion by minors*, as shown below for column 1.

$$a_1 \begin{vmatrix} b_2 & c_2 \\ b_3 & c_3 \end{vmatrix} - a_2 \begin{vmatrix} b_1 & c_1 \\ b_3 & c_3 \end{vmatrix} + a_3 \begin{vmatrix} b_1 & c_1 \\ b_2 & c_2 \end{vmatrix}$$

$$= a_1(b_2c_3 - b_3c_2) - a_2(b_1c_3 - b_3c_1) + a_3(b_1c_2 - b_2c_1)$$
$$= a_1b_2c_3 - a_1b_3c_2 - a_2b_1c_3 + a_2b_3c_1 + a_3b_1c_2 - a_3b_2c_1$$

Notice that in the second term in the original expansion above, the product of the element a_2 and its minor has been multiplied by -1. In general, if the sum of an element's row number and column number is an odd integer, multiply by -1; otherwise, do not.

a_1: 1st row, 1st column; $1 + 1 = 2$, an even integer
a_2: 2nd row, 1st column; $2 + 1 = 3$, an odd integer
a_3: 3rd row, 1st column; $3 + 1 = 4$, an even integer

Thus, only the product of a_2 and its minor is multiplied by -1.

EXAMPLE 3   Use minors to find the value of $D = \begin{vmatrix} -3 & 4 & 5 \\ 1 & -2 & -3 \\ 2 & 3 & -1 \end{vmatrix}$.

Plan   Expand the determinant using the minor of the elements of column 2.

Solution $D = -4 \begin{vmatrix} 1 & -3 \\ 2 & -1 \end{vmatrix} + (-2) \begin{vmatrix} -3 & 5 \\ 2 & -1 \end{vmatrix} - 3 \begin{vmatrix} -3 & 5 \\ 1 & -3 \end{vmatrix}$

$= -4(-1 + 6) - 2(3 - 10) - 3(9 - 5)$
$= -20 + 14 - 12,$ or -18

Thus, the value of the determinant is -18.

EXAMPLE 4 Solve the system
$$-3x + 4y + 5z = 5$$
$$x - 2y - 3z = -5$$ using minors and Cramer's Rule.
$$2x + 3y - z = -2$$

Solution The determinant of the matrix of coefficients of the system was evaluated as -18 in Example 3. Use this value and Cramer's rule.

$$x = \frac{\begin{vmatrix} 5 & 4 & 5 \\ -5 & -2 & -3 \\ -2 & 3 & -1 \end{vmatrix}}{-18}, \quad y = \frac{\begin{vmatrix} -3 & 5 & 5 \\ 1 & -5 & -3 \\ 2 & -2 & -1 \end{vmatrix}}{-18}, \quad \text{and } z = \frac{\begin{vmatrix} -3 & 4 & 5 \\ 1 & -2 & -5 \\ 2 & 3 & -2 \end{vmatrix}}{-18}$$

$$x = \frac{5\begin{vmatrix} -2 & -3 \\ 3 & -1 \end{vmatrix} - (-5)\begin{vmatrix} 4 & 5 \\ 3 & -1 \end{vmatrix} + (-2)\begin{vmatrix} 4 & 5 \\ -2 & -3 \end{vmatrix}}{-18} \quad \leftarrow \text{Numerator is expanded using Column 1.}$$

$$= \frac{5(2 + 9) + 5(-4 - 15) - 2(-12 + 10)}{-18} = \frac{55 - 95 + 4}{-18} = \frac{-36}{-18}, \text{ or } 2$$

In a similar manner, y and z can be evaluated using an expansion by minors. The values obtained are $y = -1$ and $z = 3$. Thus, $(2, -1, 3)$ is the solution of the system.

Classroom Exercises

For the determinant shown, state the value of the minor for the indicated entries.

$$\begin{vmatrix} 0 & 0 & 2 \\ 1 & 1.6 & 4 \\ 0 & 8 & 5 \end{vmatrix}$$

1. 1.6   **2.** 2   **3.** 1   **4.** 4   **5.** 5   **6.** 8

Find the value of the determinant of Exercises 1–6 using the minors of the following.

7. the elements of row 1   **8.** the elements of column 1   **9.** the elements of row 3

Written Exercises

Use the definition of a 3-by-3 determinant or the method of repeated columns to find the value of each determinant.

1. $\begin{vmatrix} -2 & -4 & 2 \\ 3 & 5 & 1 \\ 4 & -1 & -3 \end{vmatrix}$ **2.** $\begin{vmatrix} 5 & -1 & -3 \\ 2 & -4 & 2 \\ 1 & 4 & -2 \end{vmatrix}$ **3.** $\begin{vmatrix} 3 & -2 & 4 \\ 1 & -3 & 5 \\ -3 & 2 & -1 \end{vmatrix}$

Use minors to find the value of each determinant.

4. $\begin{vmatrix} -3 & 2 & -1 \\ 4 & 3 & 6 \\ -1 & 5 & 1 \end{vmatrix}$

5. $\begin{vmatrix} 4 & -2 & -4 \\ -2 & 1 & -3 \\ -3 & 3 & 2 \end{vmatrix}$

6. $\begin{vmatrix} 8 & -6 & -1 \\ 4 & 3 & -2 \\ -5 & -2 & 1 \end{vmatrix}$

Solve each system of equations using Cramer's Rule.

7. $-3x + 2y - z = 3$
$2x + 4y + 3z = 8$
$-x - 3y + 2z = 2$

8. $2x - y + 3z = 7$
$-x + 3y - 2z = -6$
$3x - 2y + 4z = 11$

9. $-x + 3y - 2z = 6$
$4x - 2y + 3z = -9$
$3x + y - z = -5$

10. $-3x + 2y - 2z = -10$
$2x + 3y + z = -1$
$-x - y + 2z = -8$

11. $-4x - y + 2z = -4$
$-3x + 2y - z = 10$
$5x - 3y + 3z = -20$

12. $3x - 2y - 4z = 10$
$4x + 3y - z = 16$
$-x - 2y + 3z = -12$

Solve each system of equations using minors and Cramer's Rule.

13. $-4x - 3y + z = 7$
$5x + y - 3z = -16$
$-x - 2y + 4z = 3$

14. $2x + 3y - 2z = 3$
$-3x + 4y - z = 12$
$-2x - y + 4z = -5$

15. $-x - 2y + 3z = 8$
$-4x - y - 2z = -22$
$2x - 3y - z = 6$

Prove that the following are true for the determinant below.

$$\begin{vmatrix} a_1 & a_2 & a_3 \\ b_1 & b_2 & b_3 \\ c_1 & c_2 & c_3 \end{vmatrix}$$

16. If the elements of a row are zero, the value of the determinant is 0.

17. If any two rows of a determinant are interchanged, the sign of the value of the determinant is changed.

18. If two rows in a determinant are the same, the value of the determinant is 0.

Mixed Review

Simplify and write each expression using only positive exponents. **5.1, 5.2**

1. $(-2a^4b^2c)^2$

2. $\dfrac{-18x^6y^4z}{-6x^4yz^6}$

3. $-3(-2x^{-3}y^2)^4$

4. $(2a^4b)(-3a^{-3}b^2)$

Brainteaser

Can an irrational number raised to an irrational power be a rational number? (HINT: Begin with $\sqrt{2}^{\sqrt{2}}$.)

16.5 Solving Linear Systems with Matrices

To solve systems of two linear equations in two variables using the inverse of a matrix

To solve systems of three linear equations in three variables using matrix transformations

The system of two linear equations $\begin{aligned} a_1x + b_1y &= c_1 \\ a_2x + b_2y &= c_2 \end{aligned}$

can be written in matrix form as $\begin{bmatrix} a_1 & b_1 \\ a_2 & b_2 \end{bmatrix} \begin{bmatrix} x \\ y \end{bmatrix} = \begin{bmatrix} c_1 \\ c_2 \end{bmatrix}$.

To see this, multiply the two matrices on the left side of the equation above.

Thus, the system $\begin{aligned} 3x - 2y &= 9 \\ -5x + 9y &= -6 \end{aligned}$ can be written in matrix form as

$$\begin{bmatrix} 3 & -2 \\ -5 & 9 \end{bmatrix} \begin{bmatrix} x \\ y \end{bmatrix} = \begin{bmatrix} 9 \\ -6 \end{bmatrix}.$$

A system of two linear equations can be solved by writing it in matrix

form and then using $\begin{bmatrix} x \\ y \end{bmatrix} = \begin{bmatrix} a_1 & b_1 \\ a_2 & b_2 \end{bmatrix}^{-1} \begin{bmatrix} c_1 \\ c_2 \end{bmatrix}$, where

$\begin{bmatrix} a_1 & b_1 \\ a_2 & b_2 \end{bmatrix}^{-1}$ is the multiplicative inverse of $\begin{bmatrix} a_1 & b_1 \\ a_2 & b_2 \end{bmatrix}$.

The above statement can be verified by first writing the following.

$$\begin{bmatrix} a_1 & b_1 \\ a_2 & b_2 \end{bmatrix} \cdot \begin{bmatrix} x \\ y \end{bmatrix} = \begin{bmatrix} c_1 \\ c_2 \end{bmatrix}$$

Then, substitute A, X, and C, respectively, for the above matrices to obtain this equation:

$$A \cdot X = C$$

Next, multiply each side on its left by A^{-1}. $A^{-1}(A \cdot X) = A^{-1}C$

Now, use the Associative Property. $(A^{-1} \cdot A) \cdot X = A^{-1}C$

Use $A^{-1} \cdot A = I$. $I \cdot X = A^{-1}C$

Use $I \cdot X = X$. $X = A^{-1}C$

Thus, $\begin{bmatrix} x \\ y \end{bmatrix} = \begin{bmatrix} a_1 & b_1 \\ a_2 & b_2 \end{bmatrix}^{-1} \begin{bmatrix} c_1 \\ c_2 \end{bmatrix}$.

Classroom Exercises

State the matrix equation for each of the following systems.

1. $3x + 4y = 6$
$-2x - y = 7$

2. $x + 2y = 3$
$x - y = 1$

3. $2x + 3y = 0$
$x = 1$

4. $x = 1$
$y = 2$

5–8. Give the augmented matrix for the four systems of Classroom Exercises 1–4.

9. Give the steps in transforming the augmented matrix $\begin{bmatrix} 2 & 4 & 6 \\ 3 & 2 & 1 \end{bmatrix}$

into the equivalent augmented matrix $\begin{bmatrix} 1 & 2 & 3 \\ 0 & 1 & 2 \end{bmatrix}$.

Written Exercises

Solve each system using the inverse of a matrix.

1. $3x - 2y = 8$
$x + 3y = -1$

2. $-2x + 4y = 2$
$3x - y = 7$

3. $5x - 2y = -8$
$-x + 3y = -1$

4. $-3x + 5y = -1$
$4x - 2y = -8$

5. $6x - 2y = 4$
$-5x + 3y = -2$

6. $-4x - 3y = 8$
$2x + y = -2$

Solve each system using matrix transformations.

7. $2x + 3y = 11$
$-x + y = 12$

8. $y - 2z = 2$
$-6y + 8z = -9$

9. $2x + 4z = 0$
$5x - 9z = -95$

10. $2x - 3y + 2z = 1$
$4x - 3y + 2z = 5$
$3x - 2y + z = -5$

11. $2x + 3y - z = -7$
$x + y - z = -4$
$3x - 2y - 3z = -7$

12. $3x - 2y + z = -5$
$-2x + 3y - 3z = 12$
$3x - 2y - 2z = 4$

13. $x + 3y - 2z = 5$
$-2x - y - 3z = 4$
$4x - 2y + z = 2$

14. $3x - y + 2z = 9$
$x - 2y - 3z = -1$
$2x - 3y + z = 10$

15. $-3x + 2y + 2z = 9$
$2x - 5y - 3z = -2$
$-6x + 3y - 4z = 4$

16. Explain in writing how to solve a system of two linear equations using the inverse of a matrix.

If $A = \begin{bmatrix} a_1 & b_1 \\ a_2 & b_2 \end{bmatrix}$ and A^{-1} exists, it can be shown that

$A^{-1} = \dfrac{1}{\text{Det } A} \begin{bmatrix} b_2 & -b_1 \\ -a_2 & a_1 \end{bmatrix}$, where Det $A = \begin{vmatrix} a_1 & b_1 \\ -a_2 & b_2 \end{vmatrix}$.

Use this formula to find A^{-1} in Exercises 17–19.

17. $A = \begin{bmatrix} 4 & -1 \\ 2 & 0 \end{bmatrix}$

18. $A = \begin{bmatrix} 6 & 3 \\ -2 & 1 \end{bmatrix}$

19. $A = \begin{bmatrix} a & -b \\ b & a \end{bmatrix}$

20. Prove the formula used in Exercises 17–19.

21. Two square matrices X and Y have the same dimensions. The matrices and their product, $X \cdot Y$, all have multiplicative inverses. Prove that $(X \cdot Y)^{-1} = Y^{-1} \cdot X^{-1}$. (HINT: Use reversible steps starting with this equality to obtain $I = I$. Then reverse the steps.)

Mixed Review

Factor each of the following.   *5.4, 5.7*

1. $x^2 - 2x - 3$
2. $6x^2 - 5x - 4$
3. $16x^2 - 1$
4. $-27x^3 + 12x$
5. $x^4 + 5x^2 + 4$
6. $x^4 - 13x^2 + 36$

Application: *Car Sales*

A foreign car company sells three different models: a 4-door car, a 2-door car, and a sports car. Each model is produced in two types: a normally equipped car and a fully equipped car. Table 1 shows the number of cars sold by model and by equipment. Table 2 shows the cost of each model based on the cost of equipment.

TABLE 1: Cars Sold

	Normal equipment	Full equipment
4-door	120	210
2-door	200	150
Sports car	70	90

TABLE 2: Cost in Dollars

	4-door	2-door	Sports car
Normal equipment	12,000	11,000	18,000
Full equipment	16,000	15,000	22,000

Exercises

1. Use matrices to show total dollar sales for each model.
2. Calculate total sales in dollars.

16.6 Problem Solving: Using Matrices

Objective

To apply matrices

Matrices can be used to represent a variety of data. They can also be used to solve a variety of practical problems.

EXAMPLE 1

A company manufactures three models of television sets and three models of videocassette recorders (VCRs) in two plants. The following tables represent the production in each plant.

PLANT 1

	Model		
	X	Y	Z
TV sets	120	72	97
VCRs	67	46	71

PLANT 2

	Model		
	X	Y	Z
TV sets	79	86	105
VCRs	75	68	53

a. Represent the data in matrix form.
b. Find the sum of the two matrices.
c. Find the difference of the two matrices (Plant 1 − Plant 2).
d. Interpret the results of Steps *b* and *c*.

Solutions

a.
$$\text{Plant 1} \quad \begin{bmatrix} 120 & 72 & 97 \\ 67 & 46 & 71 \end{bmatrix} \qquad \text{Plant 2} \quad \begin{bmatrix} 79 & 86 & 105 \\ 75 & 68 & 53 \end{bmatrix}$$

b. Plant 1 + Plant 2
$$\begin{bmatrix} 120 + 79 & 72 + 86 & 97 + 105 \\ 67 + 75 & 46 + 68 & 71 + 53 \end{bmatrix} = \begin{bmatrix} 199 & 158 & 202 \\ 142 & 114 & 124 \end{bmatrix}$$

c. Plant 1 − Plant 2
$$\begin{bmatrix} 120 - 79 & 72 - 86 & 97 - 105 \\ 67 - 75 & 46 - 68 & 71 - 53 \end{bmatrix} = \begin{bmatrix} 41 & -14 & -8 \\ -8 & -22 & 18 \end{bmatrix}$$

d. The company produces 199 Model-*X* TVs, 158 Model-*Y* TVs, 202 Model-*Z* TVs, 142 Model-*X* VCRs, 114 Model-*Y* VCRs, and 124 Model-*Z* VCRs. They also produce 41 more Model-*X* TV sets in Plant 1 than in Plant 2.

Complete the interpretation on your own.

The cost of manufacturing products is determined by the type of materials used, the quantity of materials used, and the costs, including labor, involved in processing these materials. The next example illustrates the total cost of manufacturing 100 boats.

EXAMPLE 2   A boat manufacturer builds three models of boats: Model X, Model Y, and Model Z. Each model is produced in two styles: a leisure style and a racing style. Each leisure boat is constructed with the same relative amounts of wood, plastic, fiberglass, and other materials as any other leisure boat, regardless of model. Similarly, each racing boat is constructed with the same relative amounts of materials as any other racing boat, regardless of model. The company will produce 100 boats during the month of April.

The matrix at the right represents the number of boats produced for each model and style.

$$\text{Models} \quad \begin{matrix} X \\ Y \\ Z \end{matrix} \begin{matrix} \text{Styles} \\ \text{Leisure} \quad \text{Racing} \\ \begin{bmatrix} 30 & 10 \\ 10 & 20 \\ 0 & 30 \end{bmatrix} \end{matrix} = A$$

The next matrix represents the number of units of materials used for each style of boat.

$$\text{Styles} \quad \begin{matrix} \text{Leisure} \\ \text{Racing} \end{matrix} \begin{matrix} \text{Materials} \\ \text{Wood} \quad \text{Plastic} \quad \text{Fiberglass} \quad \text{Other} \\ \begin{bmatrix} 15 & 12 & 3 & 6 \\ 1 & 4 & 25 & 15 \end{bmatrix} \end{matrix} = B$$

The next matrix represents the unit costs of the materials in dollars per unit, with the costs of the labor and processing included.

$$\text{Materials} \quad \begin{matrix} \text{Wood (unit: board feet, or bd ft)} \\ \text{Plastic (unit: square feet, or ft}^2) \\ \text{Fiberglass (unit: square feet, or ft}^2) \\ \text{Other (unit: square feet, or ft}^2) \end{matrix} \begin{matrix} \text{Costs} \\ \begin{bmatrix} 65 \\ 70 \\ 90 \\ 35 \end{bmatrix} \end{matrix} = C$$

a. Find the cost of production of each boat.
b. Find the cost of production of each model.

Solutions   **a.** Find a matrix, each element of which is the production cost of a boat.

From Lesson 7.3, (units) $\times \dfrac{\text{dollars}}{\text{unit}}$ = dollars. Therefore, find $B \cdot C$.

$$B \cdot C = \begin{bmatrix} 15 & 12 & 3 & 6 \\ 1 & 4 & 25 & 15 \end{bmatrix} \begin{bmatrix} 65 \\ 70 \\ 90 \\ 35 \end{bmatrix} \quad \leftarrow 15 \text{ bd ft of wood} \times \dfrac{65 \text{ dollars}}{\text{bd ft}}$$

$$= \begin{bmatrix} 15(65) + 12(70) + 3(90) + 6(35) \\ 1(65) + 4(70) + 25(90) + 15(35) \end{bmatrix}, \text{ or } \begin{bmatrix} 2{,}295 \\ 3{,}120 \end{bmatrix}$$

So, the cost of production is $2,295 for each leisure boat and $3,120 for each racing boat.

b. To determine the cost of production for each model, find $A \cdot (B \cdot C)$.

$$A \cdot (B \cdot C) = \begin{bmatrix} 30 & 10 \\ 10 & 20 \\ 0 & 30 \end{bmatrix} \begin{bmatrix} 2{,}295 \\ 3{,}120 \end{bmatrix} \begin{matrix} \leftarrow \ 30 \ \text{Model-}X \\ \text{leisure boats} \end{matrix} \quad \times \quad \dfrac{2{,}295 \ \text{dollars}}{\text{leisure boat}}$$

$$= \begin{bmatrix} 30(2{,}295) + 10(3{,}120) \\ 10(2{,}295) + 20(3{,}120) \\ 0(2{,}295) + 30(3{,}120) \end{bmatrix}, \text{ or } \begin{bmatrix} 100{,}050 \\ 85{,}350 \\ 93{,}600 \end{bmatrix}$$

So, the cost of production is $100,050 for 40 Model-X boats, $85,350 for 30 Model-Y boats, and $93,600 for 30 Model-Z boats.

Classroom Exercises

Refer to Example 2. Then find each of the following.

1. total production cost for all 100 boats
2. production cost for one Model-X leisure boat
3. production cost for one Model-X racing boat
4. production cost for one Model-Y leisure boat
5. production cost for one Model-Y racing boat
6. production cost for one Model-Z leisure boat
7. production cost for one Model-Z racing boat

Written Exercises

Each year each member of the Clemente family invests money in bonds, stocks, and savings certificates. The following tables represent their investments (in dollars) for two consecutive years (Exercises 1–16).

YEAR 1

	Bonds	Stocks	Certificates
Mother	400	1,200	1,000
Father	500	900	600
Paul	100	200	500
Rosa	200	300	400

YEAR 2

	Bonds	Stocks	Certificates
Mother	800	700	1,500
Father	300	900	2,000
Paul	0	800	200
Rosa	500	300	0

Use the tables above for Exercises 1–8.

1. Represent the data of Year 1 in matrix form.
2. Represent the data of Year 2 in matrix form.
3. Find the sum of the two matrices, Year 1 + Year 2.
4. Find the difference of the two matrices, Year 2 − Year 1.

5. How much did Mrs. Clemente invest in savings certificates in two years?

6. How much did Mr. Clemente invest in bonds in two years?

7. How much did son Paul invest in stock in two years?

8. How much more did daughter Rosa invest in bonds in the second year compared to the first year?

In the third year each member of the family decided to double the investments of the second year.

9. Represent the third-year investments as a product of a scalar and a matrix.

10. Find the product of the scalar and matrix.

In the fourth year, after inheriting some money, each family member was able to multiply the investments of the second year as follows: Mrs. Clemente by 3, Mr. Clemente by 2, Paul by 4, and Rosa by 5.

11. Represent the product of the above multiples and the second-year investments.

12. Find the matrix representing the fourth-year investments.

13. How much did Mr. Clemente invest in stocks in Year 4?

14. How much did Mrs. Clemente invest in bonds in Year 4?

15. How much did Rosa invest in savings certificates in Year 4?

16. How much did Paul invest in savings certificates in Year 4?

A builder builds two models of houses in three styles. Each year she builds exactly 65 houses. Table 1 represents the number of houses produced by model and style. Table 2 represents the units of materials used for each house by style and by kinds of materials used. Table 3 represents the costs of materials per unit, with labor costs included.

TABLE 1

	Colonial	Split-Entry	Ranch
Model X	20	15	10
Model Y	2	6	12

TABLE 2

	Stone	Brick	Wood	Concrete
Colonial	200	200	100	50
Split-Entry	20	120	150	300
Ranch	30	250	300	200

TABLE 3

	Costs
Stone	160
Brick	140
Wood	180
Concrete	122

17. Write Table 1 in matrix form.

18. Write Table 2 in matrix form.

19. Write Table 3 in matrix form.

20. Find the cost to build each colonial house.

21. Find the cost to build each split-entry house.

22. Find the cost to build each ranch house.

23. Find the cost to build all of the Model-*Y* ranch houses.

24. Find the cost to build all of the Model-*X* houses.

25. Find the cost to build all of the Model-*Y* houses.

26. Find the cost to build all of the houses.

Mixed Review

Let $f(x) = -3x^2 + 1$. Find each of the following. *3.2, 5.5*

1. $f(-1)$ **2.** $f(5)$ **3.** $f(a)$ **4.** $f(a + t)$

Simplify and write each expression using positive exponents. *5.2*

5. $(8p^0q)^{-1}$ **6.** $(-3a^{-2}b^3)^2$ **7.** $(-2x^4y^{-3})^3$ **8.** $(-4a^3b^{-4})(-2a^{-2}b^6)$

Application: *Networks*

A **network** is a set of points connected by arcs. Networks can represent roadways, electrical systems, and ecological relationships. A **directed network** shows the possible directions of travel or influence along each arc.

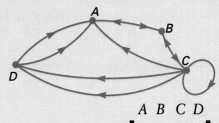

A matrix conveniently represents a network. The 0 in the first row, third column of the matrix at the right means that there are *no* paths from *A* to *C*. Likewise, there is *one* path from *C* to *C*, and there are *two* possible paths from *C* to *D*.

$$
\begin{array}{c c}
 & \begin{array}{cccc} A & B & C & D \end{array} \\
\begin{array}{c} A \\ B \\ C \\ D \end{array} &
\left[\begin{array}{cccc}
0 & 1 & 0 & 0 \\
1 & 0 & 1 & 0 \\
1 & 1 & 1 & 2 \\
2 & 0 & 0 & 0
\end{array} \right]
\end{array}
$$

1–2. Construct matrices for the directed networks at the right.

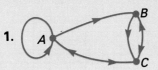

1.

2.

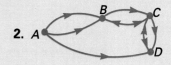

3–4. Construct directed networks for the matrices at the right.

3. $\begin{bmatrix} 2 & 1 & 1 \\ 1 & 0 & 1 \\ 1 & 1 & 0 \end{bmatrix}$

4. $\begin{bmatrix} 1 & 0 & 2 & 0 \\ 1 & 0 & 2 & 1 \\ 0 & 1 & 0 & 0 \\ 0 & 1 & 0 & 0 \end{bmatrix}$

Chapter 16 Review

Key Terms

additive identity matrix (p. 586)
additive inverse matrix (p. 586)
augmented matrix (p. 603)
Cramer's Rule (p. 594)
determinant (2-by-2) (p. 594)
determinant (3-by-3) (p. 597)
dimensions of a matrix (p. 585)

equivalent matrices (p. 604)
matrix (p. 585)
matrix transformations (p. 604)
minor (p. 599)
multiplicative identity matrix (p. 591)
multiplicative inverse matrix (p. 591)
triangular form (p. 603)

Key Ideas and Review Exercises

16.1   To add two matrices, check that their dimensions are the same. Then add corresponding elements.

Find each sum, if it exists.

1. $\begin{bmatrix} 4 & 3 & 9 \\ -5 & -1 & 4 \\ 2 & 6 & -7 \end{bmatrix} + \begin{bmatrix} 9 & 6 & -1 \\ 8 & 5 & -3 \\ -10 & 8 & -1 \end{bmatrix}$ 2. $\begin{bmatrix} 6 & -3 \\ 9 & -7 \\ -8 & 4 \end{bmatrix} + \begin{bmatrix} 8 & 3 & -5 \\ 6 & -2 & 4 \end{bmatrix}$

16.2   To multiply two matrices, check that the number of columns of the first matrix equals the number of rows of the second matrix; then multiply the row elements of the first by the column elements of the second, and add.

Find each product, if it exists.

3. $\begin{bmatrix} 8 & -2 & 3 \\ 2 & 1 & -4 \end{bmatrix} \cdot \begin{bmatrix} 4 & -1 & 3 \\ 2 & -2 & 1 \\ -5 & 3 & 2 \end{bmatrix}$ 4. $\begin{bmatrix} -4 & -2 & 3 \\ 5 & -3 & -4 \\ 6 & 9 & 3 \end{bmatrix} \cdot \begin{bmatrix} 8 & -3 & 2 \\ -4 & 9 & 8 \end{bmatrix}$

16.3   The value of a 2-by-2 determinant $\begin{vmatrix} a & b \\ c & d \end{vmatrix}$ is $ad - cb$.

To use Cramer's Rule to solve the system of two equations

$\begin{aligned} a_1 x + b_1 y = c_1 \\ a_2 x + b_2 y = c_2 \end{aligned}$, write $x = \dfrac{\begin{vmatrix} c_1 & b_1 \\ c_2 & b_2 \end{vmatrix}}{\begin{vmatrix} a_1 & b_1 \\ a_2 & b_2 \end{vmatrix}}$ and $y = \dfrac{\begin{vmatrix} a_1 & c_1 \\ a_2 & c_2 \end{vmatrix}}{\begin{vmatrix} a_1 & b_1 \\ a_2 & b_2 \end{vmatrix}}$.

Solve each system of equations using Cramer's Rule.

5. $2x - 5y = 16$
    $-3x + 2y = -13$

6. $5x - 4y = -1$
    $3y + 8 = 2x$

16.4   To find the value of a 3-by-3 determinant:
1. use the method of repeated columns, as shown on page 597, or
2. use expansion by minors, as shown on page 599.
To solve a system of three linear equations by Cramer's Rule, use 3-by-3 determinants, as shown on page 600.

Find the value of each determinant.

7. $\begin{vmatrix} 4 & -1 & 2 \\ 3 & -2 & 1 \\ -3 & 2 & 3 \end{vmatrix}$

8. $\begin{vmatrix} -3 & -2 & 3 \\ 4 & -1 & 2 \\ -1 & 3 & -2 \end{vmatrix}$

Solve each system of equations using Cramer's Rule.

9. $2x - 3y + z = -7$
$-3x + 4y + 2z = 3$
$3x - 2y - 2z = -1$

10. $4x + y - 2z = 3$
$-x + 3y + 5z = 5$
$2x - 3y + z = 9$

16.5   To solve a system like $\begin{array}{l} -3x + 2y = 2 \\ 4x - 3y = -4 \end{array}$ using the inverse of a matrix:

1. find A^{-1} where $A = \begin{bmatrix} -3 & 2 \\ 4 & -3 \end{bmatrix}$; then

2. use $\begin{bmatrix} x \\ y \end{bmatrix} = A^{-1}C$, where $C = \begin{bmatrix} 2 \\ -4 \end{bmatrix}$.

Solve each system using the inverse of a matrix.

11. $2x - 3y = -5$
$-3x + 5y = 9$

12. $3x + 4y = 6$
$-2x - y = 7$

16.6   To solve problems using matrices:
1. represent the data in matrix form,
2. determine which matrices should be multiplied,
3. multiply matrices, and
4. interpret data.

A company produced 65 Tudor-style houses and 35 Western-style houses. Table I represents the units of materials used for each style of house. Table II represents the dollar cost of materials by unit.

TABLE I

	Brick	Concrete	Lumber	Stone
Tudor	300	100	50	100
Western	100	300	200	100

TABLE II

	Cost
Brick	130
Concrete	110
Lumber	150
Stone	200

13. Find the cost of the building materials for all 100 houses.

Find each sum, if it exists.

1. $\begin{bmatrix} -5 & 8 & 7 \\ -1 & -3 & -2 \\ 6 & 9 & 10 \end{bmatrix} + \begin{bmatrix} 5 & -3 & 9 \\ -8 & 6 & 5 \\ 2 & -7 & -3 \end{bmatrix}$

2. $\begin{bmatrix} 1 \\ -3 \\ 5 \end{bmatrix} + \begin{bmatrix} 3 & 0 & -4 \end{bmatrix}$

3. $\begin{bmatrix} -8 & -3 & -9 \\ -6 & 5 & 8 \end{bmatrix} + \begin{bmatrix} -10 & 7 & 1 \\ 9 & 6 & 2 \end{bmatrix}$

Find each product, if it exists.

4. $-2 \begin{bmatrix} 7 & -6 \\ -3 & 5 \end{bmatrix}$

5. $\begin{bmatrix} -1 & 2 & 3 \\ 4 & -2 & 3 \end{bmatrix} \cdot \begin{bmatrix} 2 & 1 & 3 \\ -1 & -2 & -3 \\ 4 & 1 & -2 \end{bmatrix}$

In Exercises 6–9, $A = \begin{bmatrix} 4 & -2 \\ 3 & -1 \end{bmatrix}$, $B = \begin{bmatrix} -3 & 2 \\ -2 & 4 \end{bmatrix}$, and $C = \begin{bmatrix} 5 & -1 \\ -3 & 2 \end{bmatrix}$.

6. Find A^{-1}.

7. Prove that $A + C = C + A$.

8. Prove that $B + (-B) = O_{2 \times 2}$.

9. Prove that $AC \neq CA$.

10. Find the value of the determinant. $\begin{vmatrix} -3 & 5 \\ -1 & 6 \end{vmatrix}$

11. Use minors to find the value of the determinant. $\begin{vmatrix} 2 & -3 & 1 \\ 4 & -2 & 3 \\ -2 & 1 & 5 \end{vmatrix}$

Solve using Cramer's Rule.

12. $4x - 3y = 2$
$-2x + 4y = -6$

13. $2x - y + 3z = 5$
$-3x + 2y - z = -9$
$-x + 3y - 2z = -6$

14. $3y = x + 12$
$2x = y - 9$

15. $3x - 2y = -0.4$
$-5x - y = -1.5$

Solve each system using the inverse of a matrix.

16. $4x - 3y = 2$
$-2x + 4y = -6$

17. $2x - 5y = 15$
$3x - 7y = 22$

Solve using matrix transformations.

18. $3x - 2y + 3z = 3$
$-2x + 4y - z = -9$
$x + 3y + 2z = -9$

19. $2a - 3b + c - 2d = 1$
$-a + 2b - 3c + d = 5$
$-3a - 2b + 2c + 2d = -14$
$a + b - 2c + 3d = 4$

Choose the *one* best answer to each question or problem.

1. If $R = \begin{bmatrix} 3 & 5 \\ -2 & 0 \end{bmatrix}$ and $N = \begin{bmatrix} 1 & -1 \\ 3 & 1 \end{bmatrix}$,

then $2R - N = $ __?__

(A) $\begin{bmatrix} -2 & 6 \\ -5 & -1 \end{bmatrix}$ (B) $\begin{bmatrix} 5 & 11 \\ -7 & -1 \end{bmatrix}$

(C) $\begin{bmatrix} 2 & 4 \\ 1 & -1 \end{bmatrix}$ (D) $\begin{bmatrix} 2 & 6 \\ -5 & -1 \end{bmatrix}$

(E) $\begin{bmatrix} 4 & 12 \\ -10 & -2 \end{bmatrix}$

2. If $5\begin{bmatrix} 1 & x \\ x-y & 5 \end{bmatrix} = \begin{bmatrix} 5 & 15 \\ 20 & 25 \end{bmatrix}$,

find x and y.
(A) $x = 15, y = -5$
(B) $x = 3, y = 1$
(C) $x = 3, y = -1$
(D) $x = 3, y = -17$
(E) $x = 15, y = 5$

3. $[a \quad b \quad c] \cdot \begin{bmatrix} a \\ b \\ c \end{bmatrix} = $ __?__

(A) $[a^2 \quad b^2 \quad c^2]$
(B) $[a^2 + b^2 + c^2]$
(C) $[2a + 2b + 2c]$
(D) $[2a \quad 2b \quad 2c]$
(E) none of these

4. If $\begin{bmatrix} x \\ y \end{bmatrix} = \begin{bmatrix} 2 & 1 \\ -1 & -3 \end{bmatrix} \cdot \begin{bmatrix} 3 \\ -2 \end{bmatrix}$,

find x and y.
(A) $x = 4, y = 3$
(B) $x = 6, y = 2$
(C) $x = 9, y = 8$
(D) $x = 6, y = -6$
(E) $x = 4, y = -9$

5. If $3^{2x+1} = 27^{\frac{4}{3}}$, then $x = $ __?__
(A) $\frac{1}{6}$ (B) $\frac{1}{2}$ (C) $\frac{2}{3}$ (D) $\frac{3}{2}$
(E) 3

6. If $A \cdot B = 36$, which of the following cannot be true?
(A) $A + B > 12$
(B) $A + B < 12$
(C) $A + B = 12$
(D) $|A + B| > 12$
(E) $|A + B| < 12$

7. If $2\begin{vmatrix} x & 4 & -1 \\ 0 & 2 & 3 \\ 0 & -2 & 2 \end{vmatrix} = 60$, then

$x = $ __?__
(A) -15 (B) 3 (C) 5
(D) 10 (E) 30

8. If $ad - cb \neq 0$, then

$$\frac{\begin{vmatrix} a & b \\ c & d \end{vmatrix}}{\begin{vmatrix} -3b & a \\ -3d & c \end{vmatrix}} = \text{ __?__}$$

(A) $-\frac{1}{3}$ (B) $-\frac{1}{9}$ (C) $\frac{1}{9}$
(D) $\frac{1}{3}$ (E) none of these

9. Given the figure below, with $\overline{AB}$ parallel to $\overline{CD}$, find x.

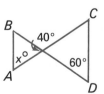

(A) 20 (B) 40 (C) 60
(D) 80 (E) 100

1. Supply the reasons to justify Steps 1–5 in the following proof.   *1.8*

$x + xy = (y + 1)x$
1. $x + xy = 1x + xy$
2. $\quad = 1x + yx$
3. $\quad = (1 + y)x$
4. $\quad = (y + 1)x$
5. $x + xy = (y + 1)x$

Solve each equation for x.

2. $2x - (6x + 8) =$ *1.4*
$\quad -3(6x - 5) + 5$

3. $|4x - 3| = 17$ *2.3*

4. $\frac{3}{x} + \frac{5}{25} = 1$ *7.6*

5. $\frac{x - a}{b} = \frac{x + c}{d}$

6. $x^2 - 6x + 25 = 0$ *9.5*

7. $x = 1 + \sqrt{2x + 6}$ *10.3*

8. Find the solution set.   *2.2*
$\quad -4x < 12 \text{ and}$
$\quad x + 5 < 2x + 3$

Find the following if $f(x) = 3x + 2$ and $g(x) = x^2 - 2$.

9. $f(-2.2)$ *3.2*

10. $g(f(a))$ *5.9*

11. $\dfrac{f(h + 4) - f(4)}{h}$ *7.1*

12. Find the slope of $\overline{PQ}$ given $P(-3,2)$ and $Q(1,-1)$.   *3.3*

13. Write an equation in standard form, $ax + by = c$, of the line through $A(2,-4)$ and *perpendicular* to the line $y = \frac{2}{3}x + 1$.   *3.7*

14. If y varies directly as x and $y = 48$ when $x = 72$, find x when $y = 60$.   *3.8*

15. Solve the system.   *4.3*
$2x + 3y + 4 = 0$
$4y - 33 = 5x$

16. Simplify $\left(\dfrac{2a^{-2}b}{3}\right)^3$ without using negative exponents.   *5.2*

17. Factor completely.   *5.7*
$5x^4 - 25x^2 - 180$

18. Find the solution set.   *6.4*
$x^2 - 6x - 16 < 0$

Simplify.

19. $(5n - 2)(3n^2 - n + 4)$ *5.3*

20. $\dfrac{x^2 - 49}{x^2 - 7x + 12} \div \dfrac{5x - 35}{4x - 12}$ *7.2*

21. $2c\sqrt{8c} + \sqrt{18c^3} - c\sqrt{32c}$, $c > 0$ *8.3*

22. $\dfrac{5}{4\sqrt{2} - \sqrt{3}}$ *8.4*

23. $\sqrt[3]{4x^2} \cdot \sqrt[3]{16x}$ *8.5*

24. $2\sqrt{-3} \cdot \sqrt{-12}$ *9.2*

25. $\dfrac{7}{5 - 2i}$

26. Copy the figure below and draw $\vec{A} + \vec{B}$.   *9.7*

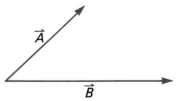

27. Use the Rational Zero Theorem to find the four zeros of $2x^4 - 5x^3 - 2x^2 + 10x - 4$.   *10.5*

28. Given $A(3,7)$, $B(-2,2)$, and $C(5,-1)$, find the distance from B to the midpoint of $\overline{AC}$.   *11.1*

29. Graph the circle described by $x^2 + y^2 - 6x + 4y + 8 = 0$. *12.1*

30. Write an equation in standard form of the ellipse with vertices $(0, \pm 2)$ and $(\pm 2\sqrt{2}, 0)$. Then graph the ellipse. *12.2*

31. A number y varies inversely as the cube of x, and $y = 9$ when $x = 4$. Find y when $x = 6$. *12.5*

32. Solve the system: $\begin{array}{c} x + y = 9 \\ xy = 18 \end{array}$ *12.9*

33. If $y = \log_2 x$, find the values of y when $x = \frac{1}{4}, \frac{1}{2}, 1, 2, 4, 8$. *13.4*

Solve. Check if necessary.

34. $3^{x+1} = 243$ *5.1*

35. $\log_b 125 = 3$ *13.3*

36. $2 \log_3 t = \log_3 4 + \log_3 (t + 3)$ *13.6*

37. Solve $5^{2x-1} = 800$ for x to 3 significant digits. *13.8*

38. How many odd 4-digit numbers can be formed from the digits 3, 4, 5, 6, 7, 8, if no digit is repeated in a number? *14.2*

39. In how many ways can a 5-member committee be selected if 8 people are available? *14.5*

40. One card is drawn from a deck of 52 playing cards. Find the probability that it is a red card or an ace. *14.8*

A box contains 4 white and 8 blue marbles. Two marbles are drawn, with the first replaced before the second is drawn. Find the probability of each event.

41. a blue, then a white *14.9*

42. a blue and a white

43. Find the two arithmetic means between 5 and 11. *15.2*

44. Find the harmonic mean of $\frac{1}{2}$ and $\frac{1}{10}$.

45. Find the sum of the first 18 terms of the arithmetic series $-7 - 2 + 3 + 8 + \dots$. *15.3*

46. Find the 10th term in the geometric sequence: 3, 6, 12 .... *15.4*

47. Find the 3rd term in the expansion of $(2x - y^2)^5$. Simplify. *15.8*

48. Find the product. *16.2*

$$\begin{bmatrix} 2 & -2 & 4 \\ -1 & 3 & 5 \end{bmatrix} \cdot \begin{bmatrix} 3 & 5 \\ 2 & -2 \\ -1 & 4 \end{bmatrix}$$

49. Find three consecutive even integers such that the square of the second integer, increased by the product of the first two, is the same as 12 more than 3 times the square of the first integer. *6.2*

50. Crew A can complete a job in 8 h and Crew B can do it in 12 h. If Crew C helps, all three together can do the job in 3 h. How long would it take Crew C working alone? *7.7*

51. Find the value of $7,500 invested at 8% compounded quarterly for 10 years. Use the formula $A = p\left(1 + \frac{r}{n}\right)^{nt}$. *13.8*

52. The arithmetic mean of two numbers is 20. One number increased by 5 is 4 times the other number. Find both numbers. *15.2*

17 TRIGONOMETRIC FUNCTIONS

Flight simulation programs use computer graphics to model the conditions of flying. Each image is generated in response to the actions of the pilot—creating a realistic panorama of an aircraft in flight.

17.1 Introduction to Right-Triangle Trigonometry

Objective

To compute the sine, cosine, and tangent of an acute angle of a right triangle

In an earlier mathematics course, you may have used properties of *similar triangles* (triangles that have the same shape) to measure indirectly the distance between two objects, such as two trees on opposite sides of a pond. It is possible to use indirect measurement in a more effective way by employing similar *right* triangles. This is one of the many purposes of trigonometry.

A second purpose of trigonometry is unrelated to indirect measurement. Trigonometry can be used to describe many phenomena that are repetitive or cyclic in nature. One example is the air-pressure vibrations that create sound, such as middle C on a piano. This tone can be described by the trigonometric formula $y = \sin 528\pi t$ and corresponds to a frequency of 264 cycles per second.

Recall that in a right triangle, the side opposite the right angle is called the *hypotenuse*. The other two sides are called *legs* of the right triangle. Each leg is opposite one of the two acute angles and adjacent to the other, as illustrated below for $\angle A$ (read "angle A") and $\angle B$.

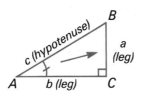
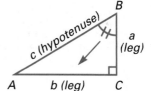

Leg a is *opposite* $\angle A$, and *adjacent* to $\angle B$.

Leg b is *opposite* $\angle B$, and *adjacent* to $\angle A$.

The two right triangles shown below are *similar* triangles.

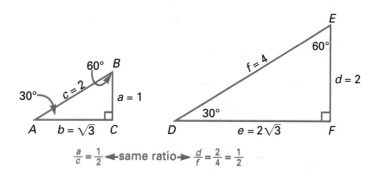

$$\frac{a}{c} = \frac{1}{2} \blacktriangleleft \text{same ratio} \blacktriangleright \frac{d}{f} = \frac{2}{4} = \frac{1}{2}$$

Two triangles are said to be **similar** if their corresponding angle measures are equal and the lengths of their corresponding sides are proportional.

By the definition of similar triangles, $\frac{a}{d} = \frac{c}{f} = \frac{1}{2}$, and $m \angle A = m \angle D = 30$. Read $m \angle A$ as "measure of angle A."

Then, by the Property of Proportions, $af = dc$, or $af = cd$, which is equivalent to the following.

$$\frac{a}{c} = \frac{d}{f} \quad \begin{array}{l}\leftarrow\text{length of leg opposite angle of 30} \\ \leftarrow\text{length of hypotenuse}\end{array}$$

This suggests that for all right triangles with a given acute angle of the same measure, the following ratio is always the same.

$$\frac{\text{length of the leg opposite the acute angle}}{\text{length of the hypotenuse}}$$

This *trigonometric ratio* is called the *sine ratio,* or *sine* of the angle.

For any given acute angle A of a right triangle, there are six trigonometric ratios. The first three of these are defined below. The remaining ratios will be defined in the next lesson.

Definitions

For all right triangles with given acute angle A:

Ratio of Lengths		**Abbreviation**
sine of $\angle A$	$= \dfrac{\text{opposite leg}}{\text{hypotenuse}}$	$\sin A = \dfrac{a}{c}$
cosine of $\angle A$	$= \dfrac{\text{adjacent leg}}{\text{hypotenuse}}$	$\cos A = \dfrac{b}{c}$
tangent of $\angle A$	$= \dfrac{\text{opposite leg}}{\text{adjacent leg}}$	$\tan A = \dfrac{a}{b}$

EXAMPLE 1   Find each trigonometric ratio to four decimal places.

a. $\sin P$
b. $\cos P$
c. $\tan P$

Plan	Use the definitions of the trigonometric ratios.

a. $\sin P = \dfrac{\text{opp}}{\text{hyp}}$ **b.** $\cos P = \dfrac{\text{adj}}{\text{hyp}}$ **c.** $\tan P = \dfrac{\text{opp}}{\text{adj}}$

$= \dfrac{15}{17}$ $= \dfrac{8}{17}$ $= \dfrac{15}{8}$

≈ 0.8824 ≈ 0.4706 $= 1.8750$

Thus, $\sin P = 0.8824$, $\cos P = 0.4706$, and $\tan P = 1.8750$ to four decimal places.

EXAMPLE 2   Find $\cos A$ to four decimal places.

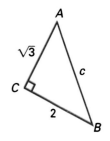

Solution   First, find c by using the Pythagorean relation.

$c^2 = 2^2 + (\sqrt{3})^2$

$c^2 = 4 + 3$

$c^2 = 7$

$c = \sqrt{7}$

$\cos A = \dfrac{\text{adj}}{\text{hyp}} = \dfrac{\sqrt{3}}{\sqrt{7}} = \dfrac{\sqrt{3} \cdot \sqrt{7}}{\sqrt{7} \cdot \sqrt{7}} = \dfrac{\sqrt{21}}{7} \approx \dfrac{4.5825757}{7} \approx 0.6546536$

Thus, $\cos A = 0.6547$ to four decimal places.

In Example 2, a calculator was used to approximate $\sqrt{21}$ as 4.5825757. However, you should follow the instructions of your teacher in deciding when to use a calculator to solve the examples and exercises of Chapters 17, 18, and 19. The final answer to an example or exercise should be equivalent no matter which method is used.

Classroom Exercises

In Exercises 1–10 refer to the figure at the right to indicate the following.

1. the leg adjacent to $\angle K$

2. the leg opposite $\angle G$

3. the leg adjacent to $\angle G$

4. the leg opposite $\angle K$

5. the hypotenuse

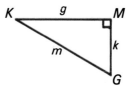

Give each trigonometric ratio as a ratio of the lengths of a pair of sides.

6. $\sin G$ **7.** $\cos K$ **8.** $\tan G$ **9.** $\cos G$ **10.** $\sin K$

Written Exercises

Find sin A, cos A, tan A, sin B, cos B, and tan B to four decimal places.

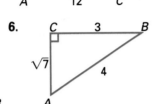

1. C 6 B
8
10

2.
13
5
A 12 C B

3. A 3 C
3√2
3
B

4. A 50
14
C 48 B

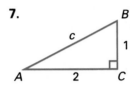

5. A
8
16
C 8√3 B

6. C 3 B
√7
4
A

7. B
c
1
A 2 C

8. A b C
√2
1
B

Find each trigonometric ratio to four decimal places. Use the figure at the right.

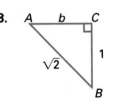

9. cos W if $t = 26, g = 10$ **10.** sin G if $t = 20, w = 16$

11. tan W if $w = 7, g = 24$ **12.** cos G if $g = 1, w = 1$

13. sin W if $w = 4, g = 6$ **14.** cos W if $w = \sqrt{5}, g = 3$

15. The lengths of the legs of a right triangle are p and f. Find the cosine of the angle opposite the leg of length p.

16. The length of a leg of a right triangle is half the length of the hypotenuse. Find the sine of the angle between that leg and the hypotenuse.

17. The lengths of the legs of a right triangle are a and b. Simplify $(\sin A)^2 + (\cos B)^2$.

18. The length of the hypotenuse of a right triangle is 1 less than twice the length of a leg. Find the sine of the angle between this leg and the hypotenuse if the other leg measures 4 units.

Mixed Review   3.5, 8.2, 9.3, 9.5

1. Simplify $\sqrt{28x^3}$.   **2.** Solve $x^2 + 4x + 6 = 0$. **3.** Graph $3x - 4y = 8$.

Brainteaser

A triangle has a right angle C, two other angles A and B, and sin A = cos A. What are the measures of ∠A and ∠B? Classified by its side lengths, what type of triangle is the triangle? Expressed in terms of the hypotenuse c, what are the lengths of sides a and b?

17.2 Using Trigonometric Tables

Objectives

To compute the cosecant, secant, and cotangent of an acute angle of a right triangle

To use a trigonometric table to find a trigonometric ratio of an acute angle of a right triangle

The three trigonometric ratios, sine, cosine, and tangent, were defined in Lesson 17.1. The *reciprocals* of these ratios are defined below.

Definitions

For all right triangles with a given acute angle A:

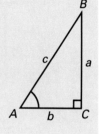

		Ratio of Lengths	Abbreviation
cosecant of $\angle A$	$=$	$\dfrac{\text{hypotenuse}}{\text{opposite leg}}$	$\csc A = \dfrac{c}{a}$
secant of $\angle A$	$=$	$\dfrac{\text{hypotenuse}}{\text{adjacent leg}}$	$\sec A = \dfrac{c}{b}$
cotangent of $\angle A$	$=$	$\dfrac{\text{adjacent leg}}{\text{opposite leg}}$	$\cot A = \dfrac{b}{a}$

EXAMPLE 1 Find $\csc T$, $\sec T$, and $\cot T$ to four decimal places.

Solution Use the Pythagorean relation to find u.

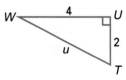

$u^2 = 4^2 + 2^2$ ← THINK: Is u greater than 4? greater
$u^2 = 16 + 4$ than 6? How do you know?
$u = \sqrt{20} = \sqrt{4 \cdot 5} = 2\sqrt{5}$

$$\csc T = \frac{\text{hyp}}{\text{opp}} \qquad\qquad \sec T = \frac{\text{hyp}}{\text{adj}} \qquad\qquad \cot T = \frac{\text{adj}}{\text{opp}}$$

$$= \frac{2\sqrt{5}}{4} \qquad\qquad = \frac{2\sqrt{5}}{2} \qquad\qquad = \frac{2}{4}$$

$$\approx \frac{2.2361}{2} \approx 1.1180 \qquad\qquad \approx 2.2361 \qquad\qquad = 0.5000$$

The tables on pages 792–796 give decimal approximations of the six basic trigonometric ratios for angles measuring from 0 to 90. Angle measure is given in multiples of ten minutes. One degree equals 60 minutes ($1° = 60'$). The values given by the table are approximate, but it is customary to use $=$ (is equal to) rather than $\approx$ (is approximately equal to).

EXAMPLE 2 Find each trigonometric ratio using the portion of a trigonometric table shown below.

a. cos 38°50′ **b.** cot 39°20′

Plan The ratio names in the top row of the page must be used with degree measures in the left-hand column (0 through 45).

x	sin x	cos x	tan x	cot x	sec x	csc x	
30′	.6088	.7934	.7673	1.3032	1.2605	1.6427	30′
40′	.6111	.7916	.7720	1.2954	1.2633	1.6365	20′
50′	.6134	.7898	.7766	1.2876	1.2662	1.6304	10′
38° 0′	.6157	.7880	.7813	1.2799	1.2690	1.6243	52° 0′
10′	.6180	.7862	.7860	1.2723	1.2719	1.6183	50′
20′	.6202	.7844	.7907	1.2647	1.2748	1.6123	40′
30′	.6225	.7826	.7954	1.2572	1.2779	1.6064	30′
40′	.6248	.7808	.8002	1.2497	1.2808	1.6005	20′
50′	.6271	.7790	.8050	1.2423	1.2837	1.5948	10′
39° 0′	.6293	.7771	.8098	1.2349	1.2868	1.5890	51° 0′
10′	.6316	.7753	.8146	1.2276	1.2898	1.5833	50′
20′	.6338	.7735	.8195	1.2203	1.2929	1.5777	40′
30′	.6361	.7716	.8243	1.2131	1.2960	1.5721	30′
40′	.6383	.7698	.8292	1.2059	1.2991	1.5666	20′
50′	.6406	.7679	.8342	1.1988	1.3022	1.5611	10′

(left labels: 38°50′ points to the .7790 / .6271 row; 39°20′ points to the .6338 / 1.2203 row)

Solutions

a. Find 38°50′ in the left column. Read across to the cos column. Thus, cos 38°50′ = 0.7790.

b. Find 39°20′ in the left column. Read across to the cot column. Thus, cot 39°20′ = 1.2203.

The ratio names at the bottom of the page must be used with degree measures in the right-hand column (45 through 90).

EXAMPLE 3 Find each trigonometric ratio.

a. cos 45°40′ **b.** sec 46°50′

Solutions

43° 0′	.6820	.7314	.9325	1.0724	1.3673	1.4663	47° 0′	
10′	.6841	.7294	.9380	1.0661	1.3711	1.4617	50′	← 46°50′
20′	.6862	.7274	.9435	1.0599	1.3748	1.4572	40′	
30′	.6884	.7254	.9490	1.0538	1.3786	1.4527	30′	
40′	.6905	.7234	.9545	1.0477	1.3824	1.4483	20′	
50′	.6926	.7214	.9601	1.0416	1.3863	1.4439	10′	
44° 0′	.6947	.7193	.9657	1.0355	1.3902	1.4396	46° 0′	
10′	.6967	.7173	.9713	1.0295	1.3941	1.4352	50′	
20′	.6988	.7153	.9770	1.0235	1.3980	1.4310	40′	← 45°40′
30′	.7009	.7133	.9827	1.0176	1.4020	1.4267	30′	
40′	.7030	.7112	.9884	1.0117	1.4061	1.4225	20′	
50′	.7050	.7092	.9942	1.0058	1.4101	1.4184	10′	
45° 0′	.7071	.7071	1.0000	1.0000	1.4142	1.4142	45° 0′	
	cos x	sin x	cot x	tan x	csc x	sec x	x	

a. Thus, cos 45°40′ = 0.6988. **b.** Thus, sec 46°50′ = 1.4617.

EXAMPLE 4   Find m ∠A if sin A = 0.8774 and ∠A is an acute angle of a right triangle. Reverse the process used in Example 2.

Solution

10'	.4720	.8816	.5354	1.8676	1.1343	2.1185		50'
20'	.4746	.8802	.5392	1.8546	1.1361	2.1070		40'
30'	.4772	.8788	.5430	1.8418	1.1379	2.0957		30'
40'	.4797	.8774	.5467	1.8291	1.1397	2.0846	20'	←61°20'
50'	.4823	.8760	.5505	1.8165	1.1415	2.0736	10'	
29° 0'	.4848	.8746	.5543	1.8040	1.1434	2.0627	61° 0'	
10'	.4874	.8732	.5581	1.7917	1.1452	2.0519	50'	
20'	.4899	.8718	.5619	1.7796	1.1471	2.0413	40'	
30'	.4924	.8704	.5658	1.7675	1.1490	2.0308	30'	
40'	.4950	.8689	.5696	1.7556	1.1509	2.0204	20'	
50'	.4975	.8675	.5735	1.7437	1.1528	2.0101	10'	
30° 0'	.5000	.8660	.5774	1.7321	1.1547	2.0000	60° 0'	
	cos x	sin x	cot x	tan x	csc x	sec x	x	

Thus, m ∠A = 61°20′ if sin A = 0.8774.

EXAMPLE 5   Find m ∠B to the nearest ten minutes if ∠B is an acute angle of a right triangle and cot B = 2.2815.

Plan   Find the number in the cot column that is closest to 2.2815. The number is 2.2817. Since "cot" is at the top of the page, read across to the angle column at the left portion of the page.

Solution

	x	sin x	cos x	tan x	cot x	sec x	csc x	
	30'	.3827	.9239	.4142	2.4142	1.0824	2.6131	30'
	40'	.3854	.9228	.4176	2.3945	1.0837	2.5949	20'
	50'	.3881	.9216	.4210	2.3750	1.0850	2.5770	10'
23°40'→	23° 0'	.3907	.9205	.4245	2.3559	1.0864	2.5593	67° 0'
	10'	.3934	.9194	.4279	2.3369	1.0877	2.5419	50'
	20'	.3961	.9182	.4314	2.3183	1.0891	2.5247	40'
	30'	.3987	.9171	.4348	2.2998	1.0904	2.5078	30'
	40'	.4014	.9159	.4383	2.2817	1.0918	2.4912	20'
	50'	.4041	.9147	.4417	2.2637	1.0932	2.4748	10'

Thus, m ∠B = 23°40′ to the nearest ten minutes.

Summary

Basic Trigonometric Ratios

$$\sin A = \frac{\text{opp}}{\text{hyp}} = \frac{a}{c} \qquad \cos A = \frac{\text{adj}}{\text{hyp}} = \frac{b}{c} \qquad \tan A = \frac{\text{opp}}{\text{adj}} = \frac{a}{b}$$

Reciprocals

$$\csc A = \frac{\text{hyp}}{\text{opp}} = \frac{c}{a} \qquad \sec A = \frac{\text{hyp}}{\text{adj}} = \frac{c}{b} \qquad \cot A = \frac{\text{adj}}{\text{opp}} = \frac{b}{a}$$

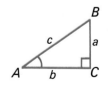

Classroom Exercises

Using the figure at the right, give each trigonometric ratio as the ratio of a pair of side lengths.

1. csc T 2. cot K 3. sec T 4. sin K

5. cos T 6. tan T 7. sin T 8. cos K

9–16. In the figure, suppose that $k = 3$, $t = 4$, and $w = 5$. Find the ratios of Classroom Exercises 1–8 to two decimal places.

Written Exercises

Find csc A, sec A, cot A, csc B, sec B, and cot B to four decimal places.

1.

2.
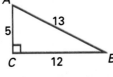

3.
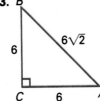

4.
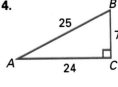

Find each trigonometric ratio. Use the tables on pages 792–796.

5. sin 28   6. tan 83   7. sec 32   8. csc 53
9. sin 60°20′   10. cot 37°20′   11. cos 39°30′   12. sin 80°40′
13. csc 19°30′   14. sec 47°10′   15. cot 87°50′   16. cos 11°40′

Find m $\angle A$ to the nearest ten minutes if $\angle A$ is an acute angle of a right triangle. Use the tables on pages 792–796.

17. sin A = 0.2079   18. csc A = 1.2361   19. tan A = 2.4751   20. sin A = 0.7934
21. sec A = 1.1471   22. cos A = 0.9492   23. sec A = 1.5721   24. cot A = 7.5958

Find each indicated trigonometric ratio to four decimal places. Use the figure at the right.

25. sec P, if $g = 6$, $p = 2$ 26. cot P, if $p = 2$, $v = 4$
27. csc G, if $v = 6$, $p = 4$ 28. cot G, if $p = \sqrt{2}$, $v = \sqrt{3}$
29. sec G, if $g = \sqrt{3}$, $p = \sqrt{2}$ 30. csc P, if $g = \sqrt{3}$, $v = \sqrt{6}$

Find m $\angle B$ to the nearest ten minutes if $\angle B$ is an acute angle of a right triangle.

31. cot B = 0.7312   32. sin B = 0.1675   33. tan B = 2.5199   34. cos B = 0.8662

35. Write how to find the cosine ratio of an acute angle A of a right triangle given the lengths of the hypotenuse and the leg opposite $\angle A$.

Simplify each expression (Exercises 36–37).

36. $(\sin 38°40')(\csc 38°40')$

37. $(\sin 23°10')^2 + (\cos 23°10')^2$

38. $\sin A = \sqrt{3} - 1$. Find $\csc A$ and simplify.

Mixed Review

Simplify.   *13.3, 8.7, 8.4, 5.2, 7.4*

1. $\log_3 27$ **2.** $32^{-\frac{4}{5}}$ **3.** $\dfrac{4}{2 - \sqrt{3}}$ **4.** $(3a - 7)^0$ **5.** $5 + \dfrac{4}{x}$

Using the Calculator

Scientific calculators use angle measures in decimals, such as 27.4 degrees, rather than in degrees and minutes. Thus, cos 38°50' is written as $\cos \left(38 + \dfrac{50}{60}\right)$. Enter the angle measure as 50 ⊝ 60 ⊕ 38 and press ⊜ to obtain 38.833333. Then press the ⌈cos⌉ key. The calculator will display 0.7789733. This is 0.7790 to four decimal places.

NOTE: Most calculators allow the use of three units of angle measure, the *degree*, the *radian* (see Lesson 18.5), and the *gradient* (a European unit of angle measure). A calculator is in the degree mode when DEG appears in the calculator display window.

EXAMPLE   Find cot 39°20' to four decimal places using a calculator.

1. Find tan 39°20'.

   $\tan 39°20' = \tan \left(39 + \dfrac{20}{60}\right)$

   Calculator steps:
   20 ⊝ 60 ⊕ 39 ⊜ ⌈tan⌉ ⇒ 0.8194625

2. Since tan and cot are reciprocals, press the reciprocal key ⌈1/x⌉. To four decimal places, cot 39°20' = 1.2203.

Exercises

Find the following ratios to four decimal places using a calculator.

1. $\sin 58.36$ **2.** $\tan 73°40'$ **3.** $\cot 42°20'$ **4.** $\csc 35°10'$

17.3 Solving Right Triangles

Objectives

To find missing measures of sides and angles of right triangles using trigonometric ratios

To solve problems using trigonometric ratios

The trigonometric ratios can be used to form equations for finding indicated side lengths or angle measures in a right triangle.

EXAMPLE 1 If m $\angle A$ = 65 and c = 140, find a to two significant digits.

Plan a is opp $\angle A$; c = 140 is the hyp.
Write an equation relating opp, hyp, and m $\angle A$ = 65.

Solution Use $\sin 65 = \dfrac{opp}{hyp} = \dfrac{a}{140}$ or $\csc 65 = \dfrac{hyp}{opp} = \dfrac{140}{a}$.

$$\sin 65 = \frac{a}{140}$$
$$140 \sin 65 = a$$
$$a = 140(0.9063) = 126.882$$

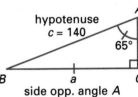
hypotenuse
c = 140
65°
B   a   C
side opp. angle A

Calculator Steps:
140 ⨯ 65 (sin) (=) 126.88309

Thus, a = 130 to two significant digits.

Recall that the acute angles of a right triangle are *complementary*; that is, the sum of their measures is 90.

EXAMPLE 2 Right triangle ABC has a right angle at C, a = 17, and b = 23. Find m $\angle A$ and m $\angle B$ to the nearest degree.

Plan First, draw and label a diagram. Then, write an equation relating a, b, and one of the two angle measures required, such as m $\angle A$.

Solution $\dfrac{17}{23} = \dfrac{opp}{adj} = \tan A$ (or $\dfrac{23}{17} = \cot A$)

$\tan A = \dfrac{17}{23} = 0.7391$
Find m $\angle A$ using a table or calculator.

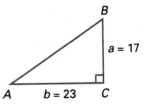
B
a = 17
A   b = 23   C

Calculator steps:
17 (÷) 23 (=) (INV) (tan) ⇒ 36.469234

Thus, m $\angle A$ = 36 to the nearest degree, and m $\angle B$ = 90 − 36 = 54.

In Examples 3 and 4 below, the **angle of elevation** or **angle of depression** is the angle between a horizontal line and the line of sight. The angles of elevation and depression are equal in measure.

EXAMPLE 3   A plane is flying at an altitude of 24,300 m. From the plane, the measure of the angle of depression of a ship is $55°30'$. How many meters, to three significant digits, is the plane from the ship?

Solution   Draw and label a diagram.
Let d = the required distance.
Write an equation with the
variable in the numerator.

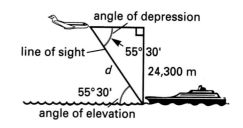
angle of depression
line of sight — $55°30'$
d 24,300 m
$55°30'$
angle of elevation

$$\frac{d}{24,300} = \csc 55°30'$$

$$\begin{aligned} d &= 24,300 \csc 55°30' \\ &= 24,300(1.2134) \\ &= 29,486 \end{aligned}$$

Calculator steps:

55.5 (sin) (1/x) (×) 24,300 (=) 29,485.776 ← The sine is the *reciprocal*
of the cosecant.

Thus, the distance is $29,500$ m to three significant digits.

EXAMPLE 4   The captain of a ship wants to determine his distance from shore. Seeking a familiar landmark, he finds a 90-ft-high lighthouse on top of a cliff. He sights both the top and bottom of the lighthouse. The measures of the two angles of elevation are 46 and 39. How far, to two significant digits, is he from the base of the cliff?

Plan   Draw and label a sketch. Two angles of elevation are given. Therefore, a system of two equations in two variables may be used.

Solution   Let d = the captain's distance from shore.
Let h = the height of the cliff.

$$\tan 39 = \frac{h}{d}$$

$$d \tan 39 = h$$

(1) $0.8098d = h$

$$\tan 46 = \frac{90 + h}{d}$$

$$d \tan 46 = 90 + h$$

(2) $1.0355d = 90 + h$

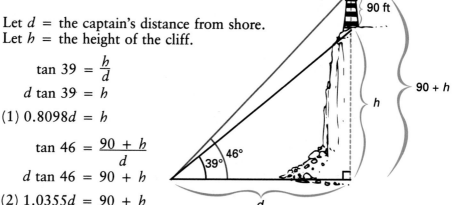
90 ft
90 + h
h
46°
39°
d

Solve the system.   (1) $h = 0.8098d$
                            (2) $1.0355d = 90 + h$

Substitute $0.8098d$ for h.   (3) $1.0355d = 90 + 0.8098d$
                            $0.2257d = 90$
                            $d = 398.7594$

Thus, the distance to the base of the vertical cliff is 400 ft to two significant digits.

The following relationships between lengths and angle measures were used to establish the number of significant digits in the answers to Examples 1–4.

Side Length		Angle measure
2 significant digits	$\longrightarrow$	nearest degree
3 significant digits	$\longrightarrow$	nearest 10 minutes or 0.1 degree

From now on, these relationships will be used in the examples and exercises.

Classroom Exercises

In right triangle ABC, give a trigonometric equation that relates the two given parts to the third part. Choose your equation so that the variable is not in a denominator.

1. m $\angle B = 40$, $c = 10$, $a = $ ___?___

2. m $\angle A = 53$, $c = 25$, $b = $ ___?___

3. m $\angle B = 32$, $a = 17$, $c = $ ___?___

4. m $\angle B = 26$, $b = 19$, $c = $ ___?___

5. m $\angle A = 39$, $c = 14$, $a = $ ___?___

6. m $\angle B = 75$, $c = 43$, $b = $ ___?___

Written Exercises

In Exercises 1–6, find lengths to two significant digits and angle measures to the nearest degree. Use the diagram above.

1. $a = 56$, $b = 21$, m $\angle A = $ ___?___, m $\angle B = $ ___?___

2. $c = 18$, $b = 13$, m $\angle B = $ ___?___, m $\angle A = $ ___?___

3. $a = 28$, m $\angle A = 39$, $b = $ ___?___, $c = $ ___?___

4. m $\angle A = 15$, $b = 30$, $a = $ ___?___, $c = $ ___?___

5. $c = 9.7$, $b = 5.3$, m $\angle A = $ ___?___, m $\angle B = $ ___?___

6. m $\angle A = 35$, $a = 19$, $b = $ ___?___, $c = $ ___?___

7. In the figure at the right, the tree casts a shadow on the ground because of the sun's rays. The length of the shadow is 34 ft. The measure of the angle of elevation is 37. Find the height of the tree to two significant digits.

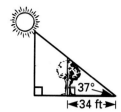

8. From the top of a lighthouse 159 ft above sea level, the angle of depression of a ship at sea is $19°20'$. Find, to three significant digits, the distance of the ship from the base of the lighthouse.

9. At a point on the ground 27.6 m from the foot of a flagpole, the angle of elevation of the top of the pole is $29°50'$. Find the height of the pole to three significant digits.

10. From a point 450 ft from the base of a building, the angles of elevation of the top and bottom of a flagpole on top of the building have measures of 60 and 55. Find the height of the flagpole to two significant digits.

11. Jane wants to find the height of a mountain. From some spot on the ground, she finds the angle of elevation to the top of the mountain to be $35°20'$. After moving 1,000 m closer to the mountain, she now finds the angle of elevation to be $50°30'$. Find the height of the mountain to three significant digits.

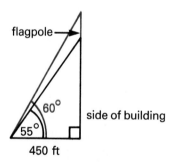

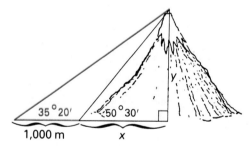

12. Two tanks on a training mission are 1,800 m apart on a straight road. The drivers find the angles of elevation to a helicopter hovering over the road between them to be 33 and 52. Find the height of the helicopter to two significant digits.

13. The length of the longer leg of a right triangle is 2 more than twice the length of the shorter leg. The length of the hypotenuse is 1 more than the length of the longer leg. Find the measure, to the nearest degree, of the angle between the shorter leg and the hypotenuse.

Mixed Review

Solve.   *6.4, 8.7, 9.5, 10.3*

1. $x^2 - 6x - 16 > 0$ **2.** $x^2 - 4x + 2 = 0$ **3.** $\sqrt{3x - 6} = 3$ **4.** $x^{\frac{2}{3}} = 4$

Application: *Horizon Distance*

If you want to see farther, you try to find a higher vantage point, perhaps a rooftop, a mountaintop, or even a space shuttle. The added height lets you see over a greater part of the earth's surface, increasing the distance at which objects disappear behind the horizon.

To find the horizon distance d, let r be the earth's radius and h the viewer's height above the earth. The line of sight to the horizon is tangent to the earth's surface, and so is perpendicular to a radius drawn to the point of tangency. From the Pythagorean relation:

$$r^2 + d^2 = (r + h)^2$$
$$r^2 + d^2 = r^2 + 2rh + h^2$$
$$d^2 = 2rh + h^2$$
$$d = \sqrt{2rh + h^2}$$

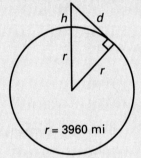

r = 3960 mi

Since ordinarily $r >> h$ ("r is much greater than h"), $rh >> h^2$. So, $2rh >> h^2$ and $2rh + h^2 \approx 2rh$. This gives the approximation

$$d \approx \sqrt{2rh} = \sqrt{2r}\sqrt{h} = \sqrt{7920}\sqrt{h} \approx 89.0\sqrt{h},$$

with d, r, and h in miles. If the height is entered as h *feet*, then

$$d \approx \sqrt{2rh} = \sqrt{(7920 \text{ mi})(h \text{ ft})\left(\frac{1 \text{ mi}}{5280 \text{ ft}}\right)} = \sqrt{\frac{7920}{5280} h \text{ mi}^2} \approx 1.22\sqrt{h} \text{ mi}.$$

The Rock of Gibralter on Spain's south coast looms 1400 ft above the Mediterannean Sea. By the formula above, a person on its peak can see about $1.22\sqrt{1400} \approx 46$ miles out to sea.

1. Leif perches 100 ft above the water in a ship's crow's nest, looking for a raft. At what distance will the raft enter his view?
2. An astronaut is orbiting 200 miles above the earth in a space shuttle. Give the horizon distance to 3 significant digits, using both the original formula and the approximation.
3. While riding a train across the Colorado plains at an elevation of 5000 ft, Trinh spots a mountain she believes to be Pike's Peak, elevation 14,110 ft. If the peak is 75 miles away, is this possible?

17.4 Angles of Rotation

Objectives
To determine the quadrant containing the terminal side of an angle of rotation

To find measures of angles coterminal with a given angle

In geometry, the measure of an angle is restricted to values between 0 and 180, including 180. However, in many practical applications, such as navigation, it is necessary to extend the concept of *angle* to *angle of rotation*.

In the figure at the right, $\angle AOB$ is an **angle of rotation** in standard position. It is formed by holding $\overrightarrow{OA}$ (read "ray OA") stationary along the x-axis, and then rotating $\overrightarrow{OB}$ from the x-axis to some position, such as 70 degrees, in a counterclockwise direction.

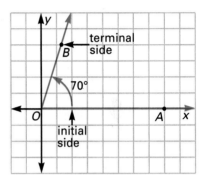

With $\overrightarrow{OA}$ (the **initial side**) fixed along the x-axis, allow $\overrightarrow{OB}$ (the **terminal side**) to continue to rotate counterclockwise. New angles are formed depending upon where $\overrightarrow{OB}$ terminates.

Some special angles of rotation, whose terminal sides coincide with a coordinate axis, are illustrated in the figures below.

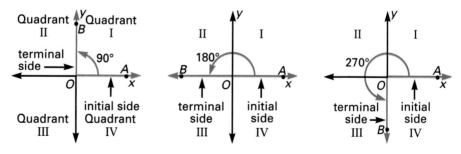

Notice that for each angle above, the initial and terminal sides lie along the coordinate axes. Such angles are called **quadrantal angles**.

It is also possible to generate angles of rotation by rotating the terminal side $\overrightarrow{OB}$ clockwise. *Clockwise* rotations correspond to *negative* angle measures.

EXAMPLE 1 Draw the angle of rotation in standard position that corresponds to each angle measure.

  a. -270 **b.** -360 **c.** -450

Solutions   **a.** Rotate clockwise.

b. Rotate clockwise until the initial and terminal sides coincide.

c. Rotate clockwise 360° and then 90° more.

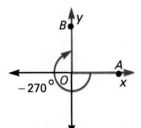   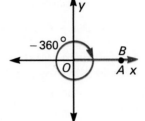   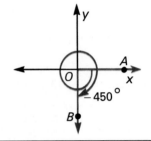

The next example involves angles of rotation that are not quadrantal angles. Note that it is customary to use the Greek letter θ (theta) to label angles of rotation.

EXAMPLE 2 Draw the angle of rotation in standard position that corresponds to each angle measure. Identify the quadrant containing the terminal side.

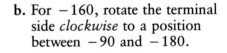

  a. $\theta = 140$ **b.** $\theta = -160$

Solutions   **a.** 140 is between 90 and 180. Rotate the terminal side *counterclockwise* to a position between 90 and 180.

b. For -160, rotate the terminal side *clockwise* to a position between -90 and -180.

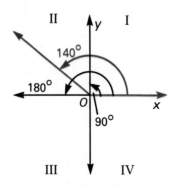   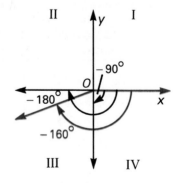

Thus, the angle with measure 140 terminates in Quadrant II.

Thus, the angle with measure -160 terminates in Quadrant III.

EXAMPLE 3   On the same set of coordinate axes, draw the angles of rotation with measures -30 and 330. State a relationship between the angles.

Solution   The two angles have the same initial and terminal sides. However, their angle measures are not equal since $-30 \neq 330$. These two angles are called *coterminal* angles.

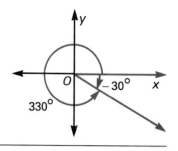

Definition   **Coterminal angles** are angles that share initial and terminal sides.

The pattern that follows suggests a formula for finding the measures of all angles coterminal with a given angle.

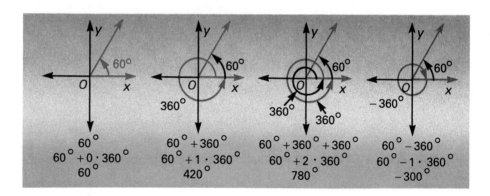

The angles with measures 60, 420, 780, and -300 share initial and terminal sides and thus are coterminal angles. Note the following.

$$60 + 0 \cdot 360 = 60$$
$$60 + 1 \cdot 360 = 420$$
$$60 + 2 \cdot 360 = 780$$
$$60 + (-1) \cdot 360 = -300$$

Notice the pattern:
$60 + n \cdot 360$,
where n is an integer.

This suggests the following formula for finding the measures of all angles coterminal with a given angle.

Coterminal Angle Formula
If angles θ_1 and θ_2 are coterminal, then $\theta_1 = \theta_2 + n \cdot 360$, where n is an integer.

EXAMPLE 4   Find two angles, one with a positive measure and the other with a negative measure, that are coterminal with an angle of measure 40.

Solution   Use the formula $\theta_1 = \theta_2 + n \cdot 360$. Assign values to n.

positive: $\theta_1 = 40 + 2 \cdot 360$ negative: $\theta_1 = 40 + (-1) \cdot 360$
$= 40 + 720 = 760$ $= 40 - 360 = -320$

Thus, an angle of measure 760 and an angle of measure -320 are coterminal with an angle of measure 40. Many other answers are also possible, depending on the choice of values for n.

EXAMPLE 5   Find the least positive measure of an angle that is coterminal with each angle. Then draw the angle.

a. angle with measure 855   b. angle with measure -320

Plan   Add integral multiples of 360, such as ± 360, ± 720, $\pm 1{,}080$, and $\pm 1{,}440$, to the given angle measure to get the least positive measure.

Solutions   a. $855 - 720 = 135$ b. $-320 + 360 = 40$

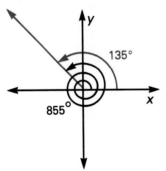   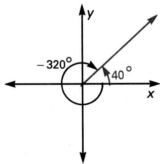

Classroom Exercises

For each angle measure, identify the quadrant in which each angle in standard position terminates, or, if the angle is quadrantal, describe the location of the terminal side.

1. 90   **2.** 160   **3.** -80 **4.** 200   **5.** -180 **6.** -320

For each angle measure, find two angle measures, one positive and one negative, so that all three angles are coterminal.

7. 35   **8.** -315 **9.** 180   **10.** -135 **11.** 160   **12.** -240

Written Exercises

Draw the angle in standard position that corresponds to each angle measure. If the angle is quadrantal, identify the axis along which the terminal side lies. If not, identify the quadrant containing the terminal side.

1. 115   **2.** -220 **3.** -90 **4.** -306 **5.** 180   **6.** 280

7. 270   **8.** -130 **9.** 350   **10.** 220   **11.** 190   **12.** -25

For each angle measure, find two angle measures, one positive and one negative, so that all three angles are coterminal.

13. 30   **14.** 205   **15.** -130 **16.** -200 **17.** -310 **18.** 90

19. -50 **20.** -270 **21.** 320   **22.** -210 **23.** -44 **24.** 290

Find the least positive measure of an angle that is coterminal with the angle of given measure. Draw the angle.

25. 405   **26.** -275 **27.** 540   **28.** 810   **29.** 1,200   **30.** $-1,260$

31. Find a general formula for the measure of any angle θ such that its terminal side will lie in the second quadrant.

32. Find a general formula for the measure of any angle θ such that an angle with measure one-fourth as large will have its terminal side in the third quadrant.

Midchapter Review

In Exercises 1–8, refer to the figure at the right. Find each trigonometric ratio to four decimal places. *17.1, 17.2*

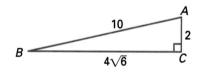

1. $\sin B$ **2.** $\cos B$ **3.** $\tan A$ **4.** $\cot B$

5. $\sin A$ **6.** $\csc B$ **7.** $\sec A$ **8.** $\cot A$

Find each trigonometric ratio. Use the tables on pages 792–796. *17.2*

9. $\sin 51$ **10.** $\cot 14$ **11.** $\csc 10$ **12.** $\cos 38°50'$

13. From a point in the middle of a park, a blimp is sighted in the distance. The measure of the angle of elevation to the blimp is $16°30'$. A range finder indicates that the blimp is 6,200 ft away. Find the altitude of the blimp to three significant digits. *17.3*

14. Find the measures of two angles, one positive and one negative, that are coterminal with an angle that has a measure of -45. *17.4*

17.5 Sines of Angles of Any Measure

Objectives

To find the sine of an angle given the coordinates of a point on the terminal side of the angle

To find the sines of angles of any measure

In the study of electricity it is necessary to work with formulas such as $E = E_{max} \sin \theta$. Here, E is the changing electromotive force (emf) created by a generator, and E_{max} is the maximum emf. However, θ is *not* the measure of an acute angle of a right triangle. In this lesson, you will learn how to define the sine ratio for angle measures that are not related to right triangles.

The sine of an angle has been defined in terms of an acute angle of a right triangle as follows.

$$\sin A = \frac{\text{length of leg opposite } \angle A}{\text{length of hypotenuse}}, \text{ or } \frac{\text{opp}}{\text{hyp}}$$

This definition can be altered to apply to an angle terminating in any quadrant.

EXAMPLE 1   Find the sine of an angle whose terminal side contains the point $P(3,4)$.

Plan

Form a right triangle by drawing both $\overline{OP}$, of length r, and $\overline{PQ}$, perpendicular to the x-axis. Use the triangle to find the sine.

Solution

$OQ = 3$ (x-coordinate of $P(3,4)$)

$QP = 4$ (y-coordinate of $P(3,4)$)

Find r, the length of the hypotenuse of the right triangle. The Pythagorean relation gives the following.

$$r^2 = 3^2 + 4^2 = 9 + 16 = 25$$

$$r = \sqrt{25} = 5$$

Thus, $\sin \theta = \dfrac{\text{opp}}{\text{hyp}} = \dfrac{4}{5}$.

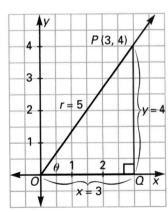

Notice in Example 1 that $\sin \theta = \dfrac{4}{5} = \dfrac{\text{opp}}{\text{hyp}}$, or $\dfrac{y\text{-coordinate of } P}{r}$.

Sine of an Angle of Any Measure

For any angle θ in standard position, with terminal side containing the point $P(x,y)$ and r = distance from P to the origin, the following is true.

$$\sin \theta = \frac{y}{r} = \frac{y}{\sqrt{x^2 + y^2}}$$

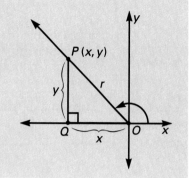

The right triangle formed by drawing a perpendicular segment from point P on a ray $\overrightarrow{OP}$ to the x-axis is called a **reference triangle**. In the figure, triangle OPQ is a reference triangle.

The sine of an angle is now no longer restricted to acute angles of right triangles. Note also that while the distance r is always positive, y and x may be positive, negative, or zero.

EXAMPLE 2 Find the sine of the angle whose terminal side contains the given point. Simplify all radicals.

a. $P(-3,4)$ **b.** $P(-1,-2)$ **c.** $P(3,-3)$

Solutions

a.

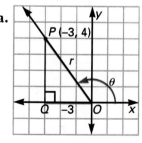

$r = \sqrt{x^2 + y^2}$
$r = \sqrt{(-3)^2 + 4^2}$
$r = \sqrt{25} = 5$
$\sin \theta = \dfrac{y}{r}$
$\qquad = \dfrac{4}{5}$

b.

$r = \sqrt{x^2 + y^2}$
$r = \sqrt{(-1)^2 + (-2)^2}$
$r = \sqrt{5}$
$\sin \theta = \dfrac{y}{r}$
$\qquad = \dfrac{-2}{\sqrt{5}} = -\dfrac{2\sqrt{5}}{5}$

c.

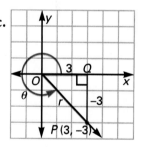

$r = \sqrt{x^2 + y^2}$
$r = \sqrt{3^2 + (-3)^2}$
$r = \sqrt{18} = 3\sqrt{2}$
$\sin \theta = \dfrac{y}{r}$
$\qquad = \dfrac{-3}{3\sqrt{2}} = -\dfrac{\sqrt{2}}{2}$

In the figure at the right below, acute angle *POQ* is called the **reference angle** of reference triangle *POQ*. Note that trigonometric ratios of angles with degree measures greater than 90 do not appear in the table of trigonometric ratios. However, the sine of an angle such as 240 can be found in terms of its reference angle as follows.

(1) An angle of measure 240 terminates in Quadrant III. (2) The reference angle has degree measure 240 − 180 = 60. (3) The *y*-coordinate in Quadrant III is *negative*.

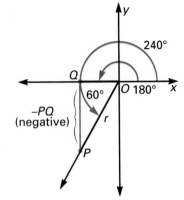

Therefore, $\sin 240 = \dfrac{-PQ}{r}$

$$= -\sin 60$$

$$= -0.8660.$$

EXAMPLE 3 Find sin 110°20′.

Solution Form the reference triangle.

Find the measure of the reference angle.

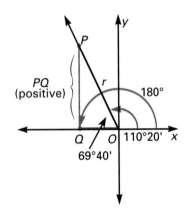

$$180° - 110°20′: \quad \begin{array}{r} 179°\ 60′ \\ -110°\ 20′ \\ \hline 69°\ 40′ \end{array}$$

Determine the correct sign.
Since *y* is positive in Quadrant II,
$\sin 110°20′ = +\sin 69°40′ = 0.9377.$

EXAMPLE 4 Find sin 310° 40′.

Solution Find the measure of the reference angle.

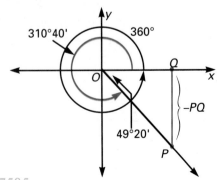

$$\begin{array}{r} 359°\ 60′ \\ -310°\ 40′ \\ \hline 49°\ 20′ \end{array}$$

Since *y* is negative in Quadrant IV,
$\sin 310°40′ = -\sin 49°20′ = -0.7585.$

Use the procedure just shown to find the sine of a negative angle measure or an angle measure that is larger than 360.

EXAMPLE 5   **a.** Find sin (-132).   **b.** Find sin 415.

Solutions   **a.**   **b.**

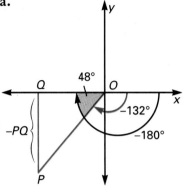
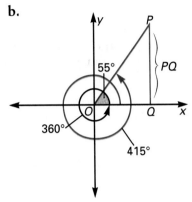

reference angle:
$|-180| - |-132|$
$= 180 - 132$, or 48
The *y*-coordinate is *negative*.
Thus, sin(-132) $= -$ sin 48
$= -0.7431$

reference angle:
$415 - 360 = 55$

The *y*-coordinate is *positive*.
Thus, sin 415 $=$ sin 55
$= 0.8192$

Focus on Reading

Indicate whether each of the following is *sometimes* true, *always* true, or *never* true.

1. The sine of an angle in Quadrant I is positive.
2. The sine of a negative angle of rotation is negative.
3. The reference angle of any nonquadrantal angle of rotation is acute.
4. The sine of an angle of rotation with terminal side in Quadrant II is negative.

Classroom Exercises

Find the sine of the angle whose terminal side contains the given point.
Simplify all radicals.

1. $P(-4,3)$ **2.** $P(6,8)$ **3.** $P(1,-1)$

Give the measure of the reference angle for each angle measure.

4. 130 **5.** 200 **6.** 350 **7.** 400 **8.** -40 **9.** -120

10–15. Find the sine of each angle with the measures given in Exercises 4–9 above.

Written Exercises

Find the sine of the angle whose terminal side contains the given point. Simplify all radicals.

1. $P(4,3)$ **2.** $P(-6,8)$ **3.** $P(8,6)$ **4.** $P(5,-12)$

5. $P(-12,-5)$ **6.** $P(-1,3)$ **7.** $P(-2,-1)$ **8** $P(5,-2)$

9. $P(-1,-1)$ **10.** $P(6,-2)$ **11.** $P(4,-4)$ **12.** $P(-2,6)$

Find the sine of each angle with the given degree measure.

13. 175	**14.** 230	**15.** 79	**16.** 329
17. 100	**18.** 129	**19.** -45	**20.** -155
21. -205	**22.** 370	**23.** -86	**24.** -350
25. 130°40′	**26.** $-100°50′$	**27.** 210°40′	**28.** 320°20′
29. 95°30′	**30.** $-170°20′$	**31.** 254°50′	**32.** $-205°10′$
33. 480	**34.** -705	**35.** $-1,130$	**36.** 1,015

37. Find the sine of an angle whose terminal side contains $P(a - 1, a + 1)$.

38. For the angle described in Exercise 37, give the restrictions on a if the angle is to terminate in the second quadrant.

Mixed Review

Find each of the following. Use the tables on pages 792–796. **17.2**

1. sec 75 **2.** cos 49 **3.** csc 82 **4.** cot 41 **5.** tan 73°40′

Brainteaser

A ship is sailing on a straight course RST. When the ship is at point R, the captain sights a lighthouse L and notes that m $\angle LRT$ is 36°30′. After sailing for 8 mi to point S, the captain then observes that m $\angle LST = 73$. What is the distance from S to the lighthouse L?

17.6 Trigonometric Ratios of Angles of Any Measure

Objectives

To find the six trigonometric ratios of an angle given the coordinates of a point on the terminal side of the angle

To find the six trigonometric ratios of angles of any measure

In Lesson 17.5, the sine ratio was defined for an angle of rotation terminating in any quadrant. The remaining five trigonometric ratios can also be defined given the coordinates of a point of the terminal side of an angle of rotation.

EXAMPLE 1

Find the six trigonometric ratios for the angle whose terminal side contains the point $P(8,6)$.

Plan

Form the reference triangle and find the lengths of its sides. Then use the definitions of trigonometric ratios.

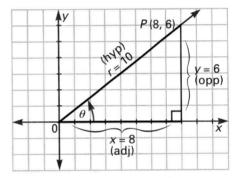

Solution

$$\text{adj} = x = 8 \text{ and opp} = y = 6$$
$$\text{hyp} = r = \sqrt{x^2 + y^2}$$
$$= \sqrt{8^2 + 6^2}$$
$$= \sqrt{100} = 10$$

$$\sin \theta = \frac{\text{opp}}{\text{hyp}} = \frac{y}{r} = \frac{6}{10} = \frac{3}{5} \qquad \csc \theta = \frac{\text{hyp}}{\text{opp}} = \frac{r}{y} = \frac{10}{6} = \frac{5}{3}$$

$$\cos \theta = \frac{\text{adj}}{\text{hyp}} = \frac{x}{r} = \frac{8}{10} = \frac{4}{5} \qquad \sec \theta = \frac{\text{hyp}}{\text{adj}} = \frac{r}{x} = \frac{10}{8} = \frac{5}{4}$$

$$\tan \theta = \frac{\text{opp}}{\text{adj}} = \frac{y}{x} = \frac{6}{8} = \frac{3}{4} \qquad \cot \theta = \frac{\text{adj}}{\text{opp}} = \frac{x}{y} = \frac{8}{6} = \frac{4}{3}$$

Definition

Trigonometric Ratios of an Angle of Any Measure
For any angle of rotation θ, with point $P(x,y)$ on the terminal side and located at a distance $r > 0$ from the origin, the trigonometric ratios of angle θ are defined as follows.

$$\sin \theta = \frac{y}{r} \qquad\qquad \cos \theta = \frac{x}{r} \qquad\qquad \tan \theta = \frac{y}{x}, x \neq 0$$

$$\csc \theta = \frac{r}{y}, y \neq 0 \qquad \sec \theta = \frac{r}{x}, x \neq 0 \qquad \cot \theta = \frac{x}{y}, y \neq 0$$

EXAMPLE 2   Find the six trigonometric ratios for the angle whose terminal side contains the given point. Simplify all radicals.

   a. $P(-4,3)$ **b.** $P(6,-2)$

Plan   Form the reference triangle. Then, use the given values of x and y and the distance formula to find r.

Solutions   **a.**

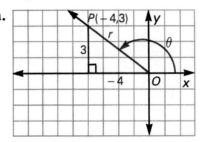

b.

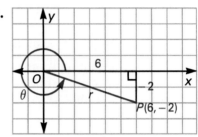

$r = \sqrt{x^2 + y^2} = \sqrt{(-4)^2 + 3^2}$
$= \sqrt{25} = 5$

$\sin \theta = \dfrac{y}{r} = \dfrac{3}{5}$

$\cos \theta = \dfrac{x}{r} = \dfrac{-4}{5} = -\dfrac{4}{5}$

$\tan \theta = \dfrac{y}{x} = \dfrac{3}{-4} = -\dfrac{3}{4}$

$\cot \theta = \dfrac{x}{y} = \dfrac{-4}{3} = -\dfrac{4}{3}$

$\sec \theta = \dfrac{r}{x} = \dfrac{5}{-4} = -\dfrac{5}{4}$

$\csc \theta = \dfrac{r}{y} = \dfrac{5}{3}$

$r = \sqrt{x^2 + y^2} = \sqrt{6^2 + (-2)^2}$
$= \sqrt{40} = \sqrt{4 \cdot 10} = 2\sqrt{10}$

$\sin \theta = \dfrac{-2}{2\sqrt{10}} = \dfrac{-1 \cdot \sqrt{10}}{\sqrt{10} \cdot \sqrt{10}}$

$\qquad = -\dfrac{\sqrt{10}}{10}$

$\cos \theta = \dfrac{6}{2\sqrt{10}} = \dfrac{3\sqrt{10}}{10}$

$\tan \theta = \dfrac{-2}{6} = -\dfrac{1}{3}$

$\cot \theta = \dfrac{6}{-2} = -\dfrac{3}{1}$, or -3

$\sec \theta = \dfrac{2\sqrt{10}}{6} = \dfrac{\sqrt{10}}{3}$

$\csc \theta = \dfrac{2\sqrt{10}}{-2} = \dfrac{\sqrt{10}}{-1}$, or $-\sqrt{10}$

As with the sine ratio, all trigonometric ratios of angles with measures greater than 90 or less than 0 can be found in terms of acute angles of reference triangles. Thus, to find a trigonometric ratio of an angle terminating in any quadrant, perform the following steps.

1. Find the corresponding trigonometric ratio of the reference angle using the tables on pages 792–796.
2. Assign the correct sign to the result; this will depend upon the signs of x and y for the given quadrant.

EXAMPLE 3   Find the six trigonometric ratios of 143°20′.

Solution   Form the reference triangle. Find the measure of the reference angle.

$$180° - 143°20': \quad 179°\ 60'$$
$$\underline{-143°\ 20'}$$
$$36°\ 40'$$

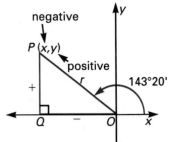

Determine the correct sign for each ratio.
In Quadrant II, x is negative and y is positive.

$$\sin 143°20' = \frac{PQ}{r} = \sin 36°40' = 0.5972$$

$$\cos 143°20' = \frac{-OQ}{r} = -\cos 36°40' = -0.8021$$

$$\tan 143°20' = \frac{PQ}{-OQ} = -\tan 36°40' = -0.7445$$

$$\csc 143°20' = \frac{r}{PQ} = \csc 36°40' = 1.6746$$

$$\sec 143°20' = \frac{r}{-OQ} = -\sec 36°40' = -1.2467$$

$$\cot 143°20' = \frac{-OQ}{PQ} = -\cot 36°40' = -1.3432$$

When the trigonometric ratios for angles of rotation are computed, it is essential first to establish the correct sign of x and y (NOTE: r is always positive.)

Sign Combinations of x and y for Angles Terminating in any Quadrant

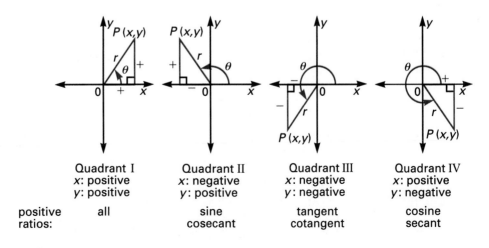

Quadrant I	Quadrant II	Quadrant III	Quadrant IV
x: positive	x: negative	x: negative	x: positive
y: positive	y: positive	y: negative	y: negative

| positive ratios: | all | sine cosecant | tangent cotangent | cosine secant |

EXAMPLE 4   Express csc 325 in terms of the cosecant of a reference angle.

Solution   Reference angle: $360 - 325 = 35$

$\csc 325 = \dfrac{r}{y} \leftarrow y$ is negative.

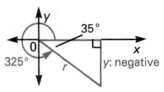

Therefore, $\csc 325 = -\csc 35$.

The six trigonometric ratios for quadrantal angles are defined in the same way as for nonquadrantal angles. However, as shown below, no reference triangles are needed for quadrantal angles.

To find sin 270 and tan 270, perform the following steps.
1. Draw the angle in standard position.
2. Choose any point on the terminal side, such as $P(0,-1)$.
3. Determine the x- and y-coordinates of P and the distance of P from the origin.
   x-coordinate of $P(0,-1)$: 0
   y-coordinate of $P(0,-1)$: -1
   Distance of $P(0,-1)$ from the origin: 1

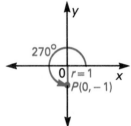

4. $\sin 270 = \dfrac{y}{r} = \dfrac{-1}{1} = -1$

$\tan 270 = \dfrac{y}{x} = \dfrac{-1}{0}$, which is *undefined*, since division by 0 is undefined.

Therefore, $\sin 270 = -1$ and $\tan 270$ is undefined.

Classroom Exercises

Identify the quadrant in which each of the following angles terminates. For quadrantal angles, identify the two quadrants that are separated by the terminal side. Also, give the signs of x and y for the angle.

1. 140   **2.** -90 **3.** 238   **4.** -50

5. 450   **6.** 65   **7.** -150 **8.** -270

9–16. Find the six trigonometric ratios for the angle measures of Exercises 1–8 above.

Find the six trigonometric ratios for each angle whose terminal side contains the given point. Simplify all radicals.

17. $P(-8,6)$ **18.** $P(-5,12)$ **19.** $P(3,3)$ **20.** $P(-2,-6)$

Written Exercises

Find the six trigonometric ratios for the angle whose terminal side contains the given point. If the ratio is undefined, so indicate. Simplify all radicals.

1. $P(6,8)$ **2.** $P(4,-3)$ **3.** $P(-5,-12)$ **4.** $P(-8,-6)$

5. $P(-12,5)$ **6.** $P(-2,2)$ **7.** $P(4,-2)$ **8.** $P(-3,1)$

9. $P(-1,-3)$ **10.** $P(2,0)$ **11.** $P(4,-1)$ **12.** $P(0,-6)$

Find the six trigonometric ratios for each angle measure. If the ratio is undefined, so indicate.

13. 105 **14.** 255 **15.** 329 **16.** -49

17. -112 **18.** 55 **19.** -235 **20.** 186

21. 313 **22.** 420 **23.** -130 **24.** -223

25. 0 **26.** 180 **27.** -90 **28.** -270

29. $112°40'$ **30.** $312°20'$ **31.** $-120°50'$ **32.** $115°30'$

Find the six trigonometric ratios for the angle whose terminal side contains the given point. Simplify all radicals.

33. $P(2,-2\sqrt{3})$ **34.** $P(-\sqrt{2},-\sqrt{2})$ **35.** $P(-\sqrt{7},-3)$ **36.** $P(\sqrt{5},-\sqrt{5})$

37. $P(\sqrt{2},\sqrt{6})$ **38.** $P(-2\sqrt{5},\sqrt{5})$ **39.** $P(-4,4\sqrt{3})$ **40.** $P(6\sqrt{2},-2\sqrt{7})$

Express each trigonometric ratio in terms of the same trigonometric ratio of a reference angle.

41. $\sin 140$ **42.** $\csc (-126)$ **43.** $\tan 336$ **44.** $\cot (-220)$

45. $\sec (-50)$ **46.** $\tan 185$ **47.** $\sin 115°40'$ **48.** $\csc 290°20'$

If $0 < \theta < 90$, express each of the six trigonometric ratios for the angles below in terms of θ.

49. $180 - \theta$ **50.** $360 - \theta$ **51.** $180 + \theta$ **52.** $90 + \theta$

53. Is the expression $2n \cdot 90$ the measure of a quadrantal angle for all positive integral values of n? Which angles? Give an expression that will define all remaining measures of quadrantal angles for n, a positive integer.

Mixed Review

Solve each equation. **8.7, 9.5, 10.4, 13.3**

1. $x^2 - 4x = -6$ **2.** $x^3 - 27 = 0$ **3.** $4^{2x-3} = 64$ **4.** $\log_8 x = \frac{2}{3}$

17.7 Trigonometric Ratios and Special Angles

Objectives

To find the six trigonometric ratios of angles of rotation whose reference angles measure 60, 30, or 45

To evaluate expressions involving trigonometric ratios

There are three special reference angles whose trigonometric ratios can be found without referring to a table or calculator.

Recall that in an equilateral triangle, the measure of each angle is 60. The altitude $\overline{AD}$ bisects both the base $\overline{BC}$ and the vertex angle A. The altitude divides the equilateral triangle into two 30-60-90 triangles. In right triangle ABD, the leg opposite the angle measuring 30, $\overline{BD}$, has half the length of the hypotenuse $\overline{AB}$.

To find AD, use the Pythagorean relation.

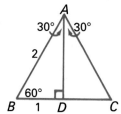

$$(AB)^2 = (BD)^2 + (AD)^2$$
$$2^2 = 1^2 + (AD)^2$$
$$3 = (AD)^2$$
$$AD = \sqrt{3}$$

Thus, if an angle in standard position has its terminal side in the first quadrant and its reference triangle is a 30-60-90 triangle, it will be convenient to use one of the following figures, for which 2 is the distance of P from the origin.

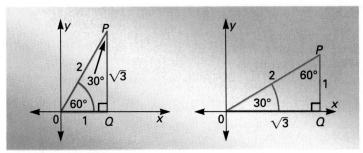

EXAMPLE 1   Find tan 30 and simplify.

Solution   Draw a reference triangle as at the right.

$$\tan 30 = \frac{y}{x} = \frac{1}{\sqrt{3}} = \frac{1 \cdot \sqrt{3}}{\sqrt{3} \cdot \sqrt{3}} = \frac{\sqrt{3}}{3}$$

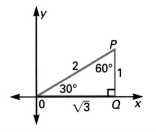

EXAMPLE 2   Find csc 120 and simplify.

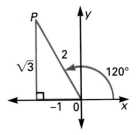

Solution   Since the standard position of 120 has its terminal side in the second quadrant, draw a 30-60-90
reference triangle in the second quadrant.

$$\csc 120 = \frac{r}{y} = \frac{2}{\sqrt{3}} = \frac{2 \cdot \sqrt{3}}{\sqrt{3} \cdot \sqrt{3}} = \frac{2\sqrt{3}}{3}$$

If a right triangle is isosceles, then the legs have equal length and each
of the acute angles has a measure of 45.

In the isosceles right triangle at the right, each leg has a
length of 1. The Pythagorean relation can be used to
find the length of the hypotenuse.

$$h^2 = 1^2 + 1^2 = 2, \text{ so } h = \sqrt{2}$$

Thus, if a reference angle has a measure of 45,
it will be convenient to use the figure at the
right for standard-position angles that terminate in the first quadrant.

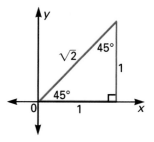

EXAMPLE 3   Find cos 45 and simplify.

Solution   The angle terminates in Quadrant I. Draw a reference triangle as shown, with legs each of length
1 and hypotenuse of length $\sqrt{2}$.

$$\cos 45 = \frac{x}{r} = \frac{1}{\sqrt{2}} = \frac{1 \cdot \sqrt{2}}{\sqrt{2} \cdot \sqrt{2}} = \frac{\sqrt{2}}{2}$$

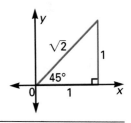

EXAMPLE 4   Find csc (−135).

Solution   Draw and label the reference triangle in
Quadrant III as shown at the right. Note
that x and y are negative in Quadrant III.

$$\csc (-135) = \frac{r}{y} = \frac{\sqrt{2}}{-1}, \text{ or } -\sqrt{2}$$

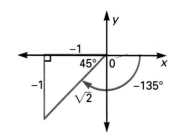

In summary, note the Special Reference Angles: 60, 30, 45.

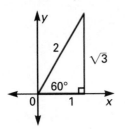 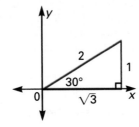 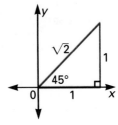

In the next example, note that it is accepted practice to write the squares of sin θ and cos θ as sin² θ and cos² θ instead of as (sin θ)² and (cos θ)². This is true for all of the trigonometric ratios.

EXAMPLE 5   Evaluate $\sin^2 \theta + \cos^2 \theta$ for $\theta = 210$.

Solution $\sin^2 \theta + \cos^2 \theta = \sin^2 210 + \cos^2 210$

$$= \left(-\frac{1}{2}\right)^2 + \left(-\frac{\sqrt{3}}{2}\right)^2$$

$$= \frac{1}{4} + \frac{3}{4} \text{, or } 1$$

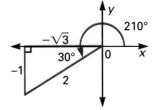

Therefore, 1 is the value of $\sin^2 \theta + \cos^2 \theta$ for $\theta = 210$.

Focus on Reading

Complete each of the following sentences.

1. In a 30-60-90 triangle, if the length of the hypotenuse is 2, then the length of the leg opposite the angle measuring 30 is __?__ .
2. In a 45-45-90 triangle, it is convenient to choose __?__ for the length of each leg and __?__ for the length of the hypotenuse.

Classroom Exercises

Give the measure of the other acute angle and the missing values.

1.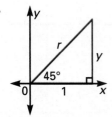

2.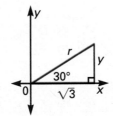

3.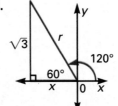

4.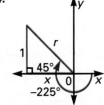

5. 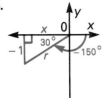   **6.** 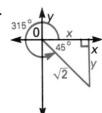   **7.** 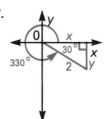   **8.**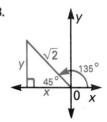

Use the figure of Classroom Exercise 8 to find each trigonometric ratio.

  9. sin 135   **10.** cos 135   **11.** cot 135   **12.** sec 135

Written Exercises

Find each trigonometric ratio. Simplify all radicals.

  1. sin 30   **2.** csc 45   **3.** cos 30   **4.** sec 30

  5. tan 60   **6.** cot 45   **7.** sin 150   **8.** cos 240

  9. sec 315   **10.** tan 120   **11.** csc 315   **12.** cot 225

13. $\sin(-225)$ **14.** $\tan(-210)$ **15.** $\sec(-45)$ **16.** $\cot(-330)$

Evaluate each expression for the indicated value of θ. Simplify all radicals.

17. $\tan^2 \theta, \theta = 60$ **18.** $2 \cos^2 \theta, \theta = 30$

19. $\cos^2 \theta - \sin^2 \theta, \theta = 240$ **20.** $\sqrt{1 - \sin^2 \theta}, \theta = 135$

21. $(2 \sin \theta)(\cos \theta), \theta = 150$ **22.** $\dfrac{\sin \theta}{\cos \theta}, \theta = 60$

Simplify each expression. Rationalize all denominators.

23. $\cos 225 - \tan(-1,050) + \sin 135$ **24.** $(\sin 30 + \cos 30)^2$

25. $\dfrac{\cot 30 - \sec 45}{\tan 60 + \csc 45}$ **26.** $\dfrac{\csc 60}{\cot 60 + \tan 60}$

Mixed Review

Find θ to the nearest ten minutes if $0 < \theta < 90$.   *17.2*

  1. $\sin \theta = 0.3746$ **2.** $\cos \theta = 0.1132$ **3.** $\tan \theta = 0.3959$ **4.** $\sec \theta = 2.430$

Brainteaser

Find the value of tan 30 · cos 45 · sin 120 · sec 150 · cot 90 · csc 225.

17.8 Finding Trigonometric Ratios and Angle Measures

Objectives   To find all angle measures between 0 and 360 corresponding to a given trigonometric ratio

To find all possible trigonometric ratios given two conditions

In the figure at the right, note that two angle measures between 0 and 360 have the same sine ratio ($\sin 210 = \sin 330 = -\frac{1}{2}$).

This happens because the sine ratio depends upon the values of y and r, which are the same for both angles.

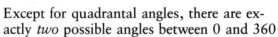

Except for quadrantal angles, there are exactly *two* possible angles between 0 and 360 that correspond to any given trigonometric ratio. The procedure for finding such angles is illustrated in Example 1 below.

EXAMPLE 1   Find the two angle measures between 0 and 360 satisfying the equation $\cos \theta = -0.9397$.

Plan   Find the measure of a reference angle and use it together with two reference triangles to locate two angles of rotation.

Solution   When x is negative, $\cos \theta$ is negative. This occurs in Quadrants II and III. Draw the reference triangles in these two quadrants. Next, find the measure of a reference angle corresponding to $|\cos \theta| = 0.9397$. From the table, the measure is 20. Then, find the measures of the two angles of rotation using the reference triangles, as shown below.

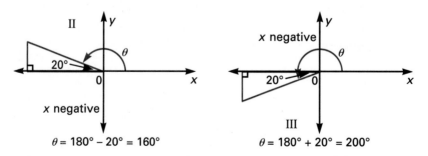

Therefore, the two angle measures satisfying the equation $\cos \theta = -0.9397$ are 160 and 200.

EXAMPLE 2   Find all values of θ, $0 < \theta < 360$, for which $\tan \theta = -0.3574$. Find the results to the nearest ten minutes.

Solution $\text{Tan } \theta$ is *negative*. The ratio $\dfrac{y}{x}$ is negative only if x and y have opposite signs ($\dfrac{+}{-}$ or $\dfrac{-}{+}$). This occurs in Quadrants II and IV (see chart at bottom of page 645). Draw the two reference triangles. Find the measure of a reference angle corresponding to $|\tan \theta| = 0.3574$. From the trigonometric table (or a calculator), the measure of each reference angle is 19°40′ to the nearest 10 minutes.

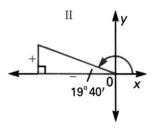

Tan θ is negative.   Tan θ is negative.
$\theta = 180° - 19° 40'$ $\theta = 360° - 19° 40'$
$\quad = 160° 20'$ $\quad = 340° 20'$

EXAMPLE 3   Find the value of each of the other five trigonometric ratios of θ if $\sin \theta = -\frac{1}{3}$ and $\tan \theta > 0$.

Plan   Locate the reference triangle and find values for x, y, and r. Do not find the measure of θ.

Solution   First, determine the quadrant of θ. The sine ratio $\dfrac{y}{r}$ is negative in Quadrants III and IV. Tan θ is positive in Quadrants I and III. To satisfy both conditions, the reference triangle must be in Quadrant III. Since $\sin \theta = -\frac{1}{3}$, or $\frac{-1}{3}$, choose $y = -1$ and $r = 3$. Use the distance formula to evaluate x.

$$r = \sqrt{x^2 + y^2}$$
$$3 = \sqrt{x^2 + (-1)^2}$$
$$9 = x^2 + 1$$
$$8 = x^2, \text{ or } x^2 = 8 \quad \text{Thus, } x = \pm\sqrt{8}.$$

Choose $x = -\sqrt{8} = -2\sqrt{2}$, since $x < 0$ in Quadrant III.

So, $\cos \theta = \dfrac{-2\sqrt{2}}{3}$

$$\tan \theta = \dfrac{-1}{-2\sqrt{2}} = \dfrac{1 \cdot \sqrt{2}}{2\sqrt{2} \cdot \sqrt{2}} = \dfrac{\sqrt{2}}{4}$$

$$\cot \theta = \dfrac{-2\sqrt{2}}{-1} = 2\sqrt{2}$$

$$\sec \theta = \dfrac{3}{-2\sqrt{2}} = \dfrac{3 \cdot \sqrt{2}}{-2\sqrt{2} \cdot \sqrt{2}} = -\dfrac{3\sqrt{2}}{4}$$

$$\csc \theta = \dfrac{3}{-1} = -3$$

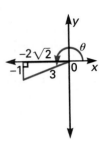

Classroom Exercises

Find the two quadrants which satisfy the given condition on θ.

1. $\cos \theta > 0$ **2.** $\sin \theta < 0$ **3.** $\tan \theta > 0$ **4.** $\sec \theta < 0$ **5.** $\cot \theta < 0$

Find the one quadrant which satisfies the given conditions on θ.

6. $\cos \theta < 0$ and $\sin \theta > 0$ **7.** $\sec \theta > 0$ and $\tan \theta > 0$

Find, to the nearest ten minutes, the two angle measures between 0 and 360 satisfying the equation.

8. $\sin \theta = 0.4723$ **9.** $\cot \theta = -3.1462$ **10.** $\sec \theta = 1.2385$

Written Exercises

Find all values of θ to the nearest ten minutes, $0 < \theta < 360$, for each given trigonometric ratio. Use a calculator to find the measure of the reference angle.

1. $\cos \theta = 0.9836$ **2.** $\tan \theta = -2.7725$

3. $\cot \theta = 1.2723$ **4.** $\sec \theta = 1.2283$

5. $\csc \theta = 1.4142$ **6.** $\sin \theta = -0.4410$

7. $\cos \theta = -0.3050$ **8.** $\sec \theta = 2.1900$

9. $\cot \theta = 3.6060$ **10.** $\tan \theta = 3.3400$

11. $\cot \theta = -9.5140$ **12.** $\sin \theta = -0.8832$

Find the value of each of the other five trigonometric ratios of angle θ. Simplify all radicals.

13. $\cos \theta = \frac{3}{5}$ and $\sin \theta > 0$ **14.** $\tan \theta = \frac{4}{3}$ and $\sin \theta < 0$

15. $\sin \theta = \frac{1}{5}$ and $\cot \theta < 0$

16. $\tan \theta = -\frac{2}{3}$ and $\sin \theta > 0$

17. $\cos \theta = -\frac{1}{5}$ and $\cot \theta > 0$

18. $\cos \theta = \frac{2}{\sqrt{5}}$ and $\sin \theta < 0$

19. $\sec \theta = 3$ and $\tan \theta < 0$

20. $\cos \theta = -\frac{1}{3}$ and $\csc \theta > 0$

In Exercises 21–22, simplify all radicals.

21. If $\cot \theta = \frac{8}{15}$ and $\cos \theta < 0$, evaluate $\sqrt{\dfrac{1 - \cos \theta}{17}}$.

22. If $\sec \theta = -\frac{5}{4}$ and $\tan \theta > 0$, evaluate

$$\frac{\tan 135 + \tan \theta}{\sin 30 \cos \theta + \sec 315 \sin 135} .$$

Mixed Review

1. $f(x) = x^2 - 2x + 4$, $g(x) = 3x - 1$. Find $f(g(x))$. **5.9**

2. Simplify. $27^{-\frac{2}{3}}$ **8.7**

3. Solve for x. **1.4**

$$-3(x - 2c) = 2 + a(bx + c)$$

4. Solve the following system.

$$3(x - 2) = 2y - 7$$
$$-2(y + 4) = -(x + 5) \quad \textbf{4.2, 4.3}$$

Application: *Tunneling*

Tunnels can be constructed by starting at opposite sides of an obstruction and meeting in the center. To make sure that the tunnel crews will meet at the center, an engineer can choose points A and B on each side of the obstruction. Then, the engineer must find a point C for which angle ACB is 90. By measuring BC and AC, the engineer can compute the measures of $\angle A$ and $\angle B$ of $\triangle ACB$. Using this information, workers starting at point A follow a line which makes the calculated angle with $\overline{AC}$. Workers at point B follow a path that makes the calculated angle with $\overline{BC}$.

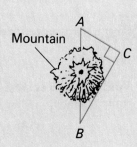

Exercises

1. Explain how the engineer in the example above uses trigonometric information to make sure that the two tunnel crews will meet.

2. If AC is 120 m and BC is 90 m, what are m $\angle A$ and m $\angle B$?

Key Terms

angle of rotation (p. 633)
angles of depression
  and elevation (p. 629)
cosecant (p. 623)
cosine (p. 620)
cotangent (p. 623)
coterminal angles (p. 635)
initial side (p. 633)

quadrantal angles (p. 633)
reference angle (p. 640)
reference triangle (p. 639)
secant (p. 623)
similar triangles (p. 620)
sine (p. 620)
tangent (p. 620)
terminal side (p. 633)

Key Ideas and Review Exercises

17.1,
17.2
To find a trigonometric ratio for an acute angle of a
right triangle, use one of the six basic trigonometric
ratios.

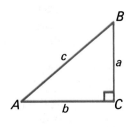

$$\sin A = \frac{\text{opp}}{\text{hyp}} \leftarrow \text{reciprocals} \rightarrow \csc A = \frac{\text{hyp}}{\text{opp}}$$

$$\cos A = \frac{\text{adj}}{\text{hyp}} \leftarrow \text{reciprocals} \rightarrow \sec A = \frac{\text{hyp}}{\text{adj}}$$

$$\tan A = \frac{\text{opp}}{\text{adj}} \leftarrow \text{reciprocals} \rightarrow \cot A = \frac{\text{adj}}{\text{opp}}$$

To find the measure of an acute angle of a right triangle, use one of the
ratios above with a table or a calculator.

Find each trigonometric ratio to four decimal places.

1. sin 42

2. tan 89

3. sec 39°20′

**Find θ to the nearest ten minutes if θ is an acute angle of a
right triangle.**

4. csc θ = 1.4530

5. tan θ = 3.5199

6. cos θ = 0.9765

Find each indicated ratio to four decimal places.

7. csc R; $s = 4$, $r = 2$

8. sin T; $s = 6$, $r = 2$

9. cot R; $t = \sqrt{3}$, $r = \sqrt{2}$

10. cos T; $r = \sqrt{7}$, $t = 3$

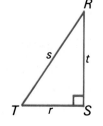

17.3 To find the measure of a side or acute angle of a right triangle, solve an equation involving a trigonometric ratio, the given data, and the missing part.

Find lengths to two significant digits and angle measures to the nearest degree. Use the figure for Exercises 7–10.

11. m $\angle T = 72$, $s = 53$,
$t =$ _____?_____ , $r =$ _____?_____

12. $s = 19$, $t = 13$,
m $\angle R =$ _____?_____ ,
m $\angle T =$ _____?_____

13. $r = 29$, m $\angle R = 42$,
$s =$ _____?_____ , $t =$ _____?_____

17.4, To find trigonometric ratios of angles of rotation, use the following
17.5, definitions, where $r = \sqrt{x^2 + y^2}$.
17.6

$\sin \theta = \frac{y}{r}$, $\cos \theta = \frac{x}{r}$, $\tan \theta = \frac{y}{x}$, $\cot \theta = \frac{x}{y}$, $\sec \theta = \frac{r}{x}$, $\csc \theta = \frac{r}{y}$

Find the six trigonometric ratios for the angle whose terminal side contains the given point. Simplify all radicals.

14. $P(-8,6)$ **15.** $P(4,-2)$ **16.** $P(3,-\sqrt{3})$ **17.** $P(-5,0)$

Find the six trigonometric ratios for each angle measure. If the ratio is undefined, so indicate.

18. 104   **19.** -48 **20.** -118 **21.** $450°50'$
22. 270   **23.** -180 **24.** 360   **25.** -270

17.7 To determine trigonometric ratios of the special angles 60, 30, and 45, refer to the top of p. 650.

Find each trigonometric ratio. Simplify all radicals.

26. $\sin 225$ **27.** $\tan 150$ **28.** $\sec 300$ **29.** $\csc (-135)$

17.8 To find all values of θ for a given ratio, $0 < \theta < 360$ (where θ is not quadrantal), or to find the remaining trigonometric ratios given two conditions, locate the reference triangle(s) and use the reference angle and values of x, y, and r.

Find all values of θ to the nearest ten minutes, $0 < \theta < 360$.

30. $\sin \theta = -0.4413$ **31.** $\cot \theta = 3.6058$

Find the value of each of the other five trigonometric ratios of θ. Simplify all radicals.

32. $\cos \theta = \frac{4}{5}$ and $\tan \theta < 0$ **33.** $\tan \theta = \frac{2}{3}$ and $\sin \theta < 0$

34. Write in your own words how to find the five remaining trigonometric ratios given one ratio and the sign of a second ratio.

Find each trigonometric ratio to four decimal places.

1. cos 79

2. csc 43°50′

3. tan 23°30′

Find m ∠A to the nearest ten minutes if ∠A is an acute angle of a right triangle.

4. sec A = 1.1575

5. tan A = 0.6445

6. sin A = 0.9686

Find each indicated trigonometric ratio to four decimal places. Use the figure at the right.

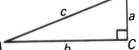

7. sin A; c = 8, b = 4

8. sec A; a = 6, b = 4

Find each length to two significant digits and each angle measure to the nearest degree. Use the figure above.

9. a = 24, m ∠A = 62, b = ___?___, c = ___?___

10. c = 8, a = 6, m ∠A = ___?___, m ∠B = ___?___

11. The angle of elevation to the top of a mountain measures 38. From a point 1,500 ft closer, the angle of elevation to the top of the mountain measures 49. Find the height of the mountain to two significant digits.

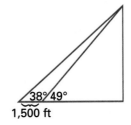

Find the six trigonometric ratios for the angle whose terminal side contains the given point. If the ratio is undefined, so indicate. Simplify all radicals.

12. $P(-4,6)$

13. $P(2, -2\sqrt{3})$

14. $(0, -8)$

Find the six trigonometric ratios for each angle measure.

15. 172

16. 220°20′

17. −65°30′

Find each trigonometric ratio. Simplify all radicals.

18. tan 90

19. sin(−180)

20. csc(−90)

21. cos 225

Find the other five trigonometric ratios of angle θ. Simplify all radicals.

22. sin $\theta = \frac{5}{13}$ and cos $\theta < 0$

23. tan $\theta = \frac{1}{3}$ and sin $\theta < 0$

24. If $\dfrac{\tan \theta_1 + \tan \theta_2}{1 - \tan \theta_1(\tan \theta_2)}$ = tan 45 and tan $\theta_1 = \frac{2}{3}$, find tan θ_2.

In each item, you are to compare a quantity in Column 1 with a quantity in Column 2. Write the letter of the correct answer from these choices.

A—The quantity in Column 1 is greater than the quantity in Column 2.
B—The quantity in Column 2 is greater than the quantity in Column 1.
C—The quantity in Column 1 is equal to the quantity in Column 2.
D—The relationship cannot be determined from the given information.

NOTE: Information centered over both columns refers to one or both of the quantities to be compared.

	Column 1	Column 2
1.	$\frac{3}{5}$ of 5	$\frac{5}{3}$ of 3
2.	$\frac{2}{3}$ of 18	$\frac{3}{2}$ of 8
3.	$x > 1$	
	$\dfrac{x + x + x + x}{x \cdot x \cdot x}$	$\dfrac{4}{x^3}$
4.	$x \neq \pm 3$	
	$\dfrac{x^2 + 6x + 9}{x + 3}$	$\dfrac{x^2 - 9}{x - 3}$
5.	Ratio of $\frac{1}{3}$ to $\frac{1}{5}$	Ratio of $\frac{2}{5}$ to $\frac{1}{3}$
6.	Ratio of $\frac{1}{4}$ to $\frac{4}{1}$	Ratio of $\frac{3}{1}$ to $\frac{1}{3}$
7.	$a > 0$	
	$\dfrac{1}{a}$	a
8.	m	$\dfrac{3m + 1}{3}$
9.	$\sqrt{48} + \sqrt{80}$	$7 + 9$

	Column 1	Column 2
10.	$\sin \theta$	1.2
11.	$\tan 45$	$\tan 315$
12.	$\sin 140$	$\cos 140$
13.	$\sin 150$	$\cos 240$
14.	$\csc 315$	$\sec(-45)$
15.		

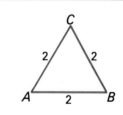

	Column 1	Column 2
	Area of $\triangle ABC$	2

16. $PR = 4$, $PQ = 4$, and $QR = 4\sqrt{2}$

Column 1	Column 2
x	45

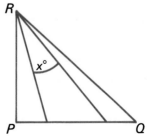

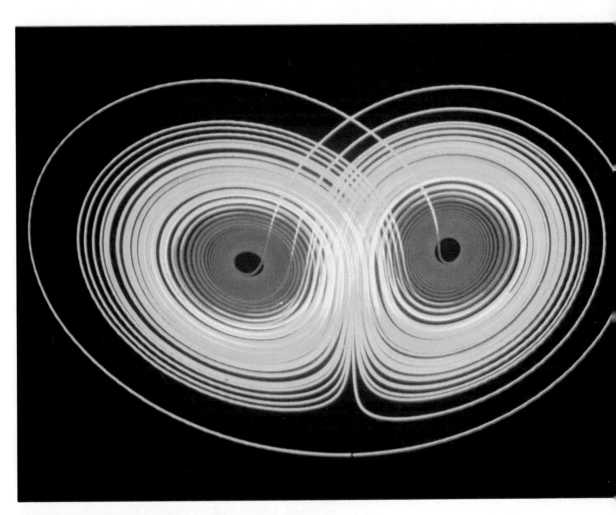

Scientists studying turbulent systems have found that there is often a surprisingly simple mathematical principle at work. The object pictured here is an example of a *strange attractor*—a graph which reveals the underlying mathematical structure of an otherwise unpredictable or "chaotic" system.

18.1 The Law of Cosines

To use the Law of Cosines to find the length of a side or the measure of an angle of a triangle

To solve problems using the Law of Cosines

It is sometimes possible to use trigonometric ratios to find the length of a side of a triangle, even if it is not a right triangle. For example, if the measures of two sides of a triangle and the included angle are known, it is possible to find the length of the third side.

In the nonright, or **oblique**, triangle at the right the measures of two sides and the included angle are 8, 10, and 58 respectively. Note (Figure 2) that h is the length of an altitude that divides triangle ABC into two right triangles, I and II. The Pythagorean relation can be used to find the length a.

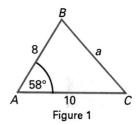

Figure 1

Triangle I: $8^2 = h^2 + x^2$, or $64 = h^2 + x^2$

Triangle II: $a^2 = h^2 + (10 - x)^2$, or

$$a^2 = h^2 + 100 - 20x + x^2$$

Subtract the equation for Triangle I from the equation for Triangle II.

$$a^2 = h^2 + 100 - 20x + x^2$$
$$\underline{64 = h^2 \qquad\qquad + x^2}$$

Figure 2

$$a^2 - 64 = 100 - 20x$$
$$a^2 = 100 + 64 - 20x$$
$$a^2 = 100 + 64 - 20(8 \cos 58) \leftarrow \text{from the diagram,} \frac{x}{8} = \cos 58$$
$$a^2 = \underline{10^2 + 8^2} - 2 \cdot \underline{10 \cdot 8} \cdot \cos 58$$

           sides   sides   included angle

This pattern suggests the following theorem, called the *Law of Cosines*.

Theorem 18.1

Law of Cosines: For any triangle ABC,
$$a^2 = b^2 + c^2 - 2bc \cos A$$
$$b^2 = a^2 + c^2 - 2ac \cos B$$
$$c^2 = a^2 + b^2 - 2ab \cos C$$

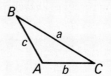

When the Law of Cosines is applied, it is important to use the correct sign for the cosine of the angle, as shown in Example 1.

EXAMPLE 1 In triangle ABC, $a = 4.0$, $b = 6.0$, and m $\angle C = 130$. Find c to two significant digits.

Plan A calculator can be used to deter-mine the cosine of 130, including the correct sign. If a table is used, recall the following.

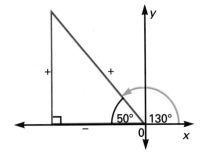

$$\cos 130 = -\cos (180 - 130)$$
$$= -\cos 50$$

This is illustrated in the reference triangle at the right.

Solution $\cos 130 = -\cos 50 = -0.6428 \leftarrow$ from the table on page 796
Use the appropriate form of the Law of Cosines.

$$c^2 = a^2 + b^2 - 2ab \cos C$$
$$c^2 = 4^2 + 6^2 - 2 \cdot 4 \cdot 6 \cdot \cos 130$$
$$c^2 = 16 + 36 - 48(-0.6428)$$
$$c^2 = 52 + 30.8544 = 82.8544$$
$$c = \sqrt{82.8544} = 9.1024392$$

Therefore, $c = 9.1$ to two significant digits.

EXAMPLE 2 Two airplanes leave the same airport at the same time. After completing their take-offs, the planes assume flight paths that form an angle of 55°. Their speeds are 450 mi/h and 600 mi/h. How far apart, to two significant digits, are the planes after 2 h?

Plan Draw and label a diagram of the flight paths. After 2 h the plane flying at 450 mi/h had flown 900 mi; the other plane had flown 1,200 mi. Use the Law of Cosines to find d.

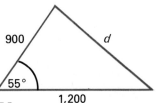

Solution $$d^2 = 900^2 + 1,200^2 - 2 \cdot 900 \cdot 1,200 \cdot \cos 55$$
$$d^2 = 810,000 + 1,440,000 - 2,160,000(0.5736) \leftarrow \text{from the table}$$
$$d^2 = 2,250,000 - 1,238,976 = 1,011,024$$
$$d = \sqrt{1,011,024} = 1005.4969, \text{ or } 1000 \text{ mi to 2 significant digits}$$

The Law of Cosines can also be used to find the measure of any angle of a triangle if the lengths of the three sides are known.

EXAMPLE 3

The lengths of the three sides of a triangular lot are 40 m, 50 m, and 80 m. Find, to the nearest degree, the measure of the largest angle of the triangular lot.

Plan

Draw and label a diagram. The largest angle is A, opposite the longest side, $a = 80$. Use the form of the Law of Cosines that involves cos A. Then solve for cos A.

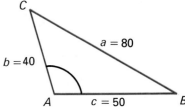

Solution

$$a^2 = b^2 + c^2 - 2bc \cos A$$
$$2bc \cos A = b^2 + c^2 - a^2$$
$$\cos A = \frac{b^2 + c^2 - a^2}{2bc}$$
$$= \frac{40^2 + 50^2 - 80^2}{2 \cdot 40 \cdot 50} = \frac{-2,300}{4,000} = -0.5750$$

Since cos $A < 0$, angle A must be obtuse. The reference triangle for angle A is illustrated at the right. Find the measure of the reference angle $180 - A$ as follows.

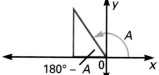

$$\cos (180 - A) = 0.5750$$
$$180 - A \approx 55 \leftarrow \text{from the table}$$
$$A \approx 125$$

The equation cos $A = -0.5750$ can also be solved using a calculator, as shown below.

Calculator Steps	Display
0.5750 $\boxed{+/-}$ $\boxed{\text{INV}}$ $\boxed{\text{cos}}$	125.09963

Thus, the measure of the largest angle is 125 to the nearest degree.

◢ *Summary*

The Law of Cosines can be used
(1) to find the measures of the angles of a triangle if the lengths of three sides are known (SSS), and
(2) to find the length of the third side of a triangle if the lengths of two sides and the measure of the included angle are known (SAS).

Classroom Exercises

Use the Law of Cosines to find an equation involving *x* or *A*.

1.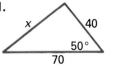

2.

3.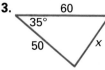

4.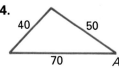

Written Exercises

In Exercises 1–4, refer to triangle *ABC* at the right. Find the lengths to two significant digits and the angle measures to the nearest degree.

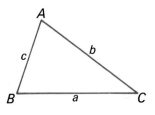

1. $a = 40$, $b = 50$, m $\angle C = 40$, $c = $ __?__
2. $a = 12$, $c = 14$, m $\angle B = 140$, $b = $ __?__
3. $a = 30$, $b = 70$, $c = 80$, m $\angle A = $ __?__
4. $a = 8.0$, $b = 4.1$, $c = 9.0$, m $\angle B = $ __?__

For Exercises 5–10, find lengths to two significant digits and angle measures to the nearest degree.

5. After two airplanes left the same airport at the same time, their flight paths formed an angle measuring 125. The first flew at 550 mi/h and the second flew at 620 mi/h. How far apart were they after 3 h?

6. A baseball diamond forms a square 90 ft on a side. The pitcher's mound is 60 ft from home plate. How far is it from the mound to third base?

7. The sides of an isosceles triangle have lengths 18, 18, and 10. Find the measure of the smallest angle of the triangle.

8. The length of the radius of a circle is 10. Two radii, $\overline{OA}$ and $\overline{OB}$, form an angle of measure 109°30′. Find the length of chord $\overline{AB}$.

9. A triangular lot has side lengths of 16 m, 26 m, and 38 m. Find the measure of the largest angle of the lot.

10. The diagonals of a parallelogram bisect each other. If their lengths are 8.0 and 10 and they intersect at an angle of 20°40', how long are the sides?

11. Prove the Law of Cosines (Theorem 18.1). Consider two cases. In the first case, no angle is obtuse. In the second case, one angle is obtuse.

12. Show that the Pythagorean relation, $a^2 + b^2 = c^2$, is a special case of the Law of Cosines.

Mixed Review

Write each trigonometric ratio in simplest radical form. *17.7, 17.8*

1. sin 270

2. tan 90

3. cos 135

4. csc 240

18.2 The Area of a Triangle

To find areas of triangles
To find areas of regular polygons

Recall that the area K of a triangle is half the product of the length of a side and the length of an altitude upon the side.

$$K = \tfrac{1}{2}bh$$

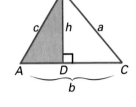

This formula can be used to derive another formula for the area of a triangle.

Write a trigonometric ratio. $\dfrac{h}{c} = \sin A$

Solve for h. $h = c \sin A$

Substitute for h in the area formula. $K = \tfrac{1}{2}b(c \sin A)$

Area of a Triangle: The area K of a triangle is one-half the product of the lengths of any two sides and the sine of the included angle.

For any triangle ABC,

$K = \tfrac{1}{2}bc \sin A,$

$K = \tfrac{1}{2}ac \sin B,$ and

$K = \tfrac{1}{2}ab \sin C$

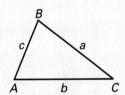

You will be asked to prove Theorem 18.2 for an obtuse included angle in Exercise 17.

EXAMPLE 1

In $\triangle ABC$, $a = 8$, $b = 10$, and m $\angle C = 120$.
Find the area of $\triangle ABC$ in simplest radical form.

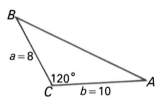

Plan

Use $K = \tfrac{1}{2}ab \sin C$.

The reference angle for 120 is one of the special angles.

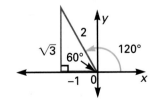

Solution

$K = \tfrac{1}{2}ab \sin C$

$= \tfrac{1}{2} \cdot 8 \cdot 10 \sin 120 = 40 \cdot \dfrac{\sqrt{3}}{2}$, or $20\sqrt{3}$

EXAMPLE 2   Find, to two significant digits, the area of a regular pentagon inscribed in a circle of radius 6.0.

Plan   Find the area K of one of the isosceles triangles in the figure at the right. Then find $5K$.

Since one complete circular rotation measures 360, each angle formed by a pair of consecutive radii has a measure of $\frac{360}{5}$, or 72.

Solution

$K = \frac{1}{2} \cdot 6 \cdot 6 \cdot \sin 72$

$\quad = 18(0.9511) = 17.1198 \leftarrow$ from the table

$5K = 5 \cdot 17.1198 = 85.5990$

The calculator steps for this example are shown below.

$0.5 \; \boxed{\times} \; 6 \; \boxed{\times} \; 6 \; \boxed{\times} \; 72 \; \boxed{\sin} \; \boxed{\times} \; 5 \; \boxed{=} \; 85.5950867$

Therefore, the area of the pentagon is 86 to two significant digits.

If the lengths of the three sides of a triangle are known, then its area can be found by using *Heron's Formula*.

Heron's Formula for the Area of a Triangle

Let $s = \frac{1}{2}(a + b + c)$, half the perimeter, or *semiperimeter* of a triangle of area K. Then $K = \sqrt{s(s - a)(s - b)(s - c)}$.

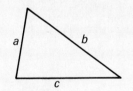

EXAMPLE 3   Find the area of a triangular lot with sides measuring 80 m, 50 m, and 70 m. Write the answer in simplest radical form.

Plan   Find the semiperimeter. Then use Heron's Formula.

Solution

$s = \frac{1}{2}(80 + 50 + 70) = 100$

$K = \sqrt{s(s - a)(s - b)(s - c)}$

$\quad = \sqrt{100(100 - 80)(100 - 50)(100 - 70)}$

$\quad = \sqrt{100 \cdot 20 \cdot 50 \cdot 30} = 1{,}000\sqrt{3}$

Therefore, the area is $1{,}000\sqrt{3}$ m^2 in simplest radical form.

Classroom Exercises

Choose an appropriate formula for finding the area of each triangle.

1. $a = 8, c = 10,$ m $\angle B = 28$ **2.** $a = 6, b = 7, c = 10$

3. $a = 9, b = 10,$ m $\angle C = 35$ **4.** $b = 9, a = 13, c = 6$

5-8. Find the area of each triangle of Classroom Exercises 1-4.

Written Exercises

Find the area of triangle ABC to two significant digits.

1. $a = 18, c = 24,$ m $\angle B = 42$ **2.** $b = 10, c = 14,$ m $\angle A = 75$

3. $c = 9.0, b = 4.0,$ m $\angle A = 67$ **4.** $a = 14, b = 16,$ m $\angle C = 60$

5. $c = 20, b = 10,$ m $\angle A = 110$ **6.** $a = 50, c = 80,$ m $\angle B = 125$

The lengths of the sides of a triangle are given. Find the area of the triangle in simplest radical form.

7. 4, 6, 8   **8.** 5, 12, 10   **9.** 5, 7, 10   **10.** 2, 4, 4

Find, to two significant digits, the area of each regular polygon inscribed in a circle of the given radius.

11. octagon (8 sides), radius 12   **12.** decagon (10 sides), radius 4.0

13. hexagon (6 sides), radius 6.0   **14.** dodecagon (12 sides), radius 10

15. Find, in simplest radical form, the area of an equilateral triangle if each side has a length of 8. Use the formula $K = \frac{1}{2}bc \sin A$, and check your answer using Heron's Formula.

16. Find, in simplest radical form, the area of an isosceles triangle if each leg has a length of 6 and the angle formed by the legs measures 120.

17. Show that if angle A is an obtuse angle of triangle ABC, then the area of the triangle is given by $K = \frac{1}{2}bc \sin A$.

18. The length of each leg of an isosceles triangle is 8. Its area is $4\sqrt{15}$ square units. Find the length of the third side.

19. For triangle ABC, $AB = 15$, $BC = 20$, and the area is 150 square units. Find the area of each of the two triangles formed by the altitude from B to $\overline{AC}$.

20. The measure of the angle formed by two congruent sides of a triangle is 120. The area is $48\sqrt{3}$ square units. Find the length of the base in simplest radical form.

Mixed Review

Graph each equation of a conic section.   *12.1-12.3, 12.5*

1. $x^2 + y^2 = 100$ **2.** $4x^2 + 9y^2 = 36$ **3.** $16x^2 - 25y^2 = 400$ **4.** $xy = 48$

18.3 The Law of Sines

Objectives
To find a length of a side of a triangle using the Law of Sines
To solve problems using the Law of Sines

In this lesson, a law is developed that allows the missing measures of a triangle to be found if two angles and a side are known (AAS or ASA). The area of a triangle (Lesson 18.2) can be found using $K = \frac{1}{2}bc \sin A$, or $K = \frac{1}{2}ac \sin B$, or $K = \frac{1}{2}ab \sin C$.

Therefore, $\frac{1}{2}bc \sin A = \frac{1}{2}ac \sin B = \frac{1}{2}ab \sin C$.

Multiply each part by 2. $bc \sin A = ac \sin B = ab \sin C$

Divide each part by abc. $\dfrac{bc \sin A}{abc} = \dfrac{ac \sin B}{abc} = \dfrac{ab \sin C}{abc}$

The result is $\dfrac{\sin A}{a} = \dfrac{\sin B}{b} = \dfrac{\sin C}{c}$.

Theorem 18.3

Law of Sines: For any triangle ABC,

$$\frac{\sin A}{a} = \frac{\sin B}{b} = \frac{\sin C}{c}.$$

EXAMPLE 1

In triangle ABC, $a = 12$, m $\angle A = 50$, and m $\angle C = 44$. Find c to two significant digits.

Plan
Use $\dfrac{\sin A}{a} = \dfrac{\sin C}{c}$, since m $\angle A$ and m $\angle C$ are given.

Solution
$\dfrac{\sin 50}{12} = \dfrac{\sin 44}{c}$

$(\sin 50)c = 12 \sin 44$

$c = \dfrac{12 \sin 44}{\sin 50} = \dfrac{12(0.6947)}{0.7660} = 10.882$

Therefore, the value of c is 11 to two significant digits.

In the Law of Sines each ratio in the proportion involves a side and the sine of the angle *opposite* that side. Therefore, when the measures of a side and two angles are given but neither angle is opposite the given side, it is necessary first to find the measure of the third angle.

EXAMPLE 2   In triangle ABC, $b = 15$, m $\angle A = 65$, and m $\angle C = 70$. Find c to two significant digits.

Plan   A proportion involving b must also include $\sin B$. To find m $\angle B$, use the fact that the sum of the measures of the angles of a triangle is 180.

Solution   m $\angle B + 65 + 70 = 180$
        m $\angle B = 45$
Apply the Law of Sines to find c.

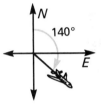

$$\frac{\sin 45}{15} = \frac{\sin 70}{c}$$

$(\sin 45)c = 15 \sin 70$

$$c = \frac{15 \sin 70}{\sin 45} = \frac{15(0.9397)}{0.7071} = 19.934$$

Therefore, $c = 20$ to two significant digits.

In ocean or air navigation, it is customary to measure angles of rotation clockwise from a north axis. In the diagram at the right, the plane is said to be flying at a **bearing** of 140.

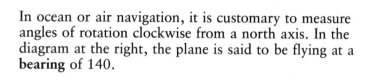

EXAMPLE 3   A ship is sailing due north. The captain observes that the bearing of a lighthouse is 40. After sailing 60 km, the captain sees that the bearing of the lighthouse has become 135. How far, to two significant digits, is the ship from the lighthouse now?

Plan   Draw and label a figure.
Find m $\angle A$ and m $\angle B$ of the triangle.

Solution   m $\angle A = 180 - 135 = 45$
m $\angle B = 180 - (45 + 40) = 95$
Apply the Law of Sines to find c.

$$\frac{\sin 95}{60} = \frac{\sin 40}{c}$$

$(\sin 95)c = 60 \sin 40$

$$c = \frac{60 \sin 40}{\sin 95}$$

$$= \frac{60(0.6428)}{0.9962} = 38.715 \leftarrow \text{Use the table or a calculator.}$$

So, the ship is 39 km from the lighthouse to two significant digits.

EXAMPLE 4   The angle of elevation to the top of a mountain from a point P on the ground is $24°10'$. The angle of elevation from a point Q directly in line with P and 1,350 ft closer is $61°40'$. Find the height h of the mountain to three significant digits.

Plan   Draw and label a figure. First find m $\angle 1$ and m $\angle 2$.

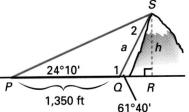

Solution
$$m \angle 1 = 180 - 61°40' = 118°20'$$
$$m \angle 2 = 180 - (24°10' + 118°20')$$
$$= 37°30'$$

(1) Use the Law of Sines to find a.

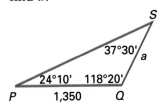

$$\frac{\sin 37°30'}{1,350} = \frac{\sin 24°10'}{a}$$

$$(\sin 37°30')a = 1,350 \sin 24°10'$$

$$a = \frac{1,350(\sin 24°10')}{\sin 37°30'}$$

$$= \frac{1,350(0.4094)}{0.6088}$$

$$= 907.8351$$

(2) Use right triangle QRS to find h.

$$\frac{h}{a} = \sin 61°40'$$

$$\frac{h}{907.8351} = \sin 61°40'$$

$$h = 907.8351(0.8802)$$

$$= 799.0765$$

The height of the mountain is 799 ft to three significant digits.

Classroom Exercises

Give a proportion for finding the indicated length. Do not find the length.

1. m $\angle A = 23$, m $\angle B = 75$, $b = 14$, $a = $ ___?___

2. m $\angle B = 25$, $b = 24$, m $\angle C = 89$, $c = $ ___?___

3. $a = 18$, m $\angle C = 45$, m $\angle A = 89$, $c = $ ___?___

4. $b = 14$, m $\angle B = 62$, m $\angle A = 13$, $a = $ ___?___

5. $a = 110$, m $\angle A = 5$, m $\angle B = 44$, $b = $ ___?___

6. m $\angle C = 100$, m $\angle B = 46$, $b = 23$, $c = $ ___?___

7–10. Find each indicated length to two significant digits in Classroom Exercises 1–4.

Written Exercises

For triangle *ABC*, find the indicated length to two significant digits.

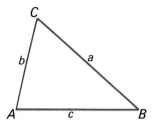

1–2. Use the data for Classroom Exercises 5–6.

3. m ∠A = 59, m ∠B = 63, b = 15, c = ___?___

4. a = 26, m ∠A = 78, m ∠C = 18, b = ___?___

5. m ∠C = 6, m ∠A = 38, c = 18, b = ___?___

6. b = 14, m ∠C = 110, m ∠B = 54, a = ___?___

For Exercises 7–12, find the indicated measure to two significant digits, unless otherwise indicated.

7. The distance between Towns *A* and *B* is 56 mi. The angle formed by the road between Towns *A* and *B* and the road between Towns *A* and *C* measures 46. The angle formed by $\overline{AB}$ and $\overline{BC}$ measures 115. Find the distance between Town *B* and Town *C*.

8. On a ship sailing north, a woman notices that a hotel on shore has a bearing of 20. A little while later, after having sailed 40 km, she observes that the bearing of the hotel is now 100. How far is the ship from the hotel now?

9. A ship is steaming south. The navigator notices that the bearing of a lighthouse is 120. After moving 8.0 mi/h for 2 h, he observes that the bearing of the lighthouse is 25. Find his distance from the lighthouse at the time of the second sighting.

10. Bill determines that the angle of elevation to the top of a building measures 40°30′. If he walks 102 ft closer to the building, the measure of the new angle of elevation will be 50°20′. Find the height of the building to three significant digits.

11. To determine the distance *AB* across a steep canyon, Megan walks 600 yd from *B* to another point, *C*. She then finds that m ∠ACB = 35 and m ∠CBA = 106. Find *AB*.

12. Find the area of a regular pentagon if each side has a length of 12.

13. Show that if *K* is the area of triangle *ABC*, then $K = a^2 \left(\dfrac{\sin B \sin C}{2 \sin A} \right)$.

Use the figure at the right to answer Exercises 14 and 15.

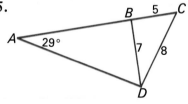

14. Find *AD* and *AB* to three significant digits.

15. Find the area of triangle *ACD*.

Mixed Review

Simplify. *7.1, 7.4, 8.4, 9.2*

1. $\dfrac{x^3 - 27}{x^2 - 9}$

2. $\dfrac{5}{x^2 - 7x + 12} - \dfrac{8}{3 - x}$

3. $\dfrac{6}{\sqrt{12}}$

4. $\dfrac{3 + i}{4 - i}$

18.4 The Ambiguous Case: Solving General Triangles

Objectives

To solve a triangle using the Law of Sines, given the measures of two sides and a nonincluded angle (SSA)

To determine whether one, two, or no triangles can be constructed when given the measures of two sides and a nonincluded angle

To use the Law of Sines and the Law of Cosines to solve triangles

In a triangle, if the measures of two sides and an angle opposite one of them are given (SSA), a value can be determined for h, the altitude to the unknown side of the triangle.

Suppose that m $\angle A$ and the lengths of sides a and b are known in the figure below. Using the sine ratio, the value of h can be found.

$$\sin A = \frac{h}{b}$$
$$h = b \sin A$$

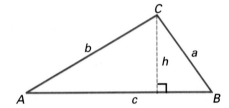

Notice that the length of side a must be greater than or equal to h. If it is not, no triangle can be formed using the given angle and sides.

The SSA case is often called the **ambiguous case**, since the number of possible triangles that can be constructed may be as few as zero and as many as two. All possibilities for constructing such triangles are summarized on this and the next page. In each possibility, the measure of $\angle A$ and the lengths of sides a and b are known.

Case 1

Angle A is an *acute* angle.

(1) If $a < h$ ($a < b \sin A$), then no triangle can be constructed.

(2) If $a = h$ ($a = b \sin A$), then one right triangle can be constructed.

(3) If $a > h$ ($a > b \sin A$) and $a < b$, then two triangles can be constructed.

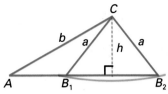

(4) If $a \geq b$, then one triangle can be constructed.

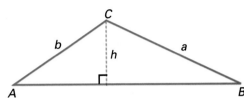

Case 2 Angle A is an *obtuse* or *right* angle.

(5) If $a \leq b$, then no triangle can be constructed.

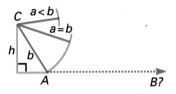

(6) If $a > b$, then one triangle can be constructed.

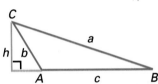

When the Law of Sines is used to find the measure of an angle for the ambiguous case, use the guidelines above to determine the number of solutions.

EXAMPLE 1 In triangle ABC, $a = 5$, $b = 20$, and m $\angle A = 40$. Find m $\angle B$ to the nearest degree.

Solution $h = b \sin A = 20(0.6428) = 12.856$
Since $a = 5$, $a < h$. Therefore, *no triangle* can be formed.

If we try to use the Law of Sines, we get an *impossible result*.

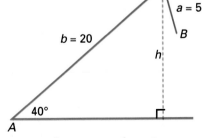

$$\frac{\sin A}{a} = \frac{\sin B}{b}$$

$$\frac{\sin 40}{5} = \frac{\sin B}{20}$$

$$5 \sin B = 20 \sin 40$$

$$\sin B = 4 \sin 40$$

$$= 4(0.6428) = 2.5712$$

There is *no solution*, since the sine ratio cannot be greater than 1.

Notice that the expression for sin B in Example 1 is as follows.

$$\sin B = \frac{b \sin A}{a}$$

Substituting h for $b \sin A$ in the numerator gives the following.

$$\sin B = \frac{h}{a}$$

Therefore, if $\sin B \le 1$, then $a \ge h$. So, it is not necessary to test if $a \ge h$ before applying the Law of Sines to find sin B.

EXAMPLE 2   In triangle ABC, m $\angle A = 40$, $a = 9.0$, $b = 12$. *Solve* the triangle. (Find the measures of all the parts of the triangle.) Find lengths to two significant digits and angle measures to the nearest degree.

Solution

$$\frac{\sin 40}{9} = \frac{\sin B}{12}$$

$$9 \sin B = 12 \sin 40$$

$$\sin B = \frac{12(0.6428)}{9}$$

$$\sin B = 0.8571$$

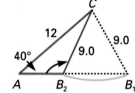

Since sin B < 1, a > h.
Therefore, two triangles can be constructed.
To the nearest degree, either m $\angle B = 59$, *or*
m $\angle B = 180 - 59 = 121$ (a Quadrant-II angle; see the illustration at the right).

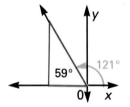

Next, find the values of m $\angle C$ and of c.

m $\angle C = 180 - 40 - 59$	m $\angle C = 180 - 40 - 121$
$= 81$	$= 19$
$c = \dfrac{a \sin C}{\sin A}$	$c = \dfrac{a \sin C}{\sin A}$
$= \dfrac{9 \sin 81}{\sin 40}$	$= \dfrac{9 \sin 19}{\sin 40}$
$= \dfrac{9(0.9877)}{0.6428}$	$= \dfrac{9(0.3256)}{0.6428}$
$= 13.8290 \approx 14$	$= 4.5584 \approx 4.6$

Thus, m $\angle B = 59$, m $\angle C = 81$, and $c = 14$.

Thus, m $\angle B = 121$, m $\angle C = 19$, and $c = 4.6$.

EXAMPLE 3   In triangle ABC, m $\angle A = 70$, $b = 2.0$, $a = 6.0$. Find the number of triangles that can be constructed.

Solution   Since $a > b$, there is only one solution.

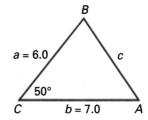

6 > 2: only *one* solution

Sometimes it is necessary to use both the Law of Sines and the Law of Cosines to solve a triangle.

For example, in triangle ABC at the right, $a = 6.0$, $b = 7.0$ and m $\angle C = 50$. How can you solve the triangle?

Note that the Law of Sines cannot be used directly, but the following sequence of steps will work.

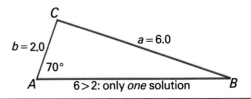

1. Since the pattern is SAS, use the Law of Cosines to find c.
2. Use the Law of Sines to find the smaller of the remaining angle measures.
3. Use the fact that the sum of the measures of the angles of a triangle is 180 to find the third angle measure.

Classroom Exercises

How many triangles can be constructed?

1. $a = 32$, $b = 21$, m $\angle A = 50$
2. $b = 14$, m $\angle A = 55$, $a = 12$ and $a > b \sin A$
3. $b = 8$, m $\angle A = 30$, $a = 4$
4. m $\angle A = 160$, $a = 20$, $b = 60$
5. m $\angle B = 70$, $b = 12$, $a = 8$

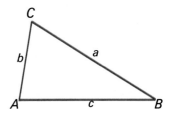

Determine which of the two laws—the Law of Sines or the Law of Cosines—can be used to find the indicated measure. If both laws need to be used, indicate which must be used first.

6. $b = 15$, $c = 14$, m $\angle A = 40$, $a =$ ___?___
7. $b = 7$, $a = 10$, m $\angle A = 80$, m $\angle B =$ ___?___
8. $a = 4$, $b = 5$, m $\angle C = 40$, m $\angle A =$ ___?___
9–11. Find the indicated measure in each of Classroom Exercises 6–8 above.

Written Exercises

In each triangle *ABC*, find the number of triangles that can be constructed. If the answer is "one" or "two," then find the indicated angle measure(s) to the nearest degree.

1. m $\angle A$ = 62, b = 14, a = 25,
    m $\angle B$ = __?__

2. m $\angle A$ = 65, b = 25, a = 20,
    m $\angle B$ = __?__

3. m $\angle A$ = 85, b = 30, a = 6,
    m $\angle B$ = __?__

4. m $\angle A$ = 100, b = 10, a = 25,
    m $\angle B$ = __?__

5. m $\angle C$ = 65, b = 14, c = 14,
    m $\angle B$ = __?__

6. m $\angle B$ = 57, a = 19, b = 3,
    m $\angle A$ = __?__

7. m $\angle B$ = 66, a = 14, b = 11,
    m $\angle A$ = __?__

8. m $\angle C$ = 43, b = 29, c = 26,
    m $\angle B$ = __?__

Solve triangle *ABC*, if possible. Find lengths to two significant digits and angle measures to the nearest degree.

9. a = 13, c = 110, m $\angle A$ = 45

10. m $\angle A$ = 42, a = 19, c = 11

11. b = 24, c = 22, m $\angle A$ = 80

12. a = 10, b = 12, m $\angle C$ = 65

13. a = 19, c = 12, m $\angle A$ = 35

14. a = 10, b = 12, c = 20

15. A triangular piece of property has sides with lengths of 40 m, 30 m, and 60 m. Find, to the nearest degree, the measure of each angle of the triangle.

16. The measures of two angles of a triangle are 50 and 70. The longest side is 6.0 ft longer than the shortest side. Find the lengths of the three sides to two significant digits.

17. In triangle *ABC*, m $\angle A$ = 40, b = 8.0, and a = 10. Find the area of the triangle.

18. The lengths of two sides of a parallelogram are 16 and 20. One angle of the parallelogram measures 120. Find, to the nearest degree, the measure of the angle formed by the shorter side and the longer diagonal.

19. The lengths of the three sides of a triangle are 8.0, 10, and 12. Find, to the nearest degree, the measure of the angle formed by the shortest side and the median to the longest side.

Mixed Review

Solve each equation. *6.6, 9.5, 10.4, 13.3*

1. $x^3 - 7x = 6$ **2.** $x^2 - 2x + 6 = 0$ **3.** $x^3 - 27 = 0$ **4.** $\log_2 (x^2 - 2x) = 3$

18.5 Radian Measure

Objectives

To convert radian measure to degree measure and vice versa
To find trigonometric ratios for radian angle measures

Until now, angles of rotation have been measured in terms of degrees. Another way of measuring angles is suggested by the development below.

In the circle at the right, the center is located at O. The arc of the *central angle AOB* is equal in length to the radius r. Angle AOB is said to have a measure of one *radian*.

In this lesson, we will use degree symbols (°) to distinguish between *degree* measurements and *radian* measurements. After this lesson, you may again assume that an angle measure is given in degrees, unless it is in terms of π or unless otherwise indicated.

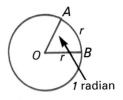

Definition

> If an angle is a central angle of a given circle and intercepts an arc of length equal to the length of a radius of the circle, then the measure of the angle is defined to be 1 **radian**.

The formula for the circumference of a circle is $C = 2\pi r$, or $C \approx 2(3.14)r$, or $C \approx 6.28r$.

In other words, there are approximately 6.28 *radii* in the circumference of a circle.

Therefore, the number of radians in a central angle that intercepts the entire circle is 2π, or approximately 6.28.

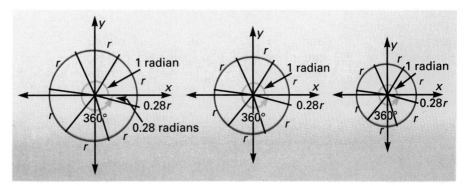

In each case above, regardless of the length of the radius, a rotation of 360° has a *radian* measure of 6.28, or 2π. When the measure of an angle is given in radians, it is customary to omit the word "radians."

The relationship $360° = 2\pi$ radians is used to convert between radians and degrees.

EXAMPLE 1   Express $-200°$ in radian measure in terms of π.

Solution $360° = 2\pi$ radians.

Therefore, $1° = \dfrac{\pi}{180}$ radians. Multiply each side by -200.

$$-200(1°) = -200 \cdot \dfrac{\pi}{180} \text{ radians} = \dfrac{-10\pi}{9} \text{ radians}$$

Therefore, $-200°$ in radian measure is $-\dfrac{10\pi}{9}$.

EXAMPLE 2   Express $\dfrac{4\pi}{3}$ in degree measure.

Solution 2π radians $= 360°$

Therefore, π radians $= 180°$. Then $\dfrac{4\pi}{3}$ radians $= \dfrac{4(180°)}{3} = 240°$.

Therefore, $\dfrac{4\pi}{3}$ radians $= 240°$.

The degree measure of one radian can be found by using
$$\pi \text{ radians} = 180°.$$

Divide each side by π.   1 radian $= \dfrac{180°}{\pi}$ ← Some calculators have a key for entering π.

$$1 \text{ radian} \approx \dfrac{180°}{3.1415927} \approx 57.29578$$

1 radian $= 57.3°$, to the nearest tenth of a degree.

EXAMPLE 3   Express the given radian measure in degrees, to the nearest degree.

a. 3   **b.** -2.6

Solutions   **a.**   1 radian $\approx 57.3°$ **b.**   1 radian $\approx 57.3°$
                  $3(1 \text{ radian}) \approx 3(57.3°)$ $-2.6(1 \text{ radian}) \approx -2.6(57.3°)$
                  $3 \text{ radians} \approx 171.9°$ $-2.6 \text{ radians} \approx -148.98°$

Thus, 3 radians $\approx 172°$ and -2.6 radians $\approx -149°$.

Radian/Degree Equivalencies

When changing degrees to radians, use

$$1° = \dfrac{\pi}{180} \text{ radians}.$$

When changing radians to degrees, use

π radians $= 180°$, or 1 radian $\approx 57.3°$.

Some frequently used angle measures are represented in both degrees and radians at the right.

degrees	30	45	60	90	360
radians	$\dfrac{\pi}{6}$	$\dfrac{\pi}{4}$	$\dfrac{\pi}{3}$	$\dfrac{\pi}{2}$	2π

EXAMPLE 4   Find the value of each trigonometric ratio in simplest radical form.

   a. $\sin \dfrac{4\pi}{3}$

   b. $\tan \dfrac{3\pi}{2}$

Solutions   **a.** $\sin \dfrac{4\pi}{3}$

   $\sin \left(\dfrac{4 \cdot 180°}{3} \right)$

   $\sin 240° = -\dfrac{\sqrt{3}}{2}$

   b. $\tan \dfrac{3\pi}{2}$

   $\tan \left(\dfrac{3 \cdot 180°}{2} \right)$

   tan 270° is undefined.

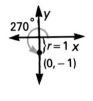

To check using a calculator, use the radian (RAD) mode and proceed as follows.

4 $\boxed{\times}$ $\boxed{\pi}$ $\boxed{\div}$ 3 $\boxed{=}$ $\boxed{\sin}$ ⇒ -0.8660254 ← Check: 3 $\boxed{\sqrt{\ }}$ $\boxed{\div}$ 2 $\boxed{=}$
   $\boxed{+/-}$ ⇒ -0.8660254

3 $\boxed{\times}$ $\boxed{\pi}$ $\boxed{\div}$ 2 $\boxed{=}$ $\boxed{\tan}$ ⇒ E Since tan 270° is undefined, the calculator displays an error message.

Classroom Exercises

Tell how to perform each conversion.

1. $\dfrac{7\pi}{6}$ to degrees   **2.** 120° to radians   **3.** 4 to degrees   **4.** $\dfrac{3\pi}{4}$ to degrees

5. π to degrees   **6.** $-60°$ to radians   **7.** $-\dfrac{\pi}{3}$ to degrees   **8.** 8 to degrees

9–16. Perform the indicated conversion in each of Classroom Exercises 1–8 above. Express degree measures to the nearest degree.

Written Exercises

Express as a radian measure in terms of π.

1. 150°   **2.** 225°   **3.** 300°   **4.** $-90°$

5. 330°   **6.** 180°   **7.** 210°   **8.** 315°

9. 160°   **10.** 80°   **11.** $-100°$ **12.** 30°

Express in degree measure.

13. $\dfrac{2\pi}{3}$ 14. $-\dfrac{5\pi}{6}$ 15. $\dfrac{3\pi}{4}$ 16. $\dfrac{11\pi}{6}$ 17. $\dfrac{7\pi}{4}$ 18. $\dfrac{5\pi}{6}$ 19. $-\dfrac{7\pi}{6}$ 20. $\dfrac{\pi}{10}$

Express the given radian measure in degree measure to the nearest degree.

21. 2   22. -4 23. 2.3   24. -5.1 25. -1 26. 1.8   27. 6   28. -3

Find the value of each trigonometric ratio. Give answers in simplest radical form.

29. $\sin \dfrac{\pi}{6}$ 30. $\cos \dfrac{\pi}{4}$ 31. $\cos \dfrac{5\pi}{6}$ 32. $\sin \dfrac{3\pi}{2}$

33. $\dfrac{1}{3} \tan^2 \dfrac{2\pi}{3}$ 34. $2 \cos \left(-\dfrac{3\pi}{4} \right)$ 35. $\cos^2 \left(\dfrac{3\pi}{4} \right) + \sin^2 \left(\dfrac{3\pi}{4} \right)$

36. If $\theta = \dfrac{\pi}{6}$ and $r = 6$, find s, the length of $\overset{\frown}{AB}$.
37. Find a general formula for s, the length of $\overset{\frown}{AB}$, in terms of r and θ.
38. If $\theta = \dfrac{\pi}{4}$ and $r = 4$, find the area of the sector of the circle formed by radii $\overline{OA}$ and $\overline{OB}$ and arc $\overset{\frown}{AB}$.
39. Write a general formula for the area A of a sector of a circle in terms of r and θ, and then in terms of r and s.

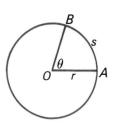

Midchapter Review

Exercises 1–4 refer to triangle *ABC*. Find lengths to two significant digits and angle measures to the nearest degree. 18.1, 18.3

1. $a = 13$, $b = 7$, m $\angle C = 47$,   2. $b = 95$, m $\angle B = 39$, m $\angle C = 65$,
$c = $ __?__ $c = $ __?__

3. $b = 12$, $c = 12$, $a = 20$,   4. $c = 5$, m $\angle A = 43$, m $\angle C = 22$,
m $\angle A = $ __?__ $b = $ __?__

Find the area of triangle *ABC*. 18.2

5. $a = 10$, $b = 14$, m $\angle C = 100$ (to   6. $a = 6$, $b = 7$, $c = 9$ (simplest radical
two significant digits)   form)

Find the number of triangles *ABC* that can be constructed. If the answer is "one" or "two," then find the indicated angle measure(s) to the nearest degree. 18.4

7. $a = 6$, $b = 12$, m $\angle A = 30$,   8. $a = 3$, $b = 12$, m $\angle A = 40$,
m $\angle B = $ __?__   m $\angle B = $ __?__

9. Express $15°$ as a radian measure in   10. Express $\dfrac{13\pi}{6}$ in degree measure. 18.5
terms of π. 18.5

18.6 Periodic Functions

To determine whether a periodic function is odd, even, or neither
To graph periodic functions
To determine the amplitude of a periodic function

A certain function f is defined for all real numbers x. Its graph is shown below. The red portion repeats itself every 8 units along the x-axis.

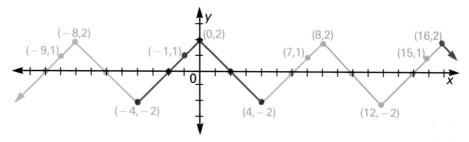

Note that the y-value for any value of x is the same as the y-value for x plus 8 or x plus an integral multiple of 8. For $x = -9$, for example,

$$f(x): \quad f(-9) = 1$$
$$f(x + 8): \quad f(-9 + 8) = f(-1) = 1$$
$$f(x + 2 \cdot 8): \quad f(-9 + 2 \cdot 8) = f(7) = 1$$
$$f(x + 3 \cdot 8): \quad f(-9 + 3 \cdot 8) = f(15) = 1$$

The basic shape of the graph of f is repeated every 8 units. Any horizontal interval of 8 units is said to be a **cycle** of the graph of function f. The function is said to be *periodic* with a *period* of 8.

Definitions

> A function f with domain D is a **periodic function** if and only if for some positive constant p, $f(x) = f(x + p)$ for every x in D. The smallest such value of p is the **period** of f.

Note that in the graph of f above, the *maximum* y-value is $M = 2$. The minimum y-value is $m = -2$. Half the difference of the maximum and minimum y-values of the function is referred to as the *amplitude* of the periodic function f. The amplitude is $\dfrac{M - m}{2} = \dfrac{2 - (-2)}{2} = \dfrac{4}{2}$, or 2.

Definition

> If a periodic function has a maximum value M and a minimum value m, then its **amplitude** is $\dfrac{M - m}{2}$.

Recall that for a function f with x in the domain D of f,

(1) f is an *even* function if $f(-x) = f(x)$ for every x in D, and

(2) f is an *odd* function if $f(-x) = -f(x)$ for every x in D.

The graph of an even function is symmetric with respect to the y-axis. The graph of an odd function is symmetric with respect to the origin.

EXAMPLE 1   Determine whether function f on the previous page is odd, even, or neither.

Solution   For every real number x, $f(-x) = f(x)$. For example, $f(-5) = f(5) = -1$ and $f(-4) = f(4) = -2$. The graph is symmetric with respect to the y-axis.

Therefore, f is an *even* function.

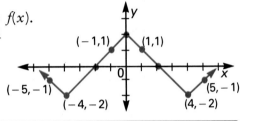

EXAMPLE 2   A portion of the graph of a periodic function f is shown at the right. The period is 6.

a. Complete the graph of f for the interval $-9 \le x \le 9$.

b. Find the amplitude of f.

c. Determine whether f is odd, even, or neither.

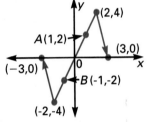

Solutions   a. Since the given part of the graph covers one cycle, repeat to the right and left until you reach -9 at the left and 9 at the right.

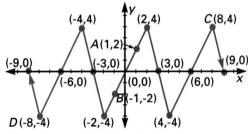

b. The maximum value of y is $M = 4$. The minimum value of y is $m = -4$. Therefore, the amplitude of f is

$$\frac{M - m}{2} = \frac{4 - (-4)}{2} = \frac{8}{2}, \text{ or } 4.$$

c. For every real number x, $f(-x) = -f(x)$. For example, $f(-1) = -2 = -f(1)$, $f(-8) = -4 = -f(8)$, and so on. The graph is symmetric with respect to the origin.

Thus, f is an *odd* function.

Classroom Exercises

The figure at the right shows the
graph of a periodic function *f*.

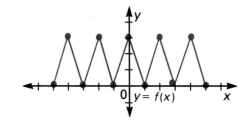

1. What is the period of function *f*?
2. What is the amplitude of *f*?
3. Is *f* odd, even, or neither?

The figure at the right shows the
graph of a periodic function *g*.

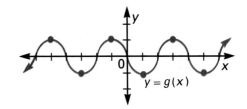

4. What is the period of function *g*?
5. What is the amplitude of *g*?
6. Is *g* odd, even, or neither?

The figure at the right shows the
graph of a periodic function *h*.

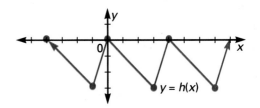

7. Tell how to graph *h* in the interval
   $4 \le x \le 12$.
8. What is the amplitude of *h*?
9. Is *h* odd, even, or neither?

Written Exercises

In Exercises 1–6, a portion of the graph of a function is shown. Use
the given period *p* to graph the function for the indicated interval. Determine whether the function is odd, even, or neither. Find the amplitude of each function.

1.

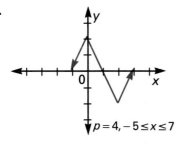

$p = 4, -5 \le x \le 7$

2.

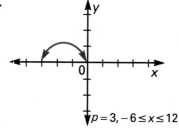

$p = 3, -6 \le x \le 12$

3.

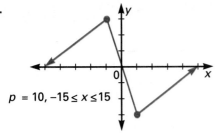

$p = 10, -15 \le x \le 15$

4.

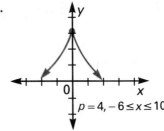

$p = 4, -6 \le x \le 10$

5.

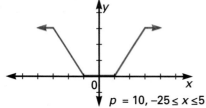

$p = 10, -25 \le x \le 5$

6.

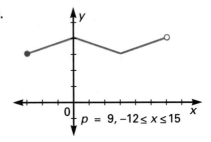

$p = 9, -12 \le x \le 15$

7. Draw the graph of $y = |2x|$, $-4 \le x \le 4$.

8. The graph of Exercise 7 represents one cycle of a periodic function f. The domain of f is the set of real numbers. What is the period of the function?

9. Graph the periodic function of Exercise 8 for $-12 \le x \le 12$. Determine the amplitude and whether the function is odd, even, or neither.

10. The graph of a periodic function g over one cycle is defined by
$y = \begin{cases} x + 1, & -1 \le x \le 2 \\ -2x + 7, & 2 < x \le 5 \end{cases}$. The domain of g is the set of real

numbers. Determine the period of the function.

11. Graph the periodic function of Exercise 10 for $-1 \le x \le 17$.

12. The graph of a periodic function h over one cycle is defined by $y = |2x - 4| + 5, 0 \le x \le 4$. The domain of h is the set of real numbers. Determine the period of the function.

13. Graph the periodic function of Exercise 12 for $-4 \le x \le 8$.

Mixed Review

Write each trigonometric ratio in simplest radical form.   **17.5, 17.7, 18.5**

1. $\sin 270$ 2. $\cos 150$ 3. $\sin 780$ 4. $\sin \pi$ 5. $\cos \dfrac{3\pi}{4}$ 6. $\sin 2\pi$

18.7 Graphing the Sine and Cosine Functions

Objectives

To graph the sine and cosine functions for given domains
To determine $\sin(-\theta)$ and $\cos(-\theta)$ for given values of $\sin\theta$ and $\cos\theta$

As shown in a trigonometry table, there is one and only one sine (or cosine) ratio corresponding to any angle measure. Therefore, the sine and cosine relations are *functions*. The graph of $T = \sin\theta$ can be drawn by setting up a table of sample values and then drawing a smooth curve through the points corresponding to these table values.

EXAMPLE 1   Graph $T = \sin\theta$ for $0 \le \theta \le 2\pi$.

Solution   Set up a table of special radian measures for θ and find corresponding values for $T = \sin\theta$. The completed table is shown below with both exact and decimal values.

θ	0	$\frac{\pi}{6}$	$\frac{\pi}{3}$	$\frac{\pi}{2}$	$\frac{2\pi}{3}$	$\frac{5\pi}{6}$	π	$\frac{7\pi}{6}$	$\frac{3\pi}{2}$	$\frac{11\pi}{6}$	2π
$T = \sin\theta$	0	$\frac{1}{2}$	$\frac{\sqrt{3}}{2}$	1	$\frac{\sqrt{3}}{2}$	$\frac{1}{2}$	0	$\frac{-1}{2}$	-1	$\frac{-1}{2}$	0
	0	0.5	0.87	1	0.87	0.5	0	-0.5	-1	-0.5	0

If a calculator is used to obtain the table values, the calculator should be in the radian (RAD) mode, as illustrated below for $\theta = \frac{7\pi}{6}$.

$7 \;\boxed{\times}\; \boxed{\pi}\; \boxed{\div}\; 6 \;\boxed{=}\; \boxed{\sin}\; \Rightarrow\; -0.5$

Plot the ordered pairs in the table. Then, draw a smooth curve through the points.

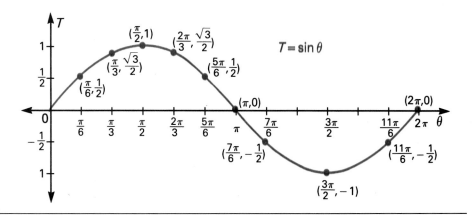

Recall that if for every θ, $f(\theta) = f(\theta + p)$, where p is constant, then f is a periodic function. The reference triangles below illustrate that the sine function is periodic.

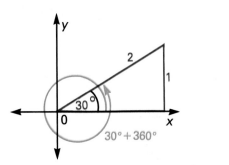 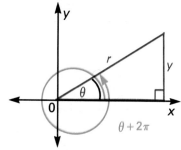

$$\sin 30 = \sin(30 + 360) = \tfrac{1}{2}, \quad \text{or} \quad \sin \theta = \sin(\theta + 2\pi) = \frac{y}{r}.$$
$$\sin \tfrac{\pi}{6} = \sin(\tfrac{\pi}{6} + 2\pi) \qquad\qquad\qquad f(\theta) = f(\theta + 2\pi)$$

Therefore, $T = \sin \theta$ is periodic. The period is 2π.

The periodicity of the sine function can be visualized more clearly by extending the graph of Example 1.

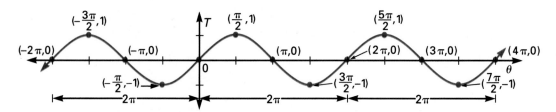

The basic shape is repeated every 2π units: $p = 2\pi$.
The maximum value M of $\sin \theta$ is 1, which occurs at $\theta = -\dfrac{3\pi}{2}$, $\theta = \dfrac{\pi}{2}$, $\theta = \dfrac{5\pi}{2}$, and so on.

The minimum value m of $\sin \theta$ is -1, which occurs at $\theta = -\dfrac{\pi}{2}$, $\theta = \dfrac{3\pi}{2}$, $\theta = \dfrac{7\pi}{2}$, and so on.

The amplitude of $\sin \theta$ is $\dfrac{M - m}{2} = \dfrac{1 - (-1)}{2} = \dfrac{2}{2} = 1$.

Notice in the graph of $T = \sin \theta$ that

as θ increases from 0 to $\tfrac{\pi}{2}$ (Quadrant I), T *increases* from 0 to 1,

as θ increases from $\tfrac{\pi}{2}$ to π (Quadrant II), T *decreases* from 1 to 0,

as θ increases from π to $\tfrac{3\pi}{2}$ (Quadrant III), T *decreases* from 0 to -1,

as θ increases from $\tfrac{3\pi}{2}$ to 2π (Quadrant IV), T *increases* from -1 to 0.

This pattern of increasing and decreasing is repeated every 2π units.

The sine function $T = \sin\theta$ is an odd function (and so is symmetric with respect to the origin), since $\sin\left(-\frac{\pi}{2}\right) = -1 = -\sin\frac{\pi}{2}$, $\sin(-\pi) = 0 = -\sin\pi$, $\sin\left(-\frac{5\pi}{2}\right) = -1 = -\sin\frac{5\pi}{2}$, and so on. In general, for any angle of measure θ,

$$\sin(-\theta) = -\sin\theta.$$

EXAMPLE 2 Find $\sin(-\theta)$ if $\sin\theta = 0.7071$.

Solution Use the property $\sin(-\theta) = -\sin\theta$.

$$\sin(-\theta) = -0.7071$$

The cosine function is also periodic, as illustrated in the reference triangles below.

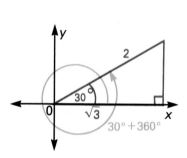 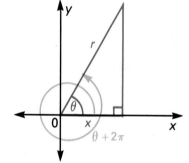

$$\cos 30 = \cos(30 + 360) = \frac{\sqrt{3}}{2}, \quad \text{or} \quad \cos\theta = \cos(\theta + 2\pi) = \frac{x}{r}.$$

Therefore, $\cos\theta$ is a periodic function with period 2π.

The periodicity is illustrated more clearly in the graph of $T = \cos\theta$. The table and graph are shown below for $-2\pi \le \theta \le 2\pi$.

θ	-2π	$-\frac{3\pi}{2}$	$-\pi$	$-\frac{\pi}{2}$	$-\frac{\pi}{3}$	0	$\frac{\pi}{3}$	$\frac{\pi}{2}$	π	$\frac{3\pi}{2}$	2π
$T = \cos\theta$	1	0	-1	0	$\frac{1}{2}$	1	$\frac{1}{2}$	0	-1	0	1

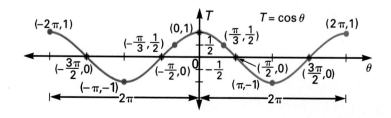

For the graph on the previous page, the basic shape is repeated every 2π units: $p = 2\pi$.

The maximum value M of $T = \cos\theta$ is 1.

The minimum value m of $T = \cos\theta$ is -1.

The amplitude of $T = \cos\theta$ is $\dfrac{M - m}{2} = \dfrac{1 - (-1)}{2} = \dfrac{2}{2} = 1$.

The cosine function $T = \cos\theta$ is an even function, since

$$\cos\left(-\frac{\pi}{3}\right) = \cos\frac{\pi}{3} = \frac{1}{2}, \cos(-\pi) = \cos\pi = -1,$$

$$\cos\left(-\frac{3\pi}{2}\right) = \cos\frac{3\pi}{2} = 0, \text{ and so on.}$$

In general, for any angle of measure θ,

$$\cos(-\theta) = \cos\theta.$$

Notice also that the cosine function, like all even functions, is symmetric with respect to the vertical axis.

Summary

$T = \sin\theta$
  period: 2π
  amplitude: 1
  $\sin(-\theta) = -\sin\theta$
  (an odd function)

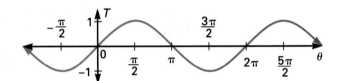

$T = \cos\theta$
  period: 2π
  amplitude: 1
  $\cos(-\theta) = \cos\theta$
  (an even function)

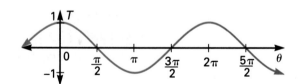

Classroom Exercises

Find each of the following.

1. the amplitude of the sine function
2. the period of the cosine function
3. the minimum value of $T = \cos\theta$
4. the period of $T = \sin\theta$
5. the maximum value of $T = \sin\theta$
6. the amplitude of the cosine function

Tell whether $T = \cos\theta$ increases or decreases as θ increases in the given interval. (HINT: Use the graph of $T = \cos\theta$ above.)

7. from 0 to $\frac{\pi}{2}$
8. from $\frac{\pi}{2}$ to π
9. from π to $\frac{3\pi}{2}$
10. from $\frac{3\pi}{2}$ to 2π

Written Exercises

In Exercises 1 and 2, complete the table of values for each function. Use the results to graph the function.

1. $T = \sin \theta$

2. $T = \cos \theta$

θ	$-\pi$	$-\dfrac{5\pi}{6}$	$-\dfrac{\pi}{2}$	$-\dfrac{\pi}{6}$	0	$\dfrac{\pi}{6}$	$\dfrac{\pi}{2}$	$\dfrac{5\pi}{6}$	π
T	?	?	?	?	?	?	?	?	?

3. As θ increases, for what values of θ is T increasing in Exercise 1?

4. As θ increases, for what values of θ is T decreasing in Exercise 1?

5. As θ increases, for what values of θ is T increasing in Exercise 2?

6. As θ increases, for what values of θ is T decreasing in Exercise 2?

7. Find $\sin(-\theta)$ if $\sin \theta = 0.5446$.

8. Find $\cos(-\theta)$ if $\cos \theta = 0.7431$.

In Exercises 9–18, use the following values of θ to make the indicated tables and graphs. Graph the indicated functions in the same coordinate plane.

$$\left\{ -2\pi,\ -\frac{11\pi}{6},\ -\frac{3\pi}{2},\ -\frac{7\pi}{6},\ -\pi,\ -\frac{5\pi}{6},\ -\frac{\pi}{2},\ -\frac{\pi}{6},\ 0 \right\}$$

9. Make a table of values for $T = \cos \theta$.

10. Graph $T = \cos \theta$.

11. As θ increases, for what values of θ is $T = \cos \theta$ increasing?

12. As θ increases, for what values of θ is $T = \cos \theta$ decreasing?

13. Make a table of values for $T = \sin \theta$.

14. Graph $T = \sin \theta$.

15. As θ increases, for what values of θ is $T = \sin \theta$ increasing?

16. As θ increases, for what values of θ is $T = \sin \theta$ decreasing?

17. For what values of θ does it appear that $\sin \theta = \cos \theta$?

18. Write briefly why the sine function is an odd periodic function.

In Exercises 19–21, $T = |\sin \theta|$ for $-2\pi \le \theta \le 4\pi$.

19. Graph $T = |\sin \theta|$.

20. Give the period of $T = |\sin \theta|$.

21. Determine whether the function is odd or even. Give a reason.

Mixed Review

Graph each function in the same coordinate plane. *11.3*

1. $y = |x|$
2. $y = -4|x|$
3. $y = 2|x - 4|$
4. $y = 3|x + 6| - 4$

Graph each function in the same coordinate plane. *11.4*

5. $y = x^2$
6. $y = 2x^2$
7. $y = \tfrac{1}{2}x^2$
8. $y = x^2 + 4$

In many real-world situations, it may not be necessary to find an exact answer to a problem. Often, an estimate may be appropriate. In such cases, however, it is important to know how much the estimate may vary from the true value.

Suppose, for example, that you want to find the distance from the Earth to Jupiter on August 19, 1990, using the table below.

SUN-CENTERED COORDINATES OF MAJOR PLANETS FOR AUG 17, 1990 IN ASTRONOMICAL UNITS

	x	y	z
Earth/Moon	0.8189001	−0.5462572	−0.2368435
Mars	1.3894497	0.1451882	0.0290001
Jupiter	−2.238098	4.328104	1.909756
Saturn	3.934523	−8.443261	−3.656253

The table gives 3-dimensional, x-y-z coordinates for the planets using a coordinate system with the Sun at the origin. The x-axis is defined as lying in the direction of the Earth's position at the first moment of Spring, the vernal equinox.

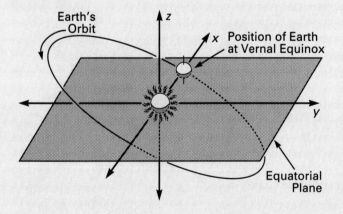

The x-y plane of the system is the **equatorial plane,** which corresponds to the plane of the equator of the Earth at the time of the equinox.

The unit of measure used in the table is the **astronomical unit (au),** which equals the mean distance of the Earth from the Sun, about 92,900,000 miles or 149,600,000 kilometers.

The values given for the Earth/Moon are actually for a point called the **center of mass** of the two bodies, which is located between their centers. This point may be used as an *approximation* of the position of the Earth, since it is not far from the Earth's center.

EXAMPLE   Estimate the distance between the centers of the Earth and Jupiter.

Solution   Use the distance formula for 3 dimensions.

$$d = \sqrt{(x_2 - x_1)^2 + (y_2 - y_1)^2 + (z_2 - z_1)^2}$$

Begin by subtracting the coordinates of Earth-Moon and Jupiter.

0.8189001	−0.5462572	−0.2368435
−2.238098	4.328104	1.909756
3.0569981	−4.8743612	−2.1465995

Substitute the results into the formula.

$$d = \sqrt{(3.0569981)^2 + (-4.8743612)^2 + (-2.1465995)^2}$$
$$= \sqrt{37.7125239} = \textbf{6.1410523 au}$$

The center of mass of the Earth/Moon system is 0.0000312 au from the center of the Earth, so this is the *maximum amount that the above estimate can be in error.* Thus, the true value of the distance, to 7 decimal places, may be as *low* as 6.1410523 − 0.0000312 = 6.1410211 au, or as *high* as 6.1410523 − 0.0000312 = 6.1410835 au.

The actual value of the distance turns out to be 6.1410804 au. This value is within the bounds of the estimate. It differs from the estimate by 0.0000281 au.

Exercises

1. Use the values given at the bottom of the previous page to express the center of mass distance 0.0000312 au in miles and kilometers. How does this compare with the radius of the Earth? (The Earth's radius = 3,959 mi or 6,371 km.)

2. Express d in the above example in miles and kilometers. Do you think that a difference of 0.0000312 au would be significant in these figures? (HINT: Do the values for the Earth–Sun distances seem to be estimates themselves?)

3. Use the values in the table to estimate the distance between the centers of the Earth and Saturn. Compare this with the actual distance of 9.1521356 au.

18.8 Changing the Amplitude and Period

Objectives

To graph functions of the form $T = c + a \cos b\theta$ and $T = c + a \sin b\theta$

To determine the amplitude, the period, and the maximum and minimum values of the above functions

Periodic functions occur in various fields, including medicine. For example, the electrocardiogram (EKG) at the right shows a heart's electrical impulses. The horizontal axis represents time in seconds, and the vertical axis represents the strength of the electrical impulses.

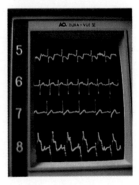

Recall that the amplitude of $T = \sin \theta$ and of $T = \cos \theta$ is 1. If $\sin \theta$ or $\cos \theta$ is multiplied by a constant a, then the amplitude will change. This can be seen for the sine function by graphing $T = \sin \theta$, $T = 2 \sin \theta$, and $T = -\frac{1}{2} \sin \theta$ on the same set of axes.

θ	$-\frac{\pi}{2}$	0	$\frac{\pi}{2}$	π	$\frac{3\pi}{2}$	2π	$\frac{5\pi}{2}$	3π
$T = \sin \theta$	-1	0	1	0	-1	0	1	0
$T = 2 \sin \theta$	-2	0	2	0	-2	0	2	0
$T = -\frac{1}{2} \sin \theta$	$\frac{1}{2}$	0	$-\frac{1}{2}$	0	$\frac{1}{2}$	0	$-\frac{1}{2}$	0

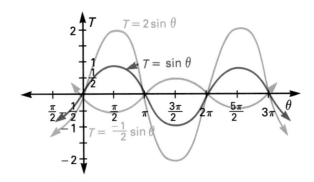

The graphs suggest that the period of $T = a \sin \theta$ is 2π regardless of the value of a.

The amplitudes of $T = 2 \sin \theta$ and $T = -\frac{1}{2} \sin \theta$ are calculated below.

$T = 2 \sin \theta$

maximum: $M = 2$

minimum: $m = -2$

amplitude: $\dfrac{M - m}{2}$

$= \dfrac{2 - (-2)}{2} = 2$

$T = -\frac{1}{2} \sin \theta$

maximum: $M = \frac{1}{2}$

minimum: $m = -\frac{1}{2}$

amplitude: $\dfrac{M - m}{2}$

$= \dfrac{\frac{1}{2} - \left(-\frac{1}{2}\right)}{2} = \frac{1}{2}$

Amplitude and Period for $T = a \sin \theta$ and $T = a \cos \theta$
If $T = a \sin \theta$ or $T = a \cos \theta$, then for each real number $a \neq 0$,
(1) the *amplitude* is $|a|$, and
(2) the *period* is 2π.

The graph of $T = \sin \theta$ or $T = \cos \theta$ is
(1) *widened* vertically if $|a| > 1$, and
(2) *narrowed* vertically if $|a| < 1$.

The graph of $T = a \sin \theta$ or $T = a \cos \theta$ can be translated (moved) upward or downward by adding or subtracting a constant. This is illustrated for the cosine function in Example 1 below.

EXAMPLE 1 Graph $T = 4 + 2 \cos \theta$ for $0 \le \theta \le 2\pi$. Find the amplitude and period.

Plan Make a table of values using multiples of $\frac{\pi}{2}$ for five sample values of θ.

Solution

θ	0	$\frac{\pi}{2}$	π	$\frac{3\pi}{2}$	2π
$\cos \theta$	1	0	-1	0	1
$4 + 2 \cos \theta$	6	4	2	4	6

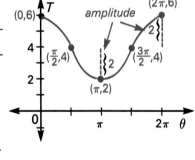

maximum value: $M = 6$

minimum value: $m = 2$

amplitude: $\dfrac{M - m}{2} = \dfrac{6 - 2}{2} = 2$

Thus, for $T = 4 + 2 \cos \theta$, the amplitude is 2 and the period is 2π.

Example 2 on the next page shows the effect of multiplying the angle measure θ by a constant. Recall that the period of $T = \sin \theta$ is 2π. Therefore, one cycle of the sine function is completed every 2π units.

EXAMPLE 2   Graph $T = \sin 2\theta$ and $T = \sin \theta$ in the same coordinate plane for $0 \le \theta \le 2\pi$. Determine the amplitude and the period of $T = \sin 2\theta$.

Plan   For $T = \sin 2\theta$, use multiples of $\frac{\pi}{4}$ to get nine sample values.

Solution

θ	0	$\frac{\pi}{4}$	$\frac{\pi}{2}$	$\frac{3\pi}{4}$	π	$\frac{5\pi}{4}$	$\frac{3\pi}{2}$	$\frac{7\pi}{4}$	2π
2θ	0	$\frac{\pi}{2}$	π	$\frac{3\pi}{2}$	2π	$\frac{5\pi}{2}$	3π	$\frac{7\pi}{2}$	4π
$\sin 2\theta$	0	1	0	-1	0	1	0	-1	0

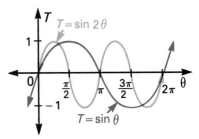

The graph of $T = \sin 2\theta$ completes 2 cycles from 0 to 2π. It completes one cycle every $\frac{2\pi}{2}$, or π units.

Therefore, the period is π and the amplitude is 1.

Notice that the graph of $T = \sin 2\theta$ above is a *horizontal narrowing* of the graph of $T = \sin \theta$. The cosine function behaves similarly.

Period for sin $b\theta$ and cos $b\theta$

If $T = \sin b\theta$, then the period is $\frac{2\pi}{|b|}$ for each real number $b \ne 0$.

If $T = \cos b\theta$, then the period is $\frac{2\pi}{|b|}$ for each real number $b \ne 0$.

The graph of $T = 4 \cos \frac{1}{3}\theta$ can be drawn without constructing a table of values. This is done by using the amplitude and the period of the graph, as illustrated in Example 3.

EXAMPLE 3   Graph one cycle of $T = 4 \cos \frac{1}{3}\theta$ in the same coordinate plane as the graph of $T = \cos \theta$. Do not use a table of values.

Plan   For $T = 4 \cos \frac{1}{3}\theta$, first determine the amplitude and the period. Then graph the function.

Solution $T = 4 \cos \frac{1}{3}\theta$: amplitude $= 4$, period $= \dfrac{2\pi}{\frac{1}{3}} = 2\pi(3) = 6\pi$

One cycle of the graph is completed every 6π units, as shown on the next page.

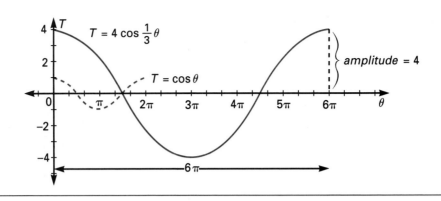

Notice that the graph of $T = 4 \cos \frac{1}{3}\theta$ is a *horizontal widening* and a *vertical widening* of the graph of $T = \cos \theta$.

EXAMPLE 4 Graph one cycle of the function $T = -1 + 2 \sin 3\theta$ without using a table of values. Determine the amplitude, the period, and the maximum and minimum values of $T = -1 + 2 \sin 3\theta$.

Solution First, graph $T = 2 \sin 3\theta$.

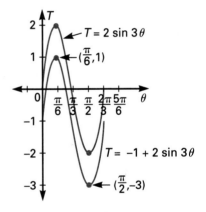

    amplitude: 2   period: $\dfrac{2\pi}{3}$

    maximum: 2   minimum: -2

Second, translate the graph of $T = 2 \sin 3\theta$ one unit downward.

    maximum: $-1 + 2 = 1$

    minimum: $-1 + (-2) = -3$

    amplitude: $\dfrac{1 - (-3)}{2} = 2$

Therefore, for $T = -1 + 2 \sin 3\theta$, the amplitude is 2, the period is $\dfrac{2\pi}{3}$, the maximum value is 1, and the minimum value is -3.

Graph of $T = a \sin b\theta$ or $T = a \cos b\theta$

Amplitude	Vertical Effect of a on Graph of $\sin \theta$ or $\cos \theta$	Period	Horizontal Effect of b on Graph of $\sin \theta$ or $\cos \theta$
$\lvert a \rvert$	widens if $\lvert a \rvert > 1$ narrows if $\lvert a \rvert < 1$	$\dfrac{2\pi}{\lvert b \rvert}$	widens if $\lvert b \rvert < 1$ narrows if $\lvert b \rvert > 1$

The graph of $T = c + a \sin b\theta$ or $T = c + a \cos b\theta$ is a translation of the graph upward c units if $c > 0$, or downward $\lvert c \rvert$ units if $c < 0$.

Classroom Exercises

Give the amplitude and the period of each function.

1. $T = 4 \cos \theta$ **2.** $T = \frac{1}{3} \sin \theta$ **3.** $T = \cos 3\theta$ **4.** $T = -\sin \theta$

5. $T = \sin 4\theta$ **6.** $T = \frac{1}{2} \cos \theta$ **7.** $T = -\sin 2\theta$ **8.** $T = 2 \cos \frac{1}{3}\theta$

9. $T = 1 + \sin \theta$ **10.** $T = 3 - \cos \theta$ **11.** $T = -4 + 4 \sin \theta$ **12.** $T = 3 + 5 \sin \theta$

13–16. Graph one cycle of each function of Classroom Exercises 1–4 without using a table of values.

Written Exercises

Graph each function for $0 \leq \theta \leq 2\pi$. Use a table of values.

1–8. Use the functions of Classroom Exercises 5–12 above.

Graph one cycle of each function without using a table of values. Determine the period and the amplitude.

9. $T = \sin 3\theta$ **10.** $T = \frac{1}{2} \sin 2\theta$ **11.** $T = 5 \cos 4\theta$

12. $T = 3 \cos 6\theta$ **13.** $T = -5 \sin 2\theta$ **14.** $T = 2 \cos \frac{1}{2}\theta$

15. $T = \frac{1}{4} \sin \frac{1}{3}\theta$ **16.** $T = 5 \cos \frac{1}{4}\theta$ **17.** $T = -6 \cos 9\theta$

Graph one cycle of each function without using a table of values. Determine the period, the amplitude, and the maximum and minimum values.

18. $T = 6 + 3 \cos 2\theta$ **19.** $T = -6 + 5 \sin 4\theta$ **20.** $T = -2\frac{1}{2} + \frac{1}{2} \sin 3\theta$

21. Graph $T = \cos \theta$ and $T = -\sin \theta$ in the same coordinate plane for $-2\pi \leq \theta \leq 2\pi$. For what values of θ does $\cos \theta = -\sin \theta$?

22. Graph $T = \theta + \sin \theta$ for $-2\pi \leq \theta \leq 2\pi$.

23. Graph $T = |3 \sin 4\theta|$ for $-2\pi \leq \theta \leq 2\pi$.

24. The graph of a sine function has a period of $\frac{\pi}{2}$, a maximum value of 6, and a minimum value of 2. Write an equation for the function.

Mixed Review

Find each trigonometric ratio in simplest radical form.   **18.5**

1. $\sec \pi$ **2.** $\tan \frac{3\pi}{4}$ **3.** $\csc \frac{7\pi}{6}$ **4.** $\cot \frac{5\pi}{3}$

18.9 Graphing the Tangent, Cotangent, Secant, and Cosecant Functions

Objective

To graph the tangent, cotangent, secant, and cosecant functions

Recall that $\tan \frac{\pi}{2}$, or $\tan 90°$, is undefined since division by zero is undefined. Similarly, $\tan \left(-\frac{\pi}{2}\right) = \frac{-1}{0}$ is undefined. More generally, the tangent ratio $\tan \theta$ is undefined for $\frac{\pi}{2}, \frac{3\pi}{2}, \frac{5\pi}{2}, \frac{7\pi}{2}, \cdot \cdot \cdot$, that is, for $\theta = \frac{\pi}{2} + k \cdot \pi$, where k is an integer.

However, as θ increases from 0 to $\frac{\pi}{2}$, $\tan \theta$ increases without bound. This is illustrated on a calculator by first selecting the radian mode (RAD), and then noticing that $\tan 1.50 = 14.10142$, $\tan 1.53 = 24.49841$, and $\tan 1.57 = 1,255.7653$.

To graph $T = \tan \theta$ for $-\frac{\pi}{2} \le \theta \le \frac{\pi}{2}$, make a table of values.

θ	$-\frac{\pi}{2}$	$-\frac{\pi}{3}$	$-\frac{\pi}{4}$	$-\frac{\pi}{6}$	0	$\frac{\pi}{6}$	$\frac{\pi}{4}$	$\frac{\pi}{3}$	$\frac{\pi}{2}$
$\tan \theta$	undef	-1.7	-1	-0.6	0	0.6	1	1.7	undef

Notice that as θ approaches $\frac{\pi}{2}$ from the left, $\tan \theta$ increases without bound. As θ approaches $-\frac{\pi}{2}$ from the right, $\tan \theta$ decreases without bound. The vertical lines $\theta = -\frac{\pi}{2}$ and $\theta = \frac{\pi}{2}$ are *asymptotes* of the graph of $T = \tan \theta$. There is no maximum or minimum value. Since there is no maximum or minimum, the amplitude is not defined.

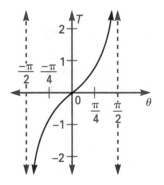

As seen in the figure at the right, the tangent function, like the sine and cosine functions, is periodic. To see this, notice that $\tan \theta = \frac{y}{x}$ and also

$\tan (\theta + \pi) = \frac{-y}{-x} = \frac{y}{x}.$

Thus, $\tan \theta = \tan (\theta + \pi)$.

Therefore, the tangent function $T = \tan \theta$ is periodic with period π.

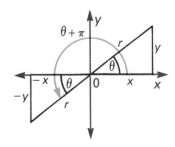

This periodicity can be visualized more clearly by extending the graph of the tangent function.

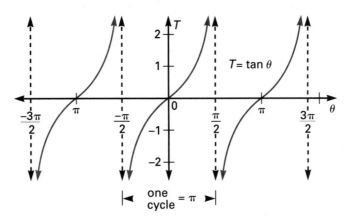

Notice that the function is an odd function because, for example, $\tan\left(-\frac{\pi}{4}\right) = -\tan\frac{\pi}{4} = -1$, $\tan\left(-\frac{\pi}{3}\right) = -\tan\frac{\pi}{3}$, and so on.

In general, for any angle measure θ for which $T = \tan(\theta)$ is defined,

$$\tan(-\theta) = -\tan\theta.$$

Recall that the period for $T = a\sin b\theta$ and of $T = a\cos b\theta$ is $\frac{2\pi}{|b|}$. However, for $T = a\tan b\theta$, the period is $\frac{\pi}{|b|}$.

EXAMPLE 1   Graph $T = \tan 2\theta$ for $-\frac{3\pi}{4} < \theta < \frac{3\pi}{4}$.

Solution   Find the period: $p = \frac{\pi}{|b|} = \frac{\pi}{2}$. Then find and draw the asymptotes.

Sketch one cycle from $-\frac{1}{2}p$ to $\frac{1}{2}p$, that is, from $-\frac{1}{2}\left(\frac{\pi}{2}\right)$ to $\frac{1}{2}\left(\frac{\pi}{2}\right)$, or from $-\frac{\pi}{4}$ to $\frac{\pi}{4}$. Repeat the cycle from $-\frac{3\pi}{4}$ to $-\frac{\pi}{4}$ and from $\frac{\pi}{4}$ to $\frac{3\pi}{4}$.

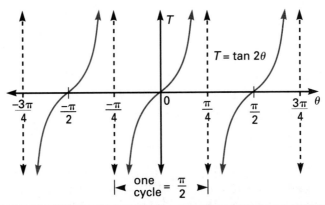

The cotangent ratio is the reciprocal of the tangent ratio. It is undefined where $\tan \theta = 0$, and 0 where $\tan \theta$ is undefined. Its period is the same as the tangent function, $\dfrac{\pi}{|b|}$, and its amplitude is not defined.

EXAMPLE 2 Graph $T = \sin \theta$ and $T = \csc \theta$ for $0 \leq \theta \leq 2\pi$ in the same coordinate plane. Determine the amplitude and the period of $T = \csc \theta$.

Plan Use the fact that the trigonometric ratios $\sin \theta$ and $\csc \theta$ are reciprocals of each other for $\sin \theta \neq 0$ to make a table of values.

Solution

θ	0	$\dfrac{\pi}{6}$	$\dfrac{\pi}{2}$	$\dfrac{5\pi}{6}$	π	$\dfrac{7\pi}{6}$	$\dfrac{3\pi}{2}$	$\dfrac{11\pi}{6}$	2π
$\sin \theta$	0	$\dfrac{1}{2}$	1	$\dfrac{1}{2}$	0	$\dfrac{-1}{2}$	-1	$\dfrac{-1}{2}$	0
$\csc \theta$	undef	2	1	2	undef	-2	-1	-2	undef

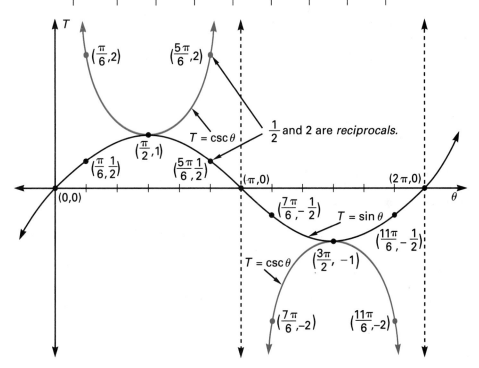

There is no maximum or minimum value for $T = \csc \theta$.
Therefore, the cosecant function has no amplitude defined.
The period of $T = \csc \theta$ is 2π, the same as that of $T = \sin \theta$.

Like sine and cosecant, the cosine and secant ratios are reciprocals of each other. The graph of $T = \sec \theta$ can be graphed on the same set of axes as the graph of $T = \cos \theta$, as done above for $\sin \theta$ and $\csc \theta$.

Classroom Exercises

Give the period of each function.

1. $T = \sec 3\theta$ **2.** $T = 4 \tan 5\theta$ **3.** $T = 2 \csc \frac{\pi}{2}\theta$ **4.** $T = 6 \cot 2\theta$

5. $T = \csc \frac{1}{3}\theta$ **6.** $T = 3 \cot \pi\theta$ **7.** $T = \sec (-\theta)$ **8.** $T = \tan 3\pi\theta$

Written Exercises

Complete the chart.

	sec θ	csc θ	tan θ	cot θ
1. Amplitude				
2. Period				
3. Maximum value				
4. Minimum value				
5. Values of θ, $0 \le \theta \le 2\pi$, for which the function is undefined				

6. Graph $T = \sec \theta$ for $0 \le \theta \le 2\pi$ using a table of values. Use multiples of $\frac{\pi}{4}$ for θ.

Graph one cycle of each of the following functions.

7. $T = \tan 3\theta$ **8.** $T = \cot \theta$ **9.** $T = \cot \frac{1}{2}\theta$ **10.** $T = \tan \frac{1}{3}\theta$

11. $T = \sec 4\theta$ **12.** $T = \csc 2\theta$ **13.** $T = \sec \frac{1}{3}\theta$ **14.** $T = \csc 3\theta$

15. Graph $T = 2 \tan 3\theta$ and $T = \tan \theta$ in the same coordinate plane for $-\frac{\pi}{2} < \theta < \frac{\pi}{2}$. Compare the graphs.

16. Graph $T = \frac{1}{2} \sec 2\theta$ and $T = \sec \theta$ in the same coordinate plane for $0 \le \theta \le 2\pi$. Compare the graphs.

17. Determine whether $T = \sec \theta$ is an odd function, an even function, or neither.

18. Determine whether $T = \csc \theta$ is an odd function, an even function, or neither.

Mixed Review

In Exercises 1–4, points $P(-2,4)$ and $Q(6,8)$ are given.

1. Write an equation of $\overleftrightarrow{PQ}$ in point-slope form.   *3.4*

2. Find the slope of a line perpendicular to $\overleftrightarrow{PQ}$.   *3.7*

3. Find the coordinates of the midpoint of $\overline{PQ}$.   *11.1*

4. Find the distance between P and Q.   *11.1*

18.10 Inverse Trigonometric Functions

Objectives
To graph inverse trigonometric functions
To determine the inverse of a trigonometric function

So far, the trigonometric functions have been defined in terms of θ and T and graphed using θ and T axes. Now, the x- and y-axes will be used.

The relation $y = \sin x$ is a function. Therefore, for each angle measure x, there is only *one* sine value. The *inverse* of the sine function is not a function, since there is not a unique angle measure for a given sine value. For example, angles of 30, 150, 390. . . . (in radians, $\frac{\pi}{6}$, $\frac{5\pi}{6}$, $\frac{13\pi}{6}$ . . . .) all have a sine value of $\frac{1}{2}$.

However, if we restrict the domain of a function so that its range is a set of unique y values, then its inverse will be a function. This is shown in Example 1 below, where the function $y = \sin x$, $-\frac{\pi}{2} \le x \le \frac{\pi}{2}$, is graphed together with its inverse. Recall the method for graphing the inverse of a function (See Lesson 13.1).

EXAMPLE 1
Graph $y = \sin x$, $-\frac{\pi}{2} \le x \le \frac{\pi}{2}$. Then graph its inverse.

Solution
(1) On the same set of axes, graph $y = \sin x$ and $y = x$ for $-\frac{\pi}{2} \le \frac{\pi}{2}$.

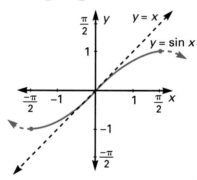

(2) Graph the mirror image of the solid portion of $y = \sin x$ with respect to the line $y = x$.

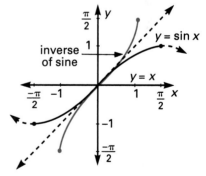

No y-value occurs more than once in the solid portion of the graph.

The domain of the inverse function is $\{x \mid -1 \le x \le 1\}$. There is just one y-value for every x-value.

Notice that the *range* of the resulting inverse function is numerically the same as the domain of the original function. The values of y in the range of an inverse function are called its **principal values**.

To indicate that the domain of the sine function is restricted to the interval from $-\frac{\pi}{2}$ to $\frac{\pi}{2}$, inclusive, write the function as $y = \text{Sin } x$ (with capital S) rather than $y = \sin x$.

The inverse of $y = \text{Sin } x$ is written $y = \text{Sin}^{-1}x$. Read "y is the angle measure whose sine is x." Note that $\text{Sin}^{-1}x$ does *not* mean $\frac{1}{\text{Sin } x}$.

The other trigonometric functions also have inverses that are functions if their domains are restricted. Two of these are graphed.

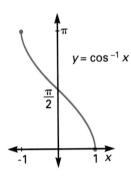

Range of Cos^{-1}
Principal Values: $\{y \mid 0 \leq y \leq \pi\}$

Range of Tan^{-1}
Principal Values: $\{y \mid -\frac{\pi}{2} < y < \frac{\pi}{2}\}$

EXAMPLE 2   Find y if $y = \text{Cos}^{-1}\left(-\frac{\sqrt{3}}{2}\right)$.

Plan   Sketch a reference triangle. Since y represents an angle measure, x and y cannot be used to label the coordinate axes of the reference-triangle diagram. Use u and v instead.

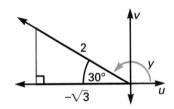

Solution $y = \text{Cos}^{-1}\left(-\frac{\sqrt{3}}{2}\right)$ means $\text{Cos } y = -\frac{\sqrt{3}}{2}$.

Principal values for inverse cosine are in Quadrants I and II. Draw the angle in Quadrant II, since the cosine is negative. The reference angle has measure 30.

Therefore, $y = 150$, or $\frac{5\pi}{6}$ in radian measure.

Often, an inverse function can be found by using a calculator.

EXAMPLE 3   Find A, to the nearest degree, if $A = \text{Tan}^{-1}(-0.9008)$.

Solution   Calculator steps: 0.9008 $\boxed{+/-}$ $\boxed{\text{INV}}$ $\boxed{\text{tan}}$ $\Rightarrow$ -42.012527
Therefore, A is -42 to the nearest degree.

It can be shown that if $x > 0$, then $\operatorname{Cot}^{-1} x = \operatorname{Tan}^{-1}\left(\frac{1}{x}\right)$. (You are asked to show this in Exercise 20.) This relationship follows from the fact that the tangent and the cotangent are reciprocals of one another. The following definitions are based on the reciprocal relationship between the secant and cosine and between the cosecant and sine. For both definitions, $|x| > 1$.

$$y = \operatorname{Sec}^{-1} x = \operatorname{Cos}^{-1}\left(\frac{1}{x}\right) \qquad \leftarrow 0 \leq y \leq \pi,\ y \neq \frac{\pi}{2}$$

$$y = \operatorname{Csc}^{-1} x = \operatorname{Sin}^{-1}\left(\frac{1}{x}\right) \qquad \leftarrow -\frac{\pi}{2} \leq y \leq \frac{\pi}{2},\ y \neq 0$$

EXAMPLE 4 Evaluate $\cos\left[\operatorname{Csc}^{-1}(-3)\right]$.

Plan Let $y = \operatorname{Csc}^{-1}(-3) = \operatorname{Sin}^{-1}\left(\frac{1}{-3}\right)$. Then $\operatorname{Sin} y = -\frac{1}{3}$.

Solution Locate y in the proper quadrant. Principal values of y are in Quadrants I and IV. Since $\operatorname{Sin} y$ is negative (as is $\operatorname{Csc} y$), angle y terminates in Quadrant IV. Find the missing leg of the reference triangle.

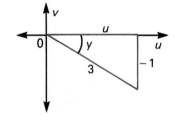

$$u^2 + (-1)^2 = 3^2$$
$$u^2 + 1 = 9$$
$$u^2 = 8$$
$$u = 2\sqrt{2} \text{ or } u = -2\sqrt{2}$$

Since y terminates in Quadrant IV, $u = 2\sqrt{2}$.

Therefore, $\cos\left[\operatorname{Csc}^{-1}(-3)\right] = \cos y = \frac{2\sqrt{2}}{3}$, or approximately 0.9428.

Calculator steps: 1 ÷ 3 = +/− INV sin cos ⟹ 0.942809

Focus on Reading

Indicate whether each statement is always true, sometimes true, or never true.

1. The inverse of a trigonometric function is a function.
2. The inverse of $y = \cos x$ is $x = \cos y$.
3. If $B = \operatorname{Tan}^{-1} 4$, then $\tan B = 4$.
4. The principal values of $y = \operatorname{Sin}^{-1} x$ are $0 \leq y \leq \pi$.
5. $\operatorname{Sin}^{-1} x$ is more conveniently written as $\dfrac{1}{\operatorname{Sin} x}$.

Classroom Exercises

Give the meaning of each statement. Is the statement true?

1. $30 = \mathrm{Sin}^{-1}\,\frac{1}{2}$ **2.** $\frac{\pi}{4} = \mathrm{Tan}^{-1}\,1$ **3.** $-90 = \mathrm{Sin}^{-1}\,1$ **4.** $180 = \mathrm{Cos}^{-1}(-1)$

Rewrite each sentence in terms of an inverse function.

5. $\mathrm{Sin}\,A = b$ **6.** $\mathrm{Cos}\,P = \frac{3}{5}$ **7.** $\mathrm{Tan}\,M = 1$ **8.** $\mathrm{Tan}\left(-\frac{3\pi}{4}\right) = -1$

Find the value of y for each of the following. Write the answer in radians.

9. $y = \mathrm{Sin}^{-1}\,0$ **10.** $y = \mathrm{Cos}^{-1}\,1$ **11.** $y = \mathrm{Tan}^{-1}\,0$ **12.** $y = \mathrm{Sin}^{-1}\left(-\frac{1}{2}\right)$

Written Exercises

Find the value of y for each of the following. Write the answer in radians.

1. $y = \mathrm{Cos}^{-1}\,0$ **2.** $y = \mathrm{Sin}^{-1}\,\frac{1}{2}$ **3.** $y = \mathrm{Cos}^{-1}\left(-\frac{1}{2}\right)$ **4.** $y = \mathrm{Tan}^{-1}\,\sqrt{3}$

5. $y = \mathrm{Sin}^{-1}\,1$ **6.** $y = \mathrm{Tan}^{-1}\,(-1)$ **7.** $y = \mathrm{Sec}^{-1}\,\frac{2}{\sqrt{3}}$ **8.** $y = \mathrm{Csc}^{-1}(-\sqrt{2})$

Find the value of each angle measure to the nearest degree.

9. $\mathrm{Sin}^{-1}\,(-0.4617)$ **10.** $\mathrm{Tan}^{-1}\,4.3635$ **11.** $\mathrm{Cos}^{-1}\,(-0.7290)$

Evaluate each expression.

12. $\cos\left[\mathrm{Sin}^{-1}\left(-\frac{\sqrt{2}}{2}\right)\right]$ **13.** $\sin\left(\mathrm{Cos}^{-1}\,\frac{3}{5}\right)$ **14.** $\tan\left[\mathrm{Sin}^{-1}\left(-\frac{5}{13}\right)\right]$

15. Graph $y = \mathrm{Cot}^{-1}\,x$ for the principal values $0 < y < \pi$.

Simplify.

16. $\sin\,(\mathrm{Cos}^{-1}\,x)$ **17.** $\csc\left(\mathrm{Tan}^{-1}\,\frac{1}{x}\right)$ **18.** $\cot\left(\mathrm{Csc}^{-1}\,\frac{\sqrt{x^2+1}}{x}\right)$

19. Graph the function $y = \mathrm{Sin}^{-1}\,(-x)$. Determine whether $\mathrm{Sin}^{-1}\,(-x) = -\mathrm{Sin}^{-1}\,x$ is true for $-1 \le x \le 1$. Explain.

20. Show that if $x > 0$, then $\mathrm{Cot}^{-1}\,x = \mathrm{Tan}^{-1}\left(\frac{1}{x}\right)$.

Mixed Review

Graph each equation of a conic section.   *11.5, 12.1, 12.2, 12.3*

1. $y = x^2 - 4x$ **2.** $x^2 + y^2 = 25$ **3.** $4x^2 + 9y^2 = 36$ **4.** $x^2 - y^2 = 25$

As a part of their records, **meteorologists** compute the daily average temperature for a given city or location by averaging the high and low temperatures for the day. The graph below shows the average temperature on the first day of each month for a given city.

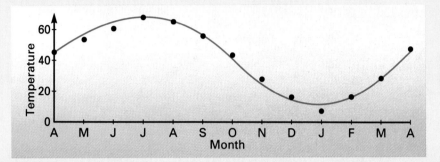

Notice that a sine curve drawn through the points models the graph. A model for a given site in the northern hemisphere can be written as

$$T(\theta) = c + a \sin \frac{2\pi}{365} (\theta - d).$$

The formula gives the average temperature for day θ, where θ is the number of days past March 21. The quantity c is the mean yearly temperature for the site, and a is the difference between the highest or lowest daily average and c. So, c translates the graph up or down and a is the amplitude. The period of the graph is $2\pi \div (2\pi/365)$ or 365 days. The quantity d is the number of days past March 21 that the daily average first exceeds the yearly mean. It translates the graph d units to the right.

EXAMPLE   In Seattle, Washington, the mean yearly temperature is 50°F, the highest average temperature is 62°F, and the daily average first reaches 50°F on April 27. What is the modeling equation for Seattle?

Solution $c = 50$ $a = 62 - 50 = 12$ $d = $ April 27 − March 21 = 37
So, $T(\theta) = 50 + 12 \sin \frac{2\pi}{365} (\theta - 37)$.

In Exercises 1–3, write the modeling equation for each city given the mean yearly temperature, the highest daily average, and the date the daily temperature first reaches the yearly mean.
  1. Detroit: 47, 71, Apr. 23   **2.** Miami: 75, 84, Apr. 25   **3.** Boston: 50, 72, Apr. 25
  4. Give Detroit's average temperature for June 21. (Use the radian mode.)

Key Terms

amplitude (p. 681)
bearing (p. 669)
cycle (p. 681)
Heron's Formula (p. 666)

Law of Cosines (p. 661)
Law of Sines (p. 668)
oblique triangle (p. 661)
period (p. 681)

periodic function (p. 681)
principal value (p. 701)
radian (p. 677)
semiperimeter (p. 666)

Key Ideas and Review Exercises

18.2   To find the area of a triangle, use $K = \frac{1}{2}bc \sin A$ for SAS, or use

$K = \sqrt{s(s - a)(s - b)(s - c)}$ for SSS, where $s = \frac{1}{2}(a + b + c)$.

1. Find the area of triangle ABC to two significant digits for $a = 12$, $c = 14$, and m $\angle B = 125$.

2. Find the area of triangle ABC in simplest radical form for $a = 4$, $b = 6$, and $c = 6$.

3. Find, to two significant digits, the area of a regular hexagon inscribed in a circle of radius 12.

**18.1,
18.3,
18.4**   To solve an oblique (nonright) triangle, use the following laws.

• Law of Cosines: $a^2 = b^2 + c^2 - 2bc \cos A$, for SSS or SAS

• Law of Sines: $\dfrac{\sin A}{a} = \dfrac{\sin B}{b} = \dfrac{\sin C}{c}$, for ASA or SSA

To determine the number of solutions for SSA, see pages 672 and 673.

Solve triangle ABC, if possible. Find lengths to two significant digits and angle measures to the nearest degree.

4. $a = 12, b = 16,$ m $\angle C = 100$

5. $a = 11, b = 12, c = 14$

6. m $\angle A = 60, c = 20, a = 18$

7. m $\angle A = 80, c = 56, a = 12$

8. Juan determines that the angle of elevation to the top of a building measures 42. At 120 ft closer to the building, the angle of elevation measures 48. Find the height of the building to two significant digits.

18.5   To convert degrees to radians, use $1° = \dfrac{\pi}{180}$ radians.

To convert radians to degrees, use π radians $= 180°$, or 1 radian $= 57.3°$.

9. Express $240°$ as a radian measure in terms of π.

10. Express $-\dfrac{3\pi}{2}$ in degree measure.

11. Express 1.7 radians in degree measure.

12. Simplify $\tan \dfrac{5\pi}{6}$.

13. Simplify $\dfrac{1}{3} \tan^2 \dfrac{4\pi}{3}$.

18.6   To determine the period of a periodic function, find the smallest value of p for which $f(x) = f(x + p)$.

To determine the amplitude of a periodic function, find $\dfrac{M - m}{2}$, where M is the maximum value and m is the minimum value of the function.

Part of the graph of a periodic function $y = f(x)$ with period $p = 6$ is shown at the right.

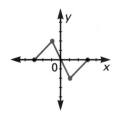

14. Graph $y = f(x)$ in the interval $-9 \le x \le 15$.

15. Find the amplitude of $y = f(x)$.

16. Determine whether f is even, odd, or neither.

18.7,   To graph trigonometric functions first determine the amplitude, if any, and
18.8,   the period using the summary table below.
18.9

	Amplitude	Period	Odd or Even	Shape of Graph
$T = \sin\theta$	1	2π	Odd: $\sin(-\theta) = -\sin\theta$	See page 685.
$T = \cos\theta$	1	2π	Even: $\cos(-\theta) = \cos\theta$	See page 687.
$T = \tan\theta$	None	π	Odd: $\tan(-\theta) = -\tan\theta$	See page 697.
$T = \cot\theta$	None	π	Odd: $\cot(-\theta) = -\cot\theta$	See Ex. 8, page 700.
$T = \sec\theta$	None	2π	Even: $\sec(-\theta) = \sec\theta$	See Ex. 6, page 700.
$T = \csc\theta$	None	2π	Odd: $\csc(-\theta) = -\csc\theta$	See page 699.

For changes in amplitude and period, see pages 695 and 697–699.

For an example of a vertical translation of a graph, see page 695.

Graph each function for $0 \le \theta \le 2\pi$ using a table of values. Give the amplitude and the period of each function.

17. $T = 3\cos 2\theta$ **18.** $T = -3 + 2\sin\theta$

Graph one cycle of each function without using a table of values.

19. $T = \frac{1}{4}\cos 3\theta$ **20.** $T = 5\sin\frac{1}{4}\theta$ **21.** $T = \tan 4\theta$ **22.** $T = 3\csc 6\theta$

23. If $\sin\theta = 0.7431$, find $\sin(-\theta)$.   **24.** If $\cos\theta = -0.8192$, find $\cos(-\theta)$.

18.10   An equation such as $y = \text{Sin}^{-1}\, x$ is read "y is the angle whose sine is x" (y is the principal value of the angle whose sine is x).

25. Find the value of y in terms of radians if $y = \text{Cos}^{-1}\left(-\dfrac{\sqrt{3}}{2}\right)$.

26. Find $\text{Tan}^{-1}(-0.3640)$ to the nearest degree.

27. Evaluate $\sin\left(\text{Cos}^{-1}\dfrac{4}{5}\right)$.

Find the area of triangle *ABC* to two significant digits.

1. $a = 12, b = 8, \text{m} \angle C = 85$

2. $a = 10, b = 8, c = 8$

3. Find, to two significant digits, the area of a regular octagon inscribed in a circle of radius 16.

In each triangle *ABC*, find lengths to two significant digits and angle measures to the nearest degree. Show all possible solutions.

4. $a = 20, b = 16, \text{m} \angle C = 105,$ $c = \underline{\ ?\ }$

5. $a = 12, b = 20, c = 12,$ $\text{m} \angle B = \underline{\ ?\ }$

6. $a = 14, \text{m} \angle A = 40, \text{m} \angle B = 80,$ $c = \underline{\ ?\ }$

7. $\text{m} \angle A = 50, b = 11, a = 10,$ $\text{m} \angle B = \underline{\ ?\ }$

8. A navigator sailing due north sights a lighthouse at a bearing of 28. After sailing another 55 km, the bearing of the lighthouse is then 115. Find his distance from the lighthouse, to two significant digits, at the time of the second measurement.

9. Express $-240°$ as a radian measure in terms of π.

10. Express $-\dfrac{2\pi}{3}$ in degree measure.

11. Simplify $\cot \dfrac{4\pi}{3}$.

12. Simplify $1 - \cos\left(-\dfrac{\pi}{3}\right)$.

Part of the graph of a periodic function $y = f(x)$ with period $p = 4$ is shown at the right. (Exercises 13–15)

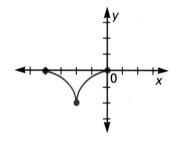

13. Graph $y = f(x)$ in the interval $-8 \le x \le 16$.

14. Find the amplitude of $y = f(x)$.

15. Determine whether f is even, odd, or neither.

16. Give the amplitude and the period of $T = 5 \cos 3\theta$.

17. Graph $T = -6 + \dfrac{1}{2} \sin \theta$ for $0 \le \theta \le 2\pi$, using a table of values. Give the amplitude and the period of the function.

Graph one cycle of each function without using a table of values.

18. $T = 3 \sin \dfrac{1}{5}\theta$

19. $T = \tan 3\theta$

20. If $\sin \theta = -0.6691$, find $\sin(-\theta)$.

21. Find the value of y in terms of radians if $y = \text{Sin}^{-1}\left(-\dfrac{\sqrt{3}}{2}\right)$.

22. The sides of a triangle are in arithmetic sequence, the shortest side being 5.0 in. long, and the perimeter of the triangle being 24.0 in. Find the length of the altitude to the 5.0-in. side to two significant digits.

College Prep Test

Choose the *one* best answer to each question or problem.

1. From which of the following statements, taken separately or together, can it be determined that m is greater than n?

I. $2m + n > 12$ II. $m + n = 7$

(A) I alone, but not II
(B) I and II taken together, but neither taken alone
(C) II alone, but not I
(D) Neither I nor II nor both
(E) Both I alone and II alone

2. Three points on a line are X, Y, and Z, in that order. If $XZ - YZ = 6$, what is the ratio $\dfrac{YZ}{XZ}$?

(A) 1:5   (B) 1:4
(C) 1:3   (D) 1:2
(E) It cannot be determined from the given information.

3. $2 \cos \dfrac{\pi}{4} =$ ___?___

(A) 0   (B) $\dfrac{1}{2}$ (C) 1

(D) $\sqrt{2}$ (E) 2

4. $\operatorname{Sin}^{-1} \dfrac{\sqrt{3}}{2} + \operatorname{Cos}^{-1} \dfrac{\sqrt{3}}{2} =$ ___?___

(A) 0   (B) $\dfrac{\pi}{6}$ (C) $\dfrac{\pi}{3}$

(D) $\dfrac{\pi}{2}$ (E) π

5. If $\dfrac{a + b}{a} = 3$ and $\dfrac{c + b}{c} = 5$, what is the value of $\dfrac{c}{a}$?

(A) 2   (B) $\dfrac{5}{3}$ (C) $\dfrac{3}{5}$

(D) 15   (E) $\dfrac{1}{2}$

6. In triangle ABC below, if $BC < BA$, which of the following is true?

(A) $x > y$ (B) $y < z$ (C) $y > x$
(D) $y = x$ (E) $z < x$

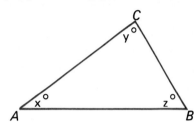

7. If $\dfrac{13a}{4}$ is an integer, then a could be any of the following except ___?___

(A) 8   (B) -32 (C) 6
(D) 0   (E) -112

8. If the area of a circle is 49π, what is the circumference of the circle?

(A) 7   (B) 7π (C) 14
(D) 14π (E) 49

9. If v, w, x, y, and z are whole numbers, then $v[w(x + y) + z]$ will be an even number when which of the following is even?

(A) v (B) w (C) x
(D) y (E) z

10. $\sin \left[\operatorname{Tan}^{-1}\left(-\dfrac{3}{4}\right) \right] =$ ___?___
(A) -1 (B) -0.75 (C) -0.6
(D) 0.6   (E) 0.75

11. If $P = \dfrac{h(a + b)}{2}$, what is the average of a and b when $P = 30$ and $h = 5$?

(A) 6   (B) 150   (C) 15
(D) $\dfrac{35}{2}$ (E) 2

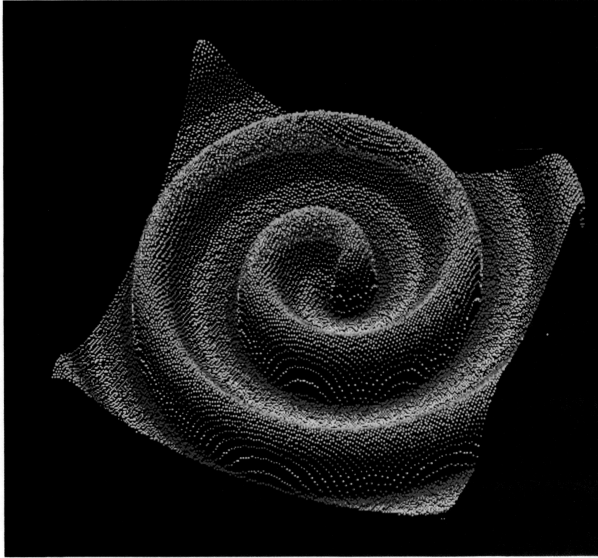

Certain systems that interact with their environment
have the ability to generate *order out of chaos.* The
image shown here is a computer-generated study of a
chemical reaction that creates a spiral within a thoroughly
mixed solution.

19.1 Basic Trigonometric Identities

Objectives

To verify the basic trigonometric identities for given values of the variables

To prove basic trigonometric identities

An equation that is true for all permissible values of its variables is called an **identity**. For example, the equation $\dfrac{4}{3x - 9} = \dfrac{4}{3(x - 3)}$ is an identity since it is true for all $x \neq 3$.

Recall that $\sin \theta$ and $\csc \theta$ are reciprocals of each other for $\sin \theta \neq 0$, as shown below.

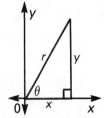

$$\sin \theta \csc \theta = \frac{y}{r} \cdot \frac{r}{y} = 1,$$

where $y \neq 0, r > 0$

Note that $y \neq 0$ if and only if $\sin \theta \neq 0$. Therefore, the equation $\sin \theta \csc \theta = 1$ is an identity since it is true for all values of θ for which $\sin \theta \neq 0$.

Reciprocal identities can be written for all of the trigonometric ratios.

Reciprocal Identities

$\sin \theta \csc \theta = 1$, for $\sin \theta \neq 0$ $\quad \left(\sin \theta = \dfrac{1}{\csc \theta} \text{ or } \csc \theta = \dfrac{1}{\sin \theta} \right)$

$\cos \theta \sec \theta = 1$, for $\cos \theta \neq 0$ $\quad \left(\cos \theta = \dfrac{1}{\sec \theta} \text{ or } \sec \theta = \dfrac{1}{\cos \theta} \right)$

$\tan \theta \cot \theta = 1$, for $\tan \theta \neq 0$, $\quad \left(\tan \theta = \dfrac{1}{\cot \theta} \text{ or } \cot \theta = \dfrac{1}{\tan \theta} \right)$

$\qquad\qquad \cot \theta \neq 0$

Recall that the acute angles of a right triangle are complementary. For example, in right triangle ABC at the right, $\angle A$ and $\angle B$ are complementary since $60 + 30 = 90$. Notice that in the triangle,

$$\sin 60 = \frac{\sqrt{3}}{2} \text{ and } \cos 30 = \frac{\sqrt{3}}{2}.$$

Therefore, $\sin 60 = \cos 30$.

This suggests that the sine of an angle equals the cosine of its complement.

The sine and cosine are said to be **cofunctions** of each other. The other cofunctions are tangent and cotangent, and secant and cosecant. The *cofunction identities* are stated below.

Cofunction Identities

$\sin \theta = \cos (90 - \theta)$ $\qquad\qquad$ $\cos \theta = \sin (90 - \theta)$

$\tan \theta = \cot (90 - \theta)$ $\qquad\qquad$ $\cot \theta = \tan (90 - \theta)$

$\sec \theta = \csc (90 - \theta)$ $\qquad\qquad$ $\csc \theta = \sec (90 - \theta)$

One of the cofunction identities was just illustrated for acute angles with positive measures. However, all of the identities are true for angles of all positive and nonpositive measures. This will be shown later in this chapter.

EXAMPLE 1   Express each function in terms of its cofunction.

$\quad$ **a.** $\cos 74$ $\qquad\qquad\qquad\qquad\qquad\qquad$ **b.** $\sec 25$

Solutions   Use the cofunction and the complement.

$\quad$ **a.** $\cos 74 = \sin (90 - 74)$ $\qquad$ **b.** $\sec 25 = \csc (90 - 25)$

$\qquad\qquad\qquad = \sin 16$ $\qquad\qquad\qquad\qquad\quad = \csc 65$

The definitions of the basic trigonometric ratios in Lesson 17.6 can be used to prove the following *quotient identities*.

Quotient Identities

$\tan \theta = \dfrac{\sin \theta}{\cos \theta}, \cos \theta \neq 0$ $\qquad\qquad$ $\cot \theta = \dfrac{\cos \theta}{\sin \theta}, \sin \theta \neq 0$

EXAMPLE 2   Prove the quotient identity $\tan \theta = \dfrac{\sin \theta}{\cos \theta}$.

Proof   From Lesson 17.6, $\sin \theta = \dfrac{y}{r}$, $\cos \theta = \dfrac{x}{r}$, and

$\tan \theta = \dfrac{y}{x}$ (see diagram).

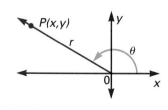

Therefore, $\dfrac{\sin \theta}{\cos \theta} = \dfrac{\dfrac{y}{r}}{\dfrac{x}{r}} = \dfrac{y}{x} = \tan \theta$.

Thus, $\tan \theta = \dfrac{\sin \theta}{\cos \theta}$ is an identity.

EXAMPLE 3   Verify that $\tan \theta = \dfrac{\sin \theta}{\cos \theta}$ for $\theta = \dfrac{5\pi}{6}$.

Solution $\dfrac{5\pi}{6}$ radians $= 150$ degrees (see diagram)

Therefore,

$$\frac{\sin 150}{\cos 150} = \frac{\dfrac{1}{2}}{\dfrac{-\sqrt{3}}{2}} = \frac{1}{2} \cdot \frac{2}{-\sqrt{3}} = \frac{1}{-\sqrt{3}} = \tan 150.$$

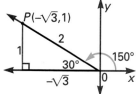

Thus, $\tan \dfrac{5\pi}{6} = \dfrac{\sin \dfrac{5\pi}{6}}{\cos \dfrac{5\pi}{6}}$.

The definitions of the trigonometric ratios can also be used to develop three special trigonometric identities called the *Pythagorean Identities*.

EXAMPLE 4   Prove that $\sin^2 \theta + \cos^2 \theta = 1$ is an identity.

Proof $\sin^2 \theta + \cos^2 \theta = \left(\dfrac{y}{r}\right)^2 + \left(\dfrac{x}{r}\right)^2 = \dfrac{y^2 + x^2}{r^2} = \dfrac{r^2}{r^2} = 1$

Therefore, $\sin^2 \theta + \cos^2 \theta = 1$ is an identity.

In Example 4, note that since $r > 0$, no denominator can be zero. Thus, all real values of the variable θ are possible values.

There are two other Pythagorean Identities. You are asked to prove one of these identities in Exercise 24. All three identities are summarized below.

Pythagorean Identities

For all values of θ for which $\tan \theta$, $\cot \theta$, $\sec \theta$, and $\csc \theta$ are defined,

$$\sin^2 \theta + \cos^2 \theta = 1$$
$$\tan^2 \theta + 1 = \sec^2 \theta$$
$$\cot^2 \theta + 1 = \csc^2 \theta$$

The Pythagorean Identities are often used in one of their alternate forms shown below.

$\sin^2 \theta = 1 - \cos^2 \theta$	$\tan^2 \theta = \sec^2 \theta - 1$	$\cot^2 \theta = \csc^2 \theta - 1$
$\cos^2 \theta = 1 - \sin^2 \theta$	$\sec^2 \theta - \tan^2 \theta = 1$	$\csc^2 \theta - \cot^2 \theta = 1$

Indicate whether each statement is true or false. If false, give a reason.

1. Trigonometric identities are equations that are true for all values of the variable θ.

2. $\sin \theta = \dfrac{1}{\csc \theta}$ is an example of a quotient identity.

3. The cofunction of $\cos \theta$ is $\sec (90 - \theta)$.

4. Another form of the Pythagorean Identity $\sin^2 \theta + \cos^2 \theta = 1$ is $\cos^2 \theta - 1 = -\sin^2 \theta$.

5. $\tan \theta = \cot (90 - \theta)$ is true only for acute angles.

Classroom Exercises

Complete each identity.

1. $\dfrac{\sin \theta}{\cos \theta} = \underline{\ ?\ }$

2. $\sec \theta \cdot \underline{\ ?\ } = 1$

3. $\cot^2 \theta = \underline{\ ?\ }$

4. $\sin^2 \theta + \cos^2 \theta = \underline{\ ?\ }$

5. $\sin (90 - \theta) = \underline{\ ?\ }$

6. $\tan \theta \cdot \cot \theta = \underline{\ ?\ }$

7. $\cot \theta = \underline{\ ?\ }$

8. $\dfrac{1}{\cot \theta} = \underline{\ ?\ }$

9. $\sec \underline{\ ?\ } = \csc \theta$

Express each function in terms of its cofunction.

10. $\cos 60$

11. $\sin 45$

12. $\cot 5$

13. $\cos \dfrac{\pi}{6}$

Verify each identity for the given value of θ.

14. $\sin^2 \theta + \cos^2 \theta = 1;\ \theta = 90$

15. $\dfrac{\sin \theta}{\cos \theta} = \tan \theta;\ \theta = 45$

Written Exercises

Express each function in terms of its cofunction.

1. $\sin 50$

2. $\tan 81$

3. $\csc 65.4$

4. $\cos \dfrac{\pi}{8}$

Verify each identity for the given value of θ.

5. $\sin \theta \csc \theta = 1;\ \theta = 135$

6. $\tan \theta = \dfrac{1}{\cot \theta};\ \theta = 135$

7. $\sin (90 - \theta) = \cos \theta;\ \theta = 60$

8. $\tan \theta = \cot (90 - \theta);\ \theta = 135$

9. $\dfrac{1}{\sec \theta} = \cos \theta;\ \theta = 30$

10. $\dfrac{\sin \theta}{\cos \theta} = \tan \theta;\ \theta = 150$

11. $\dfrac{\cos\theta}{\sin\theta} = \cot\theta;\ \theta = 240$

12. $\dfrac{1}{\tan\theta} = \cot\theta;\ \theta = 30$

13. $\cos^2\theta = 1 - \sin^2\theta;\ \theta = 270$

14. $\sec^2\theta - \tan^2\theta = 1;\ \theta = 60$

15. $1 + \cot^2\theta = \csc^2\theta;\ \theta = \dfrac{2\pi}{3}$

16. $\sin^2\theta = 1 - \cos^2\theta;\ \theta = \dfrac{5\pi}{4}$

17. $\dfrac{\cos\theta}{\sin\theta} = \cot\theta;\ \theta = \dfrac{11\pi}{6}$

18. $\cot\theta = \dfrac{1}{\tan\theta};\ \theta = \tan\theta;\ \theta = \dfrac{3\pi}{4}$

19. $1 + \tan^2\theta = \sec^2\theta;\ \theta = -\dfrac{7\pi}{3}$

20. $\cos^2\theta = 1 - \sin^2\theta;\ \theta = -\dfrac{3\pi}{2}$

Use the definitions of the trigonometric ratios to prove that each equation is an identity.

21. $\cos\theta\sec\theta = 1$

22. $\tan\theta\cot\theta = 1$

23. $\cot\theta = \dfrac{\cos\theta}{\sin\theta}$

24. $\tan^2\theta + 1 = \sec^2\theta$

Use identities from this lesson to prove the following.

25. $(1 - \sin^2\theta)(1 + \tan^2\theta) = 1$

26. $\dfrac{1}{\sec\theta - \tan\theta} - \dfrac{1}{\sec\theta + \tan\theta} = 2\tan\theta$

Mixed Review

Simplify.   *7.1, 7.4, 8.6*

1. $\dfrac{x^3 - 27}{4x - 12}$

2. $\dfrac{-x - 9}{x^2 - 9} - \dfrac{2}{3 - x}$

3. $(-32)^{-\frac{4}{5}}$

Give the amplitude and period of each function.   *18.8*

4. $T = 5\cos\frac{1}{2}\theta$

5. $T = -6\cos 4\theta$

6. $T = 3\sin 3\theta$

Brainteaser

In the figure, the measure of $\angle A$ is 30. $\angle D$ is the angle formed by the bisectors of $\angle B$ and $\angle C$. Find the measure of $\angle D$.

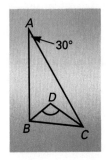

19.2 Proving Trigonometric Identities

Objectives	To simplify trigonometric expressions
	To prove trigonometric identities using basic identities

The identities of Lesson 19.1 can be used to simplify trigonometric expressions. One technique is to express all functions in terms of sines and cosines. This is illustrated in Example 1 below.

EXAMPLE 1 Simplify $\sin \theta \cot \theta$.

Plan First rewrite $\cot \theta$ in terms of $\sin \theta$ and $\cos \theta$.

Use the Quotient Identity, $\cot \theta = \dfrac{\cos \theta}{\sin \theta}$.

Solution $\sin \theta \cot \theta = \sin \theta \cdot \dfrac{\cos \theta}{\sin \theta} = \cos \theta$

Therefore, $\sin \theta \cot \theta = \cos \theta$.

You will find it helpful to make your own list of the *Reciprocal*, *Quotient*, and *Pythagorean Identities* of Lesson 19.1. Since these identities are used frequently, you should memorize them.

The Pythagorean and Reciprocal Identities are used in Example 2.

EXAMPLE 2 Simplify $\dfrac{\sin^2 \theta}{\cos \theta} + \cos \theta$.

Plan First combine the fractions by making the denominators alike.

Solution $\dfrac{\sin^2 \theta}{\cos \theta} + \dfrac{\cos \theta}{1} = \dfrac{\sin^2 \theta}{\cos \theta} + \dfrac{\cos \theta \cos \theta}{1 \cdot \cos \theta}$

$= \dfrac{\sin^2 \theta}{\cos \theta} + \dfrac{\cos^2 \theta}{\cos \theta}$

$= \dfrac{\sin^2 \theta + \cos^2 \theta}{\cos \theta}$

$= \dfrac{1}{\cos \theta} = \sec \theta$

Of the two trigonometric equations below, one is an identity.

(1) $\sin \theta \cot \theta = \cos \theta$ (2) $\sin \theta = \cos \theta$

Equation (1) from Example 1 *is* an identity. Equation (2) is false for some values of θ (such as $\theta = 30$) and thus is *not* an identity.

EXAMPLE 3   Prove that $\dfrac{\sin \theta}{1 + \cos \theta} = \dfrac{1 - \cos \theta}{\sin \theta}$ is an identity.

Plan   The right side of the equation contains only $1 - \cos \theta$ in the numerator. Get $1 - \cos \theta$ in the numerator on the left by multiplying the numerator and the denominator of the left side of the equation by $1 - \cos \theta$. Then use basic algebraic and trigonometric identities to transform the expression on the left into the one on the right side.

Solution

$$\dfrac{\sin \theta}{1 + \cos \theta} \quad \Bigg| \quad \dfrac{1 - \cos \theta}{\sin \theta}$$

$$\dfrac{(1 - \cos \theta)\sin \theta}{(1 - \cos \theta)(1 + \cos \theta)}$$

Use $(a - b)(a + b) = a^2 - b^2$. $\dfrac{(1 - \cos \theta)\sin \theta}{1 - \cos^2 \theta}$

Use the Pythagorean Identity $\dfrac{(1 - \cos \theta)\sin \theta}{\sin^2 \theta}$
$1 - \cos^2 \theta = \sin^2 \theta$.

Simplify. $\dfrac{1 - \cos \theta}{\sin \theta} = \dfrac{1 - \cos \theta}{\sin \theta}$

The original equation will be an identity *unless* the transformation steps change the restrictions on the variable θ. Since this is not the case (only integral multiples of 180 are forbidden in the original and the transformed equations), the original equation is an identity.

In Example 4, each side of an equation is transformed into a third expression that is equivalent to each of the two sides.

EXAMPLE 4   Prove that $\cos^2 \theta \tan^2 \theta + 1 = \sec^2 \theta + \sin^2 \theta - \sin^2 \theta \sec^2 \theta$ is an identity.

Plan   Use the identity $\tan \theta = \dfrac{\sin \theta}{\cos \theta}$ to simplify part of the left side. The result, $\sin^2 \theta + 1$, agrees with the right side in one term, $\sin^2 \theta$. Therefore, ignore this term and concentrate on the rest of the right side, $\sec^2 \theta - \sin^2 \theta \sec^2 \theta$.

Solution

$$\begin{array}{c|c}
\cos^2\theta \tan^2\theta + 1 & \sec^2\theta + \sin^2\theta - \sin^2\theta \sec^2\theta \\[2mm]
\cos^2\theta \dfrac{\sin^2\theta}{\cos^2\theta} + 1 & \sin^2\theta + \sec^2\theta - \sin^2\theta \sec^2\theta \\[2mm]
\sin^2\theta + 1 & \sin^2\theta + \sec^2\theta\,(1 - \sin^2\theta) \\[1mm]
 & \sin^2\theta + \sec^2\theta \cos^2\theta \\[2mm]
\multicolumn{2}{c}{\sin^2\theta + 1 = \sin^2\theta + 1}
\end{array}$$

Summary

To prove an identity,

(1) transform one side (the more complicated one, if possible) into the other side, or

(2) transform each side into the same expression.

The following guidelines should be observed when performing the above transformations.

(1) Apply the Reciprocal, Quotient, or Pythagorean Identities.

(2) Use basic algebraic identities such as $x^2 - y^2 = (x + y)(x - y)$.

(3) Combine the sum or difference of fractions into a single fraction.

(4) If all else fails, change all expressions into sines and cosines and simplify.

Classroom Exercises

Factor.

1. $\sin^2\theta - \sin\theta$ **2.** $1 - \tan^2\theta$ **3.** $\sin\theta + \sin\theta\cos\theta$

Simplify.

4. $\sin\theta \; \dfrac{1}{\sin^2\theta}$ **5.** $(1 - \cos\theta)(1 + \cos\theta)$ **6.** $\dfrac{1}{1 - \sin\theta} + \dfrac{\sin\theta}{1 - \sin\theta}$

7. $\cos\theta\tan\theta$ **8.** $\sin\theta\sec\theta$ **9.** $\dfrac{1 - \sin^2\theta}{\cos^2\theta}$

Written Exercises

Simplify.

1. $\tan\theta\cot\theta$ **2.** $\cos\theta\left(\dfrac{1}{1 - \sin^2\theta}\right)$ **3.** $\cos\theta\csc\theta$

4. $\csc^2\theta - \csc^2\theta\cos^2\theta$ **5.** $\tan^2\theta - \tan^2\theta\sin^2\theta$ **6.** $\dfrac{\cos\theta}{\cot\theta}$

7. $\dfrac{\sin \theta - \sin \theta \cos \theta}{1 - \cos \theta}$ **8.** $\dfrac{\cos \theta \sin^2 \theta + \cos^3 \theta}{\cos \theta}$ **9.** $\dfrac{2 \sin \theta}{1 - \cos^2 \theta}$

Prove that each equation is an identity. (Exercises 10–30)

10. $\dfrac{\sin \theta}{\cos \theta} + \dfrac{\cos \theta}{\sin \theta} = \dfrac{1}{\sin \theta \cos \theta}$

11. $\sin \theta + \dfrac{\cos^2 \theta}{\sin \theta} = \csc \theta$

12. $(\sin \theta + \cos \theta)^2 = 1 + 2 \sin \theta \cos \theta$

13. $\cos^2 \theta (\cot^2 \theta + 1) = \cot^2 \theta$

14. $\cos^2 \theta - \sin^2 \theta = 1 - 2 \sin^2 \theta$

15. $\dfrac{\cos \theta}{\sin \theta} + \dfrac{\sin \theta}{1 + \cos \theta} = \csc \theta$

16. $\dfrac{\cot \theta + \tan \theta}{\csc^2 \theta} = \tan \theta$

17. $\dfrac{\sin \theta + \tan \theta}{\cot \theta + \csc \theta} = \sin \theta \tan \theta$

18. $\dfrac{\cot \theta}{1 + \cot^2 \theta} = \sin \theta \cos \theta$

19. $\dfrac{1 + \cot \theta}{\cot \theta \sin \theta + \dfrac{\cos^2 \theta}{\sin \theta}} = \sec \theta$

20. $\dfrac{1 + \sec \theta}{\sin \theta + \tan \theta} = \csc \theta$

21. $\dfrac{\csc \theta - \sin \theta}{\cot^2 \theta} = \sin \theta$

22. $\dfrac{1 + \sin \theta}{\cos \theta} = \dfrac{\cos \theta}{1 - \sin \theta}$

23. $\dfrac{1}{1 - \sin \theta} + \dfrac{1}{1 + \sin \theta} = 2 \sec^2 \theta$

24. $\dfrac{\csc \theta + \cot \theta}{\tan \theta + \sin \theta} = \cot \theta \csc \theta$

25. $\sin^2 \theta \sec^2 \theta + \sin^2 \theta \csc^2 \theta = \sec^2 \theta$

26. $\sin^4 \theta - \cos^4 \theta = 2 \sin^2 \theta - 1$

27. $\dfrac{\csc \theta}{\cot \theta + \tan \theta} = \cot \theta \sin \theta$

28. $\cos \theta (2 \sec \theta + \tan \theta)(\sec \theta - 2 \tan \theta) = 2 \cos \theta - 3 \tan \theta$

29. $(\sin \theta + \cos \theta)(\tan \theta + \cot \theta) = \csc \theta + \sec \theta$

30. $\dfrac{1}{\csc \theta - \cot \theta} - \dfrac{1}{\csc \theta + \cot \theta} = 2 \cot \theta$

31. Show that the determinant equation $\begin{vmatrix} \cos \theta & -\sin \theta \\ \sin \theta & \cos \theta \end{vmatrix} = 1$ is an identity.

32. For what value of t is $t \cos \theta \sin \theta - 1 + (\sin \theta + \cos \theta)^2 = 0$ an identity?

33. Simplify $\dfrac{1 - \sin \theta + \cos \theta - \sin \theta \cos \theta}{1 - \sin^3 \theta} \div \dfrac{1 + \cos \theta}{1 + \cos \theta + \cos^2 \theta}$.

Mixed Review

Solve. *1.4, 6.1*

1. $3x - 5(2 - x) = 4 - x$

2. $7x - 4 = -2x^2$

Simplify. *17.6, 17.7, 18.5*

3. $\cos \dfrac{3\pi}{2}$

4. $\sin 120$

5. $\tan \dfrac{3\pi}{4}$

6. $\cot \dfrac{\pi}{4}$

19.3 Introduction to Trigonometric Equations

Objective

To solve trigonometric equations involving one trigonometric ratio

An equation such as $\tan \theta = \frac{\sin \theta}{\cos \theta}$ is an identity since it is true for all values of θ for which $\tan \theta$ is defined. However, most equations encountered in algebra are *conditional*, that is, true for only some values of the variable. The equations $\cos \theta = -\frac{\sqrt{3}}{2}$, $\tan \theta = 1$, and $5 \sin \theta + 6\sqrt{2} = 3 \sin \theta + 5\sqrt{2}$ are conditional equations. Solving such equations often involves recognizing special angle relationships.

Special Angle Relationships (first-quadrant position)

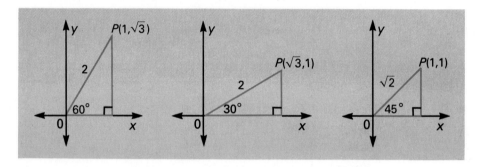

EXAMPLE 1

Solve $5 \sin \theta + 6\sqrt{2} = 3 \sin \theta + 5\sqrt{2}, 0 \le \theta < 360$.

Solution

(1) Solve for $\sin \theta$.

$$5 \sin \theta + 6\sqrt{2} = 3 \sin \theta + 5\sqrt{2}$$
$$5 \sin \theta - 3 \sin \theta = 5\sqrt{2} - 6\sqrt{2}$$
$$2 \sin \theta = -\sqrt{2}$$
$$\sin \theta = -\frac{\sqrt{2}}{2}$$

(2) Sin θ is negative only in Quadrants III and IV.

(3) The reference angle measures 45.

$(\sin 45 = \frac{1}{\sqrt{2}} = \frac{\sqrt{2}}{2})$

$\theta = 180 + 45$ or $\theta = 360 - 45$
$\quad = 225 \qquad\qquad = 315$

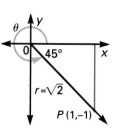

Thus, the solutions of the equation are 225 and 315.

Solutions of a trigonometric equation in a specified interval, usually $0 \leq \theta < 360$, are called *primary solutions*. Thus, 225 and 315 are primary solutions of the equation in Example 1. If there had been no specified interval, other solutions would differ from 225 and 315 by integral multiples of 360 as follows:

$\theta = 225 + k \cdot 360$ *or* $\theta = 315 + k \cdot 360$, where k is an integer.

For the next examples, the following figures review quadrantal angles.

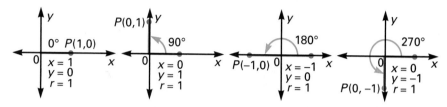

0° P(1,0)	90°	180°	270°

EXAMPLE 2   Solve $2 \cos^2 \theta = 3 \cos \theta - 1$, $0 \leq \theta < 2\pi$.

Plan   The equation is of the form $2x^2 = 3x - 1$ and is thus in quadratic form (see Lesson 9.1). Solve as a quadratic equation.

Solution
$$2 \cos^2 \theta = 3 \cos \theta - 1$$
$$2 \cos^2 \theta - 3 \cos \theta + 1 = 0$$
$$(2 \cos \theta - 1)(\cos \theta - 1) = 0$$
$$2 \cos \theta - 1 = 0 \quad or \quad \cos \theta - 1 = 0$$
$$\cos \theta = \tfrac{1}{2} \quad or \quad \cos \theta = 1$$

$\cos \theta = \tfrac{1}{2}$: $\cos \theta$ is positive in Quadrants I and IV.
        The reference angle is 60.
$\cos \theta = 1$: quadrantal angle, $\theta = 0$

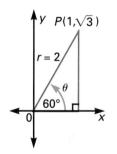
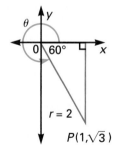
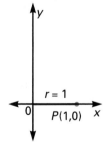

$\theta = 60$ $\theta = 360 - 60 = 300$ $\theta = 0$

Write the solutions in radians.

$$60 \cdot \frac{\pi}{180} = \frac{\pi}{3} \qquad 300 \cdot \frac{\pi}{180} = \frac{5\pi}{3} \qquad 0 \cdot \frac{\pi}{180} = 0$$

Thus, the primary solutions are 0, $\frac{\pi}{3}$, and $\frac{5\pi}{3}$.

If a trigonometric equation involves angles that are neither special nor quadrantal, then it is necessary to use a table or a calculator.

EXAMPLE 3 Solve $5 \tan^2 \theta - 4 \tan \theta = 0, 0 \le \theta < 360$. Give answers to the nearest degree.

Solution

$5 \tan^2 \theta - 4 \tan \theta = 0$
$\tan \theta (5 \tan \theta - 4) = 0 \leftarrow \tan \theta$ is the GCF.
$\tan \theta = 0 \quad or \quad 5 \tan \theta - 4 = 0$

$$\tan \theta = \frac{4}{5}, \text{ or } 0.8000$$

$\tan \theta = 0$:
quadrantal angle
Use $\tan \theta = \dfrac{0}{1} = 0$ and
$\tan \theta = \dfrac{0}{-1} = 0$.

$\tan \theta = 0.8000$:
$\tan \theta$ is positive in Quadrants I and III. From a table or a calculator, the reference angle is 39 to the nearest degree. (By calculator, 0.8 $\boxed{\text{INV}}$ $\boxed{\text{tan}}$ $\Rightarrow 38.659808$.)

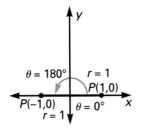
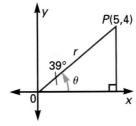
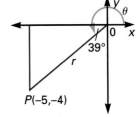

$\theta = 0 \ or \ \theta = 180$ $\theta = 39$ $\theta = 180 + 39$, or 219

Thus, the primary solutions are 0, 39, 180, and 219.

Recall that $\sin \theta$ can never be greater than 1. Thus, an equation such as $\sin \theta = 3$ has no solution.

EXAMPLE 4 Solve $-3 - 2 \sin \theta = -\sin^2 \theta, 0 \le \theta < 2\pi$.

Solution First rewrite in standard form. Then factor.

$$-3 - 2 \sin \theta = -\sin^2 \theta$$
$$\sin^2 \theta - 2 \sin \theta - 3 = 0$$
$$(\sin \theta - 3)(\sin \theta + 1) = 0$$

$\sin \theta = 3 \qquad or$
$\sin \theta$ cannot be
greater than 1. No
solution.

$\sin \theta = -1$
$\theta = 270$
$\theta = 270 \cdot \dfrac{\pi}{180}$, or
$\dfrac{3\pi}{2}$ radians

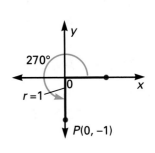

Thus, the primary solution is $\dfrac{3\pi}{2}$.

Classroom Exercises

Give the steps for solving each equation. If factoring is necessary, identify the factoring as either common-monomial factoring or factoring into two binomials.

1. $3 \sin \theta + 6 = \sin \theta + 7$ **2.** $\cos^2 \theta - \cos \theta = 0$ **3.** $3 \sin^2 \theta = 2 \sin \theta + 1$

4. $\sin^2 \theta - 1 = 0$ **5.** $4 \cos \theta - 1 = 2$ **6.** $2 \tan^2 \theta - 3 \tan \theta = 0$

Solve for θ, $0 \le \theta < 360$.

7. $\cot \theta = 1$ **8.** $2 \cos \theta = \sqrt{3}$ **9.** $6 \sin \theta + 7 = 10$

Written Exercises

Solve for θ, $0 \le \theta < 360$.

1. $\cos \theta = -\dfrac{1}{2}$ **2.** $\tan \theta = 1$ **3.** $\sin \theta = \dfrac{\sqrt{3}}{2}$ **4.** $\tan \theta = \sqrt{3}$

5. $\csc \theta = 2$ **6.** $\cos \theta = \dfrac{1}{\sqrt{2}}$ **7.** $\tan \theta = -1$ **8.** $\cos \theta = -\dfrac{\sqrt{3}}{2}$

9. $\cos \theta = -1$ **10.** $\sin \theta = 0$ **11.** $\tan \theta = 0$ **12.** $\sin \theta = 0.5299$

Solve for θ, $0 \le \theta < 2\pi$.

13. $2 \sin \theta = -\sqrt{3}$ **14.** $4 \cos \theta - 6 = -4$ **15.** $\sqrt{2} \cos \theta + 4 = 3$

16. $2 \sin \theta = 4 \sin \theta + 2$ **17.** $4\sqrt{3} \sec \theta = -8$ **18.** $5 \cot \theta - 3 = 2 \cot \theta$

Find solutions of each equation to the nearest degree, $0 \le \theta < 360$.

19. $4 \sin \theta + 1 = 3$ **20.** $3 \tan \theta - 4 = 4 \tan \theta + 2$

21. $7 \cos \theta + 3 = 2 \cos \theta - 1$ **22.** $\sqrt{2}(2 \sin \theta + \sqrt{2}) = 3\sqrt{2} \sin \theta + 1$

23. $2 \cos^2 \theta + \sqrt{3} \cos \theta = 0$ **24.** $2 \sin^2 \theta = -\sin \theta + 1$

25. $3 \cos \theta + 2 = -\cos^2 \theta$ **26.** $\tan^2 \theta = \tan \theta$

27. $\csc^2 \theta + \sqrt{2} \csc \theta = 0$ **28.** $\sec^2 \theta - 3 \sec \theta + 2 = 0$

29. $4 \cot^2 \theta - 1 = 0$ **30.** $\tan^2 \theta + 5 \tan \theta + 6 = 0$

31. $2 \sin \theta \cos \theta - 2 \cos \theta + \sin \theta = 1$ **32.** $\cos^2 \theta + 2 \cos \theta - 1 = 0$

33. $2 \cos^3 \theta - \cos^2 \theta - 2 \cos \theta = -1$ **34.** $\csc^4 \theta - 4 \csc^2 \theta + 3 = 0$

Mixed Review

Sketch one cycle of the graph of each function without using a table of values. *18.8, 18.9*

1. $f(\theta) = 3 + 2 \sin \theta$ **2.** $f(\theta) = \cos \dfrac{1}{2}\theta$

3. $f(\theta) = 3 \cos 2\theta$ **4.** $f(\theta) = \tan 3\theta$

19.4 Trigonometric Equations: Using Basic Identities

Objective

To solve trigonometric equations using trigonometric identities

In medieval times, trigonometric ratios were represented in mathematics texts by line segments, as the figure at the right illustrates. The texts, which were in Latin, referred to the line segments as *touching* (tangent) or *cutting* (secant) a circle of radius 1. The figure suggests an identity that you already know and that appears in Example 1.

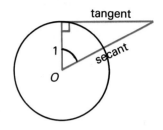

EXAMPLE 1 Solve, to the nearest degree, $\tan \theta + \sec^2 \theta = 3, 0 \le \theta < 360$.

Plan Use the Pythagorean Identity $\sec^2 \theta = \tan^2 \theta + 1$ to obtain expressions that involve only $\tan \theta$.

Solution

$$\tan \theta + \sec^2 \theta = 3$$
$$\tan \theta + (\tan^2 \theta + 1) = 3 \leftarrow \sec^2 \theta = \tan^2 \theta + 1 \text{ (See figure above.)}$$
$$\tan^2 \theta + \tan \theta - 2 = 0$$
$$(\tan \theta + 2)(\tan \theta - 1) = 0$$
$$\tan \theta + 2 = 0 \quad or \quad \tan \theta - 1 = 0$$
$$\tan \theta = -2 \quad or \quad \tan \theta = 1$$

$\tan \theta = -2$:
   $\tan \theta$ is negative in Quadrants II and IV. By calculator, the measure of the reference angle is 2 ⬚INV⬚ ⬚tan⬚ ⇒ 63.434949, or 63 to the nearest degree.
   Then, either $\theta = 180 - 63$, or 117 (Quadrant II), or $\theta = 360 - 63$, or 297 (Quadrant IV).

$\tan \theta = 1$:
   $\tan \theta$ is positive in Quadrants I and III.
   Since $\tan 45 = 1$ the reference angle measures 45.
   Then, either $\theta = 45$ (Quadrant I)
   or $\theta = 180 + 45 = 225$ (Quadrant III).

Thus, the four primary solutions are 45, 117, 225, and 297.

EXAMPLE 2   Solve and check $\cos \theta + 2 \sin \theta = 2, 0 \le \theta < 360$. Express results to the nearest degree.

Plan   The equation can be written with only sines or cosines using the Pythagorean Identity. However, since this requires that both sides of the equation be squared, the equation must be checked for extraneous roots.

To avoid a complicated solution, subtract $2 \sin \theta$ from each side before squaring.

Solution
$$\cos \theta + 2 \sin \theta = 2$$
$$\cos \theta = 2 - 2 \sin \theta$$
$$(\cos \theta)^2 = (2 - 2 \sin \theta)^2 \leftarrow \text{Square each side.}$$
$$\cos^2 \theta = 4 - 8 \sin \theta + 4 \sin^2 \theta$$
$$1 - \sin^2 \theta = 4 - 8 \sin \theta + 4 \sin^2 \theta \leftarrow \text{Use } \cos^2 \theta = 1 - \sin^2 \theta.$$
$$0 = 5 \sin^2 \theta - 8 \sin \theta + 3$$
$$0 = (5 \sin \theta - 3)(\sin \theta - 1)$$
$$5 \sin \theta - 3 = 0 \quad or \quad \sin \theta - 1 = 0$$
$$\sin \theta = \frac{3}{5} \quad or \quad \sin \theta = 1$$

$\sin \theta = 0.6$:
 sin θ is positive in Quadrants I and II. By calculator, the measure of the reference angle is 37.
Then, either $\theta = 37$ (Quadrant I)
or $\theta = 180 - 37 = 143$ (Quadrant II).

$\sin \theta = 1$:
 Then $\theta = 90$ (quadrantal angle).

Checks   In Quadrant I, $\cos 37 = 0.8$. In Quadrant II, $\cos 143 = -0.8$. $\cos 90 = 0$.

$\cos \theta + 2 \sin \theta = 2$	$\cos \theta + 2 \sin \theta = 2$	$\cos \theta + 2 \sin \theta = 2$
$\cos 37 + 2 \sin 37 \stackrel{?}{=} 2$	$\cos 143 + 2 \sin 143 \stackrel{?}{=} 2$	$\cos 90 + 2 \sin 90 \stackrel{?}{=} 2$
$0.8 + 2(0.6) \stackrel{?}{=} 2$	$-0.8 + 2(0.6) \stackrel{?}{=} 2$	$0 + 2 \cdot 1 \stackrel{?}{=} 2$
$2.0 = 2$	$0.4 = 2$	$2 = 2$
True	False	True

37 is a solution.   143 is not a solution.   90 is a solution.

Thus, the only primary solutions are 37 and 90.

Classroom Exercises

Tell how to change each equation so that only one trigonometric ratio is present.

1. $2 \cos^2 \theta = \sin \theta + 1$

2. $1 + \cot^2 \theta = 3 \csc \theta$

3. $5 \tan^2 \theta = 6 \sec \theta - 5$

4. $\sin \theta = \cos \theta$

Solve each equation, $0 \le \theta < 360$.

5. $2 \cos^2 \theta = 2 + \sin^2 \theta$

6. $\cos \theta \tan \theta = -1$

7. $\sec \theta \cos^2 \theta - 1 = 0$

8. $2 \tan^2 \theta = \sec^2 \theta$

Written Exercises

Solve each equation to the nearest degree, $0 \le \theta < 360$. If, in your solution, you square each side of an equation, check all the roots in the original equation.

1. $2 \cos^2 \theta = \sin \theta + 1$

2. $2 \cos^2 \theta - 2 \cos \theta = -\sin^2 \theta$

3. $1 + \cot^2 \theta - 3 \csc \theta = 0$

4. $2 \tan^2 \theta + 5 \sec \theta + 4 = 0$

5. $3 \sec^2 \theta - 4 \tan \theta = 2$

6. $-\cos \theta - \sin^2 \theta = 1$

7. $\cos^2 \theta = \sin^2 \theta + 1$

8. $\csc^2 \theta - 2 \cot \theta = 0$

9. $\sec^2 \theta = 2 \tan \theta + 4$

10. $6 \sin^2 \theta - \cos \theta - 5 = 0$

11. $3 \tan^2 \theta - 4 \sec \theta = 4$

12. $2 \sec^2 \theta - \tan \theta = \tan^2 \theta + 4$

13. $\sin \theta = \cos \theta$

14. $3 \sin \theta - \cos \theta = 1$

15. $\sec \theta - 1 = \tan \theta$

16. $\sec \theta + \tan \theta = \sqrt{3}$

17. $2 \sin \theta + \cos \theta = 1$

18. $1 + \cos \theta = \sin \theta$

19. $3 \cos \theta + 2 \sin \theta = 2$

20. $\sin \theta + \cos \theta = \sqrt{2}$

21. $3 \cos \theta + 4 \tan \theta + \sec \theta = 0$

22. $2 \sin \theta - \cot \theta \cos \theta = 1$

23. $\sqrt{1 - \cos^2 \theta} - \cos \theta = 1$

24. $\sqrt{3 - 3 \sin^2 \theta} + 1 = \sin \theta$

25. Solve the system $\begin{matrix} r \sin \theta = 4 \\ r \cos \theta = 3 \end{matrix}$ for r and θ, where $r > 0$ and $0 \le \theta < 360$. (HINT: Square each side and add.)

Mixed Review

Simplify. *5.3, 5.5, 8.3, 8.4, 8.6, 9.2*

1. $(x - 2)(x^2 - 5x - 2)$

2. $\sqrt{8x^3} \cdot \sqrt{2x}$

3. $(1 + \sqrt{3})(1 - \sqrt{3})$

4. $\dfrac{1 - \sqrt{2}}{1 + \sqrt{2}}$

5. $27^{-\frac{2}{3}}$

6. $\dfrac{20}{3 + i}$

19.5 Sum and Difference Identities

Objectives

To simplify trigonometric expressions using sum and difference identities

To prove trigonometric identities using sum and difference identities

Even though 15 is not a special angle measure, cos 15 can be found in terms of the sines and cosines of two special angle measures whose difference is 15. This is done by applying the following trigonometric identity for the cosine of the difference of two angles θ (theta) and ϕ (phi): $\cos(\theta - \phi) = \cos\theta\cos\phi + \sin\theta\sin\phi$.

$$\cos 15 = \cos(60 - 45)$$
$$= \cos 60 \cos 45 + \sin 60 \sin 45$$
$$= \frac{1}{2} \cdot \frac{\sqrt{2}}{2} + \frac{\sqrt{3}}{2} \cdot \frac{\sqrt{2}}{2}$$
$$= \frac{\sqrt{2}}{4} + \frac{\sqrt{6}}{4}, \text{ or } \frac{\sqrt{2} + \sqrt{6}}{4}$$

To prove the above identity, refer to the figure at the right. M, N, P, and Q are points of a circle that has its center at the origin and that has a radius with a length of one unit. A circle with a radius of one unit is called a *unit circle*. The angles of rotation for M, N, and P are θ, ϕ, and $\theta - \phi$, respectively.

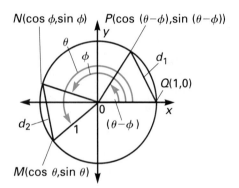

If (x,y) are the coordinates of point M, then the following are true.

$$\cos\theta = \frac{x}{r} = \frac{x}{1} = x \text{ and } \sin\theta = \frac{y}{r} = \frac{y}{1} = y$$

So, $(\cos\theta, \sin\theta)$ are the coordinates of point M of the unit circle. Similarly, it can be seen that $(\cos\phi, \sin\phi)$ and $(\cos(\theta - \phi), \sin(\theta - \phi))$ are the coordinates of points N and P, respectively.

Also in the figure, d_1 and d_2 are the lengths of chords $\overline{PQ}$ and $\overline{MN}$, respectively. Since both $\overset{\frown}{MN}$ (read "arc MN") and $\overset{\frown}{PQ}$ have $\theta - \phi$ as the measure of their central angle, these arcs are congruent. Therefore, the chords $\overline{MN}$ and $\overline{PQ}$ are also congruent. Thus, $d_1 = d_2$.

Now, using the Distance Formula, $d = \sqrt{(x_2 - x_1)^2 + (y_2 - y_1)^2}$, to express d_1 and d_2 gives the equations below.

$$d_1 = \sqrt{(\cos(\theta - \phi) - 1)^2 + (\sin(\theta - \phi) - 0)^2}$$
$$d_2 = \sqrt{(\cos\theta - \cos\phi)^2 + (\sin\theta - \sin\phi)^2}$$

Squaring and simplifying these equations gives the following.

$$d_1^2 = 2 - 2\cos(\theta - \phi)$$
$$d_2^2 = 2 - 2\cos\theta\cos\phi - 2\sin\theta\sin\phi$$

Finally, set the values of d_1^2 and d_2^2 above equal to each other.

$$2 - 2\cos(\theta - \phi) = 2 - 2\cos\theta\cos\phi - 2\sin\theta\sin\phi$$
$$-2\cos(\theta - \phi) = -2\cos\theta\cos\phi - 2\sin\theta\sin\phi$$
$$\cos(\theta - \phi) = \cos\theta\cos\phi + \sin\theta\sin\phi$$

Recall from Lesson 18.7 that $\cos(-\theta) = \cos\theta$ and $\sin(-\theta) = -\sin\theta$. These two equations can be used to derive an identity for the cosine of the *sum* of the measures of two angles. This is shown below with a direct application of the new identity.

$$\cos(\theta + \phi) = \cos[\theta - (-\phi)]$$
$$= \cos\theta\cos(-\phi) + \sin\theta\sin(-\phi)$$
$$= \cos\theta\cos\phi + \sin\theta(-\sin\phi)$$
$$= \cos\theta\cos\phi - \sin\theta\sin\phi$$

Identities for the sine and tangent of the sum or difference of two angle measures can also be derived. The sum and difference identities are listed below. You are asked to prove the last four in Exercises 24–27.

Sum and Difference Identities

$$\cos(\theta - \phi) = \cos\theta\cos\phi + \sin\theta\sin\phi$$
$$\cos(\theta + \phi) = \cos\theta\cos\phi - \sin\theta\sin\phi$$
$$\sin(\theta - \phi) = \sin\theta\cos\phi - \cos\theta\sin\phi$$
$$\sin(\theta + \phi) = \sin\theta\cos\phi + \cos\theta\sin\phi$$

$$\tan(\theta - \phi) = \frac{\tan\theta - \tan\phi}{1 + \tan\theta\tan\phi} \qquad \tan(\theta + \phi) = \frac{\tan\theta + \tan\phi}{1 - \tan\theta\tan\phi}$$

EXAMPLE 1 Simplify $\sin 40 \cos 10 - \cos 40 \sin 10$ using an identity.

Solution The expression suggests the identity for $\sin(\theta - \phi)$.

$$\sin\theta\cos\phi - \cos\theta\sin\phi = \sin(\theta - \phi)$$
$$\sin 40 \cos 10 - \cos 40 \sin 10 = \sin(40 - 10) = \sin 30 = \tfrac{1}{2}$$

EXAMPLE 2   Find tan 105 without using a calculator or a table.

Plan   Use the special angles 60 and 45 in the identity for $\tan(\theta + \phi)$.

Solution $\tan 105 = \tan(60 + 45)$

$$= \frac{\tan 60 + \tan 45}{1 - \tan 60 \tan 45} \quad \leftarrow \text{THINK: Is this number positive or negative? How do you know?}$$

$$= \frac{\sqrt{3} + 1}{1 - \sqrt{3} \cdot 1}$$

$$= \frac{(1 + \sqrt{3})(1 + \sqrt{3})}{(1 - \sqrt{3})(1 + \sqrt{3})} = \frac{1 + 2\sqrt{3} + 3}{1 - 3} = -2 - \sqrt{3}$$

Therefore, $\tan 105 = -2 - \sqrt{3}$.

EXAMPLE 3   Simplify $\cos\left(\frac{3\pi}{2} - \phi\right)$.

Solution $\frac{3\pi}{2}$ radians $= \frac{3}{2} \cdot 180 = 270$ degrees

Apply the identity $\cos(\theta - \phi) = \cos\theta \cos\phi + \sin\theta \sin\phi$.

$$\cos(270 - \phi) = \cos 270 \cos\phi + \sin 270 \sin\phi$$
$$= 0 \cdot \cos\phi + (-1) \cdot \sin\phi$$
$$= 0 - \sin\phi, \text{ or } -\sin\phi$$

Therefore, $\cos\left(\frac{3\pi}{2} - \phi\right) = -\sin\phi$.

EXAMPLE 4   Prove that $\dfrac{\sin(\theta + \phi)}{\cos\theta \cos\phi} = \tan\theta + \tan\phi$ is an identity.

Plan   First use the identity for $\sin(\theta + \phi)$ to obtain sines and cosines of θ and ϕ. Then simplify.

Solution

$$
\begin{array}{c|c}
\dfrac{\sin(\theta + \phi)}{\cos\theta \cos\phi} & \tan\theta + \tan\phi \\[2ex]
\dfrac{\sin\theta \cos\phi + \cos\theta \sin\phi}{\cos\theta \cos\phi} & \dfrac{\sin\theta}{\cos\theta} + \dfrac{\sin\phi}{\cos\phi} \\[2ex]
\dfrac{\sin\theta \cos\phi}{\cos\theta \cos\phi} + \dfrac{\cos\theta \sin\phi}{\cos\theta \cos\phi} & \\[2ex]
\dfrac{\sin\theta}{\cos\theta} + \dfrac{\sin\phi}{\cos\phi} = \dfrac{\sin\theta}{\cos\theta} + \dfrac{\sin\phi}{\cos\phi} &
\end{array}
$$

Therefore, the identity is proved.

EXAMPLE 5 Given that $\sin \theta = \frac{12}{13}$, $\cos \phi = \frac{4}{5}$, θ is in Quadrant I and ϕ is in Quadrant IV, find $\cos(\theta + \phi)$.

Plan First draw each angle in the proper quadrant. Find the missing sides of each reference triangle.

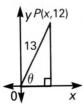

$$x^2 + 12^2 = 13^2$$
$$x^2 + 144 = 169$$
$$x^2 = 25$$
$$x = 5$$

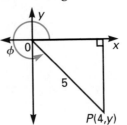

$$4^2 + y^2 = 5^2$$
$$16 + y^2 = 25$$
$$y^2 = 9$$
$$y = -3$$

$$\cos(\theta + \phi) = \cos \theta \cos \phi - \sin \theta \sin \phi$$
$$= \frac{5}{13} \cdot \frac{4}{5} - \frac{12}{13} \cdot \left(-\frac{3}{5}\right) \leftarrow \text{THINK: Can this number be greater than 1? How do you know?}$$
$$= \frac{20}{65} + \frac{36}{65}, \text{ or } \frac{56}{65}$$

Classroom Exercises

Use an identity to verify that the given equation is true.

1. $\cos 60 \cos 15 + \sin 60 \sin 15 = \dfrac{\sqrt{2}}{2}$

2. $\sin 75 \cos 45 - \cos 75 \sin 45 = \dfrac{1}{2}$

Simplify using an identity. Give your answers in simplest radical form.

3. $\sin 20 \cos 40 + \cos 20 \sin 40$

4. $\cos 70 \cos 50 - \sin 70 \sin 50$

5. $\sin(45 - 30)$ **6.** $\cos 75$

7. $\sin 105$ **8.** $\tan 15$

Written Exercises

Simplify using an identity. Give your answer in simplest radical form.

1. $\sin 80 \cos 55 + \cos 80 \sin 55$

2. $\cos 72 \cos 42 + \sin 72 \sin 42$

3. $\dfrac{\tan 17 + \tan 43}{1 - \tan 17 \tan 43}$

4. $\dfrac{\tan 155 - \tan 5}{1 + \tan 155 \tan 5}$

5. $\cos 40 \cos 50 - \sin 40 \sin 50$

6. $\cos 80 \cos 35 + \sin 80 \sin 35$

7. $\tan 75$ **8.** $\cos 195$

9. $\sin 165$ **10.** $\tan 195$

Simplify.

11. $\cos(\pi + \phi)$ **12.** $\tan(\pi - \phi)$ **13.** $\cos\left(\dfrac{\pi}{2} + \phi\right)$ **14.** $\sin\left(\dfrac{3\pi}{2} + \phi\right)$

Prove each identity.

15. $\dfrac{\sin(\theta + \phi)}{\sin \theta \sin \phi} = \cot \theta + \cot \phi$

16. $\cos(\theta - \phi) - \cos(\theta + \phi) = 2 \sin \theta \sin \phi$

17. $\sin(30 + \theta) + \sin(30 - \theta) = \cos \theta$

18. $\cos\left(\dfrac{\pi}{3} + \theta\right) + \cos\left(\dfrac{\pi}{3} - \theta\right) = \dfrac{1}{\sec \theta}$

Find $\cos(\theta - \phi)$ and $\tan(\theta + \phi)$ for the conditions in Exercises 19–21.

19. $\cos \theta = \dfrac{3}{5}$, $\tan \phi = \dfrac{5}{12}$, where θ is in Quadrant I and ϕ is in Quadrant III

20. $\sin \theta = -\dfrac{8}{17}$, $\cos \phi = -\dfrac{4}{5}$, where θ is in Quadrant IV and ϕ is in Quadrant II

21. $\cos \theta = \dfrac{1}{3}$, $\sin \phi = \dfrac{1}{2}$, where θ and ϕ are measures of acute angles

22. Show that $\cos(90 - \theta) = \sin \theta$ is an identity. (HINT: Use the difference identity for $\cos(\theta - \phi)$.)

23. Show that $\sin(90 - \theta) = \cos \theta$. (HINT: Use the identity $\cos(90 - \phi) = \sin \phi$ and let $\phi = 90 - \theta$.)

Prove the following identities. (Exercises 24–29)

24. $\sin(\theta - \phi) = \sin \theta \cos \phi - \cos \theta \sin \phi$ (HINT: Use the identities of Exercises 22 and 23.)

25. $\sin(\theta + \phi) = \sin \theta \cos \phi + \cos \theta \sin \phi$

26. $\tan(\theta - \phi) = \dfrac{\tan \theta - \tan \phi}{1 + \tan \theta \tan \phi}$

27. $\tan(\theta + \phi) = \dfrac{\tan \theta + \tan \phi}{1 - \tan \theta \tan \phi}$

28. $\sin(\theta + \phi) + \sin(\theta - \phi) = 2 \sin \theta \cos \phi$

29. $\sin A + \sin B = 2 \sin\left(\dfrac{A + B}{2}\right) \cos\left(\dfrac{A - B}{2}\right)$

30. Evaluate $\cos\left[\mathrm{Tan}^{-1}\left(-\dfrac{3}{4}\right) - \mathrm{Sin}^{-1}\dfrac{5}{13}\right]$.

Mixed Review

Solve each equation or inequality.   *5.1, 6.4, 7.6, 13.3*

1. $2^{3x+2} = 32$

2. $x^2 - 8x - 20 \le 0$

3. $\dfrac{4}{x} + \dfrac{5}{2x} = 13$

4. $\log_b 64 = 3$

Brainteaser

Equilateral triangle CDE and square $ABCD$ share a common side $\overline{CD}$. $\overline{AE}$ intersects $\overline{DC}$ at point F.

Find the degree measure of angle DFE.

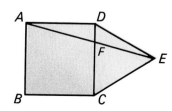

19.6 Double-Angle and Half-Angle Identities

Objectives

To simplify trigonometric expressions using double- and half-angle identities

To prove trigonometric expressions using double- and half-angle identities

The identity for sin 2θ can be derived from that for sin($\theta + \phi$) by replacing ϕ by θ as shown below.

$$\sin(\theta + \phi) = \sin \theta \cos \phi + \cos \theta \sin \phi$$
$$\sin 2\theta = \sin(\theta + \theta) = \sin \theta \cos \theta + \cos \theta \sin \theta$$
$$= \sin \theta \cos \theta + \sin \theta \cos \theta$$
$$\sin 2\theta = 2 \sin \theta \cos \theta$$

Similarly, identities for cos 2θ and tan 2θ can be developed. You will be asked to derive these identities in Exercises 24 and 25.

The double-angle identities are listed below.

Double-Angle Identities

$\sin 2\theta = 2 \sin \theta \cos \theta$

$\cos 2\theta = \cos^2 \theta - \sin^2 \theta$

$\cos 2\theta = 1 - 2 \sin^2 \theta$

$\cos 2\theta = 2 \cos^2 \theta - 1$

$\tan 2\theta = \dfrac{2 \tan \theta}{1 - \tan^2 \theta}$

EXAMPLE 1 Simplify $2 \sin 22\frac{1}{2} \cos 22\frac{1}{2}$.

Solution

$$2 \sin \theta \cos \theta = \sin 2\theta$$
$$2 \sin 22\frac{1}{2} \cos 22\frac{1}{2} = \sin 2\left(22\frac{1}{2}\right) = \sin 45 = \frac{\sqrt{2}}{2}$$

EXAMPLE 2 Find tan 2θ if $\cos \theta = \frac{3}{5}$ and $\sin \theta < 0$. Simplify the result.

Solution The cosine is positive when the sine is negative only in Quadrant IV.

$3^2 + y^2 = 5^2$ ← Solve for y.

$y = -4$ ← Quadrant IV

$\tan \theta = -\frac{4}{3}$

$$\tan 2\theta = \frac{2 \tan \theta}{1 - \tan^2 \theta} = \frac{2 \cdot \left(-\frac{4}{3}\right)}{1 - \left(-\frac{4}{3}\right)^2} = \frac{-\frac{8}{3}}{1 - \frac{16}{9}} = \frac{24}{7}$$

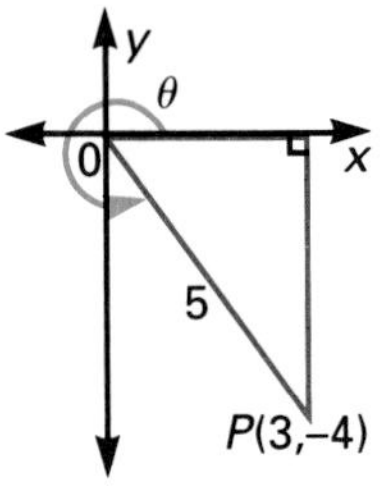

Therefore, tan $2\theta = \frac{24}{7}$.

$P(3,-4)$

EXAMPLE 3 Prove that $\cot \theta = \dfrac{1 + \cos 2\theta}{\sin 2\theta}$ is an identity.

Solution

$$
\begin{array}{c|c}
\cot \theta & \dfrac{1 + \cos 2\theta}{\sin 2\theta} \\[2ex]
\dfrac{\cos \theta}{\sin \theta} & \dfrac{1 + (2 \cos^2 \theta - 1)}{2 \sin \theta \cos \theta} \\[2ex]
& \dfrac{2 \cos^2 \theta}{2 \sin \theta \cos \theta}
\end{array}
$$

← Of the three identities for $\cos 2\theta$, choose the one that eliminates the 1 in the numerator.

$\dfrac{\cos \theta}{\sin \theta} = \dfrac{\cos \theta}{\sin \theta}$ Therefore, the identity is proved.

An identity for $\cos \dfrac{\theta}{2}$ can be derived from the double-angle identity $\cos 2\phi = 2 \cos^2 \phi - 1$. Solve for $\cos \phi$ and then replace ϕ by $\dfrac{\theta}{2}$.

$$2 \cos^2 \phi - 1 = \cos 2\phi$$

$$\cos^2 \phi = \frac{1 + \cos \phi}{2}$$

$$\cos \phi = \pm \sqrt{\frac{1 + \cos 2\phi}{2}}$$

Let $\phi = \dfrac{\theta}{2}$. $\cos \dfrac{\theta}{2} = \pm \sqrt{\dfrac{1 + \cos \left(2 \cdot \dfrac{\theta}{2}\right)}{2}}$, or $\pm \sqrt{\dfrac{1 + \cos \theta}{2}}$

An identity can be derived for $\sin \dfrac{\theta}{2}$ by solving $\cos 2\phi = 1 - 2 \sin^2 \phi$ for $\sin \phi$ and then replacing ϕ by $\dfrac{\theta}{2}$.

There are three identities for $\tan \dfrac{\theta}{2}$, one of which can be derived by using $\tan \dfrac{\theta}{2} = \dfrac{\sin \dfrac{\theta}{2}}{\cos \dfrac{\theta}{2}}$.

The half-angle identities are listed below.

Half-Angle Identities

$$\sin \frac{\theta}{2} = \pm \sqrt{\frac{1 - \cos \theta}{2}} \qquad\qquad \tan \frac{\theta}{2} = \pm \sqrt{\frac{1 - \cos \theta}{1 + \cos \theta}}$$

$$\cos \frac{\theta}{2} = \pm \sqrt{\frac{1 + \cos \theta}{2}} \qquad\qquad \tan \frac{\theta}{2} = \frac{\sin \theta}{1 + \cos \theta}$$

$$\tan \frac{\theta}{2} = \frac{1 - \cos \theta}{\sin \theta}$$

EXAMPLE 4   Find $\cos\frac{\theta}{2}$ if $\tan\theta = \frac{4}{3}$, $\cos\theta < 0$, and $0 \le \theta < 360$. Express the result in simplest radical form.

Solution   The only quadrant in which the tangent is positive and the cosine negative is Quadrant III.

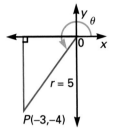

$r = \sqrt{(-3)^2 + (-4)^2} = 5$. Thus, $\cos\theta = -\frac{3}{5}$.
Since θ is in Quadrant III, $180 < \theta < 270$.

$$90 < \frac{\theta}{2} < 135$$

Therefore, $\frac{\theta}{2}$ is in Quadrant II and $\cos\frac{\theta}{2}$ is negative.

$$\cos\frac{\theta}{2} = -\sqrt{\frac{1 + \cos\theta}{2}} = -\sqrt{\frac{1 + \left(-\frac{3}{5}\right)}{2}} = -\frac{1}{\sqrt{5}} = -\frac{\sqrt{5}}{5}$$

EXAMPLE 5   Find $\tan 22\frac{1}{2}$. Express the result in simplest radical form.

Solution $22\frac{1}{2}$ is half of the special angle measure 45.

Use $\tan\frac{\theta}{2} = \dfrac{\sin\theta}{1 + \cos\theta}$, where $\theta = 45$.

$$\tan 22\frac{1}{2} = \frac{\sin 45}{1 + \cos 45}$$

$$= \frac{\frac{\sqrt{2}}{2}}{1 + \frac{\sqrt{2}}{2}} = \frac{2\left(\frac{\sqrt{2}}{2}\right)}{2\left(1 + \frac{\sqrt{2}}{2}\right)} = \frac{\sqrt{2}(2 - \sqrt{2})}{(2 + \sqrt{2})(2 - \sqrt{2})} = \sqrt{2} - 1$$

Classroom Exercises

Use an identity to verify that each equation is true.

1. $2\sin 15\cos 15 = \frac{1}{2}$

2. $\cos^2 15 - \sin^2 15 = \frac{\sqrt{3}}{2}$

Given that $180 \le \theta < 360$, find each of the following.

3. $\sin\frac{\theta}{2}$ if $\cos\theta = 0.28$

4. $\cos\frac{\theta}{2}$ if $\cos\theta = -0.28$

Use an identity to simplify each expression.

5. $\dfrac{2\tan 67\frac{1}{2}}{1 - \tan^2 67\frac{1}{2}}$

6. $1 - 2\sin^2\frac{\pi}{8}$

7. $\sin 22\frac{1}{2}$

8. $\cos 67\frac{1}{2}$

Written Exercises

Use an identity to simplify each expression.

1. $2 \sin 67\frac{1}{2} \cos 67\frac{1}{2}$ **2.** $\cos^2 22\frac{1}{2} - \sin^2 22\frac{1}{2}$ **3.** $\dfrac{2 \tan 15}{1 - \tan^2 15}$

4. $\cos 22\frac{1}{2}$ **5.** $\tan 67\frac{1}{2}$ **6.** $\sin 157.5$ **7.** $\sin \dfrac{5\pi}{8}$ **8.** $\tan \dfrac{5\pi}{8}$ **9.** $\cos \dfrac{\pi}{8}$

In Exercises 10–15, $\cos \phi = \frac{4}{5}$, $\tan \phi < 0$, and $0 \le \phi < 360$. Find each of the following.

10. $\cos 2\phi$ **11.** $\sin \dfrac{\phi}{2}$ **12.** $\tan 2\phi$ **13.** $\tan \dfrac{\phi}{2}$ **14.** $\cos \dfrac{\phi}{2}$ **15.** $\sin 2\phi$

Prove each identity.

16. $\dfrac{\cos 2\theta}{\cos \theta - \sin \theta} = \cos \theta + \sin \theta$ **17.** $\dfrac{\tan \theta - \sin \theta}{2 \tan \theta} = \sin^2 \dfrac{\theta}{2}$

18. $(\cos \theta + \sin \theta)^2 = 1 + \sin 2\theta$ **19.** $\cos^4 \theta - \sin^4 \theta = \cos 2\theta$

20. $\dfrac{1 - \tan^2 \theta}{1 + \tan^2 \theta} = \cos 2\theta$ **21.** $\sin \theta = \dfrac{\cos \theta \sin 2\theta}{1 + \cos 2\theta}$

22. $\sin 3\theta = 4 \sin \theta \cos^2 \theta - \sin \theta$ **23.** $\cos 3\theta = 4 \cos^3 \theta - 3 \cos \theta$

Derive the identities that were given in the lesson for the following.

24. $\cos 2\theta$ (3 forms)   **25.** $\tan 2\theta$ **26.** $\sin \dfrac{\theta}{2}$ **27.** $\tan \dfrac{\theta}{2}$ (3 forms)

28. The angle of elevation of a flagpole was measured at distances of 45 ft and 14.4 ft from the flagpole. The second measure of the angle of elevation was twice the first. Find the height of the flagpole to two significant digits.

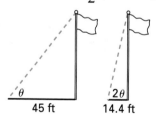

Midchapter Review

Verify each identity for the given value of θ.   **19.1**

1. $\cos(90 - \theta) = \sin \theta$; $\theta = 30$ **2.** $\cot \theta = \tan(90 - \theta)$; $\theta = 120$
3. Prove that $\tan \theta \sin \theta = \sec \theta - \cos \theta$ is an identity.   **19.2**

Solve for θ, $0 \le \theta < 2\pi$.   **19.3, 19.4**

4. $6 \csc \theta - 5 = 7$ **5.** $1 - \cos \theta = \sin \theta$

Simplify. Give your answers in radical form.   **19.5, 19.6**

6. $\cos 177\frac{1}{2} \cos 57\frac{1}{2} + \sin 177\frac{1}{2} \sin 57\frac{1}{2}$ **7.** $\tan \frac{\pi}{8}$

19.7 Trigonometric Equations: Multiple Angles

Objective

To solve trigonometric equations involving multiple angles

To solve equations involving multiples of angle measures, it is sometimes helpful first to use identities to change each multiple-angle expression to a function of a single-angle measure as shown below.

EXAMPLE 1   Solve $\sin 2\theta + \sin \theta = 0, 0 \le \theta < 2\pi$.

Plan   Use the double-angle identity $\sin 2\theta = 2 \sin \theta \cos \theta$.

Solution

$$\sin 2\theta + \sin \theta = 0$$

$$2 \sin \theta \cos \theta + \sin \theta = 0 \;\leftarrow\; \text{Factor out the common term.}$$

$$\sin \theta (2 \cos \theta + 1) = 0 \;\leftarrow\; \text{If } mn = 0, \text{ then } m = 0 \text{ or } n = 0.$$

$$\begin{array}{lll} \sin \theta = 0 & or & 2 \cos \theta + 1 = 0 \\ \theta = 0, \pi & & \cos \theta = -\frac{1}{2} \quad \text{Quadrants II, III} \\ & & \theta = \frac{2\pi}{3}, \frac{4\pi}{3} \end{array}$$

The primary solutions in radian measure are 0, $\frac{2\pi}{3}$, π, and $\frac{4\pi}{3}$.

EXAMPLE 2   Solve $\sin^2 \frac{\theta}{2} = \cos^2 \theta, 0 \le \theta < 360$.

Solution   Substitute $\pm\sqrt{\dfrac{1 - \cos \theta}{2}}$ for $\sin \dfrac{\theta}{2}$.

$$\left(\pm\sqrt{\frac{1 - \cos \theta}{2}}\right)^2 = \cos^2 \theta \;\leftarrow\; (\sqrt{a})^2 = a$$

$$\frac{1 - \cos \theta}{2} = \cos^2 \theta$$

$$1 - \cos \theta = 2 \cos^2 \theta$$

$$2 \cos^2 \theta + \cos \theta - 1 = 0 \;\leftarrow\; \text{Factor the quadratic.}$$

$$(2 \cos \theta - 1)(\cos \theta + 1) = 0$$

$$2 \cos \theta = 1 \; or \; \cos \theta = -1$$

$$\text{Quadrants I, IV} \rightarrow \quad \cos \theta = \frac{1}{2} \qquad \theta = 180$$

$$\theta = 60, 300$$

Therefore, the primary solutions are 60, 180, and 300.

Classroom Exercises

Tell what substitution is necessary to solve each equation.

1. $\sin 2\theta - \cos \theta = 0$ **2.** $\cos 2\theta + \sin \theta - 1 = 0$ **3.** $\sin^2 2\theta = \sin 2\theta$

4. $\cos^2 \frac{\theta}{2} = 1 - \frac{1}{2} \cos \theta$ **5.** $\cos 2\theta + 4 \sin^2 \frac{\theta}{2} = 1$ **6.** $3 \tan^2 \frac{\theta}{2} = 1$

7–12. Solve each of the equations of Classroom Exercises 1–6, where $0 \le \theta < 2\pi$.

Written Exercises

Solve each equation, where $0 \le \theta < 2\pi$ ($0 \le 2\theta < 4\pi$).

1. $\sin 2\theta - \sin \theta = 0$ **2.** $\cos 2\theta = \cos \theta$ **3.** $\sin 2\theta = 0$

4. $\sin 2\theta = 2 \sin \theta$ **5.** $\cos 2\theta = -\sin \theta$ **6.** $\cos \theta = \sin \frac{\theta}{2}$

Solve each equation, where $0 \le \theta < 360$.

7. $\cos 2\theta = 3 \sin \theta - 1$ **8.** $\sin 2\theta = \cos \theta$ **9.** $\cos 2\theta + 5 \cos \theta = 2$

10. $\tan 2\theta = 0$ **11.** $\cos 2\theta = 0$ **12.** $\cos \theta + \sin 2\theta = 0$

13. $\tan^2 \frac{\theta}{2} = 0$ **14.** $2 \cos^2 \frac{\theta}{2} = 3 \cos \theta$ **15.** $\sin 2\theta = \tan \theta$

16. $\cos 2\theta \sin \theta + \sin \theta = 0$ **17.** $2 \cos 2\theta + 2 \sin^2 \theta = \cos \theta$

18. $\cos \theta = \cos \frac{\theta}{2}$ **19.** $\cos(2\theta - \pi) = \sin \theta$

20. $2 \tan \frac{\theta}{2} - \csc \theta = 0$ **21.** $\sin 2\theta \cos 2\theta - 1 + \cos 2\theta = \sin 2\theta$

Mixed Review

Graph each equation.   **11.3, 12.3, 12.5, 12.7**

1. $y = |x| - 1$ **2.** $4x^2 - 25y^2 = 100$ **3.** $y = -\frac{8}{x}$ **4.** $y^2 + 8x - 6y = -1$

Brainteaser

Every triangle has three sides and three angles. Euclid proved three cases, such as side-included angle-side (SAS), in which two triangles are congruent if only three of six elements are congruent.
Yet, two triangles can have *five* of the six elements congruent and still not be congruent. Which five elements can be congruent and still produce two noncongruent triangles? Try to find two noncongruent triangles that have five congruent elements.

19.8 Vectors and Trigonometry

Objectives

To find the magnitude of a vector
To resolve a vector into two perpendicular vectors

In the figure at the right, a winch is attempting to pull a marble slab up a slippery ramp. The winch can exert a force of only 1,200 lb. Can the slab be moved up the ramp? In this lesson you will learn how to use vectors and trigonometry to answer this question.

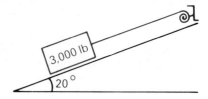

A vector $\vec{v}$ with initial point at the origin and terminal point at (a,b) is in *standard position*. It can be represented as (a,b), which is called the *rectangular form* of $\vec{v}$ (see Lesson 9.8).

In the figure below, $\vec{v_1} + \vec{v_2} = \vec{v}$. The sum, or *resultant*, of $\vec{v_1}$ and $\vec{v_2}$ is $\vec{v}$, and $\vec{v}$ is said to be *resolved* into the horizontal and vertical vector components, $\vec{v_1}$ and $\vec{v_2}$.

$$\vec{v_1} + \vec{v_2} = (-4.5,0) + (0,3)$$
$$= (-4.5 + 0, 0 + 3)$$
$$= (-4.5,3)$$
$$= \vec{v}$$

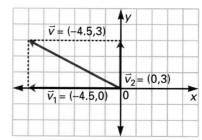

The numbers -4.5 and 3 are the *x-component* and *y-component*, respectively, of v.

Definition

If $\vec{v} = (x,y)$, in standard position, is resolved into a horizontal vector $(x,0)$ and a vertical vector $(0,y)$, then the real numbers x and y are called the **x-component** and the **y-component** of $\vec{v}$.

The **magnitude**, or length, of a vector $\vec{v}$ in standard position is represented by $\| \vec{v} \|$. The distance formula can be used to determine the magnitude of $\vec{v}$ in terms of its x- and y-components.

If $\vec{v} = (x,y)$, then $\| \vec{v} \| = \sqrt{x^2 + y^2}$.

In the figure above, $\vec{v} = (-4.5,3)$ and

$$\| \vec{v} \| = \sqrt{(-4.5)^2 + 3^2} = \sqrt{29.25} \approx 5.4.$$

The magnitude of a vector $\vec{v}$ is also called its **norm**.

EXAMPLE 1   Let $\vec{v} = (5, -3)$.

a. Draw $\vec{v}$ in standard position. Then draw $\vec{v_1}$ and $\vec{v_2}$, the horizontal and vertical vector components of $\vec{v}$.

b. Find the x- and y-components of $\vec{v}$. Then find $\|\vec{v}\|$.

Solutions   **a.**

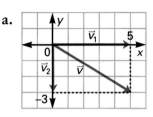

b. x-component: 5
y-component: -3
$$\|\vec{v}\| = \sqrt{5^2 + (-3)^2}$$
$$= \sqrt{34}$$
$$\approx 5.8$$

The absolute value of the x- and y-components of a vector $\vec{v} = (x, y)$ is related to the vector's magnitude $\|\vec{v}\|$ and to its reference angle α (alpha) with the x-axis.

$$\cos\alpha = \frac{|x|}{\|\vec{v}\|} \quad \text{and} \quad \sin\alpha = \frac{|y|}{\|\vec{v}\|}$$

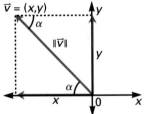

Thus, $|x| = \|\vec{v}\| \cdot \cos\alpha$ and
$|y| = \|\vec{v}\| \cdot \sin\alpha$.

In physics, it is often useful to resolve a gravitational force vector into two perpendicular vectors. For example, a force $\overrightarrow{OG}$ of 800 pounds ($\|\overrightarrow{OG}\| = 800$) can be resolved into $\overrightarrow{OA}$ and $\overrightarrow{OB}$ so that $\overrightarrow{OA}$ makes an angle of measure 25 with the horizontal and is perpendicular to $\overrightarrow{OB}$.

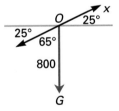

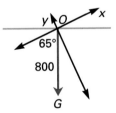

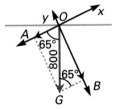

Draw an x-axis through O at angles of 25 to the horizontal and 65 to $\overrightarrow{OG}$.

Draw a y-axis through O and perpendicular to the x-axis.

Resolve $\overrightarrow{OG}$ into $\overrightarrow{OA}$ and $\overrightarrow{OB}$.

The magnitudes of $\overrightarrow{OA}$ and $\overrightarrow{OB}$ are the solutions of the equations

$$\cos 65 = \frac{\|\overrightarrow{OA}\|}{800} \quad \text{and} \quad \sin 65 = \frac{\|\overrightarrow{OB}\|}{800}.$$

EXAMPLE 2 In the figure at the right, a marble slab is being pulled up a slippery ramp by a winch. Find the number of pounds of force required to prevent the slab from sliding down the ramp.

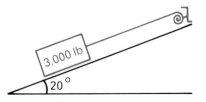

Plan Draw a force diagram with the x-axis in line with the force exerted by the winch. The slab's weight is pushing downward with a magnitude of 3,000 lb, due to gravity. Resolve this downward vector into two force vectors along the x- and y-axes as shown below.

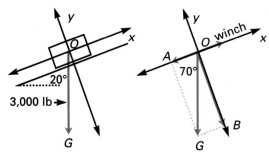

The winch must exert a force with a magnitude equal to $\|\overrightarrow{OA}\|$, since the force $\overrightarrow{OA}$ tends to pull the slab down the ramp.

Solution To find $\|\overrightarrow{OA}\|$, use right triangle AOG.

$$\cos 70 = \frac{\|\overrightarrow{OA}\|}{\|\overrightarrow{OG}\|} = \frac{\|\overrightarrow{OA}\|}{3,000}$$

$$\|\overrightarrow{OA}\| = 3,000 \cos 70 \approx 3,000(0.3420) = 1,026$$

Thus, the winch must exert a force of 1,026 lb.

Classroom Exercises

Use the figure at the right for Exercises 1–8. Express the vector in rectangular form.

1. $\overrightarrow{v}$ 2. $\overrightarrow{w}$

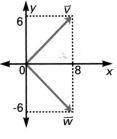

Find the x- and y-components of the vector.

3. $\overrightarrow{v}$ 4. $\overrightarrow{w}$

5. Find $\|\overrightarrow{v}\|$.   6. Find $\|\overrightarrow{w}\|$.

7. What is the magnitude of the horizontal vector component of $\overrightarrow{v}$?

8. What is the magnitude of the vertical vector component of $\overrightarrow{w}$?

Written Exercises

Draw $\vec{v}$ **in standard position. Then draw** $\vec{v_1}$ **and** $\vec{v_2}$, **the horizontal and vertical components of** $\vec{v}$.

1. $\vec{v} = (4.5, 6.0)$ 2. $\vec{v} = (-8.0, 6.0)$ 3. $\vec{v} = (-24, -10)$ 4. $\vec{v} = (15, 0)$

5-8. For the $\vec{v}$ of Exercises 1–4, find the x-component and y-component. Then find $\| \vec{v} \|$.

In Exercises 9–12, use the formulas $\cos \alpha = \dfrac{|x|}{\| \vec{v} \|}$ and $\sin \alpha = \dfrac{|y|}{\| \vec{v} \|}$ and the figure at the right to find $|x|$ and $|y|$ to the nearest whole number for the given data.

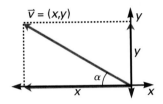

9. $\alpha = 30$, $\| \vec{v} \| = 200$ 10. $\alpha = 50$, $\| \vec{v} \| = 425$

11. $\alpha = 45$, $\| \vec{v} \| = 300$ 12. $\alpha = 62$, $\| \vec{v} \| = 510$

A mass is pulled up a slippery ramp that forms an angle α with the horizontal. For the given data, find the force required to prevent the mass from sliding down the ramp.

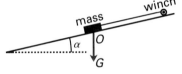

13. mass of 750 lb, $\alpha = 30$

14. mass of 2,500 lb, $\alpha = 25$

15. mass of 840 lb, $\alpha = 36$

16. mass of 1,200 lb, $\alpha = 28$

17. Prove that $\| (cx, cy) \| = |c| \cdot \| (x, y) \|$ for all real numbers c, x, and y.

In Exercises 18 and 19, use the following definition: If $\vec{v} = (a, b)$ and $\vec{w} = (c, d)$, then $\vec{v} = \vec{w}$ if and only if $a = c$ and $b = d$.

18. If $\vec{P} = (3, 5)$ and $\vec{Q} = (6m, 4 - n)$, find m and n so that $\vec{P} = \vec{Q}$.

19. Find h and k so that $(7 - h, 3k) = (k + 6, -2h)$.

20. Draw the graph of all ordered pairs (x, y) such that $\| (x, y) \| = 5$.

Mixed Review

Simplify. *5.2, 9.1, 9.2*

1. $(2a^3 b^{-2})^{-3}$ 2. $(6 - i) - (4 + 3i)$ 3. $(3 + 2i)^2$ 4. $4 \div (3i)$

5. Find two numbers whose sum is 16 and the sum of whose squares is a minimum. *11.6*

6. Is the function $f(x) = 2x^3 - x^2 + x + 4$ odd, even, or neither? *11.7*

19.9 Polar Form

Objective

To change a vector in standard position from rectangular form to polar form and vice versa

The point P in the first figure below has the **rectangular coordinates** $(-2\sqrt{3},2)$. In the second figure, the same point P is 4 units from the origin on $\overrightarrow{OP}$, which forms an angle of 150 with the positive x-axis. In the second case, P has the **polar coordinates** $(4,150°)$. The positive x-axis is sometimes called the **polar axis**.

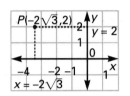
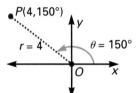

Since point P is the terminal point of $\overrightarrow{OP}$ in standard position,

(1) the rectangular form of $\overrightarrow{OP}$ is (x,y), and

(2) the **polar form** of $\overrightarrow{OP}$ is (r,θ), where
$$r = \sqrt{x^2 + y^2} = \|\overrightarrow{OP}\|,$$
$$\cos \theta = \frac{x}{r} \text{ and } \sin \theta = \frac{y}{r},$$
or $x = r \cos \theta$ and $y = r \sin \theta.$

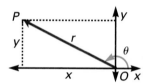

Notice that $r > 0$ and that θ is an angle of rotation.

In order to distinguish between the rectangular form and the polar form of a vector, degree symbols will be used.

EXAMPLE 1

Express $\overrightarrow{OP} = (-45, -35)$ in polar form (r,θ).

Plan

Sketch and label $\overrightarrow{OP}$ as shown. Find r and θ.

Solution

$r = \sqrt{x^2 + y^2} = \sqrt{(-45)^2 + (-35)^2} \approx 57$

$\cos \theta = \frac{x}{r} = \frac{-45}{57} = -0.7895$

Find θ by calculator as follows:

45 $\boxed{+/-}$ $\boxed{\div}$ 57 $\boxed{=}$ $\boxed{\text{INV}}$ $\boxed{\text{COS}}$ $\Rightarrow$ 142.13635

So, $\theta = 142$ *or* $\theta = 360 - 142 = 218.$

Choose $\theta = 218$ since θ is in Quadrant III.

Thus, $\overrightarrow{OP} = (57,218°)$ in polar form.

EXAMPLE 2 Express the polar form $(240, 140°)$ in rectangular form (x, y).

Solution

$$\begin{aligned} x &= r \cos \theta & y &= r \sin \theta \\ &= 240 \cos 140 & &= 240 \sin 140 \\ &= 240(-0.7660) & &= 240(0.6428) \\ &= -183.84 & &= 154.272 \\ &\approx -180 & &\approx 150 \end{aligned}$$

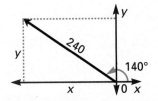

Thus, $(-180, 150)$ is the rectangular form of $(240, 140°)$.

A complex number $x + yi$ corresponds to a vector (x, y) (see Lesson 9.8). Therefore, the graph of $x + yi$ also has the polar coordinates (r, θ). Hence, $x + yi = r \cos \theta + i \cdot r \sin \theta = r(\cos \theta + i \sin \theta)$.

Definitions

The notation $r(\cos \theta + i \sin \theta)$ is called the **polar form** of the complex number $z = x + yi$, where r is the *absolute value*, or **modulus**, of z, and θ is an **argument** of z.

EXAMPLE 3 Change $-2\sqrt{3} + 2i$ to polar form, $r(\cos \theta + i \sin \theta)$.

Solution

If $z = x + yi$, then $x = -2\sqrt{3}$ and $y = 2$.
Find $r = \sqrt{x^2 + y^2}$:

$$r = \sqrt{(-2\sqrt{3})^2 + 2^2} = \sqrt{16} = 4$$

Find θ.

$$\sin \theta = \frac{y}{r} = \frac{2}{4} = \frac{1}{2}$$

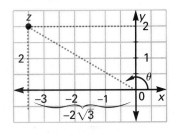

So, $\theta = 150$ and $r(\cos \theta + i \sin \theta) = 4(\cos 150 + i \sin 150)$.

Thus, $-2\sqrt{3} + 2i$ is $4(\cos 150 + i \sin 150)$ in polar form.

The polar form $r(\cos \theta + i \sin \theta)$ is sometimes abbreviated as r cis θ. Thus, $4(\cos 150 + i \sin 150)$ may be written as 4 cis 150.

The argument 150 is not unique in 4 cis 150, since multiples of 360 can be added to 150. For example, 4 cis 150 is the same number as 4 cis 510 and 4 cis (-210).

EXAMPLE 4 Change $8(\cos 240 + i \sin 240)$ to rectangular form.

Solution

$$8(\cos 240 + i \sin 240) = 8\left(\frac{-1}{2} + i \cdot \frac{-\sqrt{3}}{2}\right) = -4 - 4i\sqrt{3}$$

Classroom Exercises

Match each item at the left with one of items a–i at the right.

1. $(5, -2)$ 2. magnitude of $(5, -2)$ **a.** $\sqrt{29}$ **e.** rectangular form
3. $(5, 30°)$ 4. x-component of $(5, 30°)$ **b.** 720   **f.** argument
5. y-component of $(5, 30°)$ **c.** 770   **g.** $5 \sin 30$
6. the 315 in $3\sqrt{2}$ cis 315   **d.** polar form   **h.** $(5, -30°)$
7. $50 + k \cdot 360$ if $k = 2$ **i.** $5 \cos 30$

Written Exercises

Express each vector in polar form (r, θ) with r to two significant digits and θ to the nearest degree.

1. $(2.0, 5.0)$ 2. $(-4.0, -6.0)$ 3. $(16, -7.0)$ 4. $(-18, -15)$
5. $(-11, 12)$ 6. $(120, 200)$ 7. $(14, -130)$ 8. $(0, -140)$

Express in rectangular form, (x, y), with x and y to two significant digits.

9. $(12, 80°)$ 10. $(130, 100°)$ 11. $(8.0, 215°)$ 12. $(16, 310°)$
13. $(15, 270°)$ 14. $(260, 204°)$ 15. $(30, 290°)$ 16. $(44, 140°)$

Express each complex number in polar form.

17. $3 + 3i\sqrt{3}$ 18. $-5\sqrt{2} + 5i\sqrt{2}$ 19. $-4\sqrt{3} - 4i$ 20. $\sqrt{6} - i\sqrt{2}$
21. $4 + 0i$ 22. $-3i$ 23. -4 24. $1 + i$

Express each complex number in rectangular form.

25. $3(\cos 225 + i \sin 225)$ 26. $12(\cos 30 + i \sin 30)$
27. $10(\cos 120 + i \sin 120)$ 28. $4\sqrt{2}(\cos 300 + i \sin 300)$
29. $7[\cos(-90) + i \sin(-90)]$ 30. $2\sqrt{5}[\cos(-180) + i \sin(-180)]$
31. 2 cis 45 32. 3 cis 135 33. cis 70 34. 4 cis(-15)

Given $\overrightarrow{OP}$ and $\overrightarrow{OQ}$ in standard position, find m $\angle POQ$.

35. $\overrightarrow{OP} = (10, 35°)$, $\overrightarrow{OQ} = (8, 145°)$ 36. $\overrightarrow{OP} = (12, 70°)$, $\overrightarrow{OQ} = (15, 320°)$
37. $\overrightarrow{OP} = (4, 4)$, $\overrightarrow{OQ} = (-3, 2)$ 38. $\overrightarrow{OP} = (-\sqrt{3}, -1)$, $\overrightarrow{OQ} = (-4, 2)$

Mixed Review

Simplify.   *18.5, 19.1, 19.6*

1. $\cos \frac{\pi}{2}$ 2. $\csc \frac{4\pi}{3}$ 3. $\sin^2 \theta + \cos^2 \theta$ 4. $\dfrac{\sin 2\theta}{\cos \theta}$

19.10 Multiplication and Division of Complex Numbers: DeMoivre's Theorem

Objectives

To find products, quotients, and powers of complex numbers in polar form

To find the n different nth roots of a complex number

The polar form of a complex number has advantages over the rectangular form when finding products, quotients, and powers.

Given $z_1 = r_1(\cos \theta_1 + i \sin \theta_1)$ and $z_2 = r_2(\cos \theta_2 + i \sin \theta_2)$,

$$
\begin{aligned}
z_1 \cdot z_2 &= [r_1(\cos \theta_1 + i \sin \theta_1)] \cdot [r_2(\cos \theta_2 + i \sin \theta_2)] \\
&= r_1 r_2(\cos \theta_1 \cos \theta_2 + i^2 \sin \theta_1 \sin \theta_2 + i \sin \theta_1 \cos \theta_2 + i \cos \theta_1 \sin \theta_2) \\
&= r_1 r_2[(\cos \theta_1 \cos \theta_2 - \sin \theta_1 \sin \theta_2) + i(\sin \theta_1 \cos \theta_2 + \cos \theta_1 \sin \theta_2)] \\
&= r_1 r_2[\cos(\theta_1 + \theta_2) + i \sin(\theta_1 + \theta_2)], \text{ or } r_1 r_2 \operatorname{cis} (\theta_1 + \theta_2)
\end{aligned}
$$

Product Rule

$r_1(\cos \theta_1 + i \sin \theta_1) \cdot r_2(\cos \theta_2 + i \sin \theta_2)$
$= r_1 r_2[\cos(\theta_1 + \theta_2) + i \sin(\theta_1 + \theta_2)]$

EXAMPLE 1 Given $z_1 = 4.5(\cos 25 + i \sin 25)$ and $z_2 = 8.2(\cos 115 + i \sin 115)$, find $z_1 \cdot z_2$ in polar form and in rectangular form.

Solution
$$
\begin{aligned}
z_1 \cdot z_2 &= 4.5 \operatorname{cis} 25 \cdot 8.2 \operatorname{cis} 115 \\
&= 4.5 \cdot 8.2 \operatorname{cis}(25 + 115) \\
&= 36.9 \operatorname{cis} 140 \\
&= 36.9(\cos 140 + i \sin 140) \leftarrow \text{polar form} \\
&= 36.9(-0.7660 + 0.6428i) \\
&= -28.3 + 23.7i \leftarrow \text{rectangular form}
\end{aligned}
$$

The *quotient rule* is similar to the product rule.

Quotient Rule

$$
\dfrac{r_1(\cos \theta_1 + i \sin \theta_1)}{r_2(\cos \theta_2 + i \sin \theta_2)} = \dfrac{r_1}{r_2}[\cos(\theta_1 - \theta_2) + i \sin(\theta_1 - \theta_2)]
$$

The quotient rule can be developed as follows.

$$\frac{r_1 \text{ cis } \theta_1}{r_2 \text{ cis } \theta_2} = \frac{r_1 \text{ cis } \theta_1 \cdot \text{cis}(-\theta_2)}{r_2 \text{ cis } \theta_2 \cdot \text{cis}(-\theta_2)} = \frac{r_1}{r_2} \cdot \frac{\text{cis}(\theta_1 - \theta_2)}{\text{cis } 0} = \frac{r_1}{r_2} \cdot \text{cis}(\theta_1 - \theta_2)$$

EXAMPLE 2   Express $\dfrac{18(\cos 40 + i \sin 40)}{6(\cos 100 + i \sin 100)}$ in polar form.

Plan   Use $\dfrac{r_1 \text{ cis } \theta_1}{r_2 \text{ cis } \theta_2} = \dfrac{r_1}{r_2} \cdot \text{cis}(\theta_1 - \theta_2)$.

Solution $\dfrac{18 \text{ cis } 40}{6 \text{ cis } 100} = \dfrac{18}{6} \cdot \text{cis}(40 - 100) = 3 \text{ cis}(-60)$, or $3 \text{ cis } 300$

Thus, the quotient is $3(\cos 300 + i \sin 300)$.

Powers of complex numbers in polar form can be found as special cases of multiplication. For example, if $z = r(\cos \theta + i \sin \theta)$, then

$$z^2 = z \cdot z = r^2(\cos 2\theta + i \sin 2\theta),$$
$$z^3 = z \cdot z^2 = r^3(\cos 3\theta + i \sin 3\theta).$$

This pattern can be generalized to give *DeMoivre's Theorem*.

Theorem 19.1

DeMoivre's Theorem: For each complex number $z = r(\cos \theta + i \sin \theta)$ and for each positive integer n, $z^n = [r(\cos \theta + i \sin \theta)]^n = r^n(\cos n\theta + i \sin n\theta)$.

EXAMPLE 3   Express $(-\sqrt{2} + i\sqrt{2})^7$ in polar form and in rectangular form.

Plan   Change $z = x + yi = -\sqrt{2} + i\sqrt{2}$ to polar form.

Solution $x = -\sqrt{2}, y = \sqrt{2}, r = \sqrt{x^2 + y^2} = \sqrt{4} = 2$

$\cos \theta = \dfrac{x}{r} = \dfrac{-\sqrt{2}}{2}, \sin \theta = \dfrac{y}{r} = \dfrac{\sqrt{2}}{2}, \theta = 135$

$z = r \text{ cis } \theta = 2 \text{ cis } 135$

$z^7 = (2 \text{ cis } 135)^7$

$\quad = 2^7 \text{ cis }(7 \cdot 135) = 128 \text{ cis } 945 = 128 \text{ cis}(945 - 2 \cdot 360)$

$\quad = 128 \text{ cis } 225 \leftarrow$ polar form

$128(\cos 225 + i \sin 225) = 128\left(\dfrac{-\sqrt{2}}{2} + i \cdot \dfrac{-\sqrt{2}}{2}\right)$

$\qquad\qquad\qquad\qquad\qquad = -64\sqrt{2} - 64i\sqrt{2} \leftarrow$ rectangular form

A polar form of $-4\sqrt{3} + 4i$ is 8 cis 150. If multiples of 360 are added to the argument (150), equivalent polar forms can be written as follows.

$$8 \text{ cis } 150, \; 8 \text{ cis } 510, \; 8 \text{ cis } 870, \; 8 \text{ cis } 1{,}230, \ldots$$

If the first three of these polar forms and DeMoivre's Theorem are used, the three distinct cube roots of $-4\sqrt{3} + 4i$ can be found.

Let $z = r$ cis θ be a cube root of $-4\sqrt{3} + 4i$. Then $z^3 = -4\sqrt{3} + 4i$ and $(r \text{ cis } \theta)^3 = r^3$ cis $3\theta = 8 \text{ cis}(150 + k \cdot 360)$, where k is an integer.

Thus, $r^3 = 8$ and $3\theta = 150, 510, 870, \ldots$.
So, $r = 2$ and $\theta = 50, 170, 290, \ldots$.

The three cube roots in the form r cis θ are thus as follows.

$$z_1 = 2 \text{ cis } 50 \qquad z_2 = 2 \text{ cis } 170 \qquad z_3 = 2 \text{ cis } 290$$

The graphs of the three cube roots are on a circle of radius 2 and separated by arcs of 120°. If $3\theta = 1{,}230$, then $\theta = 410$; there is *not* a fourth distinct cube root since 410 and 50 are coterminal angles.

It can be shown that each nonzero complex number has n distinct nth roots, where n is an integer greater than 1.

EXAMPLE 4   Find the 5 fifth roots of $-2\sqrt{2} - 2i\sqrt{2}$. Express in polar form.

Plan   First, change $-2\sqrt{2} - 2i\sqrt{2}$ to polar form. Then, find five distinct numbers z such that z^5 equals the number in polar form.

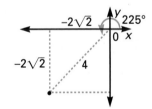

Solution $-2\sqrt{2} - 2i\sqrt{2} = 4 \text{ cis}(225 + k \cdot 360)$ in polar form.

Let $z = r$ cis θ be a fifth root of $4 \text{ cis}(225 + k \cdot 360)$.
$(r \text{ cis } \theta)^5 = r^5 \text{ cis } 5\theta = 4 \text{ cis}(225 + k \cdot 360)$

$$r^5 = 4 \; and \; 5\theta = 225 + k \cdot 360$$
$$r = \sqrt[5]{4} \qquad \theta = 45 + k \cdot 72$$

Let $k = 0, 1, 2, 3, 4$. Then $\theta = 45, 117, 189, 261, 333$.

The 5 fifth roots in polar form are $z_1 = \sqrt[5]{4}$ cis 45, $z_2 = \sqrt[5]{4}$ cis 117, $z_3 = \sqrt[5]{4}$ cis 189, $z_4 = \sqrt[5]{4}$ cis 261, and $z_5 = \sqrt[5]{4}$ cis 333.

Classroom Exercises

If $z_1 = \text{cis } 30$, $z_2 = \text{cis } 60$, $z_3 = 2 \text{ cis } 165$, and $z_4 = 3 \text{ cis } 15$, find each of the following in polar form.

1. $z_1 \cdot z_2$ **2.** $z_2 \div z_1$ **3.** $z_3 \cdot z_4$ **4.** z_4^2

5–8. For each expression of Exercises 1–4 above, find the rectangular form.

Written Exercises

Express each product in polar form and in rectangular form.

1. $3(\cos 40 + i \sin 40) \cdot 5(\cos 120 + i \sin 120)$
2. $\frac{1}{4}(\cos 100 + i \sin 100) \cdot 12(\cos 150 + i \sin 150)$
3. $5[\cos(-20) + i \sin(-20)] \cdot 4(\cos 70 + i \sin 70)$

Express each quotient in polar form with $0 \leq \theta < 360$.

4. $\dfrac{8(\cos 150 + i \sin 150)}{2(\cos 50 + i \sin 50)}$ **5.** $\dfrac{6(\cos 75 + i \sin 75)}{3(\cos 225 + i \sin 225)}$

6. $\dfrac{9(\cos 100 + i \sin 100)}{\cos(-40) + i \sin(-40)}$ **7.** $\dfrac{\cos(-20) + i \sin(-20)}{4[\cos(-220) + i \sin(-220)]}$

Express each power in polar form ($0 \leq \theta < 360$) and rectangular form.

8. $[2(\cos 50 + i \sin 50)]^3$ **9.** $[\sqrt{2}(\cos 75 + i \sin 75)]^4$
10. $(1 + i\sqrt{3})^5$ **11.** $(2 - 2i)^4$ **12.** $(-2\sqrt{3} + 2i)^3$ **13.** $(-1 + i)^8$

Express the indicated roots of each complex number in polar form.

14. $8(\cos 45 + i \sin 45)$; 3 cube roots **15.** $4 + 4i\sqrt{3}$; 3 cube roots
16. $16(\cos 80 + i \sin 80)$; 4 fourth roots **17.** $27i$; 3 cube roots
18. $-8 + 8i\sqrt{3}$; 4 fourth roots **19.** $3\sqrt{2} + 3i\sqrt{2}$; 5 fifth roots

The conjugate of $z = x + yi$ is $\bar{z} = x - yi$. Express the conjugate of each complex number in polar form.

20. $\cos 30 + i \sin 30$ **21.** $\cos 210 + i \sin 210$
22. $4(\cos 45 + i \sin 45)$ **23.** $2\sqrt{3}(\cos 300 + i \sin 300)$
24. $r(\cos \theta + i \sin \theta)$ **25.** $r[\cos(-\theta) + i \sin(-\theta)]$

Mixed Review

Solve. *2.3, 6.4, 9.5*

1. $|2x - 5| = 9$ **2.** $|2x - 5| < 9$ **3.** $x^2 - 4 < 0$ **4.** $x^2 + 2x = -4$

Application: Solar Radiation

The amount of sunlight received by any given spot on the earth's surface varies with the time of year. Let H represent the amount of solar energy intercepted by a one-meter-square patch when the sun is directly overhead. Then $H \sin \theta$ represents the energy intercepted by the same patch when the sun forms an angle of θ degrees with the horizon, as shown at the right. To understand this, notice that $H \sin \theta$ is the vertical component of the sun's light with respect to the earth. That is, it represents the amount of energy beaming down on the earth.

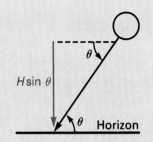

To find θ, consider the tilt of the earth. The plane containing the earth's equator forms an angle of about $23\frac{1}{2}°$ with the plane containing the earth's orbit around the sun. So, if $\propto$ is the angle of latitude of a given point in the northern hemisphere, then the angle of inclination of the sun at high noon on the longest day of the year (around June 21) equals $90° - (\propto - 23\frac{1}{2}°)$, where $23\frac{1}{2}° \le \propto \le 90°$. This is shown below. Similarly, the angle of inclination at high noon on the shortest day of the year (around December 22) equals $90° - (\propto + 23\frac{1}{2}°)$, where $0 \le \propto \le 66\frac{1}{2}°$. For example, in a city at $35°$ north latitude on December 22, $\theta = 90° - (35° + 23\frac{1}{2}°) = 31\frac{1}{2}°$. So $H \sin \theta \approx 0.52H$.

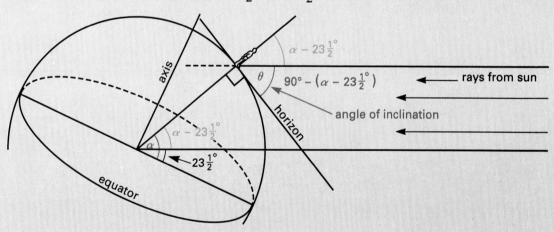

For Exercises 1–4, give the solar energy in terms of H received by a one-meter-square patch at each northern hemisphere location at the given latitude for June 21 and for December 22.

1. Houston: $30°$ **2.** Denver: $40°$ **3.** Juneau: $58°$ **4.** The Arctic Circle: $66\frac{1}{2}°$

5. Havana, Cuba sits near $23\frac{1}{2}°$ north latitude. Express the energy received on December 22 as a percent of that received on June 21.

Key Terms

argument (p. 743)
cofunction (p. 712)
components of a vector (p. 738)
DeMoivre's Theorem (p. 746)
identity (p. 711)
modulus (p. 743)

norm of a vector (p. 738)
polar axis (p. 742)
polar coordinates (p. 742)
polar form (p. 742)
rectangular coordinates (p. 742)
rectangular form (p. 738)

Key Ideas and Review Exercises

19.1   To express a function in terms of its cofunction, use the cofunction identities on page 712.

To verify an identity, show that it is true for a given value of the variable. Basic trigonometric identities are the Reciprocal Identities, Quotient Identities, and Pythagorean Identities (pages 711–713).

Express each function in terms of its cofunction.

1. $\sin 28$ **2.** $\cot 75$ **3.** $\cos \frac{\pi}{3}$ **4.** $\csc \frac{\pi}{4}$

Verify each identity for the given value of θ.

5. $\sec^2 \theta = \tan^2 \theta + 1; \theta = \frac{4\pi}{3}$ **6.** $\sin \theta = \cos \theta \tan \theta; \theta = 240$

19.2,   To simplify a trigonometric expression, use the single-angle identities of Les-
19.5,   son 19.1, the Sum and Difference Identities (page 728), and the Double- and
19.6   Half-Angle Identities (page 732–733).

To prove an identity, use the methods summarized on page 718.

Simplify.

7. $2 \sin 15 \cos 15$ **8.** $\cos 20 \cos 70 - \sin 20 \sin 70$

9. $\sin 22\frac{1}{2}$ **10.** $\sin 165$ **11.** $\tan 105$ **12.** $\sin(\pi - \theta)$

Find each of the following if $\sin \theta = -\frac{4}{5}$, $\tan \theta < 0$, $\tan \phi = \frac{5}{12}$, and $\sin \phi > 0$.

13. $\tan(\theta - \phi)$ **14.** $\cos \frac{\theta}{2}$ **15.** $\sin 2\theta$ **16.** $\cos 2\theta$

Prove that each equation is an identity.

17. $\dfrac{\sec^2 \theta}{\cot \theta + \tan \theta} = \tan \theta$ **18.** $\dfrac{\cos 2\theta}{\cos \theta - \sin \theta} = \cos \theta + \sin \theta$

19.3,
19.4, To solve a trigonometric equation, use the basic identities to rewrite the
19.7 equation in terms of one function, if possible. If both sides must be
squared, check for extraneous roots.

Solve for θ, $0 \leq \theta < 2\pi$, if θ is a special or quadrantal angle. Otherwise find solutions to the nearest degree, $0 \leq \theta < 360$.

19. $\sin \theta = -\dfrac{\sqrt{3}}{2}$ **20.** $\cos \theta = -1$ **21.** $\sin^2 \theta = \sin \theta$ **22.** $\sin 2\theta = \cot \theta$

23. $2 + 2 \sin \theta = 3 \cos^2 \theta$ **24.** $\sec \theta = \sqrt{3} + \tan \theta$

19.8 To find the magnitude of $\vec{v}$, where $\vec{v} = (x,y)$, use $\| \vec{v} \| = \sqrt{x^2 + y^2}$. To resolve a vector into two perpendicular vectors, see page 738.

25. If $\vec{v} = (-2.0, -4.0)$, draw $\vec{v}$ in standard position. Find the x- and y-components of $\vec{v}$. Then find $\| \vec{v} \|$ correct to one decimal place.

26. For $\theta = 40$ and $\| \vec{v} \| = 300$, find $|x|$ and $|y|$ to the nearest whole number.

27. A mass of 820 lb is being pulled up a ramp at an angle that measures 20 to the horizontal. Find, to the nearest pound, the number of pounds required to stop the mass from sliding down the ramp.

19.9 To convert the rectangular form $\vec{v} = (x,y)$ to the polar form (r,θ) and conversely, use $r = \sqrt{x^2 + y^2}$, and $\cos \theta = \dfrac{x}{r}$ and $\sin \theta = \dfrac{y}{r}$. The polar form (r,θ) can be written as $r(\cos \theta + i \sin \theta)$, or r cis θ.

28. Express the vector $(-15, 3.0)$ in polar form (r,θ), with r to two significant digits and θ to the nearest degree.

29. Express the vector $(14, 60°)$ in rectangular form (x,y).

30. Express the complex number $2\sqrt{3}(\cos 240 + i \sin 240)$ in rectangular form.

31. Write briefly how to express the polar form (r,θ) in the rectangular form (x,y).

19.10 For complex numbers in polar form, refer to the pages indicated below.
(1) To find products or quotients, use the rules on page 745.
(2) To find powers or nth roots, use DeMoivre's Theorem (pages 746–747).

Write each product, quotient, or power in rectangular and polar form, $0 \leq \theta < 360$.

32. $4(\cos 25 + i \sin 25) \cdot 7(\cos 150 + i \sin 150)$

33. $\dfrac{10(\cos 120 + i \sin 120)}{5(\cos 50 + i \sin 50)}$ **34.** $[3(\cos 40 + i \sin 40)]^3$

35. Express the fourth roots of $3\sqrt{2} + 3i\sqrt{2}$ in polar form.

1. Express sec 40 in terms of its cofunction.

2. Verify the identity $\cot \theta = \dfrac{\cos \theta}{\sin \theta}$ for $\theta = 150$.

Simplify.

3. $\dfrac{\sec^2 \theta - 1}{\tan \theta}$

4. $\dfrac{\cos \theta}{\sin \theta \cot \theta}$

5. $\tan 2\theta \cdot \dfrac{1 + \tan \theta}{\tan \theta}$

6. $\cos^2 15 - \sin^2 15$

7. $\tan 22\frac{1}{2}$

8. $\cos\left(\dfrac{3\pi}{2} - \theta\right)$

Find each of the following if $\cos \theta = \frac{5}{13}$, $\tan \theta < 0$, $\sin \phi = \frac{3}{5}$, and $\cos \phi < 0$.

9. $\cos(\theta - \phi)$

10. $\cos \dfrac{\theta}{2}$

11. $\tan 2\phi$

12. $\sin 2\theta$

Prove that each equation is an identity.

13. $\dfrac{1}{\tan^2 \theta} + \cot \theta \tan \theta = \csc^2 \theta$

14. $\dfrac{1}{1 - \sin \theta} + \dfrac{1}{1 + \sin \theta} = \dfrac{2}{\cos^2 \theta}$

Solve for θ, $0 \le \theta < 2\pi$, if θ is a special or quadrantal angle. Otherwise, find solutions to the nearest degree, $0 \le \theta < 360$.

15. $\sin \theta + 1 = 4 \sin \theta$

16. $\tan \theta - 2 = 4 \tan \theta$

17. $-1 + 5 \cos \theta = 2 \sin^2 \theta$

18. $\sin \theta - 1 = \cos \theta$

19. $2 + \cos 2\theta = 1 + \sin \theta$

20. $2 \sin^2 \theta - \sqrt{3} \sin \theta = 0$

21. If $\vec{v} = (-12, -8)$, draw $\vec{v}$ in standard position. Find the x- and y-components of $\vec{v}$. Then find $\| \vec{v} \|$, correct to one decimal place.

22. For $\theta = 32$, $\| \vec{v} \| = 250$, find $|x|$ and $|y|$ to the nearest whole number.

23. A mass of 650 lb is being pulled up a ramp at an angle that measures 30 to the horizontal. Find, to the nearest pound, the number of pounds required to stop the mass from sliding down the ramp.

Write each product, quotient, or power in rectangular and polar form, $0 \le \theta < 360$.

24. $3(\cos 50 + i \sin 50) \cdot 9(\cos 10 + i \sin 10)$

25. $\dfrac{15(\cos 110 + i \sin 110)}{5(\cos 40 + i \sin 40)}$

26. $[2(\cos 30 + i \sin 30)]^3$

27. Express in polar form the 5 fifth roots of $-4\sqrt{2} - 4i\sqrt{2}$.

28. Simplify $\tan(A + 2B)$ if $\cot A = \tan B = 2$.

Choose the *one* best answer to each question or problem.

1. A train traveling 90 mi/h for 1 h covers the same distance as a train traveling 60 mi/h for how many hours?
   (A) $\frac{2}{3}$ (B) 1 (C) $1\frac{1}{2}$
   (D) 3 (E) $\frac{1}{3}$

2. In a basket of 80 pears, exactly 4 were rotten. What percent of the pears were good?
   (A) 5 (B) 95 (C) 4
   (D) 96 (E) 20

3. P   Q   R   S   T

   On the line segment above, if $PQ > QR > RS > ST$, which of the following must be true?
   (A) $PR > QT$ (B) $PQ > QT$
   (C) $PS > QT$ (D) $QR > RT$
   (E) $PR > QT$

4. If $a = 7b$, then the average of a and b, in terms of b, is ___?___
   (A) $2b$ (B) $3b$ (C) $3\frac{1}{2}b$
   (D) $4b$ (E) $8b$

5. If $\frac{p}{q} = 5$ and $p = 15$, then $2p - q =$ ___?___
   (A) 10 (B) 27 (C) 33
   (D) 25 (E) -5

6. The lengths of the sides of a rectangle are 6 cm and 8 cm. Which of the following equations can be used to find θ, the angle that a diagonal makes with the longer side?
   (A) $\sin \theta = \frac{3}{4}$ (B) $\cos \theta = \frac{3}{4}$
   (C) $\tan \theta = \frac{3}{4}$ (D) $\tan \theta = \frac{4}{3}$
   (E) $\tan \theta = \frac{3}{5}$

Questions 7–8 refer to the operation represented by * and defined by the equation $x * y = xy + x - y$ for all numbers x and y. For example, $6 * 8 = 6 \cdot 8 + 6 - 8 = 46$.

7. $-5 * \frac{1}{2} =$ ___?___
   (A) -7 (B) -8
   (C) 7 (D) 8 (E) 6

8. If $5 * x = 13$, then $x =$ ___?___
   (A) 5 (B) 8 (C) $2\frac{3}{5}$
   (D) 65 (E) 2

9. $\sin \frac{\pi}{3} \cos \frac{\pi}{6} - \cos \frac{\pi}{3} \sin \frac{\pi}{6} =$ ___?___
   (A) $-\frac{1}{2}$ (B) $\frac{1}{2}$ (C) $\frac{3}{4}$
   (D) 1 (E) $\frac{5}{4}$

10. $\dfrac{2}{\tan \theta + \cot \theta} =$ ___?___
    (A) $\sin \theta$ (B) $\cos \theta$ (C) $\sin 2\theta$
    (D) $\cos 2\theta$ (E) $2 \sin \theta$

11.

    In right triangle SPQ above, if $\overline{QR}$ bisects $\angle PQS$, then $y =$ ___?___
    (A) 15 (B) 30 (C) 60
    (D) 5 (E) 65

12. $\dfrac{1}{10^{30}} - \dfrac{1}{10^{29}} =$ ___?___
    (A) $\dfrac{9}{10^{30}}$ (B) $-\dfrac{9}{10^{29}}$
    (C) $\dfrac{1}{10}$ (D) $-\dfrac{9}{10^{30}}$ (E) $\dfrac{9}{10^{29}}$

Cumulative Review (Chapters 17–19)

In right triangle ABC with right angle at C, find angle measures to the nearest degree and lengths to two significant digits.

1. $c = 6$, $b = 2$, m $\angle A = $ ___? $\quad$ *17.3*

2. $a = 20$, m $\angle A = 72$,
$\quad b = $ ___?

Find each trigonometric ratio in simplest radical form. (Exercises 3–6)

3. csc 150 $\qquad\qquad\qquad\qquad$ *17.7*

4. tan(-240)

5. $\sin\left(\frac{3\pi}{4}\right)$ $\qquad\qquad\qquad\qquad$ *18.5*

6. $\sin\left(\frac{3\pi}{2}\right)$

7. If $\sin\theta = -0.4377$, find all $\quad$ *17.8*
values of θ, $0 \le \theta < 360$, to
the nearest degree.

8. Find the values of each of the
other five trigonometric ra-
tios of θ if $\sin\theta = -\frac{5}{13}$ and
$\cos\theta < 0$.

9. Find the area of triangle ABC $\quad$ *18.2*
to two significant digits if
$a = 10$, $b = 6$, and
m $\angle C = 35$.

10. Find, to two significant
digits, the area of a regular
octagon inscribed in a circle
of radius 10.

Solve each triangle ABC. Find lengths to two significant digits and angle measures to the nearest degree. (Exercises 11–14)

11. $a = 12$, $b = 10$, $\qquad\qquad$ *18.1*
m $\angle C = 150$

12. $a = 6$, $b = 4$, $c = 8$

13. $a = 24$, m $\angle A = 70$, $\qquad$ *18.3*
m $\angle B = 50$

14. $b = 14$, $a = 6$, m $\angle A = 20$ $\quad$ *18.4*

15. Evaluate $1 - \cos\left(-\frac{2\pi}{3}\right)$. $\qquad$ *18.5*

16. Evaluate $2\sin^2\frac{4\pi}{3} - \cos\frac{4\pi}{3}$.

17. Find $\sin(-\theta)$ if $\sin\theta = $ $\qquad$ *18.7*
-0.6754.

18. Give the amplitude and the $\qquad$ *18.8*
period of $f(\theta) = 4\sin 3\theta$.

19. Sketch the graph of
$f(\theta) = -4 + \frac{1}{3}\sin 4\theta$ for
$0 \le \theta < 2\pi$, using a table
of values.

Sketch one cycle of each graph without using a table of values.

20. $f(\theta) = 2\sin\frac{1}{6}\theta$ $\qquad\qquad$ *18.8*

21. $f(\theta) = \tan 6\theta$ $\qquad\qquad\quad$ *18.9*

22. Find $\text{Sin}^{-1}\left(-\frac{1}{2}\right)$. $\qquad\qquad$ *18.10*

23. Find $\cot\left[\text{Cos}^{-1}\left(-\frac{3}{5}\right)\right]$.

Verify each identity for the given value of θ.

24. $\cot\theta = \frac{\cos\theta}{\sin\theta}$, $\theta = 240$ $\qquad$ *19.1*

25. $\tan^2\theta + 1 = \sec^2\theta$; $\theta = \frac{3\pi}{4}$

26. Express sin 20 in terms of its
cofunction.

Simplify.

27. $\dfrac{\cos\theta}{1 - \sin^2\theta}$ $\qquad\qquad\qquad$ *19.2*

28. $\sin^2\theta + \cot^2\theta + \dfrac{1}{\sec^2\theta}$

29. $\cos 70\cos 10 + \sin 70\sin 10$ $\quad$ *19.5*

30. $\sin\left(\frac{3\pi}{2} - \theta\right)$

31. $\sin 22.5$ $\qquad\qquad\qquad\qquad$ *19.6*

32. $\dfrac{\csc\theta\sin 2\theta}{\cos\theta}$

In Exercises 33–35, $\cos \theta = \frac{4}{5}$,
$\sin \theta < 0$; $\tan \phi = -\frac{12}{5}$, $\cos \phi < 0$.
Find each of the following.

33. $\sin (\theta + \phi)$ **19.5**

34. $\cos \dfrac{\theta}{2}$ **19.6**

35. $\sin 2\phi$

Prove that each equation is an identity.

36. $\cot \theta \sin \theta = \cos \theta$ **19.2**

37. $\sin^2 \theta (1 + \tan^2 \theta) = \tan^2 \theta$

38. $\dfrac{\cot^2 \theta - 1}{1 + \cot^2 \theta} = 1 - 2 \sin^2 \theta$

39. $\dfrac{2 \tan \theta - \sin 2\theta}{2 \sin^2 \theta} = \dfrac{1}{\cot \theta}$ **19.6**

Solve for θ, $0 \le \theta < 2\pi$, if θ is a special or quadrantal angle. Otherwise, find solutions to the nearest degree, $0 \le \theta < 360$.

40. $\sin \theta = 5 \sin \theta + 3$ **19.3**

41. $4 \cos \theta + 3 = 5$

42. $1 + \cos \theta = 2 \cos^2 \theta$

43. $\sin^2 \theta - \sqrt{2} \sin \theta = 0$

44. $3 - 2 \sin^2 \theta = 3 \cos \theta$ **19.4**

45. $1 + \sin \theta = \cos \theta$

46. $\sin 2\theta + \cos \theta = 0$ **19.6**

47. Draw $\vec{v} = (6, -2)$ in standard position. Find the x- and y-components of $\vec{v}$. Then find $\| \vec{v} \|$. **19.8**

48. For $\theta = 42$, $\| \vec{v} \| = 320$, find $|x|$ and $|y|$ to the nearest whole number.

49. Express the vector $(40, 220°)$ in rectangular form, with x and y to two significant digits. **19.9**

50. Change $\dfrac{1}{2} - \dfrac{i\sqrt{3}}{2}$ to polar form, $r(\cos \theta + i \sin \theta)$. **19.9**

Express each product, quotient, or power in rectangular and polar form, $0 \le \theta < 360$.

51. $4(\cos 20 + i \sin 20) \cdot$ **19.10**
    $3(\cos 70 + i \sin 70)$

52. $\dfrac{12(\cos 75 + i \sin 75)}{3(\cos 15 + i \sin 15)}$

53. $[3(\cos 40 + i \sin 40)]^3$

54. Express in polar form the 4 fourth roots of $-8 + 8i\sqrt{3}$.

55. From the top of a lighthouse 180 ft above sea level, the angle of depression of a ship at sea is $32°20'$. Find, to three significant digits, the distance of the ship from the base of the lighthouse. **17.3**

56. A 20-ft ladder is placed against a building so that the lower end is 5 ft from the base of the building. What angle, to the nearest degree, does the ladder make with the ground?

57. The diagonals of a rectangle are each 12 cm long, and they intersect at an angle of 100. Find, to two significant digits, the lengths of the sides of the rectangle. **18.1**

58. A navigator sailing north sights a lighthouse at a bearing of 36. After sailing another 67 km, he finds that the bearing of the lighthouse is 122. How far, to two significant digits, is the ship now from the lighthouse? **18.3**

Computer Investigations:
Algebra with Trigonometry

The following is a listing of the computer investigations that are included in *Holt Algebra with Trigonometry*. The related textbook page for each investigation is also indicated. For each investigation, you will need the *Investigating Algebra with the Computer* software.

Investigation: Rational Functions

Program: RATIONAL

Objectives: To use the computer to graph rational functions

To learn that the x intercepts of such a graph occur at x values which make the numerator equal to 0

To learn that vertical asymptotes occur at x values which make the denominator of the function equal to 0

To learn that a rational function often has a horizontal asymptote

1. a. With BASIC loaded into the computer and the disk in drive 1 (or drive A), run the program RATIONAL by typing the appropriate command.

Apple: RUN RATIONAL   IBM: RUN "RATIONAL"

Be sure to press the RETURN key after typing the command.

b. Read the opening messages of the program.

2. Use the program to graph the function $y = \dfrac{1}{x + 2}$.

a. Enter 0 for the degree of the numerator.

b. The numerator has only the DEGREE 0 TERM. Enter 1 for this term.

c. Enter 1 as the degree of the denominator.

d. For the denominator, enter 1 as the coefficient of the DEGREE 1 TERM, and 2 for the DEGREE 0 TERM.

e. Enter -4 for the MINIMUM X VALUE and 2 for the MAXIMUM X VALUE.

f. The computer draws the graph. Does the function have any zeros (x intercepts) over this interval?

g. For one x value the denominator of the function equals zero and therefore the function is undefined. The computer marks this point with a vertical dashed line (asymptote). What is this x value?

h. The graph also has a horizontal asymptote, which is a line the graph approaches but never touches. Which line is the horizontal asymptote?

i. What is the y intercept? (HINT: To check your estimate from the graph, substitute 0 for x in the equation defining the function.)

j. Answer N (no) to ANOTHER INTERVAL FOR SAME FUNCTION?

k. Answer Y (yes) to ANOTHER FUNCTION?

3. Have the computer graph $y = \dfrac{2x^2 + 3x - 2}{x}$.

    a. Enter 2 as the degree of the numerator. For the coefficients enter 2 as the DEGREE 2 TERM, 3 as the DEGREE 1 TERM, and -2 as the DEGREE 0 TERM.

    b. Enter 1 as the degree of the denominator; enter 1 and then 0 as the coefficients of the denominator.

    c. Enter -3 as the MINIMUM X VALUE and 2 as the MAXIMUM X VALUE.

    d. The computer draws the graph. What x value has a vertical asymptote?

    e. There is one negative zero and one positive zero. List each (to the nearest tenth).

    f. Write the factors of the numerator of the function. (?) (?)

    g. What is the relationship between the factors of the numerator and the zeros of the function?

    h. Answer N to ANOTHER INTERVAL? and Y to ANOTHER FUNCTION?

4. Graph the function, $y = \dfrac{x^2 - x - 2}{x^2 + 2x - 3}$.

    a. The numerator is degree 2 with coefficients 1, -1, and -2.

    b. The denominator is degree 2 with coefficients 1, 2, and -3.

    c. Enter -4 and 3 for the MINIMUM and MAXIMUM x values.

    d. List the zeros (x intercepts) of the function.

    e. Write the factors of the numerator of the function.

    f. What is the relationship between the factors of the numerator and the zeros of the function?

    g. List the two x values where the function is undefined.

    h. Write the factors of the denominator.

    i. What is the relationship between the factors of the denominator and the x values where the function is undefined?

    j. The function has a horizontal asymptote (not drawn by the computer). This may not be clear from the interval graphed so far. So answer Y to ANOTHER INTERVAL FOR SAME FUNCTION?

    k. Enter 1 and 8 for the MINIMUM and MAXIMUM x values.

    l. The graph now appears to get closer and closer to a horizontal line. What is the equation of this line? $y =$?

    m. Enter N to ANOTHER INTERVAL? and Y to ANOTHER FUNCTION?

5. Consider this function, $y = \dfrac{5x^2 - 11x}{2x^2 - 5x - 7}$.

    a. Before graphing this function, use algebra to make some predictions about the graph. First, factor the numerator.

b. Use the factors to predict the zeros of the function.

c. Factor the denominator.

d. Use the factors to predict the x values where vertical asymptotes will occur.

e. Now have the computer graph the function. The numerator is a polynomial of degree 2 with coefficients 5, -11, and 0. The denominator is a polynomial of degree 2 with coefficients 2, -5, -7.

f. Have the computer graph the function from $x = -4$ to $x = 6$.

g. Which of your predictions were correct?

h. It is not clear whether the function has a horizontal asymptote. Answer Y to ANOTHER INTERVAL FOR SAME FUNCTION?

i. Have the computer graph the function from $x = -8$ to $x = -1$.

j. The leftmost section of the graph may appear to be a straight line. However, it is actually a curved line that approaches a horizontal asymptote. What is the equation of this asymptote? (HINT: What fraction is formed by the first coefficient in the numerator and the first coefficient in the denominator of the function?)

k. Have the computer graph the same function from $x = 4$ to $x = 10$.

l. As x increases in the positive direction, does the graph appear to approach the same horizontal asymptote you listed in step **j**?

m. Enter N to ANOTHER INTERVAL FOR SAME FUNCTION?

n. Enter Y to ANOTHER FUNCTION?

6. a. Before having the computer graph each function below, predict the zeros and the vertical asymptotes. If necessary, factor the numerator and the denominator.

b. Graph the function over the indicated interval.

c. Check whether your predictions of the zeros and asymptotes were correct.

d. List the equation of any horizontal asymptote of the function. (To verify your estimate of a horizontal asymptote, regraph each function to the left and/or the right of the first interval listed below.)

Function	Interval	Function	Interval
i. $y = \dfrac{3}{x - 1}$	-4 to 5	**ii.** $y = \dfrac{4x}{x^2 + x - 12}$	-6 to 6
iii. $y = \dfrac{2x}{x^2 - 4}$	-3 to 3	**iv.** $y = \dfrac{4x - 3}{2x + 1}$	-5 to 5
v. $y = \dfrac{10x^2 + x - 42}{10x}$	-6 to 6	**vi.** $y = \dfrac{2x^2 - 4x - 6}{2x^2 + 3x - 5}$	-5 to 5

Investigation: Locating Real Zeros

Program: SOLVEPOL

Objectives: To use the computer to graph polynomial functions

To read the zeros of a function from its graph, and to estimate non-integral zeros to the nearest tenth.

1. **a.** With BASIC loaded into the computer and the disk in drive 1 (or drive A), run the program SOLVEPOL by typing the appropriate command.

   Apple: RUN SOLVEPOL   IBM: RUN "SOLVEPOL"

   Be sure to press the RETURN key after typing the command.

   b. Read the opening messages of the program.

2. Use the program to graph the function $f(x) = x^3 + 3x^2 - 2x - 6$.

   a. Enter 3 for THE DEGREE OF THE POLYNOMIAL.

   b. Enter the COEFFICIENTS one at a time: 1, 3, -2, -6. Press RETURN after typing each number.

   c. Enter -4 for the MINIMUM X VALUE and 3 for MAXIMUM X VALUE.

3. Use the graph on the screen to answer these questions.

   a. List one integer which is a <u>zero</u> of the function.

   b. Notice the function values for the integers -2 and -1. Is $f(-2)$ positive or negative?

   c. Is $f(-1)$ positive or negative?

   d. The function has a zero between -2 and -1. From the units marked on the x axis, estimate this zero to the nearest tenth.

4. **a.** There are two consecutive positive integers where the function changes from negative to positive. List these two integers.

   b. Estimate the positive zero of the function to the nearest tenth.

   c. Enter N (no) to ANOTHER INTERVAL FOR SAME FUNCTION?

   d. Enter Y (yes) to ANOTHER FUNCTION?

5. Have the computer graph the function $f(x) = 3x^4 + x^3 + 7x^2 + 3x - 6$.

   a. Enter 4 for the DEGREE OF THE POLYNOMIAL.

   b. Enter the COEFFICIENTS: 3, 1, 7, 3, -6.

   c. Enter -2 for MINIMUM X VALUE.

   d. Enter 2 for MAXIMUM X VALUE.

6. Use the graph to answer these questions.

    a. What negative integer is a zero of the function?

    b. List two consecutive integers which have a zero between them. ___?___ , ___?___

7. Focus on the interval between the two integers you listed in Exercise **6b.**

    a. Enter Y to ANOTHER INTERVAL FOR SAME FUNCTION?

    b. Graph the function between the two integers you listed in Exercise **7b.**

    c. Estimate the zero of the function to the nearest tenth.

    d. Enter N to ANOTHER INTERVAL FOR SAME FUNCTION?

    e. Enter Y to ANOTHER FUNCTION?

8. Have the computer graph $f(x) = 2x^3 - x^2 - 10x$.

    a. Enter 3 for the DEGREE, and the COEFFICIENTS: 2, -1, -10, 0.

    b. Enter -3 for MINIMUM X VALUE and 3 for MAXIMUM X VALUE.

    c. List two integers which are zeros of the function.

    d. List two consecutive integers which are not zeros but which have a zero between them. ___?___ and ___?___

    e. Enter N to ANOTHER INTERVAL?

9. Enter Y to ANOTHER FUNCTION?

10. a. Have the computer graph each function over the listed interval.

    b. From the graph list the zeros of each function. Estimate non–integer zeros to the nearest tenth.

Function	Interval
i. $f(x) = 3x^3 - x^2 - 6x + 2$	-2 to 2
ii. $f(x) = 8x^3 - 27x^2 + 25x - 6$	-1 to 3
iii. $f(x) = 4x^4 - 5x^2 + 2x + 1$	-2 to 2
NOTE: Enter 0 for the degree 3 coefficient.	
iv. $f(x) = 3x^5 - 8x^2$	-1 to 5
NOTE: Enter 0's for the missing coefficients.	
v. $f(x) = 2x^3 - x^2 + 2x - 1$	-2 to 2

Investigation: The Role of a in $y = ax^2$

Program: PARAB1

Objectives: To use the computer to graph parabolas of the form $y = ax^2$

To predict whether the parabola will open upward or downward

To predict whether the parabola will be more or less "steep" than the graph of $y = x^2$

1. a. With BASIC loaded into the computer and the disk in drive 1 (or drive A), run the program PARAB1 by typing the appropriate command.

        Apple: RUN PARAB1   IBM: RUN "PARAB1"

    Be sure to press the RETURN key after typing the command.

  b. Read the opening messages of the program.

2. a. Use the program to draw the graph of $y = x^2$. Type 1 for A and press RETURN.

  b. Is the graph a straight line?

  c. The graph of $y = x^2$ is a <u>parabola</u>. Does the parabola open upward or downward?

3. Have the computer draw the graph of $y = -x^2$.

  a. Answer Y (yes) to ANOTHER GRAPH? and N (no) to CLEAR THE SCREEN?

  b. Type -1 for A and press RETURN.

  c. Is the graph a parabola?

  d. Does the graph open upward or downward?

4. Have the computer draw the graph of $y = 2x^2$.

  a. Answer Y to ANOTHER GRAPH? and N to CLEAR THE SCREEN?

  b. Enter 2 for A.

  c. Does the parabola open upward or downward?

  d. Is this parabola "steeper" (more narrow) than the graph of $y = x^2$?

5. Have the computer draw the graph of $y = -.5x^2$.

  a. Answer Y to ANOTHER GRAPH? and N to CLEAR THE SCREEN?

  b. Enter $-.5$ for A.

  c. Does the parabola open upward or downward?

  d. Is this parabola "steeper" (more narrow) than the graph of $y = -x^2$?

6. Consider the equation $y = .25x^2$. Before graphing this function, answer these questions.

  a. Will the graph be a parabola?

b. Will it open upward or downward?

c. Will the graph be more steep or less steep than the graph of $y = x^2$?

7. Have the computer draw the graph of $y = .25x^2$. Leave the previous graphs on the screen. Were your predictions in Exercise 6 correct?

8. Consider the equation $y = 4x^2$. Before graphing this function, answer these questions.

a. Will the graph be a parabola?

b. Will it open upward or downward?

c. Will the graph be more steep or less steep than the graph of $y = x^2$?

9. Have the computer draw the graph of $y = 4x^2$. Leave the previous graphs on the screen. Were your predictions in Exercise 8 correct?

10. Complete each sentence below.

a. The graph of $y = ax^2$ will open upward if $a > \underline{\ ?\ }$.

b. The graph of $y = ax^2$ will open downward if $\underline{\ ?\ }$.

c. If $|a| > 1$, the graph of $y = ax^2$ will be $\underline{\ ?\ }$ (more steep/less steep) than the graph of $y = x^2$.

d. If $|a| < 1$, the graph of $y = ax^2$ will be $\underline{\ ?\ }$ (more steep/less steep) than the graph of $y = x^2$.

11. Enter Y to ANOTHER GRAPH? Then enter Y for CLEAR THE SCREEN?

12. a. As a reference, graph $y = x^2$. (Enter 1 for A.)

b. Before graphing each function below, predict whether the graph will open upward or downward and whether it will be more steep or less steep than $y = x^2$.

c. Have the computer draw the graph. Do not clear the screen.

d. Determine whether your predictions were correct.

    i. $y = 3x^2$ **ii.** $y = -2x^2$ **iii.** $y = -.8x^2$ **iv.** $y = .2x^2$

Investigation: The Role of c in $y = ax^2 + c$

Program: PARAB2

Objectives: To use the computer to graph parabolas of the form $y = ax^2 + c$
To discover that the vertex of such a parabola is $(0, c)$.

1. **a.** With BASIC loaded into the computer and the disk in drive 1 (or drive A), run the program PARAB2 by typing the appropriate command.

   Apple: RUN PARAB2   IBM: RUN "PARAB2"

   Be sure to press the RETURN key after typing the command.
   b. Read the opening messages of the program.

2. Use the program to graph $y = x^2$.
   a. Type 1 for A and press RETURN.
   b. Enter 0 for C (and press RETURN).
   c. What is the y intercept?
   d. What are the coordinates of the <u>vertex</u>? (_?_ , _?_)

3. Have the computer graph $y = x^2 + 3$.
   a. Answer Y (yes) to ANOTHER GRAPH?
   b. Keep the first graph on the screen. Answer N (no) to CLEAR THE SCREEN?
   c. Enter 1 for A and 3 for C.
   d. Does the graph of $y = x^2 + 3$ open upward or downward?
   e. Where does the graph intersect the y axis?
   f. What are the coordinates of the vertex? (_?_ , _?_)
   g. What was the effect on the graph of adding the 3 behind the x^2?

4. Have the computer graph $y = x^2 - 2$.
   a. Answer Y to ANOTHER GRAPH? and N to CLEAR THE SCREEN?
   b. Enter 1 for A and -2 for C.
   c. Does the graph open upward or downward?
   d. What are the coordinates of the vertex of this graph?
   e. What was the effect on the graph of the "-2" in the equation?

5. **a.** Before having the computer graph $y = x^2 - 5$, predict the coordinates of the vertex.
   b. Have the computer draw the graph. Do not clear the screen. Was your prediction correct?

6. Answer Y to ANOTHER GRAPH? and Y to CLEAR THE SCREEN? Then graph $y = -2x^2$.

   a. Enter -2 for A and 0 for C.

   b. Does the graph open upward or downward?

   c. What is the vertex of this graph?

7. Have the computer graph $y = -2x^2 + 5$. Do not clear the screen.

   a. Enter -2 for A and 5 for C.

   b. Does this graph open upward or downward?

   c. What are the coordinates of the vertex of this graph?

   d. What was the effect on the graph of the " $+5$" in the equation?

8. a. Before having the computer graph $y = -2x^2 - 3$, predict the vertex.

   b. Now graph the equation. Do not clear the screen. Was your prediction correct?

9. *Complete:* The coordinates of the vertex of the graph of $y = ax^2 + c$ $(a \neq 0)$ are $(\underline{\ ?\ }, \underline{\ ?\ })$.

10. Enter Y to ANOTHER GRAPH? Then enter Y for CLEAR THE SCREEN?

11. a. Predict the coordinates of the vertex of each of the two functions listed in each exercise.

   b. Have the computer graph the two functions on the same axes. (Enter 0 for C.)

   c. Check your predictions of the coordinates of the vertices.

   i. $y = 3x^2$
   $\ y = 3x^2 + 2$

   ii. $y = -4x^2$
   $\ y = -4x^2 - 3$

   iii. $y = 2x^2 - 8$
   $\ y = 2x^2 + 1$

   iv. $y = -.5x^2 + 5$
   $\ y = -.5x^2 - 4$

Investigation: The Role of b in $y = ax^2 + bx + c$

Program: PARAB3

Objectives: To use the computer to graph parabolas of the form $y = ax^2 + bx + c$

To discover that the coefficient b in the equation can move the parabola right or left and up or down

1. **a.** With BASIC loaded into the computer and the disk in drive 1 (or drive A), run the program PARAB3 by typing the appropriate command.

   Apple: RUN PARAB3   IBM: RUN "PARAB3"

   Be sure to press the RETURN key after typing the command.

   b. Read the opening messages of the program.

2. Use the program to graph the function $y = x^2$.

   a. Enter 1 for A, 0 for B, and 0 for C.

   b. The computer draws the <u>parabola</u> and the <u>axis of symmetry</u>. It also prints the equation of the axis of symmetry and the coordinates of the <u>vertex</u> of the parabola.

   c. What is the equation of the axis of symmetry? $x = \underline{\ ?\ }$

   d. What are the coordinates of the vertex? ($\underline{\ ?\ }$, $\underline{\ ?\ }$)

3. Have the computer graph $y = x^2 - 3x$.

   a. Answer Y (yes) to ANOTHER GRAPH? and N (no) to CLEAR THE SCREEN?

   b. Enter 1 for A, -3 for B, and 0 for C.

   c. Is the second parabola further to the right than the first one?

   d. What is the axis of symmetry of the second parabola (to the nearest tenth)? $x = \underline{\ ?\ }$

   e. List the coordinates of the vertex of the second parabola.

4. Repeat Exercise 3 for $y = x^2 - 5x$. (A = 1, B = -5, C = 0)

   a. Is this parabola further to the right than the previous two?

   b. What is the axis of symmetry?

   c. What are the coordinates of the vertex?

5. Repeat Exercise 3 for $y = x^2 - 6x$. (A = 1, B = -6, C = 0)

   a. Is this parabola further to the right than all the others?

   b. What is the axis of symmetry?

   c. What are the coordinates of the vertex?

6. Clear the screen and graph the following parabolas on the same axes. In each case, list the axis of symmetry and the coordinates of each vertex.

   a. $y = -2x^2$ ($b = 0$, $c = 0$)   **b.** $y = -2x^2 + 4x$ ($c = 0$)

   c. $y = -2x^2 + 7x$ **d.** $y = -2x^2 + 9x$

7. Compare the four graphs of Exercise 6. What happened to the parabolas as the coefficient of the x term became larger and larger?

8. Clear the screen and graph the following parabolas on the same axes. List each axis of symmetry and the coordinates of each vertex.

   a. $y = 3x^2 + 6x$ $(c = 0)$

   b. $y = 3x^2 + 6x - 4$

   c. $y = 3x^2 + 6x - 7$

   d. $y = 3x^2 + 6x + 4$

   e. $y = 3x^2 + 6x + 7$

9. What happened to the parabolas of Exercise 8 as the value of c changed?

10. *Complete:* For the function $y = ax^2 + bx + c$ $(a \neq 0)$, changes in the value of b move the parabola to the __?__ or __?__ as well as __?__ or __?__ .

11. Graph each set of functions on the same screen. List each axis of symmetry and the coordinates of each vertex.

Set A	**Set B**
a. $y = 4x^2$; $(b = 0, c = 0)$	**a.** $y = -.5x^2$; $(b = 0, c = 0)$
b. $y = 4x^2 - 4x$ $(c = 0)$	**b.** $y = -.5x^2 - x$ $(c = 0)$
c. $y = 4x^2 - 8x$	**c.** $y = -.5x^2 - 2x$
d. $y = 4x^2 - 12x$	**d.** $y = -.5x^2 - 4x$

Set C	**Set D**
a. $y = 5x^2 + 10x$	**a.** $y = -x^2 + 4x$
b. $y = 5x^2 + 10x - 4$	**b.** $y = -x^2 - 4x$
c. $y = 5x^2 + 10x + 3$	**c.** $y = -x^2 + 4x + 5$
d. $y = 5x^2 + 10x + 7$	**d.** $y = -x^2 - 4x - 5$

Investigation: The Circle

Program: CIRCLE

Objectives: To use the computer to graph circles of the form $(x - h)^2 + (y - k)^2 = r^2$

To discover that the center of the circle is (h, k) and the radius is r

1. **a.** With BASIC loaded into the computer and the disk in drive 1 (or drive A), run the program CIRCLE by typing the appropriate command.

   Apple: RUN CIRCLE   IBM: RUN "CIRCLE"

   Be sure to press the RETURN key after typing the command.

   b. Read the opening messages of the program.

2. Use the program to graph the equation $x^2 + y^2 = 36$. Each equation must be in the form $(X - H)\verb|^|2 + (Y - K)\verb|^|2 = R\verb|^|2$.

   a. The computer asks WHAT IS H? Type 0 and press RETURN.

   b. To WHAT IS K?, enter 0.

   c. To WHAT IS R^2?, enter 36.

   d. The graph should be a <u>circle</u>. It may not be perfectly round on your monitor. The <u>center</u> of the circle is marked by an X. What are the coordinates of the center? (_?_ , _?_)

   e. At what values does the circle intersect the x axis? _?_ and _?_

   f. At what values does the circle intersect the y axis? _?_ and _?_

   g. What is the length of the radius of the circle?

   h. Enter Y to the question ANOTHER GRAPH?

3. Have the computer graph this equation: $(x - 4)^2 + y^2 = 9$

   a. Enter 4 (not −4) for H and 0 for K.

   b. Enter 9 for R^2.

   c. The graph is a circle, although it may be distorted on your screen. What are the coordinates of the center? (_?_ , _?_)

   d. What is the length of the radius?

   e. What effect did the "−4" in the equation have on the graph?

   f. Answer Y to ANOTHER GRAPH?

4. Have the computer graph this equation: $x^2 + (y + 3)^2 = 16$

   a. Enter 0 for H.

   b. Think of $(y + 3)^2$ as $[y - (-3)]^2$. So enter −3 (not 3) for K.

   c. Enter 16 for R^2.

d. What are the coordinates of the center?

e. What is the length of the radius?

f. What effect did the " + 3" in the equation have on the graph?

g. Answer Y to ANOTHER GRAPH?

5. Have the computer graph this equation: $(x - 2)^2 + (y + 4)^2 = 25$

    a. Enter 2 for H, -4 for K, and 25 for R$\hat{}$2.

    b. What is the center?

    c. What is the length of the radius?

6. Consider this equation: $(x + 5)^2 + (y - 1)^2 = 49$
Before graphing the equation, predict the center and the length of the radius.

7. Have the computer graph the equation in Exercise 6.

    a. Enter -5 for H, 1 for K, and 49 for R$\hat{}$2.

    b. Was your prediction of the center correct?

    c. Was your prediction of the length of the radius correct?

8. Repeat Exercises **6** and **7** for this equation: $(x - 1)^2 + (y - 3)^2 = 16$
Be sure to enter the correct values for H, K, and R$\hat{}$2.

9. *Complete:* For an equation of the form $(x - h)^2 + (y - k)^2 = r^2 (r > 0)$,

    a. the graph is a __?__ ;

    b. the center is at the point (__?__ , __?__);

    c. the length of the radius is __?__ .

10. Enter Y to ANOTHER GRAPH?

11. **a.** Before graphing each equation, predict the center and the length of the radius.

    b. Have the computer graph each equation.

    c. Check whether your predictions were correct.

    d. The values of H, K, and R$\hat{}$2 are listed for the first four equations.

      i. $x^2 + y^2 = 64$ (H = 0, K = 0, R$\hat{}$2 = 64)

     ii. $(x - 3)^2 + y^2 = 81$ (H = 3, K = 0, R$\hat{}$2 = 81)

    iii. $x^2 + (y - 4)^2 = 25$ (H = 0, K = 4, R$\hat{}$2 = 25)

    iv. $(x + 3)^2 + (y - 2)^2 = 36$ (H = -3, K = 2, R$\hat{}$2 = 36)

     v. $(x - 5)^2 + (y + 1)^2 = 49$

    vi. $(x - 6)^2 + (y - 7)^2 = 4$

   vii. $(x + 8)^2 + (y + 5)^2 = 9$

Investigation: The Ellipse

Program: ELLIPSE1

Objectives: To use the computer to graph ellipses of the form $\dfrac{x^2}{a^2} + \dfrac{y^2}{b^2} = 1$, connecting each point of the ellipse to the two foci

To discover that such an ellipse is centered at the origin, that the foci are on the x axis if $a > b$ and on the y axis if $b > a$, that the x intercepts are a and $-a$, and that the y intercepts are b and $-b$

1. a. With BASIC loaded into the computer and the disk in drive 1 (or drive A), run the program ELLIPSE1 by typing the appropriate command.

        Apple: RUN ELLIPSE 1   IBM: RUN "ELLIPSE1"

Be sure to press the RETURN key after typing the command.

  b. Read the opening messages of the program.

2. Use the program to graph the equation $\dfrac{x^2}{16} + \dfrac{y^2}{4} = 1$.

  a. To the question WHAT IS A^2? type 16 and press RETURN.

  b. Enter 4 for B^2?

  c. The computer calculates the coordinates of the <u>foci</u> of the <u>ellipse</u>. It marks the foci with X's. What are the coordinates of the foci?

  d. Which axis, x or y, are the foci on?

  e. Read the messages at the bottom of the screen. Then press any key to begin graphing.

  f. As the computer draws the graph, it connects each point of the ellipse to the two foci. At the bottom of the screen it shows the distance between the point and the foci and the sum of the two distances. What do you notice about the sum of the distances?

  g. What are the x intercepts of the graph?   **h.** What are the y intercepts?

  i. Answer Y (yes) to ANOTHER GRAPH?

3. Have the computer graph $\dfrac{x^2}{25} + \dfrac{y^2}{9} = 1$.

  a. Enter 25 for A^2 and 9 for B^2.   **b.** What are the coordinates of the foci?

  c. Which axis are the foci on?

  d. Press any key to begin graphing. What is true about the sum of the distances from each point plotted to the foci?

  e. What are the x intercepts of the ellipse?

  f. What are the y intercepts?

  g. Enter Y to ANOTHER GRAPH?

4. Have the computer graph $\frac{x^2}{16} + \frac{y^2}{36} = 1$.

   a. Enter 16 for A^2 and 36 for B^2.

   b. What are the coordinates of the foci?

   c. Which axis are the foci on?

   d. Press any key to begin graphing. What is the constant sum of the distances from each point plotted to the foci?

   e. What are the x intercepts of the ellipse?

   f. What are the y intercepts?

   g. Enter Y to ANOTHER GRAPH?

5. Consider the equation $\frac{x^2}{9} + \frac{y^2}{49} = 1$.

   Before graphing it, predict the following.
   a. Which axis will the foci be on?   **b.** x intercepts?   **c.** y intercepts?

6. Graph the equation.

   a. Did you predict the correct axis for the foci?

   b. What are the coordinates of the foci?

   c. Was your prediction of the x intercepts correct?

   d. Was your prediction of the y intercepts correct?

   e. Answer Y to ANOTHER GRAPH?

7. Repeat Exercises **5** and **6** for the equation $\frac{x^2}{100} + \frac{y^2}{64} = 1$.

   a. foci on which axis?   **b.** coordinates of foci?

   c. x intercepts?   **d.** y intercepts?

8. What determines which axis contains the foci of the ellipse?

9. *Complete:* For an equation of the form $\frac{x^2}{a^2} + \frac{y^2}{b^2} = 1$,

   a. the graph is an ___?___ ;

   b. the x intercepts are ___?___ and ___?___ ;

   c. the y intercepts are ___?___ and ___?___ .

10. Enter Y to ANOTHER GRAPH?

11. **a.** Predict the x and y intercepts of the graph of each equation below. Also state which axis the foci are on.

    b. Have the computer graph the equation.

    c. Check whether your predictions were correct.

    d. List the coordinates of the foci.

    i. $\frac{x^2}{4} + \frac{y^2}{25} = 1$ **ii.** $\frac{x^2}{81} + \frac{y^2}{9} = 1$ **iii.** $\frac{x^2}{36} + \frac{y^2}{36} = 1$

    iv. In Exercise **iii**, $a^2 = b^2$. What special kind of ellipse results in this case?

Investigation: The Hyperbola

Program: HYPERB1

Objectives: To use the computer to graph hyperbolas centered at the origin

To discover that the foci of the hyperbola are on the x axis and the branches of the hyperbola open horizontally if the x term is first (the positive term) in the equation, and that the foci are on the y axis and the hyperbola opens vertically if the y term is positive

To discover that the x intercepts are a and −a if the x term is positive and that the y intercepts are a and −a if the y term is positive.

1. a. With BASIC loaded into the computer and the disk in drive 1 (or drive A), run the program HYPERB1 by typing the appropriate command.

        Apple: RUN HYPERB1   IBM: RUN "HYPERB1"

Be sure to press the RETURN key after typing the command.

  b. Read the opening messages of the program.

2. Use the program to graph the equation $\dfrac{x^2}{9} - \dfrac{y^2}{4} = 1$.

  a. Enter 1 for the form of the equation, then 9 for A^2 and 4 for B^2.

  b. The computer marks the <u>foci</u> of the <u>hyperbola</u>. Which axis are they on?

  c. What are the coordinates of the foci?

  d. Read the messages on the screen. Then press any key to begin graphing.

  e. The computer connects each point to the two foci and prints these distances and the difference between them. What is true about the difference of the distances for all the points?

  f. Do the branches of the hyperbola open vertically or horizontally?

  g. What are the x intercepts?   **h.** Does the hyperbola have any y intercepts?

  i. Press a key to see the <u>asymptotes</u>. What are their equations?

  j. The asymptotes intersect at the <u>center</u> of the hyperbola. What are the coordinates?

  k. Press a key and the computer draws a rectangle with the asymptotes as its diagonals. This rectangle contains the <u>vertices</u> of the hyperbola.

  l. Answer Y (yes) to ANOTHER GRAPH?

3. Have the computer graph $\dfrac{y^2}{16} - \dfrac{x^2}{4} = 1$.

  a. Choose 2 for the form of the equation; enter 16 for A^2 and 4 for B^2.

  b. Which axis are the foci on?   **c.** What are the coordinates of the foci?

  d. Press any key to begin graphing. What is the difference of the distances to the foci?

  e. Do the branches of this hyperbola open vertically or horizontally?

  f. Are there any x intercepts?   **g.** What are the y intercepts?

  h. List the equations of the asymptotes.   **i.** Enter Y to ANOTHER GRAPH?

4. Have the computer graph $\frac{y^2}{9} - \frac{x^2}{64} = 1$. (Form 2; $A\hat{\ }2 = 9$; $B\hat{\ }2 = 64$.)

    a. Which axis contains the foci?   **b.** What are the coordinates of the foci?

    c. Press any key to begin graphing. What is the difference of the distances from any point on the graph to the foci?

    d. Do the branches of this hyperbola open vertically or horizontally?

    e. What are the y intercepts?   **f.** What are the asymptotes' equations?

    g. Enter Y to ANOTHER GRAPH?

5. Have the computer graph $\frac{x^2}{25} - \frac{y^2}{16} = 1$. (Form 1; $A\hat{\ }2 = 25$; $B\hat{\ }2 = 16$.)

    a. Which axis contains the foci?   **b.** What are the coordinates of the foci?

    c. Press any key to see the graph. What is the constant difference between the distances to the foci?

    d. Do the branches of this hyperbola open vertically or horizontally?

    e. What are the x intercepts?   **f.** What are the equations of the asymptotes?

    g. Enter Y to ANOTHER GRAPH?

6. Before graphing $\frac{x^2}{81} - \frac{y^2}{25} = 1$, make predictions about the graph.

    a. Which axis will contain the foci?

    b. Will the branches open vertically or horizontally?

    c. x intercepts (if any)?   **d.** y intercepts (if any)?

7. Have the computer graph the equation in Exercise 6. (Form 1; $A\hat{\ }2 = 81$; $B\hat{\ }2 = 25$.)

    a. Which of your predictions were correct?

    b. What are the coordinates of the foci?

    c. What are the equations of the asymptotes?

8. Make predictions about the graph of the equation $\frac{y^2}{9} - \frac{x^2}{49} = 1$.

    a. Which axis will contain the foci?

    b. Will the branches open vertically or horizontally?

    c. x intercepts (if any)?   **d.** y intercepts (if any)?

9. Have the computer graph the equation in Exercise 8. (Form 2, $A\hat{\ }2 = 9$, $B\hat{\ }2 = 49$)

    a. Were your predictions correct?   **b.** What are the coordinates of the foci?

    c. What are the equations of the asymptotes?

10. *Complete:* For an equation of the form $\frac{x^2}{a^2} - \frac{y^2}{b^2} = 1$,

    a. the graph is a __?__ ;   **b.** the foci are on the __?__ axis;

    c. the branches open __?__ (vertically or horizontally?);

    d. the x intercepts are __?__ and __?__ ;   **e.** there are no __?__ intercepts.

11. *Complete:* For an equation of the form $\dfrac{y^2}{a^2} - \dfrac{x^2}{b^2} = 1$,

    a. the graph is a __?__ ;   **b.** the foci are on the __?__ axis;

    c. the branches open __?__ (vertically or horizontally?);

    d. the __?__ intercepts are __?__ and __?__ ;   **e.** there are no __?__ intercepts.

12. Enter Y to ANOTHER GRAPH?

13. **a.** For each equation, predict whether the foci will be on the x or y axis and whether the branches will open vertically or horizontally. Also predict the x and y intercepts (if any).

    b. Graph the equation.   **c.** Check whether your predictions were correct.

    d. List the coordinates of the foci and the equations of the asymptotes.

    i. $\dfrac{x^2}{36} - \dfrac{y^2}{9} = 1$ **ii.** $\dfrac{y^2}{36} - \dfrac{x^2}{36} = 1$ **iii.** $\dfrac{y^2}{16} - \dfrac{x^2}{64} = 1$

    iv. $\dfrac{x^2}{121} - \dfrac{y^2}{4} = 1$ **v.** $\dfrac{x^2}{144} - \dfrac{y^2}{25} = 1$ **vi.** $\dfrac{y^2}{25} - \dfrac{x^2}{49} = 1$

Investigation: The Ellipse (center not at the origin)

Program: ELLIPSE2

Objectives: To use the computer to graph ellipses not centered at the origin

               To discover that the center of the ellipse is (h, k), that the major axis has length $2a$ (where $a > b$), and the minor axis has length $2b$ (where $a > b$).

1. **a.** With BASIC loaded into the computer and the disk in drive 1 (or drive A), run the program ELLIPSE2 by typing the appropriate command.

               Apple: RUN ELLIPSE2   IBM: RUN "ELLIPSE2"

    Be sure to press the RETURN key after typing the command.

    b. Read the opening messages of the program.

2. Use the program to graph the equation $\dfrac{(x-2)^2}{25} + \dfrac{y^2}{16} = 1$.

    a. Enter 2 (not -2) for H and 0 for K.

    b. Enter 25 for the FIRST DENOMINATOR and 16 for the SECOND DENOMINATOR.

    c. The computer marks the <u>center</u> of the ellipse with an X and then graphs the ellipse. What are the coordinates of the center?

    d. Two <u>vertices</u> of the ellipse are $(-3, 0)$ and $(7, 0)$. What is the distance from each of these points to the center?

    e. What is the relationship between the answer to Exercise **2d** and the first denominator of the equation?

    f. The <u>major axis</u> of the ellipse connects the two vertices listed in Exercise **2d**. What is the length of the major axis?

g. Two other vertices of the ellipse are (2, 4) and (2, −4). What is the distance from each of these points to the center?

h. What is the relationship between the answer to Exercise **2g** and the second denominator of the equation?

i. The <u>minor axis</u> connects the two vertices listed in Exercise **2g**. The major and minor axes of the ellipse intersect at its center. What is the length of the minor axis?

j. Answer Y (yes) to ANOTHER GRAPH?

3. Graph $\dfrac{x^2}{36} + \dfrac{(y-4)^2}{9} = 1$.

   a. Enter 0 for H and 4 (not −4) for K.

   b. Enter 36 for the FIRST DENOMINATOR and 9 for the SECOND.

   c. What are the coordinates of the center of this ellipse?

   d. What are the coordinates of the end points of the major and minor axes?

   e. What is the length of the major axis?

   f. What is the length of the minor axis?

   g. Answer Y to ANOTHER GRAPH?

4. Have the computer graph $\dfrac{(x+4)^2}{49} + \dfrac{(y-2)^2}{16} = 1$.

   a. Think of $(x+4)^2$ as $[x-(-4)]^2$. So enter −4 for H and 2 for K.

   b. Enter 49 for the FIRST DENOMINATOR and 16 for the SECOND.

   c. What are the coordinates of the center of the ellipse?

   d. What is the length of the major axis?

   e. What is the length of the minor axis?

   f. Answer Y to ANOTHER GRAPH?

5. Have the computer graph $\dfrac{(x-5)^2}{9} + \dfrac{(y+3)^2}{36} = 1$.

   a. Enter 5 for H, −3 for K, and 9 and 36 for the denominators.

   b. What are the coordinates of the center?

   c. The major (longer) axis is vertical for this ellipse. What is the length of the major axis?

   d. What is the length of the minor (shorter) axis?

   e. Answer Y to ANOTHER GRAPH?

6. Before graphing $\dfrac{(x+3)^2}{64} + \dfrac{(y+1)^2}{49} = 1$, make predictions about the graph.

   a. What are the coordinates of the center?

   b. Is the major axis vertical or horizontal?

   c. What is the length of the major axis?

   d. What is the length of the minor axis?

7. Have the computer graph the equation in Exercise 6. (H = −3, K = −1, A^2 = 64, B^2 = 49) Were your predictions correct?

8. Repeat Exercises **6** and **7** for $\dfrac{(x + 7)^2}{4} + \dfrac{(y - 3)^2}{25} = 1$. (H = −7, K = 3, A^2 = 4, B^2 = 25)

9. Enter Y to ANOTHER GRAPH?

10. a. Before graphing each ellipse, repeat Exercise **6.**
   b. Graph the ellipse.
   c. Check your predictions.

   i. $\dfrac{x^2}{81} + \dfrac{y^2}{49} = 1$ (H = 0, K = 0, A^2 = 81, B^2 = 49)

   ii. $\dfrac{(x + 1)^2}{100} + \dfrac{y^2}{64} = 1$ (H = −1, K = 0, A^2 = 100, B^2 = 64)

   iii. $\dfrac{x^2}{121} + \dfrac{(y - 5)^2}{9} = 1$ (H = 0, K = 5, A^2 = 121, B^2 = 9)

   iv. $\dfrac{(x - 6)^2}{16} + \dfrac{(y + 2)^2}{25} = 1$ (H = 6, K = −2, A^2 = 16, B^2 = 25)

   v. $\dfrac{(x - 8)^2}{36} + \dfrac{(y - 1)^2}{49} = 1$

Investigation: The Hyperbola (center not at the origin)

Program: HYPERB2

Objectives: To use the computer to graph hyperbolas not centered at the origin

               To discover that the center of the hyperbola is (h, k), that the major axis has length 2a and the minor axis has length 2b

1. a. With BASIC loaded into the computer and the disk in drive 1 (or drive A), run the program HYPERB2 by typing the appropriate command.

          Apple: RUN HYPERB2   IBM: RUN "HYPERB2"

Be sure to press the RETURN key after typing the command.

   b. Read the opening messages of the program.

2. Use the program to graph $\dfrac{(x - 3)^2}{16} - \dfrac{y^2}{9} = 1$.

   a. Choose form 1. (Type a 1 and press RETURN.)

   b. Enter 3 (not −3) for H and 0 for K.

   c. Enter 16 for the FIRST DENOMINATOR and 9 for the SECOND DENOMINATOR.

   d. The computer graphs the <u>hyperbola</u>. One <u>vertex</u> is at the point (−1, 0). What are the coordinates of the other vertex?

e. Press any key and the computer draws the <u>asymptotes</u>. They intersect at the <u>center</u> of the hyperbola. What are the coordinates of this point?

f. What is the distance from each vertex to the center of the hyperbola?

g. What is the relationship between your answer in Exercise **2f** and the first denominator of the equation?

h. What are the equations of the asymptotes?

i. Press any key and the computer draws a rectangle between the branches of the hyperbola. Notice that the asymptotes contain the diagonals of this rectangle. Also, the length of the rectangle is the distance between the vertices. What is the width (the shorter dimension) of this rectangle?

j. What is the relationship of the answer to Exercise **2i** and the second denominator of the equation?

k. Answer Y (yes) to ANOTHER GRAPH?

3. Have the computer graph $\dfrac{x^2}{25} - \dfrac{(y-2)^2}{16} = 1$.

a. This is form 1; H = 0; K = 2; A^2 = 25; B^2 = 16.

b. What are the coordinates of the vertices of the hyperbola?

c. Press any key to see the asymptotes. What are the coordinates of the center where they intersect?

d. What is the distance from each vertex to the center?

e. What are the equations of the asymptotes?

f. Press a key to see the dashed rectangle. What is the width (the vertical dimension) of the rectangle?

g. Enter Y for ANOTHER GRAPH?

4. Graph $\dfrac{(y+2)^2}{4} - \dfrac{(x-1)^2}{36} = 1$.

a. Since the y term is first in this equation, choose form 2.

b. Enter 1 for H. Since $(y+2)^2 = [y-(-2)]^2$, enter -2 for K.

c. Enter 4 for the FIRST DENOMINATOR and 36 for the SECOND.

d. One vertex of the hyperbola is (1, 0). What are the coordinates of the other vertex?

e. Press a key to see the asymptotes and press again to see the rectangle.

f. What are the coordinates of the center?

g. What is the distance from each vertex to the center and what is the relationship of this number to the first denominator of the equation?

h. What is the horizontal dimension of the rectangle and what is the relationship of this number to the second denominator of the equation?

i. What are the equations of the asymptotes?

j. Enter Y for ANOTHER GRAPH?

5. Before graphing $\dfrac{(x+4)^2}{49} - \dfrac{(y+3)^2}{25} = 1$, make predictions about the graph.

   a. Will the branches of the hyperbola open vertically or horizontally?

   b. What are the coordinates of the center?

   c. What are the coordinates of the vertices?

   d. What is the distance from each vertex to the center?

   e. What is the other (vertical) dimension of the rectangle between the branches?

6. Have the computer graph the hyperbola in Exercise 5.

   a. This is form 1; H = -4; K = -3; A$\hat{\;}$2 = 49; B$\hat{\;}$2 = 25.

   b. Were your predictions correct?

   c. What are the equations of the asymptotes?

7. Repeat Exercises **5** and **6** for $\dfrac{(y-4)^2}{9} - \dfrac{(x+5)^2}{64} = 1$.

   (Form 2, H = -5, K = 4, A$\hat{\;}$2 = 9, B$\hat{\;}$2 = 64)

8. Enter Y to ANOTHER GRAPH?

9. Before graphing each hyperbola predict the following.

   a. the coordinates of the center;

   b. whether the hyperbola opens vertically or horizontally;

   c. the coordinates of the two vertices;

   d. the dimensions of the rectangle between the branches.

   e. Graph the hyperbola and check your predictions.

   f. List the equations of the asymptotes.

    i. $\dfrac{x^2}{36} - \dfrac{y^2}{25} = 1$ (Form 1, H = 0, K = 0, A$\hat{\;}$2 = 36, B$\hat{\;}$2 = 25)

    ii. $\dfrac{(x+1)^2}{81} - \dfrac{y^2}{49} = 1$ (Form 1, H = -1, K = 0, A$\hat{\;}$2 = 81, B$\hat{\;}$2 = 49)

    iii. $\dfrac{(y-3)^2}{9} - \dfrac{x^2}{100} = 1$ (Form 2, H = 0, K = 3, A$\hat{\;}$2 = 9, B$\hat{\;}$2 = 100)

    iv. $\dfrac{(y-1)^2}{16} - \dfrac{(x+2)^2}{36} = 1$ (Form 2, H = -2, K = 1, A$\hat{\;}$2 = 16, B$\hat{\;}$2 = 36)

    v. $\dfrac{(x-8)^2}{4} - \dfrac{(y+1)^2}{25} = 1$

    vi. $\dfrac{(y+5)^2}{4} - \dfrac{(x-6)^2}{64} = 1$

Investigation: The Parabola

Program: PARAB4

Objectives: To use a computer to graph parabolas of the form $y = ax^2 + bx + c$ by showing each point of the parabola as equidistant from the focus and the directrix

To discover that the focus is above the directrix when $a > 0$ and below the directrix when $a < 0$

1. a. With BASIC loaded into the computer and the disk in drive 1 (or drive A), run the program PARAB4 by typing the appropriate command.

Apple: RUN PARAB4   IBM: RUN "PARAB4"

Be sure to press the RETURN key after typing the command.

b. Read the opening messages of the program.

2. Use the program to graph the equation $y = .25x^2 + x - 4$.

a. Type .25 for A and press the RETURN key.   **b.** Enter 1 for B.   **c.** Enter -4 for C.

d. The computer draws a horizontal line called the <u>directrix</u>. What is the equation of the directrix? $y = \underline{\ ?\ }$

e. The <u>focus</u> is marked with an X. What are the coordinates of the focus?

f. Is the focus above or below the directrix?

g. Press any key to begin graphing.

h. As it plots the function, the computer connects each point to the focus and to the directrix. The segment connecting each point to the focus is the same length as the segment connecting the point to the directrix.

i. When the parabola is complete, the computer prints the equation of the <u>axis of symmetry</u>. What is it? $x = \underline{\ ?\ }$

j. What are the coordinates of the <u>vertex</u>?

k. The vertex and the focus are both on which line?

l. Is the vertex above the focus or below the focus?

m. Where is the vertex in relation to the focus and directrix?

n. Answer Y (yes) to ANOTHER GRAPH?

3. Have the computer graph $y = -.5x^2 + 3x + 2$.

a. Enter $-.5$ for A, 3 for B, and 2 for C.

b. What is the equation of the directrix?

c. What are the coordinates of the focus?

d. Is the focus above or below the directrix?

e. Press any key to begin graphing.

f. What is the equation of the axis of symmetry?

g. What are the coordinates of the vertex?

h. Is the vertex above or below the focus?

i. Answer Y to ANOTHER GRAPH?

4. Have the computer graph $y = .4x^2 - 2x - 4$.

   a. Enter .4 for A, -2 for B, and -4 for C.

   b. What is the equation of the directrix?

   c. What are the coordinates of the focus?

   d. Is the focus above or below the directrix?

   e. Press any key to start graphing.

   f. What is the equation of the axis of symmetry?

   g. What are the coordinates of the vertex?

   h. Is the vertex above or below the focus?

   i. Enter Y to ANOTHER GRAPH?

5. Make predictions about the graph of $y = -.6x^2 + 4x - 2$.

   a. Will the focus be above or below the directrix?

   b. Will the vertex be above or below the focus?

6. Have the computer graph the function in Exercise 5.   (A $= -.6$, B $= 4$, C $= -2$)

   a. Which of your predictions were correct?

   b. List this information about the parabola.
   axis of symmetry: $x = \underline{\ ?\ }$ vertex: $(\underline{\ ?\ }, \underline{\ ?\ })$
   directrix: $y = \underline{\ ?\ }$ focus: $(\underline{\ ?\ }, \underline{\ ?\ })$

   c. Enter Y to ANOTHER GRAPH?

7. Repeat Exercises **5** and **6** for the function $y = .2x^2 + 1.5x - 3$.

8. Complete: For the parabola defined by $y = ax^2 + bx + c$ ($a \neq 0$),

   a. when $a > 0$, the focus is $\underline{\ ?\ }$ (above/below) the directrix and the vertex is $\underline{\ ?\ }$ (above/ below) the focus;

   b. when $a < 0$, the focus is $\underline{\ ?\ }$ (above/below) the directrix and the vertex is $\underline{\ ?\ }$ (above/ below) the focus.

9. Enter Y to ANOTHER GRAPH?

10. For each function below,

    a. predict whether the focus will be above or below the directrix;

    b. predict whether the vertex will be above or below the focus;

    c. have the computer graph the function;

    d. check whether your predictions were correct;

    e. list the focus, directrix, axis of symmetry, and vertex.

    i. $y = x^2 - 3$ [Enter 0 for B.]   **ii.** $y = .75x^2 + 4x$ [Enter 0 for C.]

    iii. $y = -.25x^2 - 2x - 1$ **iv.** $y = -.3x^2 + 1.2x + .8$

Investigation: Systems of First- and Second-Degree Equations

Program: QUADLIN

Objectives: To use the computer to graph systems consisting of one linear and one quadratic equation

To discover that such systems may have none, one, or two solutions

1. **a.** With BASIC loaded into the computer and the disk in drive 1 (or drive A), run the program QUADLIN by typing the appropriate command.

   Apple: RUN QUADLIN   **IBM:** RUN "QUADLIN"

   Be sure to press the RETURN key after typing the command.

   b. Read the opening messages of the program.

2. Use the program to graph this system, $\begin{cases} x^2 + y^2 = 64 \\ x + y = 8 \end{cases}$

   a. For the first equation choose form 1 and press the RETURN key.

   b. Enter 1 for A (and press RETURN); enter 1 for B; enter 64 for C.

   c. Now enter the coefficients of the linear equation: A = 1, B = 1, C = 8.

   d. What figure is the graph of the quadratic equation?

   e. The straight line for the second equation intersects the first graph in how many points?

   f. List any solutions of the system: (_?_ , _?_) , (_?_ , _?_)

   g. Enter Y (yes) to ANOTHER GRAPH?

3. Have the computer graph and solve this system, $\begin{cases} 4x^2 + 25y^2 = 100 \\ x + 2y = 4 \end{cases}$

   a. Enter 1 for the form of the quadratic equation.

   b. Enter the coefficients of the first equation: A = 4, B = 25, C = 100.

   c. Enter the coefficients of the linear equation: A = 1, B = 2, C = 4.

   d. What kind of figure is the graph of the first equation?

   e. How many points do the graphs of the two equations have in common?

   f. List any solutions of the system.

   g. Enter Y to ANOTHER GRAPH?

4. Have the computer graph this system, $\begin{cases} x^2 - 9y^2 = 81 \\ 2x + 3y = -6 \end{cases}$

   a. The first equation is form 1, with A = 1, B = -9, and C = 81.

   b. For the second equation, A = 2, B = 3, C = -6.

   c. What kind of figure is the graph of the quadratic equation?

d. The line intersects the quadratic graph in how many points?

e. List any solutions of the system.

f. Enter Y to ANOTHER GRAPH?

5. Have the computer graph this system, $\begin{cases} 2x^2 + y = 5 \\ y = -3x + 1 \end{cases}$.

a. The first equation is form 2; $A = 2$; $B = 1$; $C = 5$.

b. The second equation must be converted to the form required by the computer. So add $3x$ to both sides to obtain $3x + y = 1$.

c. Enter 3 for A, 1 for B, and 1 for C.

d. What kind of figure is the graph of the first equation?

e. How many points of intersection do the two graphs have?

f. List any solutions of the system.

g. Enter Y to ANOTHER GRAPH?

6. Have the computer graph this system, $\begin{cases} 4x - 3 = y^2 \\ 9 = 6y - 3x \end{cases}$.

a. The first equation is of form 3. To match the form $AX + BY\hat{\,}2 = C$, add 3 to both sides and subtract y from both sides. This gives $4x - y^2 = 3$.

b. For the first equation, enter 4 for A, -1 for B, and 3 for C.

c. Change the second equation to the form $Ax + By = C$ and write the result.

d. Enter A, B, and C for the second equation.

e. What kind of figure is the graph of the first equation?

f. How many points of intersection do the two graphs have?

g. List any solutions of the system.

7. Enter Y to ANOTHER GRAPH?

8. a. Use the computer to graph each system. If necessary, convert each equation to the required form before entering the coefficients.

b. List the type of graph for each quadratic equation.

c. List any solutions (to the nearest tenth) of each system.

i. $\begin{cases} x^2 - y^2 = 4 \\ 3x + y = 6 \end{cases}$ **ii.** $\begin{cases} 25x^2 + 16y^2 = 400 \\ y = 5 \end{cases}$

iii. $\begin{cases} x^2 + 3y = 16 \\ 2y = 7 - 3x \end{cases}$ **iv.** $\begin{cases} y = -2x^2 + 1 \\ y = x + 5 \end{cases}$

v. $\begin{cases} y^2 = 5x \\ 10x + 3y = 5 \end{cases}$ **vi.** $\begin{cases} 81y^2 - 16x^2 = 1296 \\ 13 = 11x - 7y \end{cases}$

Investigation: Systems of Two Second–Degree Equations

Program: QUADQUAD

Objectives: To use the computer to graph systems consisting of two quadratic equations

To discover that such systems may have none, one, two, three, or four solutions.

1. **a.** With BASIC loaded into the computer and the disk in drive 1 (or drive A), run the program QUADQUAD by typing the appropriate command.

   Apple: RUN QUADQUAD   IBM: RUN "QUADQUAD"

   Be sure to press the RETURN key after typing the command.

   b. Read the opening messages of the program.

2. Use the program to graph and solve this system, $\begin{cases} 9x^2 + 4y^2 = 36 \\ x^2 - y^2 = 9 \end{cases}$.

   a. The first equation is form 1. So type a 1 and press the RETURN key.

   b. Enter 9 for A, 4 for B, and 36 for C.

   c. The second equation is also of form 1. A = 1, B = -1, and C = 9.

   d. What type of figure is the graph of the first equation?

   e. What type of figure is the graph of the second equation?

   f. The two graphs intersect in how many points?

   g. Is there any solution to the system?

   h. Enter Y (yes) to ANOTHER?

3. Use the computer to solve this system, $\begin{cases} 2x^2 + y = 4 \\ x + y^2 = 5 \end{cases}$.

   a. The first equation is form 2. A = 2, B = 1, and C = 4.

   b. The second equation is form 3. A = 1, B = 1, and C = 5.

   c. What type of figure is the graph of the first equation?

   d. What type of figure is the second graph?

   e. How many points do the two graphs have in common?

   f. List all solutions of the system.

   g. Enter Y to ANOTHER?

4. Use the computer to solve this system, $\begin{cases} 25y^2 - x^2 = 100 \\ y = 4x^2 \end{cases}$.

   a. The first equation is form 1. However, it must be transformed into the required form, which is $-x^2 + 25y^2 = 100$. So enter -1 for A, 25 for B, and 100 for C.

b. The second equation is form 2. Rearrange it by subtracting $4x^2$ from both sides to give $-4x^2 + y = 0$. So $A = -4$, $B = 1$, and $C = 0$.

c. What type of figure is the graph of each equation?

d. List all solutions to the system.

e. Enter Y to ANOTHER?

5. Use the computer to solve this system, $\begin{cases} y^2 = 8 - 4x \\ x^2 + y^2 = 81 \end{cases}$.

a. The first equation is form 3. Add $4x$ to both sides to obtain $4x + y^2 = 8$. So enter 4 for A, 1 for B, and 8 for C.

b. The second equation is in form 1, with $A = 1$, $B = 1$, and $C = 81$.

c. What type of figure is the graph of each equation?

d. List all solutions of the system.

6. Enter Y to ANOTHER?

7. a. Use the computer to solve each system. If necessary, convert each equation to the required form before entering the coefficients.

b. List the type of figure that forms the graph of each equation.

c. List any solutions of each system.

   i. $\begin{cases} x^2 + 10y = 40 \\ 25x^2 - 36y^2 = 900 \end{cases}$
       ii. $\begin{cases} x - 2y^2 = 10 \\ x = 3y^2 + 5 \end{cases}$

   iii. $\begin{cases} x^2 + y^2 = 64 \\ y^2 = x - 9 \end{cases}$
       iv. $\begin{cases} 3y^2 + 2x = 6 \\ y = 5x^2 \end{cases}$

   v. $\begin{cases} 4x^2 = 3 - 2y \\ 4 = 3x^2 - 4y \end{cases}$
       vi. $\begin{cases} 9x^2 = 25y^2 + 225 \\ 9x^2 + 144y^2 = 1296 \end{cases}$

Investigation: Exponential Functions and Equations

Program: EXPGRAPH

Objectives: To use the computer to graph exponential functions

To learn that the graph of an exponential function lies in quadrants I and II

To discover that the graph of $y = a^x$ rises from left to right if $a > 1$ and falls from left to right if $a < 1$

1. **a.** With BASIC loaded into the computer and the disk in drive 1 (or drive A), run the program EXPGRAPH by typing the appropriate command.

   Apple: RUN EXPGRAPH   IBM: RUN "EXPGRAPH"

   Be sure to press the RETURN key after typing the command.

   b. Read the opening messages of the program.

2. Use the program to graph $y = 2^x$.

   a. Type 2 for A and press RETURN.

   b. Is the graph a straight line?

   c. The graph lies in which two quadrants? __?__ and __?__

   d. Does the graph have an x intercept?

   e. What is the y intercept?

   f. As x increases (left to right), does the graph rise or does it fall?

   g. Answer Y (yes) to the question ANOTHER GRAPH?

   h. Answer N (no) to the question CLEAR THE SCREEN?

3. Have the computer graph $y = 3^x$.

   a. Type 3 for A and press RETURN.

   b. Is the graph a straight line?

   c. The graph lies in which two quadrants?

   d. Does the graph have an x intercept?

   e. What is the y intercept?

   f. For positive values of x, does the second graph rise more steeply than the first graph?

   g. What point(s) do the two graphs have in common?

   h. Answer Y to the question ANOTHER GRAPH? and N to CLEAR THE SCREEN?

4. Graph $y = .5^x$. That is, enter .5 for A.

   a. Does the graph lie in the same quadrants as the first two graphs?

   b. What point does the graph have in common with the first two graphs?

c. As x increases, does the graph rise or fall?

d. Answer Y to ANOTHER GRAPH? and Y for CLEAR THE SCREEN?

5. Consider the function $y = 4^x$. Before graphing the function, make predictions about it.
   a. Will the function rise or fall from left to right?
   b. Will the function intersect the x axis?
   c. Where will the function intersect the y axis?

6. **a.** Graph the function (enter 4 for A). Were your predictions correct?
   b. Answer Y to ANOTHER GRAPH? and N to CLEAR THE SCREEN?

7. Consider $y = .25^x$. Before seeing the graph, make predictions about it.
   a. Will the function rise or fall from left to right?
   b. What point will it have in common with $y = 4^x$?

8. **a.** Have the computer graph the function. Were your predictions correct?
   b. Enter Y to ANOTHER GRAPH? and Y to CLEAR THE SCREEN?

9. Consider these three functions.

   A. $y = 5^x$ **B.** $y = .2^x$ **C.** $y = 7^x$

   Before having the computer graph them, answer these questions. Then graph all three functions on the same screen and check your predictions.
   i. Which functions decrease (fall) from left to right?
   ii. Which functions increase (rise) from left to right?
   iii. Which function has the steepest slope; that is, which one will rise or fall the fastest?
   iv. Which point do all three functions have in common?

Investigation: The Logarithmic Function

Program: LOGGRAPH

Objectives: To use the computer to graph logarithmic functions

To learn that the graph of a logarithmic function is the inverse of the corresponding exponential function

To discover that the graph of a logarithmic function lies in quadrants I and IV

1. **a.** With BASIC loaded into the computer and the disk in drive 1 (or drive A), run the program LOGGRAPH by typing the appropriate command.

   Apple: RUN LOGGRAPH   IBM: RUN "LOGGRAPH"

   Be sure to press the RETURN key after typing the command.

   b. Read the opening messages of the program.

2. Use the program to graph $y = \log_{10} x$. ◄——This is read "log base 10 of x."

   a. Enter 10 for B and press RETURN.

   b. The computer draws the line $y = x$. Then it graphs $y = \log_{10} x$ to the right of this line and $y = 10^x$ to the left. The two curved graphs are mirror images across this line. Graphs like these are called <u>inverse functions</u>.

   c. In what quadrants does the graph of $y = \log_{10} x$ lie? __?__ and __?__

   d. What is the x intercept of $y = \log_{10} x$?   **e.** Does $y = \log_{10} x$ have a y intercept?

3. **a.** To the question ANOTHER GRAPH?, type Y (yes) and press RETURN.

   b. Have the computer graph $y = \log_2 x$. That is, enter 2 for B.

   c. The computer graphs $y = \log_2 x$ to the right of the line $y = x$ and $y = 2^x$ to the left of the line.

   d. The graph of $y = \log_2 x$ is in which quadrants?

   e. What is the x intercept of $y = \log_2 x$?

   f. Does $y = \log_2 x$ have a y intercept?   **g.** Answer Y to ANOTHER GRAPH?

4. Consider the function $y = \log_4 x$. Before having the computer graph this function and its inverse, write the equation of the inverse function.

   a. In what quadrants will the graph of $y = \log_4 x$ lie?

   b. What will be the x intercept of $y = \log_4 x$?

   c. Graph the function. Were your predictions correct?

   d. Answer Y to ANOTHER GRAPH?

5. Have the computer graph $y = \log_{.5} x$. That is, enter .5 for B.

   a. What is the inverse of $y = \log_{.5} x$?

   b. How does the graph of $y = \log_{.5} x$ differ from the earlier log graphs?

   c. What is the x intercept?   **d.** Answer N to ANOTHER GRAPH?

Computer Investigations

Formulas From Geometry

Rectangle

area: $A = \ell w$

perimeter: $P = 2\ell + 2w$

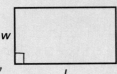

Square

area: $A = s^2$

perimeter: $P = 4s$

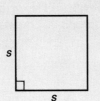

Triangle

area: $A = \frac{1}{2}bh$

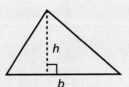

Parallelogram

area: $A = bh$

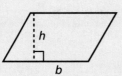

Trapezoid

area: $A = \frac{1}{2}h(b_1 + b_2)$

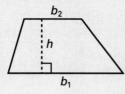

Circle

area: $A = \pi r^2$

circumference: $C = 2\pi r$

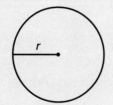

Rectangular Prism

volume: $V = \ell wh$

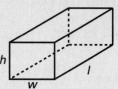

Cube

volume: $V = s^3$

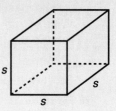

Right Circular Cylinder

volume: $V = \pi r^2 h$

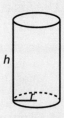

Rectangular Pyramid

volume: $V = \frac{1}{3}\ell wh$

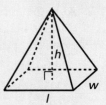

Right Circular Cone

volume: $V = \frac{1}{3}\pi r^2 h$

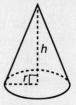

Sphere

volume: $V = \frac{4}{3}\pi r^3$

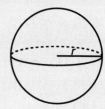

Table of Squares, Cubes, Square and Cube Roots

No.	Squares	Cubes	Square Roots	Cube Roots	No.	Squares	Cubes	Square Roots	Cube Roots
1	1	1	1.000	1.000	51	2,601	132,651	7.141	3.708
2	4	8	1.414	1.260	52	2,704	140,608	7.211	3.733
3	9	27	1.732	1.442	53	2,809	148,877	7.280	3.756
4	16	64	2.000	1.587	54	2,916	157,464	7.348	3.780
5	25	125	2.236	1.710	55	3,025	166,375	7.416	3.803
6	36	216	2.449	1.817	56	3,136	175,616	7.483	3.826
7	49	343	2.646	1.913	57	3,249	185,193	7.550	3.849
8	64	512	2.828	2.000	58	3,364	195,112	7.616	3.871
9	81	729	3.000	2.080	59	3,481	205,379	7.681	3.893
10	100	1,000	3.162	2.154	60	3,600	216,000	7.746	3.915
11	121	1,331	3.317	2.224	61	3,721	226,981	7.810	3.936
12	144	1,728	3.464	2.289	62	3,844	238,328	7.874	3.958
13	169	2,197	3.606	2.351	63	3,969	250,047	7.937	3.979
14	196	2,744	3.742	2.410	64	4,096	262,144	8.000	4.000
15	225	3,375	3.873	2.466	65	4,225	274,625	8.062	4.021
16	256	4,096	4.000	2.520	66	4,356	287,496	8.124	4.041
17	289	4,913	4.123	2.571	67	4,489	300,763	8.185	4.062
18	324	5,832	4.243	2.621	68	4,624	314,432	8.246	4.082
19	361	6,859	4.359	2.668	69	4,761	328,509	8.307	4.102
20	400	8,000	4.472	2.714	70	4,900	343,000	8.367	4.121
21	441	9,261	4.583	2.759	71	5,041	357,911	8.426	4.141
22	484	10,648	4.690	2.802	72	5,184	373,248	8.485	4.160
23	529	12,167	4.796	2.844	73	5,329	389,017	8.544	4.179
24	576	13,824	4.899	2.884	74	5,476	405,224	8.602	4.198
25	625	15,625	5.000	2.924	75	5,625	421,875	8.660	4.217
26	676	17,576	5.099	2.962	76	5,776	438,976	8.718	4.236
27	729	19,683	5.196	3.000	77	5,929	456,533	8.775	4.254
28	784	21,952	5.292	3.037	78	6,084	474,552	8.832	4.273
29	841	24,389	5.385	3.072	79	6,241	493,039	8.888	4.291
30	900	27,000	5.477	3.107	80	6,400	512,000	8.944	4.309
31	961	29,791	5.568	3.141	81	6,561	531,441	9.000	4.327
32	1,024	32,768	5.657	3.175	82	6,724	551,368	9.055	4.344
33	1,089	35,937	5.745	3.208	83	6,889	571,787	9.110	4.362
34	1,156	39,304	5.831	3.240	84	7,056	592,704	9.165	4.380
35	1,225	42,875	5.916	3.271	85	7,225	614,125	9.220	4.397
36	1,296	46,656	6.000	3.302	86	7,396	636,056	9.274	4.414
37	1,369	50,653	6.083	3.332	87	7,569	658,503	9.327	4.431
38	1,444	54,872	6.164	3.362	88	7,744	681,472	9.381	4.448
39	1,521	59,319	6.245	3.391	89	7,921	704,969	9.434	4.465
40	1,600	64,000	6.325	3.420	90	8,100	729,000	9.487	4.481
41	1,681	68,921	6.403	3.448	91	8,281	753,571	9.539	4.498
42	1,764	74,088	6.481	3.476	92	8,464	778,688	9.592	4.514
43	1,849	79,507	6.557	3.503	93	8,649	804,357	9.644	4.531
44	1,936	85,184	6.633	3.530	94	8,836	830,584	9.695	4.547
45	2,025	91,125	6.708	3.557	95	9,025	857,375	9.747	4.563
46	2,116	97,336	6.782	3.583	96	9,216	884,736	9.798	4.579
47	2,209	103,823	6.856	3.609	97	9,409	912,673	9.849	4.595
48	2,304	110,592	6.928	3.634	98	9,604	941,192	9.899	4.610
49	2,401	117,649	7.000	3.659	99	9,801	970,299	9.950	4.626
50	2,500	125,000	7.071	3.684	100	10,000	1,000,000	10.000	4.642

Table of Common Logarithms

N	0	1	2	3	4	5	6	7	8	9
1.0	0000	0043	0086	0128	0170	0212	0253	0294	0334	0374
1.1	0414	0453	0492	0531	0569	0607	0645	0682	0719	0755
1.2	0792	0828	0864	0899	0934	0969	1004	1038	1072	1106
1.3	1139	1173	1206	1239	1271	1303	1335	1367	1399	1430
1.4	1461	1492	1523	1553	1584	1614	1644	1673	1703	1732
1.5	1761	1790	1818	1847	1875	1903	1931	1959	1987	2014
1.6	2041	2068	2095	2122	2148	2175	2201	2227	2253	2279
1.7	2304	2330	2355	2380	2405	2430	2455	2480	2504	2529
1.8	2553	2577	2601	2625	2648	2672	2695	2718	2742	2765
1.9	2788	2810	2833	2856	2878	2900	2923	2945	2967	2989
2.0	3010	3032	3054	3075	3096	3118	3139	3160	3181	3201
2.1	3222	3243	3263	3284	3304	3324	3345	3365	3385	3404
2.2	3424	3444	3464	3483	3502	3522	3541	3560	3579	3598
2.3	3617	3636	3655	3674	3692	3711	3729	3747	3766	3784
2.4	3802	3820	3838	3856	3874	3892	3909	3927	3945	3962
2.5	3979	3997	4014	4031	4048	4065	4082	4099	4116	4133
2.6	4150	4166	4183	4200	4216	4232	4249	4265	4281	4298
2.7	4314	4330	4346	4362	4378	4393	4409	4425	4440	4456
2.8	4472	4487	4502	4518	4533	4548	4564	4579	4594	4609
2.9	4624	4639	4654	4669	4683	4698	4713	4728	4742	4757
3.0	4771	4786	4800	4814	4829	4843	4857	4871	4886	4900
3.1	4914	4928	4942	4955	4969	4983	4997	5011	5024	5038
3.2	5051	5065	5079	5092	5105	5119	5132	5145	5159	5172
3.3	5185	5198	5211	5224	5237	5250	5263	5276	5289	5302
3.4	5315	5328	5340	5353	5366	5378	5391	5403	5416	5428
3.5	5441	5453	5465	5478	5490	5502	5514	5527	5539	5551
3.6	5563	5575	5587	5599	5611	5623	5635	5647	5658	5670
3.7	5682	5694	5705	5717	5729	5740	5752	5763	5775	5786
3.8	5798	5809	5821	5832	5843	5855	5866	5877	5888	5899
3.9	5911	5922	5933	5944	5955	5966	5977	5988	5999	6010
4.0	6021	6031	6042	6053	6064	6075	6085	6096	6107	6117
4.1	6128	6138	6149	6160	6170	6180	6191	6201	6212	6222
4.2	6232	6243	6253	6263	6274	6284	6294	6304	6314	6325
4.3	6335	6345	6355	6365	6375	6385	6395	6405	6415	6425
4.4	6435	6444	6454	6464	6474	6484	6493	6503	6513	6522
4.5	6532	6542	6551	6561	6571	6580	6590	6599	6609	6618
4.6	6628	6637	6646	6656	6665	6675	6684	6693	6702	6712
4.7	6721	6730	6739	6749	6758	6767	6776	6785	6794	6803
4.8	6812	6821	6830	6839	6848	6857	6866	6875	6884	6893
4.9	6902	6911	6920	6928	6937	6946	6955	6964	6972	6981
5.0	6990	6998	7007	7016	7024	7033	7042	7050	7059	7067
5.1	7076	7084	7093	7101	7110	7118	7126	7135	7143	7152
5.2	7160	7168	7177	7185	7193	7202	7210	7218	7226	7235
5.3	7243	7251	7259	7267	7275	7284	7292	7300	7308	7316
5.4	7324	7332	7340	7348	7356	7364	7372	7380	7388	7396

Table of Common Logarithms

N	0	1	2	3	4	5	6	7	8	9
5.5	7404	7412	7419	7427	7435	7443	7451	7459	7466	7474
5.6	7482	7490	7497	7505	7513	7520	7528	7536	7543	7551
5.7	7559	7566	7574	7582	7589	7597	7604	7612	7619	7627
5.8	7634	7642	7649	7657	7664	7672	7679	7686	7694	7701
5.9	7709	7716	7723	7731	7738	7745	7752	7760	7767	7774
6.0	7782	7789	7796	7803	7810	7818	7825	7832	7839	7846
6.1	7853	7860	7868	7875	7882	7889	7896	7903	7910	7917
6.2	7924	7931	7938	7945	7952	7959	7966	7973	7980	7987
6.3	7993	8000	8007	8014	8021	8028	8035	8041	8048	8055
6.4	8062	8069	8075	8082	8089	8096	8102	8109	8116	8122
6.5	8129	8136	8142	8149	8156	8162	8169	8176	8182	8189
6.6	8195	8202	8209	8215	8222	8228	8235	8241	8248	8254
6.7	8261	8267	8274	8280	8287	8293	8299	8306	8312	8319
6.8	8325	8331	8338	8344	8351	8357	8363	8370	8376	8382
6.9	8388	8395	8401	8407	8414	8420	8426	8432	8439	8445
7.0	8451	8457	8463	8470	8476	8482	8488	8494	8500	8506
7.1	8513	8519	8525	8531	8537	8543	8549	8555	8561	8567
7.2	8573	8579	8585	8591	8597	8603	8609	8615	8621	8627
7.3	8633	8639	8645	8651	8657	8663	8669	8675	8681	8686
7.4	8692	8698	8704	8710	8716	8722	8727	8733	8739	8745
7.5	8751	8756	8762	8768	8774	8779	8785	8791	8797	8802
7.6	8808	8814	8820	8825	8831	8837	8842	8848	8854	8859
7.7	8865	8871	8876	8882	8887	8893	8899	8904	8910	8915
7.8	8921	8927	8932	8938	8943	8949	8954	8960	8965	8971
7.9	8976	8982	8987	8993	8998	9004	9009	9015	9020	9025
8.0	9031	9036	9042	9047	9053	9058	9063	9069	9074	9079
8.1	9085	9090	9096	9101	9106	9112	9117	9122	9128	9133
8.2	9138	9143	9149	9154	9159	9165	9170	9175	9180	9186
8.3	9191	9196	9201	9206	9212	9217	9222	9227	9232	9238
8.4	9243	9248	9253	9258	9263	9269	9274	9279	9284	9289
8.5	9294	9299	9304	9309	9315	9320	9325	9330	9335	9340
8.6	9345	9350	9355	9360	9365	9370	9375	9380	9385	9390
8.7	9395	9400	9405	9410	9415	9420	9425	9430	9435	9440
8.8	9445	9450	9455	9460	9465	9469	9474	9479	9484	9489
8.9	9494	9499	9504	9509	9513	9518	9523	9528	9533	9538
9.0	9542	9547	9552	9557	9562	9566	9571	9576	9581	9586
9.1	9590	9595	9600	9605	9609	9614	9619	9624	9628	9633
9.2	9638	9643	9647	9652	9657	9661	9666	9671	9675	9680
9.3	9685	9689	9694	9699	9703	9708	9713	9717	9722	9727
9.4	9731	9736	9741	9745	9750	9754	9759	9763	9768	9773
9.5	9777	9782	9786	9791	9795	9800	9805	9809	9814	9818
9.6	9823	9827	9832	9836	9841	9845	9850	9854	9859	9863
9.7	9868	9872	9877	9881	9886	9890	9894	9899	9903	9908
9.8	9912	9917	9921	9926	9930	9934	9939	9943	9948	9952
9.9	9956	9961	9965	9969	9974	9978	9983	9987	9991	9996

Tables

Table of Values of the Trigonometric Functions

Deg.(θ)	Rad.(θ)	Sin θ	Cos θ	Tan θ	Cot θ	Sec θ	Csc θ		
0° 00′	.0000	.0000	1.0000	.0000		1.000		1.5708	90° 00′
10′	.0029	.0029	1.0000	.0029	343.77	1.000	343.8	1.5679	50′
20′	.0058	.0058	1.0000	.0058	171.89	1.000	171.9	1.5650	40′
30′	.0087	.0087	1.0000	.0087	114.59	1.000	114.6	1.5621	30′
40′	.0116	.0116	.9999	.0116	85.940	1.000	85.95	1.5592	20′
50′	.0145	.0145	.9999	.0145	68.750	1.000	68.76	1.5563	10′
1° 00′	.0175	.0175	.9998	.0175	57.290	1.000	57.30	1.5533	89° 00′
10′	.0204	.0204	.9998	.0204	49.104	1.000	49.11	1.5504	50′
20′	.0233	.0233	.9997	.0233	42.964	1.000	42.98	1.5475	40′
30′	.0262	.0262	.9997	.0262	38.188	1.000	38.20	1.5446	30′
40′	.0291	.0291	.9996	.0291	34.368	1.000	34.38	1.5417	20′
50′	.0320	.0320	.9995	.0320	31.242	1.001	31.26	1.5388	10′
2° 00′	.0349	.0349	.9994	.0349	28.636	1.001	28.65	1.5359	88° 00′
10′	.0378	.0378	.9993	.0378	26.432	1.001	26.45	1.5330	50′
20′	.0407	.0407	.9992	.0407	24.542	1.001	24.56	1.5301	40′
30′	.0436	.0436	.9990	.0437	22.904	1.001	22.93	1.5272	30′
40′	.0465	.0465	.9989	.0466	21.470	1.001	21.49	1.5243	20′
50′	.0495	.0494	.9988	.0495	20.206	1.001	20.23	1.5213	10′
3° 00′	.0524	.0523	.9986	.0524	19.081	1.001	19.11	1.5184	87° 00′
10′	.0553	.0552	.9985	.0553	18.075	1.002	18.10	1.5155	50′
20′	.0582	.0581	.9983	.0582	17.169	1.002	17.20	1.5126	40′
30′	.0611	.0610	.9981	.0612	16.350	1.002	16.38	1.5097	30′
40′	.0640	.0640	.9980	.0641	15.605	1.002	15.64	1.5068	20′
50′	.0669	.0669	.9978	.0670	14.924	1.002	14.96	1.5039	10′
4° 00′	.0698	.0698	.9976	.0699	14.301	1.002	14.34	1.5010	86° 00′
10′	.0727	.0727	.9974	.0729	13.727	1.003	13.76	1.4981	50′
20′	.0756	.0756	.9971	.0758	13.197	1.003	13.23	1.4952	40′
30′	.0785	.0785	.9969	.0787	12.706	1.003	12.75	1.4923	30′
40′	.0814	.0814	.9967	.0816	12.251	1.003	12.29	1.4893	20′
50′	.0844	.0843	.9964	.0846	11.826	1.004	11.87	1.4864	10′
5° 00′	.0873	.0872	.9962	.0875	11.430	1.004	11.47	1.4835	85° 00′
10′	.0902	.0901	.9959	.0904	11.059	1.004	11.10	1.4806	50′
20′	.0931	.0929	.9957	.0934	10.712	1.004	10.76	1.4777	40′
30′	.0960	.0958	.9954	.0963	10.385	1.005	10.43	1.4748	30′
40′	.0989	.0987	.9951	.0992	10.078	1.005	10.13	1.4719	20′
50′	.1018	.1016	.9948	.1022	9.7882	1.005	9.839	1.4690	10′
6° 00′	.1047	.1045	.9945	.1051	9.5144	1.006	9.567	1.4661	84° 00′
10′	.1076	.1074	.9942	.1080	9.2553	1.006	9.309	1.4632	50′
20′	.1105	.1103	.9939	.1110	9.0098	1.006	9.065	1.4603	40′
30′	.1134	.1132	.9936	.1139	8.7769	1.006	8.834	1.4573	30′
40′	.1164	.1161	.9932	.1169	8.5555	1.007	8.614	1.4544	20′
50′	.1193	.1190	.9929	.1198	8.3450	1.007	8.405	1.4515	10′
7° 00′	.1222	.1219	.9925	.1228	8.1443	1.008	8.206	1.4486	83° 00′
10′	.1251	.1248	.9922	.1257	7.9530	1.008	8.016	1.4457	50′
20′	.1280	.1276	.9918	.1287	7.7704	1.008	7.834	1.4428	40′
30′	.1309	.1305	.9914	.1317	7.5958	1.009	7.661	1.4399	30′
40′	.1338	.1334	.9911	.1346	7.4287	1.009	7.496	1.4370	20′
50′	.1367	.1363	.9907	.1376	7.2687	1.009	7.337	1.4341	10′
8° 00′	.1396	.1392	.9903	.1405	7.1154	1.010	7.185	1.4312	82° 00′
10′	.1425	.1421	.9899	.1435	6.9682	1.010	7.040	1.4283	50′
20′	.1454	.1449	.9894	.1465	6.8269	1.011	6.900	1.4254	40′
30′	.1484	.1478	.9890	.1495	6.6912	1.011	6.765	1.4224	30′
40′	.1513	.1507	.9886	.1524	6.5606	1.012	6.636	1.4195	20′
50′	.1542	.1536	.9881	.1554	6.4348	1.012	6.512	1.4166	10′
9° 00′	.1571	.1564	.9877	.1584	6.3138	1.012	6.392	1.4137	81° 00′
		Cos θ	Sin θ	Cot θ	Tan θ	Csc θ	Sec θ	Rad.(θ)	Deg.(θ)

Table of Values of the Trigonometric Functions

Deg.(θ)	Rad.(θ)	Sin θ	Cos θ	Tan θ	Cot θ	Sec θ	Csc θ		
9° 00′	.1571	.1564	.9877	.1584	6.3138	1.012	6.392	1.4137	**81° 00′**
10′	.1600	.1593	.9872	.1614	6.1970	1.013	6.277	1.4108	50′
20′	.1629	.1622	.9868	.1644	6.0844	1.013	6.166	1.4079	40′
30′	.1658	.1650	.9863	.1673	5.9758	1.014	6.059	1.4050	30′
40′	.1687	.1679	.9858	.1703	5.8708	1.014	5.955	1.4021	20′
50′	.1716	.1708	.9853	.1733	5.7694	1.015	5.855	1.3992	10′
10° 00′	.1745	.1736	.9848	.1763	5.6713	1.015	5.759	1.3963	**80° 00′**
10′	.1774	.1765	.9843	.1793	5.5764	1.016	5.665	1.3934	50′
20′	.1804	.1794	.9838	.1823	5.4845	1.016	5.575	1.3904	40′
30′	.1833	.1822	.9833	.1853	5.3955	1.017	5.487	1.3875	30′
40′	.1862	.1851	.9827	.1883	5.3093	1.018	5.403	1.3846	20′
50′	.1891	.1880	.9822	.1914	5.2257	1.018	5.320	1.3817	10′
11° 00′	.1920	.1908	.9816	.1944	5.1446	1.019	5.241	1.3788	**79° 00′**
10′	.1949	.1937	.9811	.1974	5.0658	1.019	5.164	1.3759	50′
20′	.1978	.1965	.9805	.2004	4.9894	1.020	5.089	1.3730	40′
30′	.2007	.1994	.9799	.2035	4.9152	1.020	5.016	1.3701	30′
40′	.2036	.2022	.9793	.2065	4.8430	1.021	4.945	1.3672	20′
50′	.2065	.2051	.9787	.2095	4.7729	1.022	4.876	1.3643	10′
12° 00′	.2094	.2079	.9781	.2126	4.7046	1.022	4.810	1.3614	**78° 00′**
10′	.2123	.2108	.9775	.2156	4.6382	1.023	4.745	1.3584	50′
20′	.2153	.2136	.9769	.2186	4.5736	1.024	4.682	1.3555	40′
30′	.2182	.2164	.9763	.2217	4.5107	1.024	4.620	1.3526	30′
40′	.2211	.2193	.9757	.2247	4.4494	1.025	4.560	1.3497	20′
50′	.2240	.2221	.9750	.2278	4.3897	1.026	4.502	1.3468	10′
13° 00′	.2269	.2250	.9744	.2309	4.3315	1.026	4.445	1.3439	**77° 00′**
10′	.2298	.2278	.9737	.2339	4.2747	1.027	4.390	1.3410	50′
20′	.2327	.2306	.9730	.2370	4.2193	1.028	4.336	1.3381	40′
30′	.2356	.2334	.9724	.2401	4.1653	1.028	4.284	1.3352	30′
40′	.2385	.2363	.9717	.2432	4.1126	1.029	4.232	1.3323	20′
50′	.2414	.2391	.9710	.2462	4.0611	1.030	4.182	1.3294	10′
14° 00′	.2443	.2419	.9703	.2493	4.0108	1.031	4.134	1.3265	**76° 00′**
10′	.2473	.2447	.9696	.2524	3.9617	1.031	4.086	1.3235	50′
20′	.2502	.2476	.9689	.2555	3.9136	1.032	4.039	1.3206	40′
30′	.2531	.2504	.9681	.2586	3.8667	1.033	3.994	1.3177	30′
40′	.2560	.2532	.9674	.2617	3.8208	1.034	3.950	1.3148	20′
50′	.2589	.2560	.9667	.2648	3.7760	1.034	3.906	1.3119	10′
15° 00′	.2618	.2588	.9659	.2679	3.7321	1.035	3.864	1.3090	**75° 00′**
10′	.2647	.2616	.9652	.2711	3.6891	1.036	3.822	1.3061	50′
20′	.2676	.2644	.9644	.2742	3.6470	1.037	3.782	1.3032	40′
30′	.2705	.2672	.9636	.2773	3.6059	1.038	3.742	1.3003	30′
40′	.2734	.2700	.9628	.2805	3.5656	1.039	3.703	1.2974	20′
50′	.2763	.2728	.9621	.2836	3.5261	1.039	3.665	1.2945	10′
16° 00′	.2793	.2756	.9613	.2867	3.4874	1.040	3.628	1.2915	**74° 00′**
10′	.2822	.2784	.9605	.2899	3.4495	1.041	3.592	1.2886	50′
20′	.2851	.2812	.9596	.2931	3.4124	1.042	3.556	1.2857	40′
30′	.2880	.2840	.9588	.2962	3.3759	1.043	3.521	1.2828	30′
40′	.2909	.2868	.9580	.2994	3.3402	1.044	3.487	1.2799	20′
50′	.2938	.2896	.9572	.3026	3.3052	1.045	3.453	1.2770	10′
17° 00′	.2967	.2924	.9563	.3057	3.2709	1.046	3.420	1.2741	**73° 00′**
10′	.2996	.2952	.9555	.3089	3.2371	1.047	3.388	1.2712	50′
20′	.3025	.2979	.9546	.3121	3.2041	1.048	3.356	1.2683	40′
30′	.3054	.3007	.9537	.3153	3.1716	1.049	3.326	1.2654	30′
40′	.3083	.3035	.9528	.3185	3.1397	1.049	3.295	1.2625	20′
50′	.3113	.3062	.9520	.3217	3.1084	1.050	3.265	1.2595	10′
18° 00′	.3142	.3090	.9511	.3249	3.0777	1.051	3.236	1.2566	**72° 00′**
		Cos θ	Sin θ	Cot θ	Tan θ	Csc θ	Sec θ	Rad.(θ)	Deg.(θ)

Tables

793

Table of Values of the Trigonometric Functions

Deg.(θ)	Rad.(θ)	Sin θ	Cos θ	Tan θ	Cot θ	Sec θ	Csc θ		
18° 00′	.3142	.3090	.9511	.3249	3.0777	1.051	3.236	1.2566	72° 00′
10′	.3171	.3118	.9502	.3281	3.0475	1.052	3.207	1.2537	50′
20′	.3200	.3145	.9492	.3314	3.0178	1.053	3.179	1.2508	40′
30′	.3229	.3173	.9483	.3346	2.9887	1.054	3.152	1.2479	30′
40′	.3258	.3201	.9474	.3378	2.9600	1.056	3.124	1.2450	20′
50′	.3287	.3228	.9465	.3411	2.9319	1.057	3.098	1.2421	10′
19° 00′	.3316	.3256	.9455	.3443	2.9042	1.058	3.072	1.2392	71° 00′
10′	.3345	.3283	.9446	.3476	2.8770	1.059	3.046	1.2363	50′
20′	.3374	.3311	.9436	.3508	2.8502	1.060	3.021	1.2334	40′
30′	.3403	.3338	.9426	.3541	2.8239	1.061	2.996	1.2305	30′
40′	.3432	.3365	.9417	.3574	2.7980	1.062	2.971	1.2275	20′
50′	.3462	.3393	.9407	.3607	2.7725	1.063	2.947	1.2246	10′
20° 00′	.3491	.3420	.9397	.3640	2.7475	1.064	2.924	1.2217	70° 00′
10′	.3520	.3448	.9387	.3673	2.7228	1.065	2.901	1.2188	50′
20′	.3549	.3475	.9377	.3706	2.6985	1.066	2.878	1.2159	40′
30′	.3578	.3502	.9367	.3739	2.6746	1.068	2.855	1.2130	30′
40′	.3607	.3529	.9356	.3772	2.6511	1.069	2.833	1.2101	20′
50′	.3636	.3557	.9346	.3805	2.6279	1.070	2.812	1.2072	10′
21° 00′	.3665	.3584	.9336	.3839	2.6051	1.071	2.790	1.2043	69° 00′
10′	.3694	.3611	.9325	.3872	2.5826	1.072	2.769	1.2014	50′
20′	.3723	.3638	.9315	.3906	2.5605	1.074	2.749	1.1985	40′
30′	.3752	.3665	.9304	.3939	2.5386	1.075	2.729	1.1956	30′
40′	.3782	.3692	.9293	.3973	2.5172	1.076	2.709	1.1926	20′
50′	.3811	.3719	.9283	.4006	2.4960	1.077	2.689	1.1897	10′
22° 00′	.3840	.3746	.9272	.4040	2.4751	1.079	2.669	1.1868	68° 00′
10′	.3869	.3773	.9261	.4074	2.4545	1.080	2.650	1.1839	50′
20′	.3898	.3800	.9250	.4108	2.4342	1.081	2.632	1.1810	40′
30′	.3927	.3827	.9239	.4142	2.4142	1.082	2.613	1.1781	30′
40′	.3956	.3854	.9228	.4176	2.3945	1.084	2.595	1.1752	20′
50′	.3985	.3881	.9216	.4210	2.3750	1.085	2.577	1.1723	10′
23° 00′	.4014	.3907	.9205	.4245	2.3559	1.086	2.559	1.1694	67° 00′
10′	.4043	.3934	.9194	.4279	2.3369	1.088	2.542	1.1665	50′
20′	.4072	.3961	.9182	.4314	2.3183	1.089	2.525	1.1636	40′
30′	.4102	.3987	.9171	.4348	2.2998	1.090	2.508	1.1606	30′
40′	.4131	.4014	.9159	.4383	2.2817	1.092	2.491	1.1577	20′
50′	.4160	.4041	.9147	.4417	2.2637	1.093	2.475	1.1548	10′
24° 00′	.4189	.4067	.9135	.4452	2.2460	1.095	2.459	1.1519	66° 00′
10′	.4218	.4094	.9124	.4487	2.2286	1.096	2.443	1.1490	50′
20′	.4247	.4120	.9112	.4522	2.2113	1.097	2.427	1.1461	40′
30′	.4276	.4147	.9100	.4557	2.1943	1.099	2.411	1.1432	30′
40′	.4305	.4173	.9088	.4592	2.1775	1.100	2.396	1.1403	20′
50′	.4334	.4200	.9075	.4628	2.1609	1.102	2.381	1.1374	10′
25° 00′	.4363	.4226	.9063	.4663	2.1445	1.103	2.366	1.1345	65° 00′
10′	.4392	.4253	.9051	.4699	2.1283	1.105	2.352	1.1316	50′
20′	.4422	.4279	.9038	.4734	2.1123	1.106	2.337	1.1286	40′
30′	.4451	.4305	.9026	.4770	2.0965	1.108	2.323	1.1257	30′
40′	.4480	.4331	.9013	.4806	2.0809	1.109	2.309	1.1228	20′
50′	.4509	.4358	.9001	.4841	2.0655	1.111	2.295	1.1199	10′
26° 00′	.4538	.4384	.8988	.4877	2.0503	1.113	2.281	1.1170	64° 00′
10′	.4567	.4410	.8975	.4913	2.0353	1.114	2.268	1.1141	50′
20′	.4596	.4436	.8962	.4950	2.0204	1.116	2.254	1.1112	40′
30′	.4625	.4462	.8949	.4986	2.0057	1.117	2.241	1.1083	30′
40′	.4654	.4488	.8936	.5022	1.9912	1.119	2.228	1.1054	20′
50′	.4683	.4514	.8923	.5059	1.9768	1.121	2.215	1.1025	10′
27° 00′	.4712	.4540	.8910	.5095	1.9626	1.122	2.203	1.0996	63° 00′
		Cos θ	Sin θ	Cot θ	Tan θ	Csc θ	Sec θ	Rad.(θ)	Deg.(θ)

Table of Values of the Trigonometric Functions

Deg.(θ)	Rad.(θ)	Sin θ	Cos θ	Tan θ	Cot θ	Sec θ	Csc θ		
27°00'	.4712	.4540	.8910	.5095	1.9626	1.122	2.203	1.0996	63°00'
10'	.4741	.4566	.8897	.5132	1.9486	1.124	2.190	1.0966	50'
20'	.4771	.4592	.8884	.5169	1.9347	1.126	2.178	1.0937	40'
30'	.4800	.4617	.8870	.5206	1.9210	1.127	2.166	1.0908	30'
40'	.4829	.4643	.8857	.5243	1.9074	1.129	2.154	1.0879	20'
50'	.4858	.4669	.8843	.5280	1.8940	1.131	2.142	1.0850	10'
28°00'	.4887	.4695	.8829	.5317	1.8807	1.133	2.130	1.0821	62°00'
10'	.4916	.4720	.8816	.5354	1.8676	1.134	2.118	1.0792	50'
20'	.4945	.4746	.8802	.5392	1.8546	1.136	2.107	1.0763	40'
30'	.4974	.4772	.8788	.5430	1.8418	1.138	2.096	1.0734	30'
40'	.5003	.4797	.8774	.5467	1.8291	1.140	2.085	1.0705	20'
50'	.5032	.4823	.8760	.5505	1.8165	1.142	2.074	1.0676	10'
29°00'	.5061	.4848	.8746	.5543	1.8040	1.143	2.063	1.0647	61°00'
10'	.5091	.4874	.8732	.5581	1.7917	1.145	2.052	1.0617	50'
20'	.5120	.4899	.8718	.5619	1.7796	1.147	2.041	1.0588	40'
30'	.5149	.4924	.8704	.5658	1.7675	1.149	2.031	1.0559	30'
40'	.5178	.4950	.8689	.5696	1.7556	1.151	2.020	1.0530	20'
50'	.5207	.4975	.8675	.5735	1.7437	1.153	2.010	1.0501	10'
30°00'	.5236	.5000	.8660	.5774	1.7321	1.155	2.000	1.0472	60°00'
10'	.5265	.5025	.8646	.5812	1.7205	1.157	1.990	1.0443	50'
20'	.5294	.5050	.8631	.5851	1.7090	1.159	1.980	1.0414	40'
30'	.5323	.5075	.8616	.5890	1.6977	1.161	1.970	1.0385	30'
40'	.5352	.5100	.8601	.5930	1.6864	1.163	1.961	1.0356	20'
50'	.5381	.5125	.8587	.5969	1.6753	1.165	1.951	1.0327	10'
31°00'	.5411	.5150	.8572	.6009	1.6643	1.167	1.942	1.0297	59°00'
10'	.5440	.5175	.8557	.6048	1.6534	1.169	1.932	1.0268	50'
20'	.5469	.5200	.8542	.6088	1.6426	1.171	1.923	1.0239	40'
30'	.5498	.5225	.8526	.6128	1.6319	1.173	1.914	1.0210	30'
40'	.5527	.5250	.8511	.6168	1.6212	1.175	1.905	1.0181	20'
50'	.5556	.5275	.8496	.6208	1.6107	1.177	1.896	1.0152	10'
32°00'	.5585	.5299	.8480	.6249	1.6003	1.179	1.887	1.0123	58°00'
10'	.5614	.5324	.8465	.6289	1.5900	1.181	1.878	1.0094	50'
20'	.5643	.5348	.8450	.6330	1.5798	1.184	1.870	1.0065	40'
30'	.5672	.5373	.8434	.6371	1.5697	1.186	1.861	1.0036	30'
40'	.5701	.5398	.8418	.6412	1.5597	1.188	1.853	1.0007	20'
50'	.5730	.5422	.8403	.6453	1.5497	1.190	1.844	.9977	10'
33°00'	.5760	.5446	.8387	.6494	1.5399	1.192	1.836	.9948	57°00'
10'	.5789	.5471	.8371	.6536	1.5301	1.195	1.828	.9919	50'
20'	.5818	.5495	.8355	.6577	1.5204	1.197	1.820	.9890	40'
30'	.5847	.5519	.8339	.6619	1.5108	1.199	1.812	.9861	30'
40'	.5876	.5544	.8323	.6661	1.5013	1.202	1.804	.9832	20'
50'	.5905	.5568	.8307	.6703	1.4919	1.204	1.796	.9803	10'
34°00'	.5934	.5592	.8290	.6745	1.4826	1.206	1.788	.9774	56°00'
10'	.5963	.5616	.8274	.6787	1.4733	1.209	1.781	.9745	50'
20'	.5992	.5640	.8258	.6830	1.4641	1.211	1.773	.9716	40'
30'	.6021	.5664	.8241	.6873	1.4550	1.213	1.766	.9687	30'
40'	.6050	.5688	.8225	.6916	1.4460	1.216	1.758	.9657	20'
50'	.6080	.5712	.8208	.6959	1.4370	1.218	1.751	.9628	10'
35°00'	.6109	.5736	.8192	.7002	1.4281	1.221	1.743	.9599	55°00'
10'	.6138	.5760	.8175	.7046	1.4193	1.223	1.736	.9570	50'
20'	.6167	.5783	.8158	.7089	1.4106	1.226	1.729	.9541	40'
30'	.6196	.5807	.8141	.7133	1.4019	1.228	1.722	.9512	30'
40'	.6225	.5831	.8124	.7177	1.3934	1.231	1.715	.9483	20'
50'	.6254	.5854	.8107	.7221	1.3848	1.233	1.708	.9454	10'
36°00'	.6283	.5878	.8090	.7265	1.3764	1.236	1.701	.9425	54°00'
		Cos θ	Sin θ	Cot θ	Tan θ	Csc θ	Sec θ	Rad.(θ)	Deg.(θ)

Tables

795

Table of Values of the Trigonometric Functions

Tables

Deg.(θ)	Rad.(θ)	Sin θ	Cos θ	Tan θ	Cot θ	Sec θ	Csc θ		
36° 00'	.6283	.5878	.8090	.7265	1.3764	1.236	1.701	.9425	54° 00'
10'	.6312	.5901	.8073	.7310	1.3680	1.239	1.695	.9396	50'
20'	.6341	.5925	.8056	.7355	1.3597	1.241	1.688	.9367	40'
30'	.6370	.5948	.8039	.7400	1.3514	1.244	1.681	.9338	30'
40'	.6400	.5972	.8021	.7445	1.3432	1.247	1.675	.9308	20'
50'	.6429	.5995	.8004	.7490	1.3351	1.249	1.668	.9279	10'
37° 00'	.6458	.6018	.7986	.7536	1.3270	1.252	1.662	.9250	53° 00'
10'	.6487	.6041	.7969	.7581	1.3190	1.255	1.655	.9221	50'
20'	.6516	.6065	.7951	.7627	1.3111	1.258	1.649	.9192	40'
30'	.6545	.6088	.7934	.7673	1.3032	1.260	1.643	.9163	30'
40'	.6574	.6111	.7916	.7720	1.2954	1.263	1.636	.9134	20'
50'	.6603	.6134	.7898	.7766	1.2876	1.266	1.630	.9105	10'
38° 00'	.6632	.6157	.7880	.7813	1.2799	1.269	1.624	.9076	52° 00'
10'	.6661	.6180	.7862	.7860	1.2723	1.272	1.618	.9047	50'
20'	.6690	.6202	.7844	.7907	1.2647	1.275	1.612	.9018	40'
30'	.6720	.6225	.7826	.7954	1.2572	1.278	1.606	.8988	30'
40'	.6749	.6248	.7808	.8002	1.2497	1.281	1.601	.8959	20'
50'	.6778	.6271	.7790	.8050	1.2423	1.284	1.595	.8930	10'
39° 00'	.6807	.6293	.7771	.8098	1.2349	1.287	1.589	.8901	51° 00'
10'	.6836	.6316	.7753	.8146	1.2276	1.290	1.583	.8872	50'
20'	.6865	.6338	.7735	.8195	1.2203	1.293	1.578	.8843	40'
30'	.6894	.6361	.7716	.8243	1.2131	1.296	1.572	.8814	30'
40'	.6923	.6383	.7698	.8292	1.2059	1.299	1.567	.8785	20'
50'	.6952	.6406	.7679	.8342	1.1988	1.302	1.561	.8756	10'
40° 00'	.6981	.6428	.7660	.8391	1.1918	1.305	1.556	.8727	50° 00'
10'	.7010	.6450	.7642	.8441	1.1847	1.309	1.550	.8698	50'
20'	.7039	.6472	.7623	.8491	1.1778	1.312	1.545	.8668	40'
30'	.7069	.6494	.7604	.8541	1.1708	1.315	1.540	.8639	30'
40'	.7098	.6517	.7585	.8591	1.1640	1.318	1.535	.8610	20'
50'	.7127	.6539	.7566	.8642	1.1571	1.322	1.529	.8581	10'
41° 00'	.7156	.6561	.7547	.8693	1.1504	1.325	1.524	.8552	49° 00'
10'	.7185	.6583	.7528	.8744	1.1436	1.328	1.519	.8523	50'
20'	.7214	.6604	.7509	.8796	1.1369	1.332	1.514	.8494	40'
30'	.7243	.6626	.7490	.8847	1.1303	1.335	1.509	.8465	30'
40'	.7272	.6648	.7470	.8899	1.1237	1.339	1.504	.8436	20'
50'	.7301	.6670	.7451	.8952	1.1171	1.342	1.499	.8407	10'
42° 00'	.7330	.6691	.7431	.9004	1.1106	1.346	1.494	.8378	48° 00'
10'	.7359	.6713	.7412	.9057	1.1041	1.349	1.490	.8348	50'
20'	.7389	.6734	.7392	.9110	1.0977	1.353	1.485	.8319	40'
30'	.7418	.6756	.7373	.9163	1.0913	1.356	1.480	.8290	30'
40'	.7447	.6777	.7353	.9217	1.0850	1.360	1.476	.8261	20'
50'	.7476	.6799	.7333	.9271	1.0786	1.364	1.471	.8232	10'
43° 00'	.7505	.6820	.7314	.9325	1.0724	1.367	1.466	.8203	47° 00'
10'	.7534	.6841	.7294	.9380	1.0661	1.371	1.462	.8174	50'
20'	.7563	.6862	.7274	.9435	1.0599	1.375	1.457	.8145	40'
30'	.7592	.6884	.7254	.9490	1.0538	1.379	1.453	.8116	30'
40'	.7621	.6905	.7234	.9545	1.0477	1.382	1.448	.8087	20'
50'	.7650	.6926	.7214	.9601	1.0416	1.386	1.444	.8058	10'
44° 00'	.7679	.6947	.7193	.9657	1.0355	1.390	1.440	.8029	46° 00'
10'	.7709	.6967	.7173	.9713	1.0295	1.394	1.435	.7999	50'
20'	.7738	.6988	.7153	.9770	1.0235	1.398	1.431	.7970	40'
30'	.7767	.7009	.7133	.9827	1.0176	1.402	1.427	.7941	30'
40'	.7796	.7030	.7112	.9884	1.0117	1.406	1.423	.7912	20'
50'	.7825	.7050	.7092	.9942	1.0058	1.410	1.418	.7883	10'
45° 00'	.7854	.7071	.7071	1.0000	1.0000	1.414	1.414	.7854	45° 00'
		Cos θ	Sin θ	Cot θ	Tan θ	Csc θ	Sec θ	Rad.(θ)	Deg.(θ)

796

Glossary

abscissa: The x-coordinate of an ordered pair of real numbers. (p. 73)

absolute value: The absolute value of any real number x, written $|x|$, is x if $x > 0$, and $-x$ if $x < 0$. On a number line, $|x|$ is the distance between the graph of x and the origin. (pp. 1, 51)

absolute value of a complex number: The absolute value of the complex number $a + bi$, written $|a + bi|$, equals $\sqrt{a^2 + b^2}$. (p. 305)

amplitude: The amplitude of a periodic function is one-half the distance between its maximum and minimum values. (p. 681)

angle of elevation (depression): The angle between a horizontal line and the line of sight. (p. 629)

angle of rotation: An angle of rotation in standard position is the angle formed by rotating a ray counterclockwise from the positive x-axis (the *initial* side) to a new position (the *terminal* side). (p. 633)

antilogarithm: If $\log_b q = r$, then q is the base-b antilogarithm (antilog) of r. (p. 481)

arithmetic means: The terms between the first and last terms of an arithmetic sequence. *The* arithmetic mean of x and y is their average, $\dfrac{x + y}{2}$. (p. 549)

arithmetic sequence: A sequence in which the difference between successive terms (the *common difference*) is constant. (p. 545)

arithmetic series: The indicated sum of the terms of an arithmetic sequence. (p. 553)

asymptote: A line that the graph of a function approaches more and more closely without ever touching. (p. 419)

augmented matrix: A matrix representing a system of equations, in which the constants of the system form the right-hand column. (p. 603)

axiom: A statement that is assumed to be true. Also called a *postulate*. (p. 31)

base: In the power b^3, b is the base, or the repeated power. (p. 161)

bearing: An angle of rotation measured clockwise from due north. (p. 669)

binomial: A polynomial that has two terms. (p. 170)

binomial expansion: The sum of the terms of the nth power of a binomial. (p. 576)

characteristic: The integer part of a base-10 logarithm. (p. 482)

circle: The set of all points in a plane that are a fixed distance, the *radius*, from a given point, the *center*. (p. 409)

coefficient: The constant factor in a term having one or more variables. (p. 9)

combination: A subset of r objects from a set of n objects is a combination of n elements taken r at a time. (p. 511)

combined variation: A relation containing both direct and inverse variation. (p. 432)

common difference: The constant difference between successive terms in an arithmetic sequence. (p. 545)

common logarithm: A base-10 logarithm. (p. 481)

common ratio: The constant ratio between successive terms in a geometric sequence. (p. 558)

complex conjugates: Two complex numbers of the form $a + bi$ and $a - bi$. (p. 308)

complex number: A number of the form $a + bi$, where a and b are real numbers and $i = \sqrt{-1}$. (p. 303)

complex rational expression: A rational expression in which the numerator, the denominator, or both contain at least one rational expression. (p. 249)

composite of two functions: Given the functions f and g such that the range of f is in the domain of g, the composite of the two functions is the function whose value at x is $g(f(x))$. (p. 190)

compound sentence: The combination of two simple sentences by *and* or by *or*. (p. 46)

conic section: The intersection of a plane and a right circular cone. (p. 440)

conjugates: Two real numbers of the form $a + b$ and $a - b$. (p. 281)

conjunction: A compound sentence containing the connector *and*. (p. 46)

consistent system: A system of equations that has at least one solution. (p. 119)

constant of variation: The constant k in functions of the forms $y = kx$ (direct variation) or $xy = k$ (inverse variation). (pp. 105, 427)

corollary: A statement that can be inferred from a proven theorem. (p. 477)

cosecant: The cosecant of angle θ (csc θ) is the reciprocal of sin θ. (p. 643)

cosine: For an angle θ in standard position with a point $P(x,y)$ on the terminal side at a distance $r > 0$ from the origin,

the cosine of angle θ (cos θ) $= \frac{x}{r}$. (p. 643)

cotangent: The cotangent of angle θ (cot θ) is the reciprocal of tan θ. (p. 643)

coterminal angles: Angles that share initial and terminal sides. (p. 635)

degree of a monomial: The sum of the exponents of its variables. (p. 170)

degree of a polynomial: The degree of the term of highest degree. (p. 170)

dependent events: Events for which the outcome of one affects the outcome of the other. (p. 525)

dependent system: A system of equations that has infinitely many solutions. (p. 120)

determinant: A real number associated with a square matrix. (p. 594)

direct variation: A linear function defined by an equation of the form $y = kx$ ($k \neq 0$). (p. 105)

discriminant: The discriminant of the quadratic equation $ax^2 + bx + c$ is $b^2 - 4ac$. (p. 343)

disjunction: A compound sentence containing the connector *or*. (p. 46)

domain: The set of first coordinates of the ordered pairs of a relation. (p. 74)

ellipse: The set of all points in a plane such that for each point, the sum of the distances from two fixed points (the *foci*) is constant. (p. 413)

empty set: The set containing no elements. (p. 48)

equivalent equations: Equations that have the same solution set. (p. 13)

equivalent expressions: Expressions that are equal for all values of the variables for which they have meaning. (p. 10)

equivalent matrices: Augmented matrices which can be obtained from each other by any of a series of operations called *elementary matrix transformations.* (p. 604)

equivalent systems: Systems of equations with the same solution set. (p. 124)

equivalent vectors: Vectors with the same magnitude and direction. (p. 323)

even function: A function for which $f(-x) = f(x)$ for all x in the domain of f. (p. 401)

event: Any subset of the sample space. (p. 517)

exponent: In the power b^n, n is the exponent. If n is a positive integer, it tells how many times the base b is a factor in the power b^n. (p. 161)

exponential function: An exponential function with *base* b is defined by $y = b^x$, where $b > 0$ and $b \neq 1$. (p. 464)

extraneous root: A solution of a derived equation that is not a solution of the original equation. (p. 253)

factorial notation: For a positive integer n, $n!$ (read "n factorial") is defined as $n(n - 1)(n - 2) \ldots 3 \cdot 2 \cdot 1$. (p. 500)

field: A set of numbers, along with two operations, that satisfies a specified list of eleven properties. (p. 8)

frequency distribution: A display giving a set of data and the frequency of occurence of each score. (p. 529)

function: A relation in which no two ordered pairs have the same first coordinate. (p. 75)

geometric means: The terms between the first and last terms of a geometric sequence. *The* geometric mean of x and y is xy or $-xy$ ($x \neq 0$, $y \neq 0$). (p. 562)

geometric sequence: A sequence in which the ratio of successive terms (the *common ratio*) is constant. (p. 558)

geometric series: The indicated sum of the terms of a geometric sequence. (p. 566)

greatest common factor: The expression of greatest degree and greatest constant factor that is a factor of two expressions. (p. 175)

half-plane: All points of a plane that lie on one side of a given line. (p. 110)

harmonic means: The terms between the first and last terms of a harmonic sequence. (p. 550)

harmonic sequence: The reciprocal of an arithmetic sequence that has no zero terms. (p. 550)

histogram: A bar graph that displays a frequency distribution. (p. 529)

hyperbola: The set of all points in a plane such that for each point, the difference of the distances to two fixed points (the *foci*) is constant. (p. 418)

identity: An equation true for all possible values of its variables. (pp. 16, 711)

imaginary number: Any complex number, $a + bi$, for which $a = 0$ and $b \neq 0$. (p. 303)

inclusive events: Two events which can occur simultaneously. (p. 521)

inconsistent system: A system of equations that has no solution. (p. 120)

independent events: Events for which the outcome of one event doesn't affect the outcome of the other. (p. 523)

independent system: A system of equations that has at most one solution. (p. 119)

intersection: The set of all elements common to two sets. (p. 148)

inverse relations: Relations in which the order of the elements of each ordered pair is reversed. (p. 459)

inverse variation: A function defined by an equation of the form $xy = k$ or $y = \frac{k}{x}$ ($k \neq 0$). (p. 427)

irrational number: A nonterminating, nonrepeating decimal. (p. 1)

joint variation: Direct variation of a quantity as the product of two or more other quantities. (p. 431)

least common denominator: The common multiple of the denominators of two fractions of least degree and least positive constant factor. (p. 14)

linear function: A function whose graph is a line. Its equation is called a *linear equation in two variables* and has the standard form $ax + by = c$, where a and b are not both equal to zero. (p. 78)

logarithm: A logarithm is an exponent. For positive numbers b and y ($b \neq 1$) such that $b^y = x$, the *base-b logarithm* of x ($\log_b x$) is equal to y. (p. 467)

logarithmic function: The logarithmic function with *base-b* is the inverse of the exponential function, and is defined by $y = \log_b x$ ($x > 0$, $b > 0$, $b \neq 1$). (p. 469)

mantissa: The decimal portion of a base-10 logarithm. (p. 482)

matrix: A rectangular array of numbers (*elements*) enclosed by brackets. (p. 585)

mean: The number found by adding a group of scores and then dividing the sum by the number of scores. (p. 529)

median: For an odd number of scores in an ordered distribution, the middle score; for an even number of scores, the mean of the two middle scores. (p. 529)

minor: The minor of an element of a determinant is the determinant found by deleting the column and row in which the element lies. (p. 599)

mode: The score in a distribution that occurs most frequently. (p. 529)

monomial: A number, a variable, or the product of a number and one or more variables. (p. 170)

multiplicity: The number of times that a factor is repeated in the factorization of a polynomial. (p. 224)

mutually exclusive events: Events that cannot occur simultaneously. (p. 521)

normal distribution: A distribution of data with a bell-shaped curve. (p. 533)

norm of a vector: The length or *magnitude* of a vector. (p. 738)

nth root: A solution of the equation $x^n = b$, for n a positive integer. (p. 284)

odd function: A function for which $f(-x) = -f(x)$ for all x in the domain of f. (p. 402)

odds: The odds that an event will occur are defined as the ratio of the number of successful outcomes to the number of failures. (p. 519)

opposite vectors: Vectors of equal length but opposite direction. (p. 323)

ordinate: The y-coordinate of an ordered pair of real numbers. (p. 73)

parabola: The set of all points in a plane that are equidistant from a given point (the *focus*) and a given line (the *directrix*). (pp. 391, 435)

periodic function: A function such that for all x in its domain and for some positive constant p, $f(x) = f(x + p)$. The smallest such value of p is the *period*. (p. 681)

permutation: An ordered arrangement of elements of a set. (p. 503)

polar axis: The positive x-axis. (p. 742)

polar coordinates: The polar coordinates of the point P are given by the ordered pair (r,θ), where r is the distance from P to the origin and θ is the angle of rotation from the polar axis to P. (p. 742)

polar form: The polar form of the complex number $z = x + yi$ is $r(\cos \theta + i \sin \theta)$, where r is the absolute value (*modulus*) of z, and θ is an *argument* of z. (p. 743)

polynomial: A monomial or the sum of two or more monomials. (p. 170)

principal values: The range of an inverse trigonometric function when the domain is restricted to a given set. (p. 701)

probability: The number of successful outcomes divided by the total number of possible outcomes. (p. 517)

proportion: An equation of the form $\frac{a}{b} = \frac{c}{d}$. (p. 14)

quadrant: One of the four regions, labeled counterclockwise I-IV, into which a coordinate plane is divided by its axes. (p. 73)

quadrantal angle: An angle with terminal side on a coordinate axis. (p. 633)

quadratic equation: An equation of the form $ax^2 + bx + c$, where $a \neq 0$. (p. 199)

quadratic formula: The quadratic formula, $x = -b \pm \dfrac{\sqrt{b^2 - 4ac}}{2a}$, gives the solutions of the quadratic equation $ax^2 + bx + c = 0$. (p. 317)

radian: The measure of the central angle of a circle intercepting an arc of length equal to the circle's radius. (p. 677)

radical: The symbols $\sqrt{x}$ and $\sqrt[n]{x}$ are called radicals; n is the *index*, and x is the *radicand*. (pp. 267, 284))

radical equation: An equation that contains a radical with a variable as the radicand. (p. 347)

range: The set of all second coordinates of the ordered pairs of a relation. Also, in statistics, the difference between the highest and lowest values of a set of data. (pp. 74, 530)

rational expression: A quotient of two polynomials. (p. 233)

rational number: A number that can be written as the quotient of two integers, with the divisor not zero. (p. 1)

real number: Any rational or irrational number. (p. 1)

reciprocal: A number such that the product of any nonzero number and its reciprocal is equal to 1. (p. 8)

rectangular coordinates: The rectangular coordinates of the point P are given by the ordered pair (x,y), and represent the distances to P along the x- and y-axes. (p. 742)

rectangular hyperbola: The graph of the equation $xy = k$, where k is a nonzero real number. (p. 426)

reference triangle: The triangle formed by drawing a perpendicular from the terminal side of an angle in standard position to the x-axis. The acute angle formed with vertex at the origin is the *reference angle*. (pp. 639, 640)

relation: A set of ordered pairs. (p. 74)

resultant: The sum of two vectors. (p. 324)

root: Any value of a variable that satisfies an equation. (p. 13)

sample space: The set of all possible outcomes of an experiment. (p. 517)

scalar multiple: The product of a real number (*scalar*) and a vector. (p. 325)

scientific notation: The representation of a number in the form $a \times 10^c$, where $1 \leq a < 10$ and c is an integer. (p. 166)

secant: The secant of angle θ (sec θ) is the reciprocal of cos θ. (p. 643)

sequence: An ordered list of numbers. (p. 545)

sigma notation: A shorthand method of writing a series, using the summation symbol Σ (sigma). (p. 555)

simple sentence: A statement of equality or inequality of two algebraic expressions or numbers. (p. 46)

sine: For an angle θ in standard position with a point $P(x,y)$ on the terminal side at a distance $r > 0$ from the origin, the sine of angle θ (sin θ) $= \frac{y}{r}$. (p. 643)

slope: The slope of a nonvertical line containing the points $A(x_1,y_1)$ and $B(x_2,y_2)$ is given by the formula $\frac{y_2 - y_1}{x_2 - x_1}$. (p. 84)

solution set: The set of all solutions of an open sentence which belong to its domain. (p. 42)

square root: A square root of the number b is a solution of the equation $x^2 = b$. The *principal square root* of b is the positive square root. (p. 267)

standard deviation: The principal square root of the variance. (p. 530)

synthetic division: A method using only the coefficients to divide a polynomial in x by $x - a$. (p. 217)

system of equations: A set of two or more equations. (p. 119)

tangent: For an angle θ in standard position with a point $P(x,y)$ on the terminal side at a distance $r > 0$ from the origin, the tangent of θ (tan θ) $= \frac{y}{x}$, $x \neq 0$. (p. 643)

theorem: A statement that can be proved from a set of axioms. (p. 31)

translation: Moving a graph in a coordinate plane without changing its shape or rotating it. (p. 388)

trinomial: A polynomial that has three terms. (p. 170)

union: The set of all elements in either of two sets or in both of them. (p. 148)

value of a function: The range element of a function for a given value of the domain. (p. 79)

variable: A symbol that can stand for any element of a given set. (p. 3)

variance: The sum of the squares of the deviations from the mean of a set of scores, divided by the number of scores. (p. 530)

vector: A directed line segment, having magnitude and direction. (p. 323)

y-intercept: The y-coordinate of the point at which a line crosses the y-axis. (p. 91)

zero of a function: Any value of x in the domain of the function f which satisfies the equation $f(x) = 0$. (p. 234)

Answers are provided to the odd-numbered problems for the Written Exercises.

Chapter 1

Written Exercises, page 4

1. 130 **3.** −94 **5.** 88 **7.** 0 **9.** 0
11. 0 **13.** 10 **15.** −2.3 **17.** 9 **19.** 12
21. −4 **23.** Undef **25.** 33 **27.** −18
29. −210 **31.** 3,300 **33.** 625 **35.** $-\dfrac{31}{10}$
37. Because if $0 \div 0 = c$, then $c \cdot 0 = 0$, for which every number is a solution. **39.** Pos

Written Exercises, page 7

1. 8 **3.** 0 **5.** 4 **7.** 144 **9.** −60
11. −48 **13.** −217 **15.** 36 **17.** Undef
19. 165 **21.** F

Written Exercises, pages 11–12

1. $6x + 1$ **3.** $19m − 15n − 12$
5. $12x^2 − 3y^2 + 3xy$ **7.** $−10x + 2$
9. −5 **11.** −40 **13.** −31 **15.** −25.6
17. 27 **19.** Rule of Subt **21.** Distr Prop
23. Assoc Prop Mult **25.** Closure for Add
27. $9 − 7 \neq 7 − 9$ **29.** $8 \div 2 \neq 2 \div 8$

Written Exercises, pages 15–16

1. 3 **3.** $-\dfrac{1}{4}$ **5.** $1\dfrac{1}{2}$ **7.** −19 **9.** 0.84
11. 25 **13.** −4 **15.** 5 **17.** $\dfrac{c + b}{a}$
19. $\dfrac{1}{5t}$ **21.** $\dfrac{n − 4rt}{r}$ **23.** 2,500 **25.** 6.75
27. 120 **29.** $-\dfrac{1}{6}$ **31.** $-\dfrac{1}{4}$ **33.** 1 **35.** $2\dfrac{1}{2}$
37. Identity **39.** No roots **41.** No roots

Written Exercises, pages 19–21

1. $1,680 **3.** $8\dfrac{1}{4}$% **5.** 3 yr 6 mo
7. $2,400 **9.** 42 m/s **11.** 62 m/s
13. $b = \dfrac{2A}{h}$; 11.5 ft **15.** $s = \dfrac{V}{s^2}$, 3.7 m
17. 2,900 m **19.** $9\dfrac{1}{4}$% **21.** $12\dfrac{1}{2}$ yr
23. 420 m **25.** 220 m

Written Exercises, pages 24–25

1. 4 **3.** 7, 11 **5.** 9, 36, −3
7. 24, 26, 28 **9.** 3, 11, 21 **11.** 80, 81, 82
13. 45, 60, 75 **15.** May: $2,000; June:

$2,600; July: $5,200; August: $10,400
17. Any 3 consecutive even integers; {$2n$, $2n + 2$, $2n + 4$}, where n is an integer.
19. $162 − a$

Written Exercises, pages 29–30

1. $4x − 2$ **3.** $2x + 12$ **5.** $l = 18$ yd;
$A = 72$ yd^2 **7.** 4 m by 12 m **9.** 78.5 cm^2
11. 6 m, 15 m, 15 m, 21 m, 21 m **13.** 120
cm^2 **15.** 491 m^2 **17.** 13 in, 26 in, 32.5 in,
45.5 in **19.** 30 ft^2

Written Exercises, pages 34–35

1. (1) yes; (2) no: $8 \geq 6$, but $6 \ngeq 8$; (3) yes; (4)
no, not symmetric **3.** (1) yes; (2) no: $8 \nless 6$,
but $6 < 8$; (3) yes; (4) no, not symmetric **5.**
(1) yes; (2) no: 3 is a divisor of 6, but 6 is not a
divisor of 3; (3) yes; (4) no, not symmetric
7. 1. Thm 1.2; 2. Assoc Prop Mult; 3. Number
fact: $−1(−1) = 1$; 4. Mult Ident Prop; 5.
Trans Prop Eq **9.** 1. Thm 1.2; 2. Thm 1.4; 3.
Thm 1.2; 4. Rule of Subt; 5. Thm 1.6; 6.
Comm Prop Add; 7. Rule of Subt; 8. Trans
Prop Eq **11.** 1. $(−x)(−y) = (−1x)(−1y)$
(Theorem 1.2); 2, $= [−1(−1)](xy)$ (Theorem
1.7, Ex. 10); 3. $= 1(xy)$ (Number fact:
$−1(−1) = 1$); 4. $= xy$ (Mult Ident); 5.
$(−x)(−y) = xy$ (Trans Prop Eq)
13. 1. $a(−b) = (−b)a$ (Comm Prop Mult);
2. $= −(ba)$(Thm 1.3); 3. $−(ab)$ (Comm Prop
Mult); 4. $a(−b) = −(ab)$ (Trans Prop Eq)
15. 1. $(ab + c) + ad = ab + (c + ad)$ (Assoc
Prop Add); 2. $= ab + (ad + c)$ (Comm Prop
Add); 3. $= (ab + ad) + c$ (Assoc Prop Add);
4. $= a(b + d) + c$ (Distr Prop); 5. $(ab + c) +
ad = a(b + d) + c$ (Trans Prop Eq) **17.** (1)
no; (2) no; (3) no; (4) no **19.** (1) no; (2) yes;
(3) no; (4) no

Chapter 2

Written Exercises, page 44

1. $\{x \mid x \leq 5\}$

3. $\{n \mid n > 2\}$

5. $\{y \mid y > −4\}$

7. $\{x \mid x \le 4.5\}$

4.5

9. $\{x \mid x < -4\}$

−4

11. $\{x \mid x < 4\}$

4

13. $\{p \mid p > -2\}$

−2

15. $\{n \mid n < -9\}$

−9

17. $\{a \mid a < -4\frac{2}{3}\}$

$-4\frac{2}{3}$

19. $\{x \mid x \le 4\}$

4

21. $\{x \mid x > -7\}$

−7

23. $\{a \mid a < -7\}$

−7

25. $-2 < 2$, but $-\frac{1}{2} \not> \frac{1}{2}$ **27.** F; $2 < 5$ and $5 < 6$, but $2 + 5 \not< 6$ **29.** T

Written Exercises, page 50

1. $\{x \mid x < -4 \text{ or } x > 3\}$

−4   3

3. $\{y \mid y \ge 3\}$

3

5. $\{\text{Real Numbers}\}$

0

7. $\{a \mid -4 \le a < 2\}$

−4   2

9. $\{x \mid -3 < x < 12\}$

−3   12

11. $\{x \mid x < 3\}$

3

13. $\{x \mid x \ge 2\}$

2

15. $\{x \mid x \le -3 \text{ or } x > 2\}$

−3   2

17. $\{x \mid -2.5 < x < 2.2\}$

−2.5   2.2

19. Answers will vary.   **21.** $\{4, 6, 8\}$
23. $\{a \mid 2 < a < 7.25\}$ **25.** $\{x \mid -2 \le x \le 1\}$
27. $\{x \mid -1 < x < 2 \text{ or } x > 4\}$

Written Exercises, page 53

1. $-5, 13$ **3.** $-2, 14$ **5.** $-10, 2$
7. $\{x \mid x < 3 \text{ or } x > 5\}$

3   5

9. $\{y \mid -10 \le y \le 4\}$

−10   4

11. $\{c \mid -4 \le c \le 4\}$

−4   4

13. $-5.5, 3$ **15.** $1\frac{2}{3}, 5$ **17.** $-0.8, 7.2$
19. $\{y \mid -4 < y < 7\}$

−4   7

21. $\{y \mid y < -7 \text{ or } y > 3\}$

−7   3

23. $\{x \mid x < -7.5 \text{ or } x > 4\}$

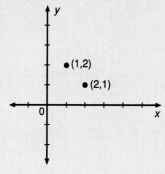

25. $-4, 10$ **27.** $-22, 8$

Written Exercises, pages 56–57

1. Between 2 yr 6 mo and 6 yr 3 mo, inclusive
3. $\{x \mid x > 12\}$ **5.** $\{w \mid 4 \text{ in} < w < 9 \text{ in}\}$
7. 6, 7, 8 and 7, 8, 9   **9.** 180 m/s $\leq v \leq$ 240 m/s   **11.** Between \$18,000 and \$24,000, inclusive   **13.** 25, 30, 35, and 30, 35, 40
15. 4 ft, 5 ft, 6 ft, 7 ft   **17.** (l, w, h):
(19,11,4), (20,12,4), (21,13,4), (22,14,4)

Written Exercises, pages 60–61

1. Frank: 10 yr, Selma: 50 yr   **3.** Matt: 8 yr, Barbara: 14 yr, Christi: 28 yr   **5.** Truck 1: 21 yr, Truck 2: 7 yr, Truck 3: 5 yr   **7.** Between 18 and 33 yr   **9.** Painting 1: less than 9 yr, Painting 2: less than 27 yr, Painting 3: less than 36 yr   **11.** $p + 8 + f$ **13.** $\dfrac{56 + 2n + p}{3}$

Written Exercises, pages 64–65

1. 2 h   **3.** $\frac{1}{2}$ h   **5.** 1:00 P.M.   **7.** 8:15 A.M.
9. 44 mi/h   **11.** 2:30 A.M.   **13.** Between 2 and $2\frac{1}{2}$ h   **15.** $\dfrac{x + y + z}{a + b + c}$

Chapter 3

Written Exercises, pages 76–77

1. D: $\{1, 2\}$, R: $\{1, 2\}$; function

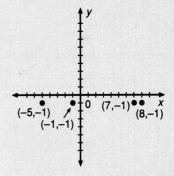

3. D: $\{-1\}$, R: $\{2, 6\}$; not a function

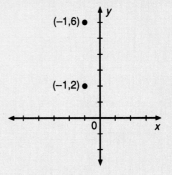

5. D: $\{-3, -2, 0, 3, 4\}$, R: $\{-3, -2, 2, 3, 4\}$; function

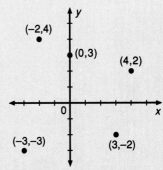

7. D: $\{-5, -1, 7, 8\}$, R: $\{-1\}$; function

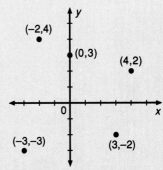

9. $\{(-2,3),(-1,-3),(1,-2),(4,1)\}$; D: $\{-2,-1,1,4,\}$, R: $\{-3,-2,1,3\}$; function
11. $\{(0,0),(1,-1),(1,1),(2,-2),(2,2),(3,-3),(3,3)\}$ D: $\{0,1,2,3\}$, R: $\{-3,-2,-1,0,1,2,3\}$; not a function   **13.** Not a function   **15.** D: $\{x \mid -3 < x \leq 4\}$, R: $\{y \mid -1 \leq y < 3\}$; function
17. D: $\{x \mid -3 \leq x \leq 3\}$, R: $\{y \mid -2 \leq y \leq 2\}$; not a function   **19.** D: $\{x \mid -3 \leq x \leq -2\}$ together with $\{x \mid -1 < x \leq 3\}$, R: $\{-1,0,2\}$; function   **21.** D: $\{\ldots-7,-4,-1,2,5,8,\ldots\}$, R: $\{\ldots-5,-3,-1,1,3,5,\ldots\}$; function

Written Exercises, page 81

1.

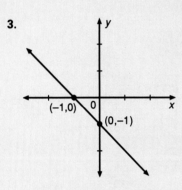

3.

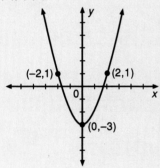

5. $(-3,6)$, $(-2,1)$, $(-1,-2)$, $(0,-3)$, $(1,-2)$, $(2,1)$, $(3,6)$

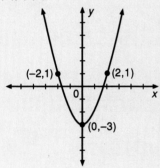

7. 32   **9.** $-4a^2 + 6$ **11.** $\{-8, -\frac{8}{3}, -2\}$

13. -5 **15.** $12a - 19$ **17.** $-4a^2 + 12a - 1$ **19.** $W(t^3) = W(ttt) = W(t) + W(t) + W(t) = 3W(t)$

Written Exercises, page 87

1. 2; up to right   **3.** 3; up to right

5. Undef; vert   **7.** $-\frac{3}{2}$; down to right

9. 0; horiz   **11.** $-\frac{1}{4}$; down to right   **13.** Yes

15. $\frac{16}{9}$ **17.** $\frac{7}{8}$ **19.** $-\frac{b}{4a}$ **21.** -3

Written Exercises, page 90

1. $-2x + 3y = 7$ **3.** $3x + 4y = 34$
5. $x + 2y = 2$ **7.** $x = -4$ **9.** $x = 2.3$
11. $9x + y = 1$ **13.** $3x - y = 1$
15. $-3x + y = 8.12$ **17.** $x = 5$
19. Answers may vary. One possible answer: For nonvertical, nonhorizontal lines, use the two points to find the slope, then use the point-slope form and one of the points to find the equation. For vertical lines, use the form $x = k$, where k is the x-coordinate of either point. For horizontal lines, use the form $y = k$, where k is the y-coordinate of either point.
21. $3x + 4y = 32$; 20

Written Exercises, page 94

1. $y = -\frac{2}{3}x + 5$ **3.** $-\frac{4}{9}$, 1   **5.** $-\frac{2}{3}$, -1
7. $-\frac{4}{5}$, $-\frac{3}{5}$

9.

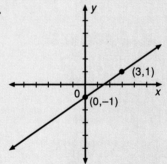

11.

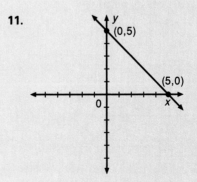

13.

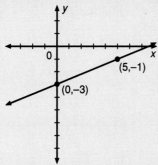

15.

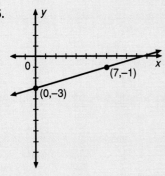

17.

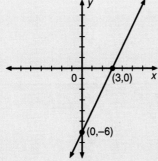

19.

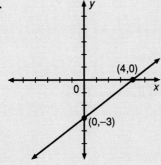

21.

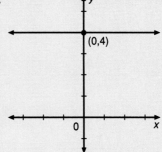

23.

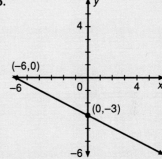

25.

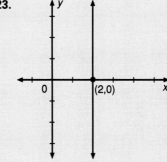

27. Yes **29.** Yes

31.

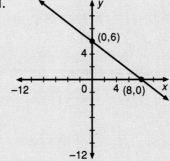

33. 2

Written Exercises, pages 98–99

1. $y = 12x + 100$ **3.** 2,600 sets
5. $y = 300x + 2,000$
7.

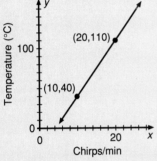

9. 250   **11.** 30°C   **13.** $7a - 30$ chirps

Written Exercises, page 102

1. $3x - y = 2$ **3.** $x + y = 1$
5. $4x + 3y = 23$ **7.** $x = 3$ **9.** $x = -5$
11. $4x - 3y = -17$ **13.** $x = 4$
15. $3x - y = 7$ **17.** parallel
19. $y = x + 3$ **21.** $x + 2y = 5$ **23.** 2

Written Exercises, pages 108–109

1. $\frac{1}{5}$, $y = \frac{1}{5}x$ **3.** $-\frac{1}{3}$, $y = -\frac{1}{3}x$

5. $1, y = x$ **7.** $\frac{3}{2}, y = \frac{3}{2}x$

9. $-\frac{5}{4}$, $y = -\frac{5}{4}x$ **11.** -1 **13.** 4   **15.** 2

17. 2   **19.** No   **21.** No   **23.** -162
25. 31.2   **27.** -36 **29.** 111   **31.** 250

33. $238   **35.** 81 ft   **37.** $133\frac{1}{3}$ cm

39. Decreases   **41.** Quadruples y

Written Exercises, page 111

1.

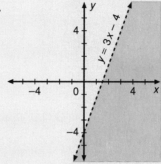

3.

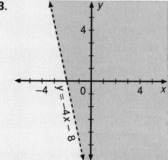

5.

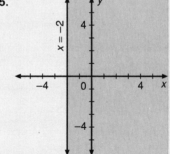

7.

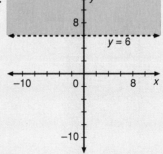

9.

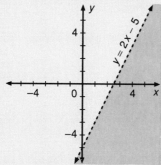

17.

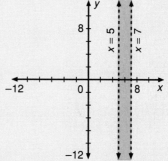

11.

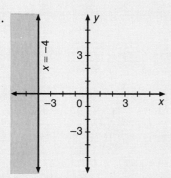

13.

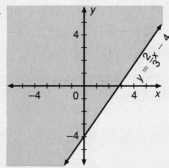

Chapter 4

Written Exercises, pages 121–122

1. Consistent, independent; $(-1,3)$

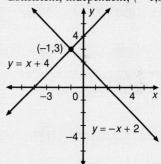

3. Consistent, dependent; $\{(x,y)\,|\,y = 3x - 2\}$

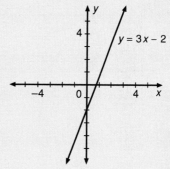

15.

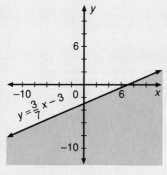

5. Inconsistent, independent; No solution

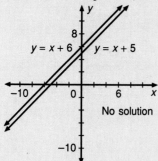

No solution

7. $\{(x,y)|y = -3x + 2\}$

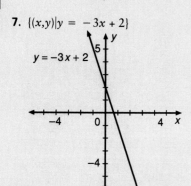

9. (1,2)

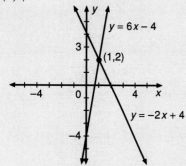

11. (1,0)

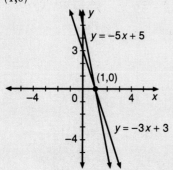

13. (2,0)

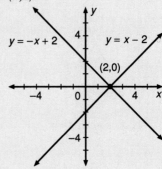

15. $k \neq -7$

Written Exercises, page 127

1. (3,6)   **3.** (3,8)   **5.** (2,−1)   **7.** $(-\frac{1}{2}, \frac{1}{2})$

9. (3,−2)   **11.** (600,400)   **13.** Martha: 16 yr; Ned: 8 yr   **15.** 84   **17.** (0,1), (3,7), (8,9)

Written Exercises, pages 131–132

1. (6,0)   **3.** (−3,−1)   **5.** (1,1)   **7.** (9,2)
9. (−3,4)   **11.** Inconsistent   **13.** (1,1)
15. (7,−4)   **17.** $(\frac{1}{3}, -\frac{1}{7})$ **19.** $(t,u) = (7,3)$
21. (7,7)   **23.** 6, 15   **25.** $l = 8$ m, $w = 3$ m
27. 1,200 people   **29.** $(1 - \frac{2a}{b}, \frac{b}{a} - 2)$

Written Exercises, pages 136–137

1. 7 dimes, 9 nickels   **3.** 200 adult, 100 student   **5.** 20%: 16 l; 30%: 64 l
7. $3,600 at $15\frac{1}{2}$%, $2,400 at $14\frac{1}{4}$%
9. $30,000   **11.** $15/day, 14¢/mi

Written Exercises, pages 140–141

1. 25 mi/h   **3.** 550 mi/h   **5.** 80, 100
7. 126   **9.** 5 mi/h   **11.** 42

Written Exercises, pages 144–145

1. (4,−2, 3)   **3.** (−2,−4,−7)   **5.** (3,2,1)
7. $(1, \frac{1}{2}, -1)$ **9.** $(-2, \frac{3}{5}, 2)$ **11.** 8, 7, 5
13. 721   **15.** (1,−1,−2)

17. $(w,x,y,z) = (4,1,2,3)$

Written Exercises, page 150

1.
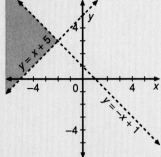

3.
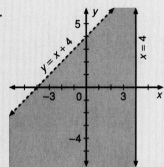

5.
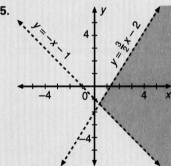

7.
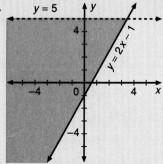

9.
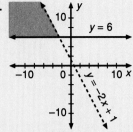

11. $(1,1)$, $(1,3)$, $(2,1)$
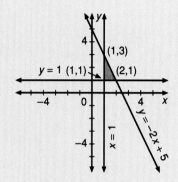

13. $y \le -4x + 7$ and $y \le x + 2$ and $y \ge -\frac{2}{3}x + \frac{1}{3}$ **15.** $y \ge 3x - 4$ and $x \le 2$ and $y \le 3$ and $y \ge -\frac{4}{3}x + \frac{1}{3}$

Chapter 5

Written Exercises, pages 163–164

1. $5x^6$ **3.** $-12yz^2$ **5.** $-22m^4n^7$

7. $-\dfrac{2b^5}{-a^6}$ **9.** $\dfrac{3a^4b}{2c^8}$ **11.** $16x^4$ **13.** $45a^2$

15. $405x^{13}$ **17.** $\dfrac{49c^6}{100d^{10}}$ **19.** 5 **21.** 2

23. 2 **25.** 2 **27.** $\frac{2}{3}$ **29.** $-10{,}000x^4y^3z^6$

31. $64a^{12}c^6d^{17}$ **33.** $\dfrac{-8m^3n^3p^9}{27x^9y^6}$

35. $x^{7a+2}y^{5b-1}$ **37.** $x^{2ac}y^{bc-c}$ **39.** $x^{5a+9}y^{4b}$

41. $x^{4a+b}y^{a-2b}$

Written Exercises, page 167

1. $\frac{1}{125}$ **3.** $\frac{2}{3}$ **5.** $\frac{1}{64}$ **7.** 0.0038 **9.** 500

11. 4,600 **13.** $\dfrac{8}{x^3}$ **15.** $\dfrac{-2}{a^3}$ **17.** x^2

19. $\dfrac{-14}{n^8}$ **21.** $\dfrac{1}{x^{12}}$ **23.** $\dfrac{c^{12}}{8}$ **25.** $\dfrac{4c^8}{5}$

27. $x^{10}y^{15}$ **29.** 20,000 **31.** 0.00005

33. 0.0069 **35.** $\dfrac{a^6c^3}{3b^5d^4}$ **37.** $\dfrac{144z^6}{x^8y^6}$

39. $\dfrac{15}{x^5y^2}$ **41.** $\dfrac{-64y^3w^{12}}{27x^9z^6}$ **43.** $\dfrac{1}{x^{n-m}}$

Written Exercises, page 173

1. 2 **3.** 3 **5.** 0 **7.** 3 **9.** $3x^2 - 4x + 1$
11. $3x^2 + 7x + 1$ **13.** $-21x^2 - 6xy + 15x$
15. $x^2 - x - 2$ **17.** $6t^2 - t - 12$
19. $10x^2 + 13xy - 3y^2$ **21.** $x^4 + 18x^2 + 80$
23. $10y^6 - 39y^3 + 14$ **25.** $8x^2 + 4x - 60$
27. $3x^3 - 5x^2 - 6x + 8$ **29.** $3xy - 2xy^2$
31. $15x^4 - 17x^2y^2 - 4y^4$ **33.** $8a^2 - 22ab + 6a + 15b^2 - 9b$ **35.** $-60y^3 + 21y^2 + 18y$ **37.** $x^{2n} + 2x^n - 8$
39. $x^{2a} + 4x^ay^{2a} + 3y^{4a}$

Written Exercises, page 177

1. $(x + 5)(x - 4)$ **3.** $(y - 3)(y - 2)$
5. Not possible **7.** Not possible
9. $(7y - 1)(y - 5)$ **11.** $(5t + 2)(2t - 3)$
13. $(3m - 4)(3m - 2)$ **15.** $3(n + 5)(n - 3)$
17. $t(3t + 2)(2t - 5)$ **19.** $2x(3x + 1)(x - 5)$

21. $(2x + y)(6x - y)$ **23.** $\dfrac{f}{m - 5}$

25. $\dfrac{d - c}{a - 4}$ **27.** $\dfrac{-9a}{6a - 2n}$ **29.** $-9, -3, 3, 9$

31. $(6y + 5)(2y - 3)$ **33.** $5xy(3x^2 + xy - 2y)$ **35.** $2y(9y^2 + 2)(2y^2 - 3)$ **37.** $(x^{3m} - 6)(x^{3m} - 2)$ **39.** $(5x^a - 3y^b)(x^a - 4y^b)$

Written Exercises, page 180

1. $n^2 - 121$ **3.** $4x^2 + 20x + 25$
5. $49a^2 - 42a + 9$ **7.** $x^2 + 2cx + c^2$
9. $9x^2 - 4a^2$ **11.** -9 **13.** $36c^2 - 10$
15. $16x^2 - 80x + 90$ **17.** $9m^4 - 16n^2$
19. $x^2 + 0.2x + 0.01$ **21.** $9x^2 + 12xh^2 + 4h^4$ **23.** 57 **25.** $-15c^2 + 62c + 168$
27. $-10a^2 + 14a + 12$ **29.** $x^{2m} - y^{2n}$

Written Exercises, page 183

1. $(2c + 5)(2c - 5)$ **3.** $(x - 4)^2$ **5.** Not possible **7.** $(y^2 + 7)(y^2 - 7)$ **9.** $(4n + 3)^2$

11. $(x + y + 4)(x + y - 4)$ **13.** $(x + y - 2)(x - y + 2)$ **15.** $(x + 3 + y)(x + 3 - y)$
17. $(a^2 - 8)^2$ **19.** $(x + 3)(x^2 - 3x + 9)$
21. $(4d + 1)(16d^2 - 4d + 1)$ **23.** 4
25. $(6x^3 + y^2)(6x^3 - y^2)$ **27.** $(5a^2 + 6b^2)^2$
29. Not possible **31.** $(m - 4n)(m^2 + 4mn + 16n^2)$ **33.** $(2t + 5v)(4t^2 - 10tv + 25v^2)$
35. $(2x + 3y + 5)(2x + 3y - 5)$ **37.** $(x^n + y^{2n+3})(x^n - y^{2n+3})$ **39.** For all real numbers x and y, $(x - y)^2 \geq 0$; $x^2 - 2xy + y^2 \geq 0$; $x^2 + y^2 \geq 2xy$.

Written Exercises, page 186

1. $4(x - 4)(x - 3)$ **3.** $3(2y + 1)(2y - 1)$
5. $4(n - 5)^2$ **7.** $(x + 2)(x - 2)(x + 3)(x - 3)$ **9.** $(x + 3)(x - 3)(x^2 + 4)$
11. $2(y + 3)(y^2 - 3y + 9)$ **13.** $-1(a - 5)^2$
15. $-1(2n - 1)(2n + 3)$ **17.** $-1(y - 1)(y^2 + y + 1)$ **19.** $(4x + 3)(2y - 5)$
21. $(4x - y)(2x + 3)$ **23.** $(x^2 + 5)(3x - 4)$
25. $(2c^2 - d)(3c - 2)$ **27.** $5a(2a - 5)(a + 2)$ **29.** $3xy(x - 1)^2$ **31.** $(3y + 2)(3y - 2)(y + 2)(y - 2)$ **33.** $(3y + 2)(3y - 2)(y^2 + 4)$ **35.** $3(2x + 5)(4x^2 - 10x + 25)$
37. $-y(y - 4)(y^2 + 4y + 16)$
39. $-2a(2c + 1)^2$ **41.** $a(a - b^2)(a - b)$
43. $(2c^2 + 3)(2c^2 - 3)(c^2 + 1)(c + 1)(c - 1)$
45. $x^3(x^{a+2} + 1)$ **47.** $y^{n+4}(y^2 + 1)$
49. $x^2(x^n - 2)^2(x^n + 2)^2$

Written Exercises, page 189

1. $3c - 2, -1$; no **3.** $8x + 10, 0$; yes
5. $4y^2 - 3y + 10, 6$; no **7.** $4n^2 + 5n - 2, 10$; no **9.** $3a^3 - 12a^2 - a + 5, 0$; yes
11. $y^2 + 2y + 4, 0$; yes **13.** $3c^2 - 2c + 12, 0$; yes **15.** $y^2 - 3, -5$; no **17.** $3x^2 + 2x - 4, 0$; yes **19.** $x^4 - x^3y + x^2y^2 - xy^3 + y^4, 0$; yes **21.** $x^{2m} + 3, 0$; yes **23.** 22

Written Exercises, pages 191–192

1. 3 **3.** 198 **5.** $1\frac{1}{3}$ **7.** 38 **9.** $3a^2 + 1$

11. $a^4 + 4a^2 + 6$ **13.** -3 **15.** -29.5
17. $-2a^2 - 8a + 3$ **19.** $-18c^2 - 24c + 3$
21. $-2c^2 + 12c - 7$ **23.** 192
25. $g(f(x)) = 3x + 24, f(g(x)) = 3x + 8$
27. $g(f(x)) = 35x - 19, f(g(x)) = 35x + 41$
29. $g(f(x)) = x^2 + 6x + 8, f(g(x)) = x^2 + 2$
31. $g(f(x)) = 6x^2 + 16, f(g(x)) = 18x^2 - 24x + 14$ **33.** $g(f(x)) = 2x, f(g(x)) = 2x$
35. $g(f(x)) = x, f(g(x)) = x$ **37.** $s(t(x)) = x^2 + 5x - 6, s(t(10)) = 144, t(s(x)) = x^2 -$

$x - 9$, $t(s(10)) = 81$ **39.** $25x^2 - 220x + 481$ **41.** $25x^2 - 20x + 1$ **43.** $-5x^2 + 17$

Chapter 6

Written Exercises, page 202

1. $-2, \frac{1}{3}$ **3.** $-5, 0$ **5.** $5, 8$ **7.** $-3, 4$
9. $0, 5$ **11.** ± 7 **13.** None **15.** 2
17. $\pm 1, \pm 5$ **19.** ± 5 **21.** $2, 5$ **23.** -5
25. $0, \pm 3$ **27.** $-5, 0, 2$ **29.** $\frac{3}{2}, 2$
31. $0, \frac{9}{16}$ **33.** $\frac{3}{2}, \frac{8}{3}$ **35.** $\pm \frac{1}{3}, \pm 2$
37. $\pm \frac{1}{3}$ **39.** $2, 5$ **41.** $-6, 0, 1$ **43.** $-2a, 3a$ **45.** $\pm a, \pm 3a$ **47.** $-3, 2$ **49.** Converse: For all real numbers m and n, if $m = 0$ or $n = 0$, then $m \cdot n = 0$. True

Written Exercises, page 206

1. $-8, -6, -4; 6, 8, 10$ **3.** 7 chairs
5. $-3, -8; 8, 3$ **7.** $-15, -10, -5; 10, 15, 20$ **9.** 10 boxes **11.** $-7, -6, -5; 5, 6, 7$
13. $-\frac{2}{3}, -6; \frac{2}{3}, 6$ **15.** 240 tiles
17. $-0.5, 0, 0.5, 1; 2.5, 3, 3.5, 4$

Written Exercises, pages 210–211

1. 6 ft, 8 ft **3.** $l = 17$ m, $w = 3$ m
5. $h = 9$ yd, $b = 6$ yd **7.** 144 m²
9. 16 ft², 64 ft² **11.** 13 m **13.** 27 m²
15. rectangle: 100 m², triangle: 50 m²
17. 5 ft **19.** 15 in. by 19 in.

Written Exercises, page 216

1. $\{n|n < 2 \text{ or } n > 6\}$
[number line: 2 6]

3. $\{c|-5 \le c \le -2\}$
[number line: −5 −2]

5. $\{x|x \le -4 \text{ or } x \ge 4\}$
[number line: −4 4]

7. $\{y|y < -5 \text{ or } y > 5\}$
[number line: −5 5]

9. $\{n|n < 0 \text{ or } n > 4\}$
[number line: 0 4]

11. $\{c|-4 \le c \le 0\}$
[number line: −4 0]

13. $\{x|-7 < x < 7\}$
15. $\{y|y < -3 \text{ or } 0 < y < 4\}$
[number line: −3 0 4]

17. $\{x|-4 \le x \le 0 \text{ or } x \ge 4\}$
[number line: −4 0 4]

19. $\{y|y < -5 \text{ or } 0 < y < 4\}$
[number line: −5 0 4]

21. $\{x|-3 < x < 2\frac{1}{2}\}$
[number line: −3 $\frac{5}{2}$]

23. $\{y|y \le 0 \text{ or } y \ge 4\frac{1}{2}\}$
[number line: 0 $\frac{9}{2}$]

25. $\{c|c < -4 \text{ or } c > 3\}$
[number line: −4 3]

27. $\{a|1 < a < 3\}$
[number line: 1 3]

29. $\{x|-3 < x < 0 \text{ or } x > 3\frac{1}{2}\}$
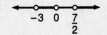
[number line: −3 0 $\frac{7}{2}$]

31. $\{a \mid -2 < a < 2 \text{ or } a < -6 \text{ or } a > 6\}$

33. Answers will vary. One possible answer: The product of two factors is negative if and only if the factors have opposite signs. Therefore, one and only one of the factors must be negative.

35. $\varnothing$

37. $\{y \mid y < -4 \text{ or } -2 < y < 2 \text{ or } y > 4\}$

39. $\{x \mid x < -5 \text{ or } -1 < x < 1 \text{ or } x > 5\}$

41. $\{n \mid -3 \le n \le 3\}$

Written Exercises, pages 221–222

1. $2x^2 - 5x + 3, -6$; no **3.** $y^3 - 3y + 8$, 0; yes **5.** $4n^4 + 4n^3 + 2n^2 + 8n - 1$, 0; yes **7.** $-21, 0; x - 5$ **9.** $17, 0; x - 2$
11. $50, 0, -12, -16, -18, 0, 80$; $x + 2, x - 2$ **13.** $c^2 - 20c + 400, 0$; yes
15. $x^4 + 3x^3 + 8x^2 + 24x + 72, 6$; no
17. 1 **19.** $2y^2 - \frac{3}{2}y + 3, 0$; yes

Written Exercises, page 225

1. $(x - 1)(x + 1)(x - 2); 1, -1, 2$

3. $(x - 1)(x + 4)(2x - 3); 1, -4, \frac{3}{2}$

5. $(x - 4)(5x + 2)(2x - 1); 4, -\frac{2}{5}, \frac{1}{2}$

7. $(x + 1)(x - 2)(x - 3)(x + 4); -1,$ $2, 3, -4$ **9.** 2 (mult 3) **11.** $-2, 4, \frac{1}{2}, -\frac{3}{2}$

13. $2x^4 + x^3 - 9x^2 - 4x + 4$
15. 13 ft by 5 ft by 3 ft

Chapter 7

Written Exercises, pages 236–237

1. $\{x \mid x \ne 3\}$ **3.** $\{x \mid x \ne 0 \text{ and } x \ne 7\}$ **5.** 5

7. $0, 9$ **9.** $\frac{1}{x^2}$ **11.** $\frac{1}{x + 5}$ **13.** $\frac{x}{x + 1}$

15. $\frac{1}{2(x - 2)}$ **17.** $\frac{3(x - 2)}{4x^2}$ **19.** $\frac{2m}{3m - 2}$

21. $-\frac{x + 2}{x - 2}$ **23.** $\frac{-(x - 6)}{3x + 4}$; $-\frac{9}{5}$

25. $2a + 3$ **27.** $\frac{-1}{m - 2}$ **29.** $(a - 2)(x + 3)$

31. $\frac{-1}{(a - 1)(a + 3)}$ **33.** 1. $\frac{a}{b} = 0$ (Given);
2. $0 \cdot b = a$ (Def of division); 3. $0 = a$ (Zero Prop for Mult); 4. $a = 0$ (Symmetric Prop Eq)

Written Exercises, pages 240–241

1. $\frac{2}{x}$ **3.** $\frac{x}{2}$ **5.** $\frac{b}{6a^2}$ **7.** $\frac{x}{12}$

9. $\frac{a}{2(a - 5)}$ **11.** $\frac{8x^2}{3}$ **13.** $\frac{3y^3(x - y)}{4x}$

15. $\frac{-(2x + 1)}{x - 1}$ **17.** $\frac{2(a + 4)}{3a^4 b^4(a - 2)}$

19. $\frac{-(3y + 1)}{x}$ **21.** $\frac{-(x^2 + 3x + 9)}{2x^2}$

23. $\frac{-(x - 10)}{4(x - 5)}$

25. $\frac{x(x^2 + 2x + 4)(y^2 + y + 1)}{x - 5 + 7y}$

Written Exercises, page 243

1. 54 ft **3.** $2\frac{1}{2}$ yd **5.** 0.3 kg **7.** 6 lb/gal
9. $11.25 **11.** $1,448.40

Written Exercises, page 248

1. $\frac{a}{5}$ **3.** $\frac{3a}{5}$ **5.** $\frac{5}{b + 3}$ **7.** $\frac{13x}{20}$

9. $\frac{3(5a + 6)}{10(a + 4)}$ **11.** $\frac{25}{18(x + 5)}$ **13.** $\frac{5}{m + 4}$

15. $\frac{x + 1}{x}$ **17.** $\frac{y + 1}{y - 2}$

19. $\frac{-2(3a^2 + 2a + 5)}{(3a + 2)(a - 5)}$ **21.** 1. $\frac{x}{z} + \frac{y}{z} = x \cdot \frac{1}{z} + y \cdot \frac{1}{z}$ (Ident Prop Mult); 2. $= (x + y)\frac{1}{z}$ (Distr Prop); 3. $= \frac{x + y}{z}$ (Ident Prop Mult);

4. $\frac{x}{z} + \frac{y}{z} = \frac{x + y}{z}$ (Trans Prop Eq)

Written Exercises, page 251

1. $\dfrac{2x+11}{11x+2}$ **3.** $\dfrac{-2(3x-14)}{5(11x-38)}$ **5.** $\dfrac{a+5}{a+2}$

7. $\dfrac{6(x+2)}{17x+14}$ **9.** $\dfrac{a}{a-1}$ **11.** $\dfrac{2x+1}{x+2}$

13. $\dfrac{-1}{3(3+h)}$ **15.** $\dfrac{-(6+h)}{9(3+h)^2}$ **17.** $\dfrac{x-y}{x+y}$

Written Exercises, pages 255–256

1. $3\frac{1}{3}$ **3.** $\frac{3}{10}$ **5.** $\frac{1}{2}$ **7.** $1\frac{9}{22}$ **9.** -6 (5 is extraneous.) **11.** 2 **13.** -19 **15.** -6 (4 is extraneous.) **17.** -6 **19.** $-\frac{1}{2}$, 5

21. No solution (5 is extraneous.) **23.** $\frac{2}{3}$, 4

25. 2, 7 **27.** $x = \dfrac{2ab}{a+b}$; 8.4

29. $b = \dfrac{a}{1+cd}$; 0.5 **31.** $\frac{3}{7}$ **33.** $\frac{4}{5}$

35. -10, 1

Written Exercises, page 260

1. $3\frac{3}{7}$ h **3.** $26\frac{2}{3}$ h **5.** Andy: 21 h; Sal: $10\frac{1}{2}$ h **7.** Florence: 30 h; Betty: 60 h

9. 6 h **11.** 12 d

Chapter 8

Written Exercises, page 270

1. 9 **3.** 1.2 **5.** Not real **7.** 4 **9.** $\frac{5}{4}$
11. $\{x \mid x \le 2\}$ **13.** ± 12 **15.** ± 6
17. ± 0.07 **19.** $\{x \mid x \le 3 \text{ or } x \ge 4\}$
21. $\{x \mid -2 \le x \le 2 \text{ or } x > 3\}$

Written Exercises, pages 274–275

1. $2\sqrt{6}$ **3.** $3\sqrt{5}$ **5.** $5\sqrt{2}$ **7.** $6\sqrt{2}$
9. $12\sqrt{2}$ **11.** $3\sqrt{5}$ **13.** $4a^2$ **15.** $6|c^5|$
17. $m^3\sqrt{m}$ **19.** $b^8\sqrt{b}$ **21.** $8|a^9|\sqrt{3}$
23. $4a\sqrt{2a}$ **25.** $2x^2y^2\sqrt{3y}$ **27.** $4x^8y^5\sqrt{5y}$
29. $-10x^7y^5\sqrt{5y}$ **31.** $12x^4y^4\sqrt{2y}$

33. Answers may vary. One possible answer: Find the greatest perfect square factor of each factor in the radicand. Then use Theorem 8.1 to simplify. **35.** x^m **37.** $x^{5m+2}y^{5n-3}\sqrt{y}$
39. Either $\sqrt{x^2+y^2} \le x+y$ or $\sqrt{x^2+y^2} > x+y$. Assume $\sqrt{x^2+y^2} > x+y$. Then $x^2+y^2 > (x+y)^2$, $x^2+y^2 > x^2+2xy+y^2$, $0 > 2xy$. But this is a contradiction. Because x and y are nonnegative numbers, $2xy$ must also be nonnegative. Therefore, the original statement is false, and $\sqrt{x^2+y^2} \le x+y$.

Written Exercises, page 278

1. $5\sqrt{3}$ **3.** $-\sqrt{2}$ **5.** $12\sqrt{2}$ **7.** $-\sqrt{11}$
9. $16\sqrt{5}$ **11.** $30\sqrt{2} - 5\sqrt{6}$ **13.** $8x$
15. $4-x$ **17.** $8\sqrt{7}$ **19.** $120\sqrt{2}$
21. $42 - 4\sqrt{6} + 4\sqrt{15}$ **23.** $46 - 14\sqrt{6}$
25. $114 - 48\sqrt{3}$ **27.** $6x - \sqrt{15xy} - 5y$
29. $10b\sqrt{ab}$ **31.** $5\sqrt{y-3}$

Written Exercises, pages 282–283

1. $2\sqrt{2}$ **3.** $2\sqrt{3}$ **5.** x^4 **7.** $2a^2\sqrt{7}$
9. $2\sqrt{2}$ **11.** $\sqrt{2}$ **13.** $\dfrac{-10\sqrt{7d}}{7d}$ **15.** $\dfrac{\sqrt{2}}{a^2}$
17. $\dfrac{2\sqrt{a}}{a^2}$ **19.** $\dfrac{x^2\sqrt{xy}}{y^2}$ **21.** $14(\sqrt{3}+\sqrt{2})$
23. $3(2-\sqrt{3})$ **25.** $\dfrac{\sqrt{2x}}{2x^2}$ **27.** $\dfrac{-2\sqrt{3x}}{xy}$
29. $5 - 2\sqrt{6}$ **31.** $\dfrac{6x + 8\sqrt{xy}}{9x - 16y}$
33. $\dfrac{27 + 10\sqrt{2}}{23}$ **35.** $3\sqrt{3} - 5$
37. $\dfrac{a + \sqrt{a^2 - b^2}}{b}$ **39.** $\sqrt{2}$

Written Exercises, page 287

1. $7x$ **3.** $-24y$ **5.** 3 **7.** -2 **9.** $2\sqrt[4]{4}$
11. $-5\sqrt[3]{2}$ **13.** x^7 **15.** a^5 **17.** $2a^5\sqrt[4]{a^3}$
19. $2b^2$ **21.** $23\sqrt[4]{2}$ **23.** $6\sqrt[6]{5}$
25. $3b^4\sqrt[3]{4}$ **27.** $2\sqrt[3]{49}$ **29.** $\dfrac{\sqrt[4]{14}}{2}$
31. $-14\sqrt[3]{3}$ **33.** $2d^2\sqrt[4]{7d^3}$ **35.** $\dfrac{\sqrt[3]{35y}}{5y}$
37. $\dfrac{\sqrt[5]{6ab^3}}{2ab}$ **39.** 1. $(\sqrt[n]{a} \cdot \sqrt[n]{b})^n =$

$(\sqrt[n]{a})^n(\sqrt[n]{b})^n$ (power of product); 2. $= ab$ (Def of principal nth root); 3. $(\sqrt[n]{ab})^n = ab$ (Def of principal nth root); 4. $(\sqrt[n]{a} \cdot \sqrt[n]{b})^n = (\sqrt[n]{ab})^n$ (Trans and Sym Prop Eq); 5. $\sqrt[n]{a} \cdot \sqrt[n]{b} = \sqrt[n]{ab}$ (If n is negative, each is the unique solution of $x^n = ab$. If n is positive, each is the unique positive solution.)

Written Exercises, pages 290–291

1. 6 3. $\frac{1}{3}$ 5. $\frac{1}{20}$ 7. $2\frac{1}{2}$ 9. -12

11. $a^{\frac{3}{5}}$ 13. $11^{\frac{4}{3}}$ 15. $23^{\frac{4}{5}}$ 17. $\sqrt[4]{5^3}; \sqrt[4]{125}$

19. $\frac{1}{\sqrt[5]{3^4}}; \frac{1}{\sqrt[5]{81}}$ 21. 8 23. 9 25. 4

27. $-\frac{4}{49}$ 29. $\frac{1}{243}$ 31. $104 33. $\frac{1}{8}$

Written Exercises, pages 294–295

1. $\frac{5^{\frac{1}{4}}}{5}$ 3. $\frac{30x^{\frac{2}{3}}}{x^2}$ 5. $4^{\frac{5}{9}}$ 7. $t^{\frac{15}{4}}$ 9. x^8y^3

11. $\frac{1}{4x^2y^3}$ 13. $a^{\frac{4}{7}}$ 15. a 17. $\frac{x^4}{y^5}$

19. $\frac{7m^2}{4}$ 21. $\sqrt[6]{x^5}$ 23. $\sqrt[12]{x}$ 25. $\frac{4}{3}$

27. $-\frac{1}{2}$ 29. $\frac{5}{4}$ 31. $-\frac{1}{3}$ 33. $\frac{x^{\frac{6}{7}}}{x}$

35. $\frac{-30a^{\frac{5}{8}}}{a^2}$ 37. $\frac{y^6}{8x^{15}}$ 39. $\frac{v^{\frac{3}{5}}t^{\frac{5}{6}}}{t}$

41. $x\sqrt{x} - \sqrt[3]{y}$ 43. $4x - 12\sqrt[6]{x^3y} + 9\sqrt[3]{y}$

45. 8

Chapter 9
Written Exercises, page 306

1. $8i$ 3. $-6i$ 5. $i\sqrt{21}$ 7. $-i\sqrt{3}$

9. $5i\sqrt{2}$ 11. $-8i\sqrt{10}$ 13. $8 - 6i$

15. $-7 + 15i\sqrt{2}$ 17. $\pm 5i$ 19. $\pm 5i\sqrt{2}$

21. $\pm 2, \pm 2i$ 23. $8 + 8i$ 25. $4 + 4i$

27. $-8 + i$ 29. 5 31. 4 33. $4\sqrt{2}$

35. $\pm 4i$ 37. $\pm\frac{2}{3}i$ 39. $\pm\sqrt{3}, \pm i\sqrt{5}$

41. $15 - 20i\sqrt{3}$ 43. 20 45. 0

47. $|a + bi| = \sqrt{a^2 + b^2} = \sqrt{a^2 + (-b)^2} = |a - bi|$

Written Exercises, page 309

1. -28 3. $6i\sqrt{6}$ 5. $-10i$ 7. $6 + 6i$

9. $15\sqrt{2} - 24i\sqrt{3}$ 11. $-2 - 14i$

13. $7 + 24i$ 15. $8 + 6i$ 17. 34 19. $125i$

21. $32i$ 23. $\frac{-5i}{4}$ 25. $\frac{7i}{2}$ 27. $\frac{-5 + 2i}{6}$

29. $-3 - 3i$ 31. i 33. $\frac{8 - 4i}{5}$ 35. $(3y + 2i)(3y - 2i)$ 37. $(a\sqrt{5} + i\sqrt{3})(a\sqrt{5} - i\sqrt{3})$

39. $-1,000,000$ 41. $4 - 16i\sqrt{3}$ 43. $\frac{i}{8}$

45. 0 47. $\overline{w + z} = \overline{(a + bi) + (c + di)} =$ $\overline{(a + c) + (b + d)i} = (a + c) - (b + d)i$ $= (a - bi) + (c - di) = \overline{a + bi} +$ $\overline{c + di} = \overline{w} + \overline{z}$ 49. $\overline{wz} = \overline{(a + bi)}$ $\overline{(c + di)} = \overline{ac + adi + bci + bdi^2} =$ $\overline{(ac - bd) + (ad + bc)i} = (ac - bd) - (ad + bc)i = ac - adi - bci - bd = (a - bi)$ $(c - di) = \overline{(a + bi)}\ \overline{(c + di)} = \overline{w} \cdot \overline{z}$

Written Exercises, page 312

1. $-3, 7$ 3. 2, 5 5. $4 \pm 3\sqrt{2}$ 7. $-3, \frac{1}{2}$

9. $-\frac{1}{3}, -\frac{1}{2}$ 11. $\frac{-7 \pm 3\sqrt{5}}{2}$ 13. $3 \pm 4i$

15. $-4 \pm 2i$ 17. $-\sqrt{2}, 5\sqrt{2}$

19. $\frac{-7 \pm \sqrt{33}}{2}$ 21. $-\frac{3}{4}$ 23. $\frac{7 \pm \sqrt{29}}{10}$

25. $-2i, 6i$ 27. $-5i, -i$ 29. $2\sqrt{2}$

Written Exercises, pages 315–316

1. 50, $x = 5$ 3. 34, $x = 3$ 5. 1, 175, $x = 5$ 7. 80 m, $t = 4$ s 9. 6, 6
11. 7,200 yd²; 60 yd by 120 yd 13. 25 trees/acre; 6,250 oranges/acre 15. $-25, x = -5$ 17. $-100, x = -4$

Written Exercises, page 319

1. 2, 4 3. $\frac{5 \pm \sqrt{33}}{2}$ 5. $-\frac{1}{2}, 1$

7. $\frac{1 \pm i\sqrt{3}}{4}$ 9. $\frac{3 \pm 2i}{2}$ 11. $\frac{1 \pm 3i}{4}$

13. $-2 \pm \sqrt{5}$ 15. $3 \pm 2\sqrt{5}$

17. 3, $\frac{-3 \pm 3i\sqrt{3}}{2}$ 19. 4, $-2 \pm 2i\sqrt{3}$

21. $-10, 5 \pm 5i\sqrt{3}$ **23.** $\dfrac{-3 \pm i\sqrt{5}}{2}$

25. $\dfrac{-\sqrt{3}}{3}, \sqrt{3}$ **27.** $-2i, -\dfrac{i}{2}$

29. $\dfrac{1}{2}, \dfrac{-1 \pm i\sqrt{3}}{4}$

9.

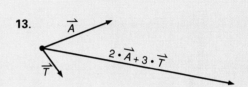

11.

Written Exercises, page 322

1. $l: 6 + 4\sqrt{3}$ ft; $A: 36 + 20\sqrt{3}$ ft^2
3. $(2 + \sqrt{5}, -2 + \sqrt{5}), (2 - \sqrt{5}, -2 - \sqrt{5})$
5. $(3 + i\sqrt{5}, -3 + i\sqrt{5}), (3 - i\sqrt{5}, -3 - i\sqrt{5})$ **7.** 1.6 yd

13.

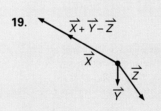

15.

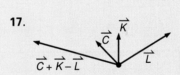

Written Exercises, page 327

1.

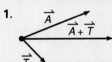

3.

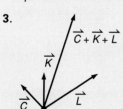

5.

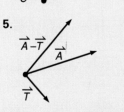

7.

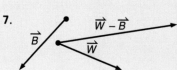

17.

19.

817

21.

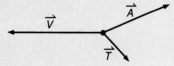

23.

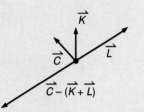

$$\overrightarrow{C} - (\overrightarrow{K} + \overrightarrow{L})$$

25.

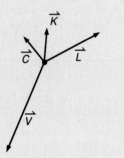

Written Exercises, page 330

1.

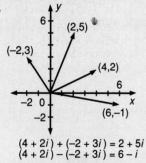

$$(4 + 2i) + (-2 + 3i) = 2 + 5i$$
$$(4 + 2i) - (-2 + 3i) = 6 - i$$

3.

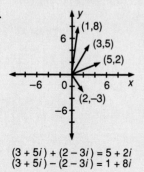

$$(3 + 5i) + (2 - 3i) = 5 + 2i$$
$$(3 + 5i) - (2 - 3i) = 1 + 8i$$

5.

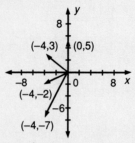

$$(-4 - 2i) + 5i = 4 + 3i$$
$$(-4 - 2i) - 5i = -4 - 7i$$

7.

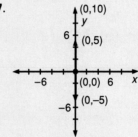

$$5i + (-5i) = 0 + 0i$$
$$5i - (-5i) = 0 + 10i$$

9.

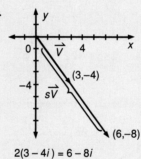

$$2(3 - 4i) = 6 - 8i$$

11.

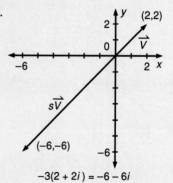

$$-3(2 + 2i) = -6 - 6i$$

13.

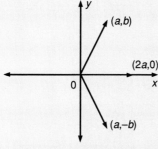

$(a + bi) + (a - bi) = 2a$

15.

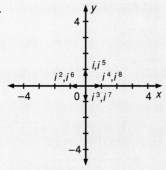

Written Exercises, page 349

1. 7 **3.** 3 **5.** 8 **7.** 10 **9.** $\pm 3\sqrt{2}$
11. $\pm 6\sqrt{2}$ **13.** 3 **15.** 9 **17.** No roots
19. 2, 6 **21.** $-12, 4$ **23.** 3, 7
25. $V = \pi r^2 h$ **27.** $-1, 6$

Written Exercises, page 355

1. $-1, 1, 2$ **3.** $-3, \pm i$ **5.** $2, \pm 3\sqrt{2}$
7. $-5, \pm i\sqrt{2}$ **9.** $U = 2, L = 0$
11. $U = 3, L = -4$ **13.** $U = 4, L = -3$
15. $-4, -1, \pm 2i\sqrt{3}$ **17.** $-1, -\frac{1}{2}, \frac{2}{3}, 1, 2$
19. $-5, 2, 2 \pm \sqrt{3}$

Written Exercises, pages 359–360

1. $\frac{1}{3}, \pm 2\sqrt{2}$ **3.** $-\frac{3}{2}, \frac{1}{3}, \frac{2}{3}$ **5.** $2, -\frac{2}{3}$,
$\pm\sqrt{5}$ **7.** $5, -\frac{2}{3}, \pm i\sqrt{2}$ **9.** $\frac{1}{2}(m = 3), -\frac{2}{3}$
11. $\pm\frac{2}{3}, \pm i\sqrt{3}$ **13.** $\pm\frac{1}{2}, \frac{3}{4}, \pm 2$

15. Answers will vary. One possible answer: Make a list of all the integral factors of the constant term and a list of all the integral factors of the coefficient of the term of highest degree. Then list all the distinct rational numbers formed by using a number from the first list for the numerator and a number from the second list for the denominator. **17.** $\pm 2\sqrt{3}, \pm\sqrt{2}$
19. $2 \pm\sqrt{5}, \pm 2\sqrt{2}$

Chapter 10

Written Exercises, page 342

1. Sum: -9, product: -22 **3.** Sum: 4,
product: 0 **5.** Sum: 2, product: $\frac{1}{2}$
7. Sum: $\frac{1}{2}$, product: -3 **9.** Sum: 0,
product: $-\frac{5}{2}$ **11.** $x^2 + 2x - 15 = 0$
13. $18x^2 + 3x - 1 = 0$ **15.** $5x^2 - 11x - 12 = 0$ **17.** $9x^2 - 12x + 4 = 0$
19. $x^2 + 16 = 0$ **21.** $x^2 + 12x + 16 = 0$
23. $4x^2 - 20x + 15 = 0$ **25.** $8x^2 + 24x + 23 = 0$ **27.** Sum: 12, product: 6
29. Sum: -3, product: -12 **31.** Sum: $3\sqrt{2}$,
product: -20 **33.** 9, 1

Written Exercises, page 346

1. 0, 1 rational **3.** 64, 2 rational **5.** 1, 2
rational **7.** 48, 2 irrational **9.** 100, 2
rational **11.** ± 4 **13.** $-14, 2$ **15.** $-3 < k < 3$ **17.** $-2 < k < 2$ **19.** Two irrational
21. One rational **23.** One irrational
25. One nonreal

Written Exercises, page 364

1.

The line passes through $(3, 0)$.

3. Between −5 and −4, between 0 and 1

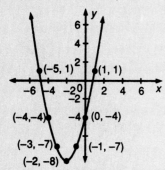

−4, between −2 and −1, between 1 and 2, 3

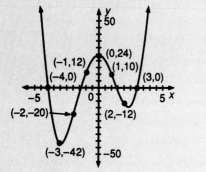

7. Between −4 and −3, between −2 and −1

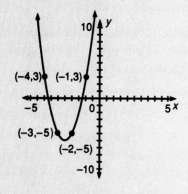

9. −2, between 0 and 1, between 2 and 3

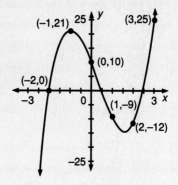

11. −4, −1, between 1 and 2, 4

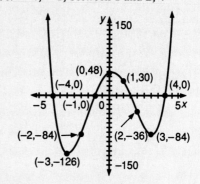

13. $\dfrac{1 \pm \sqrt{43}}{3}$; −1.9, 2.5   **15.** $0 < c < 5$

Written Exercises, pages 368–369

1. $-5i, \frac{1}{3}$ **3.** $2 + 3i, 2, -1$ **5.** 2 pos, 1 neg; 0 pos, 1 neg   **7.** 3 pos, 1 neg; 1 pos, 1 neg   **9.** 3 pos, 2 neg; 3 pos, 0 neg; 1 pos, 2 neg; 1 pos, 0 neg   **11.** 2 pos, 2 neg; 2 pos, 0 neg; 0 pos, 2 neg; 0 pos, 0 neg   **13.** $\pm 2i, 5$

Chapter 11

Written Exercises, pages 380–381

1. 4   **3.** 5   **5.** $2\sqrt{5}$ **7.** $4\sqrt{2}$ **9.** $2\sqrt{13}$

11. $\sqrt{41}$ **13.** $2\sqrt{26}$ **15.** $\dfrac{\sqrt{113}}{4}$ **17.** Yes

19. Yes   **21.** (7,4)   **23.** (−3, −5)

25. $\left(\dfrac{-11}{2}, \dfrac{-5}{2} \right)$ **27.** (−13, −3)

29. (−7, −6)   **31.** (−4, −7)   **33.** They do.

35. $\sqrt{(2c - a)^2 + (2d - b)^2}$ **37.** $\sqrt{a^2 + b^2} +$
$\sqrt{c^2 + d^2} + \sqrt{(c - a)^2 + (d - b)^2}$

Written Exercises, pages 384–385

1. Yes **3.** No **5.** $2x + y = 13$
7. $2x + 16y = 53$ **9.** Yes **11.** $2x + y = 10$
13. $3x - y = 10$ **15.** Yes **17.** $ABCD$ is a parallelogram and a rhombus. **19.** Slope of $\overline{AB}$: $\frac{1}{3}$; midpoint of $\overline{AB}$: (5,5); perpendicular bisector of $\overline{AB}$: $y = -3x + 20$; For the point $P(0,20)$: $PA = \sqrt{(2 - 0)^2 + (4 - 20)^2} = \sqrt{260}$; $PB = \sqrt{(8 - 0)^2 + (6 - 20)^2} = \sqrt{260}$. Thus, $PA = PB$. **21.** Consider the generalized equilateral triangle with vertices $A(0,0)$, $B(2a,0)$, and $C(a, a\sqrt{3})$, and the median from C to $\overline{AB}$. Midpoint M of $\overline{AB}$: $(a,0)$; Slope of $\overline{AB}$: 0(horizontal); Slope of $\overline{CM}$: undefined (vertical); So, $\overline{CM}$ is perpendicular to $\overline{AB}$, and is thus an altitude of triangle ABC.

5. $(0,-7)$;

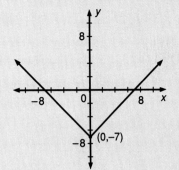

7. $(8,0)$;

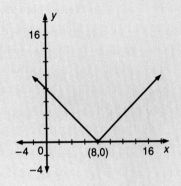

Written Exercises, page 389

1. $(0,0)$;

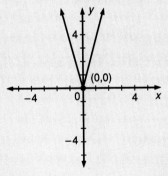

9. $(1,1)$, $x = 1$;

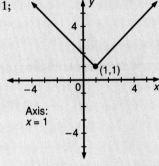

Axis: $x = 1$

3. $(0,0)$;

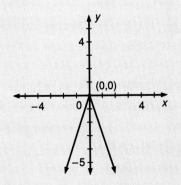

11. $(-2,2)$, $x = -2$;

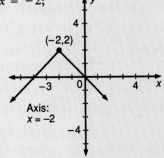

Axis: $x = -2$

13. $(-5, -4),$
   $x = -5;$

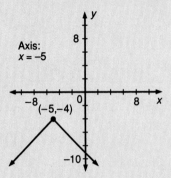

15. $(2, -3),$
   $x = 2;$

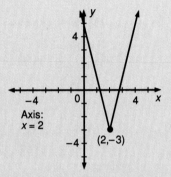

17. $(-6, -8),$
   $x = -6;$

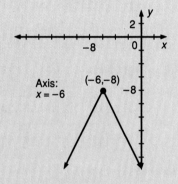

19. $(4, -5); x = 4$ **21.** $(4, -6); x = 4$

23. $\left(\frac{3}{2}, 6\right); x = \frac{3}{2}$

25.

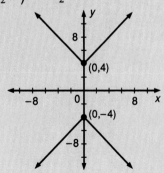

Written Exercises, page 394

1.

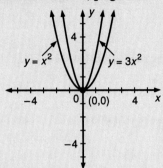

3.

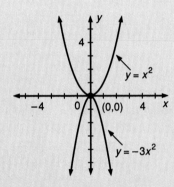

5.

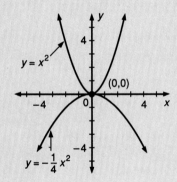

7. $(1, 0);$
   $x = 1$

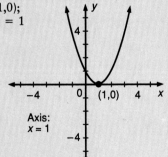

9. $(2,1)$; $x = 2$

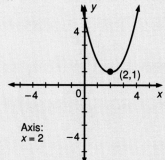

Axis:
$x = 2$

11. $(2,-5)$; $x = 2$

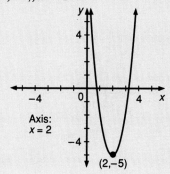

Axis:
$x = 2$

$(2,-5)$

13. $(8,0)$; $x = 8$

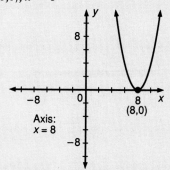

$(8,0)$

Axis:
$x = 8$

15. $(0,-7)$;
$x = 0$

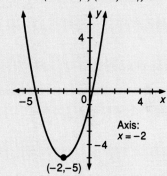

Axis:
$x = 0$

$(0,-7)$

17. $(-2,3)$; $x = -2$

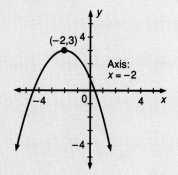

$(-2,3)$

Axis:
$x = -2$

19. $y - 1 = -\frac{1}{4}(x + 4)^2$ **21.** $y + 1 = (x - 1)^2$ **23.** $\{p \mid p > \frac{5}{2}\}$

25. $\{(h,k) \mid h < 0 \text{ and } k > -6\}$

Written Exercises, page 396

1. $y + 5 = (x + 2)^2$; $(-2,-5)$; $x = -2$

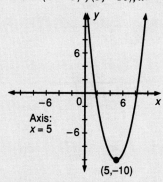

Axis:
$x = -2$

$(-2,-5)$

3. $y + 10 = (x - 5)^2$; $(5,-10)$; $x = 5$

Axis:
$x = 5$

$(5,-10)$

5. $y - 4 = (x + 4)^2$; $(-4,4)$; $x = -4$

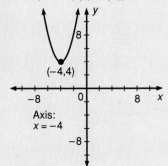

Axis:
$x = -4$

7. $y - 2 = (x + 5)^2$; $(-5,2)$; $x = -5$

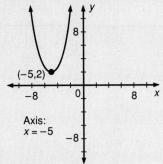

Axis:
$x = -5$

9. $y + 1 = (x - 1)^2$; $(1,-1)$; $x = 1$

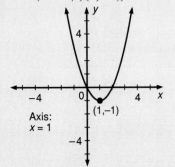

Axis:
$x = 1$

11. $y + 16 = (x - 4)^2$; $(4,-16)$; $x = 4$

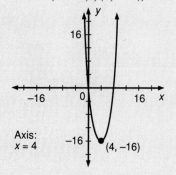

Axis:
$x = 4$

13. $y + 2 = 2(x - 2)^2$; $(2,-2)$; $x = 2$

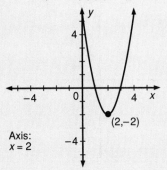

Axis:
$x = 2$

15. $y + 3 = 4(x - 1)^2$; $(1,-3)$; $x = 1$

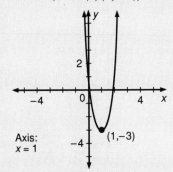

Axis:
$x = 1$

17. $y - 8 = -2(x + 1)^2$; $(-1,8)$; $x = -1$

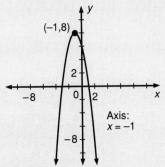

Axis:
$x = -1$

19. $y - 5 = -(x + 3)^2$; $(-3,5)$; $x = -3$

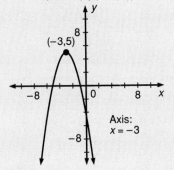

Axis:
$x = -3$

21. $y - 1 = -(x + 1)^2; (-1,1); x = -1$

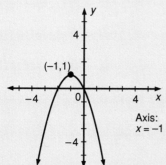

Axis:
$x = -1$

23.

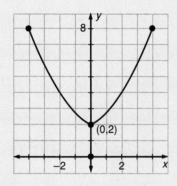

25. $b = -6, c = 13$

Written Exercises, page 400

1. -10 **3.** -8 **5.** $-\frac{167}{24}$ **7.** \$7, \$200

9. 355 **11.** 40 ft by 80 ft   **13.** 10, 10

15. 162 in^2

17.
$$y = ax^2 + bx + c$$
$$y - c = ax^2 + bx$$
$$y - c = a\left(x^2 + \frac{b}{a}x\right)$$
$$y - c + \frac{b^2}{4a} = a\left(x^2 + \frac{b}{a}x + \frac{b^2}{4a^2}\right)$$
$$y - \frac{(4ac - b^2)}{4a} = a\left[x - \left(\frac{-b}{2a}\right)\right]^2$$

Written Exercises, page 402

1. Even   **3.** Even   **5.** Odd   **7.** Odd

9. Even   **11.** Neither   **13.** ± 3

Written Exercises, pages 411–412

1. $x^2 + y^2 = 169$ **3.** $(x - 2)^2 + (y - 3)^2 = 45$ **5.** $(x + 2)^2 + (y - 8)^2 = 81$

7. $(x - 1)^2 + (y + 1)^2 = \frac{17}{4}$

9.

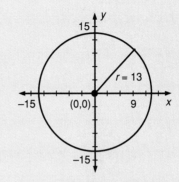

11.

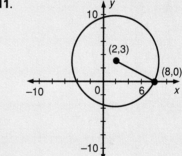

13.

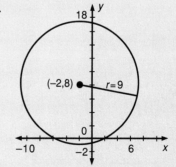

15.

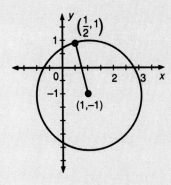

17. (0,0); 2

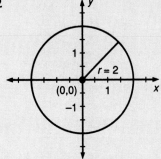

19. (0,0); 8

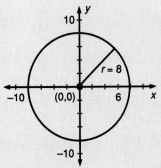

21. (0,0); $2\sqrt{2}$

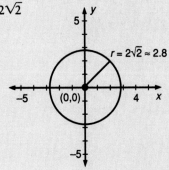

23. (0,0); $\sqrt{21}$

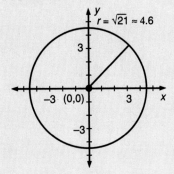

25. $(x - 3)^2 + (y - 5)^2 = 36$; (3,5); 6

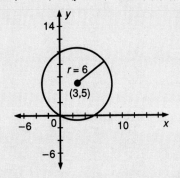

27. $(x + 3)^2 + (y - 7)^2 = 100$; $(-3,7)$; 10

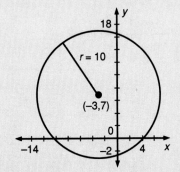

29. $(x + 1)^2 + (y - 2)^2 = 12$ **31.** By definition, $CP = r$, where $P(x,y)$ is a point on the circle, C is its center, and r is its radius. C is given as (h,k). $CP = \sqrt{(x - h)^2 + (y - k)^2} = r$. So, $(x - h)^2 + (y - k)^2 = r^2$.
33. $(x - 2)^2 + (y - 3)^2 = 11$ **35.** $x^2 + (y - 4\sqrt{2})^2 = 16$ **37.** $3x - 4y = -50$

Written Exercises, pages 416–417

1.

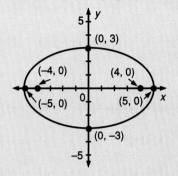

3.

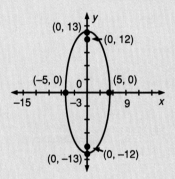

5.

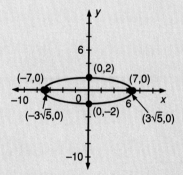

7.

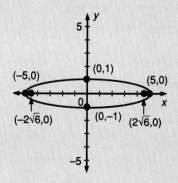

9.

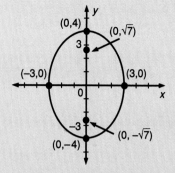

11.

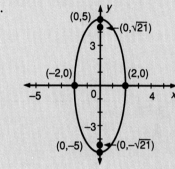

13. $\dfrac{x^2}{9} + y^2 = 1$ **15.** $\dfrac{y^2}{25} + \dfrac{x^2}{9} = 1$

17. $\dfrac{x^2}{5} + \dfrac{y^2}{3} = 1$ **19.** $\dfrac{x^2}{64} + \dfrac{y^2}{55} = 1$

21. $\dfrac{y^2}{64} + \dfrac{x^2}{55} = 1$ **23.** $\dfrac{y^2}{25} + \dfrac{x^2}{16} = 1$

25. $\dfrac{y^2}{25} + \dfrac{x^2}{16} = 1$ **27.** 17π **29.** Let $d_1 + d_2 = 2a$. Then $PF_1 + PF_2 = 2a$, and the distance formula gives the following equation.

$\sqrt{[y - (-f)]^2 + (x - 0)^2} +$
$\sqrt{(y - f)^2 + (x - 0)^2} = 2a; \sqrt{(y + f)^2 + x^2} +$
$\sqrt{(y - f)^2 + x^2} = 2a; \sqrt{(y + f)^2 + x^2} = 2a -$
$\sqrt{(y - f)^2 + x^2}; y^2 + 2fy + f^2 + x^2 = 4a^2 -$
$4a\sqrt{y^2 - 2fy + f^2 + x^2} + y^2 - 2fy + f^2 +$
$x^2; 4fy - 4a^2 = -4a\sqrt{y^2 - 2fy + f^2 + x^2};$
$fy - a^2 = -a\sqrt{y^2 - 2fy + f^2 + x^2}; f^2y^2 -$
$2a^2fy + a^4 = a^2y^2 - 2a^2fy + a^2f^2 + a^2x^2;$
$a^4 - a^2f^2 - a^2y^2 + f^2y^2 = a^2x^2; a^2(a^2 - f^2) -$
$y^2(a^2 - f^2) = a^2x^2; (a^2 - f^2)(a^2 - y^2) = a^2x^2;$
$b^2(a^2 - y^2) = a^2x^2; b^2a^2 - b^2y^2 = a^2x^2; b^2y^2 +$
$a^2x^2 = b^2a^2; \dfrac{y^2}{a^2} + \dfrac{x^2}{b^2} = 1$

Written Exercises, pages 421–422

1. Vertices: $(\pm 2,0)$; Foci: $(\pm 2\sqrt{5},0)$; $y = \pm 2x$

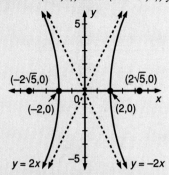

3. Vertices: $(\pm 6,0)$; Foci: $(\pm 2\sqrt{13},0)$; $y = \pm\frac{2}{3}x$

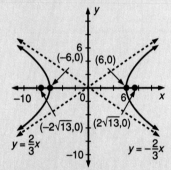

5. Vertices: $(0,\pm 8)$; Foci: $(0\pm\sqrt{113})$; $y = \pm\frac{8}{7}x$

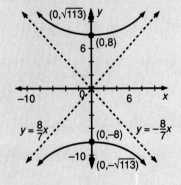

7. Vertices: $(\pm 1,0)$; Foci: $(\pm\frac{\sqrt{17}}{4},0)$; $y = \pm\frac{1}{4}x$

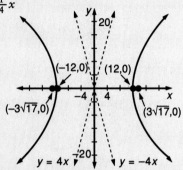

9. Vertices: $(\pm 4,0)$; Foci: $(\pm 2\sqrt{5},0)$; $y = \pm\frac{1}{2}x$

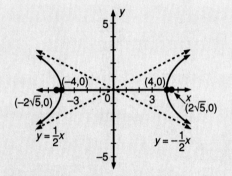

11. Vertices: $(\pm 6,0)$; Foci: $(\pm\sqrt{61},0)$; $y = \pm\frac{5}{6}x$

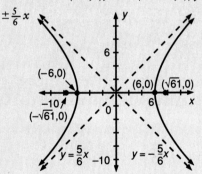

13. $\dfrac{y^2}{25} - \dfrac{x^2}{25} = 1$ **15.** $\dfrac{x^2}{16} - \dfrac{y^2}{9} = 1$

17. $\dfrac{y^2}{25} - \dfrac{x^2}{24} = 1$ **19.** $\dfrac{x^2}{100} - \dfrac{y^2}{16} = 1$

21. $\dfrac{x^2}{25} - \dfrac{y^2}{144} = 1$

23. Let $d_1 - d_2 = 2a$. Then $PF_1 - PF_2 = 2a_1$ and the distance formula gives the following equation. $\sqrt{[y - (-f)]^2 + (x - 0)^2} - \sqrt{(y - f)^2 + (x - 0)^2} = 2a$; $\sqrt{(y + f)^2 + x^2} - \sqrt{(y - f)^2 + x^2} = 2a$; $\sqrt{(y + f)^2 + x^2} = 2a + \sqrt{(y - f)^2 + x^2}$; $y^2 + 2fy + f^2 + x^2 = 4a^2 + 4a\sqrt{y^2 - 2fy + f^2 + x^2} + y^2 + x^2 - 2fy + f^2 + x^2$; $4fy - 4a^2 = 4a\sqrt{y^2 - 2fy + f^2 + x^2}$; $fy - a^2 = a\sqrt{y^2 - 2fy + f^2 + x^2}$; $f^2y^2 - 2a^2fy + a^4 = a^2y^2 - 2a^2fy + a^2f^2 + a^2x^2$; $f^2y^2 - a^2y^2 - a^2f^2 + a^4 = a^2x^2$; $y^2(f^2 - a^2) - a^2(f^2 - a^2) = a^2x^2$; $(f^2 - a^2)(y^2 - a^2) = a^2x^2$; $b^2(y^2 - a^2) = a^2x^2$; $b^2y^2 - a^2x^2 = b^2a^2$; $\dfrac{x^2}{a^2} - \dfrac{y^2}{b^2} = 1$

Written Exercises, page 425

1. Center: $(3,2)$; vertices: $(-7,2)$, $(13,2)$, $(3,-4)$, $(3,8)$; foci: $(-5,2)$, $(11,2)$

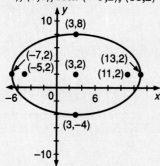

3. Center: $(3,2)$; vertices: $(-1,2)$, $(7,2)$; foci: $(-2,2)$, $(8,2)$

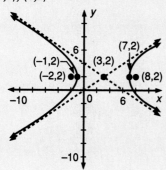

5. Center: $(-5,1)$; vertices: $(-11,1)$, $(1,1)$; foci: $(-15,1)$, $(5,1)$

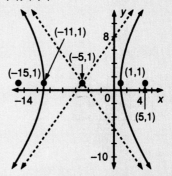

7. Center: $(3,-1)$; vertices: $(1,-1)$, $(5,-1)$, $(3,-5)$, $(3,3)$; foci: $(3,-1 - 2\sqrt{3})$, $(3,-1 + 2\sqrt{3})$

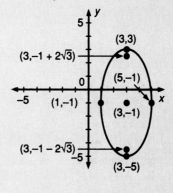

9. Center: $(1,-2)$; vertices: $(0,-2)$, $(2,-2)$; foci: $(1 - \sqrt{2},-2)$, $(1 + \sqrt{2},-2)$

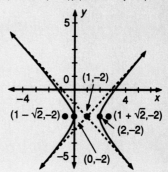

11. Center: $(-3,2)$; vertices: $(-4,2)$, $(-2,2)$; foci: $(-3-\sqrt{5},2)$, $(-3+\sqrt{5},2)$

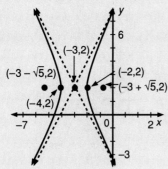

13. Center: $(3,-1)$; vertices: $(0,-1)$, $(6,-1)$, $(3,-3)$, $(3,1)$; foci: $(3-\sqrt{5},-1)$, $(3+\sqrt{5},-1)$

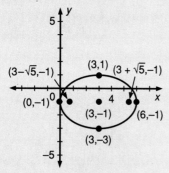

15. $\dfrac{(y-1)^2}{25} + \dfrac{(x-2)^2}{9} = 1$

17. $\dfrac{(x+1)^2}{4} - \dfrac{(y-2)^2}{12} = 1$

Written Exercises, pages 428–429

1.

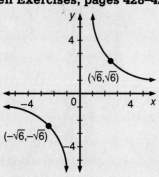

3.

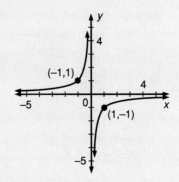

5.

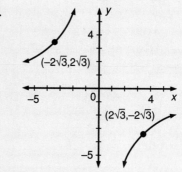

7. No   **9.** Yes; 5   **11.** Yes; 3   **13.** 9   **15.** $\frac{1}{2}$
17. 24 amps   **19.** 8   **21.** 12.5 ft-candles
23. It is halved.   **25.** 169.7 lb

Written Exercises, pages 433–434

1. 12   **3.** ± 3 **5.** ± 8 **7.** 440 cm^3
9. Multiplies y by 8   **11.** Halves y **13.** 12 ft

15. $F = \dfrac{km_1m_2}{d^2}$, 4.19×10^8 m from the earth,

or 4.10×10^7 m from the moon

Written Exercises, page 439

1. $y^2 = -32x$ **3.** $(y-8)^2 = -6\left(x - \dfrac{9}{2}\right)$

5. $(x+2)^2 = 24(y-3)$ **7.** $(x-3)^2 = 24(y-2)$

9. Vertex: $(7,2)$; focus: $(7,4)$; directrix: $y = 0$; axis: $x = 7$

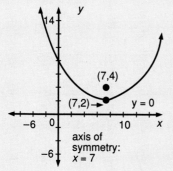

15. Vertex: $(3,2)$; focus: $(3,4)$; directrix: $y = 0$; axis: $x = 3$

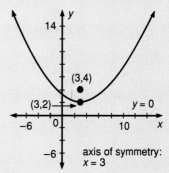

11. Vertex: $(6,-1)$; focus: $\left(6, \frac{3}{2}\right)$; directrix: $y = -\frac{7}{2}$; axis: $x = 6$

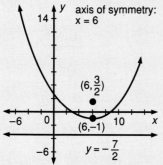

17. Vertex: $(1,-5)$; focus: $\left(1, -\frac{9}{2}\right)$; directrix: $y = -\frac{11}{2}$; axis: $x = 1$

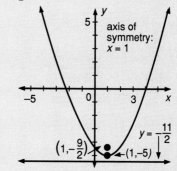

13. Vertex: $(5,6)$; focus: $(3,6)$; directrix: $x = 7$; axis: $y = 6$

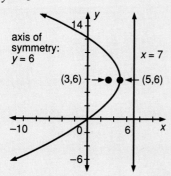

19. Vertex: $(-1,5)$; focus: $\left(-\frac{3}{2}, 5\right)$; directrix: $x = -\frac{1}{2}$; axis: $y = 5$

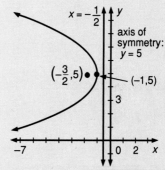

21. $y^2 = 8x$ **23.** $(x - h)^2 = 4p(y - k)$;

$y - k = a(x - h)^2$, or $(x - h)^2 = \frac{1}{a}(y - k)$,

and letting $a = \frac{1}{4p}$, $(x - h)^2 = 4p(y - k)$.

Written Exercises, page 442

1. Hyperbola **3.** Parabola **5.** Ellipse
7. Hyperbola **9.** Ellipse **11.** Ellipse
13. Circle **15.** Circle **17.** Parabola
19. Hyperbola **21.** Answers will vary. One
possible answer: The equation of a parabola has
one first-degree and one second-degree term.
No combination of real-number coefficients can
make either second-degree term a first-degree
term, so the equation cannot represent a
parabola. **23.** 3, 4

Written Exercises, pages 450–451

1. $(-3,0)$, $(3,0)$

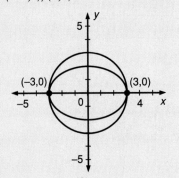

Written Exercises, pages 445–446

1. $(0,2),(2,0)$

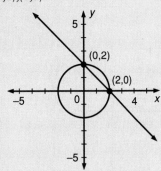

3. $(-3,-3),(3,3)$

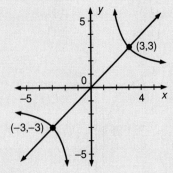

5. $(2,9), (-1,-3)$ **7.** $(3,1), (-8,12)$
9. $(3,0), \left(\frac{9}{5}, \frac{8}{5}\right)$ **11.** $(-3,5)$ **13.** No real
solutions **15.** No real solutions **17.** 8 ft, 3 ft
19. $\frac{5}{8}$ **21.** $\left(\frac{4}{5}, \frac{9}{5}\right), \left(\frac{16}{5}, \frac{1}{5}\right)$ **23.** $\left(\frac{1}{3}, \frac{1}{2}\right),$
$(2,-2)$

3. No real solutions

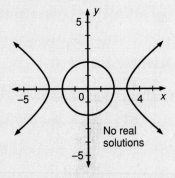

No real
solutions

5. $(\pm 2,0)$ **7.** $(-3,-2), (3,2)$ **9.** $(-2,-1),$
$(2,1)$ **11.** $(\pm 1,0)$ **13.** $(\pm 6.9, \pm 6.2)$
15. $(-3,-1), (3,1)$ **17.** 12 ft by 5 ft
19. $6 + 2\sqrt{6}$ in. by $6 - 2\sqrt{6}$ in.

21.

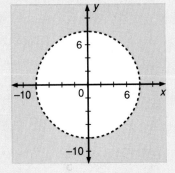

832

23.

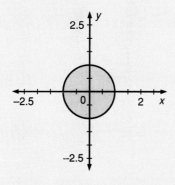

25. $(2,1), (-2,-1), (\sqrt{3}, \sqrt{3}), (-\sqrt{3}, -\sqrt{3})$

27. $\dfrac{x^2}{8} + \dfrac{y^2}{4} = 1$

Chapter 13

Written Exercises, pages 462–463

1. $\{(4,1), (4,-3), (0,2)\}$; no   **3.** $y = \dfrac{1}{4}x$; yes

5. $x = 5$; no   **7.** $y = \dfrac{3}{2}x$; yes

9. $y = \dfrac{1}{2}x + 3$; yes

11.

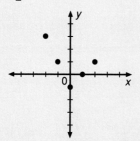

13.

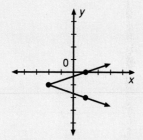

15. Yes, no   **17.** No, no   **19.** $y = \dfrac{4}{3}x + 4$;

yes   **21.** $y = -\dfrac{3}{2}x + 3$; yes   **23.** $y = \pm\sqrt{\dfrac{x}{2}}$;
no   **25.** $y = \dfrac{3}{2}x + 3$; yes   **27.** $y = -\dfrac{5}{2}x + 5$; yes   **29.** $y = \pm\sqrt{\dfrac{3-x}{2}}$; no

31. $f^{-1}(x) = \dfrac{x+3}{3}$, or $\dfrac{1}{3}x + 1$; $f(12) = 33$;
$f^{-1}(12) = 5$; $f(f^{-1}(12)) = 12$; $f^{-1}(f(12)) = 12$
33. b

Written Exercises, page 466

1. D: {real numbers}; R: $\{y|y > 0\}$;
increasing; (0,1)

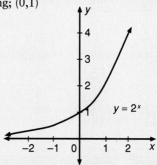

3. D: {real numbers}; R: $\{y|y > 0\}$;
decreasing; (0,1)

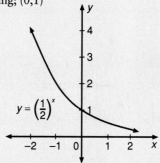

5. D: {real numbers}; R: $\{y|y > 0\}$;
increasing; (0,3)

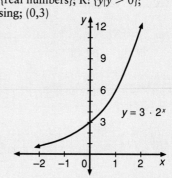

7. D: {real numbers}; R: $\{y | y > 0\}$;
increasing; (0,3)

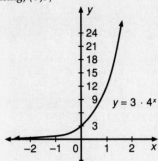

9. D: {real numbers}; R: $\{y | y < 0\}$;
decreasing; (0, −2)

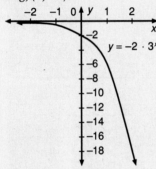

11. D: {real numbers}; R: $\{y | y > 6\}$;
increasing; (0,7)

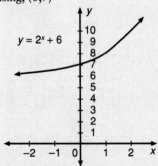

13. 100.9   15. 9.7   17. The graph of
$y = \left(\frac{1}{2}\right)^x$ is the reflection about the y–axis of
the graph of $y = 2^x$.   19. The graph of
$y = 4 \cdot 2^x$ rises more steeply than that of
$y = 2^x$.   21. The graph of $y = 2^{x-1}$ is that of
$y = 2^x$ shifted 1 unit to the right.

23.

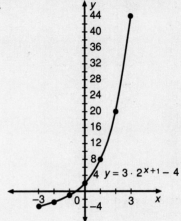

25. When $a = 1$, $y = a^x$ is just the constant
function $y = 1$.   27. The line $y = -c$

Written Exercises, page 468

1. 2   3. 1   5. −4   7. 0   9. $\frac{1}{2}$ 11. 10

13. 5   15. $\frac{1}{3}$ 17. 2   19. $\frac{3}{5}$ 21. $\frac{3}{4}$

23. −6   25. 2

Written Exercises, page 472

1. $y = \log_2 x$ 3. $y = 8^x$
5. D: $x > 0$; R: reals; incr.; (1,0)

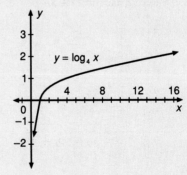

7. D: $x > 0$; R: reals; decr.; $(1,0)$

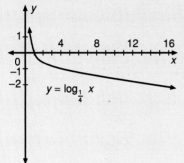

$y = \log_{\frac{1}{4}} x$

9. D: $x > 0$; R: reals; decr.; $(1,0)$

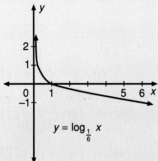

$y = \log_{\frac{1}{6}} x$

11. D: $x > 0$; R: reals; incr.; $(1,0)$

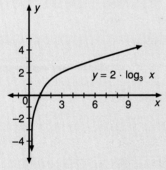

$y = 2 \cdot \log_3 x$

13. D: $x > 0$; R: reals; incr.; $\left(\frac{1}{8}, 0\right)$

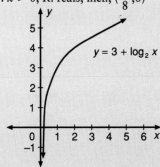

$y = 3 + \log_2 x$

15. D: $x > 0$; R: reals; decr.; $(16,0)$

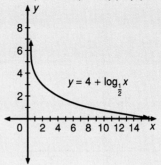

$y = 4 + \log_{\frac{1}{2}} x$

17. g: $y = \log_b x$ and g^{-1}: $x = \log_b y$. So, g^{-1}: $y = b^x$.

19.

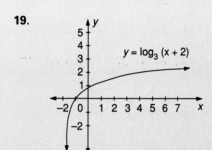

$y = \log_3 (x + 2)$

Written Exercises, page 476

1. $\log_2 6 + \log_2 c$ **3.** $\log_4 n - \log_4 9$
5. $\log_b 3 + \log_b t + \log_b w + \log_b z$
7. $\log_2 5 + \log_2 m - \log_2 c - \log_2 d$
9. $\log_2 30c$ **11.** $\log_4 \dfrac{15}{a}$ **13.** $\log_5 \dfrac{4c^2}{d}$
15. 3 **17.** 2 **19.** 2 **21.** 5 **23.** ± 5
25. 3 **27.** $\log_4 \dfrac{5c^2 - 15c}{2c - 8}$ **29.** 5

Written Exercises, pages 479–480

1. $2(\log_3 m + \log_3 n)$ **3.** $\dfrac{1}{3}(\log_5 4 - \log_5 h)$
5. $\log_2 x^7$ **7.** $\log 4 \sqrt[3]{7t}$ **9.** $\log_b \dfrac{m^5}{n^2}$
11. $\log_5 \sqrt{c^2 - 2c}$ **13.** 6 **15.** 10 **17.** 3
19. 8 **21.** 7 **23.** 7 **25.** $2(\log_5 x + 3\log_5 y - 2\log_5 z)$ **27.** $\log_3 \dfrac{6\sqrt{m}}{n^2}$

29. $\log_b \dfrac{8m^3n^6}{p^3}$ **31.** 13, 10 **33.** 0

Written Exercises, pages 484–485

1. 4.2718 **3.** 285 **5.** -2.2262
7. 7.3711 **9.** 70,100 **11.** -3 **13.** 0.1
15. -1 **17.** 1 **19.** $\dfrac{1}{2}$ **21.** $\dfrac{2}{3}$ **23.** 2.5
25. 1×10^{-5} mole/l **27.** 1.6×10^{-5} mole/l
29. 1,111.1 **31.** 28

Written Exercises, page 489

1. 1.23 **3.** 1.80 **5.** 0.712 **7.** 3.47
9. \$886.47 **11.** \$29,371.62 **13.** 1.28×10^5
15. 2.64×10^{17} **17.** 4.92 **19.** -0.678
21. 7.40 **23.** The graph of $y = 3 + \log_2 x$ is that of $y = \log_2 x$ shifted up 3 units.

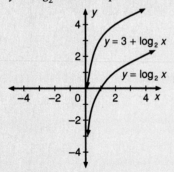

25. The graph of $y = \log_2 (x + 3)$ is that of $y = \log_2 x$ shifted left 3 units.

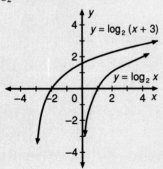

Chapter 14

Written Exercises, pages 501–502

1. 720 **3.** 40,320 **5.** 60 **7.** 26,000
9. 120 **11.** 24 **13.** 6,720 **15.** 362,880
17. 24 **19.** $(n + 3)!$

Written Exercises, pages 505–506

1. 24 **3.** 72 **5.** 105 **7.** 2,880 **9.** 138
11. 45 **13.** 48 **15.** 34,560 **17.** $\dfrac{n!}{(n - r)!}$

$$= \frac{n(n - 1)(n - 2) \cdots (n - r + 1)(n - r)!}{(n - r)!}$$

$$= n(n - 1)(n - 2) \cdots (n - r + 1) = {}_nP_r$$

19. 151,200 **21.** $n(n - 1)(n - 2)(n - 3)$

Written Exercises, page 508

1. 1,680 **3.** 364 **5.** 132 **7.** 20
9. 1,680 **11.** 90 **13.** 3,780 **15.** 138,600
17. $n^2 + 5n + 6$ **19.** $n^2 - 3n + 2$

Written Exercises, page 510

1. 3,628,800 **3.** 5,040 **5.** 144 **7.** 3
9. 72 **11.** 6

Written Exercises, page 513

1. 36 **3.** 455 **5.** 1 **7.** 66 **9.** 18,564
11. 2,598,960 **13.** 66 **15.** 12 **17.** 5
19. (9,6), (6,9), (9,3), (6,0) **21.** (8,2), (8,6), (2,8), (2,0) **23.** $\dbinom{n}{n - r} =$

$$\frac{n!}{(n - r)! \, [n - (n - r)]!} = \frac{n!}{(n - r)! \, r!} =$$

$$\frac{n!}{r!(n - r)!} = \dbinom{n}{r}$$

Written Exercises, pages 515–516

1. 6,720 **3.** 73 **5.** 10 **7.** 61,425
9. 35,035 **11.** 10 **13.** 19 **15.** 210
17. 6,300 **19.** 2,982 **21.** 1 **23.** 1,960
25. $1 + 5 + 10 + 10 + 5 + 1 = 32 = 2^5$

27. $\binom{n}{r} + \binom{n}{r+1} = \dfrac{n!}{r!(n-r)!} +$

$\dfrac{n!}{(r+1)!(n-r-1)!} = \dfrac{(r+1)n!}{(r+1)r!(n-r)!} +$

$\dfrac{n!(n-r)}{(r+1)!(n-r)(n-r-1)!} =$

$\dfrac{n!(r+1+n-r)}{(r+1)!(n-r)!} = \dfrac{(n+1)n!}{(r+1)!(n-r)!} =$

$\dfrac{(n+1)!}{(r+1)!(n-r)!} = \dfrac{(n+1)!}{(r+1)![n+1-(r+1)]!}$

$= \binom{n+1}{r+1}$

Written Exercises, page 520

1. $\dfrac{1}{8}$ **3.** $\dfrac{1}{2}$ **5.** $\dfrac{3}{4}$ **7.** $\dfrac{3}{13}$ **9.** $\dfrac{1}{4}$ **11.** $\dfrac{3}{4}$

13. 3 to 1 **15.** 3 to 1 **17.** $\dfrac{1}{8}$ **19.** $\dfrac{3}{8}$

21. $\dfrac{1}{8}$ **23.** $\dfrac{1}{4}$ **25.** $\dfrac{3}{8}$ **27.** $\dfrac{7}{8}$ **29.** 7 to 2

Written Exercises, page 524

1. $\dfrac{1}{3}$ **3.** 0 **5.** $\dfrac{2}{3}$ **7.** $\dfrac{1}{6}$ **9.** $\dfrac{5}{18}$ **11.** $\dfrac{5}{18}$

13. 1 to 5 **15.** 1 to 5 **17.** $\dfrac{7}{13}$ **19.** $\dfrac{4}{13}$

21. $\dfrac{7}{26}$ **23.** $\dfrac{3}{4}$ **25.** $\dfrac{2}{13}$ **27.** $\dfrac{3}{26}$ **29.** $\dfrac{1}{13}$

31. $\dfrac{7}{15}$ **33.** $\dfrac{7}{15}$

Written Exercises, pages 527–528

1. $\dfrac{2}{9}$ **3.** $\dfrac{4}{9}$ **5.** $\dfrac{5}{72}$ **7.** $\dfrac{1}{6}$ **9.** $\dfrac{2}{3}$ **11.** $\dfrac{1}{2}$

13. $\dfrac{25}{102}$ **15.** $\dfrac{13}{204}$ **17.** $\dfrac{13}{102}$ **19.** $\dfrac{32}{221}$

21. $\dfrac{10}{17}$ **23.** $\dfrac{1}{715}$ **25.** $\dfrac{28}{143}$ **27.** $\dfrac{3}{13}$

29. $\dfrac{33}{66,640}$ **31.** a: $\dfrac{15}{22}$; b: $\dfrac{1}{66}$

Written Exercises, page 532

1. 45, 44, 54 **3.** 21.5, 18.5, 16 **5.** 85.05, 85, 85 **7.** 55, 60, 60 and 30 **9.** 14, 3.74 **11.** 199.78, 14.13 **13.** 61 **15.** 78 lb

Written Exercises, page 535

1. $\dfrac{5}{32}$ **3.** $\dfrac{5}{8}$ **5.** $\dfrac{1}{2}$ **7.** $\dfrac{1}{2}$ **9.** 0.841

11. About 17 **13.** 66 in., 68 in., 71 in.

Chapter 15

Written Exercises, pages 547–548

1. 12 **3.** 2 **5.** $1\dfrac{3}{4}$ **7.** $-4, -7, -10$

9. $\dfrac{7}{12}, \dfrac{2}{3}, \dfrac{3}{4}$ **11.** $11, 6, 1, -4$ **13.** $4, 6.6,$ $9.2, 11.8$ **15.** 14 oz, $10\dfrac{1}{2}$ oz, 7 oz, $3\dfrac{1}{2}$ oz

17. -111 **19.** 41 **21.** $-94\dfrac{3}{5}$ **23.** 85 m

25. \$24,700 **27.** Not arithmetic **29.** $25 + 18\sqrt{3}$ **31.** 23rd **33.** 8

Written Exercises, page 552

1. 21, 28 **3.** 8.5, 8, 7.5 **5.** 0.5, 4, 7.5, 11

7. 59 **9.** 10.5 **11.** 5, 27 **13.** $\dfrac{1}{21}$ **15.** $\dfrac{3}{4}$

17. $\dfrac{4}{15}, \dfrac{4}{9}$ **19.** $\dfrac{1}{5}$ **21.** \$18,600; \$19,500; \$20,400; \$21,300; \$22,200; \$23,100; \$24,000; \$24,900 **23.** $8 + \sqrt{2}, 12 + 2\sqrt{2}, 16 + 3\sqrt{2}$

25. $2\sqrt{5}, 3\sqrt{5}$ **27.** $\dfrac{2xy}{x+y}$

Written Exercises, pages 556–557

1. 2,025 **3.** 889 **5.** 4,140 **7.** 814

9. 90 **11.** 25 **13.** $\displaystyle\sum_{k=1}^{4} 8k$ **15.** $\displaystyle\sum_{k=1}^{5} (2k+12)$

17. $\displaystyle\sum_{k=1}^{5} (2k+5)$ **19.** $\displaystyle\sum_{k=1}^{9} (19-4k)$

21. $\displaystyle\sum_{k=1}^{12} (8k+11)$ **23.** $2,418\sqrt{2}$ **25.** 307.8

27. 424 **29.** \$3220 **31.** 2

33. $\displaystyle\sum_{k=1}^{20} 5kx = 5x + 10x + 15x + \ldots + 100x$

$= 5x \cdot (1 + 2 + 3 + \cdots + 20) = 5x \cdot \displaystyle\sum_{k=1}^{20} k$

35. $\displaystyle\sum_{r=1}^{k} (2n-1) = 1 + 3 + 5 + \ldots +$

$(2k-1) = \dfrac{k}{2}(1 + 2k - 1) = \dfrac{k}{2} \cdot 2k = k^2$

Written Exercises, pages 560–561

1. 120, 240, 480 **3.** 6, -18, 54 **5.** 0.75, 1.5, 3, 6 **7.** -8, 2, $-\frac{1}{2}$, $\frac{1}{8}$ **9.** 128, -48, 18, $-6\frac{3}{4}$ **11.** $-2{,}560$ **13.** -0.0048
15. 128 **17.** $-3{,}600{,}000$ **19.** $-\frac{1}{27}$
21. $-36\sqrt{3}$ **23.** $128x^{11}$, $-512x^{14}$ **25.** $2i$, 2
27. $\frac{1}{8}$ ft **29.** 1, 4 **31.** 8

Written Exercises, pages 564–565

1. 6, 18, 54 or -6, 18, -54 **3.** 3, 1
5. -12, -36, -108 or 12, -36, 108 **7.** 1, -4 **9.** 18, 36, 72, 144 **11.** 2, $2\sqrt{2}$, 4 or -2, $2\sqrt{2}$, -4 **13.** 6 **15.** 20 **17.** 2, 18, or -2, -18 **19.** 16, 4 or -4, -16 **21.** 2,430 tons **23.** $-4\sqrt{3}$ **25.** $-5\sqrt{2}$ **27.** 0.04
29. 0.9 **31.** 8 **33.** 6 **35.** $3\sqrt{3}$
37. x, ___, y: $y = xr^2$, $r^2 = \dfrac{y}{x}$, $r = \dfrac{\sqrt{y}}{\sqrt{x}} \cdot \dfrac{\sqrt{x}}{\sqrt{x}}$
$= \dfrac{\sqrt{xy}}{x}$; $x \cdot \dfrac{\sqrt{xy}}{x} = \sqrt{xy}$

Written Exercises, page 569

1. 55,555 **3.** -726 **5.** $63\frac{3}{4}$ **7.** 44,444.4
9. $-1{,}023$ **11.** 33,333.33 **13.** 340
15. 4,004 **17.** 22,222.2 **19.** 254 **21.** $\frac{127}{64}$
23. -25 **25.** $-42\frac{5}{8}$ **27.** $319\frac{3}{8}$ in
29. $376\frac{5}{9}$ yd **31.** $2\frac{17}{24}$ **33.** $5\frac{3}{8}$
35. $\frac{14{,}701}{15{,}000}$

Written Exercises, page 574

1. Does not exist; $r = 2$ **3.** $-\frac{1}{2}$ **5.** $4\frac{4}{9}$
7. 32 **9.** 36 **11.** $5 + 0.5 + 0.05 + \ldots$, $\frac{50}{9}$
13. $8.1 + 0.081 + 0.00081 + \ldots$, $\frac{810}{99}$
15. $3.12 + 0.00312 + 0.00000312 + \ldots$, $\frac{3{,}120}{999}$ **17.** 480 cm **19.** 4, 6, 7, $7\frac{1}{2}$, $7\frac{3}{4}$, $7\frac{7}{8}$, $7\frac{15}{16}$, $7\frac{31}{32}$; converges to 8. **21.** 27, 18, 21, 20, $20\frac{1}{3}$, $20\frac{2}{9}$, $20\frac{7}{27}$, $20\frac{20}{81}$; converges to $20\frac{1}{4}$. **23.** $1 < x < 3$ **25.** $x < -1$ or $x > 1$

Written Exercises, page 577

1. $a^6 + 6a^5b + 15a^4b^2 + 20a^3b^3 + 15a^2b^4 + 6ab^5 + b^6$ **3.** $x^4 - 12x^3 + 54x^2 - 108x + 81$ **5.** $243c^5 + 810c^4d + 1{,}080c^3d^2 + 720c^2d^3 + 240cd^4 + 32d^5$ **7.** $x^{12} - 12x^{10}y + 60x^8y^2 - 160x^6y^3 + 240x^4y^4 - 192x^2y^5 + 64y^6$ **9.** $90c^2d^3$ **11.** $24x^4y^2$
13. $720c^9d^2$ **15.** $160x^9$ **17.** $243c^{15} + 810c^{12}d^2 + 1{,}080c^9d^4 + 720c^6d^6 + 240c^3d^8 + 32d^{10}$ **19.** $16p^4 - 32\sqrt{p}\,p^3 + 24p^3 - 8\sqrt{p}\,p^2 + p^2$ **21.** $x^{14} + 14x^{13} + 91x^{12} + 364x^{11} + \ldots$ **23.** $x^{20} - 10x^{18}y^3 + 45x^{16}y^6 - 120x^{14}y^9 + \ldots$

25. $\sum_{k=0}^{4} \binom{4}{k} 2^{4-k}(0.1)^k$, 19.4481

27. $\sum_{k=0}^{3} \binom{3}{k}(0.5)^k$, 3.375

Chapter 16

Written Exercises, page 588

1. $\begin{bmatrix} 6 & 11 & -9 \\ 6 & -6 & 13 \\ 4 & 6 & -10 \end{bmatrix}$ **3.** $\begin{bmatrix} -2 & 3 \\ -22 & 13 \end{bmatrix}$

5. $A + (-A) = \begin{bmatrix} 5 + (-5) & -3 + 3 \\ 7 + (-7) & 2 + (-2) \end{bmatrix}$

$= \begin{bmatrix} 0 & 0 \\ 0 & 0 \end{bmatrix} = 0_{2\times2}$

7. $A + (C + B) = \begin{bmatrix} 5 & -3 \\ 7 & 2 \end{bmatrix}$

$+ \begin{bmatrix} -1 + (-9) & (-5) + 8 \\ -7 + (-6) & (-3) + 4 \end{bmatrix}$

$= \begin{bmatrix} 5 + (-10) & -3 + 3 \\ 7 + (-13) & 2 + 1 \end{bmatrix} = \begin{bmatrix} -5 & 0 \\ -6 & 3 \end{bmatrix}$;

$(A + C) + B =$

$\begin{bmatrix} 5 + (-1) & -3 + (-5) \\ 7 + (-7) & 2 + (-3) \end{bmatrix} + \begin{bmatrix} -9 & 8 \\ -6 & 4 \end{bmatrix}$

$= \begin{bmatrix} 4 + (-9) & -8 + 8 \\ 0 + (-6) & -1 + 4 \end{bmatrix} = \begin{bmatrix} -5 & 0 \\ -6 & 3 \end{bmatrix}$;

Thus, $A + (C + B) = (A + C) + B$.

9. $B + (-B) = \begin{bmatrix} b_{11} + (-b_{11}) & b_{12} + (-b_{12}) \\ b_{21} + (-b_{21}) & b_{22} + (-b_{22}) \end{bmatrix}$

$= \begin{bmatrix} 0 & 0 \\ 0 & 0 \end{bmatrix} = O_{2\times2}$

Written Exercises, pages 592–593

1. $\begin{bmatrix} 18 \\ 0 \\ 6 \end{bmatrix}$ **3.** $\begin{bmatrix} 16 & 13 & 2 \\ 5 & -1 & -8 \end{bmatrix}$ **5.** $\begin{bmatrix} 14 & 8 & 18 \\ 29 & 15 & 39 \\ -11 & 83 & 9 \end{bmatrix}$

7. $\begin{bmatrix} \dfrac{-2}{3} & \dfrac{-1}{3} \\[2mm] \dfrac{-2}{3} & \dfrac{-4}{3} \end{bmatrix}$ **9.** $\begin{bmatrix} \dfrac{1}{23} & \dfrac{7}{46} \\[2mm] \dfrac{4}{23} & \dfrac{5}{46} \end{bmatrix}$

11. $\begin{bmatrix} \dfrac{-8}{59} & \dfrac{1}{118} & \dfrac{-13}{59} \\[3mm] \dfrac{14}{59} & \dfrac{13}{118} & \dfrac{8}{59} \\[3mm] \dfrac{-13}{59} & \dfrac{9}{118} & \dfrac{1}{59} \end{bmatrix}$

13. Not possible

15. $A \cdot C = \begin{bmatrix} 2(-2)+3(4) & 2(3)+3(2) \\ -1(-2)+4(4) & -1(3)+4(2) \end{bmatrix}$

$= \begin{bmatrix} 8 & 12 \\ 18 & 5 \end{bmatrix}$

$C \cdot A = \begin{bmatrix} -2(2)+3(-1) & -2(3)+3(4) \\ 4(2)+2(-1) & 4(3)+2(4) \end{bmatrix}$

$= \begin{bmatrix} -7 & 6 \\ 6 & 20 \end{bmatrix}$

Thus, $A \cdot C \neq C \cdot A$.
17. T **19.** F **21.** F

Written Exercises, pages 595–596

1. 17 **3.** -86 **5.** 1 **7.** -50 **9.** $-\dfrac{6}{15}$
11. $\dfrac{29}{60}$ **13.** $(-1,2)$ **15.** $(3,-1)$ **17.** $(0,-3)$
19. $(2,4)$ **21.** $(1.3,2.1)$ **23.** $(2\sqrt{2},2\sqrt{3})$

25. $|-A| = \begin{vmatrix} -a_1 & -b_1 \\ -c_1 & -d_1 \end{vmatrix} = a_1 d_1 - c_1 b_1$

$|A| = \begin{vmatrix} a_1 & b_1 \\ c_1 & d_1 \end{vmatrix} = a_1 d_1 - c_1 b_1$

Thus, $|-A| = |A|$.

27. $|A \cdot B| = \begin{vmatrix} a_1 a_2 + b_1 c_2 & a_1 b_2 + b_1 d_2 \\ c_1 a_2 + d_1 c_2 & c_1 b_2 + d_1 d_2 \end{vmatrix}$

$= (a_1 a_2 + b_1 c_2)(c_1 b_2 + d_1 d_2) -$
$\quad (c_1 a_2 + d_1 c_2)(a_1 b_2 + b_1 d_2)$
$= a_1 a_2 b_2 c_1 + b_1 b_2 c_1 c_2 + a_1 a_2 d_1 d_2 + b_1 c_2 d_1 d_2$
$\quad - [a_1 a_2 b_2 c_1 + a_2 b_1 c_1 d_2 + a_1 b_2 c_2 d_1 + b_1 c_2 d_1 d_2]$
$= b_1 b_2 c_1 c_2 + a_1 a_2 d_1 d_2 - a_2 b_1 c_1 d_2 - a_1 b_2 c_2 d_1$
$= (a_1 d_1 - c_1 b_1)(a_2 d_2 - c_2 b_2) = |A| \cdot |B|$

Written Exercises, pages 600–601

1. -70 **3.** -21 **5.** 30 **7.** $(-1,1,2)$
9. $(-2,2,1)$ **11.** $(-1,2,-3)$ **13.** $(-3,2,1)$
15. $\left(\dfrac{244}{63}, -\dfrac{38}{63}, \dfrac{244}{63}\right)$ **17.** Suppose that
consecutive rows are interchanged. If we
expand about the rows with the same elements
before and after the interchange, the same
minors are produced. However, since the rows
that are being expanded about have been shifted
up or down one step, 1 is added to or subtracted
from the row and column for each element of
the row. So, the sign of the coefficient of each
minor is reversed. For example, consider the

determinant $\begin{vmatrix} a_1 & a_2 & a_3 \\ b_1 & b_2 & b_3 \\ c_1 & c_2 & c_3 \end{vmatrix}$. Expanding about the

first row gives the following.

$D = a_1 \begin{vmatrix} b_2 & b_3 \\ c_2 & c_3 \end{vmatrix} - a_2 \begin{vmatrix} b_1 & b_3 \\ c_1 & c_3 \end{vmatrix} + a_3 \begin{vmatrix} b_1 & b_2 \\ c_1 & c_2 \end{vmatrix};$

However, if the first two rows of the
determinant are interchanged, producing

the determinant $\begin{vmatrix} b_1 & b_2 & b_3 \\ a_1 & a_2 & a_3 \\ c_1 & c_2 & c_3 \end{vmatrix}$, expanding about the

row containing the "a" elements gives the
following.

$D = -a_1 \begin{vmatrix} b_2 & b_3 \\ c_2 & c_3 \end{vmatrix} + a_2 \begin{vmatrix} b_1 & b_3 \\ c_1 & c_3 \end{vmatrix} - a_3 \begin{vmatrix} b_1 & b_2 \\ c_1 & c_2 \end{vmatrix};$

Thus, the sign of the value of the determinant is
changed.
If the first and third rows are interchanged, the
sign of each minor's coefficient remains the
same. However, the minors are inverted; that is,
the top row becomes the bottom row, and vice
versa. Thus, the sign of the value of the
determinant changes.

Written Exercises, pages 605–606

1. $(2,-1)$ **3.** $(-2,-1)$ **5.** $(1,1)$ **7.** $(-5,7)$
9. $(x,z) = (-10,5)$ **11.** $(-1,-1,2)$
13. $(1,0,-2)$ **15.** $(-3,-2,2)$

17. $\begin{bmatrix} 0 & \dfrac{1}{2} \\ -1 & 2 \end{bmatrix}$ **19.** $\begin{bmatrix} \dfrac{a}{a^2+b^2} & \dfrac{b}{a^2+b^2} \\[3mm] \dfrac{-b}{a^2+b^2} & \dfrac{a}{a^2+b^2} \end{bmatrix}$

21. Proof
1. $I = I$ Reflexive Prop Eq; 2. $(X \cdot Y) \cdot$
$(X \cdot Y)^{-1} = I$: $A \cdot A^{-1} = I$; 3. $X^{-1} \cdot X \cdot Y \cdot$
$(X \cdot Y)^{-1} = X^{-1} \cdot I$: Mult Prop Eq; 4. $I \cdot Y \cdot$
$(X \cdot Y)^{-1} = X^{-1}$: $A \cdot I = A$; 5. $Y^{-1} \cdot Y \cdot$
$(X \cdot Y)^{-1} = Y^{-1} \cdot X^{-1}$: Mult Prop Eq;
6. $(X \cdot Y)^{-1} = Y^{-1} \cdot X^{-1}$: $A \cdot A^{-1} = I$,
$A \cdot I = A$

Written Exercises, pages 609–611

1. $\begin{bmatrix} 400 & 1{,}200 & 1{,}000 \\ 500 & 900 & 600 \\ 100 & 200 & 500 \\ 200 & 300 & 400 \end{bmatrix}$ **3.** $\begin{bmatrix} 1{,}200 & 1{,}900 & 2{,}500 \\ 800 & 1{,}800 & 2{,}600 \\ 100 & 1{,}000 & 700 \\ 700 & 600 & 400 \end{bmatrix}$

5. \$2,500 **7.** \$1,000

9. $2 \cdot \begin{bmatrix} 800 & 700 & 1{,}500 \\ 300 & 900 & 2{,}000 \\ 0 & 800 & 200 \\ 500 & 300 & 0 \end{bmatrix}$

11. $\begin{bmatrix} 3(800) & 3(700) & 3(1{,}500) \\ 2(300) & 2(900) & 2(2{,}000) \\ 4(0) & 4(800) & 4(200) \\ 5(500) & 5(300) & 5(0) \end{bmatrix}$

13. \$1,800 **15.** \$0

17. $\begin{bmatrix} 20 & 15 & 10 \\ 2 & 6 & 12 \end{bmatrix}$ **19.** $\begin{bmatrix} 160 \\ 140 \\ 180 \\ 122 \end{bmatrix}$ **21.** \$83,600

23. \$1,418,400
25. \$2,088,200

Chapter 17

Written Exercises, page 622

1. 0.6000; 0.8000; 0.7500; 0.8000; 0.6000;
1.3333 **3.** 0.7071; 0.7071; 1.0000; 0.7071;
0.7071; 1.0000 **5.** 0.8660; 0.5000; 1.7321;
0.5000; 0.8660; 0.5774 **7.** 0.4472; 0.8944;
0.5000; 0.8944; 0.4472; 2.0000 **9.** 0.3846
11. 0.2917 **13.** 0.5547

15. $\dfrac{f}{\sqrt{p^2 + f^2}}$ **17.** $\dfrac{2a^2}{a^2 + b^2}$

Written Exercises, pages 626–627

1. 1.6667, 1.2500, 1.3333, 1.2500, 1.6667,
0.7500 **3.** 1.4142, 1.4142, 1.0000, 1.4142,
1.4142, 1.0000 **5.** 0.4695 **7.** 1.1792
9. 0.8689 **11.** 0.7716 **13.** 2.9957
15. 0.0378 **17.** 12 **19.** 68 **21.** 29°20′
23. 50°30′ **25.** 1.0541 **27.** 1.3416
29. 1.5811 **31.** 53°50′ **33.** 68°20′

35. Answers will vary. One possible answer:
Find the length of the adjacent side by using the
Pythagorean relation. Then divide that length by
the length of the hypotenuse. **37.** 1

Written Exercises, pages 630–631

1. 69, 21 **3.** 35, 45 **5.** 57, 33 **7.** 26 ft
9. 15.8 m **11.** 1,710 m **13.** 67

Written Exercises, page 637

1.

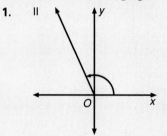

II

3. quadrantal: negative y-axis

5. quadrantal: negative x-axis

7. quadrantal: negative y-axis

9.

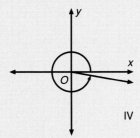

IV

11.

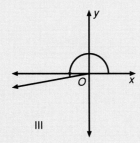

III

Exercises 13–24: Answers will vary. Possible answers are given.
13. 390, −330 **15.** 230, −490 **17.** 50, −670 **19.** 310, −410 **21.** 680, −40
23. 316, −404
25. 45

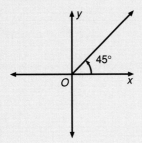

27. 180

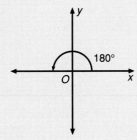

29. 120

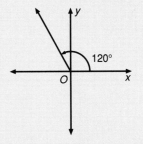

31. $\theta = x + 360n$, where $90 < x < 180$ and n is an integer.

Written Exercises, page 642

1. $\dfrac{3}{5}$ **3.** $\dfrac{3}{5}$ **5.** $\dfrac{-5}{13}$ **7.** $\dfrac{-\sqrt{5}}{5}$ **9.** $\dfrac{-\sqrt{2}}{2}$

11. $\dfrac{-\sqrt{2}}{2}$ **13.** 0.8716 **15.** 0.9816

17. 0.9848 **19.** −0.7071 **21.** 0.4226
23. −0.9976 **25.** 0.7585 **27.** −0.5100
29. 0.9954 **31.** −0.9652 **33.** 0.8660
35. −0.7660

37. $\dfrac{(a+1)\sqrt{2a^2+2}}{2a^2+2}$

Written Exercises, page 647
Exercises 1–40 are in the following order: sin; cos; tan; csc; sec; cot.

1. $\dfrac{4}{5}$; $\dfrac{3}{5}$; $\dfrac{4}{3}$; $\dfrac{5}{4}$; $\dfrac{5}{3}$; $\dfrac{3}{4}$ **3.** $\dfrac{-12}{13}$; $\dfrac{-5}{13}$;

$\dfrac{12}{5}$; $\dfrac{-13}{12}$; $\dfrac{-13}{5}$; $\dfrac{5}{12}$ **5.** $\dfrac{5}{13}$; $\dfrac{-12}{13}$; $\dfrac{-5}{12}$;

$\dfrac{13}{5}$; $\dfrac{-13}{12}$; $\dfrac{-12}{5}$ **7.** $\dfrac{-\sqrt{5}}{5}$; $\dfrac{2\sqrt{5}}{5}$; $\dfrac{-1}{2}$;

$-\sqrt{5}$; $\dfrac{\sqrt{5}}{2}$; −2 **9.** $\dfrac{-3\sqrt{10}}{10}$; $\dfrac{-\sqrt{10}}{10}$; 3;

$\dfrac{-\sqrt{10}}{3}$; $-\sqrt{10}$; $\dfrac{1}{3}$ **11.** $\dfrac{-\sqrt{17}}{17}$; $\dfrac{4\sqrt{17}}{17}$;

$\dfrac{-1}{4}$; $-\sqrt{17}$; $\dfrac{\sqrt{17}}{4}$; −4 **13.** 0.9659;

−0.2588; −3.7321; 1.0353; −3.8637;
−0.2679 **15.** −0.5150; 0.8572; −0.6009;
−1.9416; 1.1666; −1.6643 **17.** −0.9272;
−0.3746; 2.4751; −1.0785; −2.6695; 0.4040
19. 0.8192; −0.5736; −1.4281; 1.2208;
−1.7434; −0.7002 **21.** −0.7314; 0.6820;
−1.0724; −1.3673; 1.4663; −0.9325
23. −0.7660; −0.6428; 1.1918; −1.3054;
−1.5557; 0.8391 **25.** 0; 1; 0; undefined; 1;
undefined **27.** −1; 0; undefined; −1;

undefined; 0 **29.** 0.9228; -0.3854; -2.3945; 1.0837; -2.5949; -0.4176 **31.** -0.8587; -0.5125; 1.6753; -1.1646; -1.9511; 0.5969

33. $\dfrac{-\sqrt{3}}{2}$; $\dfrac{1}{2}$; $-\sqrt{3}$; $\dfrac{-2\sqrt{3}}{3}$; 2; $\dfrac{-\sqrt{3}}{3}$

35. $\dfrac{-3}{4}$; $\dfrac{-\sqrt{7}}{4}$; $\dfrac{3\sqrt{7}}{7}$; $\dfrac{-4}{3}$; $\dfrac{-4\sqrt{7}}{7}$; $\dfrac{\sqrt{7}}{3}$

37. $\dfrac{\sqrt{3}}{2}$; $\dfrac{1}{2}$; $\sqrt{3}$; $\dfrac{2\sqrt{3}}{3}$; 2; $\dfrac{\sqrt{3}}{3}$ **39.** $\dfrac{\sqrt{3}}{2}$; $\dfrac{-1}{2}$; $-\sqrt{3}$; $\dfrac{2\sqrt{3}}{3}$; -2; $\dfrac{-\sqrt{3}}{3}$ **41.** sin 40

43. $-\tan 24$ **45.** sec 50 **47.** sin 64°20′
49. $\sin\theta$; $-\cos\theta$; $-\tan\theta$; $\csc\theta$; $-\sec\theta$; $-\cot\theta$ **51.** $-\sin\theta$; $-\cos\theta$; $\tan\theta$; $-\csc\theta$; $-\sec\theta$; $\cot\theta$ **53.** Yes; 180, 360; $(2n-1)\cdot 90$

Written Exercises, page 651

1. $\dfrac{1}{2}$ **3.** $\dfrac{\sqrt{3}}{2}$ **5.** $\sqrt{3}$ **7.** $\dfrac{1}{2}$ **9.** $\sqrt{2}$

11. $-\sqrt{2}$ **13.** $\dfrac{\sqrt{2}}{2}$ **15.** $\sqrt{2}$ **17.** 3

19. $-\dfrac{1}{2}$ **21.** $\dfrac{-\sqrt{3}}{2}$ **23.** $\dfrac{-\sqrt{3}}{3}$

25. $5 - 2\sqrt{6}$

Written Exercises, pages 654–655

1. 10°20′; 349°40′ **3.** 38°10′; 218°10′
5. 45°; 135° **7.** 107°50′; 252°10′
9. 15°30′; 195°30′ **11.** 174°; 354°

13. $\sin\theta = \dfrac{4}{5}$; $\tan\theta = \dfrac{4}{3}$; $\csc\theta = \dfrac{5}{4}$; $\sec\theta = \dfrac{5}{3}$; $\cot\theta = \dfrac{3}{4}$ **15.** $\cos\theta = \dfrac{-2\sqrt{6}}{5}$; $\tan\theta = \dfrac{-\sqrt{6}}{12}$; $\csc\theta = 5$; $\sec\theta = \dfrac{-5\sqrt{6}}{12}$; $\cot\theta = -2\sqrt{6}$ **17.** $\sin\theta = \dfrac{-2\sqrt{6}}{5}$; $\tan\theta = 2\sqrt{6}$; $\csc\theta = \dfrac{-5\sqrt{6}}{12}$; $\sec\theta = -5$; $\cot\theta = \dfrac{\sqrt{6}}{12}$ **19.** $\sin\theta = \dfrac{-2\sqrt{2}}{3}$; $\cos\theta = \dfrac{1}{3}$; $\tan\theta = -2\sqrt{2}$; $\csc\theta = \dfrac{-3\sqrt{2}}{4}$; $\cot\theta = \dfrac{-\sqrt{2}}{4}$
21. $\dfrac{5}{17}$

Chapter 18

Written Exercises, page 664

1. 32 **3.** 22 **5.** 3,100 mi **7.** 32 **9.** 128

11.

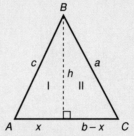

Case I:
For triangle I, $c^2 = h^2 + x^2$.
For triangle II, $a^2 = h^2 + (b - x)^2$
$$a^2 = h^2 + b^2 - 2bx + x^2$$
Subtract 1st equation from second equation.
$$a^2 = h^2 + b^2 - 2bx + x^2$$
$$c^2 = h^2 \qquad\qquad + x^2$$
$$\overline{a^2 - c^2 = b^2 - 2bx}$$
But $\dfrac{x}{c} = \cos A$, or $x = c\cos A$, so by substitution,
$a^2 - c^2 = b^2 - 2b(c\cos A)$, or
$a^2 = b^2 + c^2 - 2bc\cos A$.
Case II:

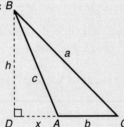

For triangle BDA, $c^2 = h^2 + x^2$.
For triangle BDC, $a^2 = h^2 + (b + x)^2$
$$a^2 = h^2 + b^2 + 2bx + x^2$$
Subtract 1st equation from second equation.
$$a^2 = h^2 + b^2 + 2bx + x^2$$
$$c^2 = h^2 \qquad\qquad + x^2$$
$$\overline{a^2 - c^2 = b^2 + 2bx}$$
But $\dfrac{x}{c} = \cos(180 - A)$, so $x = c\cos(180 - A)$,
or $x = -c\cos A$. Thus, by substitution,
$a^2 - c^2 = b^2 + 2b(-c\cos A)$, or
$a^2 = b^2 + c^2 - 2bc\cos A$.

Written Exercises, page 667

1. 140 sq units **3.** 17 sq units **5.** 94 sq units **7.** $3\sqrt{15}$ sq units **9.** $2\sqrt{66}$ sq units
11. 410 sq units **13.** 94 sq units
15. $16\sqrt{3}$ sq units

17.

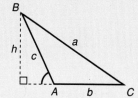

The area of any triangle is $\frac{1}{2}bh$. Now from the figure, $\frac{h}{c} = \sin(180 - A)$, or $h = c \sin (180 - A)$. But $\sin A = \sin (180 - A)$, so if the area is K, then $K = \frac{1}{2}b[c \sin (180 - A)] = \frac{1}{2}bc \sin A$. **19.** 96 sq units, 54 sq units

Written Exercises, page 671

1. 880 **3.** 14 **5.** 120 **7.** 120 mi
9. 14 mi **11.** 550 yd **13.** By Theorem 18.2, $K = \frac{1}{2}ab \sin C$. By the Law of Sines, $\frac{\sin B}{b} = \frac{\sin A}{a}$, or $b = \frac{a \sin B}{\sin A}$. Substituting for b, $K = \frac{1}{2}a\left(\frac{a \sin B}{\sin A}\right)\sin C$. So, $K = a^2\left(\frac{\sin B \sin C}{2 \sin A}\right)$. **15.** 57.2 sq units

Written Exercises, page 676

1. One; 30 **3.** None **5.** One; 65
7. None **9.** No solution **11.** m $\angle B = 53$, m $\angle C = 47$, $a = 30$ **13.** m $\angle B = 124$, m $\angle C = 21$, $b = 27$ **15.** 36, 27, 117
17. 38 square units **19.** 47

Written Exercises, pages 679–680

1. $\frac{5\pi}{6}$ **3.** $\frac{5\pi}{3}$ **5.** $\frac{11\pi}{6}$ **7.** $\frac{7\pi}{6}$ **9.** $\frac{8\pi}{9}$
11. $\frac{-5\pi}{9}$ **13.** 120 **15.** 135 **17.** 315
19. -210 **21.** 115 **23.** 132 **25.** -57
27. 344 **29.** $\frac{1}{2}$ **31.** $\frac{-\sqrt{3}}{2}$ **33.** 1 **35.** 1
37. $s = r \cdot \theta$ **39.** $A = \frac{\theta r^2}{6}$; $A = \frac{rs}{2}$

Written Exercises, pages 683–684

1. Even; 2

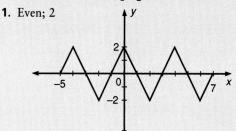

3. Odd; 3

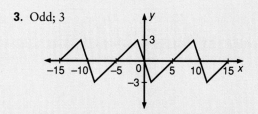

5. Even; $\frac{3}{2}$

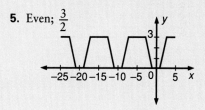

7.

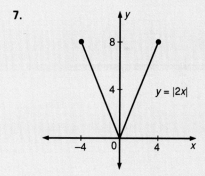

$y = |2x|$

9. Even; 4

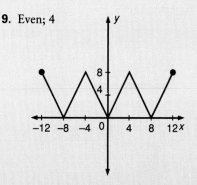

11.

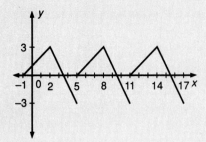

13.

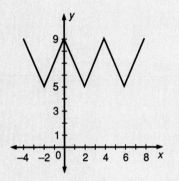

Written Exercises, page 689

1. $0, -\frac{1}{2}, -1, -\frac{1}{2}, 0, \frac{1}{2}, 1, \frac{1}{2}, 0$

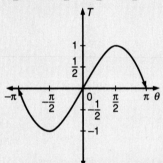

3. $-\frac{\pi}{2} \le \theta \le \frac{\pi}{2}$ **5.** $-\pi \le \theta \le 0$

7. -0.5446

9.

θ	-2π	$-\frac{11\pi}{6}$	$-\frac{3\pi}{2}$	$-\frac{7\pi}{6}$	$-\pi$	$-\frac{5\pi}{6}$	$-\frac{\pi}{2}$	$-\frac{\pi}{6}$	0
T	1	$\frac{\sqrt{3}}{2}$	0	$-\frac{\sqrt{3}}{2}$	-1	$-\frac{\sqrt{3}}{2}$	0	$\frac{\sqrt{3}}{2}$	1

11. $-\pi \le \theta \le 0$
13. $T = \sin \theta$

θ	-2π	$-\frac{11\pi}{6}$	$-\frac{3\pi}{2}$	$-\frac{7\pi}{6}$	$-\pi$	$-\frac{5\pi}{6}$	$-\frac{\pi}{2}$	$-\frac{\pi}{6}$	0
T	0	$\frac{1}{2}$	1	$\frac{1}{2}$	0	$-\frac{1}{2}$	-1	$-\frac{1}{2}$	0

15. $-2\pi \le \theta \le -\frac{3\pi}{2}$ and $-\frac{\pi}{2} \le \theta \le 0$

17. $\frac{\pi}{4} + k\pi$, where k is an integer.

19. $T = |\sin \theta|$

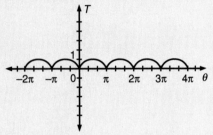

21. Even; $|\sin(-\theta)| = |-\sin \theta| = |\sin \theta|$

Written Exercises, page 696

1.

θ	$\frac{\pi}{8}$	$\frac{\pi}{4}$	$\frac{3\pi}{8}$	$\frac{\pi}{2}$	$\frac{5\pi}{8}$	$\frac{3\pi}{4}$	$\frac{7\pi}{8}$	π	$\frac{9\pi}{8}$	$\frac{5\pi}{4}$	$\frac{11\pi}{8}$	$\frac{3\pi}{2}$	$\frac{13\pi}{8}$	$\frac{7\pi}{4}$	$\frac{15\pi}{8}$	2π
T	1	0	-1	0	1	0	-1	0	1	0	-1	0	1	0	-1	0

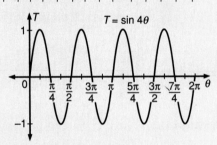

3.

θ	0	$\frac{\pi}{4}$	$\frac{\pi}{2}$	$\frac{3\pi}{4}$	π	$\frac{5\pi}{4}$	$\frac{3\pi}{2}$	$\frac{7\pi}{4}$	2π
T	0	-1	0	1	0	-1	0	1	0

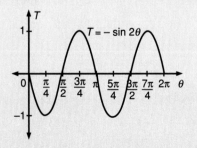

5.

θ	0	$\frac{\pi}{2}$	π	$\frac{3\pi}{2}$	2π
T	1	2	1	0	1

$T = 1 + \sin \theta$

7.

θ	0	$\frac{\pi}{2}$	π	$\frac{3\pi}{2}$	2π
T	−4	0	−4	−8	−4

$T = -4 + 4 \sin \theta$

9. $p = \frac{2\pi}{3}; a = 1$

$T = \sin 3\theta$

11. $p = \frac{\pi}{2}; a = 5$

$T = 5 \cos 4\theta$

13. $p = \pi; a = 5$

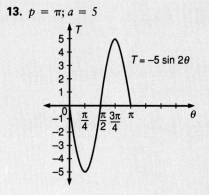

$T = -5 \sin 2\theta$

15. $p = 6\pi; a = \frac{1}{4}$

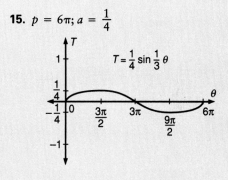

$T = \frac{1}{4} \sin \frac{1}{3} \theta$

17. $p = \frac{2\pi}{9}; a = 6$

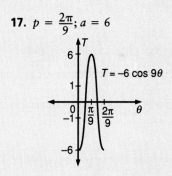

$T = -6 \cos 9\theta$

19. $p = \frac{\pi}{2}; a = 5; \text{max} = -1; \text{min} = -11$

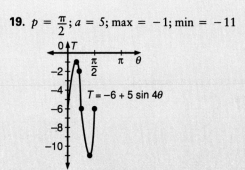

$T = -6 + 5 \sin 4\theta$

21. $-\dfrac{5\pi}{4}, -\dfrac{\pi}{4}, \dfrac{3\pi}{4}, \dfrac{7\pi}{4}$

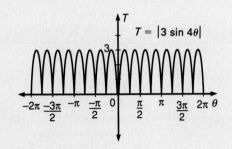

23.

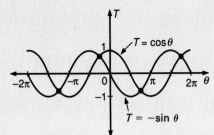

Written Exercises, page 700

1. None, None, None, None   **3.** None, None, None, None   **5.** $\dfrac{\pi}{2}, \dfrac{3\pi}{2}$; $0, \pi, 2\pi$; $\dfrac{\pi}{2}, \dfrac{3\pi}{2}$; $0, \pi, 2\pi$

7.

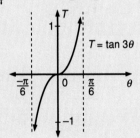

9.

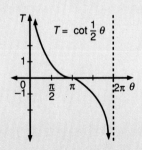

11.

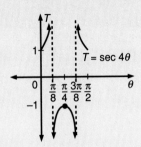

13.

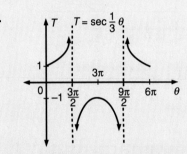

15. One possible answer: $T = 2\tan 3\theta$ is steeper, and has a period that is one-third as great as that of $T = \tan\theta$.

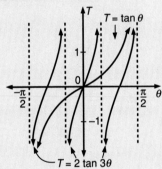

17. Even

Written Exercises, page 704

1. $\dfrac{\pi}{2}$ **3.** $\dfrac{2\pi}{3}$ **5.** $\dfrac{\pi}{2}$ **7.** $\dfrac{\pi}{6}$ **9.** -27

11. 137   **13.** $\dfrac{4}{5}$

15.

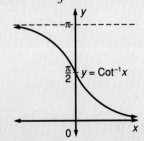

17. $\sqrt{x^2 + 1}$ **19.** True. Since $-1 \le x \le 1$, so is $-x$. Thus, the domains are the same. Let $y = \text{Sin}^{-1}(-x)$. Then $\sin y = -x$. But $\sin y$ is an odd function, so $\sin(-y) = -(-x)$, or $\sin(-y) = x$. This means that $-y = \text{Sin}^{-1} x$, or that $y = -\text{Sin}^{-1} x$. So, $\text{Sin}^{-1}(-x) = -\text{Sin}^{-1} x$.

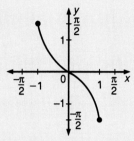

Chapter 19

Written Exercises, pages 714–715

1. $\cos 40$ **3.** $\sec 24.6$ **5.** $\sin 135 \csc 135$ $= \dfrac{1}{\sqrt{2}} \cdot \sqrt{2} = 1$ **7.** $\sin(90 - 60) =$ $\sin 30 = \dfrac{1}{2} = \cos 60$ **9.** $\dfrac{1}{\sec 30} = \dfrac{\frac{1}{2}}{\sqrt{3}} =$

$\dfrac{\sqrt{3}}{2} = \cos 30$ **11.** $\dfrac{\cos 240}{\sin 240} = -\dfrac{1}{2} \div \left(-\dfrac{\sqrt{3}}{2}\right) =$ $-\dfrac{1}{2} \cdot \left(-\dfrac{2}{\sqrt{3}}\right) = \dfrac{1}{\sqrt{3}} = \cot 240$

13. $\cos^2 270 = 0^2 = 1 - (-1)^2 = 1 -$ $\sin^2 270$ **15.** $1 + \cot^2 \dfrac{2\pi}{3} = 1 + \left(-\dfrac{1}{\sqrt{3}}\right)^2 =$ $1 + \dfrac{1}{3} = \dfrac{4}{3} = \left(\dfrac{2}{\sqrt{3}}\right)^2 = \csc^2 \dfrac{2\pi}{3}$

17. $\dfrac{\cos \frac{11\pi}{6}}{\sin \frac{11\pi}{6}} = \dfrac{\sqrt{3}}{2} \div \left(-\dfrac{1}{2}\right) = \dfrac{\sqrt{3}}{2} \cdot (-2) =$ $-\sqrt{3} = \cot \dfrac{11\pi}{6}$ **19.** $1 + \tan^2\left(-\dfrac{7\pi}{3}\right) =$ $1 + (-\sqrt{3})^2 = 4 = 2^2 = \sec^2\left(-\dfrac{7\pi}{3}\right)$

21. $\cos \theta \sec \theta = \dfrac{x}{r} \cdot \dfrac{r}{x} = 1$ **23.** $\dfrac{\cos \theta}{\sin \theta} =$ $\dfrac{x}{r} \div \dfrac{y}{r} = \dfrac{x}{r} \cdot \dfrac{r}{y} = \dfrac{x}{y} = \cot \theta$

25. $(1 - \sin^2 \theta)(1 + \tan^2 \theta) = \cos^2 \theta \sec^2 \theta$ $= \cos^2 \theta \cdot \dfrac{1}{\cos^2 \theta} = 1$

Written Exercises, pages 718–719

1. 1   **3.** $\cot \theta$ **5.** $\sin^2 \theta$ **7.** $\sin \theta$

9. $2 \csc \theta$ **11.** $\sin \theta + \dfrac{\cos^2 \theta}{\sin \theta} = \dfrac{\sin^2 \theta}{\sin \theta}$ $+ \dfrac{\cos^2 \theta}{\sin \theta} = \dfrac{1}{\sin \theta}$ **13.** $\cos^2 \theta (\cot^2 \theta$ $+ 1) = \cos^2 \theta (\csc^2 \theta) = \dfrac{\cos^2 \theta}{\sin^2 \theta} = \cot^2 \theta$

15. $\dfrac{\cos \theta}{\sin \theta} + \dfrac{\sin \theta}{1 + \cos \theta} = \dfrac{\cos \theta (1 + \cos \theta)}{\sin \theta (1 + \cos \theta)} +$ $\dfrac{\sin^2 \theta}{\sin \theta (1 + \cos \theta)} = \dfrac{\cos \theta + \cos^2 \theta + \sin^2 \theta}{\sin \theta (1 + \cos \theta)} =$ $\dfrac{\cos \theta + 1}{\sin \theta (1 + \cos \theta)} = \dfrac{1}{\sin \theta} = \csc \theta$

17. $\dfrac{\sin \theta + \tan \theta}{\cot \theta + \csc \theta} = \dfrac{\sin \theta + \frac{\sin \theta}{\cos \theta}}{\frac{\cos \theta}{\sin \theta} + \frac{1}{\sin \theta}} =$ $\dfrac{\frac{\cos \theta \sin \theta + \sin \theta}{\cos \theta}}{\frac{\cos \theta + 1}{\sin \theta}} = \dfrac{\cos \theta \sin \theta + \sin \theta}{\cos \theta} \cdot$ $\dfrac{\sin \theta}{\cos \theta + 1} = \dfrac{\sin \theta (\cos \theta + 1)}{\cos \theta} \cdot \dfrac{\sin \theta}{\cos \theta + 1} =$ $\dfrac{\sin \theta}{\cos \theta} \cdot \sin \theta = \sin \theta \tan \theta$

19. $\dfrac{1 + \cot \theta}{\cot \theta \sin \theta + \frac{\cos^2 \theta}{\sin \theta}} = \dfrac{\frac{\sin \theta}{\sin \theta} + \frac{\cos \theta}{\sin \theta}}{\frac{\cos \theta \sin \theta}{\sin \theta} + \frac{\cos^2 \theta}{\sin \theta}} =$ $\dfrac{\frac{\sin \theta + \cos \theta}{\sin \theta}} \cdot \dfrac{\sin \theta}{\cos \theta \sin \theta + \cos^2 \theta} =$ $\dfrac{\sin \theta + \cos \theta}{\cos \theta (\sin \theta + \cos \theta)} = \dfrac{1}{\cos \theta} = \sec \theta$

21. $\dfrac{\csc \theta - \sin \theta}{\cot^2 \theta} = \dfrac{\frac{1}{\sin \theta} - \frac{\sin^2 \theta}{\sin \theta}}{\frac{\cos^2 \theta}{\sin^2 \theta}} = \dfrac{\frac{\cos^2 \theta}{\sin \theta}}{\frac{\cos^2 \theta}{\sin^2 \theta}} =$ $\dfrac{\cos^2 \theta}{\sin \theta} \cdot \dfrac{\sin^2 \theta}{\cos^2 \theta} = \sin \theta$

23. $\dfrac{1}{1 - \sin \theta} + \dfrac{1}{1 + \sin \theta} =$ $\dfrac{1 + \sin \theta + 1 - \sin \theta}{1 - \sin^2 \theta} = \dfrac{2}{\cos^2 \theta} = 2 \sec^2 \theta$

25. $\sin^2 \theta \sec^2 \theta + \sin^2 \theta \csc^2 \theta = \dfrac{\sin^2 \theta}{\cos^2 \theta} +$ $\dfrac{\sin^2 \theta}{\sin^2 \theta} = \tan^2 \theta + 1 = \sec^2 \theta$

27. $\dfrac{\csc \theta}{\cot \theta + \tan \theta} = \dfrac{\dfrac{1}{\sin \theta}}{\dfrac{\cos \theta}{\sin \theta} + \dfrac{\sin \theta}{\cos \theta}} =$

$\dfrac{\dfrac{1}{\sin \theta}}{\dfrac{\cos^2 \theta + \sin^2 \theta}{\sin \theta \cos \theta}} = \dfrac{\dfrac{1}{\sin \theta}}{\dfrac{1}{\sin \theta \cos \theta}} = \dfrac{1}{\sin \theta} \cdot \sin \theta \cos \theta =$

$\cos \theta = \cos \theta \cdot \dfrac{\sin \theta}{\sin \theta} = \cot \theta \sin \theta$

29. $(\sin \theta + \cos \theta)(\tan \theta + \cot \theta) =$
$\sin \theta \tan \theta + \sin \theta \cot \theta + \cos \theta \tan \theta +$

$\cos \theta \cot \theta = \dfrac{\sin^2 \theta}{\cos \theta} + \cos \theta + \sin \theta + \dfrac{\cos^2 \theta}{\sin \theta}$

$= \dfrac{1 - \cos^2 \theta}{\cos \theta} + \cos \theta + \sin \theta + \dfrac{1 - \sin^2 \theta}{\sin \theta} =$

$(\sec \theta - \cos \theta) + \cos \theta + \sin \theta +$
$(\csc \theta - \sin \theta) = \sec \theta + \csc \theta$

31. $\begin{vmatrix} \cos \theta & -\sin \theta \\ \sin \theta & \cos \theta \end{vmatrix} = \cos^2 \theta + \sin^2 \theta = 1$

33. $\dfrac{1 + \cos \theta + \cos^2 \theta}{1 + \sin \theta + \sin^2 \theta}$

Written Exercises, page 723

1. 120, 240 **3.** 60, 120 **5.** 30, 150

7. 135, 315 **9.** 180 **11.** 0, 180 **13.** $\dfrac{4\pi}{3}$,

$\dfrac{5\pi}{3}$ **15.** $\dfrac{3\pi}{4}, \dfrac{5\pi}{4}$ **17.** $\dfrac{5\pi}{6}, \dfrac{7\pi}{6}$ **19.** 30,

150 **21.** 143, 217 **23.** 90, 150, 210, 270
25. 180 **27.** 225, 315 **29.** 63, 117, 243,
297 **31.** 90, 120, 240 **33.** 0, 60, 180, 300

Written Exercises, page 726

1. 30, 150, 270 **3.** 19, 161 **5.** 18, 45,
198, 225 **7.** 0, 180 **9.** 72, 135, 252, 315
11. 65, 180, 295 **13.** 45, 225 **15.** 0
17. 0, 127 **19.** 90, 337 **21.** 222, 318
23. 90, 180, 270 **25.** $r = 5, \theta = 53$

Written Exercises, pages 730–731

1. $\dfrac{\sqrt{2}}{2}$ **3.** $\sqrt{3}$ **5.** 0 **7.** $2 + \sqrt{3}$

9. $\dfrac{\sqrt{6} - \sqrt{2}}{4}$ **11.** $-\cos \phi$ **13.** $-\sin \phi$

15. $\dfrac{\sin (\theta + \phi)}{\sin \theta \sin \phi} = \dfrac{\sin \theta \cos \phi + \cos \theta \sin \phi}{\sin \theta \sin \phi}$

$= \dfrac{\cos \phi}{\sin \phi} + \dfrac{\cos \theta}{\sin \theta} = \cot \theta + \cot \phi$

17. $\sin (30 + \theta) + \sin (30 - \theta) =$
$(\sin 30 \cos \theta + \cos 30 \sin \theta) + (\sin 30 \cos \theta -$

$\cos 30 \sin \theta) = 2 \sin 30 \cos \theta = 2\left(\dfrac{1}{2}\right) \cos \theta$

$= \cos \theta$ **19.** $-\dfrac{56}{65}, \dfrac{63}{16}$ **21.** $\dfrac{\sqrt{3} - 2\sqrt{2}}{6}$,

$-\dfrac{8\sqrt{2} + 9\sqrt{3}}{5}$ **23.** Since $\cos (90 - \phi) =$
$\sin \phi$, letting $\phi = 90 - \theta$ gives $\cos [90 -$
$(90 - \theta)] = \sin(90 - \theta)$, or $\cos \theta = \sin (90 -$
$\theta)$. **25.** $\sin (\theta + \phi) = \sin [\theta - (-\phi)] =$
$\sin \theta \cos (-\phi) - \cos \theta \sin (-\phi) = \sin \theta \cos \phi +$
$\cos \theta \sin \phi$ **27.** $\tan (\theta + \phi) = \tan [\theta - (-\phi)]$
$= \dfrac{\tan \theta - \tan (-\phi)}{1 + \tan \theta \tan (-\phi)} = \dfrac{\tan \theta + \tan \phi}{1 - \tan \theta \tan \phi}$
29. Let $A = \theta + \phi$ and $B = \theta - \phi$. Then $\sin A$
$+ \sin B = \sin (\theta + \phi) + \sin (\theta - \phi)$ which
equals $2 \sin \theta \cos \phi$ from Exercise 28. Also, A
$+ B = (\theta + \phi) + (\theta - \phi) = 2\theta$ and $A - B =$
$(\theta + \phi) - (\theta - \phi) = 2\phi$, so $2 \sin \left(\dfrac{A + B}{2}\right)$
$\cos \left(\dfrac{A - B}{2}\right) = 2 \sin \theta \cos \phi$. Thus, $\sin A$
$+ \sin B = 2 \sin \left(\dfrac{A + B}{2}\right) \cos \left(\dfrac{A - B}{2}\right)$.

Written Exercises, page 735

1. $\dfrac{\sqrt{2}}{2}$ **3.** $\dfrac{\sqrt{3}}{3}$ **5.** $\sqrt{2} + 1$ **7.** $\dfrac{\sqrt{2 + \sqrt{2}}}{2}$

9. $\dfrac{\sqrt{2 + \sqrt{2}}}{2}$ **11.** $\dfrac{\sqrt{10}}{10}$ **13.** $-\dfrac{1}{3}$ **15.** $-\dfrac{24}{25}$

17. $\sin^2 \dfrac{\theta}{2} = \left(\sqrt{\dfrac{1 - \cos \theta}{2}}\right)^2 = \dfrac{1 - \cos \theta}{2} =$

$\dfrac{\tan \theta}{\tan \theta}\left(\dfrac{1 - \cos \theta}{2}\right) = \dfrac{\tan \theta - \sin \theta}{2 \tan \theta}$

19. $\cos^4 \theta - \sin^4 \theta = (\cos^2 \theta + \sin^2 \theta)(\cos^2 \theta$
$- \sin^2 \theta) = \cos^2 \theta - \sin^2 \theta = \cos 2\theta$

21. $\dfrac{\cos \theta \sin 2\theta}{1 + \cos 2\theta} = \dfrac{\cos \theta (2 \sin \theta \cos \theta)}{1 + (2 \cos^2 \theta - 1)} =$

$\dfrac{2 \cos^2 \theta \sin \theta}{2 \cos^2 \theta} = \sin \theta$

23. $\cos 3\theta = \cos (\theta + 2\theta) = \cos \theta \cos 2\theta -$
$\sin \theta \sin 2\theta = \cos \theta (2 \cos^2 \theta - 1) -$
$\sin \theta (2 \sin \theta \cos \theta) = 2 \cos^3 \theta - \cos \theta -$
$2(1 - \cos^2 \theta) \cos \theta = 2 \cos^3 \theta - \cos \theta -$
$2 \cos \theta + 2 \cos^3 \theta = 4 \cos^3 \theta - 3 \cos \theta$
25. $\tan 2\theta = \tan (\theta + \theta) = \dfrac{\tan \theta + \tan \theta}{1 - \tan \theta \tan \theta} =$
$\dfrac{2 \tan \theta}{1 - \tan^2 \theta}$

27. $\tan\dfrac{\theta}{2} = \dfrac{\sin\frac{\theta}{2}}{\cos\frac{\theta}{2}} = \dfrac{\pm\sqrt{\frac{1-\cos\theta}{2}}}{\pm\sqrt{\frac{1+\cos\theta}{2}}} \cdot \dfrac{\sqrt{2}}{\sqrt{2}}$

$= \pm\sqrt{\dfrac{1-\cos\theta}{1+\cos\theta}}; \tan\dfrac{\theta}{2} = \pm\sqrt{\dfrac{1-\cos\theta}{1+\cos\theta}} \cdot$

$\left(\dfrac{\sqrt{1+\cos\theta}}{\sqrt{1+\cos\theta}}\right) = \pm\dfrac{\sqrt{1-\cos^2\theta}}{(\sqrt{1+\cos\theta})^2} =$

$\pm\dfrac{\sqrt{\sin^2\theta}}{1+\cos\theta} = \dfrac{\sin\theta}{1+\cos\theta}; \tan\dfrac{\theta}{2} =$

$\pm\sqrt{\dfrac{1-\cos\theta}{1+\cos\theta}}\left(\dfrac{\sqrt{1-\cos\theta}}{\sqrt{1-\cos\theta}}\right) =$

$\pm\dfrac{(\sqrt{1-\cos\theta})^2}{\sqrt{1-\cos^2\theta}} = \pm\dfrac{1-\cos\theta}{\sqrt{\sin^2\theta}} =$

$\dfrac{1-\cos\theta}{\sin\theta}$; For the last two forms of the

identity, the "$\pm$" sign is not needed because $\tan\dfrac{\theta}{2}$ and $\sin\theta$ have the same sign.

Written Exercises, page 737

1. $0, \dfrac{\pi}{3}, \pi, \dfrac{5\pi}{3}$ **3.** $0, \dfrac{\pi}{2}, \pi, \dfrac{3\pi}{2}$ **5.** $\dfrac{\pi}{2}, \dfrac{7\pi}{6},$

$\dfrac{11\pi}{6}$ **7.** 30, 150 **9.** 60, 300 **11.** 45, 135,

225, 315 **13.** 0 **15.** 0, 45, 135, 180, 225,
315 **17.** 60, 90, 270, 300 **19.** 90, 210, 330
21. 0, 135, 180, 315

Written Exercises, page 741

1.

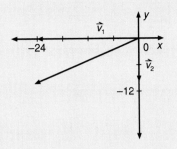

3.

5. x: 4.5, y: 6.0; 7.5 **7.** x: -24, y: -10; 26
9. 173,100 **11.** 212,212 **13.** 375 lb
15. 494 lb **17.** $\|(cx, cy)\| = \sqrt{(cx)^2 + (cy)^2}$
$= \sqrt{c^2(x^2 + y^2)} = |c|\sqrt{x^2 + y^2} = |c| \cdot \|(x, y)\|$
19. $h = 3, k = -2$

Written Exercises, page 744

1. $(5.4, 68°)$ **3.** $(17, 340°)$ **5.** $(16, 133°)$
7. $(130, 276°)$ **9.** $(2.1, 12)$ **11.** $(-6.6, -4.6)$
13. $(0, -15)$ **15.** $(10.3, -28)$ **17.** 6 cis 60
19. 8 cis 210 **21.** 4 cis 0 **23.** 4 cis 180

25. $-\dfrac{3\sqrt{2}}{2} - \dfrac{3i\sqrt{2}}{2}$ **27.** $-5 + 5i\sqrt{3}$

29. $-7i$ **31.** $\sqrt{2} + i\sqrt{2}$ **33.** $0.34 + 0.94i$
35. 110 **37.** 101

Written Exercises, page 748

1. 15 cis 160; $-14.1 + 5.13i$ **3.** 20 cis 50;

12.9 + 15.3i **5.** 2 cis 210 **7.** $\dfrac{1}{4}$ cis 200

9. 4 cis 300; $2 - 2i\sqrt{3}$ **11.** 64 cis 180; -64
13. 16 cis 0; 16 **15.** 2 cis 20, 2 cis 140,
2 cis 260 **17.** 3 cis 30, 3 cis 150, 3 cis 270
19. $\sqrt[5]{6}$ cis 9, $\sqrt[5]{6}$ cis 81, $\sqrt[5]{6}$ cis 153,
$\sqrt[5]{6}$ cis 225, $\sqrt[5]{6}$ cis 297 **21.** cis 150
23. $2\sqrt{3}$ cis 60 **25.** r cis θ

Index

Boldfaced numerals indicate the pages that contain definitions.

Index